The Fire Chief's Handbook

6th Edition

The Fire Chief's Handbook

6th Edition

Fire Engineering

PennWell Corporation
1421 South Sheridan Road
Tulsa, Oklahoma 74112 USA
1.800.752.9764
+1.918.831.9421
sales@pennwell.com
www.pennwell-store.com
www.pennwell.com

Cover design by Amy Spehar
Book design by Robin Remaley
Managing Editor: Jared Wicklund

Library of Congress Cataloging-in-Publication Data

The fire chief's handbook / by Ronny J. Coleman... {et al.}.–6th ed.
p. c.m.
Includes bibliographical references and index
ISBN 0-87814-830-2
1. Fire extinction–Handbooks, manuals, etc. 2. Fire prevention–Handbooks, manuals, etc.
I. Coleman, Ronny J.
TH9151 .F457 2003
363.37'068–dc21
2003001342

Printed in the United States of America.

1 2 3 4 5 07 06 05 04 03

Table of Contents

Section III
EQUIPMENT

Section IV
OPERATIONS

Section V
FIRE PREVENTION AND LOSS REDUCTION

Section VI
THE FUTURE

About the Editors

John M. Eversole, is Chief of Special Functions (ret.) for the Chicago Fire Department. He has a Bachelor's Degree in Management from Lewis University. Chief Eversole joined the Chicago Fire Department in February of 1969 and has served on engine companies, hook and ladder Companies, and squad companies. He is also a certified Master Instructor through the Office of the Illinois State Fire Marshal. He is an instructor teaching Fire Science programs for the Chicago City Wide Colleges and the University of Illinois.

Chief Eversole has been involved in a number of special programs such as the Deep Tunnel Project and the Hazardous Incident Team. He has coordinated the development of the Confined Space/Collapse Rescue operations and has also been working with the U.S. Department of Defense in developing a civilian emergency response program for terrorism. Chief Eversole is recently retired as the Chief of Special Functions. Under his command, he was responsible for Hazardous Materials, Technical Rescue, Specialty Apparatus, Air Sea Rescue, and the Office of Fire Investigation.

Chief Eversole is the Chairman of the Hazardous Material Committee of the IAFC and is also the Chairman of the Hazardous Materials Professional Competency Standards Committee of the NFPA. He serves as a member of the InterAgency Board for Equipment Standardization and InterOperability. He has also served on several technical committees for IFSTA. He served on the Firefighter Safety Act Panel of the U.S. Fire Administration, which resulted from the tragic Kansas City explosion. He also served as a member of *America Burning Recommissioned.*

Robert C. Barr is president of Firescope, Inc., a fire protection consultant organization specializing in fire station location and service delivery studies. He has been associated with the fire service for more than 35 years—which includes service as a volunteer and on-call firefighter, as head of a two-year fire and safety engineering technology program, as Chief Fire Marshal of Prince William County Virginia, and 15 years in the Public Fire Protection Division of the National Fire Protection Association.

Robert currently serves on the NFPA Technical Committee on Fire and Emergency Service Organization and Deployment—Career.

About the Authors

Jack A. Bennett. Retired Chief, Meno Park, California Fire Protection District. His career began in 1955 with the Los Angeles City Fire Department where he was promoted to the rank of Assistant Chief. For the past 22 years he has been a fire service instructor and writer. He is a member of the IAFC, the NFPA Fire Service and Wildland Fire Management Sections, the Fire Districts Association of California and the California State Firefighters Association.

Nick Brunacini, became a member of the Phoenix Fire Department in 1980. He was promoted through the ranks to the level of Shift Commander, the position he currently holds.

Nick has lectured internationally on fire department tactics, strategy, and firefighting operations. He is currently the production manager for the second edition of the *Fire Command* curriculum package—a project he is working on with his father Alan and brother John.

Nick lives in Phoenix, Arizona, with his wife Michele and their three daughters.

Ronny J. Coleman, is currently the president of the Fire and Emergency Services Network (FETN), and is serving as the interim fire chief for Fremont, California. Coleman previously served as fire chief for Fullerton, California, from 1985 to 1992 and fire chief for San Clemente, California, from 1973 to 1985.

Coleman has a Master of Arts Degree in Vocational Education from Cal State Long Beach, a Bachelor of Science Degree in Political Science from Cal State Fullerton, and an Associate of Arts Degree in Fire Science from Rancho Santiago College. He is a Fellow of the Institution of Fire Engineers, UK Branch. He also is a Certified Fire Chief, CFSTES; Certified Chief Officer, CFSTES; Certified Fire Officer, CFSTES; and Chief Fire Officer Designate, CFAI. Coleman also possesses a Lifetime Standardized Teaching Credential from the California Community Colleges System and is a principal with Citygate Associates, LLC in Folsom, California.

Coleman has served in many elected professional positions, including Vice President of the International Committee for Prevention and Control of Fire (CTIF); President of the International Association of Fire Chiefs; President of the California League of Cities, Fire Chiefs Department; and President of the Orange County Fire Chief's Association.

Larry Collins has been a member of the County of Los Angeles Fire Department (LACoFD) for more than 23 years. Collins is a captain, US&R Specialist, and paramedic who is assigned to US&R Company 103, a unit he helped open in 2000. He is responsible for planning, training, supervising, and performing technical rescues, helicopter high-rise operations, confined space rescue, structure collapses, helicopter swiftwater rescues, marine disaster operations, swiftwater and flood rescues, as well as other rescue operations in addition to responding to multi-alarm fires across Los Angeles County.

For a decade prior to his assignment at USAR Company 103, he was a captain of US&R Company 1, the Los Angeles County Fire Department's central Urban Search and Rescue Company. The recipient of half a dozen medals of valor for his actions while performing various rescue operations, he assisted in development of the swiftwater/flood rescue and US&R systems used in Los Angeles County, and he has served as a USAR and swiftwater rescue instructor and consultant since the early 1980s.

He is a Search Team Manager for the Los Angeles County Fire Department's FEMA Urban Search and Rescue (US&R) Task Force; and he serves as an Urban Search and Rescue Specialist on the FEMA US&R Incident Support Team. He sits on a number of local, state, and national committees dealing with diverse issues such as earthquake search and rescue, technical and canine search, terrorism response, rapid intervention, tsunami planning and response, and the use of helicopters to conduct swiftwater rescue from high-speed flood control channels and natural rivers.

Dennis Compton is a well-known speaker and the author of the *When in Doubt, Lead!* series of books, as well as many other articles and publications. His background includes a significant management, consulting, and teaching history covering a wide variety of disciplines and subjects on the public and private sectors.

Compton is currently the Fire Chief in Mesa, Arizona. He previously served as Assistant Fire Chief in the Phoenix, Arizona, Fire Department. During a career that spans more than 31 years, Chief Compton has been an active participant in the international fire service. As a result, many fire departments and other organizations have recognized his accomplishments. He is the immediate past-chair of the Executive Board of the International Fire Service Training Association (IFSTA), chair of the Congressional Fire Services Institute's National Advisory Committee, and serves on the board of the NFPA. Dennis was selected as the American Fire Sprinkler Association's Fire Service Person of the Year in 2000 and is a charter member of the Arizona Fire Service Hall of Fame. He was also selected as the Year 2001 Distinguished Alumnus of the Year by the University of Phoenix.

Jack L. Cottet graduated from Oklahoma State University with degrees in both Fire Protection and Trade and Industrial Education. He worked several years for Kemper Insurance Company as a fire protection consultant and for the Insurance Services Office completing municipal fire protection surveys. He is currently employed by Utica National Insurance as the home office Fire Protection Specialist.

He has served as a chief officer in the Cleveland Volunteer Fire Company for the past 25 years and is a New York State Fire Instructor. Mr. Cottet serves on the Board of Directors of the Certified Fire Protection Specialists Board and is a member of the NFPA technical committee on Fire Service Training. He specializes in rural water supply and conducts training programs and seminars on that subject. His rural water supply column is published monthly by *Fire Rescue* Magazine.

Thomas K. Freeman, is the fire chief/administrator of the Lisle–Woodridge Fire District in Lisle, Illinois. The district covers approximately 30 square miles, and provides fire, EMS, and related services to more than 70,000 people. The department staffs five stations with 115 fire suppression and rescue personnel. The fire district is rated ISO Class 1.

Chief Freeman has been in the Fire Service for more than 25 years. In addition to numerous state certifications, he holds an Associate's Degree in Fire Science Technology, a Baccalaureate Degree in Fire Service Management, a Master's Degree in Public Administration and Fiscal Management, and is pursuing Doctoral studies in Political Science.

Chief Freeman is a member of various fire service organizations, including the IAFC and the Illinois Fire Chiefs Association, and serves as the Vice President of the Board of Directors of the IFCA Educational and Research Foundation. Additionally, Freeman teaches throughout Illinois in the areas of Strategy and Tactics, IMS, Officer Development, and Leadership.

Barry Furey, is the executive director of the Knox County Emergency Communications District in Knoxville, Tennessee. Barry began his career in public safety in 1970, and has served as chief of the Valley Cottage, New York, Fire Department; deputy chief of the Harvest, Alabama, Fire Department, and Training Officer for the Village of Savoy, Illinois. Prior to accepting the position in Knoxville, he managed county-wide communications centers in two states, and provided consulting services to public safety agencies internationally.

He has published more than 100 articles in public safety journals, and has been involved in a variety of committee and task force assignments for the Association of Public-Safety Communications Officials (APCO). In 2000, he received the APCO Presidential Award for his work in enhanced wireless 9-1-1.

James Goodbread's professional career has encompassed more than 34 years in the United States Air Force Fire Protection—16 of which were in the position of Fire Chief at one of the largest and most complex facilities in the Air Force: Tinker Air Force Base. Goodbread managed a major Air Force fire department responsible for structural, aircraft, and EMS response. He has extensive experience with heavy industrial processes, large frame aircraft firefighting and safety, aircraft depot maintenance operations, airport emergency response, and firefighter safety and health.

Goodbread also has 12 years of experience representing the Air Force on the NFPA 1500 Firefighter Safety and Health committee, NFPA Fire Ground Operations Task Group, Chairman of the Air Force Uniform Committee, and the Air Force Breathing Apparatus Committee. He is a member of the NFPA, Oklahoma Metro Fire Chiefs, and the IAFC. He is also specialized in DOD emergency response requirements.

John Granito is a retired professor and vice president for Public Service and External Affairs of the State University of New York. He joined his first fire department in 1949 and spent 21 years as an active firefighter and officer. He was supervisor of fire training for New York State, Director of the International Fire Administration Institute for the IAFC, and an elected fire commissioner. He continues to lecture and write on fire service administration and operation, and conducts fire department studies throughout the country. He is the coordinator of the NFPA's Urban Fire Forum.

John R. Hawkins is an assistant chief with the California Department of Forestry and Fire Protection (CDF Fire) and Butte County Fire Rescue in Paradise, California. He started with CDF Fire in 1964 as a seasonal fire fighter and since has held most line positions. Currently, Chief Hawkins supervises (and is also the technical rescue coordinator for) Division 3 including three Type 1 fire crews and a training crew, Battalion 1 with five fire stations, and the Training and Safety Bureau. Chief Hawkins has an Associate of the Arts degree in Fire Science and a Bachelor of Science degree, is an EMT and rescue technician and a certified California Chief Fire Officer, and is a graduate of the NFA Executive Fire Officer Program. Chief Hawkins has more than 37 years of experience in the fire service. He previously served as a Type 1 Incident Commander and Operations Section Chief on national and state teams. Chief Hawkins instructs ICS, command, fire, EMS, and technical rescue courses and received the prestigious Ed Bent Instructor of the Year Award at the 1998 Fire Rescue West Conference.

Judy Janing, Ph.D., R.N., EMT-P is an experienced researcher and analyst, and serves as an Emergency Program Specialist with IOCAD Emergency Services Group. Janing has more than 25 years' experience in emergency response and more than 10 years' experience in education. She has served as the lead educational developer for the development or revision of several courses at the National Fire Academy and various universities. She was also involved with the Omaha, Nebraska, fire department's EMS program for many years. Her Ph.D. is in Community and Human Resources, specializing in evaluation methodology; her Master's Degree is in Adult Education. She has published several clinical, educational, and research articles in various peer journals, and has co-authored several texts. She is an adjunct faculty member at the National Fire Academy.

Jon C. Jones, a fire protection consultant from Lunenbrug, Massachusetts, holds Bachelor's and Master's degrees in science education and is a member of the SFPE, NFPA, and IAAI. Jones has been a volunteer firefighter on the Lunenburg Fire Department for 28 years, has served on the staff of the NFPA, and has served as Fire Marshal at the University of Massachusetts Medical Center.

Benjamin F. Lopes III is currently the fire chief of the Santa Clara County Fire Department in California. He began his career in the fire service in 1972 and holds a master's degree in Human Resources and Organization Development from the University of San Francisco. Lopes has been with the fire service for more than 30 years, has been a speaker at various seminars, and has served as adjunct faculty at both the National Fire Academy and the California Fire Academy. He currently is a member of the Board of Directors for FIRESCOPE.

Richard A. Marinucci has been the chief of the Farmington Hills Fire Department since 1984. The department is very active in EMS, fire prevention, public fire safety education, training, emergency management, and hazardous materials. Chief Marinucci has served as the President of the International Association of Fire Chiefs. He is the current Chair of the Commission on Chief Fire Officer Designation. In 1999, he served as Senior Advisor to Director James Lee Witt of FEMA and Acting Chief Operating Officer of the United States Fire Administration for seven months as part of a loan program between the City of Farmington Hills and FEMA. He received the Outstanding Public Service Award from the Director for his efforts. Chief Marinucci was also selected to be a panel member of the *Recommissioned America Burning* committee.

Chief Marinucci has three Bachelor of Science degrees: Secondary Education from Western Michigan University, Fire Science from Madonna College, and Fire Administration from the University of Cincinnati. He was the first graduate of the Open Learning Fire Service Program at the University of Cincinnati (summa cum laude) and was named a Distinguished Alumnus in 1995. He has since co-authored a revision of two of the open learning courses. He has also attended courses at the National Fire Academy.

He is a member of the Oakland County Fire Chiefs, Southeastern Michigan Fire Chiefs (President 1989–90), Michigan Fire Chiefs, and National Fire Protection Association. He has served as President of the Great Lakes Division of the IAFC in 1992/93. He was co-editor of the *Michigan Fire Service News* for 1989-1996.

Chief Marinucci has instructed a wide variety of training programs from basic fire fighter to chief officer training. He has been an adjunct faculty member of Madonna University, Eastern Michigan University, and Oakland Community College. He has presented programs for the International Association of Fire Chiefs, California Fire Instructors Workshop, Fire Department Instructors Conference, Fire Chiefs Association of Japan, and numerous regional, state, and local conferences and workshops across the country.

James L. McFadden, retired, is a 39-year career veteran fire chief with the California Department of Forestry and Fire Protection/San Luis Obispo County Fire Department. He has served as an adjunct instructor and course designer for the NFA, several community colleges, and the California Chief Officer curriculum. He has authored and co-authored several fire service magazine articles and a book on I-Zone and Firefighter Safety. He directed the development for most Incident Command courses and has traveled from Alaska to Australia to introduce ICS. He continues to teach and deliver presentations while consulting on a wide variety of topics.

Gregory G. Noll is a senior partner with Hildebrand and Noll Associates, Inc.; the assistant chief of Lancaster County, Pennsylvania's, Hazmat Response Team; a member of PA Task Force-1 USAR; and a member of the US Air Force Reserve Fire Protection. He has also served as the Hazardous Materials Coordinator for the Prince Georges County (MD) Fire Department. Noll has more than 32 years of experience in fire service and emergency response community, and is the co-author of seven textbooks on hazardous materials emergency response topics. He has served as subject matter expert for various NFA and DOJ projects, and is a member of the IAFC Hazardous Materials Committee, the NFPA Technical Committee on Hazardous Materials Response Personnel (NFPA 471 and 472), and the editorial advisory board for *Fire Engineering* magazine.

William C. Peters. is a 28-year veteran of the Jersey City (NJ) Fire Department and has served the past 16 years as apparatus supervisor, with responsibility for purchasing and maintaining the apparatus fleet. He is a voting member of the NFPA 1901 Apparatus Committee, representing apparatus users. Peters is the author of the *Fire Apparatus Purchasing Handbook* (Fire Engineering, 1994); two chapters on apparatus in *The Fire Chief's Handbook, 5th edition* (Fire Engineering, 1995); the instructional video *Factory Inspections of New Fire Apparatus* (Fire Engineering, 1998); and numerous apparatus-related articles. He is an advisory board member for *Fire Engineering* magazine and the Fire Department Instructors Conference (FDIC), and lectures extensively on apparatus purchasing and safety issues.

Gary Pope is a Captain with the Fairfax County Fire and Rescue Departmentin Fairfax, Virginia. Gary has been active for 27 years with the career department in Fairfax County, Virginia—19 years were spent in the Apparatus Section where he was responsible for writing apparatus specifications and managing the fire department fleet of more than 450 vehicles. Captain Pope has more than 35 years in both the career and volunteer fire service and is currently a voting member of the NFPA 1901 committee for the Standard for Automotive Fire Apparatus.

Captain Pope serves as chairman of the task group responsible for the development and maintenance of the NFPA 1915 Standard for Fire Apparatus Preventative Maintenance Program and a member of the Safety task group. Has presented several programs at the Fire Department Safety Officer's Apparatus Symposium along with programs for the Virginia Fire Chief's Association and other fire apparatus management seminars. He also served on the task group assigned to develop and implement safety inspection program for fire and rescue vehicles in the Commonwealth of Virginia. He is also certified as a Cardiac Technician in the Lord Fairfax EMS Council of Northern Virginia and serves as a volunteer with the John H. Enders Fire and Rescue Company in Berryville, Virginia.

Gary is married to a wonderful person named Pat and has two outstanding step children named Chelsea and Ammie.

Daniel Redstone, FAIA, NCARB, is president of Redstone Architects, Inc., located in Southfield, Michigan. Daniel has more than 35 years of experience in the field of architecture. He has served as Principal-in-Charge on a variety of projects including public safety facilities, government buildings, computer centers, office buildings, higher education facilities, and adaptive reuse. He is instrumental in defining project issues with his clients, and acts as a catalyst in the creation of innovative, award-winning, and cost-effective solutions through the proper use of programming methods, planning, concept analysis, project budgeting, and project design. Daniel was a member of the Michigan State Board of Architects for 10 years, and serves on the Board of Directors of NCARB (The National Council of Architectural Registration Boards). He is a member of numerous public safety associations in the state of Michigan.

Gordon M. Sachs, MPA, EFO, is chief of the Fairfield Fire & EMS and director of IOCAD Emergency Services Group in Pennsylvania. He is a former program manager with the Federal Emergency Management Agency's U.S. Fire Administration. Chief Sachs has more than 25 years of emergency response experience, particularly in the area of incident command, training, and safety. He is an adjunct faculty member in the EMS Management/Leadership, Command and Control, and Emergency Response to Terrorism curricula at the National Fire Academy. He has written many journal articles on EMS management/leadership and responder safety/health, and has authored or co-authored several books in these areas. He has a Master's Degree in Public Administration, a Bachelor of Science in Education, three Associate in Applied Science Degrees in Fire Science disciplines, and an Executive Fire Officer Certificate from the National Fire Academy.

Kathy Saunders began her fire service career with the Lawrence, Kansas, Fire Department. In her six and a half years with the LFD, she attained the uniformed rank of Major in charge of the administrative division, and was responsible for developing the annual budget, monitoring statutes, regulations and standards for compliance, acting as the department PIO, representing the department in grievance procedures as well as on the city management team, contract negotiations team, and the wellness committee. Saunders also received an Associate's degree in Fire Science, state certification as an emergency medical technician, and certifications for Firefighter I and Instructor I.

In 1997, Saunders took the position of Fire Chief of the Bloomington, Indiana, Fire Department and also served as president of the Joint Central Dispatch Policy Board during integration of the various county dispatching systems, purchased land and initiated the design of a new fire station, and assisted in reaching an agreement on a four-year collective bargaining contract.

Saunders currently serves as the Fire Marshal for Knox County, Tennessee. She leads the Knox County Fire Prevention Bureau which conducts all fire and arson investigations, public education activities, fire and life safety code inspections, and plan reviews in the unincorporated areas of the county.

Kathy Saunders is married to David Reidy, J.D., Ph.D. They reside in Knoxville, Tennessee with their daughter, Kiyoko Cecelia, and son, Kame Benjamin.

Richard A. Schnuer serves as the finance director for the city of Champaign, Illinois, where he is responsible for all aspects of financial planning, operations, and reporting. He also serves as treasurer of both the fire and police pension systems. Mr. Schnuer previously held financial management positions in other cities, the federal government, and the state of Illinois.

Mr. Schnuer is a frequent speaker to state, national, and international audiences on a variety of public finance topics, and has authored articles for professional journals. He has served as a member of several state advisory boards on topics including public employee pension systems, investment management, and revolving loan funds. He is a past President of the Illinois Government Finance Officers Association.

Mr. Schnuer holds a Masters of Business Administration degree from the University of Chicago, where he concentrated in the management of public and not-for-profit organizations. He also holds a Bachelor of Arts degree from Tufts University in Medford, Massachusetts.

Philip Stittleburg has served as the fire chief of the La Farge Fire Department in Wisconsin, since 1977 and as an assistant district attorney for Vernon County, Wisconsin, since 1974. He is also currently serving as chairman of the National Volunteer Fire Council and on the board of directors of the NFPA. He is a member of the NFPA technical committees on occupational safety and occupational medical and health.

Russell J. Strickland is a faculty member at the Maryland Fire and Rescue Institute, University of Maryland at College Park. During his more than 21 years with the Institute, he served in a number of supervisory positions, and is currently the assistant director for the Field Programs Division, responsible for development and delivery of the Institute's field programs. Active in Maryland fire, rescue, and EMS for more than 27 years, he has served with a number of fire departments and is also involved with the Maryland Fire Service Personnel Qualifications Board, the Chesapeake Society of Fire and Rescue Instructors, and the Maryland Council of the Fire and Rescue Academies.

Dr. Thomas B. Sturtevant is a program manager for the Emergency Services Training Institute (ESTI) within the Texas Engineering Extension Service, a member of the Texas A&M University System. He is responsible for the Emergency Management Administration's online Bachelor degree program, Department of Defense emergency services programs, curriculum development, and manages ESTI's accreditation with National Professional Qualification System. He was a tenured assistant professor at Chattanooga State Technical Community College, TN, where he held the positions of Dean of Distance Education and Coordinator of Fire Science Technology.

Dr. Sturtevant was a fire protection specialist with the Tennessee Valley Authority and held various firefighting positions with the San Onofre Nuclear Generating Station, CA, and the United States Air force. He is a Certified Fire Protection Specialist and has a Doctorate in Education and a Masters in Public Administration from the University of Tennessee. His research and consulting efforts focus on program evaluation and emergency service professional development.

Robert Tutterow has 25 years' fire service experience and has been the Health and Safety Officer for the Charlotte Fire Department for more than 15 years. He is an advocate for firefighter safety and is active in the NFPA standards making process. He is currently a member of the NFPA Technical Committee on Structural Fire Fighting Protective Clothing and Equipment and its oversight Correlating Committee on Fire and Emergency Services Protective Clothing and Equipment.

He chaired an ad-hoc committee that developed the *PPE Care and Use Guidelines* that was an interim document until an NFPA standard on care of PPE could be developed. In addition, Robert is the chair of the Safety Task Group on the Technical Committee for Fire Department Apparatus. He is also co-founder and officer with FIERO (Fire Industry Equipment Research Organization).

Foreword

I was very honored to be asked to write the foreword for the sixth edition of *The Fire Chief's Handbook.* I remember when I joined the fire service in 1975, my father handed me several boxes of fire textbooks that he had amassed over his career. His only comment at that moment was "this ought to keep you busy for a while." On the top was Casey's *The Fire Chief's Handbook.* Since that day, many things have changed and some remain the same. Several editions of this great book have been used as study material on promotional exams that I have taken throughout my carrier. The last edition (the fifth edition) was used on our most recent battalion chiefs exam. Along with red fire engines, Dalmatians, and the use of bagpipers, *The Fire Chief's Handbook* maintains its place along with other great traditions in the fire service.

I'm not sure if there ever was a more uncertain time in the history of the fire service than that which faces us today. This great profession will never be the same following the events of September 11th, 2001. We will continue to fight fires. Most of us will respond to calls for ill persons, heart attacks, and the occasional shooting or stabbing—but it will never be the same. Incidents such as the Oklahoma City and the Atlanta bombings have taught us to look at the way we respond differently. "Secondary devices" should be a primary—as opposed to secondary—thought at many incidents we respond to. September 11 goes beyond secondary devices; its lasting and ultimate effect, other than the immediate loss of loved ones, may not be realized for years to come.

There are many changes on the horizon for the fire service. From changes in the way we respond to fires and EMS incidents to the very makeup of departments themselves. I believe that regionalism through joint-response pacts, mutual aid, and automatic aid will be an ever-increasing way of life from now on. With this regionalism must come standardization—not only in tools and equipment—but also with procedures and protocols. Who knows, someday we may all have the same decal on the side of our apparatus. But for now, departments must begin to plan for responses to more challenging and possibly higher-magnitude incidents than the occasional second- and third-alarm fire that comes along every month or so.

There is an old adage that states "you have to learn to crawl before you can learn to walk." In the fire service, this is especially true. It is imperative that young firefighters learn the fundamental evolutions such as donning SCBA, and pulling and stretching attack lines before they learn how to run an incident utilizing the Incident Command System or set up a civilian mass decontamination corridor at a terrorism drill. This book is not a "crawl" book for the fire service. It is expected that the readers of this book are well-versed at the basic evolutions that fire fighters, company officers, and chief officers need to deliver basic services in their community. This book goes well beyond the basics. It does, however, contain fundamentals that are essential in the delivery of structural and wildland fire protection as well as Emergency Medical Service (EMS) delivery systems for both large and small departments. It is designed for fire fighters, company officers, and chief officers of all ranks and of all department types who want the latest information on the fundamentals of leadership in the fire service as well as managing the day-to-day operations of a fire department.

In reviewing the chapters for this book, I have found that that it continues on with the tradition that *Fire Engineering* has established with previous editions of this great text. Some of the previous authors have returned to update their sections of this book. Some of the previous authors who helped lay the groundwork in past editions such as Chief William E. Clark have passed away, while other previous editors and authors such as Tom Brennan, Bruce Varner, Gene Carlson, and others' busy schedules could not permit participation in this edition. However, their foundations are evident throughout this new edition. That is one of the shining aspects of our profession—we learn, develop, and then pass our knowledge and traditions on to the next generation. That is what my father did, it is what past editors and authors have done, and it is what I and the authors of this edition are attempting to do now. It is also what you are tasked with in the future; to learn, develop our profession, and then to pass your knowledge and traditions on to the next generation of firefighters.

Deputy Chief John F. Coleman
Fire Prevention/Communications
Toledo Department of Fire and Rescue
2003

Section I

MANAGEMENT

1

FIRE DEPARTMENT MANAGEMENT

Ronny J. Coleman

CHAPTER HIGHLIGHTS

- Management in the fire service is the process used to provide an orderly structure to all of the events in the life cycle of an organization while leadership is the personal skill and ability to get others to follow a direction that the leader has set out to follow.
- Basic management activities include planning, structuring, direction, conducting programs, coordinating, and evaluating.
- Structuring, which includes chain of command, unity of command, span of control, division of labor, and the use of the exception principle is a way of organizing the resources of an organization so that its work plan makes sense, work is supervised at various levels, and things get done.
- The exception principle is a concept of delegating authority so that decisions will be handled as quickly as possible from the lower levels of the organization while retaining the ability for those at lower levels to ask for help on the "exceptions" from superior officers.

INTRODUCTION

The word management is often used as a synonym for other words in fire department jargon. It is not uncommon to find the word linked with the task of supervising subordinates. Sometimes it is used to describe planning and financial activities of individuals. In other cases, the term management is used to describe leadership activities. To more fully understand what management is, fire officers must draw a distinction between it and other processes.

However, entire textbooks have been devoted to the definitions of these terms. It is not appropriate to try and duplicate the richness of those texts in only one chapter of a book that is designed to give a fire officer an overview. This chapter is designed to discuss the practical aspects of how management is expected to operate within the context of a fire organization.

Management in the fire service includes those processes used to provide an orderly structure to all of the events in the life cycle of an organization. It is a skill that can be acquired and an ability that people can learn to become better at over the years. Leadership in the fire service is the personal skill and ability to get others to follow a direction that the leader has set out to follow. It too is something that can be learned

One of the catch phrases used to describe this phenomenon is that leadership is getting the right things done and management is doing things the right way. For purposes of definition in this chapter, we are going to define management as a process and leadership as a skill. In fact, they both contain elements of each other, when viewed in the context of the leader/manager role that a contemporary fire officer often faces.

Management, when used in the context of the fire service, is a process of structuring the activities of an organization in such a way that it achieves efficiency and effectiveness in the use of human and physical resources to protect life and property.

We must also draw another distinction. Management, as used in the context of the private sector, is different from the public sector. There have been people who tried to state that fire departments should be run more like a business, but the fact is that it cannot be done in the same fashion as business.

There are many ideas from contemporary management theory offered by persons such as Drucker, Demming, Naisbitt, Covey, and others that lead fire departments to become more effective and efficient. However, there is one key distinction between fire department management and private sector management environments. A fire department is a monopoly. It does not have the ability to alter the price of its services or enter into a competitive field in which users have options to accept or reject its services in lieu of those from a competitive firm.

Having drawn that distinction, we must also add one other dimension to management in the public sector. When we are given resources by a community in order to serve a broad interest of ensuring fire safety and life safety in the community, management implies a tremendous amount of accountability and responsibility. A fire officer is a "fiduciary" or someone that has been given responsibility to serve in the interest of others. Those fire officers who do not employ management processes could be accused of wasting the community resources. If the fire service does not produce a quality service, the consequences will be felt in losses to the community. It then follows that managing public sector resources means we can use a lot of the same techniques used to manage private sector resources, but our work is performed in an environment where good management results in cost containment and an increased level of service rather than a profit.

Since this chapter was first written, the fire service has experienced a series of events that have given it monumental challenges in the field of both leadership and management. What is noteworthy is the fact that the fire service rose to the occasion of dealing with these challenges by using the same fundamentally sound processes that they were trained to use. In spite of the existence of periodic "trends" or "fads" that deviate from basic theory, the fire service tends to stay on course by its practical use of the same principles that have been used for years.

What has changed is the toolbox that today's managers need to do their jobs. In fact, the role of management information systems is much more important today than it was 10 years ago. New tools like geographic information systems (GIS) and improved communications technology is improving the ability of fire officers to accomplish their goals and objectives.

In emphasizing basics in this chapter, we are not implying simplicity. On the contrary, the more complex our jobs become, the more we need to develop tools to help the leader/manager in assessing the condition of the organization. In this way, he or she can gather and assemble information with which to make better decisions while dealing with conflicting priorities and demands.

The Basic Functions of Management

There are many textbooks to describe the recognized function activities of management. For many years there was a popular acronym—POSDCORB. This acronym stood for planning, organizing, staffing, directing, coordinating, overseeing, recruiting, and budgeting. Contemporary management literature offers many such acronyms to try to remind people of the various functions of management. The fire service does not need to invent its own management theories to support our activities. Instead, we need to understand the basic principles of management theory and then skillfully apply them to the operation of fire service agencies.

To expedite the utilization of basic management techniques for the fire service, we are going to outline an even simpler series of activities. They are: planning, structuring, direction, conducting programs, coordinating, and evaluating.

Planning

The process of planning involves looking forward to the future. The planning process used in the fire service has essentially three cycles. The first of these is a year-to-year plan. This is often reflected in the form of a yearly budget. The second level of planning entails three to five years and is commonly classified as a capital improvement or capital outlay plan. Thirdly, an even longer-range plan is called either a master plan or strategic plan.

The planning process involves identifying what needs to be done in order to achieve the overall mission of the organization. Planning has two distinct levels. The first of these are formal planning processes, which are reflected in documents that are published and adopted. In the second level are the informal planning processes used on a daily, weekly, or monthly basis in order to predict or at least anticipate events as they begin to emerge from activities.

Fire chiefs must recognize that "to plan" is a verb, meaning that action must be taken to achieve information to give a direction to the organization. A "plan" is a noun, which implies that when you take action to develop a future scenario, it needs to be documented and placed into some context in order for it to make sense. In its most simplistic form, a plan is nothing more than a vision of the future. No plan becomes a reality by itself. If a plan is not adequately followed by the individuals who conduct management processes, the plan will never become reality. In short, a plan is only a roadmap and planning is the process used to make the trip. General Dwight D. Eisenhower once stated that "Plans are useless, but planning is essential." What he was implying is that the process is more important than the product.

Structure

Structuring is essentially organizing the resources of an organization so that the work plan makes sense, work is supervised at various levels, and things get done. There are some basic definitions used in structuring an organization—chain of command, unity of command, span of control, division of labor, and the use of the exception principle. Chain of command is the hierarchical relationship between levels of authority within an organization as well as the relationship between supervisors and subordinates. Span of control is the number of people that one supervisor can effectively handle under specific circumstances. Unity of command is the definition of who works for whom, for a person should only be accountable to one boss in a system. Division of labor is the system that divides up tasks according to similarity in tasks or logic. Finally, the exception principle is a concept of delegating authority so that decisions will be handled as quickly as possible from the lower levels of the organization while retaining the ability for those at lower levels to ask for help on the "exceptions" from superior officers. Problems or exceptions can be passed upwards until they arrive on the desk of someone who has the authority to solve them. Managers must use these techniques so that they are not trying to manage everything by themselves.

The fire service has utilized the aforementioned principles extensively because they lend themselves to the creation of a "scalar" organization or an organization that is paramilitary in its structure. This is the primary reason why fire department organizations tend to look like pyramids. While this organizational pyramid concept has been useful for many years, it has not been effective in dealing with the increased complexity of fire service operations in the last few decades. So, while the management process implies that an organization must have some degree of structure if its plans are ever to be achieved, the hierarchical relationship and/or use of other relationships is equally appropriate.

Contemporary management theory includes the use of a lot of techniques that are not part of this historically scalar organizational structure. These include the use of "task force" or matrix management techniques. These concepts employ methods whereby individuals work across conventional lines of authority. There is no legitimate reason why fire service organizations cannot utilize processes that preserve traditional hierarchical relationships for day-to-day operations, but also use non-traditional "teaming" concepts to pursue organizational goals and objectives. The real test of managerial skill is the ability to accomplish the task of managing the completion of complicated tasks while at the same time maintaining the fire services desire for organizational discipline and need for expediency under emergency conditions.

However, the fire chief has a responsibility to structure an organization to achieve its plan and not necessarily conform to any artificial tradition that is counterproductive to the plan. Structure implies a table of organization. That is one of the reasons for creating a table of organization, to show that relationships can be outlined, job descriptions can be defined, and roles and relationships can be laid out in advance of activity.

The management responsibility to provide structure to an organization also implies that if plans change, the organization's structure may need to change. There is a dynamic relationship between the vision of the future and the day-to-day activities an organization engages in achieving that vision.

Establishing direction

This management process that links the plan and the various function of the organizational structure is called direction. It is the process wherein the organization defines its goals and objectives and starts marching in a specific direction to achieve them. It is also linked to the concept of the mission of the organization.

The best comparison of this concept is to relate it to a trip that a person might wish to take. A person must have a destination to plan the trip. If the location is just next door they don't spend much time thinking about the trip. However, if the trip is miles away, across the country somewhere, they must have a way of getting there. They can choose to drive, fly, or walk. Each will take them in a slightly different way, but seldom do any of the methods go directly to the destination. There are zigs and zags to the trip. What is important is the fact that you have a compass bearing for the destination.

One of the best compass bearings or way of establishing direction in an organization is the mission statement. The mission statement is the basic reason for the organization's activities. In order for the mission to be accomplished the organization must accomplish goals that take the organization in a specific direction. The mission statement of an organization coupled with goals and objectives is like a flight plan for an airplane. You might have to deviate for bad weather, but you always end up at the right airport—after a period of time.

Establishing direction involves the process of training and supervising people to create policies, practices, procedures, and even priorities to take the organization in a specific direction. It cannot be inferred that because an organization has a plan and it has organized its resources that it will achieve anything. This particular management function involves making incremental decisions on a day-to-day basis. The management function we see here is one of the proverbial crossing of t's and dotting of i's. Establishing a direction for an organization is not a one-time effort. It is a constant, ongoing struggle on a daily basis. Not unlike an automobile that requires you to constantly monitor the steering wheel as you move down the highway, an organization must be steered.

This management function uses a host of specific techniques to make sure that direction is there. These may include, but are not limited to, creating and modifying such documents as goals and objectives, priorities, work performance plans, and even preparing training and education plans for the personnel in the an organization.

If the organization is not provided with periodic adjustments it can find itself in difficulty. Not unlike the automobile analogy used before, an organization can slow down when it needs to speed up. The manager's job is to provide the necessary changes in direction or speed based on both internal and external influences.

Coordinating

Coordination activities are maintained as day-to-day working relationships of both the inner workings of the agency and the relationship with organizations who interface with the fire department. They may include, but are not limited to, interacting with water departments, building departments, police departments, dispatch communication systems, other municipal departments, neighboring fire departments, county government, state government, federal government, community organizations, etc. Typical management coordination activities consist of navigating both the organization's internal activities and monitoring the status of the organization's external relationships.

This particular management activity cannot be overestimated when there is a possibility of contradiction or conflict between one organization's activities and another. For example, a fire agency must manage its relationship with both water and building departments to assure that the fire agency has an effective code enforcement program. Failure to do this can result in conflict that can compromise the ability of a fire agency to manage the community's fire problem. Similarly, failure to maintain an effective relationship with law enforcement can result in reduced effectiveness in dealing with emergency problems in the community

Liaison also infers cooperation and communications. A key factor in maintaining balance between organizations is an understanding of the other organization's plans, structures, and directions as they impact the organization being managed. Unfortunately, many individuals do not understand that coordination is a management function, but instead feel that everyone must conform to their wishes. Nothing could be more counterproductive in terms of operation and efficiencies. The coordination function also includes an analysis of redundancy and overlap between an organization and companion or peer groups.

Program activity

The vast majority of the time taken up in the fire service is in managing programs. An organization may have an overall plan, a structure in place, direction, and be coordinating with others, but what is done on a day-to-day basis makes the organization effective and/or efficient as regards the mission statement. Therefore, the management of programs is made up of a number of essential sub-management techniques.

The first among these is the manager's responsibility to define acceptable standards of performance. The second component is to define workload. The third is to measure whether or not the organization has the ability to achieve the workload.

Another management function is to define acceptable standards in the community. Another way of saying this is that a manager's job is to set the minimum standards. What is the community standard for response time? How often should buildings be inspected for fire safety? How many hours should an individual receive training in a given year? Defining minimum standards in the management context is not entirely devoid of the impact of individual value systems. Not uncommonly, people engage in program activity driven by the individual values of the person in charge of a program. Effective managers are individuals who define standards that are appropriate for the community and are not biased by their personal values.

Once the minimum acceptable standard has been established for a particular program, then management function is to define the total workload that this standard imposes on an organization. Let's say, for example, an organization determines that all their occupancies should be inspected once a year. All they have to do is count the number of occupancies in the community for a first cut at the workload. Three or four thousand buildings may need to be inspected. The second cut in the management of this activity is to define how long it would take to conduct an average inspection. This figure multiplied by the number of buildings begins to define the workload for the entire community.

If this type of operation is conducted for each type of program conducted by a department, it can very shortly lead to the definition of the staffing requirements to achieve that workload. It follows that if a community has a certain expectation and there is a certain amount of this work to be accomplished and only so many personnel to accomplish it, one of three things will occur within the context of organizational activity. First, you will achieve everything that the community expects since you probably have an adequate staff to deal with the task. The second is that you may not achieve everything the community expects. Even though you are doing the right things, there is too much to be done since you may have inadequate staffing. Lastly, you may not achieve something that the community expects because the resources are being utilized in another fashion to work in another area. That is a problem of prioritization and not workload.

If there is any one arena that is controversial in the fire service it is the issue of staffing. Yet, staffing emerges from a management task of clearly identifying what needs to be done and how well it needs to be done. As mentioned earlier, if we were in the private sector, the ground rules would be different. For instance, if you have to produce an automobile, certain tasks have to be accomplished before that automobile is finished. You have two choices. One choice is to put a sufficient number of personnel on the assembly line to ensure the automobile's completion in a certain number of hours. The other choice is to use fewer people and take longer to produce the same vehicle. The reason this system of choices found in the private sector does not work in the fire service is that, in the past, much of the workload in the fire service has not been clearly defined.

Evaluation

Evaluation is a management task that focuses on determining whether or not the plan is being executed successfully. All too often this management task is treated superficially because it does not have a great deal of immediate response. However, lack of immediate response does not remove it from the inventory of what makes for an effective management operation. Almost all contemporary management literature implies that evaluation is equally important to the planning process. For instance, quality control gurus often point out about the fact that we should not be inspecting for defects in the end product because if we continue to inspect for defects, we will continue to find defects. Instead, it is far more advisable to focus evaluation efforts on the process. Try to determine what it takes to eliminate defects. The whole concept of total quality management and the impact of programs such as quality circles are dependent upon the use of skills in evaluating processes, not in evaluating failures.

Evaluation is also linked to accountability. The phrase, "What gets measured, gets done," means that we look at the activities of programs and the individuals executing those programs, while evaluating whether goals have been achieved or not. Everyone knows that when there has been measurable success, there is a tendency to continue to achieve. If we don't evaluate programs and activities, it is conceivable that resources will be inappropriately utilized for activities that are or become virtually nonproductive.

Management as a Cycle

Now that we have defined these events as processes, we need to reinforce the fact that they do not occur only once. Dennis Waitley, in his book, *Timing is Everything*, implies that management cycle is probably best manifested by looking at how a farmer operates. A farmer must plan his crop for the following year, plant seeds, cultivate, water; otherwise nourish his crop, and then collect it in order to achieve his reward and determine whether the yield was high enough. Waitley says this same kind of process is an important part of the management of all organizations. A fire agency does not just live for the moment. They exist to provide a service many times throughout the life cycle of a community. So, it follows that fire service agencies should spend a portion of their time on planning for the future; a portion of their time devoted to creating and implementing programs; and an appropriate amount of time devoted to nourishing growth and development of personnel. This is the only way to realize a measure of accomplishment as part of their annual activities.

Since our management lives in the fire service are not limited to one year but often involves decades of activities that inherently occur during the management cycle, it is easy to see that the cycle will repeat itself many times. There are a few dimensions of this cycle that bear some consideration by students of fire service management.

Review and justify

First, it is easy to be seduced by the fact that this year's budget needs to look a lot like last year's budget. However, it is far more importance that, as the cycle renews itself, those in managerial positions take a very close look at the decisions that have been made in the past. Failure to do this can result in an organization building up a layer of insulation between reality and organizational activity. While it is not necessary to "throw out the baby with the bath water," it is important to give the budget cycle a fresh look every year. Programs should be evaluated to see if they are accomplishing what they were supposed to accomplish. A manager should assess the expenditures in the budget cycle as if they were investments, not expenses. The attitude should be "What should you be investing in for the next year in order to accomplish the organizations mission?"

A few years ago there was a popular concept called zero-based budgeting. It was based on the notion that each and every year an agency has to re-justify its programs and activities anew. The concept created a lot anxiety among individuals that were unsure of their true purpose. On the other hand, others used it to elevate their understanding of the organization's mission, goals, and objectives to a higher state of visibility. Zero-based budgeting, in this author's opinion, was just a fad that eventually ran out of supporters. Nonetheless, this exercise forced individuals to make sure they could be confident in the basic underpinnings of their organization's reason for existence. As a result, many came out of this exercise even stronger.

Need for change

The second dimension of this cycle, from a management point of view, is the realization that if the community changes, the organization must change with it. Any time the plan, the structure, the direction, and the program activity remains almost identical for more than five to seven years, an organization may be working its way toward inefficiency and ineffectiveness.

Linkage

The last dimension of the cycle requiring attention is that all of the cycles are moving targets. In other words, five annual budgets make up one five-year plan. The minute you have completed one calendar year and it falls off the end of the cycle, there is often a need to re-evaluate and establish a new five-year plan. A series of five-year plans may make up the 20-year master plan. Yet, when you are five or ten years through that cycle the master plan should be revisited and extended into the future to assure that it remains realistic. The management task used here can best be referred to as linkage; that is to keep a perspective on what you are doing with where you are trying to take the organization.

Conclusion

Managers in fire departments are not people who spend one day planning, one day structuring, and one day doing program activity. Not unlike a gourmet cook, they know all the ingredients and utilize each of these management skills as appropriate in the context of the organization. The most important thing for fire department managers to constantly remind themselves of is the need to keep perspective. As a manager of an organization, you are 100% responsible for its accomplishments and its failures. Therefore, the degree to which you pay attention to taking care of the basics, the more likely the organization will continue to be successful. If an individual begins to focus on leading the organization, yet fails to manage it correctly, the result may be highly visible and the organization may "crash in flames." If the leader of an organization focuses entirely on giving orders and directing others to perform, the organization will lack the backbone to survive once that individual has gone on, either to retirement or for other reasons.

It is important to a draw distinction between your leading an organization and managing it. If you are leading the organization it is going somewhere. If you are managing the organization, when it gets there it will be a smoothly operating machine. One does not necessarily have to fulfill both roles of leader and manager. For instance, some of the most effective leaders are people who recognize that their management skills are lacking and surround themselves with people who are good at accomplishing the management tasks we have just mentioned. Conversely, some of the most successful managers who lack charisma and the visibility to achieve strong leadership roles have empowered people in their organizations to supply the strong leadership needed. A strong manager knows how to use a strong individual in the organization for its benefit. He or she is never threatened by persons who have greater abilities than their own.

Good managers tend to be people who are constantly alert to improving their management skills through education and training. They recognize that there is such a thing as managerial obsolescence, which is different than technological obsolescence. Effective managers endeavor to acquire new techniques to supplement the management skills that they have already practiced.

Author William D. Hitt, in his book, *The Leader-Manager* (Battelle Press, Columbus, 1988) notes, "The essence of leadership is found in a person's ability to move an organization successfully from state A to state B, that is to a higher level of performance...the leader is able to transform vision into significant actions." This suggests that one has to have vision, but must also be able to act upon it to succeed. Hitt's book demonstrates the roles of leader/manager in a view of two axes set at right angles that create four boxes. Within the boxes are four roles that persons might find themselves playing—the victim, the dreamer, the doer, and the leader-manager. The victim is low on both vision and implementation. The dreamer is high on vision and low on implementation, the doer is high on implementation and low in vision. The leader-manager is high on both dimensions.

Management is a function that can and must be learned. It is within the realm of possibility for almost everyone to acquire the skills to become an effective manager. But for someone to become a successful manager, a continual emphasis must be placed on learning how to get things done in the most effective, yet practical way.

2

OFFICE MANAGEMENT AND WORKFLOW

Benjamin F. Lopes III

CHAPTER HIGHLIGHTS

- **Fire department elements, including offices and other support services, interrelate with the local community as part of a "system."**
- **Work assignments are determined by the complexity, standardization, and level of decision-making authority in an organization.**
- **Records management requires a planned and integrated system.**
- **Information technologies are "tools" empowering the organization.**

INTRODUCTION

This chapter explores strategies for office management, including how a plan for staffing, office layout, workflow, records management, and information technology can improve how well a fire department delivers services. The information presented here applies equally to small or large, rural or metropolitan, and career or volunteer fire departments.

Fire department offices come in all shapes and sizes. Offices range from small public reception areas in fire stations to large stand-alone administrative offices located in high-rise buildings. While the office is generally regarded as the administrative place of business it is actually much more. Organizations are more complicated than the physical layout or geographic location of their offices. Organizations are composed of structures, technologies, and processes. A change to any element impacts the entire organization.

Structure refers to the relationships between people. What is the chain-of-command? Where are people assigned and located? Do reporting relationships change with specified tasks? Technology refers to the tools necessary to complete a task. Tools and their technology may range from simple to complex. Office furnishings, computers, and telephones are examples of tools. In addition, new "high tech" tools and technologies become routine and commonplace over time.

Processes are the formal and informal work practices and procedures used within the organization. What, when, and why people complete tasks within an organization is related to how and where they use the tools and technology to do it. All three elements—structures, technologies, and processes—are required to deliver fire department services.

Another way to view these organizational relationships is through a model. Many organizational theorists use the "open systems" model to view organizations and, in this way, the components and their relationships may be studied. An open system is a metaphor referring to an organism transforming inputs to outputs for survival. The change of inputs, such as people, money, or information, into outputs, such as goods and/or services, by organizations takes place in a specific environment. In the case of the fire department, that environment is the local community.

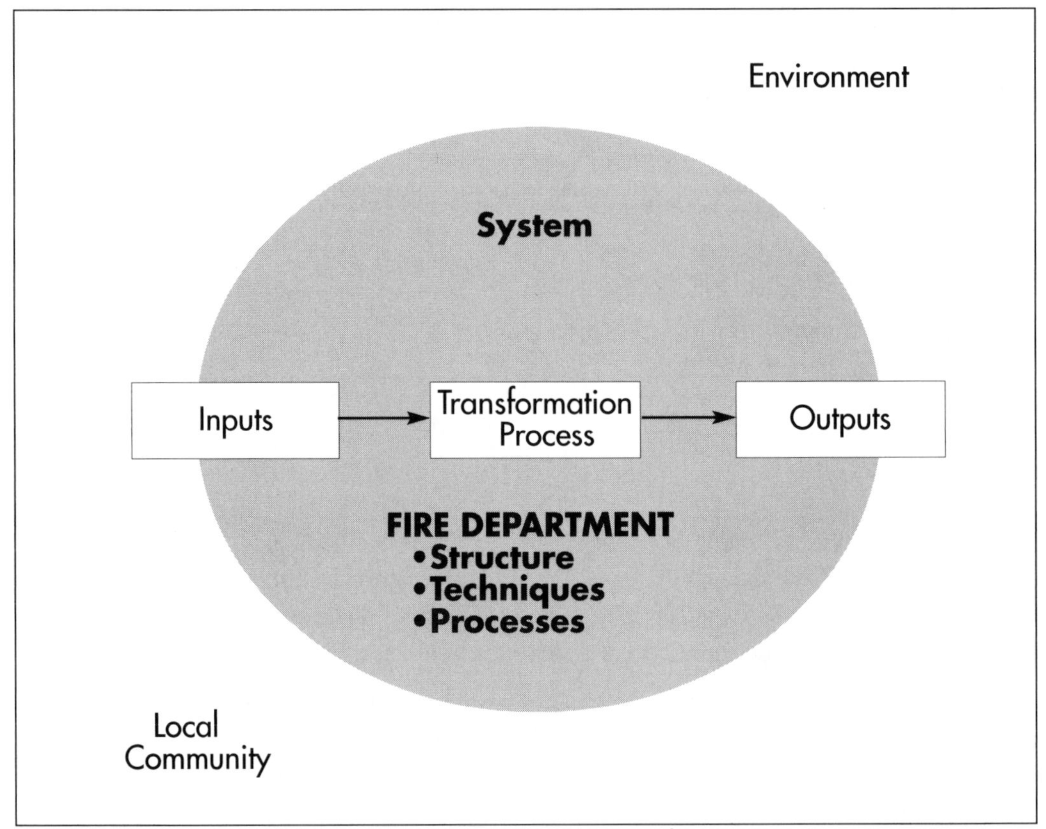

The fire department as a system: Inputs, the organization, outputs, and the environment (the community).

In the private sector, organizations compete against one another for limited resources to transform inputs into outputs that include a profit. They must be efficient and effective to survive. In the public sector, competition for resources is more limited, but the relationship between the organization, inputs, outputs, and the environment remains the same. Fire departments must also be efficient and effective to transform input resources into valued service outputs.

As part of an organization, office functions must contribute to the mission and goals of the organization. Whether supporting various divisions of the organization or providing direct service to customers, the role of the office is key to the success of the fire department.

STAFFING

Local needs and available space will determine whether the fire department offices are located in a fire station, city offices, or stand alone administrative facilities. The office provides non-emergency services to citizens and supports the internal functions of the department. Examples of non-emergency services include fire and life safety plan checks, public education, and incident record keeping. In a support role the office helps to coordinate personnel and facilities through communications and the resource allocation necessary for fire department operations.

The expectation of the modern fire department, like most government agencies, is to do more with less. Budgets, and subsequently the number of personnel, have not grown at a rate equal to local population growth and demand for services. Personnel costs comprise 80% to 90% of career fire department budgets. In volunteer organizations, where personnel costs are minimal, the budget still directly impacts the resources available. While no agency can staff at levels to handle all potential risks or demands, fire departments are measured by the ability to provide service in a timely manner. Without improvements in the tools and techniques utilized, staffing levels would become inadequate to manage the demands for service.

Determining assignments, delegating responsibilities, and follow-up

The staff that provides the services delivered through fire department offices may be civilian or uniformed fire personnel. Emergency responders may cringe at the thought of clerical support duties. In a climate where public agencies seek efficiencies, however, all employees must add value with the time available on duty. On the positive side, many human resources experts believe that job enrichment is achieved through more diverse assignments and autonomy.

Organization development experts predict managers will organize more work around teams of employees. This is due in part to the increasing difficulty for a single person to maintain the diverse skills and knowledge required for emergency work and non-emergency activities. Working on a number of teams also results in employees that are "cross-functional," that is, no one person or group is responsible for data collection, computer entry, and records storage. Instead, one person may learn and be partially responsible for a portion of all these activities. All personnel must take more responsibility for data input and completed staff work.

After human resources, the most valuable resource is information. Information is a product of data collection and analysis. Such data collection and information management must become a responsibility for all personnel. This data and information can be used in many ways such as to help educate citizens about a fire problem or to track, benchmark, inventory, record, etc. within the organization. Computer networks now facilitate data collection and ultimately distribution of information.

Staffing levels

An organization's staffing must fit the needs of the local community. At one time the office secretary held the key to important information and records, but computers and freer access to information are changing that. Information is available in "real time" on the internet. The computer allows the fire chief the freedom to draft documents and frees support staff to work on previously unrelated tasks. Fire chiefs and firefighters are thus producing and storing their own documents. Without a plan, however, an organization can duplicate effort and the only production speeded up will be that of duplicated records and wasted time.

Because of the proliferation of computer use, many fire departments have found it necessary to establish a dedicated position for the management of information and technology. The distinguishing characteristics of the position are to (1) provide technical support, and (2) to train personnel. Technical support may include selecting and purchasing hardware and software, ensuring the compatibility of components and programs, and troubleshooting system problems. Training can involve implementing new record systems and providing help to users.

A basic premise in staffing is to make assignments that enable the department to accomplish its mission and goals. In other words, the personnel strategy should be supportive of the organizational strategy. Staffing may be determined by analyzing the complexity, standardization, and decision-making requirements of an organization.

Complexity. Complexity includes the number of tasks required, the span of control, and the geographic layout of the department. Complexity issues are generally related to the size of the operation. Larger organizations generally specialize more in assignments. Are individuals assigned solely to training, supply, or other specialized tasks? Is the department composed of a single station or multiple battalions? What are the travel distances to support the operation?

What are the impacts of having support personnel perform multiple functions (reception, phones, payroll, and clerical)? How do teams come together?

While bureaucracies favor specialization, the resulting complexity generally works against excellence. Increased size and specialization slow an organization's ability to react to environmental changes. Single-purpose positions are rarely able to grasp the "big picture." It is therefore recommendable that single-purpose support staff be kept to a minimum. Large organizations should have proportionally fewer support staff positions based on economies of scale.

Standardization. A second factor in determining organizational structure is the amount of standardization required. Are goals, rules, policies, procedures, job specifications, and training standards well-documented? The more routine the work, the more it can be standardized. With a high degree of standardization personnel can complete tasks from emergency operations to clerical duties. It is recommended that the organization should have an infrastructure or conceptual framework that includes a mission statement, goals and objectives, policy manual, job specifications, and a training manual. Standardization should stimulate opportunities by clearly identifying expectations and creating a framework for participation.

Decision-making. The last factor in determining office staffing within the organization is the question "where does the decision-making (centralization) take place?" Do procedures and staff reports come from several points in the organization or just a single individual? Do support personnel serve multiple masters? Team leaders may possess the knowledge, but not necessarily the power in their position to accomplish a task. It is recommended that decisions be made by persons at the lowest level possessing the knowledge, abilities, and resources to complete the assignment. Empower employees to contribute and make a difference.

Office staffing is an obvious part of the organizational structure, but it is also part of the "system." The office serves as the point of contact for employees and citizens in non-emergency service delivery, that is, between the fire department and its environment, the community. Ultimately, how an organization is staffed tells a lot about its values, mission, and goals.

Office Design

It may not fit the public's heroic image of the fire service, but a simple fact is that many of the services provided by the fire department are delivered in an office setting.

Fire prevention personnel meet with contractors and developers. Public education can be delivered from fire department classrooms. In addition, the support functions of the department require the resources available in a typical business office. Planning to train and equip emergency responders requires access to records, files, and libraries. Designing fire apparatus can put demands on meeting space, record retrieval, and reference materials.

Any discussion of fire department office space must also include planning for auxiliary areas required for business-like conference rooms, public meeting areas, and classrooms. These and other activities will impact the requirements and layout for fire department offices from community to community. Some facilities are designed as symbols of community pride, while others take a more utilitarian approach to architecture and landscape. In either case, office layout should be examined thoroughly.

Recommendations for planning office space

Ask the following questions:

- Who will work or meet here?
- What equipment is necessary?
- How much room is needed for work? ...meetings? ...storage?
- Is there adequate privacy for meetings?
- Where is the necessary information stored?
- Is there appropriate security for confidential documents, valuable equipment and personal items?

Office layout

Written guidelines or goals will assist managers in assessing and designing office and meeting areas. The guidelines should address organization image, workflow, adequate space, and safety. The image of the fire department is defined not only by the service provided, but also by the appearance of personnel, equipment, and facilities. Employee and citizen perceptions help to establish an organization's identity in the community. How employees relate to the organization is also influenced, in large measure, by how they feel about the work environment. Citizens will form opinions of the department without ever calling for assistance. Consequently, department and public office areas should present a productive, efficient, and professional image to visitors and employees.

Well-designed, functional, safe, professional, and attractive stations, offices, and meeting spaces can contribute greatly to high morale, productivity, and efficiency. They should be consistent with the desired community image and the value placed on employees. While the resources and facilities available within each jurisdiction may vary considerably, office management and layout should receive the same consideration and planning as other service elements.

A survey of work flow is the first step in determining office layout. Work Flow and Space Requirements Checklist (See Appendix) is a tool to help you consider the relationship between work flow, space needs, and assigned functions. A much greater degree of efficiency can be achieved when you take a systematic approach to identifying the various elements of the office environment, office traffic patterns, information management, equipment, and personnel.

Work areas should provide adequate space to conduct business. Guidelines may be determined by the nature of the work conducted. A minimum of 40 $ft.^2$ to 60 $ft.^2$ is required for a single work area. Space needs also vary between free-standing and modular work-station furniture. Modular furnishings generally make better use of available space and allow easier reconfiguration of space to accommodate changing requirements. If work areas do not meet minimum established guidelines, the department should be committed to improving the work environment and look for opportunities to enhance space requirements.

Square Footage Needed for Public/Meeting Rooms

No. of Persons	Net Square Footage
4	100
6	125
8	150
12	225
20*	300*

***Add 15 $ft.^2$ for each additional person more than 20.**

In addition to basic office space needed, there is also space needed for support areas, circulation areas, and even growth factors. Square footage for file rooms, copy rooms, break rooms, libraries, bathrooms, and telecommunication rooms should be determined on a case-by-case basis. Circulation areas such as hallways, corridors, aisles, should comprise no more than 35% of usable office square footage. Growth factors should be considered when evaluating space needs, so be sure to project organizational growth for a five-year period. Apply the expected percentage of growth to the gross total of current space needs.

Too much open space can also be a problem. Vast open space can be disorienting. Architects know that proportion (the space and scale of a room) has a lot to do with the occupant's sense of security and the sense of being somewhere. Space also contributes greatly to how productive people are. The elements of the building, the furnishings, and the nature of the work must be in balance.

Office Safety

Fire department facilities should be designed to provide the highest degree of operational safety and security. Facilities should meet essential construction standards and should, at a minimum, reflect the local fire and building code requirements. Fire protection systems should mirror requirements for business occupancies of similar size and public access. Public agencies such as the fire department can easily lose credibility when they enforce standards for other structures in the community, which are not evident in their own facilities.

Building access and computer security are also emerging areas of concern. Is there a security system in place? Do office personnel have policies to prevent unauthorized persons from entering? Is visibility and lighting adequate? Are there procedures in place to prevent and/or deal with the aftermath of workplace violence?

There are several easily accessible internet websites that offer information on office air quality, ventilation, noise, workplace violence prevention, and office electrical safety. The U.S. Department of Health and Human Services, the Center for Disease Control and Prevention, the National Institute for Occupational Safety and Health, and public health departments are all good resources.

With respect to disaster safety, most communities are at risk for moderate to severe earthquakes, tornadoes, hurricanes, or similar violent catastrophic events. Fire department offices located in areas of such high risk must incorporate features that ensure the safety of employees and the public. High-rise offices may require more stringent safeguards and should be carefully evaluated.

Ergonomics. Ergonomic considerations should be an integral part of the design process, promoting efficiency and preventing work-related injuries. Ergonomics is the study of the relationship between people and the tools of their occupation. In particular, ergonomics focuses on the physical relationship of people and the way they use their equipment. A tool such as a computer workstation, for example, is said to have good ergonomic design when it can be easily adjusted to fit the user.

Office tasks often require rigid or unnatural body postures, repetitive motions, and intense concentration. With good ergonomics, the user does not have to contort their body in ways that could cause discomfort, strain, or even injury while using the tools required on the job. Good ergonomic design can help reduce tension, stress, and injury in the workplace, and have a positive effect on productivity, quality of work, and morale.

Cumulative trauma disorder (CTD). Physical disorders may develop or be aggravated by the cumulative application of bio-mechanical stress to tissues and joints over time. Acute trauma, or injury caused by a single event, does not fall within this definition. Examples of CTD include bursitis, ligament strains, muscle strains (*e.g.*, neck-tension syndrome), nerve entrapment (*e.g.*, carpal tunnel syndrome), stenosing tenosynovitis (*e.g.*, trigger finger), etc. Symptoms may include pain from exertion or pressure, change in skin color, numbness or tingling, decreased range of motion, decreased grip strength, as well as swelling of a joint or part of a limb. CTD risk is composed of the following factors:

- Frequency: The rate at which specific physical motions or exertions are repeated
- Force: Physical exertion by or pressure applied to any part of the body
- Duration: The length of any period of work activity that poses a CTD risk
- Posture: The position of a body part during work activity
- Vibration: Localized or body exposure
- Exposure: The placement of hands and feet in extreme temperature environments that cause discomfort
- Recovery time: The amount of time separating repetitive motions or exertions, or separating periods of any work activity posing a CTD risk, which is needed to prevent fatigue of the body parts performing the activity

Recommendation: Perform periodic inspections of work sites to reduce or eliminate the risk of office CTD hazards.

Inspections should include the following:

- Evaluate clerical work areas for processes, procedures, equipment, or work activities that may increase risk of repetitive stress injuries and symptoms.
- Evaluate injury records for evidence of CTD. Provide a written description of measures to control the problem.
- Implement administrative and engineering control measures.
  - Administrative controls include providing "job specific" training for awareness of symptoms and remedies, the use of rest periods, and redesign of work activities that pose a risk to employees.
  - Engineering control measures include modification to work stations, equipment, chairs, and processes.

Work station design and layout play the most critical role in eliminating sources of injury and postural problems in the office environment. Musculoskeletal effects and visual fatigue are the primary concern. These effects can usually be controlled through proper design and use of the workstation. An important point to keep in mind is that office equipment should adjust to the user, not the other way around. More active participation in the design, installation, and development of office work areas by users is likely to encourage stronger support for any changes made.

A first step in the ergonomic evaluation of a computer workstation is to establish the optimum posture of the operator. This ensures that heights and angles of the equipment (*e.g.*, chair, table, copy holder, keyboard, monitor, etc.) fit the individual, which in turn helps increase comfort and productivity.

Described below are some of the factors that affect a computer operator's performance and health. The factors are keyed to the numbered diagram to help visualize proper layout.

1. The chair should be easily adjustable to provide good support to your lower back and allow for the proper height interface with the keyboard. The seat pan should be height-adjustable to fit the operator comfortably above the floor and should be able to tilt forward or backward. The chair should be fitted with casters if the task requires the operator to get up or move around the workstation frequently. Be sure that the casters selected are appropriate for the office floor (*i.e.*, carpet versus vinyl).
2. Body posture should be as shown with right angles at the elbow, hip, and knee. The head should be held in a neutral position facing straight ahead with the eyes gazing forward or slightly down.
3. A foot rest may be needed when the operator's feet do not comfortably reach the floor, although this should not be necessary if both the table and the chair are height-adjustable. The feet should reach and touch the floor in a flat, relaxed manner. Foot and leg circulation will be affected if overextension of the feet occurs.
4. The keyboard support table can be adjustable to allow proper upper body posture, or the chair height can be adjusted to achieve the same purpose. The keyboard should be detachable to permit flexible positioning. A wrist rest should be available for those who desire it. Keeping the wrist level straight and in a relaxed position offers the worker maximum comfort for extended hours of work.
5. The document holder should be provided and adjustable in height and angle of tilt, to the same height and plane that the operator views the majority of the time. Usually the position will be at or just below screen level, allowing the operator to hold his/her head in the neutral position shown in the diagram.
6. If eye fatigue and/or vision problems are experienced by video display terminal users, they should see their ophthalmologist to rule out any need for prescription glasses. The display monitor should be positioned so that the distance from the eye to the screen can be adjustable, allowing the center of the screen to be positioned so that viewing angle is 15° to 25° below eye level. Display monitors placed too low will increase musculoskeletal tension and fatigue to the back and neck. Display monitors placed too high will increase visual fatigue. The screen should be detached from the keyboard so that each can be positioned in an optimum manner.

7. General lighting for the office and display monitor vary between operators. Light above and in front of the display monitor will glare onto the screen. Reduce glare onto the screen by appropriate placement of the display screen in the room. The display monitor should be oriented so that the operator does not face an unshielded window or a bright light source. Ideally, the screen should tilt to help eliminate screen reflections. Reduce mirror-like reflections on the screen by using an etched screen surface, a thin-film coating, or a hood. Other types of reflections on the screen can be reduced by using a neutral density, micro-mesh, or glare-resistant filter.

8. Due to the work demands put on computer operators, it is important to monitor environmental factors, such as temperature (68 °F to 78 °F) and humidity (30% to 70%). Noise should be kept within comfortable limits (<75 dbA or as low as achievable). Noise from printers and nearby equipment may be reduced with sound screens or well-placed absorbent materials such as acoustical ceiling tile, carpets, curtains, and upholstery.

Computer users have raised the issue of the various types of radiation (*e.g.*, X-ray and ELF) produced by computer equipment. Exposure levels to radiation during computer use is very low (not significantly above background) and there is currently no accepted scientific evidence that suggests radiation exposure from computer use is harmful. Adhering to the philosophy of "prudent avoidance" (if potential exposure is an issue of concern) would dictate staying about at arm's length from the monitor to eliminate almost all exposure.

Office assignments

In addition to considering image, work flow, space requirements, and safety the following issues deserve special attention.

Privacy. Many of the issues handled in the fire department require privacy. Strategic planning, personnel actions, and labor relations are just a few of the sensitive issues discussed during the course of business. Privacy is more than simply meeting behind closed doors. It also means an area where confidential topic areas can remain confidential by controlling access, visibility, and sound transmission.

Communications. Staff members should have access to the organization's common communications. Telephones, voice-mail, computer networks, email, and fire radios are all examples of the tools necessary to interact with the department and community. Radio and television also give the chief additional resources to monitor current local, state, national and international events.

Conference area. The chief's staff offices should have access to a conference area capable of seating a small group, typically four to six. The nature of business conducted will determine whether a conference table is a requirement.

Reference. Reference requirements vary with management styles. The more technical or information driven the position is, the greater the need for reference materials. Internet access has become a necessary tool for access to reference materials. Research papers, trade journals, regulations, and up-to-date news is readily available in this electronic format. Personal notes and documents, catalogs, rules and regulations, as well as state and federal administrative codes should be accessible. Some reference materials in the office may be for personal use only. Access and removal of these materials should be allowed only with the expressed permission of the owner.

Location. Offices can be grouped by function or assigned by proximity with regard to working relationships. For example, members of the training division may occupy adjacent offices. Another scenario might be to group chief officers in a particular part of the building to facilitate communications.

A fire chief's office serves as the point of origin for much of the organization's business, including planning and meetings. For this reason it is significant, not only to the individual who occupies it, but also to the organization and the community. It is a reflection of the individual and the values of the organization. The privacy, location, and available resources reflect the fact that the chief requires information for the "big picture" and places a high value on interaction with people.

Records Management

At some time every fire chief probably wonders if the fire department can keep up with the sheer volumes of paperwork required. There are federal, state, and local reporting requirements, financial records, budget documents, payroll, personnel records, codes and ordinances, apparatus records, and so on and so forth. Computers and automation can contribute to the flood of record keeping. Accreditation, performance measurement, certifications, and qualifications also drive the need for accurate and up-to-date documentation.

Even the best administrator can be intimidated when it is time to access and analyze the volumes of data collected. How can a fire chief keep ahead of this tidal wave of information? It is important to remember that expertise, in any endeavor, is the ability to pay attention to the *right* thing at the *right* time. Too much data can lead to "sensory overload." A records management system helps to reduce data into useful information. Information system management is the key to a fire department's decision-making and strategic planning effort. Consider the following strategies to attack the problem:

Identify the problem. Ask how other fire chiefs handle reports and records. Network! Network! Network! The idea is simple but powerful. Network with peers and experts. Lawyers, accountants, personnel specialists, certified mechanics, etc., are familiar with the requirements for special documentation in their respective fields of expertise. Read the city charter and national standards and professional organization publications, as well as state administrative codes governing formation and activities of the fire department. Knowledge is like a treasure hunt—you must keep digging to find it.

Plan. "No one plans to fail. They just fail to plan." A records management plan is part of the fire department's infrastructure. Without a plan, records management, like fireground activities, becomes a "happening." Everyone works hard, but the results are unpredictable.

A systems approach coordinates procedures that produce, analyze, and store records and reports. A standard approach does not mean that a department must become a "cookie cutter" bureaucracy. On the contrary, with minimal planning even the most empowered organization can be more efficient. A systems approach structures what was once left to uncertainty. It structures reporting around the "big picture" by factoring report users and report uses into the plan. It communicates expectations through a standard set of operating procedures.

Systems should not be viewed as or become obstructions to initiative, upward delegation, or endorsement for central control. Following guidelines should not become more important than getting the job done. A systems approach to reports and records gives standard, meaningful, information that can be dealt with efficiently. The process should contribute to the speed and clarity with which information will flow. Standard information should assist the organization in planning and the decision-making process.

What elements constitute a reports and records plan? The following key areas apply to both paper and electronic records:

- Accessibility: Determine whether the record is ACTIVE or ARCHIVED. Does the record require short- or long-term storage? If the record is active it must be stored for ease of access by the person or persons using the information. If the information is archived, consideration must be given to space required and retrieval. Each type of archived information should be stored in a single location.
- Accountability: Someone must be responsible to write, process, complete, act on, and store each report or record. Reference may be by individual, title, or position, but remember—groups are not accountable, individuals are. Without accountability paperwork can fall into a "black hole." The fact that a signature or initials are required will help indicate the identity of the originator, reviewer, authorization, etc.
- Format: At a minimum, fire department reports should have a letterhead or logo, form title and/or stock number, and a revision date. All official documents should reflect these basic requirements.
- Routing: A rule in effective office management is to handle each piece of paper or communication only once. Two of the most common elements in any communication are "TO:" and "FROM:." Any routing in between is either a wasted motion or due to inaccurately addressed materials.
- Security: Does the record require security? Is it for general publication or should access be limited on a need to know basis? Is there a legal requirement for security?

Ask if the record or report is really necessary. It can be very tempting to create a new form for every new situation. That's what bureaucracies do best. Every possible action is formalized and recorded. Do not require special records, reports, or forms unless they will be used frequently, something useful will be done with them once completed, or they are required by law.

Finally, communicate the plan. Employees should know the "who, what, where, when, why, and how" of communications handling in the organization. The plan should clearly state the need for records security and authorized users. An action guide or procedures manual is an excellent way to standardize and streamline paperwork. One of the advantages of written communications is that it provides consistency over time.

Reports needed

Federal, state, and local requirements should serve as the primary guidelines for the need to produce and retain reports and records. Five broad categories of reports and records generated and maintained by fire departments are financial records, personnel records, incident reports, code enforcement records, and internal documentation. A more comprehensive list is available at the end of this section. These examples are not all-inclusive but give some insight into the data generated by the average fire department. Analyzing reports and records by category helps to simplify the records and reports planning process for the fire chief.

Office communications

Many of the day-to-day communications that take place in a fire department are in written form, including personnel assignments, hydrant or street information, upcoming events logged for subsequent shifts, apparatus repairs needed, etc. Written communications have the advantage of relaying a message and lasting over an indefinite period of time. E-mailed messages closely resemble written communications, but may not have the same life span. Computer operating systems and software change at a rapid pace sometimes making historical retrieval a challenge.

Modern principles of management emphasize the importance of good communications among all levels within an organization. Sender rank may not be the determining factor in a written communication's importance. In many departments, firefighters may have key roles in budgeting and project management and therefore may have the need to communicate with others inside and outside the department. With multiple levels of communication, however, it becomes necessary to determine which records and reports are official and must be retained and which are not. Intra-office communications should also be addressed in the reports and records plan.

Depending on the size of department it may be advantageous to formalize how intra-office communications are defined, authorized, distributed, and retained. The following example covers key elements for such communications:

XYZ FD Policy: Written intra-office communications

XYZ Fire Department (FD) written communications include informal and formal messages. Only approved information shall be distributed and posted in FD facilities.

Informal: Informal communications include typed or handwritten messages and electronic mail to an individual or small group. Messages may be generated and routed by all FD personnel for approved FD business. Messages are not retained or logged in the notice and memo binder.

Formal: A notice and memo binder is required for retention of formal fire department communications. The binder provides a system for distribution and the storage of written communications distributed to all personnel, all stations, and/or special groups. It also provides hard copy portability for reference at station meetings. Company officers shall read, initial, and date new notices and memos during morning briefings. Vacation, sick, and injured personnel shall read, initial, and date appropriate communications upon return to duty.

Notices: A notice is a formal written FD communication of a specific nature distributed to all personnel. Notices require approval for distribution by the fire chief. They shall be maintained in the notice and memo binder for sixty (60) days. They shall be disposed of after sixty (60) days or as indicated in the notice.

Memorandums (Memos): A memo is a formal written FD record or reminder and is distributed to assigned personnel or small groups. Notices require approval for distribution by a program manager/chief officer. They shall be maintained in the notice and memo binder for sixty (60) days. They shall be disposed of after sixty (60) days or as indicated in the memorandum.

Announcement: An announcement is a communication of an informative nature. Announcements require approval for distributions by a chief officer. They require posting on the station bulletin board for sixty (60) days. They shall be disposed of after sixty (60) days or as indicated in the announcement. A permanent record shall be maintained by a designated individual for future FD reference.

Format

All fire department messages, notices, and memos shall contain the following items:

- Fire Department Heading
- Type of Communication: Message, memo, notice, or announcement
- Date: The date the communication is written
- To: The person or group to which the communication is directed
- From: The originator of the communication
- Subject: The topic of the communication
- Approved for Distribution: (notices and memos) Chief officer name and title authorizing distribution

Reports and records generated by the fire department

Financial Records: *(Note: Retention varies with statutory requirements.)*

1. Journals/Ledgers

Accounts receivable	Accounts payable	Disbursements
Payroll journal	Taxes receivable	Warrant register

2. Source Documents

Bank statements	Bills	Checks
Claims	Invoices	Purchase orders
Requisitions	Time slips	Vouchers
Warrants	Reimbursement requests	

3. Statements

Balance sheet	Fund balance	Revenues
Fixed assets	Cash receipts/disbursements	

4. Other

Inventory records	Budgets	Capital asset records
Schedule of investments	Lease-purchase records	Audits
Long-term debt		

Personnel Administration Records: *(Note: Rather than following statutory minimums, it is recommended that personnel records be retained indefinitely. Issues may arise long after separation.)*

1. Administrative Files

Job specifications	Labor agreement	Rules, regulations, policies

2. Confidential Personnel File (Not available to employee and/or immediate supervisor.)

Records checklist	Pictures	EEO status record
Employment tests	Medical exams	Police reports
Interviewer comments	Credit reports	Out-dated disciplinary action
Employment references	Communicable exposure reports documentation	Worker's comp.

3. Employee Action File

Records checklist	Grade changes	Payroll records
Insurance sign-up	Payroll deductions	Tax records
Application	School records	Attendance reports
Equipment issued	Certifications	Current job description
Training records	Transfers	Separation report
Current disciplinary actions	Employment agreements	Records checklist
Performance appraisals		

Incident Reports: *(Note: Incident reports are usually retained indefinitely. Documentation may be used for personnel issues or exposure.)*

Incident action plans
Incident run records
Investigation reports
Patient field records
Tactical worksheets
Customer satisfaction
Quality assurance
Public education reports
Performance
measurement

Code Enforcement Records: *(Note: Code enforcement records are generally retained for the lifetime of the property or occupancy.)*

Plan check record
Side plan and conditions
Annual inspections
Building inspection records
Certificate of occupancy
Hydrant acceptance test
Hazardous materials storage plan
Permits
Material safety data sheets

Internal Documentation: *(Note: Documentation retention varies. Some records are needed for the life cycle of the facility or equipment. Other records facilitate operations on a short term basis and should be discarded on a routine basis.)*

1. Apparatus and Equipment Records

   Annual pump tests
   Daily operator's checklist
   Specifications
   Preventative maintenance
   Fuel records
   Vehicle accident reports

2. Facility Records

   ADA compliance
   Maintenance requests
   Safety inspections

3. Supply Records

   Inventory
   Order forms

4. Other

   Contracts
   Disaster preplans
   Minutes of meetings
   SOPs
   Station log books
   Training manual
   Strategic plan

INFORMATION TECHNOLOGIES

This is the information age. At no time in history has so much raw data, covering virtually every facet of life, been available to the average person. Acquiring data is not difficult. The problem is determining how it can be used. The term "office" implies a place where data is available to be transformed into information. The fire department office serves as the "central nervous system" for information. Station and/or maintenance facility personnel or the fire inspector may be the end user of information, but generally the office helps facilitate information processing.

Fighting fire requires a particular knowledge, or information. EMS requires knowledge of local protocols. Code enforcement requires knowledge of local and state codes and ordinances. Virtually every service provided by the modern fire department has a qualified body of knowledge, that is, an information base. As stated earlier in this chapter, expertise is paying attention to the right thing at the right time. The very nature of the service is matching the correct information to the given circumstances. Information technologies include the data, equipment, techniques, and processes that are converted into information for the organization.

Technology is defined as a scientific method for achieving a practical purpose. Technology includes tools and techniques (or scientific methods) to gain power or leverage, making a task easier (practical purpose). In physics Power = Work/Time. More work in less time gives more power. Leverage produces an output force greater than the input force.

Technology does not mean that you need the latest or most powerful gizmo, but instead technology management includes a philosophy that advocates the use of tools to augment human effort. Technology becomes even more powerful when applied to ideas in achieving a goal. The "systems model" also promotes the concept that technology can have a significant influence on the organizational structure of the fire department. Tools designed for speed and efficiency can have significant cascading effects on the organization's social system. With the power of technology it's not who you know but how you know.

Organizations should be organized around processes instead of tasks. Processes cover a range of activities when compared to "compartmentalized" tasks. This means people must cross traditional barriers or "turf" to deliver service. Technology can help overcome three facts of organizational life"people are busy; people are in different places; and information is not always where it is needed."

Benefits

Automated systems assist with repetitive tasks, report compilation, record updates, and report generation. The significant benefits of an automated system to the fire chief are leveraging time, communicating, information access, and decision making.

Leveraging time. Computers do not decrease workloads. Instead, they have increased the capability to collect data and turn it into information. As previously stated, more data is now available. The speed of retrieval, sorting, and selecting allows more data to be "crunched."

Communicating. In the past the "grapevine" was the best method to get information out fast. "Telephone, tell a friend, tell a fireman," has been an axiom for the speed of informal fire department communications. With communication technologies, information can be distributed over unlimited distances in real time to multiple users. The person generating a record can create, distribute, and store information in a few keystrokes, fax, or conference call. Electronic communication is not a substitute for face-to-face communications. It does, however, allow a fire department, particularly with multiple facilities, the ability to link ideas and expertise regardless of location.

Information access. The real strengths of computing are the speed of information storage, retrieval, and manipulation. Days and weeks of search and calculation have been compressed to nanoseconds. People who are held accountable must have access to plans, procedures, and other people. Computer technologies, particularly networking, allow greater opportunity for personnel to participate. Networking allows employees to "cross barriers of space, time, and social category to share expertise, opinions, and ideas."

Decision making. Information is power. To draw an analogy, some economists focus on the distribution of wealth and note that the trend appears to be a shrinking middle class. Conversely, with information technologies, more people have access to information; therefore, more people can make informed decisions. Access to computing tools will radically shape a department's culture.

Limitations

Technology can be seductive. Speeding up existing processes may only result in marginal improvements in effectiveness. Traditional organizational structures are built around efficiency and control, whereas the new paradigm is service and quality. The information age requires an organization built around outcomes, not tasks.

There is a difference between automating and informating. "Automating tends to concentrate on the smart machine and to cut out or reduce people. Informating organizations also use smart machines but in interaction with smart people." Smart machines need smart people to work with them. People who work with these tools need to consider:

- ***It's a tool for the job, not the job itself.*** Information technology is like a command system, tactics, and other "guidelines" used in the fire service. It is a strategy, or approach, to processing information. How we use tools can sometimes become more important than getting the job done. If the technology becomes the focus of our labor it can obstruct what it was intended to do, which is to assist us in our tasks. Flexibility and initiative should always be valued in the fire service, on the emergency scene, and in the office. A degree of "looseness" is implied in empowerment. Technology is not the end product, service is. Technology should be transparent. It is merely a tool.
- ***Garbage in equals garbage out.*** Since the inception of computerized systems the rule remains the same, "garbage in equals garbage out." Outputs (information) are only as good as the inputs (data entered). A seemingly small error in data entry

can distort the "reality" of the information. Users have to understand the purpose or audience and the implications of the data used.

- ***Information technology is impersonal.*** People have more trouble sensing what others are feeling when communicating electronically compared to face-to-face communications. Non-verbal and other communication subtleties do not lend themselves to electronic communications. "Flaming" (rude and impulsive behavior) is more common in electronic communications. Tendencies to be outspoken in electronic communications may lead to increased group conflict.
- ***Information overload.*** More and more information moves faster and faster. It is often difficult to distinguish priority data from other data, thereby making answering email very time consuming. Learning about new software and systems takes time. Manipulating more data requires time.
- ***Control.*** "Not surprisingly, people often resist changes in information control that diminish their position." The access to information is a great equalizer in the work force. "Often information overload is really an argument about control. Typically the problem is not too much information literally, but a lack of control over information exchange." Sharing information and employee empowerment does not fit everyone's leadership style, fortunately only the unenlightened few.

Plan your information system to address users, needs, and processes. If information is stored in the "wrong" format, or is not accessible by the end user, the system becomes inefficient. Policies should encourage information exchange and define parameters clearly. A written policy similar to the following example may help to establish a degree of standardization for the system.

XYZ FD policy: Computer use

Purpose: The XYZ Fire Department (FD) operates a computer network and stores data on behalf of the citizens we serve. Computers operating in FD facilities shall be considered as FD owned tools. The FD computer system will be continuously evaluated for security, access, and applications. There may be unforeseen circumstances requiring further refinement of this policy. As such, certain restrictions for protection from inappropriate use, accidental or intentional destruction, unauthorized modification, and/or disclosure of information will apply.

Code of ethics: All data, files, and software stored, maintained, or placed on any FD computer or computer media, including transferable media such as diskettes, are FD property, without exception. The fact that individual items or collections of data or software are public in nature, or actually are public records, does not diminish the "proprietary" aspects of FD ownership.

- Use of computers is limited to FD business.
- Use of an assigned computer requires permission from the responsible owner.
- Only authorized software is allowed. At no time shall software be added to or removed from the network without the expressed permission of the system manager.
- Unauthorized pirating of software may result in severe fines and disciplinary actions against the responsible employee. Some data security violations can carry criminal punishment; some misuse can be deemed a misdemeanor.
- Do not knowingly omit data, falsify records, make misleading entries, or destroy official records.
- Do not interfere with or snoop around in other people's computer work.
- Do not download or store any materials that may be considered offensive in the work environment.

Public interest must always precede self interest. Individual actions must uphold and safeguard the FD's ability to operate efficiently, confidentially, and accurately.

Essentials for success

A study on managing information technologies was completed by the Syracuse University School of Information Studies. The study examined how county governments used technology to collect, organize, manipulate, and disseminate information. The Syracuse study highlights several factors that are essential to the successful use of information technologies. To extract the maximum value from information technologies, an organization must have the following:

- Leadership: Top officials must establish a supportive environment, through formal policies, visible support and their own use of information technologies. Not only must they be interested in and conversant with information technologies, they must also champion a commitment to technology at all levels.
- Practical Vision: There must be a vision of the organization's future complete with an understanding of the role technology can play in realizing it and a clear information technology plan to guide the organization toward that vision.
- Valued Tools: Information technologies must be integrated into government as tools that enable it to carry out administrative functions, deliver services, comply with mandates and respond quickly to new demands. Efficiency is measured; effectiveness is valued.

- Ongoing Training: Training of end-users at all levels is critical to realize the maximum return on an investment in information technologies, to ensure productivity and to encourage creative solutions.
- Adequate Resources: The organization must allocate people, facilities and money to information technologies; verbal support is not enough. Know what you have and ask for what you need.
- Teamwork: A project team including technical experts, managers and end-users is vital. All players should be involved, informed, and heard.

In addition to all of this, officials should remember to:

- Be creative. Encourage innovation.
- Focus on doing it better; improvement is always possible.
- Make the technologies "invisible" to the end-users; they should concern themselves with their productivity, not how the technologies work.
- Look outside your own territory; benchmark your practices against the best.
- Focus on useful information: technology is an enabling tool. Treat it that way.

As technology expands so too does our view of information itself. Traditionally, information in the fire service meant data collected as historical records, a past-tense view of what has already taken place. Forms and records were stored in cabinets and retrieved for analysis. Now, however, office-related information is becoming increasingly available as emergency operations information. Today's technology allows "real time" remote access to critical decision-making information. For example, weather related information is now available to predict fire behavior on wildfire incidents. In addition to real time information, digitized information now allows access to maps and preplans by responding units. Incident commanders are able to locate and track on-scene resources while dispatchers are taking on an increasing role in incident management.

The next generation of information technology will combine information with emergency scene tasks to help us in incident decision-making. The information support may be provided remotely, possibly from the fire department's offices. On-board video monitors and sensors could simultaneously feed information on existing conditions back to the fire department office. Computerized occupancy files integrated in geographic information systems (GIS) will increasingly be accessed for plot plans, hazardous storage, and their relationships to surrounding neighborhoods. Support (office) personnel, located miles away, will take an active role in suppression and enforcement efforts.

Conclusion

Many of the functions necessary to deliver fire department services are provided through, or supported by, the administrative office. The fire department operates as part of an "open system" within the community. The office helps the organization transform people, information, and money into services.

Fire departments must function as part of "the information society." "The essential requirement, therefore, for all its workers is that they are able to read, interpret, and fit together the elements of this currency, irrespective, almost, of what the data actually relates to. That is a skill of the brain." Fire department personnel, including the fire chief, must become "cross-functional," increasingly filling more of an "office" role. Individuals must take more responsibility for data input and completed staff work which will result in information for decision-making.

More "horsepower" does not necessarily lead to improvements in service. Records management must be planned thoroughly. With thoughtful planning by fire department personnel, a records system can streamline input, use, manipulation, storage, and recall of documents. Without a plan, electronic communications and automated processes can actually increase inefficiency.

Information technologies are empowering tools, allowing more of the organization to communicate issues, ideas, problems, and solutions. Yet caution is also in order because technology can be seductive. It is not an end result. It is a means to provide better results and better service. Information technology will change the way we live and work. The fire department of the future will be measured by the ability to use information to its advantage in managing fire service activities.

References

Sproul, L. and Kiesler, S., *Connections,* p.19. MIT Press, Cambridge, Massachusetts, 1993.

Sproul, L. and Kiesler, S., *Connections,* p.105. MIT Press, Cambridge, Massachusetts, 1993.

Fletcher, Bretschneider, Marchand, Rosenbaum, and Bertot, *Managing Information Technology: Transforming County Governments in the 1990's.* Richter, *Governing.* August 1992.

APPENDIX

Workflow & Space Requirements Checklist

Name: ____________________ **Title:** __________________ **Date:** _______

Job Description (briefly describe your job function and responsibilities)

____
____
____
____
____

I. OFFICE EQUIPMENT: Identify the office equipment used in your work area.

Equipment	Dimensions	Electrical Requirement	Shared?
			❑Yes ❑No
			❑Yes ❑No
			❑Yes ❑No
			❑Yes ❑No
			❑Yes ❑No

Telecommunications: Do you: ❑ have your own phone? ❑ share a phone?

If shared, with whom? ______________________________________

What is the percentage of time that you spend on the phone daily? ______________

Are conversations confidential? ❑ Yes ❑ No ❑ Sometimes

Computer: Do you: ❑ have your own computer? ❑ share a computer?

If shared, with whom? ______________________________________

What is the percentage of time that you spend on the computer daily? ___________

Is data confidential? ❑ Yes ❑ No

Personal Materials: Identify all the materials that must be stored in your work space (*i.e.*, books, binders, files). Include the size of the material and the linear inches of space required for each type. Note materials that require security.

Material Size (length x width) Linear Inches Required

____

____

____

____

Reference Materials: Identify all the materials that must be stored in your work area such as catalogs, phone books, mailing lists, etc. Include the size of the material and the linear inches of space required for each type.

Material Size (length x width) Linear Inches Required

____

____

____

____

____

Furniture: Identify any specialized furnishings—such as drawing boards, file cabinets, tables, maps, charts, etc. Include dimensions.

Material Quantity Dimensions

____

____

____

____

____

Special Design Considerations: Are you: ❑ right-handed ❑ left-handed

Do you have special needs (lighting, special accommodation, etc.): ❑ Yes ❑ No
If yes, please explain:

____

____

____

____

____

II. INTERACTION AND WORK FLOW: Prioritize the individuals you work closest with inside your department. Give the reason for the interaction, such as work flow, shared files, shared equipment, or common projects.

Name Reason

____

____

____

____

____

Is work completed, then filed? ❑ Yes ❑ No

If so, where?

____

____

Is work sent to another individual or department for completion and/or storage?

❑ Yes ❑ No

If so, where?

____

____

Special Areas and Requirements: Identify and describe any special areas (*e.g.*, reception area, storage, etc.) necessary to accommodate functional requirements.

Special Area Description

____

____

____

____

Are there needs for confidentiality? ❑ Yes ❑ No

Conferences: Does the job require conferences in the assigned work space?

❑ Yes ❑ No

On the average, how often are conferences held per week? ________________

On the average, how many individuals attend these conferences: ________________

How long do the conferences usually last? ________________________

Are these conferences confidential? ❑ Yes ❑ No

Who is responsible for scheduling conference areas?

Name(s): ______________________________________

List any requirements (including materials, audio/visual, writing, display surface, etc.) used during the conferences.

____

____

____

____

III. SKETCH YOUR WORK AREA: Does the area address WORK FLOW and SPACE REQUIREMENTS?

3

FINANCIAL MANAGEMENT

Richard A. Schnuer

The author would like to acknowledge the following for their thoughtful comments on various sections of this chapter: Brett Christensen of Marquette Associates; Stephen T. McElhaney, FSA, of William M. Mercer, Inc.; Stan Helgerson of the Village of Carol Stream, Illinois; Pat Krolak of Marquette Associates; Joanne Malinowski of Kane, McKenna Capital, Inc.; Dave Richardson of the Village of Streamwood, Illinois; William Stafford of the City of Evanston, Illinois; James Tuton of the University of Illinois; and of the City of Champaign, Illinois, Chief John Corbly, Elizabeth Hannan, and Colleen Livermore; also Linda Bamber of Tufts University for coaching in writing; and my family for their support in this and other endeavors.

CHAPTER HIGHLIGHTS

- An overview of budget development
- Covers basic tools that allow the fire manager to manage the budget effectively
- The importance of accurate accounting and comprehensive financial reporting
- Describes sources used to fund fire services
- Discusses the most widely used forms of debt financing and the expert services required to issue debt
- Provides a primer to purchasing methods
- Covers financial principles that underlie investing funds for both operations (short-term) as well as pension benefits (long-term)
- Describes the major components of a public sector investment policy
- Covers common types of pension plans, although most of the section describes "defined-benefit" plans

Introduction

Financial management is the art of directing the acquisition and judicious use of money to accomplish an end, and providing stewardship over financial resources entrusted to you. This chapter attempts to provide fire managers with the basic concepts and practices essential to sound financial management of the fire service.

Almost all fire managers, regardless of the governance system that they operate within, will find many of topics in this chapter useful. However, this chapter also covers topics such as pension fund management that are typically more applicable to managers of independent districts.

Most of the chapter sections can be read independently. However, some sections discuss closely related concepts that will be better understood by reviewing all relevant sections. In particular, the sections on Budget Development, Budget Administration, and Accounting and Financial Reporting present related concepts. Also, both the section on Investment Policies for Operating Funds, and the discussion of pension investments in the section on Pension System Management build on the section on Investment Principles.

Obviously, comprehensive treatment of public financial management is beyond the scope of this publication. Fire managers who hold these responsibilities can further their education through a variety of professional development opportunities. One excellent service provider, particularly for state and local government officials, is the Government Finance Officers Association of the United States and Canada (GFOA). It can be found online at www.gfoa.org. This organization is considered by many to be the leading membership organization for public finance professionals in the U.S. and Canada. The organization's services include publications, training programs (both traditional and videoconferences), and development and promulgation of guidelines on a variety of public finance issues. The names and Internet addresses of more specialized membership organizations in the field of public finance are provided in the relevant sections of this chapter.

For the long-term management of a fire service organization—achieving its mission, creating its future, and ensuring the fire, life, and environmental safety future of the community itself—not many things are more important than giving professional attention to the concepts presented in this chapter, and integrating this knowledge with the other information in this handbook.

Budget Development

Governmental budgeting is perplexing to many. Common complaints include that budgets are too focused on minor details, that budget decisions are not rational, and that budget processes promote "gamesmanship" such as "use it or lose it" attitudes. Unfortunately, some of these complaints are true. However, some result from lack of understanding of the budget process and its goals.

This section attempts to assist both fire chiefs in independent districts, who have some control over the budget process, as well as fire managers in departments that are part of general-purpose governments. The latter may be tempted to skip some of the discussion that appears targeted toward managers of independent districts. However, fire managers in general-purpose governments should find that the discussion of broader budget issues will help them to understand the budget processes used by their governments, and become more effective in using those processes to achieve their objectives.

Perspectives on budgeting

To some extent, confusion over the budget process results from failure to understand the different perspectives that various players have of the process:

Fire chief. Your primary goal is meeting the objectives outlined in your strategic and/or operational plans, or more specifically, obtaining funds to provide services.

Financial officer. His primary goal is keeping the government fiscally healthy. In short, that means balancing revenues and expenditures, which requires fiscal control. However, the financial officer's concern with control does not mean he is unconcerned about services. Granted, his focus on long-term fiscal health may lead the financial officer to oppose an expenditure that the fire chief desires. But the ability of the fire service to operate in the long term depends on the fiscal health of the governmental entity.

Professional chief executive officer, if different from the fire chief* (e.g., *mayor or city manager). His primary focus is on services, but he must consider all of the services offered by the governmental entity. Therefore, he views the budget process as a means to allocate resources appropriately to meet the government's many objectives. The chief executive officer will evaluate the needs of the community in making budget decisions. In addition, he will evaluate the performance of the fire department (and other services), desiring to allocate resources where they will be used most effectively.

Elected officials. Cynics might say that that the primary objective of elected officials is getting reelected, so their perspective of the budget process is simply a means to that end. Others would say that many elected officials act in the best interests of citizens. They may take equal concern in all the services of the government, but may, to some extent, have more concern for particular services and/or constituent groups. Elected officials' knowledge of budget issues varies as well. While they often have good insight into public needs, they may (especially if new to elected office) have little knowledge of fire operations or the government's fiscal affairs and practices.

Meeting these various goals actually entails several cyclical activities including:

1. Gathering input, which includes any goals the elected officials may have established, plans (including the fire service's strategic plan), community values, staff recommendations, evaluation of past fire service efforts, and technical analysis
2. Decision-making, which requires establishing priorities in light of available resources, while balancing the community's immediate needs with its long-term fiscal health
3. Developing outputs, which include establishing policies, programs, and operational procedures to meet service needs, and the budget document itself
4. Budget administration, which entails carrying out services and expending resources, and
5. Measuring results of the activities funded through the budget process, which has obvious usefulness for operations; in addition, evaluation information is also fed back (becomes inputs) to the next year's budget process.

This chapter primarily discusses decision-making and budget administration, while the remaining topics are discussed in other chapters. The chapter on strategic planning covers gathering inputs, and another chapter covers performance measurement. Developing outputs (policies and programs to meet service needs) is discussed in several chapters of this handbook.

One output, however, is covered in this chapter—the budget document. The budget document should reflect the various activities above. In fact, the Government Finance Officers' Association, in its Distinguished Budget Awards Program, characterizes the budget document as a:

- Policy document
- Financial planning document
- Operations plan (for services)
- Communications tool

This chapter should help fire managers understand budget processes and the documents that result from them.

Forms of budgeting

Unlike annual financial reports, there are no mandated national standards for budget processes or budget documents. (However, some states prescribe certain standards and/or practices for their local governments. In addition, several national organizations recently developed a set of recommended budget practices, which are described below.) Since budget processes can take many forms, this document cannot specifically address your entity's budget process. However, to better understand your entity's process and use it to meet your goals, it may be helpful to review the major types of budget processes that have developed over the years.

Before doing so, however, an important distinction must be understood—the difference between a line-item accounting system and a line-item budget. Every accounting system has a "chart of accounts" that organizes accounting information into various categories. The categories allow users of the systems to find particular transactions, summarize information, and gain an overall understanding of the entities' financial position. (See the section on Accounting and Financial Information for a discussion on how a typical chart of accounts is organized.) For example, a fire manager wants the purchase of medical supplies for EMS put in the correct "object code" so that he can tell at the end of the year the total amount that was purchased. Line-item accounting systems support, and will always be part of, budget systems.

A line-item budget system refers to one in which line item expenditures are the focus of budget development and decision-making. The budget document presents line-item detail, and elected officials review and approve (or revise) line-item expenditures. An EMS manager, for example, might have to justify the amounts recommended for medical supplies, office supplies, vehicle operations and repairs, etc.

In the other budget systems described below, managers and elected officials focus the budget process on the programs provided, and their costs and benefits. During the budget development process, the EMS manager and elected officials would focus on the quality and quantity of service compared to previously established goals and objectives. They would also review the total cost of their EMS operation and how it compares with similar jurisdictions, but not focus on particular cost items.

To support budget processes that look beyond the detail, accounting systems must provide information on the cost of various programs in addition to line-item information. Once the budget is approved, at whatever level, the fire manager still has to allocate the budget for each program (*e.g.*, EMS, fire response, fire prevention, etc.) to various object codes (*e.g.*, salaries, medical supplies, vehicle operation and repairs, etc.).

Lump sum budget. Legislative bodies initially appropriated single lump sums for broad purposes. For example, during the Civil War, the U.S. Congress simply allocated large amounts of money "for the Army of the Potomac." Of course, when the funds ran out and more was requested, Congress had little information on how the funds were spent.

Line-item budget. Developed in New York City in the early 20th century, these budgets provide much more information and control to fiscal officers and legislators than lump-sum budgets. Most major U.S. cities adopted line-item budgets in the 1920s.

Program budget. Line-item budgets provide little information on the overall uses of public funds. Therefore, program budgets were created in the 1940s to allow upper managers and legislators to focus on the objectives for which public funds are allocated. Use of program budget methods, broadly defined, has become increasingly widespread. This increase has been facilitated by modern financial management systems that make it easier to categorize revenue and expenditure information by purpose, in addition to capturing object code information.

Performance budget. Also developed in the 1940s, these budgets focus on the results achieved from the allocation of resources to various purposes. Performance can be evaluated in many ways, including units of work accomplished (workload), efficiency, and effectiveness. Emphasis on the latter was an important new development in budgeting.

Planning-program-budgeting (PPB). Rather than a single approach, PPB is a collection of methods developed in the 1960s that attempt to bring better long-range planning and improved analytical techniques into the budget process. While PPB is rarely used formally and in its entirety today, many of its methods are practiced. For example, the concept of "life-cycle costing" (evaluating the total cost of a piece of equipment over its useful life, including costs of operation, repair, etc.) is an aspect of PPB.

Management by objectives (MBO). Many governments adopted this method in the 1960s to focus on the results achieved through budget allocation. Organizational units, and in some cases individuals, set goals and objectives. MBO is not a budget process *per se,* but can be related to the budget process by providing information on the amount of funds allocated to each objective through the various organizational units.

Zero-based budgeting (ZBB). This method became well known when President Jimmy Carter made it mandatory for federal agencies during his administration in the 1970s. It had been used in many states and the private sector in previous decades. It was developed in response to the criticism that most budget methods tend to assume continuation of the current year's programs and services, and merely focus on relatively small changes for the next year ("incremental budgeting"). In ZBB, the entire agency is evaluated every year. Each program is described and priced, resulting in a "decision package." The program could be continued or dropped in favor of a higher-priority program. While this technique forced a useful examination of old practices that may have outlived their usefulness, most governments found it impractical and unnecessary (*e.g.*, is it truly useful to thoroughly re-examine, each year, the level of fire response needed in a particular jurisdiction?) While few governments practice formal ZBB, many conduct a thorough review of agency programs and budgets on a cyclical basis such as once every five years. Other governments require agencies to submit budgets at different levels, such as 2001's level of a 5% increase and a 5% decrease. Examination of possible increases and decreases is intended to force evaluation of the effectiveness of services provided by the agency, and articulation of relative priorities.

Recent budget trends

Recent budget reform efforts have focused on the outcomes of budget allocation decisions, and how the budget process helps or (all too often) hinders management. In their popular book *Reinventing Government,* David Osborne and Ted Gaebler issued several criticisms of governmental budget practices, including:

- Inappropriately tight expenditure controls that waste managers' time and rob them of flexibility
- A focus on accountability for detailed matters (*e.g.*, line items) rather than focusing on results
- Lack of information on the cost of services
- Failure to incorporate the profit motive into budgeting, where possible
- Planning periods that are too short (*i.e.*, the annual budget) to allow managers to plan and budget effectively

A number of state and local governments have begun to address these problems through a variety of techniques. These include: allocating lump sums for defined programs with clearly articulated goals, paying contractors based on outcomes rather than unit costs, allowing managers to carry over funds from one fiscal year to the next, adopting multi-year budgets, developing performance measures, and determining the cost of their services. The last two methods are discussed below.

Performance measures. In recent years, the Governmental Accounting Standards Board (GASB, the standards-setting agency for governmental accounting) has promoted the use of "service effort and accomplishment measures" (SEAs) in year-end financial reports. Membership organizations representing government managers, including the GFOA and the International City/County Management Association (ICMA), have strenuously objected to GASB's suggestion that it might mandate the use of SEAs through accounting standards.

Partly in response to GASB, the ICMA has launched a major effort to establish performance measures for common municipal services, including fire services. (More information can be found at *www.icma.org*.) As a result of these and other efforts, many governmental agencies have incorporated performance objectives and measures into their budget processes. One aspect of many performance measures is comparing your agency's performance to other comparable agencies, known as "benchmarking."

Fire managers might, at first, view such efforts as another administrative chore. It's better, perhaps, to view them as an opportunity. Despite the criticisms of budgeting and recent reforms noted above, too many public managers suffer budget processes that focus on anything but the job that the department is doing and the resources it needs to do its job better. Budget processes too often entail higher-level managers and elected officials spending hours questioning fire managers as to why they need $X for office supplies, $Y for small tools, etc. By developing performance measures and including them in their budget documents and presentations, fire managers have a chance to steer their budget processes in a more useful direction.

Some managers might protest, saying that the governing board and/or chief executive dictate the budget process and discussion. Of course, that can be true. Nevertheless, any budget document can be transmitted with a cover memo that articulates major goals and measures of accomplishment in achieving those goals. Further, some elected officials may be more willing to rid themselves of line-item detail than public managers may think.

Costing services. In addition to establishing performance measures, many governments have begun to determine the total cost of providing various services, which is referred to as "costing" services or "activity-based costing." Although common in the private sector, most governments cannot state their cost to provide individual services because their accounting systems were not set up to do so. The systems often centralize many expenses rather than allocating them to the appropriate service areas. For example, all support costs (*e.g.*, facilities maintenance) may be accounted in one fund and all pension costs in another.

The effort to "cost" services has two objectives. First, policy-makers can make more informed decisions when evaluating whether to fund one service or another. Second, managers can compare the cost of providing services with comparable agencies, and in some cases, with the private sector. Agencies with higher-than-average costs could evaluate their management practices (and those of less-costly agencies) to determine whether they might become more efficient, and whether contracting privately for some services might make better use of taxpayer money.

Recommended budget practices

In the late 1990s, eight national associations representing government administrators and elected officials joined together to develop recommended budget practices. The organizations formed the National Advisory Council on State and Local Budgeting, to which they also appointed representatives of academia, labor unions, and private-sector financial services providers. According to Jeffrey Esser, Executive Director of the Government Finance Officers Association, the recommended budget practices represent "an unprecedented cooperative effort by several organizations with diverse interests to examine and agree on key aspects of good budgeting."[1]

Just what is "good budgeting?" The Advisory Council recommended that budget processes incorporate several principles. In its report, the Council stressed that application of these principles should be guided by a long-term perspective and involvement of stakeholders in the budget process. The principles and their elements are listed below.

1. *Establish broad goals to guide government decision making.* A government should have broad goals that provide overall direction for the government and serve as a basis for decision-making. Related elements:
   - Assess community needs, priorities, challenges, and opportunities
   - Identify opportunities and challenges for government services, capital assets, and management
   - Develop and disseminate broad goals
2. *Develop approaches to achieve goals.* A government should have specific policies, plans, programs, and management strategies to define how it will achieve its long-term goals. Related elements:
   - Adopt financial policies
   - Develop programmatic, operating, and capital policies and plans
   - Develop programs and services that are consistent with policies and plans
   - Develop management strategies
3. *Develop a budget consistent with approaches to achieve goals.* A financial plan and budget that moves toward achievement of goals, within the constraints of available resources, should be prepared and adopted. Related elements:
   - Develop a process for preparing and adopting a budget
   - Develop and evaluate financial options
   - Make choices necessary to adopt a budget
4. *Evaluate performance and make adjustments.* Program and financial performance should be continually evaluated, and adjustments made, to encourage progress towards achieving goals. Related elements:
   - Monitor, measure, and evaluate performance
   - Make adjustments as needed

Readers will note that only the third principle addresses activities that are typically thought of as "budgeting." The first principle essentially entails strategic planning, while the fourth entails performance evaluation, both of which are discussed in separate chapters of this book. The second chapter entails operational planning, discussed throughout this book. Fire managers seeking to examine the Advisory Council's full recommendations may obtain its report from the GFOA.

Policy framework

If the budget is to reflect a long-term view, fire managers and elected officials should consider long-term issues in developing a budget. Developing long-term goals and policies is best accomplished before the budget process, for two reasons. First, it is a lot of work that is difficult to complete while putting together a budget with all of its detail and complexity at the same time. Second, policy-makers can take a more comprehensive and long-term view when they do not face immediate decisions, especially since policy-makers may have considerable pressure from various constituent groups when they face those decisions. Key components of long-term policies that affect budget decisions are addressed in the following paragraphs.

Strategic plans. Since the budget process allocates resources to competing objectives, you must first define your objectives. Fire service managers without clearly defined objectives will be at a disadvantage when competing with other services for scarce resources. More importantly, defining objectives will help ensure that appropriated funds are used to the best advantage. Strategic planning is discussed in another chapter of this book.

Service plans. Sometimes referred to as "business plans," these are narrative descriptions of the services that will be provided with the funds appropriated through the budget process. Keep the audience in mind and avoid jargon when writing these plans. Ideally, service plans will include performance indicators.

Long-term financial plan. Ideally, the jurisdiction should project its financial condition for the next three years (or perhaps five). This can be a difficult task, especially if the jurisdiction's revenues and expenses are complex. However, a long-range financial plan will help the jurisdiction identify opportunities and challenges that can be met most effectively with advance planning, through the budget process and other means.

Financial policies. This refers to sound financial management practices that generally vary little from year to year. Many may seem obvious, such as not spending "one-time" funds (such as proceeds from sale of equipment) on recurring expenses such as salaries. However, financial policies can provide a check on risky practices that might be tempting in the desire to provide services. Ideally, the governing body should review and adopt the financial policies.

One policy topic that should be discussed is the amount of reserve funds ("fund balance") that should be maintained to help the jurisdiction respond to an unexpected drop in revenues or a catastrophe. The policy should state the minimum and maximum amount of reserve, and when it would be appropriate to use the reserve. The Government Finance Officers Association has developed a "recommended practice" regarding fund balance policy.

You should also consider the jurisdiction's practice for replacing capital items such as fire trucks. Meet with your financial officer to discuss vehicle replacement schedules. Having determined appropriate replacement schedules for various types of vehicles, you can then approach the governing body with a long-term funding plan. For example, the jurisdiction could set aside funds each year to purchase apparatus with cash, or use lease-purchase financing (discussed under the Debt Financing section of this chapter). Dealing with these issues through a policy avoids raising them in the middle of putting together the budget. You should avoid surprising the financial officer and governing body during the budget process with a large, unexpected budget request such as the need to replace a number of vehicles (or even a very expensive single vehicle). You are likely to find yourself having to justify each aspect of the vehicle's performance including its mileage, downtime for repairs, conformance with current technology, etc. (Remember that adopting a funding plan does not mean that you will simply replace a vehicle at the scheduled year with no review of its performance. If a vehicle could last a year or two longer, advise the fiscal officer to put off replacement. You will look like a model of fiscal restraint, and build trust for the inevitable day that you recommend replacing a vehicle before its scheduled date.)

Other areas that should be considered are:

- Policies for wages, wage increases and benefits. (These may be established as part of a separate human resources policy.)
- Revenue policies, which should cover the jurisdiction's major revenues, the circumstances under which the jurisdiction may wish to increase or decrease the revenues, and the type of revenues (*i.e.*, fees versus taxes) that should be used to fund particular services. (Funding Sources is also covered in a separate section of this chapter.)
- The uses of grant revenues.
- Funding items that might be difficult to estimate and/or do not occur on a regular basis, such as liability claims.
- Use of budgeted funds that have not been expended at the end of the year.
- The relationship between recurring revenues and expenses. Of course, budgeted expenditures for recurring costs (such as salaries) should be less than budgeted recurring revenues. The jurisdiction may also want to consider establishing a cushion in the event that revenues do not meet expectations. For example, the jurisdiction may keep budgeted revenues for recurring expenses at least 2% under expected revenues. (The need for a cushion will depend upon the predictability of each jurisdiction's funding sources.)

How to develop a budget

At this point, readers may be saying, "Understanding the concepts that underlie the budget process is interesting, but exactly how do I put together a budget?" As noted above, budget processes vary considerably from place to place in response to local preferences and state laws. However, it is possible to describe, in broad terms, the major steps to develop a budget.

Update and review relevant plans and policies. Fire managers should identify issues arising from the review and seek guidance from upper management and/or the governing body prior to "putting together the numbers" in the budget. Conducting formal updates of these plans every other year may be adequate in some entities, but they should at least be reviewed annually. Examples of issues that managers might raise following plan review include:

- Should certain revenues be increased due to declines in other revenue sources?
- Is it an appropriate time to add new services (or expand existing services) that are high priorities in the strategic plan? If so, how will they be funded? (In addition to new programs, service issues also include significant strategic and tactical delivery changes.)
- Could opportunities to cooperate with other fire services allow the jurisdiction to cut costs?

For each of these issues, outline the concept and identify its impact. Make initial estimates (not detailed) of the resources needed for new program, including level of effort, costs, etc. Following preliminary approval, proceed to the second stage where you complete the proposal, including justification based on sound objectives, and detailed estimates of cost and/or staff time.

Estimate expenses. If you have established plans and policies for major costs, as recommended above, you will derive a large portion of cost estimates from those sources. Still, even with policies a great deal of technical work may be necessary.

Provide detailed guidance to people who put together budget information. Even with clear policies, if you don't provide detailed instructions you will receive disparate responses. For example, you may want division managers to give you their priorities for new expenditures. Inevitably, one manager will recommend several items, listing the priority of each as "high." So, you will have to include specific guidelines such as "rate each priority high, medium, or low, with one-third of the priorities in each category." Ideally, you should provide "fill in the blank" worksheets and/or examples with the instructions. (An example is like the picture that "says a thousand words.")

Focus on the larger costs. The "80–20" rule applies to most fire services—they spend 80% of their budget on 20% of the items. The smaller items should be reviewed, but don't spend too much time on them. Variances from year to year in those items will probably cancel one another out. Only a very small fire service should worry whether it will spend $80 or $100 on a particular supply item.

Plan on variations expected beyond the budget year. Perhaps you with to add a fire station so you need to hire twelve firefighters. Their salaries for the next budget year may total $XX, but is that a fair representation of your costs? In many jurisdictions, firefighters receive significant increases when they complete their probationary period. It may be appropriate to budget $YY next year, but the true ongoing cost of hiring the 12 firefighters may be 15% greater than $YY. When you put together the proposal to open the station, you must identify sufficient funding, on a long-term basis, to paying the employees after they complete probation. Table 3–1 shows how to explain an activity that will cost one amount in the first year (the budget year) but another amount on an ongoing basis. (These variances over time also demonstrate the usefulness of a long-term financial plan.)

Don't budget for every contingency. Many jurisdictions experience large payments to retiring employees for unused vacation leave. Perhaps you have five employees at retirement age. But is it reasonable to budget payments for all five? It is not likely that all five will leave in one year, and budgeting the expense for all will leave you with substantial unspent funds. Perhaps you could use those funds elsewhere. Further, budgeting significant amounts that are not needed may cause you to loose credibility with your financial officer and governing body. Better to budget for those employees who you expect will actually retire, or better yet, suggest that your financial officer establish a contingency for this purpose.

Table 3–1 Proposed Company Inspection Program

Budget	Startup Costs	Recurring Costs	Total–First Year Cost
Personnel			
Two new FTE	$ 00	$100,000	$100,000
Existing shift personnel	00	140,000*	140,000
Materials	10,000	5,000	15,000
Capital Assets:			
Two inspectors' cars	40,000**	5,000**	45,000**
Total Program Costs	50,000	250,000	300,000
Less: Opportunity costs*	00	140,000	140,000
New funding request	$50,000	$110,000	$160,000

**Opportunity costs are the value of existing rsources that would be allocated to the new program. These costs would not be eliminated if the new program were not approved.*

***These costs would be funded through the equipment replacement fund.*

Estimate Revenues. Avoid estimating them too low. If your revenues significantly exceed your estimates every year, you will lose credibility and your governing body will be tempted to overspend. Better to budget revenues closer to expectations, but plan to not spend all of it. If the revenues do exceed projections, you can always use the money to fund one-time expenses or add to reserves that may have fallen over the years.

General guidance in budget preparation

Push to have actual needs definition, implementation planning, and cost calculations made as close to actual delivery responsibility as possible. Develop standards for proposal writing and train all personnel in their application. Establish reasonable deadlines for budget submissions, publish them well in advance, and adhere to them. Require that all proposals for increased costs, even small ones, be submitted in writing. It's easy for people to say, "We need," but if they care about the idea, they will put it in writing! Use the department's strategic plan as an initial "test" for appropriate consideration of added costs. Encourage and train proposal writers to develop objectives stating the outcome for the community, avoiding the use of undefined fire service acronyms and jargon. Develop an atmosphere that both supports ideas, yet allows personnel to feel comfortable enough to play "critic" for one other. Test proposals on non-fire service trial audiences (spouses, secretaries, and friends might serve this purpose).

The major components of a good budget document are as follows:

Transmittal memorandum. Your intended audience may not read the entire budget document, so consider the messages that you consider most critical for them to receive. The transmittal memorandum should highlight your fiscal condition and any major changes proposed in the new budget.

Relevant policies and plans. Consider including a summary of your strategic plan, long-term financial plan, and financial policies.

Summary information. In addition to traditional tables summarizing revenues and expenditures, use charts and graphs to summarize them. With improved microcomputer software, you can obtain many good financial packages at a reasonable cost that can be used by people without specialized skills. Good microcomputer charting programs include primers ("tutorials") on designing good graphics for data display. In addition, cost-effective training is available through many sources including the Internet, on CD-ROMs, and at local community colleges. Therefore, only some key tips are provided here:

- Use one primary message per chart.
- The message should be quickly understood. Charts and graphs should stand on their own rather than requiring extensive interpretation.
- If multiple related messages are needed to show correlations, display ideas separately first, and then combine them on one chart.
- Use color, but use it judiciously and always for a reason. Use hot colors for negative issues, and cooler colors for positive aspects.
- Choose the right type of graph:
  - Pie charts to show proportional pieces of a whole.
  - Bar or column graphs to make comparisons.
  - Line graphs for a running value over a timeframe.

- Mix graph types if your data allow it—three or more charts with the same general format will begin to run together in the reader's mind.

Service plan and performance indicators. You should always include your service plans, since these relate directly to the level of funding requested in the budget document. Service plans for all the entity's divisions or organizational unit may be placed at the beginning of the budget document, or incorporated in the separate sections for each program or organizational unit.

Fund statements. These are traditional statements showing the revenues, expenditures, and year-end financial position (fund balance) for each fund.

Expenditure information. Include the information that most people think of as "the budget," which is the expenditures for each program and/or organizational unit (depending on the format used by the particular entity). Even at this level, the information should be summarized rather than line-item detail, unless required by law. Of course, some elected officials view it as their job to review line-item detail, at times due to mistrust of public employees, and will not easily give it up. Summary personnel information is often included as well. (See Appendix 1)

Other information required by state law. A local government's authority to spend money flows from state law, which often includes specific requirements to include certain information in the budget document and/or ordinance adopting the budget.

Prepare for the presentation of the budget proposal. Do your homework. Regardless of the type of legal entity or source of funding, keep the following points in mind—they can make or break your budget justification efforts:

- Know your community, its population, and its hazards
- Know your customers' special needs
- Know your programs, objectives, and budget
- Know your current performance
- Know your jurisdiction's budgeting system
- Know your political system
- Know your audience

Know your community, its population, and its hazards. There isn't much a chief can do that would be more embarrassing, or potentially damaging to his department's budget sales effort, than to misstate some community detail that is held as common knowledge or make the chief appear out of touch. Examples of this might include drastically under- or overstating some statistic like total population, or land area in the city limits, or making an offhand comment about ethnic populations. Have a general working knowledge of your community's demographics, and, more importantly, know how fire and life safety are affected by those conditions. General economic condition, age and upkeep of structures, unemployment rate, literacy levels, age, incidence of drug abuse, and blight all have a major impact on emergency service demands and fire problems. Be realistic about the local economic condition; presenting a grossly expansionary budget when the economy is in the pits and tax revenue is down may hurt credibility, unless those programs can help the community recover from the economic downturn.

Don't overlook major incident or disaster potentials, even if your history hasn't included such events. Every part of the country has the potential to suffer natural disasters. Being unprepared for the inevitable could end your career, and be catastrophic for the community. A major fire (not even a conflagration) could have a long-lasting negative impact on the department's credibility and budget if the chief and department had done nothing to prepare for it.

Know your customers' special needs. Every community has special-needs populations, such as the elderly or large numbers of non-English-speaking residents. Remember, these are your customers. Design programs to meet their needs and concerns. That might include safety education of the elderly or a daily phone check for shut-ins. Some may say phoning shut-ins is "not our business." But if the department provides EMS, has dispatchers that are not busy 100 percent of the time and a computer to track the shut-ins, it may be a good fit, provide familiarity between department staff and its customers, and score some points at budget time.

Know your program, objectives, and budget. Mission, strategic plan, objectives, and program initiatives are the basis of budget justification. It is much easier to play on the political field when you can use these items to provide focus. Approving expenses is more palatable for elected officials when the benefits to the community are clear.

Consider the following examples of program objectives:

A. Perform preventive maintenance on all motorized apparatus and equipment every 25 to 30 hours of operation by hiring two additional full-time mechanics, upgrading tools and diagnostic equipment, and expanding spare parts inventory.

B. Ensure that no deaths, injuries, or property damage will result from on-scene equipment failures attributable to insufficient maintenance.

Here is another set of objectives:

C. Replace all breathing apparatus with positive pressure equipment that meets OSHA standards and NFPA 1500.

D. Reduce firefighter smoke inhalation injuries (and resulting Workmen's Compensation and insurance expenses) by 30% and avoid potential liability created by providing breathing apparatus that meets current standards.

Both A and C are written to explain the activities from an internal perspective, that of service personnel. However, the audience for the budget is primarily people outside the fire services. From an external perspective, A could easily sound expensive and downright unnecessary; after all, no one does preventive maintenance on his automobile every 25 hours of operation. C most likely would defy understanding; are those "oxygen" bottles not already pressurized?

Non-fire service people can understand B and D. Preventing deaths, injuries, and property damage are words right from the mission statement, and, in an elected official's mind, they are the fundamental purpose of funding a fire department. Even the most callous councilperson can understand the reason for preventing firefighter smoke inhalation injuries, even if it's only to prevent liability to the government entity.

Know your department's current performance. Do you know the specific progress that your department is making toward its major goals and objectives (as defined in the strategic plan)? Take this little open-book quiz. Don't worry, you'll grade it yourself and you don't have to show it to anyone else! Seriously, do this. You might find it quite enlightening. Use a separate piece of paper. As you work, also include the time it would take to research the answers you can't give from personal memory.

1. What is your approximate budget?
2. What is the population protected: is it growing or shrinking?
3. How many total responses were made last year: Fire? EMS?
4. What were the fire losses in dollars last year?
5. What was the motor fuel budget for the last budget year?
6. How many civilian fire fatalities did the jurisdiction suffer last year? The last three years? Is the per capita trend up or down from five years ago? Same questions for fire injuries.
7. If you charge for EMS, what was your collection rate last year? If you didn't charge for EMS, did the department produce any income? How much?
8. What were the top three structure fire causal factors, in descending order, last year?
9. Question number 6, but for fire personnel?
10. How many firefighter injuries caused lost time?
11. How many firefighter injuries occurred during non-emergency activities?
12. What were the three leading causes of fires where there were injuries or deaths?
13. What was the code blue (clinically dead at some point) save rate in your jurisdiction last year?
14. How many civilians did department personnel rescue from structure fires last year?
15. How many prevention inspections did your agency conduct last year? How many violations were discovered?
16. How many violations were cleared by the first re-inspection? How many went to court?

17. What were the three most common violations?
18. How do the causal factors from questions 8 and 12 compare with violations found during inspections? How much training has the department conducted in hazard recognition?
19. How many total public education contacts did the department make last year? How many of those contacts focused on the causal factors identified in questions 8 and 12?
20. What was the total cost of your fire prevention efforts, including opportunity costs?

If you were able to answer all these questions accurately, immediately, and without referring to your annual report (or other sources of information), you are rare indeed. (Or your department has no alarms and no programs.) If you were able to answer all questions by devoting just five minutes to each, or you found questions 8 and on just as easy to answer as questions 1 through 4, you are also in a rare department. If you found the questions getting progressively harder to answer, you're not alone. If you had questions you couldn't answer at all, you're also not alone. Maybe it's time to get a cup of coffee, sit back for a few minutes, and ponder just how seriously your department takes fire prevention—and then consider whether you should make any changes in emphasis, budget, or both.

If you don't run an ambulance or have a formal EMS program, is that a valid reason for not knowing the answer to question 13? Maybe you don't have a definition for a "code blue save," but given the significance in impact of that performance measure, maybe it's time to sit down with your medical director and some of the paramedics and have them develop one.

Maybe there are other holes in your information system that need to be filled. Every year, as objectives are identified, make sure the information system provides for data collection that will measure achievement. Also, if possible, gather baseline data about current performance. (Information management, planning, and financial management are very closely related. The National Fire Academy offers courses on these subjects.)

If you knew all the answers to all the questions, or at least knew where to look to quickly get the answers, wouldn't you feel much better prepared to enter the budget justification arena?

Know your jurisdiction's budgeting system. Know the levels and timing of administrative reviews before the budget goes to the governing body. As stated earlier in this chapter, understand the personal agendas of those who can affect what gets to the council, and the light in which it is presented. Especially in larger jurisdictions, these levels of bureaucratic review can be extensive, as budget analysts, assistant finance directors, budget directors, assistant city managers, legislative aides, and the city manager may all review budget proposals before elected officials gets to see (what's left of) the proposal.

Have a working understanding of the major thrusts of other departments' budgets. Be prepared to give insight into interdepartmental dependencies and complementary programs.

Know your political system. Know the local rules of the political game. Know whom you could lobby to influence the decision, whom you can lobby without breaking organizational and/or community norms, and when. Some of the rules may be unwritten. The word "lobby" may be taboo. Department members, even the chief, may be prohibited from initiating contact with elected officials. Understand the latitude of your response if an elected official initiates the contact with a question or concern.

Make sure department personnel know the boundaries, and inform them about major department program features, goals, and objectives, since it is unrealistic to think that no elected official will ever ask a firefighter what the rank and file think about some budget issue. A firefighter could be an elected official's long-time friend or neighbor. Also make sure you know where the community power centers are. Sometimes they are civic organizations; in other communities it may be a local restaurant where the movers and shakers meet for breakfast.

Get to know the personal agendas of the elected officials—what are their "hot buttons?" One elected official was always concerned about the upkeep and appearance of the city's cemetery. The fire department proposed "Cleanup, Fix-up Week" designed to remove fire hazards from in and around buildings throughout the city wide, launched with a ceremony at the city cemetery where new landscaping and fencing could be shown off.

Know your audience. When you give presentations on budget proposals or department activities, know your audience. Know their general education levels and political leanings. It may not be very effective to give a folksy, "good old boy" presentation to a group of engineers and business managers, just as a technical discussion of fire flow analysis and sprinkler head loading would not go over very well with a neighborhood group of homeowners.

When you speak to Chamber of Commerce business owners, highlight how your programs will enhance their marketability, insurance rates, liability management, and business in general. Point out that fire department programs and services which they must use (plan review, permit management, inspections) are designed to maximize their access to your programs, will be completed in a timely fashion, and will minimize business interruptions. Then, make sure you can deliver on your promises. Explain how, if they experience a loss or fire, you can help them get up to speed with as little business interruption as possible. These assurances will interest business people more, and do more for your budget, than pictures of chrome and expensive fire stations.

Homeowners will appreciate knowing that your operations and training are aimed at, first and foremost, protecting their loved ones, and, second, protecting their property, (especially the irreplaceables—pictures, mementos, and valuable papers), rather than at breaking windows, cutting holes, squirting water, and getting back to the station ASAP. Think about the media coverage of residential fires. The public sees piles of burned rubble and personal belongings being shoveled out of broken windows. Have you ever seen a television or newspaper picture of furniture neatly protected by salvage covers? The most common interior news footage is of firefighters pulling lath and plaster and making a bigger mess. Sure it's a messy job, but try to understand the public's ignorance of the technical aspects, and their sensitivities about the destruction or rough handling of personal belongings.

Knowing your audience takes us to the next major focus of our justification efforts—the formal budget presentation before the funding body.

The formal budget presentation

Generally the real work in "selling" the budget is accomplished before it goes before the elected body in a formal presentation. Even so, using the following practices will also help gain budget approval:

- Focus the presentation on demonstrating that specific positive outcomes will result from the funding you have proposed. Don't overload the document or your presentation with too much detail. Keep the focus on objectives, risk, and service levels.
- Respond to questions directly rather than trying to "dodge" a tough question. Have reference materials available for the unexpected question. However, you can't anticipate everything. It's better to be honest and admit that you don't know the answer than resort to "smoke and mirrors." Smoke eventually clears and mirrors can break. Don't risk hurting your future credibility.
- Realize that the elected officials are also on a stage, and may ask questions for which they know the answer. This may be done for the public's benefit, or to get answers officially in the record, or to show other officials, the media, or the public that they are doing their job of critical review. Rather than becoming annoyed by this type of question, keep in mind that by asking it of you, they are endorsing your credibility. Be careful how you answer, especially when your first impulse is to say, "If you had read the materials provided…." or, "As I already explained in your work session…."
- Be aware of local identity sensitivities. Especially in smaller communities on the fringes of a major city, it may not be wise to say, "This is the way City Z does it." You may quickly be reminded, "We're not City Z and that's the way we like it!"
- Be aware of body language. Build into your presentation several opportunities to jump to a summary. Have some alternate tacks in mind if you read negative reaction.
- Have an articulate program manager make the presentation for a major new program. This must be coordinated with the upper management (*e.g.*, the city manager), if applicable, but the chief of department should not feel that he personally must present every program or detail. Allowing others to make a presentation demonstrates your belief that the department has more than one highly competent individual. If compatible with local protocol, put together a

team presentation; let the chief set the tone and introduce other staff and program managers, who often can provide better answers to detailed questions.

- Be concise. Elected officials' time is important so get to the point. Respect the meeting agenda, even if others have not. If you run long, make sure it was their doing.
- There is no better way to make a concise presentation and focus your audience on key points than to use visuals. A good graphic will convey its primary message in seven seconds or less, and that's without an oral presentation. Charts and graphs can make complex data much more understandable. (Tips on preparing charts and graphs are provided in the section above on preparing the budget document.)
- Mere pictures can add interest and impact to a verbal presentation. The equipment required to display pictures—a digital camera, computer and projector—has dropped in price significantly in recent years. However, don't let the flashiness of visuals (particularly special effects) detract from your message or steal the show. A little bit of flash can add interest; use too much and the audience may begin to wonder how much your presentation cost or how much time you had available to devote to it.
- Use valid comparative data. Make comparisons to communities that are similar in size, demographics, and geography. Refer to national statistics, but be careful to refer to communities of similar size and within your area of the country.

Finally, know when to shut up. If you've got them sold, don't feel obliged to give your full presentation, or you may just talk them out of it! If you read negative body language or sense a negative tone in their questioning, don't get argumentative. If the vote goes against you, take a deep breath, bite your tongue if you have to, and sit down and ponder any lessons learned. Don't argue - you will have other chances. Don't take it personally. If you presented a solid budget based on community needs and programs to meet them, you did your job.

When a budget proposal is rejected. When a proposal is not approved during the budget process, whether by a city manager or governing body, give feedback to the person who originated it, who may want to know that:

- Given the fiscal constraints and agency priorities, the proposal didn't survive the internal cut.
- It was an excellent proposal, but in the current political environment, it wasn't the right time. However, if politics change, it's ready to go. (And we all know that sometimes that change is brought about by a disaster…ours, a neighbor's, or the nation's.)
- The concept needs to be more fully developed (*i.e.*, more clearly defined objectives, more research to tie outcomes to process, better needs analysis, etc.).

Whatever the reason the proposal was rejected, the originator showed initiative and positive risk taking in its development. Reward and recognize that, don't punish.

SUMMARY

Following these principles in a dedicated, professional way (which does not necessarily mean paid) will allow fire managers to build on each past effort, increasing the following with each budget cycle: credibility, the accuracy and completeness of budget projections, the usefulness and ease of revising plans and performance measures, analysis skills, and application of the foregoing procedures. Problems will arise, of course, but the information and skills gathered by applying these principles consistently will help in earlier problem identification and process improvement.

BUDGET ADMINISTRATION

Regardless of whether your budget is tattered or intact after final adoption, there are a number of tasks required to properly manage what has just been passed. This section outlines those tasks.

Before the new fiscal year begins

First, redistribute to the program budgets the actual amounts funded. Few things are more frustrating to a program manager that to have $X removed from a line-item budget without specifics as to what program(s) the reduction is supposed to affect. The financial officer will probably make those changes, but if not, the chief must do so. Alternatively, if the governing body adopted a program budget, you will have to determine the line items to reduce.

Develop a spending plan if one is not already in the budget submittal. This is more than taking each line-item amount and dividing by twelve months, although for some types of expenditures that is appropriate. Estimate procurement timing and fluctuations in cash flow by month, or at least by quarter, for major types of expenditures, particularly those which can fluctuate greatly such as overtime.

These projections serve two purposes. First, they allow you to know whether expenditures are in control. This may not be apparent without careful projections. For example, many agencies provide more paid holidays during the second half of the year due to the Thanksgiving and Christmas holidays. An agency with a January 1 fiscal year could find itself short of funds if one-half its overtime pay were expended in the first half of the year, leaving insufficient funds for overtime associated with holidays. Second, the projections will help the fiscal officer invest idle funds for maximum return, yet have them available when projected.

Train personnel in the policies and procedures related to spending authority. In particular, involve program managers and others who deal directly with money handling, spending, and requisitions. Who is authorized to initiate procurement, for what items and up to what amounts, and under what circumstances? Someone needs to have emergency spending authority. In some departments that might be the chief only. In most departments the on-duty individual who has shift command responsibility is appropriate. (See the section on Purchasing for more discussion of this area.)

Establish transfer authority. From time to time, object code line item account balances may need to be adjusted, moving excess funds from one account to make up for shortages in another. Typically public entities will restrict transfers for some items such as salaries for full-time employees. Depending on the budgeting system used, this transfer authority typically rests in the hands of the program manager for transfer within the program, the division or functional area manager for transfers between the programs of the division, and the chief for transfers between divisions. The procedure for petitioning restricted transfers should be made clear.

Finally, establish position controls. Since personnel costs often constitute from 60 to 90 percent of a public entity's budget, controlling the number of employees is essential. Position controls are often included in the legislation that adopted the budget, or in separate but related legislation. Regardless of whether the governing body enacts position control, administrative systems must be in place to ensure that employment practices are consistent with the positions anticipated when the budget was developed. Position control usually includes the number of positions and their pay classifications. (Part-time positions should be indicated by "full-time equivalent positions" or "FTEs.") Many modern computerized financial management systems include position control systems as part of (or related to) the payroll systems. The authorized positions are set up on the computer system at the beginning of the fiscal year, and an employee cannot be paid unless he or she fills a designated position.

Ongoing budget administration

Periodically (at least monthly), review and report on the condition of program and line-item accounts. Such reviews will allow managers to identify problems as soon as (or even before) they arise. These reports should show total funds budgeted for the year, what has been budgeted, spent, and committed by requisition, and variance (percentage high/low according to spending plan). In this age of computer networks, it is possible to get account status on-line, in real time, and get up-to-the-minute account information. On-line entry of purchase requisitions and vouchers keeps the accounts up to date.

When reviewing expenditures, don't be too concerned about minor variances, particularly in relatively small line-items. As noted in the section on budget development, it is impossible to accurately predict each item. However, you may want to make transfers from some accounts to others, even of relatively small amounts, to adjust the spending plan to changed conditions.

Major revenues should be monitored regularly (ideally monthly) for early identification of any fiscal problems. This is less important in jurisdictions whose revenues do not typically fluctuate greatly from year to year, such as property taxes. Monitoring is critical in jurisdictions where a large portion of their revenues fluctuate significantly with economic conditions (called "cyclical revenues") such as sales taxes and taxes on natural resource production (*e.g.*, oil and natural gas). Major revenues should at least be monitored at the end of each quarter.

Management should report the agency's fiscal condition to upper management and the governing body at least quarterly. If conditions differ significantly from expectations, it may be desirable to adjust the fiscal plan and formally amend the budget. Many agencies have a formal review of program performance at the six-month point of the budget year. That period is long enough to get most new programs up and running. Even if there are not measurable service impacts at that point, there will at least be actions to implement the programs. Are these things on track? If not, the six remaining months are time enough to adjust major programs, possibly to see results by the end-of-year review. The six-month review also provides an opportunity to respond to any fiscal problems that might have arisen.

Fiscal practices to avoid

In the long run, both the fire department and the public are better served in an environment that encourages responsible use of community resources, including fiscal resources. Certain common fiscal practices handicap such an environment. One is the "legal spender concept." It says, "Spend all of your budget! If we don't, we'll not only lose the amount we didn't spend this year, we'll lose it next year as well. City Hall will say, 'You didn't need it last year, so we won't give it to you this year!'"

Some jurisdictions try to short-circuit the rash of spending at the end of the year by freezing procurement during the last quarter of the fiscal year. This only works once! In subsequent years managers simply spend more sooner. The danger of this approach is that the resources won't be available if last quarter expenses are higher than the manager anticipated. Also, departments may spend the money on lower-priority needs early in the year, rather than risk "giving it up" at the end.

Another solution that hasn't worked well was exemplified by a fire protection board that gave the chief a bonus each year, based substantially on the amount of the budget that was unspent. Things finally came to a head, and the chief "resigned," when the board discovered the firefighters were sharing rubber turnout coats among shifts.

To a large extent, the "use it or lose it" attitude stems from the belief that funds not used by the fire service will not be used to full benefit. While prevalent, and perhaps understandable, this attitude demonstrates the presumption that staff are in the best position to determine how to best use community resources. Firefighters, believing that the fire service is the most important program of the local government, will want to use community resources there. But the same holds true for police officers, road crews, and all other units of government. The most important way that a fire chief can foster responsible fiscal management is to promote the idea that all government units serve the public, and that in a democracy, elected officials should have the final say in determining funding levels.

In addition, there are several more specific ways to minimize the effects of the "legal spender" attitude. The simplest is for the jurisdiction to allow unspent funds to carry over, either in the same programs/accounts, or within the budget at the chief's discretion. There still may be a perception that the jurisdiction will diminish the subsequent budget by a similar amount, but at least managers won't feel the need to argue for money they already have. Some jurisdictions split any savings, with half allocated to the individual departments and the other half returned to fund balance. Another approach is to have unspent monies routinely rolled into the department's apparatus replacement fund.

Accounting and Financial Reporting

Fire service managers may feel that accounting and financial reporting are, at best, necessary but nonproductive activities, so the less time spent on them, the better. It is hoped that this section will allow fire service managers to develop an understanding of accounting and financial reporting, and an appreciation of their role in the financial management system. At the same time, this section will not attempt to turn fire service managers into accountants!

Overview of accounting systems

Accounting systems may seem to serve the needs of only "bean counters" and regulators who enjoy such matters. But imagine that you had no accounting system, instead simply recorded the money you took in and the money paid to vendors and employees. At the end of the year, what would you know about your fiscal affairs?

You would (hopefully) know how much money you had taken in and disbursed, and to whom. But that's it. You wouldn't know how much you had spent on particular products, classes of employees, or programs. You couldn't say how much money you generated from particular revenue sources. An accounting system will tell you that, and more. You would know how much money you owed to vendors for products that they had delivered but you hadn't paid for yet. You would also know how much money was owed to you for taxes or services that you hadn't received yet. An accounting system is a system to classify and categorize financial transactions so that you can look at them comprehensively and make sense of them.

To make it easier to manage the myriad of financial transactions, accounting systems generally keep each type of transaction in a subsystem called a "subsidiary ledger." Examples include payroll, cash receipts, and accounts payable ledgers. Data in the subsidiary ledgers are "rolled up" into a central system called the "general ledger" (or "GL"). For example, a payroll system keeps track of employees' wages, other benefits such as medical insurance and pensions, and employee deductions for taxes, pension contributions, etc. Once the account staff has confirmed that a particular week's payroll transactions have all been recorded properly, they "post" information from the payroll ledger to the GL.

Chart of accounts. A "chart of accounts" allows accounting systems to classify financial data so that they can answer questions such as "what were our revenues from ambulance billings?" and "how much money did we spend on medical supplies?" Most fire service managers are familiar with the string of numbers, called "line items," which make up a chart of accounts.

A simple chart of accounts might have three sets of data such as XX-XXX-XXXXX, where the first set of numbers is the fund, the second is the department, and the third is the "object code." So, a fire department might have a line item such as 02-040-01110 used for regular salaries, where:

Fund: 02 General Dept: 040 Fire Object Code: 01100: Regular Salaries

However, a more complex chart of accounts can provide additional information such as the organizational division and program. Let's say that you have a Fire Prevention Division program headed by a full-time Fire Prevention Manager. In addition to fire safety education and other duties, he coordinates the fire inspection program and personally performs some building inspections. Your chart of accounts includes sections for divisions and programs, so the hours that the Fire Prevention Manager devotes to building inspections are paid through line item 02-040-30-003-01110, where:

Fund: 02 General Dept: 040 Fire Division: 30 Fire Prevention Program: 003
Building Inspection Object Code: 011000: Regular Salaries

The hours that the fire response units devote to fire response (and related preparation activities) are paid through 02-040-20-002-01100, where:

Fund: 02 General Dept: 040 Fire Division: 20 Suppression Program: 002
Fire Response Object Code: 011000: Regular Salaries

However, the hours that the fire response unit devotes to building inspection are paid through 02-040-20-003-01100, where:

Fund: 02 General Dept: 040 Fire Division: 20 Suppression Program: 003
Building Inspection Object Code: 011000: Regular Salaries

With a good computer-based financial management system, you can "slice and dice" the financial information in a variety of useful ways. In this example, you can determine the entire cost of the Building Inspection program, including all wages for employees performing inspections, regardless of the divisions to which they are assigned. You can then present a budget based on expenditures for each program, and hopefully elevate the focus of budget reviews on the services provided rather than the expenditures for each line item. (See the discussion on "program budgeting" in the section on Budget Development.) You can also tell the total wages paid to employees in each division, such as suppression.

Basis of accounting. The simplest accounting systems merely record cash transactions; an accounting entry is made only when cash is received or paid. This is called the "cash basis" of accounting. But such systems provide a limited picture of the agency's financial status. They do not tell you how much money is owed for goods and services that you have received but not paid for, or how much is owed to the agency.

These differences can be substantial! For example, there is typically a considerable lag between the time that property taxes are levied and received. As a result, many agencies might have little cash in the bank at certain times in the year, making their financial conditions appear dire. But they might be owed millions in property tax revenues that will be collected in several months. Under the "accrual basis" of accounting, the amount of property taxes actually expected (not merely levied) would be recorded as a "receivable" in the accounting system. While the agency might still have little cash, its accounts would present a more accurate picture of its financial condition.

Most government agencies maintain their accounting records in accordance with a third basis of accounting, called "modified accrual." As the name implies, it's a mixture between the cash and accrual bases of accounting. Revenues are recorded only when they will be available to pay "current" liabilities (those due in the near future). Expenditures are recognized only when they will be paid with "current" assets (cash and assets that can be quickly converted to cash). Most governmental budgets are presented on the modified accrual basis of accounting.

Annual financial reports differ, however. Most states require that local governments issue financial reports consistent with generally accepted accounting principles, or "GAAP." Many people have heard this term but may wonder exactly what that means. A national standard-setting body, the Government Accounting Standards Board (GASB), establishes GAAP. (There is a similar standard-setting body for private-sector organizations.)

As a practical matter, it is not easy to maintain an entity's accounts in accordance with GAAP on a day-to-day basis. So most governments, as well as private-sector organizations, make end-of-year adjustments to their accounting records to bring them into compliance with GAAP. For example, during the year an agency may not keep track of how much money it owes vendors until the vendors bill the agency. However, at the end of the year, to comply with GAAP, the agency would review its orders to find out how much it owes vendors who haven't billed the agency yet, and record that amount as a "payable" in its accounting system.

Financial reporting. As noted above, GASB governs financial reporting for governmental agencies. Exactly what does GASB require? The following are the key elements of annual financial reports that meet new GASB guidelines issued in the late 1990s (*GASB Statement 34*):

1. "Management Discussion and Analysis" which includes:
   a. Condensed financial information allowing readers to compare the current and prior fiscal periods
   b. Analysis of the City's overall financial position and results of operations
   c. Analysis of the balances and transactions for major individual funds
   d. Analysis of variations from the original and final amended budget for the General Fund
   e. Description of significant capital asset and long-term debt activity during the year
2. "Government-Wide Financial Statements" presented on the accrual basis of accounting, including the following statements:
   a. Statement of Net Assets, similar to a Balance Sheet, which reports assets and liabilities for the entity as a whole. Assets will include infrastructure assets (*e.g.*, roads and bridges), since the entity uses them to provide services.
   b. Statement of Activities, which focuses on the cost of programs, such as Public Safety, Public Works, General Government, etc. It is intended to identify the extent to which each program draws from general revenues or is self-supporting.
3. "Fund Financial Statements" that will include the balance sheet and statement of revenues, expenditures and changes in fund balance for each "major fund." (This section of the annual financial report is the most similar to the required reports prior to the issuance of GASB 34.) The "non-major" funds will be shown in this section of the financial statements in total only, by fund type. Unlike the government-wide financial statements, the fund financial statements will be presented either the accrual or modified accrual basis of accounting, depending on the particular fund type.

External audits. Most organizations are required to have their financial reports audited by an independent audit firm. What does the auditor do?

Opinion Letter. The auditor's primary job is to determine whether the financial report fairly presents the financial position of the entity in accordance with generally accepted accounting principles. "Fairly" does not mean "perfectly." Even small entities have thousands of transactions, and some small mistakes will certainly go uncorrected. The goal of the auditor is to determine whether the financial statements are "materially" correct, as defined by audit standards.

The auditor does that by reviewing the overall accounting system and "testing" a certain number of transactions in accordance with auditing procedures, which defined by standard-setting bodies. The auditor's work usually results in a simple one or two-page "opinion letter" stating that the report meets applicable accounting standards, known as a "clean audit opinion." If the auditor did not find that the report meets GAAP, the auditor's opinion would note areas of exception.

Single-audit Report. In addition, entities receiving federal funds must retain an audit firm to perform certain other audit procedures at the same time that the standard audit is performed. These procedures are intended to allow the auditor to determine whether the agency complied with grant requirements, which go beyond financial requirements. The auditor will issue a "single-audit report" in conformance with federal law and applicable audit procedures.

Management Report. Another common audit report in the public sector is the "Management Report." In this report, the auditor suggests changes to the entity's financial practices to improve financial reporting and/or controls. The items noted in the management report do not indicate significant weaknesses. If they were, they would be listed in the opinion letter. Managers may not have sufficient staff or other resources to implement the auditor's suggestions, or may simply disagree with them. However, managers should carefully review the management report, discuss it with the auditor, and consider putting the suggestions into practice.

Fraud Report. A standard financial audit is not intended to find fraud. Doing so would require a much more detailed review and additional audit procedures. Remember that the auditor's job is to determine whether the financial statements are "materially" correct. A misstatement of thousands of dollars in a financial report would generally not be material for a multi-million-dollar entity, so significant theft might go undetected by an auditor. Fire service managers concerned about fraud should discuss that with their auditors and engage them to undertake additional audit procedures.

A word on auditor selection: government accounting standards are becoming increasingly complex. When selecting auditors, managers of public sector entities should inquire into the number of governmental audits performed by the firms under consideration, and the time that members of the firms devote to training in government accounting standards. (Several organizations of finance professionals have model "requests for proposals" for audit firms.) As discussed in the section on purchasing, simply selecting the low-cost firm is usually not appropriate when employing professional services. Unfortunately, many audit firms (especially smaller ones) do not have sufficient public sector business to make it cost effective to develop the required expertise. Therefore, some government entities will need to look outside their borders for a qualified audit firm. The consequences of engaging an unqualified auditor can be significant, including the potential loss of grant funds and issuing debt at higher interest rates.

Specialized accounting and reporting requirements. Many fire departments receive grants from the federal and/or state government. The grantor will require certain reports, sometimes periodically and always when the grant terminates. Before your agency begins to receive the grant funds, consider how to meet the reporting requirements. At the very least, you will need to track all grant revenues and expenditures. If the grant is from the federal government, you will need to meet the "single-audit" requirements mentioned above. And note: if grants from the state government are "pass-through" federal funds, they will be subject to single-audit requirements as well. Discuss the source of the funds with the grantor agency. Advise your external auditor of any grant funds received and special requirements related to them. Finally, most grants require some amount of local "match." Be sure to identify the source of the matching funds prior to accepting the grant. Make sure that the source meets the requirements of the grantor agency, and track the expenditure of matching funds as you do the grant funds.

Most states require that local governments provide them some type of financial report. Some states require the agency's standard year-end financial report, with the opinion of an external auditor. Other states require reports in unique formats, although in many cases the numbers can be pulled fairly easily from the standard year-end financial report. If you are responsible for financial reporting, contact the agency of your state government that deals with local government affairs to ensure that you are aware of the reporting requirements.

Funding Sources

No discussion of financial management would be complete without information on funding sources. While this has always been a concern for volunteer and fire protection district organizations, historically it has not been a pressing concern for city and county departments. In those areas, funding has always been seen as a problem for the jurisdiction's finance department, mayor or county executive, and the elected body. This has changed. As budgets of public entities become tighter, fire managers may wish to suggest revenue source when they make spending proposals.

In this section, the first topic discussed is the beneficiary of public services and the purposes that governments can achieve with various revenues, aside from funding services. Next, certain characteristics of various revenue sources will be discussed. While property tax has been the primary revenue source for local government entities, it continues to decrease in popularity. As a result, public entities are making more use of other sources, particularly fees. Therefore, the final section will discuss most of the available funding sources, ending with a case study in setting user fees.

Of course, the availability of particular revenue sources will vary with state law, and usually varies between independent fire districts and general-purpose municipal governments (*e.g.*, cities and counties). Fire districts have had some success obtaining approval of additional revenue sources by their state legislatures. Approval is generally easier when revenues are dedicated to particular services and are contingent on voter approval within each fire district. Managers of fire districts with limited funding sources may wish to consider, as they review the list of revenue sources below, whether their state legislature might approve new revenue sources.

Evaluate beneficiaries and the governmental purpose

For fire managers, the purpose of any revenue is obvious—funding services and related facilities and equipment. However, the particular beneficiaries of each service can play a factor in determining its funding source. Public managers and elected officials also consider two other purposes (or consequences) of public revenues, which are to modify behavior and to redistribute wealth. These issues—the beneficiaries of the service and the purpose of the revenue mechanism—both have a part in determining the appropriate funding method.

Service beneficiaries. When a public entity wishes to pay for services (including infrastructure), it should identify the beneficiaries to determine the degree of public subsidy versus user fees. The benefit to be received (often referred to as the "good") can be categorized as a:

- *Public Good.* These benefit primarily the community as a whole. Environmental protection is an example. On the local level, streets, bridges, and water and sewer systems benefit virtually all members of the community.
- *Merit Good.* Both specific users and the community as a whole benefit from such services. Programs to provide prenatal care to low-income mothers are an example. While the mother and child are direct beneficiaries, society benefits by avoiding the greater costs of providing health care for infants born with preventable deficiencies. Many public entities place improvements that serve specific properties, but are also available to the general public (such as sidewalks) in this category.
- *Private Good.* The benefit is primarily for the specific user. Examples include public golf courses, community swimming pools, and other recreation facilities. Generally, public policy holds that users should pay for such services and facilities.

Into which categories does the fire department fall? All of them. The availability of services to prevent conflagration, and manage mass casualty and community-wide disaster incidents, is generally considered to be for the public good. Public education programs are also consistent with the concept of the public good. Merit good activities would include some inspection services that not only benefit the specific occupants, but also help protect jobs and the tax base. Some response activities also qualify as merit goods to the degree that they protect the environment and also protect the community's neighborhoods from damage. Private good services include most responses to private business and personal properties, and individual EMS calls.

Clearly, the lines between the categories are fuzzy, and reasonable people will view them differently. While most would agree that reviewing building plans primarily benefits the developer and property owner (private good), most also would agree that the community benefits from having structures that conform to building, fire, and life safety codes (merit good). Where does fire suppression stop benefiting the private property owner and begin to benefit the community?

Answering these questions locally is fundamental in adopting service fees. It may seem radical, even unacceptable, to charge a fee for fire suppression or other traditionally "free" services provided by the fire department. But in these times of shrinking revenues, it more often is a choice between charging actual users a fee for the service, and removing or cutting back the service for all. (See "Setting User Fees, Impact Fees, and Benefit Assessments," below, for additional information on this subject.)

Modify behavior. In addition to raising revenues to fund services, governments can modify behavior by charging taxes and/or fees. For example, many government entities impose so-called "sin" taxes to discourage certain types of behavior or activities such as cigarette smoking. More pertinent to fire services are fines (which are a type of fee) for negligent false alarms, which encourage better maintenance and management of the systems.

Since taxes and fees raise revenues as well as modify behavior, it is important to recognize both of these consequences when establishing them. It would be undesirable to apply a "behavior-modifying" fee to a desirable activity. For example, governments should encourage installation of sprinkler and other private protection systems, rather than discourage their installation through unreasonably high fees. Some cost-of-service recovery may be warranted, as is collecting information about the systems (controlling behavior). Therefore, a permit with a small fee to help the agency keep track of sprinklered properties may be appropriate, as long as the fee does not prevent the community from receiving the merit good of sprinklered buildings.

Redistribute wealth. Finally, a government can redistribute wealth through its choice of revenues. That is why many income tax systems increase the tax rate with higher levels of income. Redistribution of income also takes the form of free or reduced-cost services or grants to lower income families and individuals. At first glance this may not seem relevant to fire departments. However, this principle is used when fire agencies establish sliding-fee scales, based on the ability to pay, for services such as EMS, or when bills are routinely written off for those unable to pay. Providing smoke detectors to poor families is another example of redistributing wealth.

Revenue characteristics. Once the purposes of revenue collection are established, each source may be evaluated against several characteristics.

- Equity
- Efficiency
- Elasticity
- Diversification

Equity has three components: horizontal equity, vertical equity, and equity based on use. Most people agree on the definition of horizontal equity: households with the same resources (*e.g.*, income or wealth) should pay the same amount of tax.

Vertical equity, however, generates more debate. It refers to the extent to which one's tax payments increase as the income of the taxpayer increases. For many years, the majority of the public considered a tax fair if higher-income people paid a higher percent of their income as taxes. A revenue system structured to achieve that is referred to as "progressive." The opposite—paying a lower percent of one's income as tax, as income increases—is called "regressive." Recently, the third situation, a "flat" tax, has gained in popularity; all persons pay the same percent of tax regardless of income. For example, a household with an income of $30,000 might pay state income taxes of $1,500 (5%) while a household with an income of $130,000 would pay $6,500—the same 5%.

Many people also consider fees equitable, since they vary with the extent to which the service is used; in other words, equity is based on use. Others disagree. For example, many governments charge fees for basic services such as water. Lower-income people tend to use a higher portion of their income on basic services than higher-income people. Is that equitable? That is a policy judgment.

In Table 3–2, which summarizes the characteristics of various revenues, equity is evaluated both ways—by the extent to which payments of taxes or fees vary with income, and the extent to which the amount of fee or tax paid relates to the amount of services received.

Efficiency can actually be evaluated on the following two planes: 1) whether the tax or fee can be collected efficiently, and 2) if the purpose is to modify behavior or gain control, how well that objective is met.

Elasticity is whether or not the revenue source will expand (and contract) with costs and demand for services. The amount of fees collected increases with demand if there is a good link between fees and demand. However, due to wage increases and inflation, costs may increase more than the increase in service demand. On the other hand, no fee increase may be needed if the service is able to enjoy some economy of scale as the amount of use increases (unit costs go down as output goes up). The amount of tax revenues has only an indirect relation to the amount of services provided. For example, building construction leads to more services but also more property tax revenues. Unfortunately, service demand can also increase unrelated to the level of tax revenues. Therefore, the issue of elasticity with respect to taxes is often whether revenues expand due to economic growth and/or inflation.

Diversification cannot be applied to any particular revenue source. Rather, it applies to the agency's revenues as a whole. To the extent possible, agencies should not "put all of their eggs in one basket." Over-reliance on any particular revenue source can lead to problems when economic or political conditions change. For example, revenues on natural resources (*e.g.*, oil and natural gas) are obviously tied to the market prices for those resources, which can fluctuate significantly. State-shared revenues may fluctuate with the preferences of the elected officials in office at any given time. Even a jurisdiction with many revenue sources could find its income fluctuating considerably if a large part of the revenues are tied to the same underlying factors. Fire service managers are advised to evaluate their revenues as a whole, to achieve adequate diversification to ensure against potentially devastating income swings.

Summary. Table 3–2 lists the most common forms of revenue collected and evaluates them using the above criteria. It is important to remember that actual evaluation must occur locally, since local conditions affect the evaluation of each revenue source. The ratings on the table below reflect the most common circumstances, which may not apply to a particular jurisdiction.

Table 3–2 Equity Efficiency

Revenue Source	Equity re: Level of Income	Extent of Service Use	Easy to Collect	Achieves Behavior Control	Elasticity
Property Tax	Medium	High	High	Low	Medium
Personal Property Tax	Low	Low	Low	Low	Low
Sales Tax	Low	Low	High	Medium	High
Income Tax	High	Low	High	Low	High
Franchise Tax	Low	Low	High	Medium	Medium
Consumption Tax	Medium	Low	High	Medium	High
Insurance Tax	High	High	High	Low	High
Fire Tax	Low	High	High	Low	High
Use Tax	Low	Low	Low	Low	High
Lienses/Permits	Medium	High	Medium	High	High
Service Fees	Low	High	Medium	High	High
Subscription Fees	Low	High	Medium	High	High
Finds & Penalties	Low	High	Medium	High	Low
Contributions	High	High	Low	N/A	High

Revenue sources

Property tax. Known technically as *ad valorem* (in proportion to the value) taxes, property taxes are the largest source of revenue for most local governments. The tax is on real property (defined as land or anything fixed to it), based on a percentage of assessed valuation. (In some areas, the resulting tax rate is known as "millage.") Assessed valuation bears a relation to market value, with the ideal relationship usually set by the state. Property tax payments tend to increase as household income increases, since wealthier households own more expensive real estate. Therefore, while the property tax is not as equitable as income taxes, it is much more equitable than most other revenue sources available to local governments.

Unfortunately, inconsistent assessment practices diminish the equity of property taxes. In some states, property taxes are administered in such as way as to produce fair results. That is, a property with a certain market value in one place is assessed at the same amount as another property with the same market value in another location. In other states, however, assessment practices can differ from one county to another, or even from one city to another.

Elasticity of property taxes may be either high or low depending on local conditions. Property tax revenues often do not suffer greatly from temporary economic declines, since property values rarely respond quickly to changes in economic conditions. That can provide much-needed stability for governments during recessions, which might lead to an immediate drop in income and sales tax revenues but little change in property tax revenues.

On the other hand, a long-term change in economic conditions can affect property tax revenues significantly. This is a major problem where urban flight has resulted in inner city decline. Properties are abandoned or devalued, resulting in declining revenues at the same time service demands are increasing. If the public entity is "landlocked," its property tax base may not be expanding.

Another factor affecting growth in property tax revenues is the many taxpayer revolts that have led to passage of legislation to constrain tax increases. In some states, revenues are constrained even when building construction and annexation have expanded the geographic area and population served by the public entity.

Finally, local jurisdictions must often consider the interests and practices of other taxing bodies (*e.g.*, cities, counties and schools) when setting property tax rates. A fire district might have a low property tax rate compared with other fire districts, but the overall tax rate for all governmental entities in the area might be relatively high. The governing board of the fire district might be reluctant to raise property taxes due to the high overall tax rate.

Sales tax. The largest revenue source for many states—sales tax—is also a major revenue source for many municipalities. The term "sales tax" refers to a tax on the purchase of general merchandise, based on a percentage of the cost of goods purchased. It is low on equity, since lower-income people spend a higher percentage of their income on basic commodities than higher-income people, who use more services. Certain items considered necessities, such as grocery food and medicine, often are not taxed to make the tax more equitable. The amount of revenue generated by sales taxes fluctuates with the economy and with seasonal business cycles. The extent of fluctuation, however, can vary from one place to another, depending on the characteristics of the local economy. Sales taxes have the advantage of raising revenues from non-residents who use the services of the local government but do not pay other revenues (such as property taxes) to the local government. This is particularly true for urban centers that provide a high level of services to non-residents, particularly streets and related transportation systems.

Personal property tax. Annual taxes on items such as boats, cars, airplanes, major appliances, furniture, and jewelry, a personal property tax is hard to assess because the items are movable, and other than vehicles, are usually not matters of public record. Therefore, while personal property taxes were intended to be somewhat equitable, they are much less so than intended.

Income tax. Imposed by the majority of states and some cities on income earned in the jurisdiction, income taxes are calculated like federal income tax, usually on a progressive scale. If a flat rate is imposed, there is usually a provision for forgiving taxes for low-income families; otherwise, the high equity effect is lost.

Franchises. Typically required of public utilities (water, sewer, electric, gas, telephone, cable TV, and Internet access) that use public rights-of-way, these are agreements with local government entities. Under the agreements, local governments receive fees in exchange for the following:

- Granting the utilities permission to operate in the jurisdiction
- Granting use of the local government's rights-of-way (in effect, a rental fee)
- Compensation for damage to due to the utilities' cuts in the entities' infrastructure (studies have shown that even if the utilities make repairs, the infrastructure suffers long-term damage)

The landscape for franchises has changed considerably in recent years. Many utilities have become much more competitive, with multiple companies providing telephone, cable, and even electric service. Most of the companies require access to the public rights-of-way. At the same time, the federal government has restricted local governments' ability to regulate telephone, cable TV, and Internet service providers.

Despite these changes, local governments can still receive fair compensation for use of their rights-of-way by private companies. Many are experiencing increased revenues, particularly from providers of telecommunications and Internet services. Similarly, many local governments are leasing their properties for companies that site cell phone towers on the properties. As an added advantage, the local governments can negotiate for placement of communications equipment used by public safety agencies and public works departments.

Franchise and right-of-way fees have the effect of an operation expense passed on the consumer. They achieve high equity to the extent that the consumer has discretion over the level of his or her use of the utilities' services. For example, use of cable television service is highly discretionary, whereas electric service generally is not.

"Sin" tax. Imposed on sales of specific commodities or services with the objective of modifying behavior as well as raising revenue, "sin" taxes include taxes on tobacco products, liquor and gambling. While generally viewed as equitable by society, they often draw higher percentages of revenues from low-income individuals. When they are adopted for specific purposes, and publicized as such, they can achieve high "control" efficiency. Tobacco taxes earmarked for fire protection, and liquor taxes earmarked for EMS and rescue services, both make sense as service demand causal factors.

Consumption tax. Also imposed on sales of specific items or services, consumption taxes are unlike "sin" taxes in that consumption taxes are not levied with the intent of modifying behavior. Rather, these taxes apply to items that are considered more discretionary such as prepared food and beverages, entertainment (*e.g.*, movies), and hotel and motel use. In urban centers, these taxes may also result in drawing revenue from nonresident populations that use the services provided by the local government in the urban center. While not levied for the purpose of modifying behavior, these taxes can modify behavior if consumers perceive the tax rate as sufficiently high.

Utility tax. A selective sales or excise tax on consumption of natural gas, electricity and/or telecommunications services, this tax may be based on the cost of goods or services sold. Alternatively, the tax may be based on the volume of product sold, particularly in the case of natural gas (therms sold) and electricity (kilowatt hours sold). That provides stability of revenues during period of volatile prices, but provides growth only if consumption is increasing or the tax rate is increased.

A set amount for each telephone line, such as $1.00 per month, is a variation on a utility tax. Many jurisdictions have used such taxes to fund emergency services dispatch, especially if dedicated to enhanced 9-1-1 capabilities. Due to their dedicated purpose, and the low monthly amount for residences (although high for some commercial enterprises), these taxes have often won voter approval when required by state law.

Insurance tax. Based on the dollar amount of premiums paid to insurance companies, some states earmark at least a portion of property insurance premium taxes for fire training, local department equipment, or operating expenses, thereby achieving a high degree of connection between the revenues paid by each property and the fire services provided. Without earmarked funds, the connection is lost.

Fire tax. Imposed by a fire protection district pursuant to state laws, this tax obviously connects property to the service provided. Like property taxes, this tax correlates with income to some extent, but much less so for retirees and others on low or fixed incomes unless provisions are made for "ability to pay" adjustments. A fire tax is low in behavior modification unless incentives are established for built-in protection, and it potentially suffers from the same low elasticity as property taxes if the jurisdiction is land-locked and already substantially developed.

Use tax. Imposed on the use of certain tangible goods, auto licenses provide an example of a use taxes. They often are based on vehicle weight (low income-equity but high use-equity) or value (high income-equity). Usually earmarked for transportation systems, some are used for EMS. Some states affect behavior modification by imposing higher fees for low-gas-mileage vehicles.

License and permit fees. Issued for a variety of activities, the primary purpose of licenses and permits is to control and modify behavior. Examples include licensing dogs and services such as plumbers, electricians, exterminators, and taxis. Examples of activities requiring permits include gun ownership, building construction, hazardous materials use or storage, installing signs, operating massage parlors and selling liquor. Of course, licenses and permits generate fees, so they can help offset the costs of delivering some related services. However, they are not usually a major contributor of revenue. Specific to the fire department, most building and fire codes provide for the establishment of reasonable fees for such items as materials storage, transportation, and use (explosives, gaseous and liquefied fuels, and other hazardous materials), process control (refinishing, welding, construction, and demolition), and standby operation services (fire watch, blaster, fire safety director, etc.).

Fees for service. Often called "cost recovery" fees, these are intended to offset some or all of the costs of delivering a service. As noted above, they may be viewed as equitable because the amount of fees that a household pays is dependent upon the household's use of the services. However, when equity is viewed as households paying for services in proportion to household income, the equity of fees are low unless there are provisions for ability-to-pay adjustments. From a public safety and public relations standpoint, fire agencies should be careful not to discourage low-income families from using emergency services appropriately because of high fees (or even the perception of high fees).

Subscription fees. Fees or dues for annual service, subscription fees have been used by some volunteer fire organizations for years. The concept has spread to other types of organizations for a variety of services. Property owners pay (or make a donation) annually for the right to use the service in case of an emergency. Over the years, there have been several widely publicized incidents in which departments have responded to calls for structure fires, but watched the buildings burn (after verifying that no lives were involved) once it was learned that the owners had not paid subscription fees. A fire department in this situation is faced with an obvious dilemma. Refusing to fight a fire appears cold-hearted, yet if word spreads that members will fight the fire even if the fee has not been paid, many residents will not pay. Some residents choose to roll the dice, especially if the department has a provision allowing nonsubscribers to pay a higher fee (which may still be a bargain) at the time of service.

The level of income-equity depends on whether special provisions are made for low income households. Higher fees scaled to the nature of the hazard presented by each property can potentially modify behavior by encouraging private protection systems.

The subscription concept also has been applied to EMS. Commonly, EMS agencies bill Medicare, Medicaid, and private insurance at a standard rate. The district forgives the amount of the bill that insurance doesn't cover, and forgives the entire bill for the uninsured. This has not been without its problems, as some insurance companies have refused to pay their regular benefit, saying that the bills aren't real because the uninsured are forgiven, or that the whole subscription plan represents co-insurance. One possible way around these problems is to make the program official (*i.e.*, a co-op co-insurance plan). Subscription rates would need to be higher for those without private insurance. The unpaid portions then would be paid "on the books" with monies from the dues pool.

Perhaps fire departments need a lesson in insurance billing. Many homeowner policies have a provision to pay up to a certain amount (*e.g.*, $500) for fire suppression services. This provision is often valid only where there is no tax-supported fire department, just the type of environment in which the volunteer or subscription service department operates.

Fines and penalties. Not a major source of income except perhaps for agencies that enforce property maintenance codes, fines and penalties are assessed for traffic, health, building, fire, life safety, false alarm etc., violations of the law. The major purpose of fines and penalties is to modify behavior rather than raise revenues. Typically the funds are not earmarked to fire services. However, doing so can add further justification to such fines and fees.

Contributions. The lifeblood of most volunteer organizations, contributions are a minor source of income for most paid departments, which use contributions for special projects. Recent examples include public fire safety education trailers, smoke detector giveaways, fire prevention educational materials geared toward children, and purchases of special rescue equipment.

There are a few other sources of income that don't lend themselves to the above evaluation criteria, but nonetheless are worth exploration:

Intergovernmental revenue. Almost always transferred from a higher level of government, these can take many forms. Agencies may win grants from the federal and state governments through competitive processes. Two publications that list sources of grants are the *Federal Domestic Assistance Program* and *Foundation Register.* (Both are available from the Government Printing Office.) While many restrictions apply, agencies with unique ideas that have the potential for bettering their communities should consider applying for grants. Writing successful grant proposals takes special skills and experience, so some jurisdictions and larger departments have specialists in grantsmanship. Other agencies can consider contracting with a grant-writing firm, some of which receive compensation only when grants are received.

Many municipalities can receive some federal revenues without entering into a competitive grant program. These revenues are also restricted to specific uses such as low-income housing, transportation, and urban development. These funds have diminished due to a shift in public opinion regarding the role of government, and the perception that some of the programs have limited effectiveness. However, some areas of funding have increased, such as a federal program adopted in the late 1990s that provides funding for local police departments. Many states have increased their funding for fire services somewhat in the aftermath of the attacks on September 11, 2001, and the federal government may do so as well.

Investment income. Resulting from the investment of idle general funds, investment income can generate funds for operating purposes. However, governments should be careful to not become too dependent on investment earnings. Doing so can lead entities to invest funds with the primary objective of earning income rather than preserving the funds. Investments are discussed extensively in later sections of this chapter.

Rental income. Usually not a major source of income, jurisdictions often rent out public facilities on a short- or long-term basis. In some cases, the number of properties and volume of income may justify establishing an enterprise fund to manage the properties and account for the expenses and income. This is a special fund that encapsulates all expenses and income of a specific program to account for it as a business.

Marketing income. Earnings from sales of products, services, or "intellectual property," marketing revenues are not usually a major source of government income. However, some jurisdictions sell products or services that are not provided adequately by the private sector. (Before embarking on such an enterprise, a government jurisdiction should examine closely the issue of competition with private enterprise and its own mission.) Various fire departments have, at one time or another, been involved in fire extinguisher sales and service, smoke alarm sales, dive bottle filling, etc. Potential liability has caused many of those same departments to drop such sales activities. Fire departments with significant public education resources have sold videotapes and other educational materials. If the marketing income project is significant in dollars or level of effort, it may warrant setting up an enterprise fund.

Many jurisdictions have missed potential income from the sale of intangible assets created in the course of business. This might include patents on processes and inventions, and copyrights on training materials and software. Contrary to common belief, such items are not automatically in the public domain nor are they the jurisdiction's property—except for the federal government. Without an agreement assigning "intellectual property" rights, copyrights and patents legally "belong" to the individual(s) who provided "creative supervision" to the project.

Many public entities require employees, as a condition of employment, to assign rights resulting from their work for the entity. However, many universities that did not allow professors to share intellectual property rights found the quality of research decline and many of the best professors leave. Now rights-sharing agreements are the norm at research universities, which once again derive a significant portion of their revenue from patents and copyrights. Consult your own legal authorities for legal provisions applicable to your area and situation.

Setting user fees, impact fees, and benefit assessments

While they are different approaches to achieve cost recovery, user fees, impact fees, and benefit assessments are sufficiently related that one discussion will serve all three.

User fees are charges made only to the actual recipient of some service. The most common in the fire service has been ambulance or EMS charges. Others include inspection fees, plans review fees, hazardous materials incident "cost-recovery" fees, event standby fees, dispatch fees (charged to other agencies, private ambulances, alarm monitoring firms, etc.), confined space operations for industry, and fire suppression charges.

Impact fees are generally one-time charges to the developers or owners of new properties. These fees allow agencies to accrue revenues for facilities and equipment required to serve the new customers, since the agencies' other revenues may cover operating costs but not major capital costs. (Impact fees may also recover costs for facilities already constructed.) State laws generally regulate impact fees. Generally, multi-purpose public entities (such as cities) must dedicate the impact fees to the specific services for which they are charged. In addition, the way in which the fees are calculated must bear a reasonable relation to the use of the services for which they are charged.

Sometimes impact fees are applied across the board—by some formula—to all new properties. In other jurisdictions, the fees are applied to only "extraordinary" hazards (defined by size, occupancy type, process, or some combination thereof), or to properties built outside primary response zones. One example of an impact fee formula would be the standard fire flow formula, but instead of calculating gallons per minute the formula might use dollars. The formula should also take into account the size and type of construction. The inclusion of a factor for fixed use hazards is good in concept. However, uses change, and unless the impact fee is reapplied with each change of use it may not be adequate. The formula also applies penalties for exterior combustible finish and gives credit for automatic sprinkler systems, with the intent of encouraging fire prevention measures.

Benefit assessments are usually charged annually, for a specified number of years, to all properties based on a formula designed to calculate the relative benefits and costs of providing public fire protection to the property to which the charges are assessed. Such fees are in essence special taxes to replace or supplement revenue from traditional tax sources.

In each of the approaches above, one of the goals is to more closely match benefits received to financial responsibility. With user fees, the cost to general taxpayers is kept at a minimum by shifting costs to those who actually use the service. In the latter two, the costs of making resources available to all properties are distributed based on the relative shared costs of resources needed for the particular property (type, size, etc.).

Setting user or impact fees for heretofore "free" (totally tax subsidized) services is an economic, political, and public relations decision. It may be wise to test the political waters before seriously considering such fees. On the other hand, in some jurisdictions the push toward fees comes from the political arena—or from the public—to limit general tax rates.

Another concern with respect to service fees is organization morale and public reaction. Some personnel may feel that imposing fees will tarnish the fire service in the public's eye. Some readers may feel that the concept is nearly immoral, or at least politically unthinkable in their jurisdiction. Many will fear losing the public's respect, or, in the case of inspection fees, increasing adversarial relations with the business community. All of these concerns must be discussed and worked through. Experience has shown that in many jurisdictions, including those that held the fears mentioned above, there was a year or two of minor grumbling from the business community when inspection fees were adopted. Usually these came from same occupancies that grumble when they see the inspectors now. By the third year, the fees were generally accepted as just another cost of doing business.

The following provides the general steps in establishing service fees. (Items 3, 4, and 6 ideally would include the participation of the elected body.)

1. Identify the resources used or needed, and determine the actual costs of delivering the service. Program budgeting makes cost determination much easier. Some agencies will choose to include administrative costs. These can include department administration (*e.g.*, the chief's salary) and central administration such as financial and personnel services (if provided outside the department). Other agencies do not include administrative costs, primarily to make the fees more palatable to elected officials and the public.
2. Determine the statutory authority and requirements for adopting fees for service.
3. Determine the beneficiary distribution of the service (*i.e.*, the extent to which it is a private good, merit good, or public good as discussed previously).
4. Determine the goal of the fee program. Is it to recover all costs including overhead and administrative costs, or part of the program's costs? Do you wish to modify behavior or redistribute wealth through the fee?
5. Given the information gained in the steps above, determine the method to calculate the proposed fees. Design the information system required to collect, track, follow up, and renew the fees. Include these in costs as appropriate.
6. Seek approval from your legislative body to proceed to investigate the fee.
7. Obtain input from employees and representative fee-payers, marketing the proposal to the parties that will pay the fee. A basic recommended approach is to

highlight the mission of the service to be funded and major program objectives, using a primary "selling line." One example is, "To keep tax rates as low as possible without reducing services, actual users of XX service will be charged a fee for the service."

8. To the extent possible, modify the proposal to reflect the input received from these parties.
9. Adopt the fee as required by law (*e.g.*, hold public hearings, pass ordinances, and hold elections).
10. Implement and test the information system as needed.
11. Train personnel.
12. Implement the fee.

Debt Financing

Debt is not a revenue source. Rather, it is a means of financing a project by spreading the cost over a period of time. Therefore, any manager proposing to finance a project with debt should also propose the revenue source to repay the debt.

In most cases, investors who purchase debt instruments (*e.g.*, bonds) from state and local governments do not pay federal taxes on the interest payments from the debt. That makes the income more valuable than income from the debt of private entities. As a result, the debt holders are willing to accept lower interest rates from debt of public agencies than from private entities. For example, if debt issued by a private entity paid 7% interest, a public agency with similar creditworthiness could issue debt paying an interest rate of about 5%–5.5%.

Debt is a contract between the borrower and lender. This contract can take several forms, some more appropriate for certain purposes than others. All states regulate local governments' use of debt, generally including the amount of debt that they may issue. In addition, the federal government, which previously had few regulations regarding state and local debt, adopted significant regulations in the 1980s. Thus, debt presents challenges in two areas. The first is structuring the debt to meet the agency's financial purposes while also making the debt marketable to potential bondholders. ("Debt structure" refers to issues such as the source of money to repay the debt, the period of time until the debt is fully paid, and whether the debt service payments are level over time, increase or decrease.) The second challenge is ensuring that the debt agreement will meet various legal tests.

Major players in debt issuance

- *Chief Financial Officer* (CFO). The public agency's CFO should lead the debt issuance. Since issuing debt for one purpose will affect the agency's ability to issue debt for another purpose, the CFO will have two goals. One is to fund the particular project in question, and the other is to maintain the agency's sound financial position.
- *Underwriter.* This is the financial institution that purchases the debt.
- *Financial advisor.* This term refers to an individual or firm that performs a number of functions on behalf of the agency issuing the debt, including:
  - Advising the agency on the amount of debt it can issue
  - Advising the agency how to structure the debt to match the project purpose and to ensure that it will attract financial institutions to purchase the debt
  - Conducting the sale of the debt
  - Preparing the document explaining the project purpose and nature of the debt (known as an "official statement," the document is similar to a "prospectus" in the private sector)
  - Assisting in obtaining a credit rating and/or bond insurance
  - Establishing a timetable for issuing the debt, including legally required steps (*e.g.*, public hearings and publication of legal notices)
- *Bond counsel.* This is an attorney specializing in the issuance of public debt. The attorney makes sure that the public agency has followed the steps necessary to issue debt in accordance with state and federal law. Bond counsel prepares the ordinance that the governing body adopts to issue the debt, and prepares other documents as necessary. Few entities' corporate counsels have the expertise to serve as bond counsel, but an entity's corporate counsel will work with its bond counsel.

Selling debt is very complex, requiring specialized expertise that only the very largest public entities have on staff. Therefore, entities should retain a financial advisor and bond counsel when issuing (or even considering) debt.

Primary forms of debt

Short-term operating loans are usually available from financial institutions, and are secured by income expected from other sources. For example, a fire agency may receive property tax revenues to fund its services, but the revenues are not received until several months into the fiscal year, leaving the agency with a temporary cash-flow problem. The agency could issue "tax anticipation notes (TANs)," a form of debt in which the property taxes are pledged as repayment. The agency's bond rating may be affected by excessive short-term borrowing, making any future debt issuances (whether short- or long-term) more expensive.

Public agencies commonly use *bonds* to finance capital improvements and major equipment purchases. The advantages include the ability to borrow from a nation-wide market, ensuring a competitive interest rate, and to secure large sums of money. With a good bond rating and/or bond insurance, a jurisdiction can secure an interest rate below that available in the commercial market. There are disadvantages. Issuing bonds can entail fees for bond counsel, financial advisor, underwriter, rating agency and/or bond insurance, printing, accounting, and payment fees. However, unless the amount of the debt issuance is relatively small, these out-of-pocket costs would be offset by the lower interest rates compared to other financing methods.

Bond issuance also entails considerable time of the public agency's staff (and sometimes elected officials) to retain the various professional service providers, decide on the specific bond issue, prepare necessary documents, and seek voter approval of the bond issue (required in most states if the debt is secured by a separate property tax levy—see *GO* bonds below). Bonds should be used judiciously, to finance equipment or facilities that have life expectancies longer than the lives of the bonds. Otherwise, jurisdictions would find themselves paying for assets after they are no longer in service.

Generally, bonds issued for fire services are repaid with property tax revenues. This makes them very secure for bondholders, unless the entity has serious financial problems. The public entity secures these bonds with its "full faith and credit," so these bonds are referred to as *general obligation* (or *GO*) bonds. However, bonds may be secured with other taxes or with fees. Such bonds are referred to as "revenue bonds" because they are secured by a designated revenue source rather than by the entity's full faith and credit. Purchasers of such bonds will require a higher interest rate to offset the higher risk entailed in holding bonds with a limited security pledge. Revenue bonds are common for municipal enterprises such as water and sewer utilities, but are not typically applicable to fire services.

General Obligation Debt Certificates are a general obligation of the agency, payable from the general funds of the agency and such other sources of payment as are otherwise lawfully available. There is no levy of a separate property tax to pay the amounts due on the debt certificates. In a resolution issuing debt certificates, the agency would covenant to appropriate funds annually, and in a timely manner, to fund the debt certificate payments when due. Debt certificates do not provide the same high degree of security to investors as general obligation bonds. As a result, they carry interest rates that are slightly higher than general obligation bonds. Because they are not payable from a separate property tax levy, debt certificates do not require voter approval.

Lease-purchase agreements should not be confused with rental agreements, because the terms are often used interchangeably, especially in reference to real estate (*e.g.*, "leasing an apartment"). Lease-purchase agreements are forms of debt, but differ from bonds in several ways. The most important is that generally the government need not pledge its property taxes to repay the debt. As a result, most states exempt lease arrangements from the requirement that voters approve the debt issuance. Also, public entities pay a higher interest rate under lease-purchase financing as compared to general obligation debt, since the debt is not secured as well.

Generally, leases are not secured by the public agency's "full faith and credit." Rather, the entity must repay the debt as long as the governing body appropriates sufficient funds. (While the entity may technically "walk away" from the debt, it would have difficulty borrowing money again. Therefore, this provision provides an "out" for the entity only in the direst financial conditions.) The entity pledges, as collateral, whatever it purchased with the funds that it borrowed. Apparatus, communication and computer equipment, buildings, and other major purchases are candidates for lease-purchase agreements. At the end of the lease period, the equipment either belongs to the jurisdiction, may be purchased at some prearranged prices, or may be turned over to the leaser, depending on the original contract.

Lease-back agreements are a variation of lease-purchase agreements, with some unique and interesting features for a jurisdiction. In this arrangement the jurisdiction sells a current asset—land, building, or apparatus—to an investor, who in turn leases it back to the jurisdiction at an agreed-upon rate. This offers several potential advantages to the parties. The jurisdiction receives a major infusion of cash with no restriction on its use, and usually with no adverse effect on its bond rating. If the investor is a profitable corporation, it may be able to use the purchase to provide a tax shelter for other earnings. In any case, the investor is able to declare depreciation on the property for tax purposes (something the jurisdiction has no need to do) potentially sufficient to render the lease payments tax-free.

Merely reviewing the forgoing description of major public sector debt instruments may leave some fire managers feeling out of their depth. Determining the most effective method for each purchase or project, within the context of the agency's overall debt management plan, can be a formidable task. That is why this writer strongly recommends that public agencies retain independent financial advisors to assist them in meeting their financing needs.

Methods of sale

Debt may be sold competitively. In the case of loans, bonds and debt certificates, the agency's financial advisor solicits bids from numerous underwriters. The winning underwriter is the one that proposes to lend money to the agency at the lowest interest rate. The calculation of the lowest interest rate takes into account the fee that the underwriter proposes to charge for buying the debt—known as the *underwriter's discount*—which is typically quoted as a percentage of the dollar amount of debt sold.

Alternatively, loans, bonds and debt certificates can be sold through negotiated sales. To do so, the entity selects the underwriter in advance and then negotiates the terms of the sale, based on the nature of the debt issuance and prevailing interest rates at the time of sale. The agency may engage the underwriter to perform many of the duties of the financial advisor, leading some public entities to enter into a debt issuance without a financial advisor. However, this author believes that is not wise. The underwriter has a strong incentive to purchase the debt at a higher interest rate, making it easier to sell to the ultimate holders of the debt. Only the very largest public agencies have staff with the required expertise and experience to ensure that the debt is sold at a fair price. The remaining agencies should use independent financial advisors to protect the agencies' interests. When a financial advisor represents an agency in a negotiated sale, the financial advisor evaluates the underwriter's discount (an amount the agency pays the underwriter "up front" from the proceeds of the bond sale), and the proposed interest rates on the debt, to make sure they are comparable to current market rates and conditions. The cost to employ a financial advisory is easily offset by savings in interest payments over the life of the debt issue.

Similarly, the public agency should could seek competitive bids for lease agreements from several finance companies, or at least negotiate with a number of qualified companies. Having selected a leasing company, a public agency could enter into an agreement to finance one project, or enter into a "master lease agreement." Such agreements spell out the terms by which the two parties will do business, including an interest rate (that may be tied to a published interest rate index). The public agency can then lease-finance a number of projects through the master lease agreement, without having to select a leasing company each time. An example of the usefulness of this arrangement is financing the purchase of vehicles that will be replaced at various points in time.

PURCHASING

Unfortunately, governmental purchasing procedures receive many of the same complaints as governmental budgeting procedures—that they are too focused on control, are overly complex, and provide managers insufficient latitude—in short, that they detract from the ability of government managers to respond effectively and efficiently to citizen needs. While these complaints are often true, most fire managers can understand why these systems have developed. Government purchasing activities still suffer, at times, from the cronyism or outright corruption that characterized many local governments' activities in earlier years. Tight controls were developed in response to those problems.

Fortunately, the pendulum is swinging back, as many governments seek to achieve balance between control and flexibility. This section will briefly outline the major steps in traditional, centrally controlled purchasing systems. It will go on to discuss some key purchasing issues so that fire managers can design systems to suit their needs.

Traditional purchasing systems

In many medium and large organizations, purchasing functions are centralized in the finance department or other administrative department. Traditionally, acquiring goods or services entails the following steps:

1. Requisition: The program manager (*e.g.*, fire chief) sends a "requisition" (a form stating the items needed, quantity, etc.) to the purchasing department.
2. Purchase order: The purchasing department checks that the program manager has sufficient funds in his budget to purchase the items, and then determines how the goods or services should be acquired. If the purchasing manager has standing contracts with suppliers of certain products, the purchasing process is relatively simple. The purchasing department issues a "purchase order" (PO) requesting the supplier to provide the items. If a contract for the required items is not in place, the purchasing department prepares specifications for the items and selects a vendor to supply them through one of the purchasing procedures discussed below.
3. Delivery: The supplier delivers the items in accordance with the terms on the PO. To protect against fraud, the items might be delivered to a central receiving unit, which would confirm that the items were received consistent with the PO. The receiving unit then sends the items to the department requesting them, and sends the purchasing department paperwork (such as a packing slip that came with the product, or a receipt) confirming that the items had been received.
4. Payment: The vendor sends an invoice to the purchasing department, which matches it up to the packing slip or receipt, and then issues payment to the vendor.

Note the advantages of the procedure above:

1. Control: The system provides budgetary controls. In addition, the process protects against fraud by separating duties—recall that one group of people requested the items, another ordered it, and another received it. Fraudulent actions such as diverting the items to personal use or paying a vendor for items not received (with a kickback to the public employees) would require collusion of at least two separate departments.
2. Efficiency: The purchasing department can order similar items at once, resulting in better pricing, fewer deliveries, and less staff work.
3. Effectiveness: The purchasing department develops knowledge of the products needed, so it can write specifications to fit those needs. (Fire managers may argue that they know their own needs best, and that is certainly true for turnout gear, but how many fire managers can write good specifications for copiers?)

Of course, the disadvantages are also obvious:

1. The purchase requires work by several organizational units, resulting in inefficient use of staff time.
2. The process creates delay in acquiring the items needed. This is particularly true if, to receive quantity discounts, the purchasing department waits for many requisitions of the same item before making the purchase. (Perhaps the fire department should have requested the items earlier, but that adds another administrative task to plan in advance.)
3. The purchasing department may not fully understand the needs of the fire department, and acquire items that do not perform as expected.

Increasingly governments are revising their purchasing procedures to provide more latitude to program managers while maintaining fundamental controls and efficiency. The following sections will discuss some of the key issues in purchasing, and how governments are approaching these to find the balance between controls and efficiency that fits their needs.

Differentiation by item cost

Governments have traditionally prescribed different purchasing procedures for items of different cost. For example, a fire department requiring a small tool could simply pick it up from a local hardware store. However, such latitude was often limited to relatively inexpensive items, say, under $100.

Recently, many government agencies have begun to allow program managers to purchase more expensive items without the extensive procedures outlined above. It is not uncommon for medium or even small governments to give program managers discretion to purchase items costing up to $5,000 or $10,000. However, that does not mean that the governments have no policies or procedures with respect to such purchases. Rather, the procedures are appropriate to the cost of the items. An example of a policy that attempts to balance controls and flexibility is as follows:

- Purchases under $1,000: No specific procedures specified; program managers are authorized to make such purchases using good judgment.
- Purchases over $1,000 and under $10,000: Department managers may purchase the items after receiving at least three bids or quotes (in accordance with procedures discussed below) from independent vendors. The program manager might be required to document the bids or quotes for review by the finance department, either when the program manager requests payment to the vendor, or through a periodic audit by the finance department.

- Purchases over $10,000: The program manager might be authorized to purchase the items following review of the proposed purchasing method and budget availability by the finance department. Alternatively, a central purchasing department (if the government agency has one) would purchase these items.

A related issue is the level at which approval from the governing body is required. Often this is established as a set amount, for example, $20,000. Some governments authorize purchases through their budget processes. Their budget documents contain a list of items that staff plans to purchase during the year, and adoption of the budget authorizes staff to purchase the listed items.

Control items

In addition to basing the level of control on the cost of the items, the government might desire high levels of control for particular goods and services, regardless of cost. Some of these controls entail involvement by specialized staff, such an agency's information technology (IT) department. These departments would work with the purchasing department, if the government has one. In addition to specialized items, the government might have higher-level controls on politically sensitive purchases. So-called "control items" might include:

- Computer and Telecommunications Equipment: The government's information technology (IT) staff might acquire and install all hardware and software to ensure that such equipment will interface properly with other equipment (such as the government's computer network). IT staff also wants to ensure that it has the skills and tools to maintain the hardware and software.
- Vehicles: If the government has a central garage, it might purchase all vehicles.
- Audio-Video Equipment: This can be politically sensitive, as some elected officials may take a dim view of the value of such items to citizen service.
- Travel: Since travel for training and other purposes can be politically sensitive, some jurisdictions require review of these requests by central authorities such as a finance director, city manager, or mayor.
- Memberships in Professional Organizations and Memberships: While important, expenditures for such purchases can be politically sensitive. At times, certain controls are put in place in response to that sensitivity.

Basic fraud control

Regardless of the particular policies in place, all governments should adhere to basic internal controls to protect against fraud in purchasing. The most fundamental fraud protection is to make no payment until the person (or department) making the payment has matched a vendor invoice to a receipt provided by the staff member who acquired the goods or services. The person making payment should also have verification that the good or purchase was for public purposes. Such verification could entail prior approval of the expenditure, and/or approval by the employee's supervisor.

In addition to putting basic fraud protection in its purchasing procedures, fire services should consider ethical issues. Vendors should not be favored for any reason other than their ability to provide quality products and services at a fair price. This may seem obvious. However, many fire chiefs have been offered discounts from local companies when purchasing items for their personal use. Could you truly say that accepting a discount would not influence your decision, the next time that you have to select a store to provide that product or service to the fire service?

In addition, in public service it is important to consider the appearance of ethics as well as its fact. How many fire departments routinely contract for goods or services from companies owned or managed by relatives of firefighters, especially for relatively small jobs where the firms are not selected through open, competitive selection processes? The companies may perform quality work at a fair price, but how does it look to the other contractors who would like an opportunity to compete for the government's business? How does it look to the taxpayer who sees relatives continually obtaining government contracts?

Purchase orders, e-procurement, and P-cards

"Open" or "blanket" POs are another means to provide flexibility to program managers while achieving some of the benefits of centralization. Under this procedure, a central authority determines the common needs of departments in the governmental entity and contracts with a vendor (or vendors) to provide the goods and/or services throughout the year. Procedurally, this is often accomplished by issuing a purchase order, but since it is for "open" purchases rather than a single purchase, it is referred to as an open or blanket PO. The government achieves some of the benefits of centralized purchasing: lower prices by contracting for large quantities, and ensuring objectivity in the selection of vendors. The quantity pricing can be achieved even without accepting delivery of all the items at once. Rather, agencies of the government can order relatively small quantities as needed throughout the year, usually with delivery to their sites, and still obtain the quantity discount. Obviously, that provides flexibility to program managers.

E-procurement is essentially a form of blanket PO, combined with the benefits of internet technology. To use this technology, a government agency would conduct a bid process to contract with a vendor for a set of products such as office supplies or vehicle parts. The vendor that wins the bid would program the offered prices into its computer. The government agency would provide to the vendor a list of employees authorized to make purchases, often along with their department names, account codes to which products should be charged, etc. These would be programmed into the vendor's computer as well. The authorized employees would then access the vendor's products through its on-line store (web site), which would record the items purchased and match it with the billing information.

Purchasing cards (or "P-cards") are a relatively recent development that creates another opportunity to achieve control and cost-effectiveness while providing greater flexibility to program managers. P-cards are credit cards issued to staff of the government agency, although the agency is responsible for all payments on the P-cards. The unique aspect of P-cards is the control they provide to the government agency. The agency can limit the use of each credit card to specific items and/or specific dollar amounts of purchases. The bank issuing the P-card puts that information in its data system, and the store or service provider would approve only purchases within the policy limits.

For example, an agency that wished to provide P-cards for travel purposes only could limit use of the card to hotels, rental car companies, and airlines, with no individual purchase exceeding $1,000. Staff members with P-cards would have the flexibility to purchase travel services as needed without obtaining cash advances. Another agency might authorize use of P-cards for purchase of supplies and small equipment from particular vendors with whom the purchasing department had established contracts. That would allow firefighters to purchase needed items if supplies became depleted during the weekend.

Managers should be aware, however, that P-cards are still subject to abuse. Simply restricting the items purchased does not ensure that the items were used for the public good rather than for personal use. In addition to implementing the "basic fraud controls" described above, governments should require staff members using P-cards to 1) provide a receipt for every good or service purchased with the card, and 2) sign a document acknowledging that misuse of the card can and will result in disciplinary action, and authorizing deduction of the purchase from the employee's paycheck if the employee does not reimburse the fire department on a timely basis.

Cooperative purchasing

Many government entities have found that they can reduce their costs substantially by entering into cooperative purchasing agreements with other governmental entities. In some cases these agreements are relatively *ad hoc*. In other words, two or more governments work together to purchase a limited number of items to achieve quantity discounts. On the other hand, some governments have established freestanding non-profit organizations with professional purchasers who procure a wide variety of goods on behalf of the member entities. The member governments pay a fee and may have to commit to participating in the organization for a designated period, so that the other members can be assured that the organization will be able to achieve significant quantity discounts, offsetting the administrative costs. (Of course, while the organization requires a fee, it frees staff of participating governments from purchasing the items themselves.) This may replace the need for the government to employ a purchasing agent, or at least, allow management staff to devote more time to citizen service.

Some states offer another form of cooperative purchasing to their local governments. When the state contracts to purchase goods, it may require that the vendors supply the goods to local units of government at the same cost. Local governments receive the same prices as the state, and in some cases the state's contract conditions apply as well, which may include better terms than individual local governments could negotiate on their own.

Such arrangements can provide an excellent means to purchase a wide variety of items, from office supplies to tires to computers. If the local government feels assured that the state officials have conducted a fair and competitive process to select vendors, the local governments can be reasonably assured that the state list offers the best available pricing. Still, local governments may want to occasionally solicit bids for some products on the state list to make sure they are competitive.

Purchasing methods

The discussion above included several references to bidding for goods and services. However, bidding is merely one method to select vendors to provide desired goods or services. This section describes various selection methods and the circumstance under which they are most appropriate for a particular purchase.

Readers should bear in mind that the goal is to procure products and services that meet their needs at the lowest price. It is rarely possible to receive both the highest quality and the lowest price. Agencies have to exercise judgment in evaluating the relative benefits of price and quality. Certain purchasing methods are more appropriate, in certain situations, to provide the information required to make that evaluation.

Fairness to vendors who compete for government contracts is critical in all of the following selection methods. Government agencies should ensure that information provided to one is provided to all. Specifications should not be narrowly written to create advantage for one vendor, unless the agency truly requires the particular specification. Since state law often regulates these and other purchasing procedures, readers should consult appropriate sources for guidance in this area.

Bids. Most readers are familiar with this process. It is appropriate when a government agency can develop unambiguous specifications for a particular product or service. Having done so, the agency then invites bids from vendors. Usually the agency sends the bid documents to qualified vendors of whom they are aware. In addition, the agency may advertise for bids in relevant trade publications and a local newspaper.

The agency's bid documents include detailed instructions regarding the form of response, including the date, time and place to which vendors must submit bids. The bids are often opened publicly by the purchasing department or other official such as a City Clerk. The contract is awarded to the lowest responsive and responsible bidder.

"Responsive" means that the vendor provided all of the requested information and met all specification in the bid documents. For example, the agency may require that the vendor have certain types and levels of insurance. (This is common, and protects the agency in the event that the vendor causes injuries or damage to property in the course of doing work for the agency.) A vendor failing to provide documentation of the required insurance would not receive the contract, even if the vendor met the specifications for the product or service. "Responsible" essentially means qualified. For example, a construction contractor who had never performed a large job might be disqualified from receiving a contract for a very large project, even if the contractor bid the lowest cost. Obviously, such decisions require judgment on the part of the agency, and legal consultation is recommended for any agency that considers rejecting the low-cost bid.

Request for Proposals (RFP). This method is appropriate when the government agency cannot write clear specifications for the work to be performed, and when the vendor's qualifications are paramount. In addition, this method should be used when the agency wants the opportunity to evaluate vendors' offers that have different price/quality mixes.

The RFP method is typically used to procure professional services. The agency will prepare an RFP that describes:

- The work to be performed (often called the "scope of work")
- The date by which the work must be completed (often with intermediate milestones)
- The criteria by which proposals will be evaluated, and the information that an interested firm must include in its proposal, which generally includes:
  - The firm's past work on similar projects
  - Qualifications of the firm's staff, particularly key personnel who would provide services to the agency
  - If appropriate, the firm's technical and support services (for example, the agency may require that an engineering firm have certain computer resources)
  - An outline of the firm's approach to the work
  - References for similar work
  - The estimated cost to perform the work, and hourly rates and other fees (although some governments receive this information later in the selection process)

The RFP is sent to qualified firms and may be advertised in trade journals; governments generally do not advertise RFPs in newspapers. Similar to bids, responses are due on a certain date, but the government generally uses more latitude in accepting a proposal that might be considered unresponsive in a bid situation.

Having received proposals, the government agency evaluates the firms' ability to perform the work, using the criteria specified in the proposal. Some agencies use fairly informal review processes, while others use more formal systems. In those, each criterion is given a certain number of possible points, and each proposal is then reviewed and given points on each criterion. The points are then added up to determine the most qualified firm.

Various government agencies also approach the issue of cost in different ways. Some review the fees in all proposals to determine a "competitive range." They enter into negotiation (described below) with firms in the competitive range. Other agencies simply select the vendor with the best qualifications, regardless of cost. (Even in that case, they might ensure that the firms' fees are not far out of line with the fees of other firms with similar qualifications.) The fees may entail an hourly rate, sometimes with a clause that total fees will not exceed a set amount. (Generally, firms will not agree to such clauses unless they are confident that the scope of work is clearly defined so that they will not have to devote significant hours on unanticipated tasks.) Costs such as document reproduction may be reimbursed separately from fees for staff services.

Having selected a firm, the agency enters into contract negotiations. At this time, the firm and agency define the work process and product more clearly. This allows the agency to take advantage of the firm's expertise in the type of work required. The firm might suggest methods to perform some tasks more efficiently and/or effectively than the agency expected. Or, the firm may believe that the agency failed to include some necessary tasks in its scope of work. Fees are negotiated as well, as they depend on the job definition.

While the RFP process is typically used to select firms to provide professional services, it can be used to procure tangible products as well. This method is appropriate when the agency would like the opportunity to consider the cost/benefit trade-off of various products, without being tied into particular specifications (as in the bid method).

Request for Qualifications (RFQ). This is a variation of an RFP, typically used when seeking professional services or other specialized services. The agency would state its objective and solicit statements of qualifications from firms that provide services necessary to achieve the objective. Upon receiving responses to the RFQ, the agency would negotiate the specific services and terms of compensation with the firm that has the best qualifications, or perhaps with the top two firms (in accordance with the process stated in the RFQ).

An RFQ is particularly appropriate for a project where the agency is uncertain of the specific steps necessary to meet its objective, and therefore cannot prepare a scope of services as is typical for an RFP. The RFQ can also be used to select a firm to provide services on a number of projects over a period of time. For example, an agency might wish to engage a firm to advise it on multiple bond issues, with somewhat different characteristics, over the course of several years. The agency would select the firm with the best overall qualifications to perform financial advisory work, rather than the firm with the best approach to a particular project.

Request for Information (RFI). This process is appropriate when an agency believes it has insufficient information to adequately specify the product or services. For example, an agency might be aware that some vendors have updated their products using newer technologies. However, the agency is not certain that it is aware of the specific technology used and how it enhances the products' functionality. The agency could prepare an RFI stating its functional objective and asking vendors to respond with the best product to achieve the objective. The RFI itself would be similar to an RFP, but without a detailed product specification.

When issuing an RFI, an agency has two options with respect to vendor selection. The agency could select a vendor based on the responses to the RFI alone. For instance, in the example above, after reviewing responses to the RFI the agency might conclude that one (and only one) firm had made technological improvements to its product that made it uniquely suited to the agency's use. Alternatively, the agency could use the information gained through the RFI process to develop bid specifications (or an RFP). That would be appropriate if the agency found that several manufacturers had added desired features to their products, all of which suit the agency's needs. Whichever process is used, the agency must state the process in the RFI.

Sole-source Purchasing. Fire managers should use this approach when only one vendor supplies the desired product or service. The practice is clearly warranted when necessary to achieve equipment compatibility is required (*e.g.*, a part for a computer or a vehicle is available from only the manufacturer or authorized dealer). Agencies might also wish to standardize on one make of equipment to achieve compatibility. Perhaps only one engineering firm has the particular expertise to perform a certain service (*e.g.*, a structural engineer needed to inspect a damaged building for structural integrity).

Fire managers should use sole-source purchasing judiciously. Obvious pitfalls can include paying a higher price due to lack of competitive procurement, and accusations from citizens or elected officials that you favor the particular vendor. Sometimes these problems can be avoided, even when purchasing options are limited. Perhaps you need a part produced by a particular manufacturer, but there may be several authorized dealers. In that case, there is no reason not to seek bids from multiple dealers. In the case of the engineering firm, how do you know that only one firm has staff with the requisite qualifications? You could issue an RFI to verify that, indicating that you will select a qualified vendor and contract with that party for the following three years. Doing so would provide reasonable defense as to why you did not issue an RFP each time you needed to engage a structural engineer. And it lets the qualified firm know that you are always on the outlook for competitors, so they should not take advantage of their unique position by charging excess fees. (See Appendix 3)

Additional sources of information

Managers wishing to pursue further information on the topic of purchasing may wish to use the resources of the following organizations:

- The National Institute of Governmental Purchasing, a non-profit organization, located at: *http://www.nigp.org/index.htm*
- The National Purchasing Institute, which is the Public Sector Affiliate of the Institute for Supply Management (formerly known as the National Association of Purchasing Management). This organization can be found at: *http://www.nationalpurchasinginstitute.com/*

INVESTMENT PRINCIPLES

Many issues confront fire managers whose responsibilities include investment management. The field of investment management is complex and the stakes are high. Understandably, many fire managers have had little formal training in this area. Therefore, this section will provide fire managers with concepts that are fundamental understanding investment issues. These concepts apply to funds invested to meet operational needs and capital requirements, as well as to pension funds.

Types of investments

Various types of investments are referred to as "asset classes," of which there are two basic types. The first is "fixed-income" securities, such as bonds, which are debt instruments. These are similar to your home loan, in which the financial institution loaned you the money in return for your promise to pay, each month, a certain amount of money (hence the term "fixed-income" security) that includes interest and a portion of the principle. The financial institution does not own your house, although it has certain rights to it in the event that you default on your loan.

When investors first deal with fixed-income securities, they will likely encounter some unfamiliar terms such as "coupon rate," "coupon payment," and "par value." The coupon rate is simply the interest rate that the issuer of the bond will pay to the investor. The payment itself is known as the coupon payment; it is a function of the par value (the value of the bond declared by the issuer, usually $1,000 for corporate bonds), and the coupon rate. For example, a company that issues $1,000 par value bonds, and makes coupon payments semi-annually at a 12% coupon rate, pays the investor $120 each year in coupon payments (two $60 payments).

The second asset class is "equity," which is direct ownership of a company. Most major companies raise funds by selling ownership to the public. Their stocks are traded openly in financial markets. When the company makes money, it is obligated to first pay any holders of bonds. Stockholders receive profits only after the company has met its obligations to bondholders. The profits may be reflected in payments made to stockholders, called "dividends." More importantly, profits are reflected in increased stock prices.

Thus, fixed-income securities are less risky, while equities are usually more profitable over long periods of time. This is an important concept that you should become familiar with; it will be discussed in more detail later in this article. Investors must balance the level of risk they are willing to assume against the rate of return they wish to earn. Assuming more risk usually means a greater return; playing it safe with less risky investments necessarily means lower returns.

While equities and fixed-income securities comprise the two major asset classes, others exist as well. These include real estate, private equities, and hedge funds that many pension funds use.

There are various sub-categories within these broad asset categories. Within fixed-income securities, major sub-classes, in order of least-risk to more-risk, include:

- Securities issued by the federal government and secured by its full faith and credit
- Securities issued by agencies of the federal government, backed by the federal government's full faith and credit
- Securities issued by agencies of the federal government but not backed by the federal government's full faith and credit
- Mortgage-backed securities
- Securities issued by major domestic corporations with established products
- Securities issued by foreign governments
- Securities issued by smaller corporations with products that are not well established, or perhaps are merely in the development stage

Within equities, major asset classes include:

- Stocks of major corporations without established products
- Stocks of smaller corporations without well-established products
- Stocks of foreign corporations, particularly those in developing nations

While it is easy to classify investments into groups, it is less easy to put the equities into simple categories according to their relative risks. A small company with well-established products might be less risky than a large corporation whose heyday has passed. Similarly, companies based overseas can be very stable or risky.

Time horizon

It is critical to understand the period of time for which you are investing, known as the "time horizon," and invest accordingly. This is one of the most fundamental concepts in finance. Most people intuitively understand this principle with respect to their own money. They invest one way with money that they are saving to purchase a vehicle (short-term investments), and another way with their retirement savings (long-term investments). Governmental entities have cash reserves for operating needs such as paying employees and purchasing supplies. Many also have separate reserves for pension requirements. Clearly, the time horizon for each of these reserves is different, and for reasons discussed below, the investment strategies should be different.

Closely tied to the concept of the time horizon is the "yield curve" for fixed-income securities. The yield curve is the shape of a line on a graph, where the x-axis represents time and the y-axis is the interest rate (or "yield"). Under most financial conditions, the yield curve slopes upward as shown on Exhibit YY, below. That is, the longer period of time to which you commit your investment, the higher interest rate you will get. The investor expects a greater reward for giving up his or her money for a longer period of time, since the investor won't be able to take advantage of other opportunities that may come along. The investor is also subject to greater risk, since there's a greater chance over a long period of time that conditions could change for the worse. Investors are willing to pay more interest to obtain funds for a longer period of time. They can use the funds to develop products and bring them to sale, which might take some time. They don't have to worry about paying back the principle in a short period of time and having to find other money to borrow. So, investors are willing to pay higher interest rates for long-term securities.

Exhibit YY

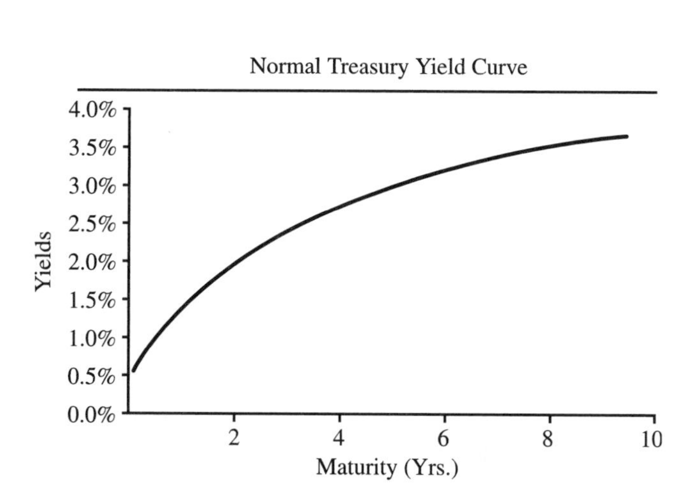

Risk

Risk versus reward. All forms of investment entail some risk; the only question is the degree of risk. At one end of the scale are securities issued by the U.S. Treasury maturing in one year or less, known as Treasury bills (or "T-bills"). Treasury bills are commonly used in financial analysis as a proxy for a "risk-free" investment. While T-bills are not entirely risk free, we've probably got much bigger problems to worry about if the federal government cannot make its debt payments!

On the opposite scale of T-bills are investments considered "highly speculative." An example is bonds issued by companies with obvious financial problems. That could include large, formerly well-established companies that have run into difficulty more recently, or smaller, "start-up" companies. There is a significant chance that an investment in a start-up company will not pay off. On the other hand, when an investment such as this does pay off, it can pay off big. Tupperware and Hewlett-Packard come to mind as good examples of investments that were initially considered highly speculative.

This relationship between risk and reward is not a fluke. Investors putting their money in high-risk investments will demand a high return if the investment pays off. Otherwise, they could simply put their money in Treasury bills.

The time horizon is also important in equity investments. Over the short term, investments in stocks are relatively risky. They can go up one year and down the next. However, stock investments become much less risky when kept for longer periods of time. The chance of a loss becomes increasingly lower, and the chance of a gain becomes much greater. In fact, there has never been a twenty-year period when an investment in U.S. equities, as a whole, lost money. That is why equities are particularly appropriate for investments with long-term time horizons, while short-term investments generally include little or no equities.

Risk tolerance. Firefighters are very familiar with this concept, although they may not call it "risk tolerance." Each time that firefighters respond to an incident, command personnel must evaluate the benefits of various courses of action against the risk entailed in those actions. A low-risk course of action may be appropriate in one situation while a higher-risk course of action may be taken in another situation.

Similarly, it is critical to determine your tolerance for investment risk. That means to evaluate the consequences of a gain or loss on your investments and decide how much risk you should take. Generally, governmental entities take a low-risk approach when investing "operating" funds that are used to pay employees and purchase supplies. The consequences of loss of principle can be great in comparison to the relatively small earnings gains that might be achieved. On the other hand, pension funds usually warrant higher-risk investments. Why? Because investment earnings play a significant role in pension system funding. Failure to achieve reasonable earnings could cause significant tax increases to fund benefits, diminishing the funds available for direct services. Further, most pension funds can be invested for long periods of time. That allows pension funds to realize the higher earnings that equities usually offer, while weathering the ups and downs of the stock market.

In addition to evaluating the practical effects of gaining or losing funds, investors must also consider the effect of a gain or loss on their reputations. This is particularly true in the public sector, where people who lack in-depth knowledge of the subject at hand are often in a position to evaluate managers' decisions. A fire manager might have an excellent track record for investing public funds that goes unnoticed for many years because the governing board is more interested in issues that directly affect citizens' safety. Yet a relatively small loss on one investment could bring the manager's overall capabilities into question, and become the subject of a newspaper headline. As with much of the fire manager's work, good communications with elected officials and other key players, in advance, can prevent having events taken out of proportion. Nevertheless, fire managers must recognize that risk tolerance is a subjective matter, and that the decision concerning risk tolerance should take public sentiment into account.

Volatility. The extent to which a security (or a group of securities such as a mutual fund) fluctuates in price on the open market is referred to as its "volatility." Investors should be aware of several principles with respect to volatility. First, when determining the appropriate asset classes for a portfolio, the volatility of one asset class relative to another is important. For example, since fixed-income securities are generally less volatile than equities, they are more appropriate for a portfolio where preservation of capital is paramount. Second, having decided to invest in a particular asset class, the process of choosing a particular security should include a comparison of volatility with securities in the same asset class. Only an "apples to apples" comparison will provide useful information to guide investment decisions.

Investment professionals use the term "beta" to describe a security's volatility in comparison to similar securities. A beta of 1 means that the security fluctuates in price to the same extent as comparable securities. A beta higher than 1 indicates higher than average volatility, and a beta below 1 indicates lower volatility than average.

Another important principle is that volatility decreases over time, since the law of averages comes into play. For example, fixed-income securities fluctuate in price in relation to general interest rates. Since general interest rates rise and fall quite a bit over long periods of time, the volatility of a particular security diminishes over time. This principle explains why it is appropriate to invest pension funds in more volatile securities if they are expected to have higher returns over long periods of time.

Components of risk

Market risk. This is the risk that the value of a security will rise or fall due to changes in the financial markets. This could be due to overall economic conditions such as a recession that causes the value of most corporate stocks to decline. Fixed-income securities are subject to a particular type of market risk called "interest rate risk." When interest rates rise, the value of existing fixed-income securities falls. Why? Because investors could purchase new securities offering higher interest rates. Since older securities pay less interest, investors will want to be compensated by paying less for them. Of course, the reverse is also true—the value of existing securities increase when interest rates drop.

The amount of risk is tied to the maturity of the security, meaning the time that the issuer will fully pay off the principle to the purchaser along with the final interest payment. The length of a fixed-income security is referred to as its "maturity." Securities with longer maturities are riskier than those with shorter securities. For example, if interest rates rise, what is the risk to the owner of a bond maturing in one month? Not much. In one month the investor will get all of his money back and can re-invest it in a higher-rate security. The investor only loses a lower rate of interest for one month. But what if the investor held a bond due to mature in ten years? If interest rates were to rise, the investor would receive relatively low interest rates on his bond for ten years, until the bond matured and the investor could purchase a higher-rate bond. If the investor sold the bond, he or she would get less than the amount he or she would have received when interest rates were lower.

For a group of fixed-income securities, a measure of its overall market risk is its "average maturity." As the term implies, it is the average of the maturities of all securities in the group. If interest rates were to fall, a group of securities with an average maturity of five years will gain more in value than a group of securities with an average maturity of two years. The reverse is also true. In other words, the longer the average maturity of a group of securities, the more sensitive the group is to changes in interest rates. This is the same idea mentioned earlier regarding the time horizon of investments—longer-term fixed-income investments generally provide greater long-term earnings, but have higher short-term risk of fluctuation in value.

A more precise measure of interest rate risk for a group of securities, commonly used by investment professionals, is "duration." Duration measures the change in market value for the securities for each one percent of interest rate changes. For example, if interest rates were to fall 1%, the value of a group of securities with a duration of two years would gain 2% in market value.

Credit (default) risk. This is the risk that an investor will lose principal due to default by the issuer, or default by another financial institution (*e.g.*, a broker) selling the security. Credit risk can be largely eliminated by carefully screening financial institutions with which you do business, and by implementing certain investment practices, such as delivery-versus-payment purchases. (This and other practices are discussed below.)

Liquidity risk. This is the risk that your funds will not be available when you need them because you have invested them in a security that cannot be easily sold, or not easily sold at a fair price. Billions of dollars of many fixed income securities, such as most government bonds and bonds of major corporations, are purchased and sold every business day. The holder of one of these securities could sell it within a day to meet a cash need. However, other securities are not actively traded. The owner might have difficulty selling it quickly, or would have to sell it at a lower price than its usual market value due to lack of buyers on the particular day that the owner needed the cash.

Political risk. This is generally associated with investments in overseas companies, some of which are subject to greater chances of war and political instability than domestic (U.S.) companies. However, domestic companies are subject to some political risk. For example, a decision by the federal government to greatly increase or decrease defense spending could affect the future revenues of defense industries, resulting in a change in the values of their stocks.

Currency risk. This type of risk is also associated with investments in overseas companies. It is the risk that you will gain or lose money due to changes in the value of the U.S. dollar relative to the value of the currency of a company whose stock you own. Many consumers confront currency valuation when they purchase foreign products. For example, if the U.S. dollar increases in value relative to the Japanese yen, the Japanese car that you want to purchase becomes less expensive. Why? Because the car is priced in yen. When you purchase the car, you trade your dollars for yen. If the dollar becomes stronger, it takes fewer dollars to get the amount of yen that the car is selling for. At the same time, the yens that you get when you sell your stock in a Japanese company are worth fewer dollars.

Inflationary risk. The discussion of risk, especially risk related to equities, may lead some investors to ask why they should purchase riskier securities. The answer is inflationary risk. Very safe, short-term investments will earn so little earnings, compared to inflation, so that the buying power of your money may hardly increase at all. In some situations it could even decrease. Therefore, taking on greater risk is prudent in many situations.

Asset allocation

Asset allocation refers to the decision regarding the type of investments to purchase. For example, you could put all of your assets in fixed income investments, all in equities, or half in one and half in the other. Here is a critical point: this decision will affect investment earnings more than any other. Many investors focus on the question of which particular securities to purchase (*e.g.*, GM stock versus Ford stock). That is not the most important decision. The most important decision is the type of investment. The decision concerning the amount of each type of investment to include in the portfolio is referred to as "asset allocation." (A collection of securities is referred to as an "investment portfolio.") Studies have shown that the asset allocation decision contributes over 90% of the total return to the portfolio. The choice of specific securities, therefore, plays a relatively small role.

Diversification

Diversification is another key concept. It refers to the practice of purchasing various types of securities for an investment portfolio. Diversification allows you to gain the benefits of each type of security, while mitigating the risk of each type. Some of the key ways in which portfolios can be diversified include the following:

- *Asset class.* For a variety of reasons, some asset classes tend to perform well when others perform poorly. For example, stock prices may fall during a recession, diminishing earnings from equities, but interest rates may also fall, increasing the value of bonds. Therefore, diversification by asset class is critical for certain investment portfolios, particularly those designed for longer-term investments such as pensions. That does not mean that all asset classes belong in all portfolios. In particular, equities are generally not appropriate for operating fund investments.
- *Maturity date.* Purchasing securities of various maturities is referred to as "diversifying across the yield curve." The practice allows you to obtain higher interest rates that are generally associated with securities with longer maturities, while also keeping some shorter-term investments to meet cash needs (diminish "liquidity risk") while militating against the loss of portfolio value that would accompany a rise in interest rates ("interest rate risk").
- *Issuer.* Securities issued by one organization will carry more or less credit risk than those issued by another. An investor can achieve a mixture of risk levels in his portfolio by holding securities issued by a variety of organizations. This is particularly important because levels of risk cannot always be determined in advance, as demonstrated by the corporate bond defaults in the early part of this century. Firms thought to have excellent prospects failed quite quickly. As the saying goes, "Don't put all of your eggs in one basket."

Often the best way to achieve diversification is through pre-established investment portfolios. These can take various forms. Mutual funds, which many individuals invest in, are also available to institutional investors such as public agencies. Some companies specialize in institutional mutual funds. Institutions (such as a fire agency) that can invest a minimum of perhaps $500,000 or $1,000,000 can obtain lower costs and a higher level of service than individuals. Available mutual funds cover the gamut of asset allocations, from short-term fixed-income securities to "balanced funds" to highly risky equity funds. Fire agencies can achieve similar results by using the services of an investment management firm. Such firms can tailor investment portfolios to each investor's specifications. Investors must have sufficient funds to make such services cost-effective, but many medium fire agencies would fall in that category.

Most states operate "local government investment pools." These usually operate similarly to private "money market mutual funds," a special category of mutual funds that are highly regulated by the Securities and Exchange Commission (SEC). The funds hold very short-term securities. When investors need funds to meet payments, they can usually get their cash out on a day's notice, referred to as "daily liquidity."

Indices

An index is a device that indicates a value or quantity. Most people are familiar with the term "consumer price index," which is a measurement of the price changes faced by average consumers. Similarly, many indices have been constructed to indicate the value of various investments. This is accomplished by including certain securities in a mock portfolio, called an index, and measuring and publishing the performance of the index. Most people are familiar with the "Dow Jones Average." It is a measurement of the value of the U.S. stock market, weighted toward certain types of stocks. Many companies publish indices to measure the value of certain types of stocks (*e.g.*, high-tech stocks, foreign stocks) and certain types of bonds (*e.g.*, short-term government bonds, corporate bonds, etc.)

Indices serve a variety of purposes to investors. One is measuring the performance of your investments. Looking at the absolute return (*e.g.*, 7%) is not sufficient. You should measure the performance of your investments against comparable investments. Also, some investment managers construct actual portfolios that replicate particular indices. These are called "index funds." Investors can purchase the index fund that meets their particular needs. The uses of indices will be discussed further in later sections.

Active versus passive management

"Active" management refers to the practice of buying and selling securities with some regularity. This might be done with the expectation that interest rates will rise or fall in the future. Active management is also pursued to exploit other perceived opportunities in the financial markets, such as when one type of security is currently selling at a low price (when viewed on an historical basis) compared to other types of securities.

"Passive" management refers to investing without respect to the particular market conditions at the time. Some common passive management techniques are discussed next.

Tie investment maturities to expected cash outflow. For example, let us assume a fire district pays employees and vendors bi-weekly on alternating weeks, with disbursements of $100,000 each week. The district could purchase securities of $100,000 that mature each week, ensuring sufficient funds to make payments. The district would simply purchase securities for as many dates as possible. If we assume that the district in our example has $2,600,000 to invest, it would purchase one security for each of the next 26 weeks.

Build a "ladder" of securities. Let us assume, for the moment, that the district in our example has a regular source of revenues that is often sufficient to meet payments every week. If the district were to invest funds to mature each week, it would usually have too many funds on hand, and have to reinvest the funds. It might choose instead to purchase twenty-six $100,000 securities, maturing each month for the next 26 months. By extending its investments to over two years it should achieve higher interest rates than the fire district that invested for only 26 weeks.

While interest rates would go up and down over the 26-month period, the district will "ride out" the ups and downs in interest rates, gaining the average of those rates over time.

Purchase an "index" fund. The goal of an index fund is to meet, not beat, a portfolio of comparable securities. That might appear to set your sites low. However, consider that many portfolio managers under-perform indexes of comparable securities, so matching the performance of the index is reasonable. Also, since the manager of an index fund simply seeks to replicate the index, investment decisions and operations are easier, resulting in lower management fees than actively managed portfolios. Many mutual fund companies and other financial institutions offer a variety of index funds.

There are two primary reasons to adopt passive investment management. First, these forms can be executed without expertise or special financial analysis tools. Second, many economists believe that no person, regardless of expertise or tools, can consistently predict interest rates or otherwise determine how financial markets might perform in the future. Essentially, they believe the concept of active management is flawed.

Few small or even mid-sized public entities will find that they have sufficient expertise and tools to successfully engage in active management. An entity that believes that active management has a place in its investment strategy should consider employing a professional investment advisor. The legal relationships with such advisors can vary. In some cases, the public official may simply take suggestions from a broker to purchase or sell a particular security. In other cases, the public entity may turn over all or a portion of its funds to an "investment advisor" or "investment manager." The manager is given full discretion over the funds within predetermined guidelines. Whichever option the public entity might choose, it should use the following practices with respect to investment managers:

- Select managers in accordance with the procedures discussed under the section on Investment Policies below.
- Enter into a written contract with managers, also discussed under Investment Policies.
- Give each manager specific guidance regarding the investments that he or she can make or recommend to you. At a minimum, the manager should stay within the boundaries of your investment policy. Many entities provide more specific direction such as the specific asset sub-class in which the manager should invest. In either case, the manager should confirm receipt of the guidance in writing.

Investment Policies for Operating Funds

This section describes the key components of an investment policy for operating funds. The most important reason to develop a policy is to promote good decisions and actions. However, there are also political reasons for adopting a policy. It will educate the governing body on investment issues and processes. Should your investment decisions come into question, your defense will be improved by showing that they were made in accordance with prior policy. This helps to demonstrate that you made reasonable decisions in light of the circumstances, even if the particular decisions did not pan out as you would have liked.

By developing an investment policy, you will broadly define your entity's investment practices. Therefore, this section will describe an investment policy, but at the same time will discuss practical issues concerning each policy area.

Keep in mind that the policy described below pertains to operating funds. That refers to funds designated for ongoing operational needs such as paying employees and purchasing equipment. In many governments without large cash reserves, all of the securities in operating funds have maturities no greater than one year, because the funds are needed for payments in the near future. Other governments are fortunate to have more significant cash reserves. Their portfolios may have average maturities of one or two years. Despite the exact time horizon, the goals of most operating portfolios are fairly similar to one another. However, they are quite different from pension funds portfolios, which have much longer time horizons and different objectives. Some policy components discussed blow, such a collateralizing bank deposits, apply to pension funds as well as to operating funds. Other components apply to both types of funds in principle, but the specific application will vary between pension funds and operating funds.

The Government Finance Officers Association (*www.gfoa.org*) and the Municipal Treasurer's Association (*www.mtausa.org*) both publish model investment policies and related materials. These provide a good basis for your policy—modified, of course, to meet local laws and circumstances. Since these materials are available, this document will merely highlight the major components of an investment policy.

Objectives

As with any enterprise, a clear understanding of your objectives is fundamental to establishing a course of action. Most operating funds are invested with objectives of legal compliance, safety, liquidity, and yield. These will be discussed in turn.

Legal compliance. All states have legislation governing the investment of public funds. While you can obtain a copy of the legislation itself, often it is more helpful to obtain materials that put the legislation in laymen's terms, and include interpretations through case law and sometimes other sources. The professional associations mentioned above have state chapters that develop such materials. Also, many states provide materials for local governments that explain their state's laws. The materials may be provided by the State Treasurer, or by the department that deals with local government affairs.

Safety. Most public managers and governing bodies agree that, aside from legal compliance, the primary objective of an investment portfolio is to preserve the funds, because they have been reserved for governmental services. While the portfolio should yield investment earnings, the primary objective of the portfolio is not to maximize earnings. Rather, safety of the investments is the primary objective. This does not mean that public funds should be invested in only investments that hold the least risk (*e.g.*, U.S. Treasury securities). Some other investments with slightly increased risk of loss are appropriate for public operating funds.

Liquidity. As noted above, "liquidity" refers to the readiness with which the investments can be turned into cash. This is the next most important objective, since operating funds are, by definition, held to meet ongoing needs. This does not mean, however, that all the securities must be highly liquid. Some governmental entities have sufficient cash reserves so that some of it can be invested in highly liquid investments, while some can be invested in less liquid securities paying higher yields.

Yield. Investment earnings can provide significant funds for services. More important than the exact amount of the earnings, however, is the fact that citizens expect public managers to put idle public funds to good use. Few fire managers went into the fire service to manage money. However, once there, they have a responsibility to earn a reasonable rate of return, subject to the objectives of safety and liquidity. That responsibility, as many others, just goes with the turf.

Standards of care

The investment policy should provide general guidance to the administrator regarding the execution of his or her responsibilities. The first statement is some form of the "prudent person" policy, such as the following two paragraphs:

Prudence. Investments shall be made with judgment and care, under circumstances then prevailing, which persons of prudence, discretion and intelligence exercise in the management of their own affairs, not for speculation, but for investment, considering the probable safety of their capital as well as the probable income to be derived.

Staff acting in accordance with written procedures and this investment policy, and exercising due diligence, shall be relieved of personal liability for an individual security's loss in value incurred by market price changes, provided deviations from expectations are reported in a timely fashion and the appropriate action is taken to control adverse developments in accordance with the terms of this policy.

Investment policies should also address ethical standards, even if the organization has a more general ethics policy. A statement such as the following paragraph is recommended:

Ethics and conflicts of interest. Staff involved in the investment process shall not conduct personal activities that could conflict with the management of the investment program or could impair their ability to make impartial decisions. Staff shall disclose to (the chief executive or appropriate governing body) a material interest in or material business relationship with financial institutions with which they conduct business.

Authority and separation of duties

The policy should specify who has the authority to make investment decisions. While the number of people should be limited to protect against fraud, adequate backup is needed as well. Also, to protect against fraud, the policy should state that the people who perform certain accounting functions should not also be authorized to make investment decisions.

Separating duties can be difficult in small organizations that have only a few finance staff members. Secretarial staff or other administrative staff members, who are not generally involved in financial operations, can perform some duties to provide that separation. In many cases the duties may not be time-consuming or require financial expertise.

In any organization there is a trade-off between efficiency and protection against fraud. Your agency's auditors can provide guidance in this regard.

Operational procedures

Many governments include certain operational procedures in their policies, governing such matters as wire transfers and recording investments. In fact, some state laws require local governments to adopt operational procedures as part of their investment policies. Alternatively, the investment policy can charge a specific position (*e.g.*, the fire chief or chief financial officer) with responsibility for developing written procedures covering specified operational practices. This avoids putting operational matters in the policy, and leaving them at the operational level where they can more easily be adopted and revised as needed. A typical statement in an investment policy follows:

> *The Fire Chief shall establish written procedures and internal controls for the operation of the investment program consistent with this Investment Policy. The Fire Chief is responsible for all investment transactions. No person may engage in an investment transaction except as provided under the terms of this policy and the procedures established by the Fire Chief.*

Selection of financial institutions

This section of the policy will discuss how to select financial institutions with which you will do business.

Broker/dealers. These are parties from whom you can purchase securities. If you have the responsibility to invest your agency's funds, you know that you get many calls from brokers who would like to sell you securities. *Caveat emptor* ("Let the buyer beware")! Most brokers who call you are reputable. However, many will offer what appear to be great securities, with rates of interest well above others available. If it sounds too good to be true, it probably is.

You should select brokers using the same type of careful analysis that you would put into purchasing fire apparatus. Some basic steps are:

- Talk to governments in your area, and to your state professional association (fire chiefs and/or finance officers associations), to get the names of reputable brokers who have been in business for some time.
- Ask selected brokers to complete a form that provides you with relevant information (samples available from professional associations).
- Get and call brokers' references. Some finance officers consider this the most important step in the broker selection process.
- Check the history of complaints against the brokers on the Central Registration Depository, which is a database maintained jointly by the North American Securities Administrators Association and the National Association of Securities Dealers.
- Review the qualifications of the brokers and their staff members.
- Review the brokers' financial statements.
- Make sure that the brokers are registered with your state's regulatory body.
- Once you have selected brokers, send them your investment policy, and have them send you a written confirmation that they have received and read it.

Investment advisors. This term refers primarily to firms that manage a portfolio of funds on your behalf, but may also include fee-only consultants. You should require your investment advisors to

- have registered with the SEC and with your state's regulatory agency
- have executed industry-standard agreements for investment advisory services (such as the Sample Agreement for Investment Advisory Services developed by the GFOA)
- have their accounts audited by an independent public accounting firm

- prohibit self-dealing
- have third-party custody arrangements
- use delivery vs. payment in securities transactions

Depositories. This term refers to financial institutions in which you deposit funds. All time and demand deposits shall be made with financial institutions that have executed industry-standard collateral agreements that specify collateral requirements, designate independent third-party custodians, and are executed under the terms of FIRREA.

Competitive purchasing

This section of the policy will spell out how to purchase securities. Most public entities state that they will purchase investment securities competitively, just as they make other purchases. For example, if you wish to invest $200,000 in a six-month bank Certificate of Deposit, you would call several banks to obtain their interest rates. (Of course, the banks should previously have executed a collateral agreement with you.)

If you wish to purchase a security such as a T-bill, you would contact two to three brokers and tell them a specific time to call you back with bids on a particular security. (Due to the nature of financial markets, where interest rates can change in minutes, it is not feasible to obtain bids from more than a few brokers.)

U.S. Treasury securities can also be purchased directly from the federal government. This method has the advantage of low cost, as it avoids profit taken by a broker when he sells a security to you. There is no need to take bids for securities purchased from the Treasury, since you know that you are getting the best price available. However, the Treasury will not provide services that you may desire, such as suggesting particular securities that might fit your portfolio at a particular point in time.

The policy itself does not have to describe the specific investment steps. It should, however, indicate that your agency would purchase securities and make deposits competitively.

Authorized investments

Many think of this as the "guts" of the investment policy. This section will specify the types of investments that you may make, including the portion of the portfolio that can be invested in each type. State law often limits the specific types of investments, but less often specifies the percent allocations.

Some managers may believe that their state laws are fairly restrictive, so all they have to do for this section of the policy is copy the state law. However, that is not always appropriate. For example, one state allows local governments to purchase 30-year U.S. Treasury

Bonds for their operating portfolios. While Treasury securities carry virtually no credit risk, such long-term securities carry significant interest rate risk, and are rarely appropriate for an operating portfolio.

A typical investment policy for operating funds would authorize investment in the following securities:

- U.S. Treasury instruments
- Agencies of the federal government no longer than four years
- Collateralized bank deposits no longer than one year
- Money market mutual funds regulated by the SEC, often limited to funds that have received the highest rating from at least one of the major rating companies
- The state's local government investment pool

Governmental entities should generally not invest in securities with maturities longer than five years, due to the increased risk of fluctuation in value. Exceptions are appropriate where the government has a specific cash requirement for a date later than five years, such as a scheduled debt payment.

Many public investment policies also allow funds to be placed in other securities such as bankers' acceptance notes, corporate bonds, and repurchase agreements. However, purchases of such securities are best left to managers with significant education and experience in finance, and should be made only after developing policies and procedures that address the use of such securities.

Obviously, the policies will vary with local conditions. For example, an agency with relatively low cash reserves should not keep a large part of its investments tied up in longer-term securities. That agency's investment policy should limit the portion of investments that could be placed in long-term securities and increase the portion that could be invested in liquid investments, particularly money market mutual funds and/or a local government investment pool.

Unsuitable securities

An investment policy should specifically prohibit investment in risky securities that have been problematic for public sector investors, due to loss of principal. For operating funds, the policy will typically prohibit the purchase of mortgage-backed security derivatives, or of all derivatives.

Weighted average maturity

The policy should state the maximum weighted average maturity (WAM) for the portfolio as a whole. As a general rule, 90 days to 3 years is an appropriate WAM for operating funds (*e.g.*, "the maximum weighted average maturity of the portfolio shall not exceed two years").

Diversification

As noted in the discussion of investment principles, diversification allows your portfolio to benefit from the advantage of various types of investments and protect against their weaknesses. The following are typical diversification guidelines:

Type of Security	Max. % in Portfolio
Collateralized demand deposits	25%
Collateralized certificates of deposit	25%
Maximum per institution	15%
U.S. Treasury obligations	75%
Full faith and credit in U.S. government agency obligations	20%
U.S. agencies and instrumentalities	50%
Investment pools	40%
Money market mutual funds	50%
Maximum per fund	25%

Diversification guidelines typically do not state minimums. However, a minimum guideline is appropriate in one instance. Since future cash requirements are difficult to predict, the policy should require that a certain portion of the portfolio be kept in liquid funds such as money market mutual funds and/or a local government investment pool. The exact portion will depend on each entity's particular cash flow characteristics.

Collateralization of deposits

"Collateralization" refers to a method to secure bank deposits. Say you deposit $200,000 in a bank, which is $100,000 greater (as of this writing) than the amount insured by the Federal Deposit Insurance Corporation (FDIC). To "collateralize" the second $100,000, the bank would take a security worth $100,000 and send it to another, independent ("custodial") bank to be held in your name. If the bank with your deposits were to fail, you would receive $100,000 from the FDIC. You would also instruct the custodial bank to sell the security and give you the second $100,000.

Most states, and most public agencies' investment policies, require collateralization of all deposits over the FDIC insurance limit. Insuring that your deposits are properly collateralized requires several steps, the first of which is to execute a collateral agreement with your bank. Fortunately, the Government Finance Officers Association has developed a model collateral agreement, referred to as a "securities service agreement." You should work with your legal counsel to tailor that agreement to meet your state's laws.

Method of sale and custody of securities

Gone are the days when securities were in paper form, and loss of the paper meant loss of thousands (or millions) of dollars. Almost all securities are now in electronic form. However, that does not mean that you should have no concerns about how the securities are kept.

Many governments purchase securities by simply sending funds (via a check or wire transfer) to the selected broker. This method has two drawbacks. First, you have provided payment but have no assurance that you will receive the security. While problems are rare, they do happen. The security may not be credited to your account for some reason. Fault may or may not lie with your broker. Or, your funds may be wired to the wrong place. While a good broker will rectify the situation, including paying you any lost interest (unless the situation is clearly your fault), there is no guarantee of that.

The method to get around this problem is called "delivery versus payment," or DVP. You designate a third party (such as a bank trust department) to handle the trade. You give your money to the third party, who does not release it until the broker has delivered the security. This protects both the broker and the public entity.

The third party can also provide custodial services, which addresses the second weakness of sending your funds to a broker. When you do that, the security is "held" by the broker in the broker's name. While the security has a designation that it belongs to another party, the security is not held in your name.

An independent custodian will hold ("safekeep") securities in your name. The fee is usually based on the value of the securities held. However, such fees often assume that you purchase and sell securities frequently, which creates more work and expense for the custodian. Therefore, ask the custodian to base its fee on your level of investment activity. You should select a custodian through a competitive process, just as you would select any service provider. However, some State Treasurers have made arrangements for custodians to safekeep securities for local governments of their states, often at very reasonable fees.

Reporting

Finally, the policy should require that administrators provide regular reports (at least quarterly) on the status and performance of the investment portfolio. Often a finance committee of the governing body reviews the reports. Reports should include the following:

- A listing that fully describes each security held at the end of the quarter, including its book and market value, type of security, and maturity date
- The percentage of the total portfolio in each asset sub-class
- Earnings for the period
- The performance for each portion of the portfolio on a total return and yield basis for the past quarter, year, and five years, compared to a benchmark for each portion of the portfolio as well as for the portfolio as a whole

The information concerning the market value of securities and collateral should be obtained from sources independent of the broker or depository institution with which the public entity purchased the security or placed the deposits.

"Market value" refers to the price that the security would actually sell for on the reporting date, rather than the cost of the security when you purchased it, or the "book value," which is based on a calculation that might be recorded in your entity's accounting system.

"Benchmark" refers to a performance standard that the portfolio should meet or exceed over a reasonable period of time (about three to five years). The benchmark should be included in the policy, and is usually an established index, as discussed above under Investment Principles.

Accurate and complete performance reporting requires some financial tools and expertise, and therefore is beyond the capacity of many smaller public entities. Such entities should calculate performance to the best of their ability, and strive to improve reporting over time. You should require any professional investment advisor to provide performance reports that meet the above requirements completely.

Pension System Management

There are essentially two types of retirement systems, "defined benefit" and "defined contribution." (The primary purpose of such systems is to provide retirement benefits, but disability benefits are often incorporated into them as well. For simplicity, this section will refer to combined systems as "pension systems.") Pension system managers are generally known as "trustees;" so that term will be used in reference to fire managers and others responsible managing pension systems. Also, the term "participants" will be used at times. This refers to both employees and retirees. While some pension systems may have little contact with retirees (simply sending the benefit checks), others have closer ongoing relationships, as discussed below.

Defined benefit (DB). This is the type of system traditionally found in the public sector. The employee contributes a set percentage of his or her pay, and in return the employer promises to provide specific benefits. Retirement benefits are typically calculated by a formula that multiplies the years of employment by a certain figure, and the result is the percent of the employee's salary that he or she receives upon retirement. For example, benefits might equal 2.5% of regular pay for every year of employment. The plan is set up so that the employee and employer contributions should gain sufficient investment earnings over time to fund the benefits when needed. If benefits exceed predictions or earnings fall short, the employer must make up the difference.

Defined contribution (DC). This type of pension plan often provides the primary pension benefit for private sector employees. Such plans are also offered by most public entities as optional additions to their DB pension plans. With defined contribution plans, the employer, the employee, or both contribute a certain amount of money to the pension system. (This might be a set percentage of pay, or the employee can choose a contribution level, up to a maximum percent of salary, which the firm matches.) The contributions are invested, and at retirement the employee receives the contributions plus their full earnings. The investments may have grown in value substantially or little. In a DC plan, the contribution is defined, whereas in a DB plan, the benefits are defined.

Trustees have certain legal responsibilities under both types of plans. These are referred to as "fiduciary" responsibilities. Since a full discussion of these is beyond the scope of this publication, fire managers who serve as pension trustees should undertake extensive education with respect to their fiduciary responsibilities, both generally and with respect to the specific laws (state, federal, and/or local) that govern their particular pension plans. Trustees' responsibilities are more extensive for defined benefit plans than for defined compensation plans, since their features differ as discussed below. Nevertheless, trustee's responsibilities under both types of plans are significant. Each type of plan has advantages and disadvantages, some of which are as follows:

Defined benefit plans. They provide the participant with a guaranteed amount of benefits, generally for retirements as well as disabilities. While the amount of benefits does not vary with investment performance, which is viewed as positive by many participants, the systems typically lack flexibility in many ways. The pension benefit is normally paid as a monthly amount, which may not meet the participant's particular needs. (For example, an employee might desire cash at retirement to start a business or purchase a vacation home.) From the employer's perspective, these plans have a major disadvantage in that the cost is unknown. The employer must contribute an annual amount determined by an actuary to be necessary to fund the system over the long-term. (The calculation of this amount is discussed later.) Despite actuaries' best efforts to predict the future, many employers find the cost of their defined pension plans exceed expectations, due to various factors such as higher-than-predicted employee disability rates and retiree longevity.

Defined contribution plans. They provide participants with much greater flexibility, but much less certainly regarding the amount of the benefits. At retirement, a participant can typically remove some or all of the funds, or convert some or all of the funds into an annual annuity to be paid for the remainder of his or her life, similar to a DB plan. Another important advantage is the plan's "portability." The participant can typically combine the assets from one DC plan with another. This is very useful to the firefighter who continues to work upon retirement. If the retiree does not withdraw all of his money from the pension plan, the system will continue an ongoing relationship with the retiree for record keeping and participant education (discussed below), just as with current employees.

However, many employees are uncomfortable with investing and may not make good investment decisions, despite education provided by the plan managers. Also, regardless of the amount in the plan at retirement, an employee could outlive his accumulation if he doesn't draw it down responsibly. Defined contribution plans often provide disability benefits as well. The disability provisions of some plans are similar to those of defined benefit plans, while with other defined compensation plans, the amount of benefit varies (to some extent) with investment performance.

From the employer's perspective, a DC plan provides a predictable cost. The employer pays the percentage of pay specified in the plan's provisions, and is ensured of no additional assessments due to factors such as higher rates of retirement, longer participant longevity, or lower than predicted investment earnings.

Managing defined benefit plans

Trustees of defined benefit plans have two basic functions. One is to provide funding for the benefits. This entails determining the annual funding level needed to provide future pension benefits, and investing funds for a long-term time horizon, which is much more difficult than investing operating funds. The second function is making decisions regarding benefit claims, including disability claims. Since these issues have many variables and complexities, trustees are well advised to employ professionals to assist them in performing many aspects of their duties. Discussion of these professionals is included under the appropriate topics below.

Considering applications for benefits. In most cases, determining normal retirement benefits for an individual is straightforward. Each pension plan's provisions, which are usually adopted by state or local law), generally provide clear formulas for calculating retirement benefits. There are usually no questions as to whether a participant who has applied for benefits meets the eligibility criteria, which are generally the participant's age and years of employment.

Determining benefits can be more difficult, however, when considering applications for disability benefits. Generally, trustees have to make two determinations. The first is whether the applicant is disabled. The second is whether the disability is duty-related (as defined in the plan provisions), which is usually compensated with higher benefits. Definitions of disability often vary from state to state, and may differ from the state's laws concerning workers compensation.

In considering whether to grant a disability pension, trustees should require that a physician suitable to the pension board examine the applicant. Many states require multiple examinations, and that is a good practice. The examination should be performed by a physician specializing in the particular condition that the applicant purports to have. Some medical practices specialize in assessing (and treating) claims of duty-related injuries. The physician should be provided with information regarding the duties of a firefighter and asked to give an opinion as to whether the applicant has the physical capability to perform those duties, and if not, whether the physician believes that the disability was caused (as defined in the plan's provisions) by the applicant's work as a firefighter.

Unfortunately, despite the trustees' attempt to obtain a clear opinion from a doctor or doctors, lay people will find it difficult to read some physicians' reports and make a conclusion as to whether the applicant is fit for duty as defined in the applicable governing law. Therefore, it is common for a pension system to employ a qualified individual (perhaps referred to as a "case manager") to obtain the records of the doctors' examinations and render an opinion to the trustees. The case manager may recommend additional examinations or tests, and may ask the doctors to clarify aspects of their opinions. Case managers often have formal medical training.

Interpreting pension plan provisions regarding benefits for duty-related injuries adds another layer of difficulty. For example, the existence of previous medical conditions contributing to the applicant's current medical condition may or may not affect the determination of whether the applicant is eligible for duty-related benefits. In most states these issues have been litigated, providing substantial case law, but clear precedents may not have been established. Therefore, trustees should consider seeking the opinion of an attorney familiar with their particular plan's provisions and case law regarding those provisions. The job of the attorney is not to determine whether the applicant is eligible for a duty-related pension, but rather, how the plan provisions, as defined by case law, relate to the particular fact situation at hand.

In addition, an attorney can provide useful guidance on how the pension system managers should proceed in their consideration of the application for disability benefits. The applicant is entitled to "due process" in the consideration of his or her claim. Essentially, the applicant must be given an opportunity to fully present the claim, and the pension system must consider it fairly. In addition to the general principles of due process, which have been well developed in case law, state law and/or other governing statutes often require that consideration of a pension application follow certain steps. For example, the body that would consider the request might be required to provide notice to the applicant a certain number of days prior to the hearing on the application.

If the trustees do not follow due process and the specific requirements of governing legislation, their decision could be successfully appealed. (The appeals process itself is defined in the governing legislation.) If an appellate body were to decide that the trustees did not follow due process in making its decision, the trustees would likely have to reconsider the application for benefits, at the very least. Depending upon state law, the appellate body might grant the pension that the trustees had denied, to be paid with interest. For that reason, if there is any chance that the pension system's determination might be challenged, trustees should obtain legal guidance on how to proceed in considering benefit applications.

Actuaries and estimating funding levels. Funding defined benefit pension plans is one of the more complex issues in public finance management. The reason is that a pension plan promises to provide certain benefits in the future. Under good financial management practices and most state laws, pension systems cannot simply pay benefits with cash on hand as needed. Each year, pension systems should set aside sufficient money to pay the benefits promised to the employees for working that year. Then, funds to pay the benefits will be available when needed, regardless of the employer's financial condition at the time. Funding the pension system in any other way is a recipe for financial trouble.

Unfortunately, many variables affect the amount of future pension payments, so the amount that the pension system should set aside in any one year must be estimated. The variables include participants' life spans (called mortality), when participants will retire, how many will become disabled (both on the job and off), the extent of future salary increases, and how much earnings the fund's investments will accrue until needed to pay benefits. These variables are referred to as "assumptions."

The job of estimating the amount of pension funding needed each year, given all of these variables, falls to an actuary. The actuary should first review all of the key assumptions with the pension managers. The selected assumptions may reflect each entity's experience and/or experience for a larger population. For example, the number of years of employment at which firefighters retire may be based on data for your entity, because that is influenced by your entity's particular employment practices. On the other hand, mortality is most often set using "mortality tables" based on large numbers of people and used by actuaries throughout the county.

Preferably, trustees should retain an actuary who has experience developing assumptions for plans such as yours. If no suitable actuary were available, an actuary familiar with public safety pension systems would be a good second choice. That actuary could learn your particular plans' provisions. Finally, the trustees and actuary should consult with the system's investment consultants (discussed below) when making an assumption regarding the annual rate of earnings on investments.

As noted above, pension systems should set aside sufficient money in one year (as determined by the actuary) to pay the benefits promised to the employees for working that year. (Most governing laws mandate that practice.) That amount of money is referred to as the "normal cost" of the pension system. Theoretically, the pension system that sets aside the normal cost each year should have all the money it needs to pay benefits.

Unfortunately, the actuary's estimates are never 100% accurate! While actual experience and assumptions can vary for many reasons, some factors are common to most funds. Many pension systems are experiencing higher rates of disability than in previous years, and the higher rates may not be reflected in actuarial assumptions. On the other hand, many defined pension systems became over-funded by the large stock market gains of the mid to late 1990s.

Another reason that funding levels may be insufficient is that the governing body increased benefits but did not provide funding for those benefits. In many states, the state government sets the amount of public pension benefits. Benefit increases are very tempting politically because they do not have to be paid until future years. Legislators can promise future benefits but do not immediately have to raise taxes or cut other services to provide those benefits. Too often, that is left to future generations. When recommending or considering increases to pension benefits, responsible public managers will ask their pension actuaries to estimate the annual cost to fund those benefits, and consider the budget impact of setting aside that amount each year. Unfortunately, some pension systems may have insufficient funds simply because governing bodies knowingly failed to fund the normal cost determined by the actuary.

As a result of all of the above issues, the actuary must determine whether prior years' funding was actually sufficient to pay the benefits promised to current participants (both active and retired). The actuary does that by answering the following question: "What is the value, in today's dollars, of the future benefits that the entity will pay to current participants?" That amount is called the "pension obligation" or "pension liability." That figure is compared to the amount of assets that the pension system currently holds. If those figures are equal, the pension system is "fully-funded." If assets exceed the pension obligation, the system is "overfunded." If assets are short of obligations, the system is "underfunded," and the difference between assets and the pension obligation is described as the "unfunded liability." The following outline may make this easier to visualize:

Comparison of Assets to Liabilities	Funding Status
Assets = Pension Obligation	Fully Funded
Assets > Pension Obligation	Overfunded
Assets < Pension Obligation	Underfunded

A pension system that is underfunded or overfunded will adjust its future funding so that, if everything goes as planned, assets will equal liabilities at a later point in time. Due to the sometimes large amounts of unfunded liabilities, it is simply not feasible for most employers, public or private, to cover an unfunded liability in just one year.

Instead, an unfunded liability is paid off as the mortgage on a house. This is referred to as "amortizing the unfunded liability." Similarly, any amount of overfunding would generally be amortized over a number of years. The current year's amortization amount is combined with the normal cost to make up the system's total annual funding requirement for that year.

This method of determining the annual funding requirement is required for private-sector defined-benefit plans and is typical for most public sector plans. However, some states provisions for funding public sector plans that vary from the description above.

Overview of pension investing for defined benefit plans. As noted above, investing pension funds is much more difficult than investing operating funds, due to the longer time horizon and expanded of investment opportunities. However, the basic principles that apply to investing operating funds hold true for pension funds. Pension managers should first develop an investment plan, consisting of an investment policy and asset allocation. Then, managers will implement the plan.

Investment consultants, policies, and asset allocation. Pension policies for investments should generally contain the same components as policies for operating funds. Since investment policy components were discussed in the section on Investment Policies for Operating Funds (above), they will not be discussed here with one exception, and that is the issue of asset allocation. Asset allocation is essential, because as discussed above, studies have shown that the asset allocation decision contributes to over 90% of the portfolio's total return. While asset allocation is important for operating funds, it is difficult for an investment manager to "get in too much trouble" investing operating funds if the manager follows a prudent investment policy. Investment earnings might fall somewhat short of reasonable expectations, but the fund is unlikely to lose any of the principal amount of the investment.

On the other hand, asset allocation is even more critical for pension funds. The longer time horizon affords the opportunity to invest in riskier securities, and investment income plays a considerable role in funding pension benefits. To achieve investment goals, more risky securities are appropriate for pension funds. Even relatively small pension funds should generally have some assets in domestic equities, and larger funds will want to consider international equities, private equities, real estate and other asset classes. In fact, investing in only very safe securities might be considered imprudent, if an investment manager failed to take reasonable risks to earn funds needed to pay future pension benefits.

Fortunately, the wider variety of investments appropriate for pension funds also allows trustees to diversify the portfolio to control swings in the asset value (volatility). Just as the asset allocation decision plays the key role in determining investment returns, well-documented studies have concluded that achieving appropriate diversification through asset allocation is the primary factor in controlling volatility.

Obviously, investing in a wider range of securities requires careful consideration of their risks and benefits. For that reason, persons with considerable expertise and experience in should guide pension system investment programs. Pension systems that do not have persons with those qualifications on staff should engage independent investment consultants. They will work with the trustees to develop the investment plan (an investment policy and asset allocation) and then work with the system to execute the program, as described below.

Pension systems can employ investment consultants in many different ways. The author strongly prefers a fee-only investment consultant. Such a firm receives compensation from only its clients, insuring its independence from other financial services providers. Other forms of payment may appear to reduce fees, but they only hide fees and bring into question the consultant's independence. For example, a consultant may receive fees for selling securities owned or managed by another arm of its firm, and/or fees from investment management firms to whom it refers the pension system's business. Clearly, such payment systems bring into question the consultant's ability to provide unbiased advice. When choosing a consultant, ask whether the firm receives any form of compensation from financial services providers. If the answer is yes, look into such relationships carefully, and avoid them if other qualified consultants are available.

Due to the many variables to which pension systems are subject, it is not possible to suggest asset allocation guidelines. However, nationwide practices provide a benchmark. On average, public pension systems have about 40% of their assets in fixed-income securities. The remaining 60% are in equities, although some systems (typically the larger ones) also hold a small amount of other relatively risky investments such as real estate. Therefore, managers of fire pension systems should not be surprised should their consultants recommend similar asset allocations. (On average, private-sector pension systems have somewhat higher allocations to equities.)

Investment managers. Once the pension system has determined its asset allocation, it must select specific investments within each asset class. This can be done in many ways. For example, a pension system seeking to invest in equities of large companies could choose among the many well-known and well-managed low-cost, no-load mutual funds. "No load" means that the investment firm charges no fees to put your money into the fund or take it out of the fund (called a "back-end" load). All funds charge fees (a percent of invested assets) to cover management and operating expenses, but paying loads is almost always an unnecessary expense, as you can find comparable no-load funds.

Mutual funds are particularly appropriate for smaller pension systems. A larger pension system, on the other hand, may have sufficient assets to hire an investment management firm to purchase stocks specifically for that pension system. Such firms are referred to as "investment managers" or "investment advisors." Of course, larger systems should consider mutual funds as well; they simply have more options.

These various options are different legal mechanisms that a pension system can use to invest its fund. The options are available for both fixed-income securities and equities. Each option has certain benefits and drawbacks, a discussion of which is beyond the scope of this chapter. (State law may also require, or prohibit, certain of these options.) For simplicity, all vehicles though which pension funds can be invested will be referred to as "investment managers" in the following discussion.

Regardless of the specific vehicle, an independent investment consultant can guide you through the process of selecting the investment manager. Let us say that the pension system wishes to invest $1 million in stocks of medium-sized U.S. companies. Investment companies typically have a database in which they track hundreds of investment managers. The consultant will first "screen" the database to find investment managers who will work with $1 million (which is too low for some investment managers) and who meet your initial criteria, which might include the total value of assets that the firm manages, the period of time that the firm has managed mid-sized domestic stock portfolios, etc. The consultant will then look at the past performance of each firm's investments, and at issues that could affect future investment performance such as:

- Has the firm's staff been stable, particular key investment officers?
- Has the firm's ownership been stable (or has it recently been sold to another firm), which may result in a change in personnel and/or investment methods?
- Does the firm have up-to-date technology?
- Has the firm's investment performance been balanced by reasonable risk levels?

The consultant will typically suggest three to five investment managers to the trustees. All should be well-qualified. The trustees may wish to interview one or more of the firms prior to select the manager that they believe will best meet their needs.

Custodians and accounting. Since pension funds can be invested in a wide variety of asset classes, via a wide variety of investment vehicles, proper accounting for the funds becomes more difficult. For that reason, most pension funds employ a third-party custodial bank. All investment transactions go through the custodian, who can then provide comprehensive reporting of the system's assets. The custodian might hold assets of the pension fund that are managed by its investment managers. This protects the pension system against the possibility of fraud. The custodian accepts all cash, interest, dividends, prepayments, etc., on behalf of the pension system. By sweeping the money immediately into an interest-bearing account (such as a money-market mutual fund), the custodian maximizes investment earnings on idle funds. Also, having a custodian provides a third party to transact security purchases and sales on behalf of the pension fund on a DVP basis (discussed in the section on Investment Policies, above).

Custodial banking services have become much more cost effective as a result of technology. At the same time, it is becoming increasingly specialized and fewer banks provide these services than in previous years. The investment consultant can assist the pension system in selecting a qualified custodian.

Reporting investment performance. Finally, the defined benefit pension system should have a comprehensive quarterly report on the performance of its investments. It is important to remember that the pension system will have investments that can fluctuate widely and can lose money, especially over relatively short periods of time. Therefore, it is critical to measure performance against the selected benchmark, rather than an arbitrary standard. For example, if a pension system's large-company U.S. stock investments lost 5% during the past year, the investments would be performing very well if the benchmark (such as the S & P 500 Index) lost 10%! The benchmark should be clearly defined in the investment policy guidelines.

The trustees should not merely accept the investment reports provided by its investment managers. The system should hire a firm to produce independent performance reports. (If the trustees have engaged an investment consultant, that firm will provide such reports.) The pension system will require the investment managers to provide data on investment activity directly to the firm consultant or other that will develop the performance report. (If all financial activity of the pension system goes through a custodial bank, that bank can provide the data for performance reporting.) The performance reporting service will calculate the increase or decrease on investments for each investment manager, and compare that figure to the benchmark established in the investment policy. The report should also analyze the investment holdings of each investment manager and compare it to the directives provided by the pension system.

For example, let's say the pension trustees engaged Investment Manager A to invest in large-company stocks. The trustees must ensure that the investment manager does just that, and does not deviate into another asset type (for example, small company stocks). Why? The trustees might have engaged Investment Manager B to purchase small-company stocks. If Manager A invested in those as well, the system would have asset duplication (into small-company stocks) instead of the desired diversification.

The investment report should also compare the fund's overall investment portfolio with the asset allocation and performance benchmark stated in the investment policy. Due to gains and losses over time, the system's actual asset allocation may deviate from the investment policy. When that occurs, the system will have to sell assets of one class and purchase another. For example, take a system with $4 million, with 50% allocated to fixed-income securities and 50% to equities. Since stocks have better gains than bonds, over time you would expect the allocation to develop a larger weighting toward stocks, say 55%. The system should then sell 5% of its equity investments and put them in fixed-income investments to "rebalance" to its target asset allocation of 50/50. This can be hard to do, because the system will have to sell its better-performing assets and put them into an asset class that has performed less well! Nevertheless, since asset allocation is essential to investment earnings and reduction of volatility, rebalancing is a critical component of investment management.

Managing defined contribution plans

In the public sector, most employers offer defined contribution (DC) plans as voluntary auxiliaries to their defined benefit plans, rather than as their primary retirement plans. In these cases, the employers typically make no contribution to the plan. Employees may contribute as much as they like, subject to the limits under federal law. While the use of DC plans is becoming more prevalent as the primary retirement plan in the public sector, it is seldom the primary plan for public safety employees. Nevertheless, DC plans provide an important adjunct to defined benefit plans.

DC plans are also referred to as "deferred compensation" plans. They have that name because under federal tax law, the employee's contributions are considered "deferred income" when they are deducted from the employee's paycheck. As a result, the contributions are not subject to federal tax during employment, but are taxed when the employee withdraws the funds.

Deferred compensation plans are often referred to by the sections of the U.S. tax code that authorized them (*e.g.*, 401(k) plans for private-sector employees, 457 plans for public-sector employees, and 403(b) plans for employees of educational institutions). These sections of federal law specify limits on contributions, minimum ages at which contributions may be withdrawn without penalty, and other provisions. In 2001, the federal government adopted legislation making the provisions of these deferred compensation laws very similar.

Managing defined contribution plans is easier for employers in two ways. First, once the employer makes its contribution to the plan, it has fulfilled its obligation to fund the plan. The employer contribution goes into an account for each specific employee, with any employee contributions. In most plans the employee can choose among several investment options, and the employee's account may grow rapidly or slowly, or lose money, depending on the performance of the employee's investments. (The trustees of some private sector defined compensation plans manage the investments, rather than giving each employee a choice of investment options, but that practice is very rare in the pubic sector.)

Regardless of the investment performance, the employer has fulfilled its funding obligation. The trustees do not have to conduct an actuarial study to determine whether the assets of the plan are sufficient to provide the promised benefit payments, because no specific benefit payments have been promised. DC plans are also easier for employers because the trustees (who usually include at least some employer representatives) typically do not manage the investments.

However, the employer has other responsibilities, described below. These are:

- Providing a diversified set of investment options
- Participant education
- Record keeping

Employers can provide some of these services in-house, and/or hire private firms to provide the services. Separate firms could be employed to provide each service, or one firm could provide all of the services. (This method is called "bundled" services, whereas contracting for each service separately is called "unbundled.") Generally, only large employers provide the services in-house and/or use unbundled services. Small and medium-sized entities often lack the expertise to contract for and manage the various services required. Also, service providers may be able to provide unbundled services to small entities at a reasonable cost. Fortunately, several firms specialize in providing bundled defined compensation services to pubic sector employers.

In selecting service providers, trustees should consider fees in addition to quality of services. There are a variety of ways in which fees may be assessed to a DC plan. The employer will usually incur fees for participant education, although these may be "rolled into" other fees. In general, there is a per participant fee for record keeping. These may be paid by the employee or by the employer. Then, each particular investment option has investment management fees that vary with the type of investment. These are almost always paid by the employee. The management fees charged by different investment management firms for identical or similar investment vehicles can vary greatly. Higher costs do not always mean higher quality; many investment companies provide similar products at markedly different costs. Plan trustees will want to make prudent decisions to offer quality investment options at reasonable costs. When selecting service providers, trustees must remember that the any fees paid by employees will reduce the growth of their retirement funds.

Investment options. As noted above, employers usually select the investment options available to participants. These options should be sufficiently large to allow each employee to select investments that are appropriate for his or her retirement needs and stage in life. Typically, younger participants should select investments that have more volatility in the short-term, but a better chance of higher yield in the long-term. Less volatile investments are generally a better choice for older investors. The options should allow employees to find investments that match the levels of risk that they are willing to take. Also, some employees prefer to invest in "socially-conscious" funds, which consider only the stocks of companies with good practices with respect to issues such as environmental protection and employment, in addition to the performance of the companies' stocks.

More investment options are not always better. With more options, participant education becomes more difficult and participants can become confused by a bewildering array of options. That could lead participants to make any number of errors. For example, a participant could select a variety of stock funds in an attempt to diversity his or her investments. However, the funds might invest in many of the same types of stocks, so the investor's assets could be much less diversified than hoped.

Many DC plans now offer four or five "model portfolios" in an attempt to make the investment decision easier. Each model portfolio provides an appropriate array of investments for participants at typical stages of life. For example, a young employee might choose a "growth portfolio" that has 80% of its assets in equities and 20% in fixed-income securities. Another model portfolio, appropriate for retirees who rely on income from savings in the DC plan and cannot risk much loss, might have the opposite asset allocation.

Participant education. While the trustees do not generally invest the assets of deferred compensation plans, they must provide education to participants so that they can make reasonable decisions. Legally, this is a fiduciary responsibility of the trustees, as much as investing the assets themselves. The educational program should include mailings to participants, allow each participant the opportunity to meet individually with representatives of the service provider to discuss his or her investments and savings plan, a web site with financial information and planning tools, and perhaps group classes on such topics as financial planning for retirement.

Most employers meet this obligation by engaging private firms. Employers should consider the extent to which the firm's compensation, or the compensation of individual representatives of the firm, depends upon participants' investment choices. If the firm or its representatives receive larger fees when participants choose one investment option over another, it will be difficult for the firm to remain unbiased when it educates employees regarding investment options. Many firms provide good services at a reasonable cost and do not "push" participants into inappropriate investment decisions.

Record keeping. DC plans must account for the contributions, withdrawals, and investment gains and losses of each employee. This is relatively straightforward, but can be costly since relatively small amounts of sums must be recorded within each employee's account. For example, take an employee who earns $40,000 per year, is paid twice each month, and puts 3% of his or her earnings in a deferred compensation plan, split equally into three different investment vehicles. The employee's contribution to the deferred contribution plan would be $50 per pay period, with about $17 put into each investment vehicle.

Firms that specialize in accounting for deferred compensation plans, including the bundled service providers, can put more of the employees' contributions into retirement savings and less into accounting fees. However, managers should not necessarily select the low-cost service provider! Aside from receiving the incorrect amount of pay, few experiences are more upsetting to employees than having their retirement savings recorded improperly.

Hybrid retirement programs

Finally, a word about "hybrid" retirement programs—these attempt to combine the best of DC and DB systems. One method gaining popularity is the so-called Deferred Retirement Option Program, or "DROP." In these, employees participate in traditional defined benefit plans for a certain period of time. Often that is the period required to retire with "full" benefits, such as 20 years of service. At that point in time, the employee can elect to participate in the DROP program.

Such programs work as follows. First, in many cases the employee must commit to retire in a specified period of time; say, within five years. Having done so, the employee and employer cease contributions to the DB program. The employee's DB benefits are frozen at the level that the employee would receive if they were to retire at that time. The employee will receive that level of benefits upon retirement.

During the DROP period, the pension payments are not paid to the employee, but are held within the pension fund, where they accumulated with interest. Upon retirement, the employee receives a regular pension from the DB plan, and receives a lump sum payment for the amount that has accumulated in his or her DROP account. The advantage to the employee is that he or she receives the fixed, guaranteed income of the DB plan. The fixed pension is computed at a time that it might have the greatest present value. In addition, the DROP provides the employee with the lump sum benefit that is typical of a DC plan. That provides flexibility; upon retirement, the employee can withdraw all or some of the DROP assets to purchase a fishing boat or start a new business, or leave the assets in the plan. There, the assets will (hopefully) accumulate more earnings, and the retiree can withdraw the assets at a later date.

The advantage to the employer is that the employee's benefits from the DB plan stop growing when the employee enters the DROP program. The employer also knows when the employee will retire, and can establish recruitment and training plans accordingly, which provides some of the flexibility of a DC plan. The extent of benefit to the employees and the employer depend on the exact structure of each DROP program.

DROP programs are often touted as cost neutral to employers. However, as with any component of a retirement system, the reality is complex and "the devil is in the details." Most DROP programs add cost to the pension plan, which is paid by the employer unless the DROP is put in place with offsetting employee contributions. Trustees considering such programs should obtain guidance from professionals (including an actuary) who have the capability to correctly assess the benefits and risks of various plan options.

CONCLUSION

The foregoing discussion may seem daunting to fire chiefs whose financial systems are not well developed. They might ask, "Where do I start?" That is a fair question. A sound, comprehensive financial management program cannot be put into place any more easily than a comprehensive fire service including prevention, disaster preparedness, response to hazardous waste events, etc. So where should you start?

Remember the definition of financial management provided at the beginning of this chapter: "Financial management is the art of directing the acquisition and judicious use of money to accomplish an end, and providing stewardship over financial resources entrusted to you." To determine the most important areas of improvement in your agency's financial management, you must assess your current financial management system. This need not be a formal process. Perhaps some readers of this chapter made a mental checklist, margin notes, or a list, indicating the areas in which their agency does well and areas that need attention.

Go back to that list or notes, and ask yourself where red flags went up. Perhaps you're not sure that banks collateralize your bank deposits properly. Or, your purchasing process doesn't include basic fraud controls. If so, some relatively straightforward actions could improve the safety of agency funds entrusted to you.

You should also consider the areas of improvement that would pay the greatest dividends. Do you have difficulty receiving approval from your governing board to replace large-ticket apparatus? Instituting an annual set-aside, or a lease-purchase program, could alleviate that problem. Or perhaps your budget presentation needs to focus more closely on the benefits of the equipment from the citizen's point of view. Is the board of trustees of your pension fund investing assets similar to the way in which your agency invests operating funds? If so, retaining an independent investment consultant could guide you to a sound investment program that will reap benefits for years to come.

In short, you should consider financial management issues the same way that you consider fire service issues. You can't do everything at once, so determine the areas that will provide the greatest benefit to your community. Through continual improvement to one area after another, you will develop a strong financial management system.

References

[1] *Recommended Budget Practices: A Framework for Improved State and Local Budgeting,* p. vii. Government Finance Officers Association, Chicago, IL, 1998.

APPENDIX 1

A Case Study in Budget Reform

The managers of one government agency were able to get around this problem when they proposed a number of improvements to the budget process and documents to their elected board. This included the addition of useful information such as fiscal policies and program goals and objectives. At the same time, the managers proposed removing line-item detail from the budget document, indicating that it would become quite big, with increased production costs. The managers offered to make the line-item detail available in the chief executive's office for any elected official who wanted to examine it. The managers reminded the elected officials that anyone wishing to discuss the detail was welcome meet with the fiscal officer or chief executive officer.

The elected officials agreed to the change. In the first year, just one or two elected officials decided to examine the line-item detail. Over time, even they declined to do so. In the meantime, the agency managers were able to focus the governing board's review on more meaningful information such as the agency's fiscal position, success of the various departments in meeting their service objectives, and recommended programs for the coming fiscal year.

APPENDIX 2

A Case Study in Establishing User Fees

The political directive was for two EMS fee schedules: one for residents, which recognized that while taxpayers pay for availability, actual users should assume the costs of use; the other for nonresidents, who should assume the entire cost of service. The chief developed two cost models: 1) a program budget with appropriate overhead, divided by actual response unit-hours produced; and 2) total hourly unit availability costs. The governing body adopted the fees.

Each year the new figures for personnel salaries, units produced, unit hours available, and adjusted overhead were plugged into the computer spreadsheet and new EMS rates were set.

One evening a petroleum company outside the jurisdiction's boundaries dumped a tractor-trailer load of gasoline in the median strip of the interstate. The fire department and other local agencies were on the scene for days. The morning after the incident was terminated, a bill was in the mail to the petroleum company for $37,557 in fire department service fees. Ten days later a check for that amount arrived. Other responding agencies prepared invoices for the incident, but it took them weeks.

APPENDIX 3

A Case Study in the Procurement of Equipment Apparatus

A fire department in Illinois combined many of the purchasing methods discussed in this chapter to acquire two engines at a reasonable cost that also met the department's needs well. First, the department conducted research to determine the specifications and reliability of engines currently produced. (It did not issue a formal RFI, but the process that the department used was similar to it.) Based on the responses, and with approval from its governing body, the department engaged in a competitive negotiation process with two manufacturers. Their selection was based on the following criteria: the manufacturers produced apparatus that would meet the department's needs, and they had good long-term track records in the industry.

The fire department then prepared and issued a formal RFP to the manufacturers that required a technical response (proposed vehicle that met the department's specifications) as well as a cost proposal. The department opened the proposals in a formal manner, similar to a bid, allowing each vendor to know the cost proposed by the other vendor.

Following that, the fire department engaged in negotiations with the manufacturers, allowing the fire department to gain a better understanding of the costs and benefits of various options for the fire truck. For example, the department found that the particular compartments it had initially specified were an "option" for one manufacturer. The specified compartments cost several thousand dollars more than the manufacturer's standard compartments, which were equally functional. Therefore, the department modified its specifications to allow the standard compartments.

The department then required both manufacturers to submit final proposals. Both proposed engines that were well suited to the department's needs, at competitive prices. The department selected the low-cost engine. It also reviewed the prices of similar engines purchased by neighboring communities through bid processes. By doing so, the department confirmed that the competitive negotiated process had saved tens of thousands of dollars compared to the bid process.

4

MANAGEMENT AND THE LAW

Philip Stittleburg

CHAPTER HIGHLIGHTS

- Fire chiefs who understand the basic concepts of the American legal system (federal, state, and local), will be prepared to deal effectively with continuing legal challenges confronting the American fire service.
- Fire chiefs who understand in general terms the path of a civil lawsuit will be prepared to deal effectively with lawsuits that arise.
- Fire chiefs who have a working knowledge of negligence law will be prepared to deal effectively with the challenges raised by new legal responsibilities.

INTRODUCTION

The concept of law has been around as long as people (and fire) have existed. It is not far from the imposition of rules of conduct and behavior that governed a family, to the development of rules that governed a tribe or a society. Over time, those rules, written and unwritten, provided the means by which acts that were a detriment to the community ("an eye for an eye" for example) were controlled by that community.

In some societies, elected or appointed bodies of elders or other highly regarded members of the community wrote down those rules in the form of laws. In other societies, those rules were developed on a case-by-case basis by judges as they dealt with problems. In Great Britain, the principal parent of the American legal system, both written law (statutory law) and judge-made law (common law) governed the conduct of British citizens.

The Anglo-Saxon approach to law also distinguished between a legal wrong that affected one individual and a legal wrong that injured a society as a whole. If one person suffered an injury, the Anglo-Saxons provided a civil legal system through which the injured person could recover damages. If an act or injury hurt society as well as an individual, as in the case of a murder, for example, the society would act on behalf of the individual through the criminal legal system. The goal of the criminal legal system was often to protect society from further criminal acts by punishing the criminal, rather than helping a victim recover losses.

Foundations of American Law

While the American legal system was built primarily from the British legal system, Americans added a unique feature of their own, a constitution. In states that had been territories under France or Spain, state legal systems still reflect their French or Spanish roots. For example, in Louisiana, traces of the French "Napoleonic Code" are still very much in evidence, and California land law had its roots in the Spanish legal system.

Although the Americans won the Revolutionary War, their first efforts at organizing a new country were not successful. Since each state was unwilling to grant meaningful power to the Continental Congress, the states collectively could do very little. So each state sent representatives to a "Constitutional Convention" to try again to develop a system that would allow the states to work as a unit, but at the same time preserve the fundamental rights of each state to govern its own affairs.

This time, the states were more successful. Their representatives agreed on a basic formula: the states would create a federal government and give that federal government certain powers; all other powers would be in the hands of each state. The details of this formula were embodied in the United States Constitution, which was adopted by all the states in 1787.

The adoption of the Constitution marked the true end of the American Revolution. While the United States' legal system was based on a large part of Great Britain's legal system, Great Britain has no constitution. Over the succeeding two centuries, the Constitution itself, and the struggle between and among the states and the federal government about what the Constitution says and means, have shaped the American legal system.

The American Legal System

Law in the United States depends on two sources for development and explanation: legislative bodies (Congress and state legislatures) and courts, both state and federal.

Legislatures and statutes

Legislatures create what lawyers call *statutory law*. A statutory law is adopted by a legislative body, written in the form of a statute, and then signed by the president or a state's governor (or passed again over the president's or a governor's veto). The intent of statutory law is to describe, as clearly as possible, conduct or behavior that is acceptable or unacceptable, and the consequences of violating those standards. In cases where the statute is not clear, lawyers will look at the legislative history of a statute (as chronicled in the *Congressional Record*, for example) for clarification and guidance.

Within constitutional limits, legislative bodies also can bestow some of their powers on other bodies. A state legislature can, for example, delegate to a state agency the power to create, enforce, and amend a state fire code. Such a code would not necessarily itself be a statute, but because it has the power of statutory law behind it, the state fire code would have the power of statutory law. Congress often passes broad statutes and then grants federal agencies the power to adopt the specific rules necessary for enforcement of those statutes. Since most county, city, town, and other local governments were created by state statutes, the powers held by those local governments are also delegated powers.

Courts and common law

Courts create *common* (or case) *law* through decisions made in connection with a particular legal controversy; the decision applies to any future case involving the same (or similar) facts. Not all courts have the power to create or change common law. Generally, state and federal trial courts cannot issue statements of law that are binding on other courts. Courts that have the power to create or change common law are called *appellate courts* or *courts of record*. The decisions of these courts are published in the form of "opinions" that contain a statement of the facts of a case, a discussion of how those facts relate to existing common or statutory law, and a legal conclusion or "finding."

Appellate courts have a great deal of respect for *precedent*, or law set forth in earlier court decisions. Because common law is such an important part of the American legal system, judges are very reluctant to destabilize the legal system by frequent changes in common law. It may take a decade of discussion at the appellate level before an earlier case is overruled.

Jurisdiction

Statutory law and common law are not always distinct. A legislative body may, by statute, overrule common law. An appellate court may declare a statute unconstitutional (a power attributed to the federal courts as early as 1803 in the case of *Marbury v. Madison*).

Other aspects of jurisdiction are regulated by statutory law or rule. A decision by the North Carolina Supreme Court does not establish new law in California. A decision by the Federal District Court in Dallas does not bind a Federal District judge in Boston. (In matters involving fire departments and firefighters, only the U.S. Supreme Court has truly national jurisdiction.) Furthermore, a resident of Florida cannot file a lawsuit against his local fire department in a Louisiana court. (Note—The concept of jurisdiction is considerably more complex than this summary suggests. The summary is intended as a brief explanation of why a court decision or trial verdict in one state does not automatically apply in another.) Here are some general rules of jurisdiction in the United States:

- State laws apply only in each respective state.
- State appellate court statements of common law apply only in the state of origin.
- Federal courts have jurisdiction when federal laws or constitutional rights have been violated, or when there is "diversity of jurisdiction" (a case that involves people and laws of two or more states).
- Federal courts generally respect state common law in "diversity of jurisdiction" cases.

The Law and the Fire Service

The year 1964 marked a fundamental change in the relationship between the American fire service and the law. Before, a fire chief's encounter with the American legal system likely was associated with a fire code enforcement matter, or, occasionally, a grievance. But in 1964, Congress passed the first civil rights act since Civil War times. The *Civil Rights Act of 1964* stimulated enormous interest in individual rights, including the right to obtain (and retain) a job as a firefighter.

Also by 1964, erosion of the concept of "sovereign immunity," the absolute immunity from negligence lawsuits customarily enjoyed by local governments, was well underway, spreading from west to east, state by state. The decline of sovereign immunity was more than matched by a marked increase in litigation, and by expansion of the damage awards juries were likely to grant a plaintiff who had suffered some kind of injury.

The convergence of the *Civil Rights Act of 1964* and the "negligence revolution" meant that fire chiefs found themselves suddenly thrust into legal controversies that had little to do with fire codes or grievances. This trend continued as complex environmental and safety laws were passed, and as Congress and legislatures began including local governments (and thus fire departments) in coverage of statutes like the *Fair Labor Standards Act (FLSA)*.

Due Process

Fire chiefs sometimes run afoul of the Constitution in connection with disciplinary actions, particularly those that involve loss of pay, rank, or job. For employment law purposes, this simple checklist of the basic elements of due process will help the fire chief maintain organizational discipline without violating rights.

- Provide clear notice of specific charges or allegations.
- Provide enough notice to enable the accused employee to defend himself.
- Provide a hearing where the employee can present his side of the story.
- Make sure the employee knows that he can bring legal help (the employer is not bound to pay for such help).
- Give the employee (or representative) the opportunity to question witnesses.
- Make sure that some kind of record of the hearing is maintained.
- Make a decision based on the evidence, not on circumstances not directly related to the charge or accusation.
- Impose disciplinary action that is proportionate to the offense.

Fire chiefs should remember that "due process" does not require disciplinary hearings to be conducted with the same formality as a criminal trial (courts consider disciplinary hearings to be civil, not criminal, even if the result of a hearing is loss of a job), but "due process" does assume fair treatment.

Federal Statutes and Cases

Literally hundreds of federal laws affect fire chiefs throughout the U.S. Since these laws are implemented by federal agencies, they are accompanied by thousands of pages of federal regulations. Of all those laws and rules, a relative handful are critical to fire department managers. The following are summaries of the most influential of the federal laws.

Civil rights acts

Congress has passed four civil rights acts since the end of the Civil War. The *Civil Rights Act of 1866* (42 U.S. Code, Section 1981) provides that "all persons...shall have the same right...to make and enforce contracts...as enjoyed by white citizens." The word "contracts" has been interpreted by courts to include the "contract" implied in hiring someone; therefore, Section 1981 has been applied to governmental discrimination in hiring (the *Civil Rights Act of 1991* broadened the concept of "contract" to include discrimination on the job after hiring). The *Civil Rights Act of 1871* (42 U.S. Code, Section 1983) prohibits "...the deprivation of any rights, privileges, or immunities secured by the Constitution and laws...." The application of Section 1983 is limited by other language to local governments, but Section 1983 is only a means by which "deprivation" of a specific constitutional right at the local level can be punished by lawsuits.

The *Civil Rights Act of 1964* (42 U.S. Code, Section 2000) has probably done more to affect the public workplace than any other law. This act, in effect, codified the Supreme Court's finding in *Brown v. Board of Education* (1954) that segregation on the basis of race is inherently discriminatory, and thus a violation of the Constitution. Title VII of the act prohibits employer use of race, color, religion, sex, or national origin in connection with any hiring, promotion, dismissal, or other employment decision, unless the employer can demonstrate that use of race, color, religion, sex, or national origin in such decisions is connected to a "bona fide occupational qualification" (BFOQ). Very few employers have ever been able to establish a BFOQ exemption under Title VII.

Starting in the second half of the 1980s, the U.S. Supreme Court handed down a series of rulings that provided a more restrictive interpretation of the civil rights acts of 1866 and 1964. After two years, and a presidential veto, the *Civil Rights Act of 1991* was passed. The 1991 civil rights act is summarized in Table 4–1 in a "before-and-after" comparison.

The *Civil Rights Act of 1991* also affects discrimination in multinational U.S. corporations, expert-witness fee recovery, *Americans with Disabilities Act* "good-faith" defenses, and the availability of jury trials for intentional discrimination cases. Finally, the 1991 act allows plaintiff recovery of compensatory damages as well as out-of-pocket expenses from the employer, including local government employers. Compensatory damages are "capped" from $50,000 to $300,000, according to the size of the employer.

Table 4–1 Civil Rights Act of 1991

Before 1991	After 1991
Section 1981 (Civil Rights Act of 1866) covers discrimination in hiring, promotion in some cases, but *not* harassment, discrimination in firing, etc. [*Patterson v. McLean Credit Union*, 491 U.S. 164 (1989)]	Section 1981 covers all forms of racial discrimination in employment.

Discriminatory seniority rules can only be challenged when adopted. [*Lorance v. ATUT Technologies,* 490 U.S. 900 (1989)]	Discriminatory seniority systems can be challenged when adopted and when they affect employees in the future.
White firefighters could challenge an affirmative action consent decree (an agreement between parties approved by the court) several years after the decree was approved by the court. [*Martin v. Wilks,* 490 U.S. 755 (1989)]	Consent-decree challenges are prohibited if the challenger had a "reasonable opportunity to object" at the time or their interests were "adequately represented by another party" (*e.g.,* a union).
Employers are not liable for discrimination regarding actions if the employer can demonstrate that the action would have been taken without a discriminatory motive. [*Price Waterhouse v. Hopkins,* 490 U.S. 228 (1989)]	Intentional discrimination is unlawful, even if the same action would have been taken without evidence of discrimination.
Employers do not have to prove *business necessity* to defend against employment practices that have a "disparate impact" on a protected group. Employers only have to provide a business *justification* for those practices. [*Wards Cove Packing v. Atonio,* 490 U.S. 642 (1989)]	Employment practices that have a disparate impact on a protected group require "...demonstra[tion] that the practice is job-related for the position and consistent with business necessity." Plaintiffs must specify the particular practices they are challenging.

Age Discrimination in Employment Act (ADEA). *ADEA* (29 U.S. Code, Section 621) was passed in 1967 and has enjoyed an interesting history since then. Originally, it prohibited discrimination on the basis of age against all persons 40 years of age or older, and it covered private employers and local governments employing 20 or more employees. However, in 1986, Congress amended *ADEA,* creating an exemption to allow state and local governments to retain mandatory retirement age requirements for firefighters, police, and correction officers until the amendment expired in 1993. Then in 1996, Congress reinstated the exemption as it appeared prior to its expiration, and directed the U.S. Department of Health and Human Services and the National Institute of Occupational Safety and Health (NIOSH) to study the issue of firefighter fitness.

Throughout *ADEA*'s history, debate continued over whether it also applied to local governments that employ fewer than 20 employees until the U.S. Supreme Court rendered the issue irrelevant in January 2000 in the case of *Kimel, et al v. Florida Board of Regents, et al* [120 SupCt 631 (2000)] when it ruled that *ADEA* did not apply at all to state or local government employees in states that have retained sovereign immunity doctrines. There are 23 sovereign immunity states, those being: Connecticut, Delaware, Georgia, Hawaii, Idaho, Kansas, Louisiana, Maine, Michigan, Mississippi, Montana, Nebraska, Nevada, New Jersey, Ohio, Oklahoma, Oregon, Pennsylvania, Rhode Island, Tennessee, Utah, Vermont, and Virginia.

Given the fact that Congress reinstated the mandatory retirement age exception in 1996, *Kimel* didn't really change anything in that regard. Furthermore, those states that had already abolished sovereign immunity have frequently been dealing with age discrimination claims on a state level for some time. The end result of all this is that governments still are not free to discriminate at will, but these discrimination claims will more frequently be dealt with at the state level.

Fair Labor Standards Act (FLSA). When first enacted in 1938, *FLSA* (29 U.S. Code, Section 201) did not apply to government employees. Over the years, the U.S. Supreme Court rebuffed suits brought by public employees to obtain *FLSA* coverage until *Garcia v. San Antonio Metropolitan Transit Authority* [469 U.S. 528 (1985)]. In the wake of *Garcia,* Congress amended *FLSA* to include public employees and charged the U.S. Department of Labor (DOL) with the development of rules on how *FLSA* was to be applied in local government workplaces.

FLSA governs work schedules; specifically, it establishes the right of employees to time-and-one-half compensation after a minimum number of hours worked in a defined work period. In public agency fire departments, a "defined work period" ranges from 7 to 28 consecutive days, while hours-worked thresholds vary according to a particular fire department's work schedule. For fire departments that are not public agencies, such as those that are private non-profit corporations, a "defined work period" is 7 consecutive days. It is important to remember that a work period is not necessarily the same as a pay period and is established by the employer.

Public agency fire service employees who work more than 212 hours in a 28 consecutive day work period are entitled to time-and-one-half compensation for the excess hours worked. For those with shorter work periods, the hours are computed on a *pro rata* basis and average out to about 7.57 hours per consecutive day. Non-public-agency firefighters must be paid overtime for hours worked over 40 in one week.

The administration of *FLSA* presents some major problems. For one thing, the law provides that a career firefighter working for a fire department subject to *FLSA* cannot volunteer the same services that he or she is paid to perform to the same department. This provision would certainly seem to be clear enough on its face, but can be difficult to apply. The issue of who is the employer can become quite confused when different organizations provide various services in the same geographic area.

More problems arise when applying the act to volunteers, but the difference is critical to an employer because volunteers are not subject to the act. The first challenge usually is determining who is a volunteer. While it might seem that anyone who receives some sort of financial benefit from an employer is not a volunteer, this is not the case. In fact, an individual performing volunteer services for a municipality will not be considered an employee (and thus not subject to the act) if (1) the individual receives no compensation or is paid expenses, reasonable benefits, or a nominal fee, and (2) such services are not the same type of services that the individual is employed to perform for such public agency.

Permissible expenses could include such things as a uniform allowance for the purpose of providing and maintaining a uniform; out-of-pocket expenses incurred during volunteering; and tuition, transportation, and meal costs involved in attending classes to learn how to provide the services. Reasonable benefits can mean inclusion of volunteers in group health insurance plans, pension plans, or length-of-service awards. However, such benefits must be paid to all persons who volunteer in the same capacity.

Determining what is a "nominal fee" is much more difficult (kind of like trying to decide when a hill becomes a mountain). A nominal fee cannot be a substitute for compensation and cannot be tied to productivity, but unfortunately neither the DOL nor the courts have given us any bright line to use. For example, one federal district court held in 1997 that firefighters who were regularly paid between $5.05 and $9.00 per hour for their duties were employees, not volunteers, and therefore covered by the act. Yet in a 1987 ruling, the DOL held that volunteer firefighters paid from $8.00 to $25.00 per assignment, with a maximum of two weekly assignments, received nominal pay and remained in the "volunteer" category.

If there is a question as to whether someone is a volunteer or an employee within the meaning of the act, it is a good idea to ask the DOL for a Wage and Hour Opinion letter. You can also find quite a lot of *FLSA* information on the DOL web site at *www.dol.gov*. However, ignoring the issue is probably not an option, as the DOL is aggressively pursuing wage-and-hour violations.

Equal Pay Act of 1963. The *Equal Pay Act of 1963* (29 U.S. Code, Section 206) is an amendment to *FLSA* that, in a phrase, requires "equal pay for equal work." In other words, an employer cannot pay employees of one gender less than those of the opposite gender for doing the same work if the difference in pay is based on gender. The act includes limited exceptions and a prohibition against reducing the salary of any employee to meet the "equal-pay" requirements of the act.

After the U.S. Supreme Court handed down its decision in *Kimel* (see previous *ADEA* section above), the applicability of this act to sovereign immunity states is questionable. It would seem that the arguments that led the Supreme Court to conclude that *ADEA* does not apply to those states would also apply here.

Americans with Disabilities Act (ADA). Although the *ADA* (42 U.S. Code, Section 1201) became law in 1990, many of its requirements had first been applied to many local governments as part of the *Rehabilitation Act of 1973.* The difference between *ADA* and the "Rehab Act" is that the 1973 act applied to federal fund recipients, while *ADA* initially applied to all state and local governments, regardless of the number of employees.

ADA protects "qualified individuals with disabilities" from discrimination in public service or employment. A "qualified individual" is defined as a person who can perform the "essential functions" of a job "with or without reasonable accommodation." A "disability" includes:

- A physical or mental impairment that substantially limits one or more of the major life activities of the individual
- A record of such an impairment
- Being regarded as having such an impairment

Since its inception, the *ADA* has proven difficult to administer. Deciding if an individual has a disability under the act and what constitutes a reasonable accommodation has resulted in many lawsuits. A few examples will illustrate the problem.

In 1992, Dennis Henderson applied for a position as a firefighter with the Pontiac (Michigan) Fire Department. He placed seventh among 107 applicants and had 19 years of previous experience, but had sight in only one eye. The fire department, relying on a provision in NFPA 1582, *Standard on Medical Requirements for Firefighters,* which requires firefighters to have binocular vision, declined to hire him. The U.S. Department of Justice then filed a lawsuit against the fire department, claiming that it had violated Henderson's rights under the *ADA.* The fire department ultimately settled the claim by hiring Henderson and giving him five years of seniority and more than $100,000 in back pay and benefits.

Next came the case of *Aka v. Washington Hospital Center* [156 S3rd, 1284 (1998)] decided by the District of Columbia Circuit Court of Appeals in 1998. In that case, Mr. Aka was told by his doctors after a heart-bypass operation that he could not return to his former job as a hospital orderly. He then asked his employer to transfer him to a job that involved a light or moderate level of exertion. When he was unsuccessful in securing a transfer to any such position, despite the fact that openings did exist at the hospital and were filled with other people, he filed a lawsuit under the *ADA* as a qualified person with a disability. The hospital relied on the argument that it had no obligation to prefer job applicants with disabilities to other applicants on the basis of disability. The court differentiated here, saying that Aka was not a job applicant under the *ADA,* but rather was an incumbent employee seeking reassignment, that reassignment was a reasonable accommodation under the *ADA,* and ruled in Aka's favor.

The *Henderson* and *Aka* cases seem to be the high water mark in cases that apply the *ADA* either directly or by analogy to the fire service. In 1999, the U.S. Supreme Court decided the case of *Sutton, et al v. United Air Lines Inc.* [119 Sup. Ct 2139 (1999)], which although not a fire service case, basically eliminated the Henderson-type argument. In *Sutton,* twin sisters applied for jobs as United airline pilots. United required its pilots to have uncorrected visual acuity no worse than 20/100. While the Suttons had uncorrected vision of 20/200 or worse, they had corrected vision of 20/20. Consequently, they argued that a reasonable accommodation would be to allow them to wear glasses to correct their vision.

The Supreme Court ruled against the Suttons, holding that they were not disabled under the *ADA.* It stated that "the use or nonuse of a corrective device does not determine whether an individual is disabled; that determination depends on whether the limitations an individual with an impairment actually faces are in fact substantially limiting." The Court also noted that when Congress adopted the *ADA,* it believed that approximately 36 million people would be covered by it, while the Suttons' argument would place more than half of the entire U.S. population in the disabled category. Clearly, if *Sutton* had been decided before Dennis Henderson applied for a job, the Pontiac Fire Department could have saved a lot of money.

The U.S. Supreme Court further limited the application of the *ADA* in February 2001 when it handed down its decision in *University of Alabama Board of Trustees v. Garrett* [121 Sup. Ct 955 (2001)]. In that case, the U.S. Supreme Court held that state employees are not protected by the *ADA* in those states that retained sovereign immunity. (For a list of sovereign immunity states, see the previous section on the *ADEA*.) It should be noted, however, that this decision does not apply to local governments, such as counties, towns, cities, and special districts in those states.

As mentioned above, a "qualified individual" must be able to perform the "essential functions" of a job "with or without reasonable accommodation." It is up to the employer to define these "essential functions." This requires an analysis of what a job really requires, not just what it would be helpful for the employee to be able to do. The frequency with which an employee may be called upon to perform a particular function does not determine whether the function is essential though. For instance, requiring that a firefighter be capable of removing a victim from a burning building can be a legitimate essential function, even though a firefighter may only rarely, if ever, actually be required to make such a rescue. Close attention to the drafting of the essential job functions will help prevent confusion by establishing exactly what is required as a firefighter.

Family and Medical Leave Act (FMLA). *FMLA* (29 U.S. Code, Section 2601) established a right for employees to have unpaid medical leave when employed by employers with 50 or more employees, including local governments. *FMLA's* provisions include the following:

- Up to 12 weeks of unpaid leave in any 12-month period for childbirth, adoption, the care of a child, spouse, or parent with a serious health problem, or for an employee's own serious health condition when that employee has exhausted sick leave or other benefits.
- An employee on *FMLA* leave is guaranteed health care benefits and his or her own job (or an equivalent job) on return from leave. Seniority does not automatically accrue unless the employer so chooses.
- The employee must have worked for the employer for at least one year and cannot collect unemployment or other government benefits during the leave period.
- The employer can ask for medical opinions and other verification of the need for *FMLA* leave and can require that the employee who does not return to work after

the leave repay the employer the cost of the health care benefits that are guaranteed during the leave.

- An employer can deny leave to any of the highest-paid 10% of its employees if allowing the leave would cause "substantial and grievous injury" to the employer's operations. In a typical city, this means that the fire chief, police chief, and similarly compensated staff could be denied family leave under *FMLA*.

Note that in the case of fire departments that are part of a town, city, or county government, "employer" is that town, city, or county, not the fire department. This means that *FMLA* applies to a 35-member career fire department if the department's parent government has 50 or more total employees. The "ten-percent" rule applies to the highest-paid 10% of the employees of the town, city, or county. A fire district that is not a part of another government unit would be covered by *FMLA* only if it had 50 or more employees.

Occupational Safety and Health Act (OSHA). *OSHA* was enacted by Congress in 1973. Under the act, the Occupational Safety and Health Administration (*OSHA*) was created and empowered to develop workplace safety standards. *OSHA* also created a research office called the National Institute of Occupational Safety and Health (NIOSH).

OSHA coverage of local government employees generally (and firefighters in particular) varies according to whether a given state conducts no occupational safety and health actions of its own, thereby leaving such actions to federal *OSHA*), or operates its own occupational safety and health program, with standards at least as stringent as federal *OSHA* standards ("state plan" states). Federal *OSHA* usually does not cover local government employees; local government regulation is imposed by states.

Even though federal *OSHA* standards may not apply in a given jurisdiction, they can be introduced as evidence of a "community standard" in a civil lawsuit.

There are several *OSHA* standards that are of special interest to fire chiefs. One sets out the regulations relating to bloodborne pathogens (29 CFR 1910.1030). This provision requires employers to furnish their employees with the necessary protective clothing and equipment to protect them if they come into contact with certain human bodily fluids. It also provides for the care of personal protective clothing that may have become contaminated, and sets out training requirements. It further requires employers to develop written infection-control plans.

Another important *OSHA* regulation is the *Respiratory Protection Standard* (29 CFR 1910.134). This standard governs SCBAs and deals with such topics as medical information, fit testing, and use. Of particular importance are the provisions of paragraph (g)(4) of this standard that set out the "two-in/two-out" rule. This rule gained considerable prominence during a revision of NFPA 1500, *Standard on Fire Department Occupational Safety and Health Program,* some years ago. It provides that when performing interior structural firefighting, at least two members must operate together when in an immediate danger to life or health (IDLH) atmosphere. These members must remain in visual or voice contact with each other at all times (radios are not sufficient to satisfy this, although they obviously can and should be used). Furthermore, at least two additional employees must remain outside of the IDLH atmosphere, though this does not necessarily mean outside of the building, since particularly in large buildings, not all parts thereof will have an IDLH atmosphere.

Other *OSHA* regulations cover such subjects as industrial fire brigades and hazardous materials scene workers. They also deserve your attention.

Hazardous materials

Although federal laws intended to protect the environment from hazardous wastes and materials were first passed over 20 years ago, only recently have federal laws included provisions dealing with firefighter exposure to the risks of hazardous materials incidents.

The *Federal Water Pollution Control Act* (1974) established the National Response Center and development of a "National Oil and Hazardous Substances Pollution Contingency Plan," but the focus of this act was environmental protection, not hazardous materials emergency response. Therefore, after the "Love Canal" scandal of the late 1970s, Congress passed the *Comprehensive Emergency Response, Compensation and Liability Act of 1980* (also known as "*CERCLA*" and "Superfund"). The focus of *CERCLA* was on hazardous materials site cleanup; Congress estimated 400 such sites, but, at last count, some 10,000 sites had been identified.

When more than 2,500 people near a chemical plant in Bhopal, India were killed by a methyl isocyanate release, Americans became more concerned about the potential for a major release in the U.S. When the same substance was released in Institute, West Virginia (with no injuries), Congress responded in 1986 with the *Superfund Amendments and Reauthorization Act (SARA). SARA* specifically addressed training for hazardous materials responders; Title III of *SARA* (the *Emergency Planning and Community Right-to-Know Act*) provided for local and regional planning for chemical emergencies and reporting of chemical releases.

The latest addition to the *SARA* family of hazardous materials statutes is the *Hazardous Materials Transportation Uniform Safety Act of 1990.* This act provides a funding mechanism, financed by registration fees from hazardous materials shippers and carriers for state and local hazardous materials response training.

Federal Case Law

In several areas, the federal government has addressed fire-related issues through case law, or a combination of case law and statutes. Below are some major examples.

Arson investigation

Arson is a crime. As such, arson investigation is treated by the courts the same as investigation of any crime; therefore, the rights of arson suspects are protected in the same manner as the rights of any other criminal suspects.

The first landmark arson investigation case was handed down by the U.S. Supreme Court in *Michigan v. Tyler,* 436 U.S. 499 (1978). In *Tyler,* the fire department responded to a fire in a furniture store shortly before midnight. Around 2:00 a.m., the officer in charge of determining origin and cause arrived and was informed by the suppression crew that two plastic containers of flammable liquid had been found in the building, obviously causing him to suspect arson. After looking around a bit, he suspended his investigation because of smoke and steam and left the store. He, or other investigators, returned at 8:00 a.m. and 9:00 a.m. the same day and also four, seven, and twenty-five days later to continue their investigation.

Tyler, ultimately convicted of conspiracy to burn real property, appealed on the basis that the investigators were required to obtain a criminal search warrant before re-entering the store. The Supreme Court held that the re-entries on the same day as the fire did not require a search warrant because they were really just continuations of the entry at the time of the fire, which had been suspended due to the smoke and steam. However, the subsequent entries over the next several days and weeks could not be sustained on this basis and any evidence gathered in those searches must be suppressed because they violated the Fourth Amendment prohibition against warrant searches and seizures without a warrant.

Because *Tyler* did not set out a specific time by which "re-entry" investigations must be completed, it fostered confusion rather than clarity. This confusion was soon made manifest in *Michigan v. Clifford,* 464 U.S. 287 (1984), which involved a fire in a personal residence, an important distinction from *Tyler.* In *Clifford,* a fire occurred at the Clifford home at about 5:40 a.m. while the Cliffords were away on a camping trip. The fire was extinguished and all officials left the house about 7:00 a.m. At that time, evidence discovered during the suppression activities led the fire department to suspect an arson fire.

The Cliffords were notified of the fire by a neighbor and they, subsequently, instructed the neighbor to secure the house. When the fire department investigator arrived about 1:00 p.m. the same day, he found a work crew pumping water out of the basement and boarding up the house. Probably relying on the *Tyler* case, he entered the house without either permission or a search warrant and proceeded with his investigation, discovering and seizing a crock-pot, electric timer, and two Coleman® fuel cans at the point of origin. All of this led to the Cliffords being charged with arson and their subsequent challenge of the 1:00 p.m. warrantless and consentless search of their residence.

The Supreme Court found that this set of facts differed from the *Tyler* case in two important respects. First, the challenged search could not be considered a continuation of the search conducted at the time of the fire because in the interval the Cliffords had taken steps to secure their home from further intrusion. Second, the privacy interests in a personal residence are significantly greater than those in a furniture store. Consequently, the evidence discovered in the 1:00 p.m. search (with the exception of a Coleman® fuel cans discovered but not seized during suppression) was held to be inadmissible.

The lesson learned from *Tyler* and *Clifford* is this: if in doubt, get a warrant or permission to enter the premises. If the purpose of entering the premises is simply to determine origin and cause, an administrative warrant of the type discussed under the following topic, Inspections, will suffice. If there is already suspicion of a crime, a criminal warrant will be necessary.

Inspections

Until 1967, the U.S. Supreme Court had consistently ruled that a fire inspection did not require a warrant. But the Supreme Court changed course that year in the cases of *Camara v. San Francisco,* 387 U.S. 523 (1967) and *See v. Seattle,* 387 U.S. 541 (1967).

Both cases dealt with the circumstances under which a warrant is required to carry out an inspection of private property, and the type of warrant needed. The Supreme Court noted that when an inspector demanded entry without a warrant, the occupant had no way of knowing whether enforcement of the code involved required inspection of his premises, no way of knowing the lawful limits of the inspector's power to search, and no way of knowing whether the inspector himself was acting under proper authorization. All of these questions are ones to be dealt with by a neutral magistrate when reviewing an application for a warrant.

Applications for criminal search warrants must be based upon probable cause to believe that evidence of a crime will be discovered on the premises to be searched. However, the need to enter premises to conduct a routine periodic inspection is based upon an appraisal of conditions in the area as a whole, not on knowledge of conditions on the particular premises to which entry is sought. Consequently, the criminal warrant procedure does not work in this setting.

The Supreme Court went on to say that area code enforcement inspections are reasonable for several reasons. First, they have a long history of judicial and public acceptance. Second, public interest demands that all dangerous conditions be prevented or abated. Finally, such inspections are neither personal in nature nor aimed at the discovery of evidence of a crime and involve a relatively limited invasion of the citizen's property. For these reasons, it is appropriate for a magistrate to issue an administrative search warrant (as distinguished from a criminal search warrant) simply upon a showing that access to the property in question has been denied and that the property is due for a routine periodic inspection. Needless to say, no warrant of any kind is required if there is consent to the inspection.

Free speech

Courts are continually confronted with a First Amendment issue of whether or not free speech has proceeded beyond Constitutional limits. This issue is particularly pointed in fire departments, where the right of firefighters to complain is limited by a court-recognized need for "esprit-de-corps" and efficient and safe operation. When courts consider free speech cases in this setting, they are compelled to weigh the speech in question to determine whether it addressed a matter of public concern. If so, they must determine whether the First Amendment interest in that speech outweighed any injury that the speech might cause to the government's interest in promoting the efficiency of the public service it performs through its employees.

Courts are likely to review some or all of the following questions in arriving at their decision:

- Does the speech create potential problems in maintaining discipline or harmony among co-workers?
- Is the employment relationship one in which personal loyalty and confidence are necessary?
- Did the speech impede the employee's ability to perform his or her responsibilities?
- What were the time, place, and manner of the speech?
- In what context did the underlying dispute arise?
- Was the matter one in which debate was vital to informed decision-making?
- Is the speaker regarded as a member of the general public?

Drug testing

The Constitution's Fourth Amendment prohibits unreasonable searches and seizures, and courts have agreed that drug testing by a public employer is a search within the meaning of the Fourth Amendment. But the Fourth Amendment is not absolute. In the case of public employee drug testing, courts continually work to balance an individual's interest in protection of his or her rights with the government's interests, particularly workplace and public safety interests.

The U.S. Supreme Court established the framework for balancing individual and governmental interests in two cases—*Skinner v. Railway Labor Executives Association,* 489 U.S. 602 (1989) and *National Treasury Employees Union v. Von Raab,* 489 U.S. 656 (1989). *Skinner* challenged Federal Railroad Administration regulations requiring drug testing after rail accidents, and in cases of violations of certain safety rules. In *Skinner,* the Court balanced the compelling government interest in rail safety against the "limited" intrusion of a blood or urine test, and held in favor of the Federal Railroad Administration.

Von Raab was handed down the same day as *Skinner,* and relied on the same "balancing act;" the personal rights of U.S. Treasury agents were held to be less compelling than those of the Treasury in maintaining confidence in agents who carried guns and had access to classified materials.

Since *Skinner* and *Von Raab* were decided, the greatest challenges have arisen in the area of suspicionless drug testing. Suspicionless drug testing is not necessarily unreasonable under the Fourth Amendment, but will only be considered reasonable when the government can demonstrate compelling special needs beyond enforcement of the law that outweigh the privacy expectations of the individuals subject to testing. In an effort to protect personal privacy, employers may provide assurance that test results will not be turned over to law enforcement and will be used only for purposes relating to the subject's employment. With persons engaged in safety-sensitive activities, it is usually possible to establish these compelling special needs.

To establish a suspicionless drug-testing program likely to withstand Constitutional challenges, employers must ensure that employees understand the program and measures must be taken to ensure the accuracy of the testing. It is important that the program clearly identify the following:

- Employee groups (categories) subject to testing
- Events that trigger testing (*e.g.*, pre-employment, transfer, or promotion; post-accident; return to duty; random, etc.)
- Related behaviors that are prohibited
- Consequences for violating drug and alcohol regulations (Violations occur when employees test positive, or when they refuse to submit to testing.)

In deciding whether to institute a suspicionless drug testing policy, an employer should consider whether such a program is justified in light of the costs of such a program, the likely effect on employee morale and the nature of the perceived problem in the community. It is also important to remember that for those municipalities with unionized employees, the National Labor Relations Board has ruled that drug and alcohol testing for current employees is a mandatory subject of bargaining.

STATE LEGAL SYSTEMS

Although each state has its own legal system, all 50 state systems share many common elements. Generally, state civil law is based on a combination of case law and statutory law, and state criminal law is based largely on statutory law. Here are typical areas of state legal control that affect fire chiefs on a daily basis.

Workers' compensation

All 50 states have adopted workers' compensation laws. The principle behind workers' compensation is simple; it is essentially a "no-fault" insurance system in which the injured employee trades the right to file a lawsuit against the employer for guaranteed benefits. The state-by-state reality however is complex. A growing number of states allow employees to collect workers' compensation benefits and also to file a lawsuit against the employer. In these cases, the employee must prove that the employer was grossly negligent, or knew of a dangerous condition in the workplace.

A few states have their own unique variations on the workers' compensation pattern. In South Carolina, for example, an injured employee can choose between filing a lawsuit against the employer or collecting workers' compensation benefits. In some jurisdictions in New York, the local government employer can be sued if that employer failed to provide safety training in an area of known risk.

Since firefighting is an inherently dangerous occupation, firefighters often use the workers' compensation system. When disputes arise, they usually involve questions regarding a condition, such as a chronic back problem that may have resulted from one or more workplace injuries, or the extent of a disability.

Personnel relations

Many states have established government employee rights by statute. These include rights to a hearing before disciplinary actions are taken, and bargaining rights (and rights to arbitration) in states that authorize recognition of employee bargaining units, as well as antidiscrimination laws.

For those fire departments that have unionized personnel, the collective bargaining agreement will usually provide for a disciplinary procedure, often that includes a graduated scale for increased severity of discipline for frequent or more serious infractions. The fire department is legally obligated to follow this procedure; therefore, it is important to be familiar with its terms.

Proper documentation is essential in sustaining the imposition of discipline. Simply saying that an employee's performance has been unsatisfactory in the past probably won't be enough to support disciplinary action, particularly if the employee has consistently been given pay raises and favorable performance reviews over the years.

It is also important to be consistent in imposing discipline. A sure way to have a disciplinary action overturned, and perhaps to end up in a discrimination suit in the bargain, is to impose different penalties on persons who have committed similar infractions and have similar records.

Whenever possible, avoid imposing discipline at times when your motives may appear suspicious. A transfer to a different fire station, a change in work hours, or assignment of new or different duties may all appear retaliatory if they occur at or near the time when discipline is being considered.

The collective bargaining agreement will also set out what matters are subject to bargaining. Be sure to examine the agreement before establishing policies or standards that may require bargaining.

Codes and code enforcement

State law approaches fire inspections in one of two ways. The state can adopt a code and require statewide enforcement, or the state can authorize local governments to adopt their own codes and enforce them. There is a wide range of variation in-between. Some states establish a minimum fire code, but allow local governments to adopt more stringent amendments. Others give counties the choice of adopting a county-wide fire code, but limit a county to adopting only one specific model code if that county chooses to adopt a code.

Fire inspections tend to be maintenance inspections of buildings that already exist. States also regulate construction of new buildings through codes. In some states, the fire code can be a construction code as well as a maintenance code by including provisions for installation of sprinkler systems in new and existing construction. In others, a separate building code is adopted. As in the case of fire codes, some states allow local amendments to building codes, while others require use of a single code statewide. These statewide codes are known as "mini-maxi codes," because they establish minimum and maximum standards in the same code document.

The legal authority for code enforcement at the state and local level is the Constitutionally recognized concept of "police power," that is, the power of a government to regulate in the interests of the health and safety of its citizens. Courts give a government's policing power great respect when a government's right to regulate is called into question. However, because police power is so broad, courts have also imposed limits on police power. One limitation on fire inspection power is the requirement of either permission from an occupant to enter an occupancy to inspect it or an administrative search warrant.

The legal theory behind this requirement (established in the *Camara v. San Francisco* and *See v. Seattle* cases discussed previously under "Inspections") is that an inspection is an administrative "search" within the meaning of the Fourth Amendment. In the wake of the *Camara* and *See* cases, the states adopted statutory systems for procurement of administrative search warrants.

Firefighter training and certification

Although few states have developed statutory minimum training requirements for firefighters, most states have provided voluntary firefighter training standards, either by statute or administrative rule. Knowledge of these standards is important for fire chiefs because such standards would probably be admissible as evidence of an *expected* standard for local firefighters in a negligence lawsuit (see the "Negligence" section of this chapter).

Criminal laws

Every state has adopted misdemeanor laws that regulate the behavior of the public at fire or emergency scenes (accident scene traffic regulations, fire station and hydrant parking regulations, etc.) Most states have adopted laws that provide greater criminal penalties for someone who assaults a police officer. Many states now have added firefighters and rescue workers to these peace officer assault statutes. There is a wide range of state criminal law that deals with the crime of arson and related crimes.

Limitations on liability

In response to case law that limited or eliminated "sovereign immunity" (see the "State Case Law" section below), many states adopted statutes that defined and limited local government (and firefighter) immunity from negligence suits. While each such statute is different, there are common elements, including:

- Scope of employment—Liability-limitation law coverage is generally restricted to employees "acting within the scope of their employment," in other words, doing what they're authorized to do. A firefighter who performs field surgery at an auto accident scene is probably acting beyond the scope of his employment.
- Level of negligence—Liability-limitation laws usually do not cover reckless behavior or negligence that borders on intentional behavior. Some statutes exclude "willful or wanton negligence," "gross negligence," or "malicious and corrupt behavior" from protection.

- Type of act or action—Some liability-limitation laws restrict their protection to actions that require judgment (*i.e.*, "discretionary" acts), as opposed to acts that do not (*i.e.*, "ministerial" acts). In one recent state case, *Invest-Cast v. City of Blaine,* 471 NW 2d 368 (1991), the Minnesota Supreme Court demonstrated its reluctance to let a discretionary-act defense bar a lawsuit. The court decided that the question of whether or not a fire-attack decision was "discretionary" within the meaning of the law was one for a jury to decide. (Note that when the case was retried, the jury ruled in favor of the fire department and firefighters.)

Occupational safety and health

All states have laws that deal with occupational safety and health. Some states deal with safety and health in connection with workers' compensation. Others rely totally on federal *OSHA* standards or inspectors.

Since the federal *OSHA* applies in all 50 states, its standards represent the minimum standards for most employees. Because of a historical (and Constitutional) reluctance on the part of the federal government to impose regulations on state and local employees, it is presently up to individual states to determine whether *OSHA* regulations apply to local government employees like firefighters, or to volunteer as well as career firefighters.

States that have chosen to include firefighters in *OSHA* coverage either have their own enforcement system and state safety rules that are at least as strict as federal *OSHA* rules (so-called "state plan states"), or have no enforcement mechanism, and leave *OSHA* enforcement up to the U.S. Department of Labor's *OSHA* inspection and enforcement staff.

"Good Samaritan" laws

As malpractice lawsuits became more frequent (and more costly) in the '60s and '70s, health care professionals began calling for limitations on their liability when they acted as "Good Samaritans" at accident scenes. Many states responded by adopting "Good Samaritan" statutes that limited health care professional liability at accident scenes. Some states included firefighters and rescue workers in "Good Samaritan" coverage, but in other states, those "Good Samaritan" laws do not cover firefighters, but rather only protect volunteer firefighters.

Several states also have adopted "Good Samaritan" laws for those who help at hazardous materials incidents. In general, these laws provide that if a hazardous materials incident "Good Samaritan" was not responsible for the release, and does not act in a grossly negligent manner, such "Good Samaritans" have liability protection (the level of protection varies from state to state).

State Case Law

Case law at the state level is developed by courts of record-appellate courts. In some states, only the state supreme court can establish precedent through case law. In others, an intermediate court, often called a court of appeals, can also establish or clarify state law in its decisions.

Case law can serve to clarify or define statutory law, or it can address legal issues not addressed in statutes. In some states, limitations on local government liability are set forth in statutes; in others, the only limitations on local liability are found in state supreme court cases.

Outlined below are some significant areas of law that are most often established (or changed) by case law.

The "Firemen's Rule"

When someone is injured by another person's action (or by a failure to act), that injured person can to file a lawsuit against the other for negligence. But in many states, if a property owner violates the fire code and a firefighter is injured fighting a fire in that occupancy, the injured firefighter cannot file a lawsuit against the negligent owner. That is because of a longstanding statement of case law called the "Firemen's Rule." The theory behind the Firemen's Rule is, simply, that when someone becomes a firefighter, that person assumes the risks of the job. Those risks include exposure to the results of someone else's carelessness. In some states, the Firemen's Rule is not applied when the conduct that brought about the firefighter's injury was not "reasonably foreseeable," that is, no firefighter could have anticipated it.

In a number of states, the Firemen's Rule no longer exists, having been abolished by either court decision or legislative action. There seems to be a nationwide movement toward disposing of the Firemen's Rule and allowing the claims of injured firefighters to be decided under traditional negligence law.

Personal and governmental liability

While many states deal with a fire chief's personal liability and a local government's general liability with statutes, other states set standards for liability protection in case law. In North Carolina, for example, there are statutes that authorize a local government to reimburse an employee who loses a lawsuit, but the rules regarding when an employee can be sued are found in case law, not statutory law.

The standards that courts use to limit local government and employee liability are the same, or similar to, those used in statutes. As long as employees are acting "within the scope of employment" and not acting in a "grossly negligent," "willfully and wantonly negligent," or "malicious and corrupt" manner, the employee avoids personal liability. If the act or decision is ministerial (as opposed to discretionary) or governmental (as opposed to proprietary), case law can protect local governments from liability as effectively as can statutory law.

"Duty" and inspections

To prove negligence, an injured person must demonstrate that the person who caused the injury owed a "duty of care" to that injured person. In most cases, that duty of care is assumed. But some state courts have created a special rule to deal with fire inspector liability.

With fire codes running hundreds of pages, few courts would expect a fire inspector to do a perfect job. If fire inspectors were held strictly liable for perfect inspections, many courts argue, no community would ever have an inspection program. In these jurisdictions, the duty to enforce the fire code is a duty to the community at large. In legal terms, this means that the fire inspector owes no duty to any individuals in the building. This "duty doctrine" is not absolute; a fire inspector can, for example, create a duty to an occupant by assuring that occupant that the building is safe. However, in states where it is followed, the duty doctrine can be a powerful defense.

Local Ordinances and Rules

Local government organizations and special districts each have their own rules and regulations, usually in the form of ordinances, a charter, or both. In some states, any local government action must be based on authority from the state government. In others, the local government may exercise any power that is not limited by the state. Sometimes, local governments operate under a combination of both state law and a local charter.

For fire chiefs, the most important ordinance or charter provision is that which establishes the fire department and describes its mission. This is because the success of a governmental immunity defense in a lawsuit often depends on whether firefighters or fire officers were "acting within the scope of their employment," that is, doing what they were authorized to do by law. Such ordinances also help define "scope of duty" for workers' compensation purposes.

Other local ordinances regulate subjects as diverse as salary plans, personnel procedures, fire station parking, use of fire hydrants for emergency and nonemergency purposes, and presentation of annual reports by the fire chief.

Key Legal Questions for the Fire Chief

The time for a fire chief to make the acquaintance of the department's, town's, or city's attorney is not just after a lawsuit has been filed. Every fire chief should meet with the attorney who will represent him, and ask the following questions:

- What is the status of sovereign or governmental immunity in our state? In other words, how much liability protection do my department and I have?
- What are the major governmental immunity cases in our state? Do any of them deal with fire departments? Ask for copies. Most legal decisions are not that difficult to read and understand.
- Could you get me copies of any governmental immunity statutes that apply in our state? Ask for "annotated statutes," which include brief summaries of cases that interpret those statutes.
- What is the status of the "duty doctrine" for fire inspections in our state? The "duty doctrine" is a complex legal argument that says, in effect, that fire code enforcement is for the good of the public at large, rather than individuals; unless an individual can establish a duty, no negligence case will stand.
- What other case law or statutory defenses do we have in our state against a negligence claim? Assumption of risk? Discretionary versus ministerial? Governmental versus proprietary activities?
- What do our courts or laws say about indemnifying an employee who loses a lawsuit arising out of that employee's work? What do they say about the town, city, county, or district furnishing legal support or legal fees?
- Is there a Good Samaritan law in our state for rescuers, or for helpers at hazardous materials incidents? Does it cover firefighters? (Many Good Samaritan laws cover passersby, but not firefighters.)
- Is the Firemen's Rule in effect in our state? What are the exceptions? (The Firemen's Rule limits the ability of firefighters to recover damages for line-of-duty injuries from a negligent property owner.)
- What right do public employees in our state have to sue for work-related injuries beyond workers' compensation benefits? When would I be liable as a supervisor or manager? (In some states, workers' compensation benefits are the only benefits an injured employee can get from an employer. In other states, under certain circumstances, an injured employee can sue the employer.)

NEGLIGENCE

Over the past several hundred years, British and American courts developed the concept of *tort,* injury to a person or that person's property. One type is an *intentional tort,* which includes defamation (libel and slander), malicious prosecution, and civil battery. Another, and by far the most common tort, is *negligence*.

The basic assumption behind negligence law is that each person owes a duty to others to act reasonably. An injury to a person or property is assumed to be the result of a failure to act reasonably, and unless the injured person was the negligent party, that injured person is entitled to "be made whole," that is, to be fairly compensated by the party responsible for the injury.

To prove negligence, an injured person must be able to demonstrate the following conditions:

- The responsible person owed a duty of care to the injured person.
- That duty was breached through an action or failure to act.
- The act (or failure) was the proximate cause of the injury (the injury could reasonably have been foreseen).
- The injured party must be able to prove actual injury (negligence law does not apply to a hypothetical situation or to someone who fears a future injury).

Governmental negligence liability and immunity

For centuries of tort law evolution, governments were held immune from tort liability. This immunity was called "sovereign" immunity, from the early days when kings or queens made the laws. According to theory, since the sovereign created courts and judges, the sovereign could not be sued in their own courts unless they agreed to allow such lawsuits. Although history records some unusual kings and queens, no records show that any ever agreed to be sued. When kings and queens disappeared from the scene, sovereign immunity became governmental immunity.

Through the first half of the 20th century, absolute governmental immunity remained the rule of law. As long as government employees acted within the scope of their employment, that is, did what they were authorized to do, those employees were protected by the same doctrine. But in the early 1960s, state supreme courts began to take a closer look at the governmental immunity doctrine.

One reason for this development was a sense that justice was not serving people injured by governmental action or failure to act. An injured plaintiff who could prove a very good negligence case could never win if the government was negligent.

A second reason was that the federal government already had defined limited situations where it would accept liability in negligence cases. Therefore, in state after state, courts either abandoned the governmental immunity doctrine altogether or put severe limits on its application.

This trend brought immediate reaction from state and local governments. The principal arena for debate over the decline of governmental immunity was in state legislatures. Since governmental immunity was a case law rule, legislatures could act independently of the courts to write statutes reinstating governmental immunity. Most state legislatures opted for limited governmental immunity, which meant that state and local governments could be sued, but only under certain circumstances. Those circumstances were based primarily on two concepts.

The first concept involves *ministerial acts versus discretionary acts.* If the act that led to injury required *discretionary judgment,* a government was held immune from liability. Legislatures and courts reasoned that government employees needed to be unafraid to make judgments in difficult situations, and so provided immunity in those cases. But for acts that did not require special judgment, or *ministerial acts,* immunity was withheld. One state court held that failure on the part of a housing inspector to correctly count electrical outlets to see if a house met a two-outlet-per-room requirement meant liability for that inspector's employer. Counting electrical outlets, the court reasoned, was ministerial, since counting required no discretionary judgment.

Most states hold fire apparatus drivers to the same standard of care and liability, as drivers of any other vehicles. This holding is a variation on the ministerial-versus-discretionary theory. Since anyone with a driver's license can legally operate a vehicle, no single class of vehicle operators receives any more legal protection than any other.

The second is *governmental versus proprietary,* a liability limitation concept based on government functions. According to this theory, governments should have liability protection for functions that are unique to governments. Otherwise, if governments are held liable and decide to abandon those "governmental" functions, there is no one else who will act in a government's place. Examples of governmental functions in different jurisdictions include fire protection, law enforcement, and public assistance.

However, governments also compete with the private sector, and, in those areas, should receive no immunity. The idea is that if government abandons one of these "proprietary" functions in the face of negligence lawsuit losses, the private sector could step in. Examples of proprietary functions in different jurisdictions include providing bus service, airports, hospitals, and water service.

The extent of governmental immunity varies from state to state; in general, within those immunity areas courts will dismiss negligence lawsuits. But if a court rules that a local government is not immune, that lack of immunity does not mean that the plaintiff has won, only that the plaintiff will have his day in court.

Local governments have all the defenses that any individual would have in a negligence suit, and at least one not generally available to defendants other than local governments. These defenses are outlined below.

Contributory (or comparative) negligence. Courts have long held that an injured plaintiff should not recover damages for his own negligence. The modern-day legal translation in most states is that a damage award will be reduced to the extent that the plaintiff is responsible for injury. In some of those states, that allocation is based on a jury determination of percentage of responsibility. If, for example, a jury holds a plaintiff to be 35% responsible for his own injuries, any award of damages against the defendant will be reduced by 35%. But in a handful of states, any liability on the part of a plaintiff will bar recovery of damages. In North Carolina, for example, a plaintiff who is only 5% responsible for his injuries will lose the case.

Assumption of risk. If an office building tenant is put on notice that his occupancy is in violation of the fire code, and that tenant is subsequently injured by a fire caused (or supported) by that violation, it can be argued that the tenant "assumed the risk" of that injury by staying in the building or not eliminating the violation. More common is the bystander at an emergency scene who moves inside the barricade tape and is injured by an action connected with the emergency. The fact that the bystander has made his way to the wrong side of the barricade tape is evidence of assumption of risk.

Duty. The question of whether a defendant owed a duty of care to the plaintiff is at the heart of negligence law. In some states, courts of record have established a "duty doctrine" that is applied in code enforcement situations.

This "duty doctrine" presumes that fire and other safety codes were enacted for the good of the community, not for the benefit of any particular person. Therefore, a fire inspector who fails to notice a fire code violation during an inspection owes no duty on account of that failure *to any particular person* who is injured as a result. As one court stated that doctrine, "a duty to all is a duty to no one," therefore, no individual can establish the duty relationship that a negligence action requires.

However, an inspector can create a duty to an individual. If an inspector tells an occupant that the building is safe, for example, and that occupant is injured later as a result of the inspector's failure to detect a violation, that injured occupant can argue that he relied on the inspector's representation of safety. That inspector's representation would establish a "duty relationship" with the occupant; the duty doctrine defense would subsequently be lost.

A fire code requirement that protects particular classes of persons, (for example, occupants of daycare centers) creates a duty relationship with those occupants. Courts have ruled that the duty-doctrine defense does not apply in these cases.

Personal negligence liability and immunity

The same state laws (or court cases) that define or limit governmental liability and immunity generally deal with the liability of individual government employees. Those laws and cases approach personal liability in a variety of ways.

In some states, an employee can be found negligent, and so will share the cost of negligence with the governmental employer (unless that employer is otherwise immune). Those same states generally authorize employer provision of legal counsel to an employee who has been sued, and employer reimbursement of an employee who loses a suit in some circumstances.

In other states, employee liability is limited as long as that employee's act was within the scope of employment, that is, the employee was doing what he was authorized to do. Using this approach, a firefighter who is negligent in rescuing someone from a hazardous materials emergency may be protected, but a firefighter who grabs a gun to shoot a fleeing felon is doing something firefighters typically are not authorized to do and may be personally liable.

A third approach centers on just how negligent the act was. This approach draws a distinction between "ordinary" negligence and "willful or wanton," "gross," or "malicious" negligence. Ordinary negligence can merit legal protection; gross negligence can leave the employee unprotected. What is the difference between these levels of negligence? Foreseeability is one way that courts draw the line. If an injury resulting from an act was virtually certain, and a "reasonable person under the same or similar circumstances" would have thus avoided the act, that act can be characterized as "willful." Another distinction is based on the concept of "recklessness"; if the circumstances surrounding the act indicate that only someone acting in a reckless manner could have brought about the injury, the negligence-meter needle moves into the red zone.

Aside from statutory or case law protection, local government employees can invoke customary negligence defenses (comparative/contributory negligence, assumption of risk, etc.).

It is essential to note that legal rules governing personal negligence of public employees are different in every state. Fire chiefs should talk with their department's, city's, town's, county's, or district's attorneys for information on where fire chiefs in their community stand in regard to personal negligence exposure.

Legal Trouble Spots

When fire departments (or fire chiefs) are sued, the lawsuit usually occurs as a result of one or more of the actions discussed below.

Doing something unauthorized. If a fire department is not authorized by ordinance, charter, or other official statement to deal with a problem, for example a hazardous materials incident, the issue undoubtedly will be raised in the wake of a legally "messy" incident. Lawyers do not read fire service books or periodicals to keep up with trends in the fire service. They always start their research with dustier sources (*i.e.,* city ordinances). That's a good reason for fire chiefs to review the ordinances or charters that empower the fire department to do what it does, and to ask for changes to cover things that they are doing but are not empowered to do.

Ignoring legally mandated procedures. There are some things that laws require a fire chief to do, whether he wants to do them or not. Signing documents that require the fire chief's signature is one example. Fire chiefs should not delegate legally mandated procedures if the law does not allow it.

Departing from rules, regulations, and standard operating procedures. If someone is injured as a result of something a firefighter does (or does not do), the first place a plaintiff's lawyer will look for a "standard of care" is in the department's own rules, regulations, or standard operating procedures (SOP). If a fire chief issues an order, establishes a rule, or authorizes an SOP, the fire chief needs to know that, like it or not, a standard is being set, and department members will be held to that standard.

Violating civil rights. By now, every fire chief should know the dangers of discrimination, but there are enough civil rights complaints and lawsuits on file to show that not every chief has learned this lesson. The astute fire chief knows the difference between "legal" discrimination (in some jurisdictions, refusal to hire a smoker is perfectly legal) and illegal discrimination, which involves constitutionally protected classes like race and gender.

Denying due process. Even in "right-to-work" states, local government employees like firefighters are still entitled to a hearing before dismissal, even for cause. The American sense of fair play has been translated into a legal expectation that before a public employee loses something—a job, pay (through disciplinary action), or promotional opportunity—that employee is entitled to a hearing.

Failure to document. As long as several years after a fire, most states allow someone who alleges injury as a result to file a lawsuit. How good is the typical fire chief's memory of an incident after several years? If that incident involved death, serious injury, or major property loss, will a copy of the incident report be enough when that chief is on the witness stand? Lawyers are great believers in "diaper documentation," or covering one's backside with paper.

The Lawsuit Process

Each state (and the federal government) has its own specific rules of civil procedure, or the rules that govern how a lawsuit is handled. There are variations from state to state, but the fundamental structure of lawsuit procedure is consistent.

A lawsuit begins with a *complaint.* The complaint, filed by the *plaintiff* (the injured person), is essentially a story of the incident that caused the alleged injury, allegations of the defendant's negligent behavior, and a request for *damages* (*compensatory* damages, which cover the actual loss suffered by the plaintiff, and, in some cases, *punitive* damages, which are intended to punish the defendant's negligent behavior to deter others).

Once the complaint is filed in court and served on the *defendant* (the person accused of negligence), the defendant responds with an *answer.* This answer reviews the story of the case presented in the complaint and includes denials when the defendant's story differs from the plaintiff's version.

While plaintiffs and defendants can serve as their own attorneys, most hire an attorney to represent their interests and lead them through the legal system. It is very important to note that, in the American legal system, the primary responsibility of an attorney is to represent the interests of the client. So, the plaintiff's attorney will work very hard to question (and in the process cast doubt upon) any statement, evidence, or testimony that the defendant offers. The defense attorney will do the same on behalf of the defendant. The American legal system relies on the judge and jury to sort the evidence, apply case law or statutory law to that evidence, and arrive at the correct decision.

Once the *complaint* and *answer* preliminaries are over, the process of collecting evidence begins. This process is called *discovery.*

Contrary to what happens in television courtrooms, the American legal system does not encourage "surprise" witnesses or evidence. Because our legal system actually encourages settlements in civil cases, it also encourages the revelation of all evidence before trial or during the discovery process, in the hope that the plaintiff and defendant will be able to settle their dispute without a trial.

The discovery process includes *depositions* (statements taken outside court, under oath), *written interrogatories* (questionnaires completed by witnesses or others with an interest in the case), and *requests for production of documents* (lists of reports, memos, and other documents that relate to the lawsuit). When the discovery process is complete and there is no sign of a settlement, the suit proceeds to trial. A civil lawsuit may be heard by a judge and jury, or (most often at the plaintiff's option) a judge alone, who also acts as the jury.

The respective roles of the judge and jury are at the heart of the American legal system. The judge is the *trier of law,* that is, he applies the rules of civil procedure in court, and tells the jury what the law is as the jury hears testimony and examines evidence. The jury is the *trier of fact*; the jury decides what the facts are in a case, based on testimony and evidence, and then applies the law (as defined by the judge) to those facts in coming to a verdict. With few exceptions, a jury's decision on the facts is the final decision. Courts of appeal (appellate courts) regularly review judicial decisions and statements on what the law is, but are reluctant to interfere with a jury's findings of fact.

During the trial, the plaintiff's and defendant's attorneys continue in their roles as advocates for their clients. They present their cases through testimony by *witnesses* who have some direct knowledge of the events that brought about the lawsuit (including the defendant and plaintiff), and by *expert witnesses,* who are authorized by the judge to testify about matters not generally known or understood by the public. The attorneys may introduce documentary or physical evidence to demonstrate facts about the case. They also may compare testimony at the trial with what they learned during the discovery process, pointing out inconsistencies to the jury.

In their roles as advocates, attorneys will seek to call into doubt, or discredit, testimony offered by witnesses for "the other side." This process is called *impeachment of witnesses.* But attorneys will not call witnesses to tell a jury what the law says about a case. Again, declarations of law come only from the judge.

Trials are governed by complex rules of procedure. Many of those rules deal with what can be offered to a jury as *evidence.* A fundamental aspect of the American legal system is that jurors should always see and hear the best, most reliable, evidence. Direct testimony by a witness about statements which that witness made will be admitted into evidence; however, as a general rule, what a third person reports someone to have said *(hearsay)* will not be admitted as evidence. Other rules are designed to keep a trial on track. If testimony seems to have nothing to do with the issues at trial, the judge may exclude that evidence because it is irrelevant.

The goal of each attorney is to demonstrate by a *preponderance* (or a greater weight) of the evidence that his client is right. Preponderance of the evidence can be a tricky concept. One way to think of preponderance is a tilt of the scales of justice in favor of the client. Another is to visualize a football game where the client scores as soon as he crosses the 50-yard line. Preponderance of the evidence is *not* the same as the criminal standard of "beyond a reasonable doubt" (which, using the football analogy, requires a trip all the way to the end zone).

At the close of the trial, the judge instructs the jury as to the law they must apply to the facts they have heard. A jury's deliberations on the law and the facts lead to a *verdict.* If the verdict is in favor of the defendant, the plaintiff can appeal; likewise, the defendant can appeal a verdict in favor of the plaintiff.

Standards and the Law

In negligence lawsuit situations, a critical question for judges and juries is whether a "standard of care" was met. In the eyes of the law, each of us owes the other a duty to behave as a "reasonable person," to exercise "reasonable care." How is "reasonable care" defined when it comes to fire fighting, fire safety, hazardous materials incident management, or training? Generally, courts look to what a "reasonable fire officer" or a "reasonable fire chief" would have done under the same or similar circumstances.

If a fire department has its own rules, regulations, or standard operating procedures, these are what a court would examine to find evidence of a standard of behavior or care. If the legal issue goes beyond a local fire department's practices, courts will look to other standards.

National Fire Protection Association (NFPA) 1710 Standard on Organization and Deployment of Fire Suppression, Emergency Medical Operations, and Special Operations to the Public by Career Fire Departments is a good example of such a standard. It is not unlawful to operate outside the requirements of 1710 (assuming that the appropriate authority having jurisdiction has not adopted it), but this doesn't mean that it couldn't be evidence of "reasonable care" at trial. A person arguing that he or she suffered a loss because of inadequate or inappropriate deployment would be very likely to present the standard to the jury as evidence that the fire department failed to meet a nationally recognized standard of care. The fire department, on the other hand, would probably present evidence that it followed its own procedures or standards. The jury would then have to decide which standard really defines "reasonable care."

There are many places where evidence of "reasonable care" may be found. For instance, in addition to NFPA, there are National Fire Academy courses and International Fire Service Training Association (IFSTA) manuals. Since all of these organizations are national in scope and fire departments throughout the nation use their materials, their documents frequently serve as benchmarks for rules, regulations, and standard operating procedures.

When such standards exist, the best strategy a fire chief can undertake is to read and understand the standard and develop a plan on how each part of a standard is to be addressed (*not* necessarily adopted) in that chief's department. A decision on how a standard is to be addressed implies that the standard was studied carefully and the fire chief's response takes into account not only the standard, but local conditions. *The worst strategy is to ignore such standards.*

Tips for Witnesses

Chief officers can expect to be called on to testify in court at least once in their careers. Because they have knowledge about fires, fire fighting, and fire prevention that the general public does not have, fire chiefs generally will be qualified to testify as experts (whether they feel like experts or not, and whether they are chiefs of career or volunteer departments).

Here are a few important tips for fire chiefs to remember about being a witness:

- **Keep your résumé up-to-date.** The "friendly" lawyer (assuming you are not testifying as a hostile witness) will need it to ensure familiarity with your qualifications, and the jury will want to know about your background.
- **Wear your uniform.** The American fire service is an honorable institution and its members are entitled to the respect that a fire service uniform commands.
- **Do not be late.** Although the justice system is often slow, individual judges have little patience for anyone who believes that anything is more important than their courtrooms and cases.
- **Listen to the entire question.** Do not try to answer a question before the lawyer finishes asking it. Also, think before you begin to answer.
- **Answer only what you are asked.** Once the first round of questioning (direct examination) is out of the way, witnesses are fair game for leading questions. Do not try to speed up the process. One wise old lawyer once said, "There are only four good answers to any question a witness is asked. They are 'Yes,' 'No,' 'I don't know,' and 'Would you repeat the question?'"
- **Do not joust with the lawyers.** A good lawyer for the opposing side will do as much as the rules of procedure allow to poke and prod witnesses into losing their tempers or arguing. After all, angry witnesses have been known to self-destruct. Be patient with lawyers and do not get into arguments or debates. Do not forget that you are as sophisticated on their turf—the courtroom—as you would expect one of them to be on yours.
- **Never lie.** As bad as the truth may sound, it is never as bad as an untruth. Lawyers are trained in law school and by experience to pounce on inconsistencies that indicate the truth is not being told. A witness caught lying is destroyed as a witness and may be guilty of the crime of perjury.

Tips for Avoiding Legal Problems

Criminal activity

From time to time, it will be necessary to decide whether to deny an offer of employment to someone with a criminal conviction, or to terminate the employment of a current employee who has engaged in criminal activity. Obviously, there are competing interests here. The individual is clearly concerned with obtaining or continuing employment in a field where the jobs are few and the openings are coveted. On the other hand, the employer has to be concerned with the effect on the department's public image of employing someone with a criminal record. Furthermore, any future misdeeds, depending on their circumstances, may become the employer's responsibility.

The determination as to whether employment can be denied or terminated because of a criminal conviction usually turns on the issue of relevance. If the offense is substantially related to the job duties the person will be performing, then the employer is on solid ground in denying or terminating employment. Thus, refusal to hire or dismissal based on commission of property crimes may be the easiest to justify because of the unique nature of the fire service. Firefighters enter people's homes at unexpected times and on short notice, thus giving homeowners little or no time to secure their valuables. Furthermore, they routinely work amongst these valuables while the owner has no opportunity to observe them, leaving the owner no alternative but to rely on their honesty. Be certain to document the circumstances of the offense and its relevance to the employee's potential or current job assignment. Also, be alert to any state laws on the subject, as many states have criminal conviction discrimination statutes.

Motor vehicle laws

Many people have the perception that motor vehicle laws do not apply to emergency vehicles when they are operating with emergency lights and siren. Nothing, of course, could be further from the truth.

All states have laws relating to the operation of emergency vehicles. While these laws vary from state to state, they carry a common theme. It is that emergency vehicle operators cannot be cited for violating certain provisions of the motor vehicle code while operating with emergency lights and siren (note that in many states, *both* lights and siren are required to take advantage of this law). This exemption from enforcement of the motor vehicle code usually applies to those provisions relating to speed, stopping at stop lights and signs, one-way travel, turning in specified directions, and parking in no-parking areas.

It is critical to understand that this is not at all the same as immunizing the emergency vehicle driver from the consequences of an accident that occurs while taking advantage of these exemptions. Many statutes emphasize that a driver can violate these portions of the motor vehicle code only when it can be done safely. In other words, an emergency vehicle operator using emergency lights and siren who proceeds through a red light without stopping can not be issued a ticket for failing to stop at the light. However, if the driver is involved in an accident while running the light, they may still be sued for the damage caused. In this sense, the law setting out the obligation to operate safely may actually impose a greater burden on the driver than if there were no exemption at all. Furthermore, if the emergency vehicle operator's conduct is truly outrageous, it might result in criminal charges, such as recklessly endangering another's safety, for which there is no exemption.

Privately owned vehicles equipped, or partially equipped, with emergency warning devices present a special problem. First, if such vehicles are to be used, they should be specifically authorized by the appropriate fire department authority and care should be taken to determine that they are properly equipped so as to meet the statutory definition of "authorized emergency vehicle." Next, the department should determine clear guidelines as to the circumstances under which such private vehicles may be used.

Since the motor vehicle code does not regulate many of the driver's activities, it is critical for the fire department to do so. Set clear policies regarding operating speed, observance of traffic control signs and signals, and operation of warning devices. Be sure that drivers understand and observe the rules.

Scope of employment

Scope of employment consists of the boundaries or limits of one's duty to their employer. This might seem as simple as "just doing my job," but it is not. The challenge often becomes defining exactly what "my job" is. The case of *Carrell, et al v. City of Portage, Indiana* 609 F.Supp. 314 (D.C.Ind. 1985) is a good case in point.

In *Carrell,* two Portage, Indiana firefighters were returning to quarters after responding to a fire when they saw a man in a drunken condition stumbling along the side of a busy highway. One of them radioed the police for assistance, but before the police could arrive, the man walked onto the highway in the path of an oncoming automobile. The firefighters shined their vehicle spotlight in the direction of the pedestrian and the oncoming motorist in an effort to warn the driver. Unfortunately, the motorist struck and killed the pedestrian anyway, and his widow subsequently sued both the motorist and the firefighters. One of her allegations was that when the firefighters shined their spotlight, they may have distracted or blinded the oncoming driver, thus contributing to or causing the accident.

The firefighters argued that they should be dismissed from the lawsuit because Indiana law granted them immunity from liability when performing a discretionary function within the scope of their employment. The court then had to decide what the scope of their employment really was.

The court noted that neither Indiana law nor their individual contracts with the city of Portage imposed any general duty to maintain public safety on the firefighters. Since they had no duty to police the highways, they were not acting in the scope of their employment when they attempted this rescue, and therefore could not claim immunity. Consequently, they were required to remain in the lawsuit.

This case highlights the importance of defining one's scope of employment. Remember that ultimately this is the employer's decision. Often this may mean seeking written direction from the governing body (*i.e.*, the city council, town board, fire district board, or the equivalent). When drafting a scope-of-employment statement, pay particular attention to response to non-emergency (good-will) calls and clearly define who can authorize such responses.

Sexual harassment

As more women enter the fire service, circumstances giving rise to sexual harassment claims are likely to increase. This situation will probably continue until all personnel understand and appreciate the boundaries of acceptable conduct.

Sexual harassment is unwelcome sexual advances, requests for sexual favors, and other verbal or physical conduct of a sexual nature. It falls into two categories, *quid pro quo* and *hostile work environment.*

Quid pro quo sexual harassment consists of an employee's supervisor using their authority to obtain sexual favors from a subordinate. It is derived from a Latin term meaning "this for that." For instance, a male supervisor who tells a female employee that he will see that she gets a promotion if she has sex with him is engaging in quid pro quo sexual harassment.

Hostile work environment cases may occur when inappropriate remarks of a sexual nature are made, pictures of nudes or persons engaging in sex acts are displayed, or inappropriate touching takes place, to mention only a few instances. It is important to note that, unlike quid pro quo cases, hostile work environment cases can occur between coworkers rather that just between supervisors and subordinates. However, in either instance, the employer can be held liable for the employee's acts.

It is very important to have a clear, written policy prohibiting sexual harassment and to assure that all employees are familiar with it. The policy should educate employees about the nature of sexual harassment and direct them as to what constitutes acceptable and unacceptable conduct. It should also give clear instructions on what to do if harassment occurs. The policy should also include directions on how to report harassment, and care should be taken to provide alternate means of reporting. This is important, since a policy that requires reporting to one's immediate supervisor would be useless in a quid pro quo situation where the immediate supervisor is the offending party.

If a complaint is filed, act on it. Arrange to have it properly and fully investigated by someone capable of gathering the facts, being careful to select an investigator whose impartiality will not be questioned. Find out early on what the victim wants to see as an outcome. Is it just cessation of the offending behavior or is something more being requested? Interviews should be conducted with the alleged victim and perpetrator, as well as any witnesses.

When the investigation is completed, determine whether or not the complaint is founded. If it is, take the necessary follow-up action, which may include disciplining the offending employee or transferring them to a different work location. Use the investigation to determine whether present policies and procedures are working. If they are not, fix them. Use the experience to further educate employees on the subject.

Mutual aid contracts

The practice of fire departments helping one another is nothing new. However, often such arrangements are informal in nature, and not infrequently are even unwritten contracts. While it may seem nice to operate with nothing more than a handshake, it is nevertheless unwise. A written contract forces the parties to consider what their expectations are of the other party, and memorializes those expectations for the benefit of those who may not have been around when the original agreement was made.

A mutual aid contract is a contract like any other. It sets out certain rights and obligations of the parties and liability can flow from failure to abide by its terms. As the demand for more efficient delivery of government services increases, mutual aid contracts will proliferate. Furthermore, as the need to deliver more complex and costly services increases, fire departments will be forced to make greater use of pooled resources, once again encouraging the creation of new mutual aid agreements and expansion of existing agreements. Responses to hazardous materials and terrorism incidents are prime examples of calls where mutual aid is particularly advantageous.

Unfortunately, events frequently overtake contracts rather quickly, causing them to become outdated and inaccurate. It is important to regularly review mutual aid contracts to assure that they still correctly recite the terms that everyone thinks they should. When performing the review, be sure that it covers such topics as:

- An accurate description of services to be provided
- Circumstances under which a request for assistance can be refused
- Circumstances under which resources can be withdrawn after response, but while the incident continues
- Liability coverage, including possible hold harmless and indemnification provisions
- Workers' compensation coverage
- Whether the terms comply with applicable laws
- Integration of command and accountability functions

Risk assessment

Much of what we do in the fire service relates to identifying and managing risk. This practice is as applicable in assessing exposure to legal liability as in any other context. There are four generally accepted methods of managing risk.

Eliminate the risk. One obvious way to avoid a risk is to simply eliminate the activity that creates the risk. Naturally, it would not make sense for a fire department to eliminate the risk associated with fire suppression by eliminating response to fire calls. However, there may be other activities that can be eliminated. For instance, perhaps the department should consider allowing privatization of a portion or all of its EMS activities if the demand for service has outrun its ability to meet that demand. Other types of specialized response (*e.g.*, high-angle rescue, underwater rescue) may likewise be candidates for review and possible elimination. Since good-will calls lead to a high proportion of claims, this type of response should also be reviewed.

Elimination of an exposure can also involve addressing a particular hazard rather than eliminating an entire type of response. For instance, using a slide pole rather than a staircase may reduce station house injuries. Properly securing equipment that must be carried in the passenger compartment can eliminate the risk of tools turning into missiles that injure occupants when dislodged.

Assume the risk. This is usually not an option for most fire departments, as it amounts to self-insuring. However, there are instances where the governing body (municipality or corporation) may want to consider this option.

Reduce the risk. This is the most viable option for the fire service, since its work is risky by nature. This approach involves identifying, addressing, and reviewing risks. As a result of the review, the fire department should institute loss-control and accident-prevention programs. Reduction of risk also includes the identification of fire risks and the creation of appropriate pre-plans.

As applied to vehicle operation, the program would include accident investigation, as well as the reporting of near accidents. Information gained from the investigations would be used to create or revise training programs to avoid recurrences.

Uniformity of fire department operations can reduce risk by having everyone act in a predictable manner. Consequently, a goal of this approach is to create forms and procedures that enable all members to follow established procedures.

Transfer the risk. This means buying insurance. When doing so, it is critical that the policy limits are adequate and that the appropriate risks are covered. No one is likely to miss the obvious exposures such as motor vehicle liability. However, be alert for potential problem areas, such as good-intent calls and fundraising activities. If your fire department is going to detonate fireworks or sponsor a circus, for instance, be sure that your insurance agent knows about it and arranges for the proper coverage. Functions where alcohol is served present special risks and coverage for such exposures must be properly addressed.

Employment relations is another frequently overlooked area. Since managing people is difficult and frequently results in serious disagreements, proper coverage in this area must be provided.

Conclusion

A fire chief's authority to make personnel decisions, to assume command at an emergency incident, or to order withdrawal of firefighters in the face of a major hazardous materials incident is based on the law. The recent evolution of American common and statutory law has resulted in the imposition of new legal *responsibilities* for the fire chief, alongside the chief's traditional authority. This evolution does not mean that fire chiefs need law degrees.

In the 1960s, many fire chiefs could not have foreseen the proliferation of hazardous materials in their communities and were not equipped by training or experience to deal with LPG (liquid propane or butane gas) explosions or major anhydrous ammonia leaks. But once those chiefs learned the potential harm of hazardous materials, they understood the need to adapt. Today, the fire chief who *has not* learned the basics of dealing with hazardous materials is rare indeed.

By applying the same familiarization process to legal issues, today's fire chief can prepare to deal more effectively with the challenges raised by new legal responsibilities. As in the case of hazardous materials, there are a few basic concepts to master (concepts discussed in this chapter), and there are more and more frequent training opportunities. Finally, there is "mutual aid" in the form of fire chiefs who have had legal experiences (good and bad) and who are willing to share those experiences with other chiefs through organizations like the International Association of Fire Chiefs (IAFC) and through fire service books and periodicals.

Fire chiefs cannot immunize themselves against lawsuits and other outcomes of these legal responsibilities. But by treating the legal aspects of fire protection as a new feature of the fire service environment to be studied and "preplanned," fire chiefs can successfully meet the challenges of the fire service and the law.

A fire chief who understands the broad concepts of the American legal system (federal, state, and local), who has a working knowledge of negligence law, and who understands in general terms the path that a civil lawsuit takes is prepared to deal effectively with the continuing legal challenges confronting the American fire service.

5

INSURANCE GRADING OF FIRE DEPARTMENTS

Thomas K. Freeman

CHAPTER HIGHLIGHTS

- The first municipal fire protection surveys were initiated in 1889 by the National Board of Fire Underwriters (NBFU) who hired a former fire department officer to examine conditions and evaluate the needs of fire departments and fire facilities throughout the country as a way to assist cities with their fire protection problems.
- Formed in 1971, the Insurance Services Office, Inc. (ISO), Jersey City, NJ, is the prime agency used nationwide to perform specific property surveys and public protection surveys and is the leading supplier of statistical, actuarial, and underwriting information for and about the property and casualty insurance industry.
- In 1980, the Fire Suppression Rating Schedule (FSRS) was released. It lists ten different protection classifications of which Class 1 areas receive the lowest insurance rates and Class 10 areas the highest (or no recognition).
- A city's class is determined by considering the needed fire flow, receiving and handling fire alarms, fire department, water supply, and divergence.

INTRODUCTION

Because you'll probably have at least one encounter with an insurance rating survey during your fire service career, it is important to have a basic knowledge of what a survey is, how it works, and how to prepare for one. The intent of the public protection classification survey is to help insurance companies establish appropriate fire insurance premiums for residential and commercial properties. Understanding these basic principles and concepts should make your first encounter with insurance surveys less intimidating.

What is the ISO?

Today, the Insurance Services Office, Inc., (ISO) is a nationwide for-profit service organization that provides services to the property and casualty insurance industries. It was formed over the course of many years as more than 20 different insurance-related organizations merged; it now employs thousands of people. The ISO is the leading supplier of statistical, actuarial, and underwriting information for and about the property and casualty insurance industry.

The ISO is actually much broader in scope than just that part we in the fire service normally see and with which we interact. Located in Jersey City, New Jersey, the ISO provides various services to the following lines of insurance: boiler and machinery, commercial automobile, commercial inland marine, commercial multiple line, crime, dwelling fire and allied lines, farm and farm owners, general liability, glass, homeowners, nuclear energy liability, personal inland marine, personal insurance coverage, private passenger automobile, and professional liability and flood, in conjunction with the National Flood Insurance Program of the Federal Insurance Administration.

The ISO performs the functions of specific property surveys and public protection surveys that were previously conducted by the ISO and its predecessors. The ISO provides full services in 43 states and limited services in other states. The ISO does not grade municipalities in Washington, Idaho, Hawaii, Mississippi, North Carolina, Texas, Louisiana or the District of Columbia, because these states have their own rating organizations, for example, the Washington Surveying and Rating Bureau in Washington state. Most of these state bureaus do, however, apply schedules that are the same as or similar to the schedules currently used by ISO (only two states, Mississippi and Washington, do not use the current ISO grading schedule).

ISO's own statistics show that its representatives visit more than two million commercial buildings in the U.S. and make information on those buildings available to more than 1,500 affiliated insurance companies and their agents. In addition, ISO surveys more than 43,000 response jurisdictions in the 44 states in which it operates. Buildings are inspected every 10 years or every 15 years, depending on the survey schedule, unless a special request is made.

The History of Fire Department Surveys

The first municipal fire protection surveys were initiated in 1889 as a way to assist cities with their fire protection problems. At that time, the National Board of Fire Underwriters (NBFU) hired a representative, a former fire department officer, to examine the conditions and to evaluate the needs of fire departments and fire facilities throughout the country. At first, the degree of public fire protection available was evaluated subjectively, based on the representative's judgment. Later, to standardize public fire protection evaluation, a committee of the National Fire Protection Association was formed to develop a public fire protection "rating" schedule. This schedule provided for five classes of protection. For each class, a few specifications were given for the water system and the fire department.

Keep in mind that at the beginning of the 20th century the urban centers of most of America's large cities consisted mostly of wood frame or wood joist masonry multiple-story buildings. With little space between them there were significant exposure problems. The lack of good transportation systems at this time made it economically expedient to concentrate diverse mixtures of business in small quadrants.

Complicating this early urban scenario were relatively new water supply systems. These systems were expensive to construct and, to hold costs down, were frequently undersized, unstable, and unreliable. They were no match for the potential suppression demands that these dense and potentially catastrophic urban centers posed. Early firefighting forces sometimes had to rely only on hydrant pressure, occasionally augmented by hand or steam-operated pumps, which produced fire streams that were too weak to reach and penetrate the significant fire loads they often encountered.

It was only a matter of time before a number of severe conflagrations occurred, demonstrating the need for sweeping changes in how fire protection was assessed and provided. The Baltimore conflagration in 1904 focused national attention on the vulnerability of many of America's cities to widespread, devastating conflagrations. This conflagration also alerted the nation's insurers to the significant financial exposure to their industry as a whole of uncontrolled and apparently unchecked fire risk, for the Baltimore conflagration caused more than \$50 million in damage (the equivalent of \$688 million in today's dollars). Cities were growing rapidly, most with little or no advance planning and seldom with building or zoning laws. For insurance companies, the message was clear—they needed advance information on fire loss characteristics of cities to conduct their business prudently and efficiently.

After the Baltimore conflagration the National Board of Fire Underwriters (NBFU) assembled an engineering staff whose sole purpose was to survey the fire conditions and fire susceptibility of metropolitan U.S. cities. The reports they generated were designed to include a wide variety of information, including the fire department, alarm systems, water supply, fire loss, fire prevention related codes (along with their enforcement), streets, buildings, and conflagration hazard areas (their probability and potential). In these early reports, even the city's police department was surveyed.

After these reports were compiled, recommendations were made to the city regarding improvements that could be made in each of the areas reviewed and underwriting information was furnished to the insurance companies about the fire risks inherent in each locality. The intention was that these teams would revisit the areas periodically to see whether their advice was being heeded and improvements made. It is interesting to note that many of these early reports pointed out significant failure rates in both fire apparatus and equipment, as well as inadequacies in water systems which could have or did contribute to fires that outpaced local resources.

In 1905, the NBFU developed a model building code that could be adopted by those cities which wanted to begin controlling hazards which had, for a long period of time, gone relatively unchecked. In October 1905, the NBFU released a report on San Francisco that said, in part:

> *Not only is the hazard extreme within the congested value district, but it is augmented by the presence of a compact surrounding great-height, large-area frame residential district, itself unmanageable from a firefighting standpoint by reason of adverse conditions introduced by topography. In fact, San Francisco has violated all underwriting traditions and precedent by not burning up. That it has not done so is largely due to the vigilance of the fire department, which cannot be relied upon indefinitely to stave off the inevitable.*

While perhaps not seen as prophetic when it was written, the NBFU's prophecy was in fact fulfilled on April 18, 1906, when the Great San Francisco earthquake caused fires that resulted in an estimated $350 million worth of damage (the equivalent of more than $5 billion worth of damage today).

In 1909, contrary to the program's original intent (that of making the surveys "one-time shots" with follow up inspections to detail progress) it was decided to make the program permanent. Thus, in 1916, the first grading schedule was released. It included seven features to be reviewed and the corresponding points that could be assigned to each. The format of the 1916 grading schedule was used for the next 64 years. The 1916 schedule established criteria that identified which communities were well protected. Deficiency points were assigned whenever a community was unable to meet a portion(s) of a given criterion. Deficiency points also could be assigned for significant effects of an area's climate, as well as for what would become known as "divergence," that is, the difference between a fire department's capabilities and the usability of the water supply. Perhaps the most significant difference between the 1916 schedule and later versions was the apparent intent of the 1916 schedule to look at fire protection as it applied to the central business core. The 1916 schedule evaluated a whopping total of 236 items and sub-items. Today's schedule reviews only 119 items.

Changes in the grading schedule after 1916 were linked primarily to changes in society and technology and to how cities were developing. Changes in the 1920s and 1930s mirrored the changes in the fire service, which progressed from horse-drawn apparatus to motorized apparatus. Changes in the 1940s emphasized protection beyond the central core as businesses moved to the outskirts of cities, whereas the 1950s and 1960s continued the emphasis on protecting cities as a whole. During this period significant improvements also were made to municipal water systems.

It was in 1971 that the ISO was formed. The organization's main interest in public protection was to recognize the impact that effective public protection had on individual property fire rates, given that public fire protection could affect the percentage of loss (value) that could be expected in a fire situation.

The 1974 schedule, referred to as the "Grading Schedule for Municipal Fire Protection," contained modifications that continued to recognize changes in society and technology and in the fire service. The point value assigned to water supply reliability was reduced by giving equal weight to the water supply and the fire department; the point value of fire alarm box systems was reduced and the structural conditions element was eliminated. The central business district was de-emphasized; instead, it was evaluated like any other part of a city. Rather than concentrating on the central business district, attention was given to the built-up areas of the city. The focus of the schedule changed, commencing with the 1974 revision from conflagration-type fires and the level of protection needed to contain them, to concern for fires in individual buildings. In order to emphasize the new focus on individual building ratings, this schedule included a caveat in the introduction that stated that the schedule was to be used as a fire insurance rating tool, not to analyze all aspects of a public fire protection program, and that it should not be used for purposes other than insurance rating. In other words, the new system was not intended to be used to rate public fire protection needs, city programs, or both, and should not be used for such! It was in 1980 that the sweeping changes were made to the schedule that would result in the system still in use today.

Modern Insurance Grading—the Fire Suppression Rating Schedule

1980 ISO Fire Suppression Rating Schedule

FIRE ALARM	
❑ Receipt of Fire Alarms	2%
❑ Operators	3%
❑ Alarm Dispatch Circuit Facilities	5%
	10%
FIRE DEPARTMENT	
❑ Pumpers	11%
❑ Ladder / Service Companies	6%
❑ Distribution of Companies	4%
❑ Pumper Capacity	5%
❑ Department Staffing	15%
❑ Training	9%
	50%

WATER SUPPLY	
❑ Supply Works; Fire Flow Delivery; and Distribution of Hydrants	35%
❑ Hydrants: Size, Type and Installation	2%
❑ Hydrants: Inspection and Condition	3%
	40%

In 1980, the Fire Suppression Rating Schedule (FSRS) was released. It lists ten different protection classifications of which Class 1 areas receive the lowest insurance rates and Class 10 areas the highest (or no recognition). The FSRS simply identifies varying levels of fire suppression capabilities that are applied to the individual property fire insurance rate relativities.

In developing a modified insurance classification system for cities, the emphasis had been on the objective analysis of evaluating suppression features, of measuring major differences among cities, and of recognizing the potential for interface between the ISO's individual building survey and rate-making function and its city-wide classification function. The current classification system is not intended to present a complete analysis of the public fire protection needs of a city and should not be used for such an evaluation.

The ISO reduced the strong emphasis which had been placed on reliability by redundancy, and, instead, measures existing performance. The ISO no longer evaluates street box alarm systems because these systems have been abused extensively in recent years by persons reporting false alarms and because only a small percentage of alarms are received from street alarm boxes versus telephone alarms.

Because of the direct interface with another of its functions, individual building surveys and rate-making, the ISO no longer considers it necessary to evaluate building, electrical, and fire prevention laws. Enforcing these laws has its greatest impact on conditions in individual buildings which are surveyed and rated separately.

Insurance rates for those buildings are evaluated using a commercial rating schedule after a field survey has been conducted by an ISO representative. Individual rates consider many of the same factors controlled by these municipal laws and also reflect the level of enforcement if unsatisfactory conditions are permitted to exist. The impact of laws and enforcement on residential property is measured by the influence on losses for that general class of property.

The new classification system for cities is a credit-type schedule, as opposed to a deficiency-type schedule like its predecessor, although the new schedule can "take points away" in circumstances where actions or activities occur that are deemed improper. The system is objective in that each item can be evaluated mathematically and a corresponding amount of credit calculated. This system evaluates a city's ability to suppress fires in buildings of "average" size once they are actually burning. It avoids penalizing cities for the effects of fire in large buildings by creating a separate section that establishes individual classifications for buildings that require large fire flows to suppress much larger fires.

Overview of the Fire Suppression Rating Schedule (FSRS)

The Fire Suppression Rating Schedule (FSRS) is divided into two sections. Section I is a Public Protection Classification (PPC) which is an indication of an entity's ability to handle fires in buildings of small to moderate size. These are defined as buildings which have a needed fire flow (NFF) of 3,500 gpm or less. Section II of the FSRS consists of individual public protection classification numbers for larger properties that have NFF greater than 3,500 gpm.

Because most communities design their fire protection based on normally expected fires, this design is recognized in the different concepts of these two sections. The public protection classification (PPC) number or class determined in Section I applies to average-size buildings with a NFF less than 3,500 gpm, whereas the aspect of the fire protection demands for larger buildings (those with a NFF of more than 3,500 gpm) has been removed from that evaluation. Section II is applied individually to each building with a NFF greater than 3,500 gpm in order to develop an individual classification number that reflects the available protection for that specific property.

The FSRS establishes a NFF in gallons per minute for suppression of a fire in a building. Representative building locations are selected throughout a city, and a basic (mean) fire flow is determined. All properties that exceed a NFF of 3,500 gpm are reviewed separately because fire protection control for these larger buildings is considered to be more the responsibility of the individual property owner. Sprinkler systems, smoke detectors, construction upgrades, and other fire protection improvements therefore become more of an incentive to these property owners.

The FSRS has three major features: fire alarm, fire department, and water system, all of which directly affect the measurement of fire suppression insofar as their city-wide effect is concerned. The fire alarm section examines how the public reports a fire and how the fire department receives that report. In a typical alarm received by telephone, the call taker will receive the call and alert firefighters, advising them of the location of the emergency. Because different cities receive fire alarms in different ways, the FSRS attempts to review all possible variations and assigns points to indicate equivalencies. The fire department section considers apparatus, equipment, staffing, automatic and mutual aid, pre-fire planning, and training. The interrelationships of engines, truck companies, minor equipment, paid and volunteer firefighters, and department training are all evaluated using a point system to relate equivalencies. The water system section considers the supply works, the main capacity to deliver fire flow, distribution of hydrants, hydrant size, type and installation, hydrant inspection and condition, and alternative water supplies.

The major differences in the FSRS versus the 1974 Grading Schedule for Municipal Fire Protection, as shown in the previous charts, can be summarized into several broad conceptual changes. First, the FSRS attempts to take a view of fire protection that is macroscopic as opposed to microscopic. The FSRS attempts, on a relative scale of 1 to 10, to quantify or assess the capability of a community to control and suppress fires when they occur and thereby limit fire loss. Remember that limiting fire loss is what is most important to ISO; fire protection and suppression are merely mechanisms to limit that loss. Therefore, the only items reviewed are those which directly affect and assist (or hinder) the suppression of a fire.

Second, the FSRS operates on a credit basis rather than on a deficiency basis (as did its predecessors). The 1974 schedule was based on a theoretically perfect community with hypothetically perfect scores in each area being graded. Every community evaluated was compared to the "perfect" score and was given "deficiencies" for each area in which it did not match up. Taking points away for not doing certain things implied that communities were penalized for not having excellent fire protection and tended to imply that the ISO was "grading" fire protection, a notion, as previously stated, that the ISO has tried to dispel.

The FSRS has inverted this process by setting minimum criteria; the conditions found in a certain community are "credited" from that minimum. The ISO philosophy is that such an approach helps to offset the misrepresentation that it is setting standards for fire protection.

In addition, the FSRS has had all "subjective" criteria removed (it is hoped that this has resulted in an objective survey) and has concentrated on significant differences among communities and de-emphasized the time spent on those items that seem to be fairly uniform among communities. It is a "performance schedule" as compared to the "specification" type schedule that preceded it. Rather than setting out specific criteria as to how each area reviewed is to be structured, the FSRS looks more at whether the criteria are met than at how they are met. For example, where previous schedules required that a city have a water system to receive anything other than a Class 9 rating, the FSRS allows alternative methods of water delivery, such as tanker shuttles and large diameter hose to be used, as long as the appropriate quantity of water can be delivered. Finally, with the exception of the 1974 schedule, which allowed the needed fire flow in a sprinklered building to be reduced by as much as half, predecessors to the FSRS did not recognize the importance of fire sprinkler systems. The FSRS recognizes the importance of sprinklers by excluding properties that are fully sprinklered and are "graded" as sprinklered as determined by ratings developed through a separate grading schedule, from the development of needed fire flows. It is important that a building be graded as "sprinklered" because buildings that are not so graded (*i.e.*, buildings that are only "partially" sprinklered) are not exempted from the flow requirements.

The Contents of the Fire Suppression Rating Schedule

Contained within the grading survey are numerous areas that are reviewed and scored, and then used cumulatively, along with the other areas reviewed, to determine a city's rating. What follows is an overview of the contents of the survey, with particular attention given to the areas reviewed and *how* they are reviewed.

The introduction contains the background material necessary to properly apply the schedule. It defines "city" as including everything from cities to districts and it explains how numbers are rounded and decimal points dropped for the purposes of computing calculations. Several portions warrant a closer look. Section 101, entitled "Scope," states that the schedule measures the major elements of a city's fire suppression system and contains the following disclaimer:

> *"The schedule is a fire insurance rating tool, and is not intended to analyze all aspects of a comprehensive public fire protection program. It should not be used for purposes other than insurance rating."*

Item 106 addresses the minimum facilities and conditions that are required to get any rating other than a 10 (municipality with less than recognized protection). These requirements include:

- The fire department is a permanently organized entity under state or local law
- It has a person in charge (fire chief)
- It serves an area with definite boundaries and is either legally or contractually obligated to protect same
- Membership shall ensure the response of at least four people to structure fires
- Training must be conducted two hours every two months for each member
- There shall be no delay in receipt of alarms and dispatch of equipment
- There shall be at least one piece of apparatus meeting NFPA 1901
- Apparatus meet the standard for automotive fire apparatus and all apparatus shall be protected from the weather

(Note: this is a partial list only)

Application of particular section to types at minimum facilities

Classes 1–8: This section tells under what circumstances different portions of the schedule shall be applied. For instance, if a city has both a piece of apparatus with a pump that has a rated capacity in excess of 250 gpm at 150 psi *and* a water system (alternatives will be discussed later) capable of delivering 250 gpm or more for a period of two hours plus consumption at the maximum daily rate, then Sections 300–301 are applied (*i.e.*, the city can obtain between a Class 1 and a Class 8).

Class 9: If the city lacks the water system but has a piece of apparatus with a pump that has a capacity of 50 gpm or more at 150 psi and at least a 300-gallon water tank, then the city can obtain a Class 9 (Items 800–802).

Class 10: If the city lacks all the above, then the schedule doesn't apply to them. They automatically get a Class 10 (*i.e.*, less than minimum recognized protection).

Needed fire flow (NFF)

This section discusses how NFF is determined and specifies what kinds of buildings should not be used in determining NFF. The factors used to determine NFF are: (1) construction class; (2) occupancy class; and (3) exposure factor. Normally, a specified number of buildings that are suspected to have a large NFF (based on findings by ISO representatives) are selected to determine the NFF.

It is important to note that buildings graded sprinklered by the ISO are not subject to being included in the group of buildings used to determine NFF. Therefore, it is to a community's advantage to fully sprinkler as many buildings as possible (and assure that they are graded sprinklered). A list of buildings graded as "sprinklered" by the ISO should be obtained by the municipality. The fewer buildings that require a larger NFF (and are used in calculating a city's NFF), the lower the NFF; this results in a potentially better classification. It is important for a city to carefully review any building's used NFF calculations. There have been instances of fully sprinklered buildings used for NFF because the ISO had not graded them as sprinklered. This occurred because it lacked information or lacked sprinkler tests for the buildings in question. NFF is also used for various other calculations in the schedule.

Receiving and handling fire alarms

The first area of review is receiving and handling fire alarms; it represents 10% of the total grade. Included under this section are areas concerning telephone service, operators (call takers), and dispatch circuitry involved in the receipt and dispatch of emergency calls.

Telephone service is reviewed to ascertain general accordance with NFPA 1221 *Standard for the Installation, Maintenance and Use of Public Fire Service Communication Systems.* The number of telephone lines needed (both emergency and business) is based on the size of the population served, and range from one fire line and one business line for a population up to 40,000, to four fire lines and three business lines for populations of 300,000 or more. Be aware that certain phone service situations can create a substantial loss of credit. For example, if non-fire emergency calls are received on fire emergency lines, then the number of lines needed has to be doubled. Automatic telephone dialing equipment used to report alarms requires separate lines. If only one phone number is listed in the telephone directory for both fire and business purposes, no credit is given for the fire line(s). Finally, even though the 1980 schedule is a "credit system," points are actually deducted if information concerning a fire is received by one call taker who then must pass the information on to another communications center. If the original call taker transfers the actual caller (patches the call) to another communications center which takes the information, then no points are deducted. In addition to these requirements, there must also be a device that permits immediate playback of calls received.

The telephone directory is also reviewed for the following areas of compliance:

1. The fire emergency number is printed on the inside front cover of the white pages (blank lines for the customer to fill in the fire emergency number do not comply).
2. Both the emergency number and the business number are listed under "Fire Department" in the white pages.
3. Both the emergency number and business number are listed under the name of the city in the white pages. *(Note: If the individual fire station phone numbers are listed, additional points are deducted from the overall points credited.)*

The next area covers operators. The number of operators needed, as defined by NFPA 1221, is determined by the number of calls received, that is, all calls required to be handled by the operator. It is important to note that this includes all calls...not simply "9-1-1" calls. According to the current edition of NFPA 1221, for a jurisdiction that receives fewer than 600 calls annually the schedule allows the following variations in operator requirements. If the jurisdiction receives from 600 to 2,500 alarms per year, at least one specially trained operator shall be on duty at all times. If more than 2,500 calls per year are received, then at least two fully trained operators shall be on duty at all times, with more as required by actual traffic. If a jurisdiction receives more than 10,000 calls per year, then three on-duty operators are required with two backup operators, for a total of five, although the two backups may be doing other work or resting. To assess whether a municipality complies with this section, use the most current edition of NFPA 1221 date requirement.

The final communications area reviewed is dispatch circuits, which includes a requirement for two separate dispatch circuits (*i.e.*, radio and telephone, radio and microwave, etc.) of which one circuit must be one of the following:

- A supervised wire circuit
- A radio channel with duplicate base transmitters, receivers, mikes, and antenna (if the primary transmitter fails, switchover to the backup must be automatic with visual and audible indication to the operator, unless the controls are located where someone is always on duty; then manual switchover is permitted if it can be accomplished within 30 seconds of failure)
- A microwave-supervised carrier channel
- A polling of self-interrogating redundant radio or microwave radio system
- A properly arranged, supervised phone circuit

Note: If fewer than 600 calls are received annually, then only one circuit is required.

The second dispatch circuit does not have to be supervised and can be either a wired circuit or radio channel; if the second circuit is a radio channel, it does not require duplication as would be required for the primary circuit. When two dispatch circuits are required, all alarms for fires in buildings must be transmitted from the communications center to the fire stations by two means (*i.e.*, radio and printer, telephone and radio, etc.). The following types of dispatch circuits are credited under the schedules:

- Radio, voice-amplification, facsimile or teletype; visually recorded (facsimile and teletype) devices shall be accompanied by an audible alerting device to alert personnel
- Radio receivers carried by firefighters, and a transmitter at the communications center
- Outside sounding device
- Voice receivers at firefighters' homes or businesses and a transmitter at the communications center
- Group alerting telephone circuits

Circuits and other system components are required to be monitored so that defects and faults that would affect system performance can be detected rapidly. Circuits also are required to be recorded, either taped or hard-copied, depending on the type of dispatch circuits being used (*i.e.*, radio, teletype, etc.).

Supervision of the primary power supply is not required. An emergency power supply is not required, but may receive credit if provided at the site of transmitters (communications center) and receivers (fire stations). Acceptable emergency power supplies are batteries that last for four hours, an automatically started generator, manually started generators, wet cell batteries or dry cell for radio receivers of the voice-amplification type. It is important that these power supplies are tested in accordance with NFPA criterion in order to receive credit form the ISO.

Total credit for receiving and handling fire alarms includes the sum of "credit for telephone service" plus "credit for operators" plus "credit for dispatch circuits."

Fire department

The fire department, the next major area to be reviewed, represents 50% of the total grade. The review begins with "needed" engine companies, based on maximum number needed for basic fire flow, distribution and/or operations.

Basic fire flow (BFF) is the fifth largest needed fire flow (NFF) of all the NFF which are calculated in Items 310–340, with the maximum BFF being 3,500 gpm (*i.e.*, if the NFF calculated were 500 gpm, 750 gpm, 1,000 gpm, 1,500 gpm, 2,000 gpm, 2,500 gpm, 3,000 gpm, 3,500 gpm, 4,000 gpm and 4,500 gpm, then the BFF would be 2,500 gpm). The number of needed engine companies for BFF is one engine for 500 gpm to 1,000 gpm BFF, two engines for 1,250 gpm to 2,500 gpm BFF, and three engines for 3,000 gpm to 3,500 gpm BFF.

By distribution, an additional engine company is needed for each area where a company that is required by BFF will not meet the first due response distance to 50% of the built-upon, (*i.e.*, hydranted area that is within the satisfactory response travel distance). Travel distance is defined as 1½ miles as measured on "all-weather" roads for engine companies. In addition, if responses outside the city deplete resources available to the city, then additional engine companies may be required. The total number of engine companies needed by distribution is the number of needed existing engine companies plus the number of additionally needed engine company locations. For operations, the standard response is two engine companies, except when only one engine is required by BFF. The number of needed engine companies is the greatest number of engines needed based on either BFF distribution or operations, plus any additionally required companies.

In contrast to needed engine companies is the credit given for existing engine companies. Engines that are staffed on first alarms are given credit as existing engine companies if, based on a certification test or three-hour acceptance test, the pump meets all the following requirements:

- 100% of rated pressure at 150 psi net pump pressure
- 70% of rated pressure at 200 psi net pump pressure
- 50% of rated pressure at 250 psi net pump pressure

A pump must be permanently mounted and must have a minimum rating of 250 gpm at 150 psi in order to qualify an apparatus for credit. There are two additional ways to receive engine company credit. The first is that apparatus carrying engine and ladder or service company equipment will be credited as existing engine companies, if needed. Second, automatic aid engine companies that are within five miles of a city and that respond according to a plan are credited if they replace the need for engine companies.

While the formula is somewhat complex, if a staffed engine company responds on a first alarm and is considered "extra," it can be counted either as a ladder or service company when it carries the appropriate equipment, or it may be credited as part of a two-piece engine company. In this case, the credit for the equipment it carries can be combined with the equipment credited on another engine in order to gain the maximum credit available under engine equipment.

For each apparatus that meets the criteria for an existing engine company, the following additional items are reviewed:

- Pump capacity up to 500 gpm
- Hose 2½" up to 400 ft., plus an additional 800 ft. or longer of 2½"
- Pumper equipment and hose
- 300-gallon booster tank
- 200 feet booster hose at 1½" or 1¾"
- 400 feet 1½" or larger hose
- 200 feet 1½" or larger hose spare (or carried). *Note: 1¾" hose is acceptable in lieu of 1½" hose*
- One master stream device (1,000 gpm) *Note: This is not needed for BFF of less than 1,500 gpm*
- One distributing nozzle
- One foam nozzle
- Ten gallons foam carried
- Fifteen gallons foam spare or carried
- Two 2½" play pipes with shutoff
- Two 2½" straight stream and spray shutoffs
- Two 1½" straight stream and spray shutoffs
- Four SCBAs (30-minute minimum)
- Four additional SCBA spare cylinders (carried)
- Two 12 ft. x 18 ft. salvage covers
- Two handlights
- One each: hose clamp, 2½" hydrant hose gate, 2½" hose jacket
- 2½" x 1½" x 1½" gated wye, portable and mobile
- Radio, roof ladder and 24-ft. extension

Many of the needed equipment items that are listed have equivalencies acceptable to ISO. The ISO should be contacted to determine if a needed piece of equipment has an acceptable equivalent that would receive credit in the grading schedule.

Equipment credit is then prorated depending on quantity possessed versus quantity required. In addition, each engine must be pump tested annually (similar to the NFPA 1901 certification test) with decreasing credit for less frequent tests down to once every five years. Hose also must be tested annually as described in NFPA 1962 *Standard for the Care, Use and Maintenance of Fire Hose, Including Connections and Nozzles* with credit decreasing at lower psi achieved (250 psi, 200 psi, or 150 psi) and for less frequent tests down to once every five years. If no records of pump tests are maintained, then the credit is reduced by 20%. If the equipment is carried on an automatic aid engine, then based upon such variables as common communications, interdepartmental training, etc., the maximum credit than can be derived for that equipment is 90% of the total credit awarded.

One reserve engine is required for every eight engine companies needed. Pump, hose, and equipment on reserves are credited just as they are on existing engines. Pump capacity is also reviewed with the requirement that the available pump capacity of all existing engines be sufficient to meet BFF.

Ladder or service companies are reviewed next. The number needed here is higher than the number needed for distribution or operations. From a distribution perspective, every protected area must have a ladder or service company response and if any protected area is beyond 2½ miles of an existing ladder or service company, then additional ladder or service companies may be needed. In these cases, the need may be met by ladder or service companies at existing or needed engine company locations. This obviously prevents requiring locations just for ladder or service companies.

From the operational perspective, any standard response on first alarms for building fires should have a ladder or service company. If a ladder or service company does not respond, an engine company responding with any ladder or service company equipment should be considered as an engine/ladder or engine/service company.

Whether a ladder or service company is needed depends on the type of area protected. Response areas with five or more buildings that are three stories high or are 35 ft. or more in height, or that have five buildings that have a NFF exceeding 3,500 gpm, or any combination of these two scenarios, should have a ladder company. In consideration of building height, all buildings, including those with sprinklers, are used. If no individual response area needs a ladder company, but the buildings in the city as a whole meet the above requirements, then at least one ladder company is needed. Response areas that do not need a ladder company should have a service company.

Companies that respond to first alarms carrying any of the equipment required for ladder companies will be considered existing ladder companies when ladder companies are needed. While the rules are quite complex, generally an existing engine company that carries any ladder company equipment is considered an engine/ladder company and gets credited as one half of an existing ladder company. If it is not credited as an engine company, but it carries ladder company equipment, an engine company will be credited as one ladder company. The same general rules apply to engine/service companies. The following equipment is required on a service company:

- One large spray nozzle (500 gpm minimum)
- Six SCBAs (30-minute minimum)
- Six spare cylinders
- Ten 12 ft. x 18 ft. salvage covers
- One electric generator
- Three 500-watt floodlights
- One smoke ejector
- One oxyacetylene cutting unit
- One power saw
- Four handlights
- One hose roller
- Six pike poles (6 ft., 8 ft., and 12 ft.)
- Two radios (1 mounted, 1 portable)
- Two ladders (10-ft. attic, 14-ft. extension)

In addition to the service company equipment, the following equipment is required for ladder companies:

- One 16-ft. roof ladder
- One 20-ft. roof ladder
- One 28-ft. extension ladder
- One 35-ft. extension ladder
- One 40-ft extension ladder
- One elevated stream device (able to reach the lesser of 100 ft. or the height of any building protected)

As with engine companies, many of the listed needed equipment items for service and ladder companies have equivalencies acceptable to ISO. The ISO should be contacted to determine if a needed piece of equipment has an acceptable equivalent that would receive credit in the grading schedule. An annual test of the aerial ladder/elevating platform also is required; variable credit is given depending on the frequency of tests—from a high of 100% credit for annual tests to a low of 0% credit for a test frequency of five years or more. In addition to the annual test, a nondestructive test (NDT), as defined in NFPA 1904, is needed every five years. *Note: If there are no records of the tests which an agency claims to have conducted, then the points awarded are reduced by 20%, emphasizing the importance of proper record keeping.*

For every eight ladder or service companies needed, one reserve ladder or service company is needed. If one of the eight needed companies is a ladder, then the reserve should be a ladder as well, as opposed to just a service company. The equipment on the reserve ladder or service companies shall be credited according to previous equipment schedules for ladder and service companies.

Part of the formula for fire department credit includes a review of company distribution, which requires an engine company within 1½ miles of every built-up area of the city and a ladder or service company within 2½ miles of every built-up area of the city.

The next area examined under the fire department is existing company personnel, that is, the average number of firefighters and company officers on duty for existing companies as determined by certain criteria. The total number of members on duty shall be the yearly average of on-duty personnel, including all time off. Chief officers and non-suppression personnel are not included in computing on-duty strength, except when more than one chief officer responds; then, those who perform company duties may be credited as firefighters. While the ISO does not discuss minimum staffing, this is the one area of the survey where a city can receive unlimited points (*i.e.*, the more on-duty personnel that respond, the more points credited).

Personnel on apparatus not credited as existing engine, ladder, or service companies but who regularly respond to first alarms to aid existing companies are included for the purpose of increasing total company strength. Personnel on units such as ambulances may be credited if they are involved in firefighting operations, depending on the extent to which they are available and used for first-alarm response.

On-call, volunteer and off-duty paid members responding to first alarms are credited based on the average number staffing apparatus on first alarms. Call and volunteer shall be credited the same as on-duty paid personnel proportionately for the time they spend sleeping at stations; otherwise every *three* volunteers or call personnel credited as responding count for *one* on-duty person (3:1). The importance of keeping good records comes into play here again, because if good records are not kept to document response, the credit ratio of volunteers to on-duty increases from 3:1 to 6:1 (*i.e.*, six volunteers equal one paid person).

The last area reviewed under the fire department is training. Facilities, aids, and actual training provided are examined. Facility and aids credit is given for drill towers, fire buildings, including smoke rooms, flammable liquid pits, library and training manuals, slide and movie projectors, pump and hydrant cutaways, and training areas, which may include streets and open areas when no other training facility is provided. Videos on fighting combustible/flammable liquid fires could replace the need for flammable liquid pits. The points credited for facilities and aids are then prorated for use with credit being given for the following:

(a) Eight half-day company drills (3 hours) per year.

(b) Four half-day multi-company drills (3 hours) per year.

(c) Two night drills (3 hours) per year.

The fewer of the above drills held, the less credit is given for facility and aids. Some drills may qualify for multiple credit. For instance, a single company drill held for three hours at night can credit as (a) and (c); a multi-company drill held for three hours at night can get credits as (a), (b) and (c). The following additional training also is examined and credited:

- Up to 20 hours per member per month of company training at the station
- Up to two days per year for all officers
- Up to four half-day sessions for driver/operators
- Up to a 40-hour class for all new driver/operators
- One-half day per member per year on radioactivity (hazardous materials training could replace the need for separate training on radioactivity)
- Up to 240 hours per new recruit

In addition, "pre-fire" planning inspections, including updated notes and sketches of all commercial, industrial, institutional, and similar buildings should be made twice a year. In order to receive full credit, all firefighters must participate in training that includes reviews of the department's preplans. The total points given for training are reduced 10% for incomplete records and 20% for no records.

Credit for fire department is the sum of the total credits given for "engine companies" plus "reserve pumpers" plus "pumper capacity" plus "ladder service" plus "reserve ladder service" plus "distribution" plus "company personnel" plus "training." The following can be used as a mental proxy for calculating fire department credit:

CREDIT = Engine companies + reserve pumpers + pumper capacity + ladder service + reserve ladder service + distribution + company personnel + training

Water supply

The third and final major area to be reviewed is the water supply, which counts for the remaining 40% of the grade. This item reviews the water supply that is available for the city's fire suppression. If it is determined that more than 85% of the community being graded is not within 1,000 feet of a recognized water system, the area devoid of fire hydrants may get no better than a Class 9 rating.

Several elements go into determining the "supply works capacity" which is one of the four factors used in calculating the credible rate of flow at each test location. Maximum daily consumption is the average rate of consumption on the maximum day. The maximum day is that 24-hour period with the highest consumption in the last three-year period, excluding highs caused by unusual operations (*i.e.*, major fires) or that won't occur again due to system changes. A water system is reviewed at a residual pressure of 20 psi. The fire flow duration should be two hours for NFFs up to 2,500 gpm and three hours for those of 3,000 gpm to 3,500 gpm. The ability of the water supply system to deliver the NFF is measured at representative locations throughout the city and at each location. The supply works capacity, main, and hydrant distribution are reviewed separately.

The supply works capacity is determined by subtracting the maximum daily consumption from the sum of the average maximum water storage plus the effective pumping capacity of pumps (expressed in gpm) when delivering at normal operating pressures and the delivery capabilities of filters plus emergency supplies (*i.e.*, supplies not normally used). To this amount is then added: (1) suction supply (static supplies from which a fire department can draft to supply water); and (2) fire department supply (water delivered by fire department vehicles). This last area, fire department supply, is perhaps the most significant change in the grading schedule and represents a major philosophical shift from the method of water delivery to how much water can be delivered with little emphasis on its delivery method. This change has had a major beneficial effect on those communities that are not connected to large municipal water systems, for it has allowed them, through tanker shuttles and other alternatives, to establish deliverable flows that meet or exceed the 250 gpm required for two hours under Item 201; this enables a city to obtain a rating better than a Class 9. Many communities formerly rated a "9" have improved two to three classes under this new alternative. This alternative requires a ISO representative to witness each tanker's fill time, dump time and set-up time. This information then is converted into a gpm flow for each tanker by dividing the amount of water carried (less 10% for spillage) by a combination of the dump and fill time and travel time. Also recorded is the pump capacity, travel and setup time at the fire and supply point, as well as folding tank and fire site pumper tank capacities. All this is then processed through computer software which computes the theoretical flow available. Considering that 40% of a city's total grade is water related, this alternative was not only long overdue, but a welcome relief to much of the nation's fire service.

Main capacity is also reviewed at the same test locations that are considered for supply works capacity. The results of actual flow tests at these locations indicate the ability of mains to carry water to those locations.

The final element used in determining the creditable flow rate at each test location is hydrant distribution. Each hydrant within 1,000 feet of a test location (measured as hose can be laid) is reviewed to determine if it can satisfy the NFF at that location. Gallons-per-minute credit for hydrants is as follows:

For each hydrant within 300 feet of test site, measured to the nearest corner of the test site/building credit, 1,000 gpm; if within 301 feet to 600 feet, credit 670 gpm; and if within 601 feet to 1,000 feet of the test site, credit 250 gpm. If, for instance, there were two hydrants within 300 feet and one hydrant within 600 feet, the total maximum flow by distribution would be 2,670 gpm (1,000 gpm + 1,000 gpm + 670 gpm).

Gallons-per-minute credit for hydrants

Distance from test site	Credit given
Up to 300	1,000 gpm
301 to 600	670 gpm
More than 600	250 gpm

Hydrant credit can be further limited according to size and number of ports as follows:

(a)	At least one pumper outlet	1,000 gpm max
(b)	Two or more hose outlets, no pumper outlet	750 gpm max
(c)	One hose outlet only	500 gpm max

Hydrants that are in another city but within measurable distances of the test site are credited as any other hydrant. If a fire department can demonstrate by the use of large diameter hose, that it can flow greater quantities than allowed above, then the actual flow shall be used for calculating flow by distribution. This is based on whether the department's standard operating procedure requires a pumper at the hydrant and the large-diameter hose between the pumper and the fire.

The capability of the water system at the test site will be calculated using the lesser of the following measurements:

- NFF at that site
- What the building requires
- The supply works capacity at the site (*i.e.*, how much water is available)
- The main capacity at the site (*i.e.*, how much water can actually be flowed)
- Hydrant distribution at the site (*i.e.*, number and distance of hydrants)

Credit for the supply system is determined by factoring the capability of the water system (at the test locations) and the NFF at the test locations.

Credit for hydrants is given based on points assigned to various types of hydrants with maximum points being given to hydrants with a 6" or larger branch and a pumper outlet with or without 2½" outlets. The fewest points are given to hydrants with: (a) only one 2½" outlet; and (b) smaller than a 6" branch, as well as to hydrants that are flush types, to cisterns, and to suction points. Points are prorated according to the number of hydrants of each type compared with the total number of hydrants.

Inspection and condition of hydrants is the last area reviewed under water supply. Inspection and condition of hydrants should be in accordance with American Water Works Association (AWWA) Manual M-l7. The credit given for this area is based on frequency of inspection and condition of the hydrants. The frequency of inspection is the average time interval between the three most recent inspections, with maximum credit given for half-year cycles and minimum credit for cycles of five years or more. Points are deducted for incomplete or no records, if hydrants are not subject to full system pressure during tests, and if inspection of cisterns and suction points does not include actual pumper drafting. Condition is reviewed by giving credit for three categories: (a) standard (no leaks, well located, operates easily); (b) usable; and (c) not usable.

Credit for water supply is determined by the total of the "credit for supply system" plus the "credit for hydrants" plus the "credit for inspection and condition."

The Public Protection Classification

The Public Protection Classification (PPC) is determined by adding the points credited to "receiving and handling fire alarms," up to 10 points, plus those credited to the fire department, up to 50 points, plus those credited to water supply, up to 40 points. From this total is deducted what's called "divergence," which is 50% of the difference between 80% of the fire department credit and 100% of the water supply credit. The divergence represents the disparity, if any, between the fire department's capabilities and the ability to supply water.

Divergence = *50% (100% water supply credit -80% of fire department credit)*

"Percentage to class correlations" are as follows:

Percentage to class correlations chart

Class	Percentage
1	90.00 or more
2	80.00 to 89.99
3	70.00 to 79.99
4	60.00 to 69.99
5	50.00 to 59.99
6	40.00 to 49.99
7	30.00 to 39.99
8	20.00 to 29.99
9	10.00 to 19.99
10	0.00 to 9.99

As an example, if a city received the following credits:

(a) Receiving and handling fire alarm — 9 points (out of 10 possible)

(b) Fire department — 35 points (out of 50 possible)

(c) Water supply — 31 points (out of 40 possible)

Divergence would be "minus 1.5 points" (*i.e.*, 50%[(31) – 80%(35)]).

The resulting PPC would be a Class 3; (9 + 35 + 31 = 75 – .5[(31) – .8(35)] = 73.5; 70.00 to 79.99 = Class 3).

Cities which cannot meet the requirements in Item 201 (previously discussed) to obtain at least a Class 8 may obtain a Class 9 if under Items 800–802 they can meet all of the following minimum criteria:

1. One piece of apparatus, which is NFPA 1901 compliant, with a permanently mounted pump capable of delivering 50 gpm or more at 150 psi and a 300 gallon tank
2. Records which indicate date, time, and location of fires, the number of responding members, meetings, training, maintenance of equipment and apparatus, and an up-to-date member roster
3. And a minimum of the following equipment:
   - At least two 150-ft. lengths of ¾" or 1" booster
   - 1½" pre-connect or equivalent each with a nozzle that can discharge both a straight and a spray stream
   - Two portable extinguishers for use on Class A, B, or C fires, with minimum rating of 20-BC in dry chemical, 10-BC in CO_2 and 2-A in water
   - One 12-ft. ladder with hooks
   - One 24-ft. extension ladder
   - One pick head ax
   - Two handlights
   - One each—pike pole, bolt cutter, claw tool, crow bar

Note: Out of 80 possible points for 1, 2, and 3 above, a city must get at least 70 in order to get a Class 9; otherwise it will receive a Class 10.

Individual properties

Section II of the FSRS deals with public protection classifications for specifically rated properties that have a NFF of more than 3,500 gpm. While specific details of how those classifications are arrived at will not be reviewed in this chapter, suffice it to say that the protection class of an individual property is the lower of either of two protection factors–fire department companies or water supply. The protection class of a subject building will be the same as that for the city unless the individual public protection classification (PPC) indicates a poorer class in which case the poorer class (but not less than Class 9 when the city is a Class 9 or better) will apply to the subject building.

The Survey Process

There are several methods by which a survey can be initiated. First a community, because of what it perceives to have been significant changes or improvements to its water system or fire department, may request a regrading. This request, which details the changes or improvements, is normally made in writing from the chief executive officer of the city to the ISO. The ISO will do an office review to determine if the changes to the fire protection would alter the community classification. If it is determined that it is probable, ISO will schedule an appointment with the community for a regrading. This will largely depend on how "backed up" the particular ISO office is and could run anywhere from three months to a year or longer, unless the ISO already plans to be in the area.

Another way that a regrading may occur is if a major annexation occurs which significantly changes the boundaries of the city or a new city incorporates with its own fire department. If either one of these situations is brought to the attention of the ISO, it will probably initiate a regrading.

Finally, the ISO has a generally established resurvey cycle of its own that it uses to attempt to regrade communities with a population of fewer than 25,000 every 15 years and those with populations more than 25,000 every 10 years. A resurvey may also be initiated by the ISO as a result of the ISO Community Outreach Program. The Commission on Fire Accreditation (CFAI) and the Insurance Services Office, Inc., have partnered in an information gathering effort through this valuable survey program. The program is designed to collect information on essential protection features within a community. This information can then be used by the insurance companies to reflect the most current conditions in their insurance rating and underwriting programs.

In the case of a field survey, the ISO field office will notify the city several weeks prior to the survey. At that time it will also request certain information from the city. This information may either be requested in advance or the city may be advised to have it ready for the field representative upon their arrival. Materials requested may include, but not be limited to the following:

- A listing of all the significant representatives of the city (*i.e.*, city manager, fire chief, public works director, mayor, etc.)

- A map of the city showing all fire station locations
- A map showing the location of all hydrants in the city
- Pertinent information concerning the city's water system (*i.e.*, maximum daily consumption, pump capacity, etc.)
- Pertinent information regarding the city itself (*i.e.*, population, etc.)
- An information sheet for each piece of fire apparatus that the fire department operates including information about the pump, the tank, and all other equipment carried on board

In addition to the above material, it would be extremely helpful to have the following information ready in a user-friendly form for the ISO representative:

1. Staffing records for the last year, including minimum and maximum levels if applicable
2. Records of all time-off during the last year, including vacation, sick leave, Kelly days, etc.
3. A training record for each member of the department, broken down by category of training
4. A list of all training facilities and aids
5. Information regarding the communications center, including number of calls received, number of operators on duty, and type and numbers of dispatch circuits
6. Copies of all preplans completed, including sketches showing frequency and dates completed
7. A complete set of water system maps, including schematics of all plants and distribution sites
8. Copies of all formal written automatic aid agreements
9. A complete record of all hours spent in stations and all responses made if volunteers are part of the system
10. A record of all structural responses for the last year, including the number of personnel responding to each

Once the ISO representative arrives, he or she will become familiarized with the material that has been presented to him or her and then proceed to set out a schedule of activities to be accomplished during the inspection. This will include station visits to spot check facilities, equipment, and apparatus listed on the information sheets discussed above, inspecting the communications center, and identifying the buildings that will constitute the test locations. The department will need to have the necessary personnel available to assist the representative with the flow tests that will be conducted at each test location. It would also be wise to review the batch list compiled of properties in the city and make sure that all buildings that are sprinklered are so identified so that sprinklered complexes do not enter into the computation of BFF; this may help reduce the number of needed engine companies.

After the ISO representative has completed all tasks necessary on site, he or she will return to the field office where all the final computations will be done. Once the grading is completed, it will be forwarded for review. After review, the final survey results will be released during the next reporting cycle.

While the ISO normally only informs the city of its classification, if the city so requests, the ISO will furnish a copy of the classification details showing the number of points awarded in each of the subcategories under the three main categories as compared to the maximum number of points available. This is very helpful in determining areas for improvement.

If a city's classification improves, the implementation will be promptly processed. If, however, the city regresses (*i.e.*, the classification becomes worse), then implementation will normally be delayed and the city will be given an opportunity to make any necessary improvements or changes to avoid the regression. This will normally require city personnel to communicate a plan of action in writing to the ISO, along with a timetable for implementation.

Conclusion

While it is true that a reduction in the classification of a city can reduce the amount of insurance premiums paid by its residents and businesses, it is equally true that the cost of making the improvements necessary to obtain that reduction (which translates into higher taxes paid by those same residents and businesses) can exceed the insurance savings realized. All too often a fire chief, tempted to want the "bragging rights" which go along with lower classifications, seeks to make improvements, citing a better rating as the reason, without comparing the cost of improvements to the benefit to the taxpayer.

Very often a property owner can realize just as much in savings to his own particular property by making fire safety improvements therein (*i.e.*, fire alarm or extinguishing systems). This will result in as much of a reduction in his own premium as could have been realized by a reduction in overall city classification. This is particularly true for homeowners insurance. In many states, once a community reaches a certain classification, Class 3-4 in many cases, the homeowner's premium may not change, even if there is further improvement in the city's rating. This is precisely the reason that the ISO stresses the fact, as previously mentioned, that the FSRS is an insurance rating tool and not intended to be used as a means of evaluating the quality of fire protection provided by a community.

6

LEADERSHIP FOR TODAY AND TOMORROW

Dennis Compton

CHAPTER HIGHLIGHTS

- Leaders have to be able to effectively lead themselves if they expect to effectively lead others.
- Basic leadership traits and behaviors that form a strong leadership foundation include commitment to and passion for the mission, fairness in decision-making, loyalty to others, respect for all people in the group, and the ability to enjoy one's work.
- Leaders must provide motivational opportunities for each individual and must upgrade and maintain job content to create meaningful and challenging situations for employees so that everyone can realize their potential while being included and valued.
- Organizational structure helps define the chain of command, clarify the span of control of various managers and supervisors, and align areas of work and responsibility throughout the organization.

INTRODUCTION

The leadership skills required to effectively operate a fire department are very similar to the requirements for any organization. Fire departments offer their chief executive officers (CEOs) some unique challenges and the quality of basic leadership throughout a department has a significant impact on the performance of the people who make up the organization.

This chapter will explore leadership from various perspectives. These include the internal and external customer needs, a unique systems perspective, and the personal leadership necessary to sustain the leaders themselves and keep them positive, productive, and healthy. An executive level manager once commented in a meeting that he was so busy that he didn't have time to lead. What he failed to realize is that leadership manifests and presents itself in everything those in formal positions of leadership say and do. Good or bad, fire officers are always leading, they are always teaching, and others in the organization are always watching and learning from them.

Fire service leaders use a number of techniques to provide the vision, guidance, direction, and inspiration necessary to keep members of the organization aligned with the mission. There is no magic formula that works for every leader in every situation. Capability, flexibility, and adaptability are keys to a fire service leader's success. However, there are basic, timeless leadership behaviors that, if not practiced, will result in some degree of organizational and personal dysfunction. Fire officers cannot lead effectively without being cognizant of their own behavior, nor can they lead successfully in spite of the workforce. Effective leadership does not occur by chance.

Much of what makes leaders effective and separates exceptional leaders from the others is related to behaviors that can be practiced and improved. Exceptional leaders infuse positive leadership practices into the core of the organizations activities. A person's leadership approach is directly related to their philosophical beliefs about people in general and what motivates people towards individual and organizational goals. This chapter will explore leadership concepts and behaviors that are progressive and timeless to enhance the performance of fire service leaders today and tomorrow.

A Strong Leadership Foundation

The quality of leadership in an organization is key to the quality of supervision and management in that organization. It is impossible to completely separate the characteristics of the leadership, supervision, and management that exist. They are connected to each other in many ways. Together with the level of quality and commitment of the employees in general, they significantly regulate the effectiveness of people's performance. What follows are some basic leadership concepts that tend to set the stage for a leader's overall effectiveness.

Leadership by example

In 600 BC, Lao Tsu wrote ten concepts of leading people. In his first concept, Lao professes that one leads primarily by example. The concept of leading by example has been stressed in every class we have ever taken on effective leadership. The inability or unwillingness for leaders to translate this concept into action undermines a person's ability to lead others perhaps more than any other factor. "Do as I say, not as I do" did not get its start in the fire service, nor in any recent generation of leaders. The leader who lacks the ability to behave the way they expect others to behave is undermining their ability to lead and has a negative impact on other's performances.

Treating people with respect

Two key elements of any positive, constructive relationship are the levels of trust and respect that exists between the people involved in the relationship. These are critical to the quality of outcomes achieved as a result of the relationship, whether at work or in any other aspect of life. One significantly affects the other. People tend to be inspired to behave and perform in a particular way when they feel that there is a sense of mutual respect between them and the leader. Mutual respect has a direct impact on mutual trust. Simply stated, people don't trust leaders they don't respect, and they do not respect leaders that are disrespectful towards them.[1]

Personal leadership

The challenges a leader faces during a long career can take a toll on a person. It is difficult for a leader to help others be positive, productive, and healthy in their work if they cannot provide those things for themselves. Over time, people can become unhappy with work in general and with the organization as a whole. It is difficult to be around a worker with this approach because their negative outlook and actions can be contagious to others. It can be devastating to a work group when the person demonstrating this negative behavior is the formal leader of the group.

Leaders have to be able to effectively lead themselves if they expect to effectively lead others. This requires strict attention to some important things that will help them stay the course of excellent leadership:

- Pay attention to their level of physical, emotional, and psychological wellness.
- Practice ethical personal values and self-discipline. When in doubt, do the right thing.
- Maintain a plan for the future to help sustain confidence.
- Value the things in life that no amount of money can buy.
- Act as if they really believe that people are the most important resource in the organization.
- Stay connected to a mentor.

- Maintain open lines of communication with others.
- Stay competent.
- Recover and rebound from the bad things and avoid living in the failures of the past.

Personal leadership is key to a person's ability to effectively lead others. It is critical to realize and respect this fact. The way we feel about ourselves and the way we, as well as others, view our character is very important to effective leadership. William Hersey Davis once wrote:

> *The circumstances amid which you live*
> *Determine your reputation.*
> *The truth you believe determines your character.*
> *Reputation is what you are supposed to be,*
> *Character is what you are.*
> *Reputation is the photograph,*
> *Character is the face.*
> *Reputation comes over one from without,*
> *Character grows up from within.*
> *Reputation is what you have when you come to a new community,*
> *Character is what you have when you go away.*
> *Your reputation is made in a moment,*
> *Your character is built in a lifetime.*
> *Your reputation is learned in an hour,*
> *Your character does not come to light for a year.*
> *Reputation grows like a mushroom,*
> *Character lasts like eternity...*

Other traits and behaviors

There are a few other basic leadership traits and behaviors that form a strong leadership foundation. These include:

- Displaying a commitment and passion for the mission
- Valuing fairness as a consideration in decision-making
- Showing loyalty to others
- Respecting equality among the people in the group
- Displaying a sense of joy from work (in other words, enjoy your work)

The effectiveness of leaders directly relates to the presence of each of these basic elements that together form a strong leadership foundation. They are "the right of passage" to everything else about leadership that will be discussed in this chapter.

Ethics

Ethics are measured based on a person's words and behavior as they carry out their personal and organizational responsibilities and commitments. Living up to everyone's ethical expectations of a leader can be difficult because of all the variables and different perspectives from which ethical behavior is evaluated and measured. Nobody can meet everyone's expectations every day, but being mindful that people are watching their leaders, and making ethical assumptions based on what they see and hear, will help guide us.

Ethical responsibilities for fire service leaders, as well as all leaders in general, fall into five major areas:

1. The commitment we make to our external customers through the mission
2. The commitment, support, and treatment leaders display towards the internal customers (each other)
3. The leadership and followership behaviors acted out daily by leaders
4. The degree of professionalism and honesty that is evident in the leader's conduct
5. The commitment that leaders make to maintaining themselves as positive, productive, and healthy contributors to the organization

The discussion of concepts and guidelines that follows can help leaders be more successful in each general area of ethical behavior.

The primary reason fire department leaders exist is to guide the delivery of services. These services are delivered to our external customers and, to the best of one's ability, the leader supports those who are in place internally to deliver that service. The fire service exists to prevent harm, by building a safe environment, educating people to avoid a variety of dangerous behaviors or conditions, responding to emergencies, and operating a full range of support programs designed to ensure the competence of the overall system's performance. Remembering this mission-driven reality will help prevent leaders from getting distracted and falling short on this critical measurement of ethical behavior.

The level of respect people have for us is directly related to the level of respect we show them. The following list of behaviors will help leaders define key elements of respectful treatment of others. Ignoring the importance of these behaviors can damage relationships, reduce effectiveness, compromise the mission, and make the leader's ability to work with others difficult at best.

- Be considerate of other people's feelings and property. Be kind to each other.
- Practice a high degree of discretion in the things we say and do to (or in the company of) others.
- Accept the differences that diverse people bring to a group and respect the importance of diversity in organizational effectiveness.
- Emphasize the importance of unity as a measure of organizational strength.

People expect certain things from their leaders. We all know that leaders in one setting are almost always followers in another. The versatility required to lead or to follow is a key element in a person's success. The way a leader leads and the way a leader is able to be led, teaches others a lot about how they themselves are expected to act in the roles of leader and follower. Remember that as leaders, we are always teaching others through our own actions whether good or bad.

Honesty, integrity, loyalty to others, and a respect for the public trust bestowed upon members of the fire service will always be critical to our success as leaders. All the Boy Scout and Girl Scout behaviors we learned as kids will be important measures of our degree of ethics throughout our lives.

Leaders must also be fair to themselves and work to maintain their own sense of well-being. As leaders, it is difficult to get others to do what we are not willing to do ourselves. This applies to every behavioral expectation we have of members of the organization. To be viewed as ethical, leaders must model the behavior they expect of others. It relates directly to the basic need to lead by example. There is no escaping this requirement in the ethics equation.

It may be impossible for a leader to measure up to everyone's ethical standards every day, but people must be mindful of their own conduct as leaders. If trust and respect are to be earned and maintained, fire service leaders must pay attention to basic areas of ethical conduct.

Coaching, Directing, and Motivating

A key element of effective leadership is the ability to inspire people by creating an environment that others find motivating and stimulating. Abraham Maslow and Frederick Herzberg both created motivational theories that identify basic human needs in a hierarchical framework that propose to drive human behavior and motivation.

Maslow's hierarchy consists of five levels of motivational factors beginning with the need to satisfy lower level needs before one can focus on higher level needs. They are as follows:

Maslow's Hierarchy of Needs

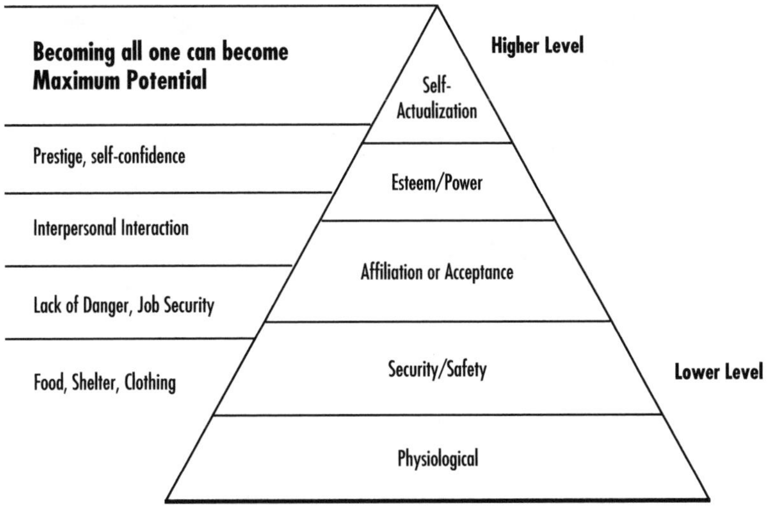

Higher Level

Becoming all one can become Maximum Potential	Self-Actualization
Prestige, self-confidence	Esteem/Power
Interpersonal Interaction	Affiliation or Acceptance
Lack of Danger, Job Security	Security/Safety
Food, Shelter, Clothing	Physiological

Lower Level

A satisfied need is no longer a motivator. Therefore, leaders must provide motivational opportunities for each individual based on their current position in the needs hierarchy.

Herzberg proposed a two-factor theory of maintenance needs and motivational needs. His maintenance needs correspond to Maslow's physiological, security, and social needs. Herzberg's motivational needs correspond to Maslow's identified needs for esteem and self-actualization. According to Herzberg, the maintenance factors of physiological, security, and affiliation needs must be fulfilled before a worker can even begin to be motivated. Herzberg further states that the true motivators are the need for esteem and self-actualization. This being true, leaders must give considerable attention to upgrading and maintaining job content and creating meaningful and challenging situations for employees.

There are many other motivational theories that leaders can research and study. These include but are not limited to:

- Robert Fulmer's Process Theory of Motivation
- Victor Vroom's Expectancy Theory
- Situational Leadership
- Fred Fiedler's Contingency Approach to Leadership
- Robert House's Path-Goal Approach to Leadership

One thing that these theories all have in common is the concept that people are individuals. Each person's particular needs and makeup, as well as the situation at hand, have significant impact on their level of motivation and what might be the appropriate leadership approach. Effective leaders must be people centered, mission focused, and flexible in their approach or they will not be successful.

In their book, *Global Leaders for the 21st Century,* Michael Marquardt and Nancy Berger predict that for companies to survive in the future, they must become learning and teaching organizations. Leaders must pass along their learning to others and they must also teach, coach, and mentor so that people are developed to apply what they learn to the business of the organization. The leader motivates, implores, inspires, and promotes each team member so that they perform with minimal supervision or interference. In doing so, they are also teaching leadership.

Steven Edwards, in his book *Fire Service Personnel Management,* said, "Creating an environment in which everyone can realize their potential while being included and valued is the ultimate goal." Considering the following factors will contribute to the efforts of the fire service leader in creating a positive environment:

Fairness: Ensure that access to training and career opportunities, promotions, and employment are based on job-related knowledge, skills, and abilities. The principles of equity should guide all actions.

Respect: Demand a work environment in which employees are treated with dignity and respect, free from mistreatment and harassment, such as unwelcome remarks, jokes, sexual innuendo, and the like.

Trust: Maintain a work environment where employees can raise issues and make suggestions without fear of reprisal, one in which they feel fully comfortable in contributing at all times.

Flexibility: Consider the changing and different needs of employees as balanced against the mission of the department. Be flexible where appropriate and it is warranted.

Sensitivity: Foster awareness of the differences in people and situations. Active efforts to ensure inclusion of people of all races, sex, and cultures are welcomed and appreciated.

Edwards believes that a key to fire service leadership is the understanding and utilization of five basic skills:

1. Empower others. As a fire officer you share power and information, solicit input, and reward people based on performance. You attempt to manage more as a colleague and encourage participation by maintaining accountability and providing a framework for success of those who work with you.
2. Develop others. A supervisor's main tasks in development include coaching, setting the example, mentoring, and providing opportunities for growth for all employees. A supervisor conscious of development will delegate freely and

counsel employees when necessary, as well as provide training and educational opportunities that are individualized if necessary.

3. Value diversity. As a manager, understand your own strengths, weaknesses, and biases. Diversity is seen as an asset in attempting to understand different cultures, people, and situations. You readily help others identify their needs and options.
4. Work for change. Through supporting employees by adapting policies, systems, and practices when practical to help meet their needs, a supervisor is a key player in identifying and influencing organizational change.
5. Communicate responsibility. Clear communication of expectations, asking questions for clarification, listening, and demonstrating empathy when appropriate help a supervisor develop clarity across cultures and languages and provide sensitivity toward individual differences.

Direction and empowerment

The art of providing adequate direction and coaching to an empowered work group is critical to the successful implementation of an enlightened approach to leadership. How do leaders provide appropriate direction to a workforce in a way that encourages innovation, empowerment, and change? How do leaders communicate the organization's expectations of each member and generally outline how they are expected to function as a member of the team? These are accomplished by providing organizational anchors that define expectations, provide clarity, and keep the group focused and moving forward together. The four organizational anchors are:

1. A clearly defined mission and clear customer expectations.
2. Shared organizational values and an understanding of the organizational culture.
3. Well managed financial resources.
4. Appropriate levels of training; organizational policies; realistic and complete standard operating procedures (SOPs); and definitive strategic and operational plans.

These anchors require constant management and attention to develop them and keep them current. It is this work (building and maintaining the anchors) that blends sound management practices and enlightened leadership behaviors.

Imagine the employees in an organization occupying the inside of a circle. They sometimes work as individuals; in organizational groups such as functions, divisions, or sections; in teams of two or more; or periodically on special project teams. Employees are configured in many different ways, and each employee comes to work each day with their own separate "to do" list. The reality is that employees are going many different directions, all at once, together and separately, everyday. The anchors keep the individual members from operating without focus and provide necessary direction and expectations from which leaders can coach. They define the outer perimeter of the circle. Rather than controlling people by keeping them in hierarchical boxes and lines, the leader's role is to keep people in the circle. There is a big difference and this difference substantially changes and enhances the leadership culture of an organization.

To ensure that employees function on the inside of the circle, it is important that leaders have the skills necessary to coach, counsel, and nudge people as needed on a daily basis. This is most effective when it comes from a member's first level supervisor. When second level supervisors (or greater) must get involved in employee behavior or performance issues, the situation is usually past the stage where counseling, coaching, or nudging a person back into the circle can be effective. In fact, when first level supervisors fail to help their employees stay inside the circle, the employees can drift so far outside acceptable norms that returning them to the organization may not be possible—and that's a shame. When an employee is operating outside the circle, they are operating without the organizational anchors that are in place to support and assist all members. This is not healthy for the organization as a whole, nor for the worker who is in this uncomfortable situation. Should they fail, they could personally have a very negative impact on the rest of the organization including the overall organizational image.

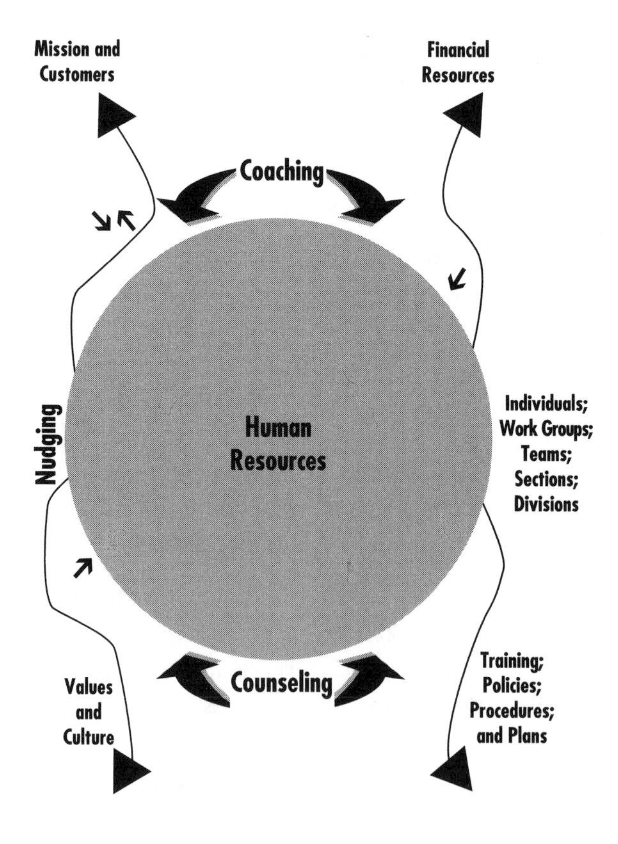

Leaders must be willing to help employees function within the direction and expectations established by the organizational anchors. This translates into helping them stay inside the circle. Leaders must be willing to coach, counsel, and nudge on a regular, daily basis to accomplish this. The organization as a whole depends on focus and direction being provided so that empowerment and innovation can thrive and serve the internal and external customers needs in an exceptional way.

Leadership Through Progressive Employee Relations

There is not a more important leadership measurement in the workplace than employee relations. The way that the workforce and management interact has a direct impact on the quality of the service or product, and thus, the organization's bottom line. This provides more than ample reason for leaders to pay a lot of attention to employee relations, whether you work in a fire department or other organization.

In many organizations, this worker/management relationship is less than positive and perhaps even ineffective. Consistent, positive employee relationships require significant commitment and the setting aside of traditional roles and approaches to dealing with each other. It also includes finding commonalities in agendas and a leadership commitment to work through issues rather than arrive at conflict-guided stalemates on a regular basis. This requires a relationship geared toward mutual trust, mutual respect, problem solving, and joint planning. This model requires strong leadership, hard work, and a willingness to take risks with each other.

Basic organizational structure

The structure of the organization sets the tone for the way the people inside the system (workers and managers) view their roles, the roles of others, the focus of their collective efforts, and how they each individually fit into the overall picture. We send a lot of messages through the way we organize our resources. In progressively led organizations, the way we organize ourselves structurally and the way we want the system to operate structurally become two different models.

Basic organizational structures are utilized to display our resources. They are typically organized in a hierarchical framework that includes:

- The CEO
- Functions that report to the CEO
- Major divisions that report to the CEO
- Sections that form divisions and report directly to division heads
- Work units that form sections and report directly to section managers

Sample Organizational Structure

CEO
FUNCTION
DIVISION
SECTION
WORK UNIT

This traditional organizational structure is important to the effectiveness of the people trying to manage, lead, and work inside the system as well as customers and others trying to access the organization from the outside. The organizational structure also does the following:

- Defines reporting relationships and the chain of command.
- Clarifies the span of control of various managers and supervisors.
- Identifies divisions and sections that may serve as sub-units (or accounts) in the overall budget. (In fact, the organizational structure usually accompanies the budget to display these financial and organizational relationships.)
- Aligns areas of work and responsibility throughout the organization so that it is clear who is responsible for what work.

The problem isn't when we organize ourselves in a standard hierarchical framework—the problems surface when we operate the organization in that way day after day. Some of the common disadvantages that surface by operating in a strict hierarchy include:

- Focusing on primarily top-down communications, which creates difficulty getting ideas and issues "up" the organization. It can be very frustrating to function at the "bottom" layers where the unheard majority of the workforce exists in most organizations.
- Creating rigid organizational lines within which territorialism prevails in many decisions.

- Lack of team orientation and a blurred organizational focus. This can result in excessive competition among members of the functions, divisions, and sections.
- An unspoken understanding that boxes on the organizational structure translate to status or stature rather than role or responsibility.
- An organizational climate could be created that leaves the impression that the resources of the organization are in place to serve the "top boxes" of the structure rather than to deliver service or products to the internal and external customers.

There are times when a standard, hierarchical structure must be used to direct work, especially in times of crisis. However, these times are the exception. The fact is, most people perform more effectively when the resources are formed into an operational model that is intended to guide and support the mission, thus focusing the leadership, management, and resources on the customers.

The operational structure displays how we should expect the various organizational components to operate on a day-to-day basis. The sample operational structure focuses the CEO, Functions, and Divisions directly on the services provided to customers and the effectiveness of the members who provide those services.

The circular structure shows the relationships and interdependence of the organizational elements to one another. It retains areas of responsibility for specific managers and workers; however, it communicates a supportive environment rather than one based upon hierarchy, authority, or control.

Sample Operational Structure

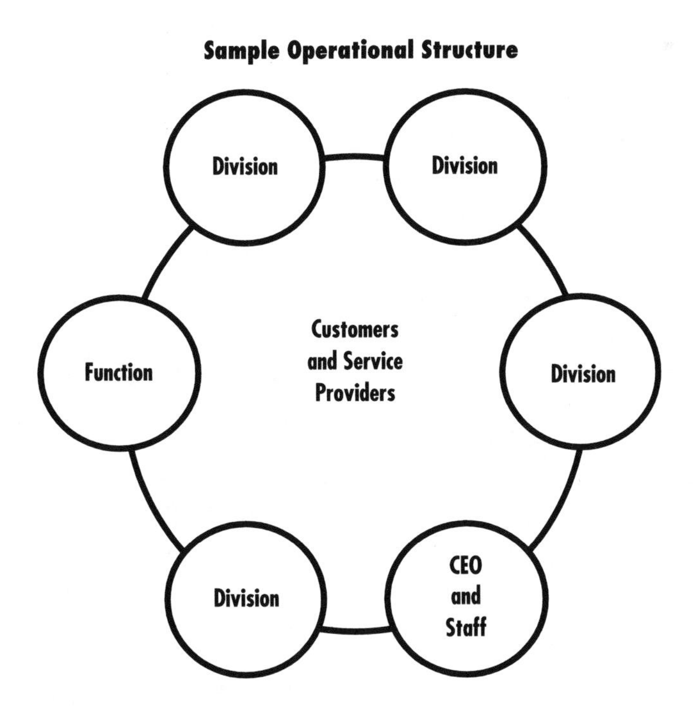

It's one thing to draw this operational structure and distribute it to the workforce, but it's another thing to lead that workforce towards bringing this model to life. Advantages of the operational structure include:

- Focusing on customer service and making the members as successful as possible in delivering that service
- Creating a sense of partnership among managers and workers throughout the organization
- Improving organizational communications by causing multi-directional routes of communicating—workers with customers and among each other; managers with workers; managers with each other; all focusing on improving external and internal customer service
- Easing the ability for members to get ideas and issues into the system for consideration and for the organization to process and implement change
- Enhancing the concept and the expectation that the workforce is empowered to make decisions within their specific organizational position and role

Instilling such an operational structure can be difficult. Doing so requires leaders and others in the organization to accept a different role than what might have been expected in the more traditional hierarchy of a standard organizational structure. Some members of the organization will adapt quickly and easily to these different expectations and roles, others will need more time, and some may not be able to make a complete transition and will require on-going coaching and more direct supervision.

As previously mentioned, the important thing is that the CEO clearly communicates this operational expectation and that leaders reward the workforce at every opportunity for acting within these redefined roles. There will be immediate and short-term benefits, but the long-term improvement in the collective performance of the system as a whole will be nothing short of spectacular. Implementing changes more readily, creating an atmosphere of organizational focus and individual empowerment, and opening lines of communications are critical benefits to the leader who masters this approach.

Organization and operational structures set the tone for how the people within the organization interact with customers and each other irrespective of stature, rank, or role within the system. Both structures are important to the success of the organization and complement each other very well. It's important that people possess a sense of ownership and responsibility for the area to which they are specifically assigned within the organization, but it is just as important that we not let those specific assignments isolate the various organizational areas from each other. The organizational structure has its place, but the operational structure provides important guidance as well. This concept can set the stage for, and be the cornerstone of, enlightened leadership and positive employee relations in an organization.

In addition to dealing with organizational design, there are specific steps that can be taken that will improve the level of employee relations in an organization. These can only be effective when the organization's leadership takes the initiative to bring them to life. They include:

- Develop descriptive literature that defines certain internal (cultural and philosophical) expectations of all members.
- Develop a plan for improvement for the internal environment in the organization.
- Ensure that the culture and values that guide your fire department are known to the membership.
- Minimize the difference in the leadership philosophy between good times (when things are going well) and bad times (when things are not going well).
- Develop a process to enhance the quality of employee relations and involvement in the department and include specific ongoing goals to do so.
- Encourage people throughout the organization to supply input into the system.
- Emphasize the importance of developing effective, positive, lasting relationships (of all types) in the organization.
- Identify and communicate the general expectations of supervisors and other leaders within the department. Do not allow supervisors to mistreat or be disrespectful to employees, nor employees to be disrespectful to supervisors or each other. It is not appropriate behavior as it interferes with the mission.
- Place emphasis on individual expectations and accountability in the department.

Empowerment must be accompanied by accountability for one's actions and decisions.

Employee relations is not a stand-alone program; it is connected to every managerial and leadership issue in the system. Employee relations must receive appropriate attention as a significant leadership issue in the organization.

A Systems Approach to Fire Department Leadership

There are many separate (yet interrelated) component parts that comprise a systems approach to guide leaders in providing effective fire and life safety services. Leaders today must pay attention to all components of the fire and life safety equation, avoiding the temptation to focus only on areas within which they have expertise or interest, or that are politically popular.

One of the arts of leading a fire department, or for that matter any organization, is coming to the realization that everything in the system is connected to everything else.

It is impossible to address or ignore any one aspect of the system without impacting one or more other component parts. This applies whether the work group is career, volunteer, or a combination of both. Exceptional fire service leaders know that they are guiding a system with the whole of the process being only as strong as the weakest part. Fire departments provide a variety of line services and support functions within the overall mission. Each element is important and requires attention when building an exceptional fire and life safety system and leading that system towards the future.

Everything we read or hear today about leading organizations tells us that to be effective, we must know what business we are in, then become the best we can be at that business. When asked about their experience receiving emergency services, many customers tell us that the day they had to call the fire department for an emergency response was either one of the worst days of their life or the worst day of their life. That is a unique relationship to have with one's customers. According to our customers, we (the entire fire service) are in the "worst day of their life" business. Fire service leaders who have come to accept this fact realize that they must build and lead each component of the fire and life safety system as if it were in place to address the worst day of someone's life because our customers believe that it is. Any weak component of the system can produce a negative outcome for customers on what they perceive to be the worst day of their life.

Emphasis today is placed on building sound infrastructures within our communities. It is understood that infrastructures addressing transportation, utilities, communications, parks, etc., are critical to a community's quality of life and overall viability. This same concept applies directly to the fire and life safety infrastructure of a community. Fire service leaders are responsible for building a community infrastructure that is directed toward preventing the worst day of someone's life from occurring in the first place and teaching people how to prevent it, or how they might survive it should it occur. However, fire emergencies will occur and an emergency response component designed to provide fast, skillful, and caring service, as well as support systems (internal and external) designed around maintaining quality service delivery programs must be in place. Fire service leaders must additionally develop positive relationships and thus enhance the human and physical resources that are required for smooth operations. Unless each of these system components receives adequate, ongoing attention from fire service and community leaders, the fire and life safety infrastructure of the community is compromised and may not be effective in addressing a particular customer's "worst day."

We can use a common three-legged stool to illustrate how the elements and components of this infrastructure work. To be useful at all, the stool must be made of a substantial material. Each part of the stool—the seat, legs, and braces—are totally dependent on each other for strength. If one of the parts is weak, the stool is unstable and cannot be used. It will not support the weight being applied to it, and will fail at some point. In an organizational application, the stool represents the most important resource in the organization—the people. The quality of the material from which the stool is made is reflected in these people's individual and collective levels of commitment, competence, ethics, diversity, and compassion for the external customers and for each other. The collective of each person's capabilities determine the basic strength of the stool. The stool must be strong enough to support the entire weight of the mission and this requires capable, well-led people. If not present, the system (stool) will fail at some point.

The three legs of the stool represent the primary line (external) service delivery programs of a fire department: fire prevention, public education, and emergency response. Each of these three primary line services to the effectiveness of the system (none of which are an option) is like one leg on the stool. Strong prevention through true consensus-based fire and building codes, including requirements for built-in protection and effective fire investigation is a requirement. An all-risk approach to public education must be in place. Finally, an emergency response component designed to deliver a full range of emergency services, including response to disasters and acts of terrorism, is an absolute necessity. Each of the three legs of the stool is equally important and represents a critical component of the fire and life safety infrastructure of a community.

The braces between the legs represent the staffing (internal) elements of the system. They are in place to support the effectiveness of the line services and the members who provide those services. The braces are critical to the strength, stability, and effectiveness of the system (the stool). Without the braces, the legs do not provide enough stability to support the full weight of the fire and life safety mission.

Fire prevention

Fire prevention services are line services provided to customers through the application of true consensus-based fire and building codes and standards that incorporate current performance requirements and practices. These form the foundation for maintaining a built environment that minimizes the negative effect of fires and other events that occur. Built-in protection, such as automatic fire sprinklers, is also a key to protecting life and property, as well as to maintaining the economic stability of our communities and our nation. The future must include a greater emphasis on the installation of fire sprinklers in homes. Fire service leaders must make this a priority in the future and set the leadership example by installing fire sprinklers in their own homes.

Public education

All-risk public education programs represent another critical line service provided to customers by fire departments. These include school-based programs, programs targeted at high risk groups within the population, and general education programs provided for the community in general. It also includes activities such as the NFPA sponsored Fire

Prevention Week (FPW), health fairs, and other similar events geared to deliver a full range of public education messages. When implementing an all-risk public education program in a school setting, the most comprehensive delivery tool that exists is the National Fire Protection Association's (NFPA) Risk Watch® Program. Risk Watch® is readily incorporated into school curriculum and has a very positive overall effect on teachers, the fire department, other partners, and people's behavior.

The Risk Watch® model is based on leadership that results in building community coalitions and does not rely solely on the fire department for its success. The fire department is a key partner and in many cases the lead agency for the program, but others with related missions should also be involved. Risk Watch® can serve as the centerpiece for a fire department's community involvement and community leadership efforts.

Public education programs designed for the general population and groups at highest risk are also critical program components. The NFPA's Center for High Risk Outreach is a resource to guide fire department leaders in these efforts.

Emergency response

Emergency response is also a critical line service delivery component of the system. When customers dial "9-1-1" they expect service that is fast, performed skillfully, and provided by firefighters who care about them, their loved ones, and their property.

Whether the emergency response workforce of the system is career, volunteer, combination, industrial, public, or private, accessing the resources must be quick and easy for the customer. Leaders must ensure an emergency response system that is properly located, staffed, trained, and equipped, so that it will not fail to perform at its utmost level of skill anytime an individual or the community as a whole experiences a fire or other emergency situation.

Support components

The support components of the system brace the legs of the stool and are critical to its overall stability. These typically make up the staff responsibilities of a fire department that include the following items.

Training and preparation. The level of training, the organization's ability to prepare members to perform their roles, is critical to meeting the daily requirements of the external and internal customers, and to a large extent, drives the organization's ability to change and develop. As leaders, we must ensure that this training and development includes all members of the department—those in prevention, public education, emergency response, and those who work in support of those who provide the line services.

Member and system support. Adequate support from the standpoint of basic human resource management is critical to "bracing" the legs of the stool and enhancing the stool's stability and the system's effectiveness. Fair compensation, an emphasis on safety, medical and chemical exposure management, an incident management system (IMS), standard operating procedures (SOPs), and incorporating humane supervisory, management, and leadership practices are some of the more common elements of this important brace, but there are many more.

System support includes a clear understanding of the organization's vision, mission, and values by all the members. It also includes short and long range planning, as well as the organization's ability to bring those plans to life by leading their implementation. These are critical leadership responsibilities that must be addressed if we expect our people to stay committed and to function in a positive, productive manner. This single component of the stool is a key leadership issue that impacts the strength and stability (the effectiveness) of the other legs and braces more than any other single factor.

Partnerships, relationships, and politics. It is difficult for organizations to achieve anything that is community-based without forming partnerships with others who share their vision or have similar missions. Community coalitions and partnerships that are formed for advocating or implementing specific programs can be used for many other purposes as well.

As leaders, the relationships we have internally and externally have a tremendous impact on the organization's overall performance. Personal and organizational relationships can and unavoidably do impact the mission. This impact must be considered in any leader's behavior.

Community involvement is critical to a fire department's community image and standing. Exploring mission-related ways to positively and actively participate with others has a significant impact on the effectiveness of a fire department leader.

If fire department leadership lacks the ability to be politically effective, the department has difficulty getting or keeping required resources. Additionally, leaders that continually struggle with politics usually struggle with other basic areas of management and leadership as well.

Infrastructure and equipment. This final brace of the stool addresses the infrastructure and equipment necessary to successfully deliver the mission. This is the hardware side of the system rather than the human side. It includes acquiring, building or maintaining a myriad of resources such as facilities, real estate, apparatus, equipment, tools, communications and dispatch systems, public education displays, safe houses, fire sprinkler mobile education units, commodities, supplies, technology, E-Business, etc. This brace of the stool includes the things the organization needs to effectively deliver the mission.

Conclusion

This chapter addresses fire department leadership from traditional, non-traditional, and progressive perspectives. A leader cannot achieve organizational or personal success in spite of the people they are in place to lead. There are basic leadership behaviors that are the "right of passage" to everything else about leadership. The need to inspire and coach others, create a positive, productive work environment with positive relationships, understand the entire system, and realize that the organization is successful only when each component of the system is effective are keys to exceptional fire service leadership.

The stool illustration demonstrates the overall investment a community must make to the fire and life safety mission of their fire department. Everything a fire department does fits into this systems model that the stool illustrates. Everything is connected to everything else; no element of the system functions totally independent of the other parts. Although it is not possible for leaders to give each leg and brace 100% of the attention needed at all times, it is important that each receives sufficient attention from leaders to ensure that the stability of the stool is not compromised to the point of structural (or system) failure. After all, in a given situation, who can predict which leg will save a life? Will lives be saved due to a specific fire or building code feature that was incorporated into a structure that subsequently prevented a fire from extending to an area of the building where lives would certainly have been lost? Will a life be saved because a 10-year-old child was taught a Risk Watch® class that resulted in her administering the Heimlich maneuver to her 7-year-old brother when he was choking? Will lives be saved due to physical rescues from burning buildings or skillful and timely extrications of people from car crashes that enabled them to quickly receive appropriate emergency medical treatment? The fact is that each of these line service delivery components (legs) save lives and the staff support areas (braces) ensure that members of organizations are properly trained, supported, positioned, connected, and equipped to do so.

As fire departments continue to expand prevention, public education, and emergency response services, our support programs must be developed as well. The fire and life safety mission of fire departments will continue to change as we move to the future. As fire department leaders, we should not reduce our emphasis on fire safety. We should instead increase the emphasis on the other life safety pieces of the all risk equation. We should not de-emphasize or reduce our firefighting capabilities. We should instead increase our effectiveness in the full range of emergency response services. It is not an either/or choice. It is a system of component parts that we, as fire service leaders, have been entrusted with building, supporting, and leading, now and into the future. Fire departments and other fire service organizations are in place to prevent, educate, or respond to the worst day of a customer's life, and must do so effectively every time. If the customers we serve were our own loved ones, the least we would expect or accept are only the most effective service levels.

Finally, the stool represents a system designed to help leaders build and maintain the fire and life safety infrastructure of a community. The stool is always under construction just as it should be. It's hard to be effective leaders if we are not clear about what we are suppose to be leading our people and other resources toward. As fire service leaders, the structural integrity of the stool is our primary responsibility. Remember that nothing in the organization stands alone because everything is connected to everything else. Realizing this and incorporating this concept into your leadership philosophy and approach will serve you well over the short and long term.

References

[1] *Compton, Dennis.When In Doubt, Lead. Fire Protection Publications, Oklahoma State University, OK, January 2002, Part 3.*

[2] *Ibid,* Part 2

[3] *The Fire Chief's Handbook, 5th Ed. PennWell Publishing Company, Tulsa, OK, 1995, pp. 215-280.*

[4] Marquardt, Michael J. and Berger, Nancy O. *Global Leaders of 21st Century, State University of New York Press, NY, September 2000, pp.* 20 & 21

[5] Edwards, Steven T. *Fire Service Personnel Management, Prentice Hall, Inc., NJ, April 2000, pp.* 30 & 31

[6] *When In Doubt, Lead,* Op. Cit., Part 3.

[7] *When In Doubt, Lead, Op. Cit.,* Part 1, Introduction.

[8] *When In Doubt, Lead, Op. Cit.,* Part 3, Chapter 4.

[9] *When In Doubt, Lead, Op. Cit.,* Part 3

Section II

HUMAN RESOURCES

7

PERSONNEL ADMINISTRATION

Kathy J. Saunders

CHAPTER HIGHLIGHTS

- **Briefly outlines a few of the items in the effective personnel manager's toolbox.**
- **Identifies some of the educational resources available to the fire officer to learn how to effectively manage and lead personnel.**
- **Reviews the topics pertinent for efficiently and legally administering a fire department's personnel.**

PERSONNEL

In most fire departments, the amount of time normally spent on the fireground or even on emergency medical responses is miniscule compared to the time spent in the firehouse, at training, on inspections, and on all the other myriad tasks during a firefighter's average shift. The type of leadership skills needed for incident command is vastly different than the leadership skills necessary to deal with personnel during the rest of the day. All too often mid-level fire officers are selected and promoted based on their ability at emergency incidents. As individuals move up the promotional ladder, they are expected to learn how to manage and lead firefighters through trial and error on the job. The ability to fight fire and to manage an emergency incident through strong working knowledge of the incident command system does not teach new fire officers at any level how to lead and manage people, except on the fireground. In order to be an effective leader in a fire department, the knowledge and skills necessary to manage personnel when NOT at emergency incidents must be learned. If the fire department does not provide this education, it is imperative that fire officers take the initiative to attain this information on their own.

The Toolbox

A fire officer needs to compile a full array of relevant knowledge, skills and abilities in the area of personnel management. Although some of the knowledge might be considered theoretical, it has direct applications in leading and administering a fire department. This topic is no doubt covered in more depth in other chapters, but a brief example here is beneficial in order to stress the importance of this facet of personnel administration.

One example of a basic management tool is behavioral science theories. In this area, the works of Abraham Maslow and Douglas McGregor are central. Maslow developed a "hierarchy of needs" both to describe the issues important to employees at any given time, and to predict successful motivational factors. He categorized needs from basic to complex, with the theory that basic needs must be met before the higher needs become important. Maslow's hierarchy of needs is listed as follows:

1. Physiological needs (food, shelter)
2. Safety needs
3. Social needs (friends, positive interpersonal interaction)
4. Ego needs (recognition, self-esteem)
5. Self-actualization needs (inner growth, creativity)

Douglas McGregor developed the X and Y theories of motivation. Theory X assumes people do not want to work. To motivate employees, the threat of discipline must be constantly applied, managers must closely supervise personnel, and each task must be rigidly defined. Theory Y assumes that employees are willing to work and will be loyal and dependable if they are given sufficient responsibility to complete their tasks, their accomplishments are acknowledged, and they believe they are contributing to the good of the organization.

The work of Maslow and McGregor are considered rudimentary today, and have been significantly refined and modified. For example, a "Theory Z" was developed a number of years ago that goes beyond employee participation to employee-driven contribution, decision-making and control. Out of this arose the total quality management (TQM) and kaizen (constant improvement) movements.

The evolution of management theories, from the traditional approach, characterized by McGregor's Theory X, into more participative styles, which encompass Maslow's hierarchy and McGregor's Theory Y, and further into empowering approaches, such as TQM and kaizen, give fire officers an important base of knowledge from which to work. Although the study of management theories has largely occurred in the private sector, adaptation to the public sector is easily accomplished.

Other examples of organizational principles about which fire officers should be knowledgeable include change management (also known as organizational development); the elements of fire department culture; development of organizational plans which includes strategic planning; and the types of organizational power, to name a few.

Another important item that should be present in every personnel manager's toolbox is a thorough understanding of workforce demographics. This includes factors such as the number of women, age distribution, ethnic and racial makeup, percentage of employees with generational roots in the area versus recent transplants from other parts of the country, and economic status of the firefighters as it relates to the economic strata of the community.

To use age demographics as an example, with the advent of the term "baby boomer" a whole new vocabulary was created, concurrent with an expanded area of study in employment management circles. "Baby boomer" was originally used simply to describe a large, age-based demographic group born between 1946 and 1962. However, the study of age demographics has expanded to include identifying the society in which each significant age group matured and as a result, identifying the common characteristics of the group.

The generations currently represented in the workforce today start with the "Silent Generation." Members of this group were born during the Great Depression and entered the workforce in the 1950s and '60s. Economic growth meant job security and steady economic advancement. Young people starting out with a company expected to retire from the same company. "The Greatest Generation" by Tom Brokaw provides an interesting look at this group.

The "Baby Boomers" are next. They are still the largest demographic age group and make up the majority of fire service employees. They experienced the social and cultural shifts of the 1960s and '70s. They entered the workforce during a period of economic growth and were central to the "me decade" of the 1980s. However, the economic downturn in the late '80s and into the '90s caused many of the boomers to lose their jobs and their worldview reflects that.

"Generation X" (born between 1963 and 1980) was so named because they feel as though they are a lost or forgotten generation. They are also known as the "Baby Busters" because they are the smallest demographic age group in the workforce, which also contributes to the feeling of being forgotten. They are cynical about the future and bitter towards the Boomers. They believe their generation will suffer the consequences caused by Boomer overindulgence, both economically and environmentally. The information revolution shaped this group, for which the term latchkey kid was created.

Generation Y members were born since 1981 and are the biggest demographic bulge since the Boomers. After the latchkey experience of the "X"-ers, "Y"-ers grew up in the "Decade of the Child" as humanistic theories of child psychology permeated counseling, education and parenting. Fathers become more involved in child rearing and businesses started recognizing family needs. This is the first generation completely immersed in the information age, which tends to encourage a global perspective as well as an emphasis on education.

The study of age demographics is critical to fire officers for a variety of reasons. As mentioned earlier, motivational factors are often driven by the employees' world view. Fire officers must understand what motivates others and just as importantly, they must know what motivates themselves. As resources for technological advances are allocated and programs are implemented, it is important to understand the comfort levels and the knowledge base of different age groups. In the area of training, learning styles and techniques are directly affected by the age and world view of the employees. Finally, as the "X"-ers and "Y"-ers enter the workforce and move up the promotional ladder, the successful fire officer must know how to create effective teams with members of all generational groups, whether the team is a firefighting crew or the executive staff of the fire department.

There are many other examples of tools the effective personnel manager must have. This was intended only to highlight briefly the importance of attaining the knowledge, skills and abilities necessary. The next step is to identify the resources available for the fire officer to get the necessary knowledge, skills and abilities.

Resources

The easiest self-help resources to obtain are books. Many excellent books are available that include information on fire personnel management and administration, one of which you are holding in your hands. *Managing Fire Services,* edited by Ronny Coleman and John A. Granito (affectionately called the green book), is one of the seminal works in the area. Other helpful books are *Recreating the Fire Service* by William J. Hewitt, and *Fire Services Today,* edited by Gerard J. Hoetmer. The quest to learn should not be limited only to books pertaining to the fire service. Myriad books have been written on the general topic of management and administration. The trouble may well be to narrow the number down to a reasonable amount of reading.

Another self-help resource is the Internet. The information explosion has not passed the fire service by. Trade journals such as *Fire Engineering* and *Fire Chief,* which are online, often include excellent fire management articles. Many journal sites also include lists of links to other fire-related sites. Fire-related websites can often be added by anyone hitting the page, so the quality of linked sites range through the entire spectrum. Websites such as for the National Fire Academy and the National Fire Protection Association (NFPA) not only provide updated information in their areas of interest, but also make on-line learning opportunities available. Not only is the Internet a vital source for information and education, it is absolutely critical for anyone serious about furthering their career to be comfortable with computers, and the research and informational resources of the Internet.

At the local level, the obvious place to start is with the local fire departments. Some departments offer excellent management and leadership classes in house. Members from surrounding departments may be allowed to sit in on the classes on a limited basis. Larger fire departments may have a community access cable channel dedicated to fire training. Some may even offer leadership academies, and may allow certain personnel from outside the department to participate.

Another good local resource is community colleges and vocational schools, many of which have fire science programs. Not many universities confer four-year degrees in fire administration or a related area. The Big Two are Oklahoma State University and the University of Maryland. Several other universities have developed respectable four-year degrees including Eastern Kentucky University. Making the commitment to attain an associate's degree or the longer commitment required of a four-year degree is not necessary to take advantage of the course offerings at a local university or community college. Again, as with books, any business administration program—whether in a university or a community college—will probably offer courses in organizational behavior, resource management, change management, etc.

State and regional fire schools often offer several courses beyond the hands-on, technically oriented ones. It is also advisable to learn where the major fire service organizations hold their annual conferences. For example, the Fire Department Instructors Conference (FDIC) is held annually in Indianapolis, Indiana. Anyone living within a reasonable distance of Indianapolis would be well advised to attend FDIC when possible for classes on a full range of fire service related topics. The location of Fire-Rescue International changes each year, giving fire officers across the nation who are interested in furthering their knowledge, access to some excellent opportunities. Other conferences providing access to personnel management and administration classes include Firehouse Expo, FDIC West, and the NFPA conferences—to name a few.

Of course, one of the best national resources available at an unbelievably good price is the National Fire Academy (NFA), located in Emmitsburg, Maryland. As a result of the historic document, *America Burning: The Report of the National Commission on Fire Prevention and Control,* Congress passed a law creating the U.S. Fire Administration (USFA) and its delivery vehicle, the NFA. One of the directives of *America Burning* was the need for comprehensive training for fire personnel, especially in the areas of incident and department management.

The Federal Emergency Management Agency (FEMA) purchased St. Joseph's College in 1979 for use as a training facility and renamed it the National Emergency Training Center (NETC). The 100+ acre campus currently houses the USFA that includes the NFA, the Emergency Management Institute (EMI), and the National Fallen Firefighter's Memorial.

In Emmitsburg, the NFA offers courses on a wide variety of topics including both specialized training courses and advanced management programs in a residential setting that encourages intensive learning. The resident curriculum is intended to target, among others, middle- and top-level fire officers. The crown jewel of the NFA's management courses is the Executive Fire Officer Program (EFOP). As stated on the NFA's web page, the purpose of EFOP is "to provide senior officers and others in key leadership positions with:

- An understanding of:
  - the need to transform fire and emergency service organizations from being reactive to proactive; with an emphasis on leadership development, prevention, and risk reduction.
  - transforming fire and emergency service organizations to reflect the diversity of America's communities; the value of research and its application to the profession.
  - the value of lifelong learning.
- Enhanced executive-level knowledge, skills and abilities necessary to lead these transformations, conduct research, and engage in lifelong learning."

EFOP consists of four graduate or upper-division-baccalaureate level courses, to be attended over four fiscal years. An applied research project (ARP) must be submitted within six months after completing each course before attending the next course. Only chief officers of local or state fire service agencies who have attained an associate's degree or higher from an accredited institution of higher learning are eligible for EFOP.

The NFA has also developed the Volunteer Incentive Program (VIP), which is specifically designed for volunteer emergency service providers. The VIP condenses a variety of two-week resident courses down into intensive six-day courses and specially tailors them to the volunteer experience. Included in the VIP are courses designed to enhance the volunteer fire officer's ability to manage and lead a fire service organization. For example, courses entitled Leadership and Administration and Fire Service Planning Concepts for the 21st Century are currently being offered during the 2001–2002 academic year.

As mentioned earlier, the costs for attending resident courses are minimal. The Academy does not charge tuition costs; and all instruction and course materials are free. "Dorm" rooms that often look more like hotel rooms are provided at no charge. The USFA reimburses all eligible students for the cost of an economy-class nonrefundable airline ticket, for one course per fiscal year. Mileage up to a certain amount will also be reimbursed. The only cost to attend the Academy is a meal ticket that provides three meals a day in the campus cafeteria. At the time of this writing, the total cost for attending a two-week resident course in Emmitsburg is less than two hundred dollars.

For those who cannot attend on-campus courses, the NFA also offers distance delivery training. Direct delivery two-day courses are available due to the partnership of the NFA with state and local fire training programs. Courses are selected and co-sponsored by the NFA and the state or local training program.

Regional delivery courses are the result of the demand for resident courses. Regional delivery courses are the same courses offered at the Emmitsburg campus. Minor modifications may be required but course content, training materials, and instructor ratios are exactly the same as the resident courses. The Training Resources and Data Exchange

(TRADE) network coordinates and implements the regional delivery program. TRADE is composed of state and local training and education administrators who are in a good position to know the training needs of their region. Students are responsible for all costs associated with attending a regional delivery course; however the NFA does provide a small stipend to eligible students to help offset the travel and lodging costs.

The NFA only recently started offering self-study courses on-line through FEMA's Virtual Campus. Several courses are offered at any given time, and the offerings change over time. Currently a 13-hour course called Fire Service Supervision is being offered, covering topics such as stress management, time management, interpersonal communications, motivation, counseling, conflict resolution, and group dynamics.

To sum up, resources abound for the fire officer looking for good educational tools in the area of personnel administration. All it takes is a little perseverance and patience to find the information.

Administration

Recruitment

Aggressive recruitment through a wide variety of sources is a must. The most effective traditional recruitment sources for the fire service are newspapers, employee referrals, colleges and universities, trade journals and magazines, and fire service professional societies. However, the need to diversify the work force has resulted in the creation of innovative recruitment strategies, since the above referenced sources have not been very successful in attracting minority and female candidates.

Minority leaders connected with churches and community centers should be contacted to assist with minority recruitment. Local role models can help create public service announcements for local community access channels. Mentoring programs can be established, so that minorities and women interested in the profession have an avenue to meet and learn from other minorities and women. The physical agility test can be offered to interested candidates prior to the actual date. For women, the physical requirements of the job are often an issue. Mini-academies can be conducted to help educate interested women in the specifics of firefighting.

However, to be successful at broadening the outreach of recruiting, a commitment must be made to conduct educational activities year around, not just when open positions are about to be posted. The underlying problem may be based on the fact that most minorities and women never had the dream of being a firefighter when they were a kid. So as adults, it is not a profession in which they naturally envision themselves. Developing that vision often takes personal, one-on-one contact. And that takes time.

Engine companies can be sent to high school sporting events in diverse and minority neighborhoods, as well as to high school women's sporting events in every neighborhood. In cooperation with religious leaders, church events could be attended regularly. A fire engine should regularly be seen in front of the Boys & Girls Club, and the YMCA. The purpose of all these visits is two-fold. Not only are these a natural for public education, but they are great opportunities to talk to people about the job of the firefighter. Which means engine companies cannot be assigned to this kind of event haphazardly or without training.

Fire departments historically do a great job of training firefighters in the technical aspects of the job, but do not rate well in the "soft" parts of the job. Very few fire departments are recruiting from among their own ranks and then training the volunteers to be good ambassadors for their department. But that is exactly what is needed. Rank and file firefighters need training so that they understand the absolutely crucial role of ambassador every one of them plays when they climb into that uniform every third day. Based on that minimal training, those interested in pursuing it can be further trained to attend high school basketball games and church ice cream socials.

Developing a diverse work force means having a realistic long-term plan. With each recruit class, the lessons learned for what was effective and what was not needs to become incorporated into the plan, keeping in mind that in the recruitment and selection process, a community should not vary its standards. The key is to have all candidates "playing the same game on the same playing field," just with a better cross-section of candidates.

Selection and testing. Few things are as critical to an organization as the selection of new employees. Many employers rely solely on interviews and resumes, but these subjective criteria can lead to unpleasant surprises. As a result, many employers have turned to tests for objective, cost-effective ways of identifying the best candidates. Properly validated tests are a useful supplement to other selection devises such as interviews and reference checks. It is imperative that fire departments have a comprehensive job description that is issued to all applicants when they begin the job application process. At a minimum, the job description should include:

- Starting salary
- General statement of duties
- Supervision received
- Minimum qualifications
- Character requirements
- Physical requirements

Additional information should be provided to all applicants:

- Last date for filing the application
- The selection process
- The purpose of each step in the selection process

- Dates and times of all tests
- Who will be conducting all tests
- How tie scores will be resolved
- Whether tests are numerically scored or of a pass/fail variety

Since passage of the Americans with Disabilities Act (ADA) in 1990, fire departments have struggled with physical agility tests. Although "physical agility" tests are specifically allowed by the ADA, "physical examinations" are not allowed until after a conditional offer of employment is extended. Partly as a response to ADA requirements, and partly because the time was ripe to address the issue of firefighter wellness and fitness in a comprehensive, valid, fair manner, the International Association of Fire Chiefs (IAFC) and the International Association of Fire Fighters (IAFF) teamed up with ten pairs of municipalities and their local unions to form the Fire Service Joint Labor Management Wellness-Fitness Task Force. The Task Force has dedicated itself to developing a total approach to wellness and fitness in the fire service.

The Task Force has developed several different programs; the one of particular interest here is the Candidate Physical Ability Test (CPAT). The CPAT was developed to be a fair and valid evaluation tool to help in selecting firefighters and to ensure that all candidates have the physical ability to complete critical tasks safely and effectively. The CPAT program covers every aspect of administering the CPAT, including recruiting and mentoring programs, providing recruits with fitness guidance to help prepare them for the CPAT and setting up and administering the test. For more information or to purchase the CPAT, contact the IAFC or IAFF.

Needs assessment. There are many kinds of tests, each of which provides different information. Before using tests, first determine what the purpose of the test is. This will determine the appropriate tests. Some are designed for employers to administer and score themselves. Other require a skilled professional to ensure proper use. It is important to know how the publisher expects the test to be used. Otherwise it may not be valid.

- *Intelligence* testing is really a test of learning ability, particularly the ability to learn through the use of printed material.
- *Aptitude* testing attempts to determine what an applicant may be able to do. It may evaluate physical dexterity or test for mechanical, clerical or similar forms of comprehension.
- *Achievement* testing determines the degree of knowledge in specific fields such as electrical or mechanical theory, bookkeeping expertise, and other skill areas.
- *Personality* tests attempt to assess applicants in terms of "traits," such as introversion versus extroversion, drive, decision-making style, and temperament. They can be useful in predicting whether an applicant has the traits that are usually associated with success in particular jobs, such as sales or supervision.

- *Honesty* tests, sometimes called integrity tests, are designed to determine the integrity of people who take them by measuring attitudes toward dishonesty and propensity for theft-type behavior. Some employers say they can reduce theft and related problem behaviors by using the results of honesty tests.

EEOC guidelines for testing. Under the Civil Rights Act of 1964, employers may use "any professionally developed ability test," provided that the results are not used as a basis for discrimination. Tests that have an adverse impact on minorities and are not justified by business necessity are discriminatory, according to Equal Employment Opportunity Commission (EEOC) guidelines that have been upheld by the U.S. Supreme Court.

The federal Uniform Guidelines on Employee Selection Procedures have been adopted by the U.S. Departments of Labor and Justice, the EEOC, and the Office of Personnel Management (civil service). They apply to all employers covered by Title VII of the Civil Rights Act. Originally developed as technical standards for validating formal tests, the guidelines have gradually been expanded to cover all aspects of employee selection. This includes application blanks, oral interviews, experience and skill requirements, plus any other factor used by employers in selecting applicants. The main principles underlying the guidelines are described here.

Adverse impact. Employer policies and practices that have an unfavorable impact on any race, gender, or ethnic group are illegal unless justified by business necessity. "Business necessity" means that there must be a clear relationship between what is evaluated by the selection procedure and the performance required by the job.

Do not start a testing program without professional advice. Procure tests from established sources well versed in the management process. Keep a complete record of all tests given both to employees and rejected applicants.

Probation

The candidate, once hired, is generally on probation. This probation period is usually one year and is often covered by state law or the labor contract. The purpose of having a special review program for newly hired employees is to enable supervisors to make corrections in work habits and to reach a decision about whether or not to keep the employee. It is unlikely that a new employee who avoids work, arrives late, or just does not seem to care will somehow get better over time.

The purpose of the probationary period is to provide guidance to new employees and to have company officers work with them to make them into efficient firefighters. For a probation program to be effective the worker's immediate supervisor should be required to approve, in writing, the retention or termination of the employee before the probationary period ends. If the employee is to be retained, policy also may provide for an increase in wages based on satisfactory completion of the probationary period. An ongoing review process should be used periodically during the probationary period and should include periodic appraisals and counseling.

If possible, do not refer to employees who complete probation as permanent employees. That expression implies that the worker has a right to lifetime employment with the organization, and it can be used in a wrongful discharge case.

Most departments use the probationary period as a tool for introducing the employee to the organization and for getting the new worker's full attention turned to performing the duties of the job well. This gives the firefighters a greater sense of belonging to the organization.

One important point to remember is that all new employees should go through an initial review period and be evaluated to determine if they should be retained, whether or not they serve out a formal probationary period. Another is that generally any form of employee review after the probationary period passes is considered a condition of employment and, as such, may have to be negotiated if the department falls under the purview of a labor instrument.

Promotions

Promotions within a department are one of the most important functions the fire chief must administer. The officers in the department are the leaders, the decision makers, the ones who will shape the culture of the department for years to come. The selection process can also raise morale, or if there are no promotions, or if they are tainted, they can cause morale to plummet. A fair, consistent, valid promotional process is crucial to keeping a department operating smoothly.

Departments conduct promotional exams in a variety of ways. They may be conducted within the fire department by department personnel or by the municipal human resources department. The general scope of advancement is usually from firefighter to company officer, lieutenant, or captain, then on to the various grades of chief officer. In most cases, a certain time in grade must be served before an individual can advance to a higher position or rank.

Once a need for a promotional exam arises within a department, a notice is promulgated informing those eligible of the pending examination. This should be done well in advance of the anticipated vacancy, to give those eligible as much time as possible to prepare for the exam. At this time, the department should provide a list of study materials from which the examination will be drawn.

A job description and an appropriate application form should be provided before a predetermined deadline. Exam procedures can include a written test, oral test, assessment center, hands-on evolution, or any combination thereof. In any event, the various components of the exam generally are weighted. For example, the oral test is 50% and the written test is 50%. In most cases, the written portion of the exam is given first and the candidate must receive a passing grade on it before proceeding to the next phase.

If an outside contractor is used, the conductor of the written portion of the exam should ensure that the exam is validated, that is, that the exam will stand up in a court of law and that the firm will support the exam in any litigation process. Oral boards are usually composed of officers of ranks higher than the position being tested, and often are chosen from other departments. They frequently are "drafted" from professional organizations within the fire service.

Many departments now are using the assessment center as part of the examination process, sometimes in conjunction with a written exam. The rationale is that this type of exam is more thorough and demonstrates candidates' ability to "think on their feet" more than the traditional oral examination. The assessment center may consist of demonstrating the ability to establish priorities via an "in-basket" process, providing concise written reports, and participating in real-world situations via the role-playing process.

It is imperative that when a department conducts this type of exam, it selects an experienced firm or in-house group to conduct the process. The examination administrator is responsible for briefing the candidates; procuring, training and briefing the assessors; becoming familiar with the job description and various standard operating procedures (SOPs); and for being an efficient facilitator of the entire process.

Once the candidates are ranked (again, the selection process varies according to state law, the labor contract, and the promotional policies of the department), the administrator charged with selecting the candidates (it is hoped this will be the fire chief) may, by regulation, have to select the top candidate, one of the top three candidates, or use some other selection process. In addition, there may be an interview process that involves the department's top administration, depending on the size of the department and the promotional rank involved.

One example of how a promotional examination process might be conducted is described here. Once the need arises for a promotional examination, a copy of the job description is sent to the examiner and a bibliography is promulgated. The exams are sent to the moderator who brings the unopened box of exams to the exam site and opens them in front of the candidates. The exam booklet number, along with the candidate's name, is given to the moderator. The moderator conducts the exam. Upon completion of the exam, a copy is made of the exam answer sheet and is given to the candidate, thereby eliminating the possibility of any exam tampering. The moderator then mails the exams back to the examiner for correction. When the exams are corrected, the candidate's test results and his/her number are sent to the moderator and to the department. The department then must meet with the moderator to match the score and numbers to a name. Seventy percent is a passing grade and accounts for 50% of the total score. Only those candidates who pass the written portion of the exam are eligible to move on to the oral phase. If the second portion of the exam is strictly an oral exam, a three-member oral panel is selected from other departments. No member of this panel should be from a neighboring department.

If an assessment center type exam is conducted, it must be used consistently to avoid having one captain selected by an assessment center and another later candidate selected by a traditional oral panel. In both types of exam, a moderator should be present to resolve any problems that may arise. After the oral portion, the assessors turn their scores in to the moderator and the written and oral marks are added to establish the final score. Candidates are notified by mail of their test results, the list is approved by the authority having jurisdiction, and the appointment is made. The appointment should be made within a few weeks after the oral portion to avoid any demoralizing time lapses.

The examination process may be negotiable. If so, union and management must agree on the entire process. This includes the number and types of examinations to be conducted, who will conduct the exams, whether interviews will be conducted and if so by whom, the weight of each step in the process, and once the candidates are ranked, the discretion afforded the fire chief in the final selection.

Affirmative action. Affirmative action means taking positive steps to recruit, hire, train, and promote individuals from groups that have been subject to discrimination based on race, gender, and other characteristics. Affirmative action goes beyond "equal employment opportunity," which requires employers to eliminate discriminatory conditions and to treat all employees equally in the workplace.

Affirmative action requirements may be imposed by state or federal regulations for government contractors, subcontractors, and loan recipients as part of conciliation agreements by state and federal agencies, by court order and under employers' voluntary affirmative action programs (AAPs). Basically such programs are required to provide the following:

- An analysis of all major job categories together with statistical information on the minority and female population of the surrounding labor area for comparison with the total work force in the facility.
- Goals, timetables and commitments designed to correct identifiable deficiencies.
- Support data for the above analysis compiled and kept up to date, together with seniority rosters and applicant flow data.
- Special attention to the upper categories, from craftsworkers to officials and managers, as defined in the EEO-1. Analyses must be made for all job categories for which there are available statistics, and employers must declare any underuse, even for a figure as low as one percent. In some cases, separate AAPs also are required for disabled individuals and veterans.

Reverse discrimination. On one hand each employer is required to hire, promote, develop, and train its work force in a nondiscriminatory fashion. Yet if imbalances develop in the representation of minorities, then cities that are required to have affirmative action plans must try to eliminate the under-representation. Similarly, cities that do not fall under affirmative action requirements may want to correct the under-representation so as to have a work force more representative of the population. Be advised that denying job opportunities to any class of worker, including majority workers, simply because the management of an organization "thinks" it should hire or train more minorities is an invitation to be sued for reverse discrimination.

Nevertheless, in more than one case an employer has survived challenges to voluntary AAPs when its plan included four basic elements:

1. The employer made a detailed and thoroughly documented self-analysis to determine whether minorities were under-represented.
2. Finding that there were fewer minority members in the work force than would normally be expected, the employer took reasonable action, relative to the degree of under-representation, to correct it. This included establishing specific but reasonable written goals and time tables for correcting the under-representation. In one case, the employer set a 50% hiring ratio and implemented it by selecting applicants alternately from two applicant lists: one for minorities and one for non-minorities; this way, no group was totally excluded from opportunities.
3. The affirmative action plan was for a limited period of time (for example, 12 months) and was suspended when goals were met.
4. It did not result in displacement of non-minority employees, nor did it unduly restrict opportunities for non-minorities.

It is clear, based on Supreme Court decisions, that the courts are going to uphold any reasonable AAP, at least for the time being, based on widespread support for affirmative action. Keep the following caveats in mind:

- Organizations should not enter into voluntary AAPs unless there is a real and verifiable imbalance in the employer's work force compared to the percentage of women and minorities in the labor market.
- No AAP should be used to completely deny opportunity to a class of worker; rather, it is expected that goals will be flexible, not rigid quotas, and that the AAP will move slowly but deliberately toward achieving its objectives.
- Factors such as gender and race may be considered in making hiring and promotion decisions, for example, to place more women in positions where they are under-represented, providing that those are not the only criteria used. This could mean, for example, that an organization can select a female or minority worker based on a lower but passing test score within a pool of qualified workers, without fear of losing a reverse discrimination suit if the selection meets 1 and 2 above.

It should be noted that the affirmative action policies outlined above should be administered by the town or city of which the fire department is an integral part.

Equal employment opportunity

Equal employment opportunity gives employees rights under a number of laws:

- The Civil Rights Act of 1964 (Title II) prohibits discrimination in employment because of "race, color, religion, sex or national origin." It covers all employers with 15 or more employees, including state and local governments. Title VII is administered by the federal Equal Employment Opportunity Commission (EEOC). It can investigate and act in cooperation with state and other federal agencies and can bring suit for enforcement of its decisions. Where there is an approved state or local agency, complaints must be filed with that agency, which then has 60 days before the EEOC assumes jurisdiction.
- The Age Discrimination in Employment Act (ADEA) prohibits discrimination against persons between the ages of 40 and 70. It covers all private employers of 20 or more employees, members of labor unions with 25 or more members, local and state governments, and employment agencies that serve covered organizations. The ADEA is administered by the EEOC.
- The Equal Pay Act of 1963 is an amendment to the Fair Labor Standards Act (FLSA) and therefore covers all employees engaged in interstate commerce. State and local government employees (except elected and policy-making officials) also are covered. This law forbids employers from discriminating by paying lower wage rates or providing a diminished benefit package to employees of one sex versus the other when the work they do is equal and is performed under similar working conditions. The EEOC is responsible for enforcement and can conduct investigations and bring suit for enforcement of its decisions. The Act is enforced by the federal government without the assistance of state and local agencies.
- The Rehabilitation Act of 1973 deals with government contractors and employees who receive federal assistance.
- The Americans with Disabilities Act of 1990 (ADA) affects all employers covered by the Civil Rights Act. It applies to employers of 15 or more employees as of July 26, 1994. This Act enacted sweeping changes in many facets of public life, from employment to facility accessibility. The major provisions of this act, as it applies to personnel, are discussed below. Although there are several acts that deal with the protection of persons with disabilities, following the guidelines for the ADA meets most of the requirements for the other acts as well.

Definitions

Disabled: A disabled person is one who:

- has a physical or mental impairment that substantially limits one or more major life activities
- has a record of such impairment (*e.g.*, a former drug addict or a recovered heart attack victim)
- is regarded as having such an impairment

This includes any physiological disorders such as retardation, emotional illness, and specific learning disabilities. The words "substantially" and "major" were inserted to eliminate minor, irrelevant claims of discrimination. When the courts first started interpreting the ADA, the courts found a variety of characteristics, including bad backs, shortness of stature, and obesity to be disabilities under certain circumstances. However, over the past several years, the Supreme Court has issued a series of rulings on the ADA that have for the most part narrowed the broad terms of the law.

Recent Supreme Court cases have focused on giving precision and content to the term "major life activity." Examples have included household chores, bathing, and brushing one's teeth. And although it has not yet been definitively decided, it has been suggested that working at a specific job is not a major life activity.

Several interesting ADA cases are currently pending before the Supreme Court. One will answer the question whether employers must accommodate the needs of disabled workers even if the accommodation overrides the seniority rights of other workers. Another asks whether an employer can refuse to hire someone whose disability would make the job a threat to his own health or life.

Disease. Employers who receive federal funds cannot discriminate against employees disabled by contagious diseases unless they currently pose a real risk of infection to others or cannot perform their jobs. Although temporary illnesses do not qualify as disabilities, Supreme Court decisions indicate that anyone with a continuing medical problem may be covered.

Drug and alcohol abuse. Individuals who have serious debilitating addictions to drugs or alcohol are protected under the law, provided they are able to perform the job safely and effectively. Practically speaking the law would probably protect most recovered alcoholics and former drug addicts—including those currently undergoing treatment—but few if any current users of illegal substances.

Reasonable accommodation. The ADA protects individuals who, with "reasonable accommodation," can perform the "essential functions" of the job. If they cannot perform these essential functions—even with reasonable accommodation—then they may legally be rejected. Accommodation may include altering facilities, restructuring jobs, and revising schedules. It is not necessary to start remodeling the work area or making other "prospective" changes to the facility. None of this must be done until the employer is actually faced with accommodating an individual with a disability and then only if the accommodation is reasonable.

Disabled individuals may be rejected only if, with reasonable accommodation, they still are unable to perform the job's essential functions. They cannot be rejected because they are unable to perform occasional or marginal assignments. Therefore, a person who, with reasonable accommodation, cannot operate a typewriter need not be hired as a clerk-typist. However, to reject an applicant for a typist's position because they cannot drive to the bank once a week would likely be a mistake. This means that it is necessary to be able to distinguish between the essential and non-essential functions of each job. The best way to do this is with job descriptions.

While the law does not require job descriptions, they are valuable for many reasons. They can give essential guidance to interviewers, supervisors, and physicians. If job descriptions already exist, make sure they are current. Consider adding an "essential functions" section to the descriptions, or otherwise identifying those aspects of the jobs that are truly essential to their performance.

Physical examinations. The ADA permits physicals to be administered after an offer of employment has been made but before the individual actually starts work. The offer may be conditional on passing the exam if (1) the employer requires all entering employees in the same classification to take an exam, (2) the results are confidential, and (3) the exam is job-related. The act also permits drug testing. It does not protect individuals currently using illegal drugs or alcoholics who are unable to do their jobs.

Employee medical information must be kept in confidential files separate from the general personnel folder; access to medical files should be severely restricted. The act does not specify what should be held in the file, but it certainly should include any physical examination results, information related to worker's compensation claims, and any details of an employee's illnesses or injuries that are not job-related.

Be certain that all interviewers, first-line supervisors, and members of management are familiar with the requirements of these laws.

Employers are required to screen applicants in a non-discriminatory manner, which means that inquiries that disproportionately screen out women and minorities are prohibited. The EEOC considers inquiries about an applicant's race, color, religion, or national origin to be completely irrelevant in terms of job ability. Such inquiries, whether direct or indirect, may be regarded as evidence of discrimination.

There are exceptions. Pre-employment inquiries made pursuant to the requirements of a local, state, or federal fair employment practices law are permissible. Also excepted are those infrequent instances where religion and national origin are "bona fide occupational qualifications," or where the employer can prove that the inquiry is justified by business necessity that is job-related.

Before rejecting a disabled applicant, determine what accommodation is needed. This is best done by discussing possible accommodations with the applicant. Decide whether the disability can be reasonably accommodated. If not, ask the applicant whether he/she can provide some or all of the accommodation. Evaluate whether all possible accommodations have been considered. Make sure the rejection is based on the applicant's inability to perform an essential job function. Document the accommodations considered and the reasons for their rejection.

The law requires affirmative action only for applicants or employees who are qualified to perform the job with reasonable accommodation. However, the employer can no longer make broad assumptions about a disabled individual's ability to perform. If an employment decision is made based on a disability, the employer must be able to prove that even after reasonable accommodation the individual cannot satisfactorily accomplish the duties of the job.

Sexual harassment. Sexual harassment of any employee or job applicant is prohibited. Such harassment is defined as any unwelcome sexual advance or request for sexual favors or any conduct of a sexual nature when (1) submission is made explicitly or implicitly a term or condition of employment, (2) submission or rejection is used as the basis for employment decisions, or (3) such conduct has the purpose or effect of substantially interfering with an individual's work, or creates an intimidating, hostile, or offensive working environment.

Employers are liable for acts of sexual harassment by their employees and may be held liable for such acts of non-employees where the employer knows of the conduct or should have known about it, unless the employer can show it took immediate and appropriate corrective action. The employer may be accused of sex discrimination if a person who qualified for an employment opportunity was denied that opportunity because it was given to one who submitted to an employer's sexual advances.

Establish a policy stating sexual harassment is strictly prohibited and appropriate disciplinary action will be taken against violators. Cover the following points:

- Definition. Specify what types of acts constitute sexual harassment.
- Complaint procedures. State to whom the individual should report alleged harassment. This is typically the supervisor, fire chief or the human resources department. Include women in the list of those to whom complaints can be reported.
- Supervisor's responsibilities. State to whom supervisors should report complaints and what action they should take.
- Investigation. List who will conduct the investigation of the alleged harassment.

- Disciplinary action. Describe what types of action may be taken depending on the severity of the harassment and other factors.
- Prevention. Spell out management's responsibility to notify workers of the sexual harassment policy and its enforcement procedures.

Communicate the policy to all employees, informing them of their right to bring complaints to management and to appropriate agencies. Immediately investigate and take action on all claims of sexual harassment. Finally, maintain thorough records of all actions taken under this guideline.

Sexual orientation. Discrimination against individuals based on their sexual orientation is prohibited. This includes having a past or current preference for heterosexuality, homosexuality, or bisexuality—or being identified with such preference. This applies to all public and private employers except religious entities where the work is connected to the purpose of the entity.

Employers cannot discipline or discharge, discriminate in wages or conditions of employment, or refuse to hire or promote on the basis of sexual orientation. Professional associations cannot refuse membership, employment agencies cannot fail to refer individuals for employment, and labor unions cannot exclude persons from full membership based on their sexual orientation.

Some employers have expressed fear that individuals may now be more open about their sexual preferences. But employers still have the right to set standards of conduct. Employers can prohibit any activities that disrupt the work place, require everyone to wear standard attire for the work being performed, and can control any form of harassment—sexual or otherwise.

PENSIONS

A firefighter's pension plan is a major benefit of the job. In the fire service, most provisions of the pension plan are either covered under a union contract and thus negotiable, or defined by state law. Even when the pension plan is dictated by state law, the local municipality often has an important part to play. The fire chief is usually on the pension board or in some way administers the pension. Therefore, it is important that the chief understands not only the department's pension plan but also the laws and regulations that apply to pensions.

All pension plans fall under the jurisdiction of the Employee Retirement Income Security Act of 1974 (ERISA). ERISA provides mandatory rules on all pension, profit sharing, and stock bonus plans as well as some welfare plans. State laws must coincide with ERISA, so in states that by law establish firefighters' pension plans, it is best to become familiar with the state statutes, to learn the specifics of the plan. If the pension plan is covered by the union contract, the fire chief must become familiar with ERISA to ensure compliance with the law.

The basic purpose of the act is to protect the rights of employees in terms of participation and vesting under pension plans, to set standards for management of the funds involved and to require reports to be filed with the government and given to each participant in summary form. The U.S. Departments of Labor and Treasury jointly administer ERISA.

There are essentially two types of pension plans:

- Defined-Benefit plans. The employer agrees to pay retirement benefits based on a formula that is typically some percentage of salary for each year of service.
- Defined-Contribution plans. The employer agrees to contribute to the employee's account a specific amount of money, typically a percentage of pay.

In addition, for reporting purposes, ERISA treats as pension plans any form of deferred compensation such as deferred profit-sharing, stock purchases or savings and thrift plans as well as pension plans. A bonus plan under which payments are systematically deferred and paid out over several years is also considered a pension plan for reporting purposes.

A cash bonus plan, a cash profit-sharing plan pertinent to the fire service and severance pay of less than two years are considered forms of compensation and are not regulated by ERISA.

The contributions made by the employer and the rights obtained under such plans by employees involve substantial sums. Therefore, no plan should be set up or amended without the advice of an attorney, accountant or benefits specialist who is experienced in this field.

Subsidized Education Programs

Many fire departments provide subsidies for educational advancement, most notably in the field of fire science or administration. Stipends may be paid for obtaining an associate, bachelor's or master's degree in fire science or fire administration. Stipends may be paid for a certain number of credits obtained toward the above degrees. Some departments pay for all or for a portion of the cost of this education, provided the firefighter receives a certain grade. In addition, monies may be paid for additional technical knowledge and skills such as paramedic or emergency medical technician (EMT) certification, fire investigator certification, and other advanced technical skills that may be acquired.

Some departments that are fortunate to have an educational institution within their jurisdiction may have a cooperative agreement in place that allows firefighters to attend a university, college, or community college at a reduced rate. For example, the University of New Haven, located in West Haven, Connecticut has a policy under which police and firefighters within the city may enroll in the fire science program or criminal justice program at one-half the tuition rate. This is certainly beneficial for those who want to take advantage of the educational opportunity.

In addition, the West Haven Fire Department has a live-in program under which fire science students are selected to live in the local fire stations. They must have some prior experience, but are trained accordingly and perform the same firefighting functions as members of the department.

The subsidizing of educational programs and pay incentives for educational achievement described above should lead to a better trained, more highly motivated department, provided the leadership of the department encourages the program and provides guidelines to firefighters in their quest for educational advancement.

Occupational Safety and Health Act of 1970

The Occupational Safety and Health Act of 1970 (OSHA) governs safety and health in the private sector. Many states, in addition to meeting federal OSHA requirements, must adhere to their own state's occupational safety and health plan. The federal law is enforced by the Occupational Safety and Health Administration within the U. S. Department of Labor.

The major provisions of the act are:

- Inspection without advance notice. Investigations may be made on either a routine basis or because of a specific complaint. The courts have sustained arguments that inspection without notice is unconstitutional, but as a practical matter, the government can obtain a warrant without proof of any violation. Under these circumstances, there isn't much point to refusing access to an OSHA representative.
- Representative employees must accompany inspectors during a visit. The U.S. Department of Labor amended regulations in 1977 to require that employers must compensate employees for time spent in "walk-around" inspections and for any conferences arising from the inspections. Since then, however, OSHA has withdrawn this regulation because the majority of employees on walk-arounds are already compensated by their employers or unions.
- Employees are entitled to file a complaint confidentially.
- Protection is afforded for employees who exercise their rights under the law. Retaliation is forbidden.

General duty clause

The law includes an extremely broad "general duty clause" that requires employees to comply with safety rules. In some states that have a state OSHA, the state inspectors use the general duty clause to apply NFPA standards on the fire service, even though the state or local municipality never formally adopted the NFPA standards.

There are also OSHA regulations that cover specific areas, which include:

- Hazard communication—the "right-to-know" laws
- Walking and working surfaces—guarding openings, ladders, scaffolds, power lifts, and housekeeping
- Exits—emergency exits, emergency plans, and fire prevention plans
- Environmental control—ventilation, noise exposure, radiation, hazard signs, and tags
- Hazardous materials—gases, flammable liquids, explosives, hazardous waste, emergency response
- Personal protective equipment—protection of eyes, face, head, and feet, and respiratory system
- Fire protection—extinguishers and sprinkler systems, detection and alarm systems
- Materials handling and storage
- Guarding and operating machinery
- Training in and operation of welding, cutting, and glazing
- Electrical wiring and electrical equipment
- Operating hand-held equipment

Physical fitness. Firefighting is one of the few professions that requires employees to be physically fit due to the extremely hostile environment in which they work. Many departments, both career and volunteer, maintain some type of voluntary physical fitness program. Others have requirements or provisions in the labor contract for maintaining a certain type of physical fitness program. In many departments, the firefighters and the administration work hand-in-hand to provide physical fitness equipment.

As mentioned earlier, the Fire Service Joint Labor Management Wellness-Fitness Task Force, a joint venture of the IAFF and IAFC in conjunction with ten pairs of participating municipalities and their locals, also developed the Fire Service Joint Labor Management Wellness/Fitness Initiative. The purpose of the Initiative was to develop an overall wellness/fitness system that was holistic, positive, rehabilitating, and educational, to incorporate the following points:

- Overcome the historic fire service punitive mentality of physical fitness and wellness issues
- Move beyond negative timed, task-based performance testing to progressive wellness improvement
- Require a commitment by labor and management to a positive, individualized fitness/wellness program
- Develop a holistic wellness approach that includes medical, fitness, injury/fitness/medical rehabilitation, and behavioral health

The Initiative is designed for incumbent fire service personnel. It requires mandatory participation by all uniformed personnel in the department, and allows for age, gender, and position in the department. The program calls for on-duty time participation using facilities provided or arranged by the department. It provides rehabilitation and remedial support for those in need, contains training and education components, and is reasonable and equitable to all participants. For more information, contact the IAFF or IAFC.

EMPLOYEE BENEFITS

The terms "benefit" or "fringe benefit" have historically been used to cover all paid time not worked and all additions to pay for the benefit of employees. Benefits accounted for only about four percent of payroll costs prior to World War II. At that time, the federal government instituted wage controls that froze the pay of many workers. One way these "frozen" workers could get a raise was to quit and be hired by another employer; this led to employers developing enhanced benefit programs as a means of keeping employees from leaving. This, combined with collective bargaining, recognition of employees' needs for security, and the desire to provide some form of tax-free reward to employees has led to major expansions of benefit programs. Today, benefits have become so important, and so costly, that referring to them as "fringe" benefits has become outdated.

Employee health care benefits is one of the most important benefits available to firefighters and their families. Fire chiefs must be familiar with the benefits provided and the process of handling such claims. Problems with health care benefits, even though they may be administered by a human resources director, will eventually end up on the fire chief's desk for assistance in adjudicating the problem. Poor handling of employee health benefits can lead to mistrust and lowered morale in the department.

There are many types of health care plans currently available and in use. Some of them include:

- Commercial policies that insure on an indemnity basis, paying a specific amount toward each covered expense. Most medical care givers agree to accept a lesser amount than their full charge when the patient is covered by a health care plan.

- Health Maintenance Organizations (HMOs) provide service directly either through their own staffs of medical professionals or by contracting with groups of professionals who spend most of their time caring for HMO patients. HMO plans are prepaid, that is, the cost of all services is covered by the premiums (sometimes with a small co-payment) and no bill is issued to the patient.
- Self-insurance is practical for large employers who play claims directly through their own insurance plans. Some small firms self-insure but limit their exposure by having major medical coverage. Self-insured plans, for the most part, are regulated by the federal government under ERISA.
- Major medical insurance supplements the basic hospital/surgical/medical coverage so that when the basic coverage has been exhausted the major medical takes over. Major medical coverage initially was developed as a separate policy. More recently, however, "comprehensive" medical plans have developed that combine the features of both the hospital/surgical/medical and major medical policies into one package.

Cafeteria plans. Cafeteria or flexible benefits plans can be used in conjunction with any of the above types of health care. The terms "flexible benefits," "flex plan," and "cafeteria plan" are used interchangeably. Strictly speaking, flexible benefit plans, or flex plans, are any type of plan that provides employees a choice of benefits. Cafeteria plans offer employees a choice of both the type of benefits and the extent of coverage. Cafeteria plans can be established as tax-qualified under Section 125 of the IRS code, allowing employees to reduce their taxable income by using pretax dollars to pay for benefits. Many IRS and ERISA regulations affect cafeteria plans so an employer that wishes to establish such a plan should do so only under the guidance of an experienced CPA or benefits consultant. The following are the various types of cafeteria plans.

Premium Only Plan (POP). Also known as premium-conversion accounts, POPs are the least complex type of flex plan. Employees pay for their health, life, and disability insurance coverage with pre-tax dollars by reducing their salaries (through payroll deduction) by the exact amount of their insurance premiums. Since this reduces taxable income, employees then save on federal and state income taxes, while both employees and employers save on Social Security taxes.

Flexible Spending Account (FSA). Also referred to as reimbursement accounts or salary reduction plans, FSAs permit employees to use pre-tax dollars to pay for health care costs not covered by the employer's insurance plan. Employees must estimate these expenses in advance for the year. This amount is deducted from their pay. Employees then pay out of pocket as expenses are incurred, and are later reimbursed by the employer. Flexible spending accounts also can be used to finance up to $5,000 annually in dependent care or elder care expenses. Although FSAs can be used independently, they are typically set up in conjunction with POPs.

Full Flex/Cafeteria plans. Employees are given a set dollar amount to spend on a "cafeteria menu" of benefits: life, health, dental, vision, disability, 401K, and even vacation time. Expenses beyond the allotted dollar amount are covered through a POP or FSA. Employees make their own choices for types of coverage and deductible amounts. Full cafeteria plans, the most complex type of flexible benefit plan, are normally feasible only for larger employers and must typically be managed either in-house using computer software packages or by an outside firm that specializes in flexible benefit administration. Some health care benefits are paid entirely by the municipality, although the trend today is to institute some form of co-pay to offset the rising costs of health care benefits.

Due to the rapid increase in medical expenses, employers have been seeking ways to reduce costs, including shifting some of the expense to the employees. To do this, both supplemental plans and comprehensive plans often make use of four features:

1. Employee Contributions. Surveys show that a slight majority of employers pay 100% of the insurance premium for employees, but only about one-third pay the full premiums for dependents.
2. Deductibles. With a deductible feature, the insured employee will pay a set amount of expense before the plan takes over; this might be $100 or $200 per individual, and $200, $400, or more for an entire family. The deductible may be paid on a per year or per event basis.
3. Co-Insurance. The plan pays some set percentage of the medical expense, say 80 percent, while the worker pays the balance.
4. Maximum Limits on Benefits. Where they are imposed, maximum limits on benefits are usually very high, for example, $1,000,000 for a lifetime of coverage.

The purpose of these limitations, of course, is not only to reduce costs directly, but also to give employees some incentive to control medical expenses by having the insured employee bear part of the cost.

Consolidated Omnibus Budget Reconciliation Act of 1985 (COBRA)

The Consolidated Omnibus Budget Reconciliation Act of 1985 requires that individuals who lose coverage under group plans because of termination must be given the opportunity to continue the coverage. Employers of fewer than 20 workers need only to comply with applicable state laws. However, COBRA clearly applies to businesses with 20 or more workers. Federal COBRA covers hospital, medical, surgical, dental, prescription drugs, hearing and vision, and group insurance plans.

Upon termination of employment, the former employee has the right to continue coverage for a maximum of 18 months, unless the termination was for gross misconduct. In layoffs, the coverage is 18 months. In the case of divorce or legal separation, the covered spouse and dependents may continue coverage for 36 months. Qualified individuals must elect to continue coverage within 60 days of notification of the qualifying event. The covered employee or dependent must make the first premium payment to the employer within 45 days of their election to take the COBRA benefits.

Employee Assistance Programs (EAPs)

EAPs are confidential counseling programs designed to help workers and sometimes their families with personal problems. Employers establish EAPs to provide employees with affordable access to treatment. In turn, they reap savings by reducing the costs associated with poor productivity, on-the-job injuries, excessive absenteeism, and turnover.

Originally established to handle the problems associated with alcohol and substance abuse, EAPs now also deal with emotional, mental health, financial, and family/marital problems. They sometimes are operated through the organization's own medical or health department, or they may be provided by an outside firm on a contract basis, either in house or at the outside firm's facility. Employees are often permitted to use the services of the EAP voluntarily and without the employer's knowledge of the nature of the problems being handled. On the other hand, employees may be required to make use of the EAP as part of a rehabilitation program, after having shown up for work in an unfit condition, for example.

Labor Relations

Fire chiefs who manage unionized fire departments face a challenge that in many cases, can be very rewarding, provided the fire chief is aware of relevant labor laws, maintains integrity in dealing with firefighters and, above all, takes part in labor-management negotiations. Any administrator who negotiates a contract or settles a grievance without the fire chief being present and taking part in the process is courting disaster. If a fire chief has well-trained employees, has established SOPs and practices unity of command, treats all personnel in a fair and equitable manner including issuing disciplinary action, and has compassionate personnel, then generally, the labor relations problems are greatly reduced.

There will always be some problems, conflicts and differences in the interpretation of the contract. The key to settling the differences is the willingness of both sides to sit down and discuss the problem. Fire chiefs should not be afraid to express their honest opinions. There will be times when an agreement cannot be reached. In these cases, the key is for both sides to disagree in an amicable manner and let the grievance procedure established in the contract take its course.

Unfortunately, at times when the problem reaches this stage, conflicts may result in some form of retaliation. As the conflict escalates, innocent firefighters can be "hit with shrapnel" and the problem escalates further. The key here is for the fire chiefs to maintain their personal integrity, not retaliate for the sake of "getting even," issue discipline fairly if there are violations of rules and regulations, stand firm, and above all be willing to discuss the issue further if necessary. Conflict management benefits no one—not the union, the department, and not the citizens and taxpayers.

If the fire chief leads the department, and is honest in all dealings with both subordinates and superiors, then conflict management will be greatly reduced and the citizens and firefighters will have a department that benefits from community involvement, a department in which all concerned can be proud.

8

SAFETY AND OCCUPATIONAL HEALTH

Dr. Thomas B. Sturtevant

CHAPTER HIGHLIGHTS

- The National Institute for Occupational Safety and Health (NIOSH) and the Occupational Safety and Health Administration (OSHA) are two different agencies who, while having different responsibilities, often work together toward the common goal of protecting employees' safety and health.
- The National Fire Protection Agency (NFPA) is a non-profit membership organization with a mission similar to the fire service that is to reduce the burden of fire and other hazards on the quality of life through the development of codes and standards, research, and education.
- The minimum requirements for a Fire Department Occupational safety and health program are outlined in NFPA Standard 1500 (1997 Edition).
- The classic risk management model involves risk identification, risk evaluation, establishment of priorities, determination of risk control techniques, and program monitoring (NFPA 1021 Fire Officer IV, 5-7.1).

INTRODUCTION

It is an understatement to say that the fire service is a hazardous occupation. According to the National Fire Protection Association (NFPA), since 1997 the fire service has experienced more than 2,700 on-duty firefighter fatalities. Although emergency activities pose serious risks to firefighters, not all fatalities occur on the fireground. During 2000, only 38% of all fatalities occurred on the fireground while 24% occurred when responding to or returning from an alarm and 14% occurred during training activities. Progress toward reducing the number of fire service fatalities and injures has occurred over the years (see figures 8–1 and 8–2). However, this progress is not as evident with fatalities as it is with injuries. The reduction in firefighter fatalities appears questionable over the past 10 years while a continued reduction in firefighter injuries is obvious over the same time period. In fact, the NFPA reports that the estimated number of injuries for the year 2000 was the lowest since 1977. Many fire chiefs agree these numbers are still high and through hard work continued progress can be made.

The role and scope of the fire service has changed over the years and continues to change. These changes, more often than not, have increased the risk to emergency responders. Thirty years ago, fire departments primarily responded to structural fires. Over the

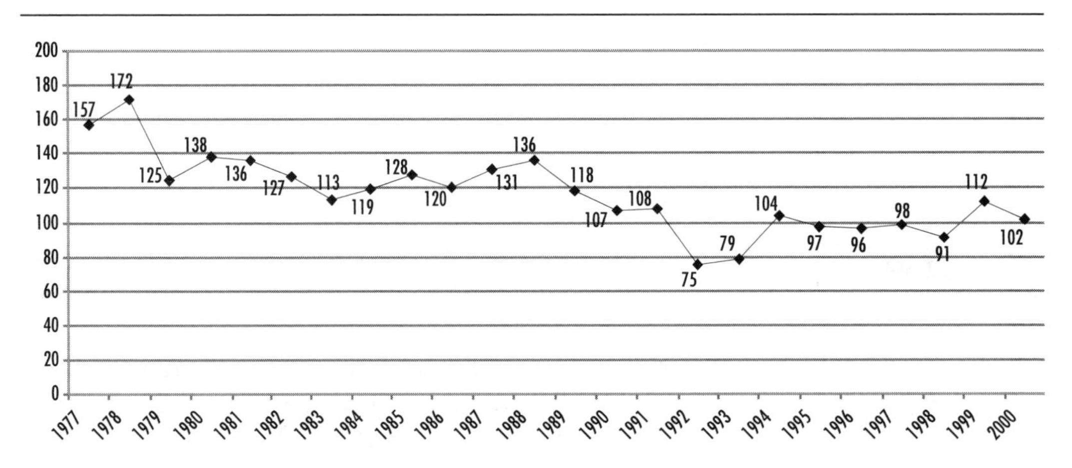

Figure 8–1 Total number of on-duty firefighter fatalities from 1977 to 2000.
Source: NFPA JOURNAL, July/August 2001.

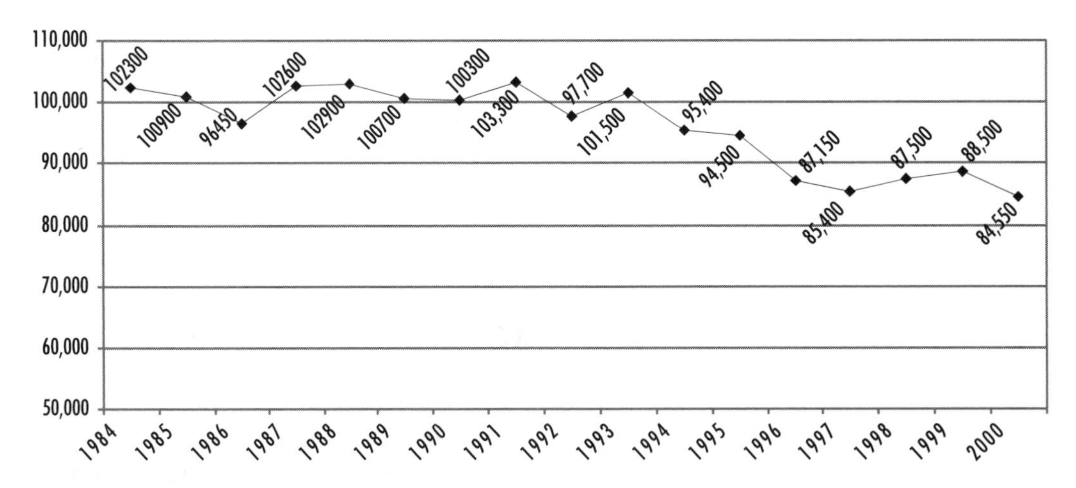

Figure 8–2 Total number of on-duty firefighter injuries from 1984 to 2000.
Source: NFPA JOURNAL, November/December 2001.

past 15 years, fire departments have increasingly responded to medical, hazardous materials, and advanced rescue emergencies. The tragic events of September 11, 2001, further changed the role and scope of the fire service. According to the United States Fire Administration (USFA), a total of 343 firefighters died as the result of the terrorist attack on the World Trade Center. The terms "home front protection" and "weapons of mass destruction" (WMD) have become part of the fire service vocabulary and responsibility. Today, citizens rely upon and expect the fire service to respond instantly to a wide range of emergencies.

Increasingly, comprehensive occupational safety and health programs are used as a means to address the hazardous nature of the fire service. There are several factors that motivate or influence the development of such programs. One factor relates to the ethos of the fire chief. That is, many fire chiefs believe that protecting their employees from harm is simply the morally right thing to do. Another factor relates to the potential for legal issues such as negligence and liability. Although the courts appear to take into account the hazardous nature of emergency response when deciding on response-related cases, it can be very difficult, not to mention embarrassing, to explain why a fire department did not establish what is considered nationally accepted minimal safety policies and procedures. Financial reasons have also motivated the development of occupational health and safety programs. Simply put, losses can be very expensive. Finally, regulations, such as those authorized by the Occupational Safety and Health Act (OSHA), and standards, such as those developed by the NFPA, motivate the development of these programs.

As the risk to the fire service increases, it becomes important to ask, "What level of risk is acceptable to you and your department?" The development of an occupational safety and health program helps answer this question. This chapter presents a brief overview of relevant safety regulations and standards and discusses the salient issues and concepts associated with the development of a department's occupational safety and health program.

Safety and Occupational Health

OSHA was signed into law by President Nixon on December 29, 1970, and became effective on April 28, 1971. Section 2(b) of the act indicates the purpose is "to provide for the general welfare, to assure so far as possible every working man and woman in the Nation safe and healthful working conditions..." This act authorizes enforcement efforts to private sector employers specifically excluding state and local governments. Included in this act are responsibilities for both employers and employees. Employers are required to 1) provide employees with a work place free from recognized hazards that are likely to cause death or serious physical harm to employees, 2) comply with occupational safety and health regulations, and 3) maintain recording keeping and reporting requirements. Employees are required to conduct their own actions in a safe manner and to comply with

OSHA regulations and standards. Finally, this act established two important agencies, the National Institute for Occupational Safety and Health (NIOSH) and the Occupational Safety and Health Administration (OSHA). Although they are two different agencies with different responsibilities, they often work together toward the common goal of protecting employees' safety and health.

National Institute for Occupational Safety and Health (NIOSH)

NIOSH is part of the Center for Disease Control and Prevention (CDC) under the U.S. Department of Health and Human Services. This federal agency is responsible for research activities related to occupational disease and injury. In addition, NIOSH investigates potentially hazardous working conditions when requested by employers or employees, makes recommendations, and disseminates safety information. Within these responsibilities, NIOSH provides valuable assistance, both directly and indirectly, to the fire service.

One example that illustrates each of these responsibilities of NIOSH is the Fire Fighter Fatality Investigation and Prevention Program. The stated goals of this program are to 1) better define the magnitude and characteristics of on-duty fatalities and severe injuries among firefighters, 2) develop recommendations for the prevention of these injuries and deaths, and 3) implement and disseminate prevention efforts. NIOSH has also provided assistance to fire departments through pre-incident inspections. For example, the fire departments can request a NIOSH inspection of their self-contained breathing apparatus (SCBA) maintenance program. The benefits of doing so include validating an already good program and/or providing valuable suggestions and recommendations to improve the SCBA program.

Additional information on NIOSH, fire service related NIOSH activities, and available reports, alerts, and other information can be found on the NIOSH Internet Home page (*http://www.cdc.gov/niosh/homepage.html*). A small example of information available on the NIOSH web site includes the following:

- NIOSH Firefighter Home page (*http://www.cdc.gov/niosh/firehome.html*) featuring Fire Fighter Fatality Data, Fire Fighter Fatality/Injury Investigation Reports, SCBA information, and NIOSH Publications and Alerts on Fire Fighting Hazards
- A Curriculum Guide For Public-Safety And Emergency-Response Workers (*http://www.cdc.gov/niosh/89-108pd.html*)
- Traumatic Incident Stress: Information For Emergency Response Workers (*http://www.cdc.gov/niosh/unp-trinstrs.html*)
- NIOSH ALERT: Preventing Allergic Reactions to Natural Rubber Latex in the Workplace (*http://www.cdc.gov/niosh/latexalt.html*)

Occupational Safety and Health Administration (OSHA)

OSHA is part of the U.S. Department of Labor (DOL) and focuses on creating and enforcing workplace safety and health regulations. The stated mission of OSHA is to prevent work-related injuries, illnesses, and deaths. To accomplish this mission, OSHA conducts inspections, investigates complaints and accidents, provides safety related training, partners with federal, state, and local entities, and establishes and enforces safety regulations. The regulations promulgated by OSHA carry the weight of law and are contained in the Code of Federal Regulations (CFR). Most of the regulations related to the fire service can be found in CFR Title 29, Parts 1901 and 1910.

All states are subject to enforcement of OSHA regulations. However, the Occupational Safety and Health Act of 1970 encourages states to establish their own occupational safety and health standards. A state can elect to establish its own occupational safety and health plan as long as the plan includes coverage of public employees and is at least as effective as the federal OSHA regulations. OSHA state regulations have the same effect of federal regulations. As of August 2001, the following states and territories have approved OSHA state plans:

Alaska	Arizona	California
*Connecticut	Hawaii	Indiana
Iowa	Kentucky	Maryland
Michigan	Minnesota	Nevada
*New Jersey	New Mexico	*New York
North Carolina	Oregon	Puerto Rico
South Carolina	Tennessee	Utah
Vermont	Virgin Islands	Virginia
Washington	Wyoming	

**These state plans cover state and local government (public sector) employment only.*

Application of OSHA standards to fire departments. As stated previously, the Occupational Safety and Health Act of 1970 includes enforcement of OSHA regulations for private sector employers. Consequently, some municipal fire departments are not subject to Federal OSHA regulations. However, OSHA regulations do apply to private fire departments such as industrial fire departments or brigades. In addition, OSHA regulations apply to fire departments wherein employer-employee relationship can be established. For example, if a volunteer firefighter receives monetary compensation for training, responding to an emergency, or other activity, an employer-employee relationship might exist. In turn, the volunteer fire department would be subject to OSHA regulations. In general, state law provides the basis for establishing employer-employee relationships. States that develop their own occupational safety and health program are required to extend enforcement to state and local government employment. Fire departments within the states and jurisdictions listed above must comply with their state OSHA plan that, in turn, must meet or exceed federal OSHA regulations. Finally, OSHA regulations on hazardous materials (29 CFR 1910.120) apply to all fire departments through legislation promulgated by the Environmental Protection Agency (EPA). The EPA legislation applies to municipal and volunteer fire departments in states that do not have an OSHA-approved plan.

As a Fire Chief, you should be familiar with applicable federal or state OSHA regulations. The OSHA web site (*http://www.osha.gov*) contains a wealth of information including the Occupational Safety and Health Act of 1970, a directory of OSHA regulations with search capability, and local OSHA office information. Several important OSHA regulations you should be familiar with include:

29 CFR 1910.120—Hazardous Waste Operations and Emergency Response (HazWoper)

29 CFR 1910.134—Respiratory Protection

29 CFR 1910.146—Permit Required Confined Spaces

29 CFR 1910.156—Fire Brigades

29 CFR 1910.1030—Occupational Exposure to Blood-borne Pathogens

National Fire Protection Association

Fire chiefs should be familiar with the NFPA as an organization, NFPA codes and standards, and the open, public process NFPA uses to develop codes and standards. The NFPA is a non-profit membership organization with a mission similar to the fire service—to reduce the burden of fire and other hazards on the quality of life. The NFPA attempts to accomplish this mission through development of codes and standards, research, and education. Membership in NFPA is open to all fire service professionals and other interested individuals and organizations. The standards produced through the NFPA process are considered minimum consensus standards. The standards contain language to this effect as well as a statement allowing jurisdictions to exceed the minimum requirements.

Application of NFPA standards to fire departments

Unlike OSHA and the regulations they enforce, NFPA cannot enforce its own standards. In addition, NFPA standards do not carry the weight of law unless they are officially adopted as such by a local or state government entity. Once a standard is officially adopted, the governing entity can enforce compliance. However, even when not officially adopted in a jurisdiction, any existing national safety standard developed from a consensus system can be introduced in court to establish reasonable behavior. It should be noted that NFPA standards are sometimes used as the basis for OSHA regulations.

Local governments can adopt standards in several ways. First, they can adopt only specific objectives within a standard. Next, they can adopt specific individual standards or all of them at the same time. Finally, governments can adopt a standard or standards by referencing name or by name and edition. To that end, it becomes important to maintain awareness of changes within the standards, especially if the standards are adopted by name only. Conceivably, the department would be required to maintain compliance with standards as they change.

NFPA standards making process

A consensus-based process involving public review and comment is used to develop and revise NFPA standards. Because of this, fire service members have several opportunities to participate in the standard making process. One way to participate is by volunteering to serve on a technical committee. Although anyone can volunteer to serve, NFPA regulations prevent any one special-interest group from holding more than a third of any committee membership. Large fire service organizations can have membership on a committee and represent many organization members in this way. Another way to participate is to submit a proposal for a new project. If a new project is initiated, or whenever a standard is revised, interested individuals have several additional opportunities to participate. In each case, NFPA publishes announcements with timelines and asks for comments and suggestions. Proposals for existing standards can be submitted to the NFPA at any time and must be considered by the appropriate technical committee in the next revision cycle. If the committee disagrees with a public proposal, it must provide a technical reason for not accepting the proposal. Finally, fire service members can participate in the final approval process. During annual NFPA meetings, proposed standards are discussed in open debate prior to a final vote. Although anyone can attend and comment during the debate, only NFPA members are allowed to vote. After a standard is accepted by the NFPA membership, it is given an edition and a revision cycle is established. Sometimes changes are required to the standard prior to the schedule revision date and occur by someone proposing that a Tentative Interim Amendment (TIA) be added to the standard. Although included with the standard, TIAs are "tentative" because they have not been processed through the entire standards-making procedures and are "interim" because they are effective only between standard editions. All TIA automatically become a proposal for the next edition of the standard. When NFPA standards are used, you should ensure you have the correct edition of the standard and should check for any TIAs. Both can be found within the standard. Many standards follow a five-year revision cycle.

The NFPA web site (*http://www.nfpa.org*) is an excellent location to visit to learn more about the NFPA. In addition, the web site contains a complete list of all NFPA standards along with detailed information on revision cycles, comments, and reports.

NFPA safety related standards

A large number of NFPA standards focus directly or indirectly on fire service safety. The NFPA standards within the 1500 numbering sequence relate directly to occupational safety and health. These standards will be discussed briefly here and in more detail later in this chapter. Many of the other NFPA safety related standards are listed here and discussed further in other chapters of this text.

NFPA 1500 Standard on Fire Department Occupational Safety and Health Program. In 1983 the NFPA began the process of forming a technical committee to consider a standard on fire department occupational safety and health programs. This committee included firefighters (both career and volunteer), fire chiefs, and equipment manufacturers. It also included attorneys, physicians, risk management professionals, and representatives from other associated fields. Among the organizations represented were the International Association of Fire Chiefs (IAFC), International Association of Fire Fighters (IAFF), National Volunteer Fire Council (NVFC), International Society of Fire Service Instructors (ISFSI), the Fire Marshals Association of North America (FMANA), and associated NFPA sections. This committee has the largest fire service representation of NFPA committees.

The committee began work on developing a document to address the health and safety concerns of municipal fire departments. There was no differentiation among career, volunteer, or combination departments. Initially, the only document available for the fire service was OSHA Regulation CFR 1910.156 Fire Brigades which originally was designed for industrial fire brigades. At the time, the standard was being used for enforcement purposes and in some states for municipal fire departments, yet it did not address some of the important issues of firefighter health and safety.

Over a four-year period, the committee worked with other NFPA technical committees: protective clothing and equipment, fire apparatus and equipment, training, and the professional qualifications committees. All addressed areas in other standards that affected firefighter health and safety. Standard 1500 was first issued by the NFPA in the summer of 1987 amid controversy over its impact on the fire service. Perhaps the most popular argument was that new standard placed too great of a financial burden on fire departments to the extent that some departments would be forced out of the fire suppression business. Since then, through the standards-making process, the committee has revised the standard on a five-year cycle, first in 1992 and then in 1997. The 1997 standard is now the current edition.

The stated purpose of NFPA Standard 1500 is to set forth minimum requirements for a fire department's occupational safety and health program and specify safety guidelines for emergency response personnel. As such, it provides the framework for emergency service related safety and health programs. The requirements of NFPA Standard 1500 are applicable to all types of fire departments providing emergency response services, excluding industrial fire brigades or fire departments meeting the requirements of NFPA 600, Standard on Industrial Fire Brigades. Although the standard is comprehensive, it encourages exceeding the standard and provides for meeting the standard in a variety of ways.

The original concept was for Standard 1500 to be an umbrella document: that is, its chapters ultimately would become specific stand-alone documents. This process grew to the point where these additional documents evolved into specific standard numbers overseen by their own task groups. The task groups in turn reported to the full technical committee. The task groups included fire department incident management (1561), infectious disease control for the fire service (1581), medical requirements for firefighters (1582), and fire department safety officer (1521). In addition, the 1997 edition of the standard references 38 additional NFPA documents as mandatory requirements.

NFPA 1521 Standard for Fire Department Safety Officer (1997 edition). This standard was changed from Standard 1501 to NFPA Standard 1521 in 1992 to reflect the numbering system for the occupational safety and health documents. The purpose of this standard is to establish minimum requirements for a fire department's or emergency service organization's health and safety officer and incident safety officer. Specifically, it includes qualifications, authority, and functions of both positions.

NFPA 1561 Standard on Emergency Services Incident Management Systems (2000 edition). The 2000 edition of this standard saw a name change from "Standard on Fire Department Incident Management Systems" (IMS) to its current title. With the understanding that the safety and health of those operating at an emergency scene depends upon how the incident is managed, this standard contains the minimum requirements of an IMS. Specifically, it defines and describes the essential elements of an IMS, such as the system structure, components, roles, and responsibilities.

NFPA 1581 Standard on Fire Department Infection Control Program (2000 Edition). This standard was developed to establish minimum criteria for a fire department infection control program while in the fire station, in emergency apparatus, and/or in any location where routine emergency operations are taking place. Designed to be compatible with applicable federal regulations and guidelines, the standard includes program components; fire department facility requirements; emergency operations protection; and cleaning, disinfecting, and disposal requirements.

NFPA 1582 Standard on Medical Requirements for Fire Fighters and Information for Fire Department Physicians (2000 edition). The purpose of this standard is to specify minimum medical requirements for both new and existing members. Designed primarily for fire department physicians, the standard includes requirements on medical evaluation and medical conditions by organ system. To better assist physicians, the standard provides additional information on department activities. Because of this, the title was changes from "Standard on Medical Requirements for Fire Fighters" to its existing title.

NFPA 1250 Recommended Practice in Emergency Service Organization Risk Management (2000 Edition). The purpose of this standard is to outline recommended practices for developing, implementing, and evaluating emergency services risk management programs. These recommended practices expand on requirements in chapter 2 of NFPA Standard 1500 and outlines an entire risk management program.

Table 8–1 NFPA documents referenced in NFPA 1500.
Source: NFPA 1500 Standard on Fire Department Occupational Safety and Health Program (1997 Edition).

NFPA 10, Standard for Portable Fire Extinguishers, 1994 edition.
NFPA 101®, Life Safety Code®, 1997 edition.
NFPA 472, Standard for Professional Competence of Responders to Hazardous Materials Incidents, 1997 edition.
NFPA 473, Standard for Competencies for EMS Personnel Responding to Hazardous Materials Incidents, 1997 edition.
NFPA 600, Standard on Industrial Fire Brigades, 1996 edition.
NFPA 1001, Standard on Fire Fighter Professional Qualifications, 1997 edition.
NFPA 1002, Standard for Fire Department Vehicle Driver/Operator Professional Qualifications, 1993 edition.
NFPA 1003, Standard for Airport Fire Fighter Professional Qualifications, 1994 edition.
NFPA 1021, Standard for Fire Officer Professional Qualifications, 1997 edition.
NFPA 1041, Standard for Fire Service Instructor Professional Qualifications, 1996 edition.
NFPA 1051, Standard for Wildland Fire Fighter Professional Qualifications, 1995 edition.
NFPA 1403, Standard on Live Fire Training Evolutions, 1997 edition.
NFPA 1404, Standard for a Fire Department Self-Contained Breathing Apparatus Program, 1996 edition.
NFPA 1405, Guide for Land-Based Fire Fighters Who Respond to Marine Vessel Fires, 1996 edition.
NFPA 1521, Standard for Fire Department Safety Officer, 1997 edition.
NFPA 1561, Standard on Fire Department Incident Management System, 1995 edition.
NFPA 1581, Standard on Fire Department Infection Control Program, 1995 edition.
NFPA 1582, Standard on Medical Requirements for Fire Fighters, 1997 edition.
NFPA 1901, Standard for Automotive Fire Apparatus, 1996 edition.
NFPA 1906, Standard for Wildland Fire Apparatus, 1995 edition.
NFPA 1911, Standard for Service Tests of Pumps on Fire Department Apparatus, 1997 edition.
NFPA 1914, Standard for Testing Fire Department Aerial Devices, 1997 edition.
NFPA 1931, Standard on Design of and Design Verification Tests for Fire Department Ground Ladders, 1994 edition.
NFPA 1932, Standard on Use, Maintenance, and Service Testing of Fire Department Ground Ladders, 1994 edition.
NFPA 1961, Standard on Fire Hose, 1997 edition.
NFPA 1962, Standard for the Care, Use, and Service Testing of Fire Hose Including Couplings and Nozzles, 1993 edition.
NFPA 1964, Standard for Spray Nozzles (Shutoff and Tip), 1993 edition.
NFPA 1971, Standard on Protective Ensemble for Structural Fire Fighting, 1997 edition.
NFPA 1975, Standard on Station/Work Uniforms for Fire Fighters, 1994 edition.
NFPA 1976, Standard on Protective Clothing for Proximity Fire Fighting, 1992 edition.
NFPA 1977, Standard on Protective Clothing and Equipment for Wildland Fire Fighting, 1993 edition.
NFPA 1981, Standard on Open-Circuit Self-Contained Breathing Apparatus for Fire Fighters, 1997 edition.
NFPA 1982, Standard on Personal Alert Safety Systems (PASS) for Fire Fighters, 1993 edition.
NFPA 1983, Standard on Fire Service Life Safety Rope and System Components, 1995 edition.
NFPA 1991, Standard on Vapor-Protective Suits for Hazardous Chemical Emergencies, 1994 edition.
NFPA 1992, Standard on Liquid Splash-Protective Suits for Hazardous Chemical Emergencies, 1994 edition.
NFPA 1993, Standard on Support Function Protective Clothing for Hazardous Chemical Operations, 1994 edition.
NFPA 1999, Standard on Protective Clothing for Emergency Medical Operations, 1997 edition.

Occupational Safety and Health Program

Developing the program

Preparation and planning are key elements in the development of an occupational safety and health program. Preparation involves becoming familiar with related regulations and standards. The previous discussion on regulations and standards provides only a general overview. During preparation, fire chiefs must determine specifics about federal, state, and local requirements related to regulations (Federal v. State OSHA) and standards (adoption of NFPA standards). The worksheets provided in the back of NFPA Standard 1500 can be very helpful in gaining an overview of the depth and breadth of the program. Planning involves thinking through the development, implementation, and evaluation aspects of the occupational safety and health program. The plan should include time lines, funding, and specific elements that will be included in the occupational safety and health program. At a minimum, NFPA Standard 1500 should be used as a guide in the development of the plan. Specifically, the plan should include organizational issues that help integrate safety into every aspect of department operations and activities. In addition, the plan should include focus or target areas such as emergency operations, training and education, and facility safety. Specific elements of an occupational safety and health program are provided in the next section.

After preparation and planning, work can begin on the development of the program. The first step is to conduct a detailed comparison between identified requirements and existing practices. Some suggest this step should be completed in two phases–an initial assessment to gain a quick overview followed by a more detailed review. Regardless of the method used, the end result should be a very detailed review that documents the extent of compliance as well as an itemized listing of documents, procedures, and policies. The next step is to prioritize those areas where compliance is required. This step includes identifying areas of non-compliance, evaluating the extent of non-compliance, and exploring alternative methods to attain compliance. Several considerations used to select among identified alternatives include the cost for compliance, the estimated time to implement, the estimated time for results, the ease of implementation, and the estimated effectiveness. The next step is to develop an implementation plan for non-compliance areas. The plan should be detailed and contain responsibilities, costs, time frames, and goals. In the end, the process should provide a document that clearly identifies compliance and non-compliance areas as well as a plan to achieve compliance. Finally, the plan should be reviewed and then approved or adopted. The review process can be internal, by an individual or committee within the department, or external, through local resources such as a city's risk management department or though a consulting service. The approval or adoption of the plan can occur using informal means such as a cover letter from the chief or through formal means such as an ordinance, law, or statute.

Administering the program

Efficient and effective administration of the program requires that safety be integrated into every decision and activity to the extent that it becomes part of the organization's culture. This process is not easy and takes time. However, with proper focus and effort a healthy safety culture can be realized. Several elements used to assist in this effort include (1) written safety policies, goals, and objectives, (2) active participation at all levels within the organization, and (3) clearly defined roles and responsibilities. Although the fire chief bears ultimate responsibility for the program, the health and safety officer (HSO) should be responsible for the day-to-day administration of the program. In addition, a health and safety committee (HSC), composed of representatives from all levels and areas within the organization, should actively assist the HSO. For example, including a labor representative as an important member of the HSC may help reduce labor and management issues while increasing the chance of success through buy-in by union members. Finally, data is used on the front end to establish objectives, during implementation to determine net effects, and on the back end to determine the overall effectiveness of the program. In that light, maintaining adequate data collection procedures and a record keeping system becomes a critical component of administering the program.

Evaluating the program

The purpose of evaluating the occupational safety and health program is to determine the overall effectiveness of the program. If conducted properly, the evaluation should be able to determine what parts of the program are effective and what parts are not effective. Perhaps the most difficult question is how to measure effectiveness. Although there are no hard and fast rules, several indicators of effectiveness might include changes in (1) the number of injuries and accidents, (2) the cost associated with the injuries and accidents, (3) compliance to regulations and standards, (4) overtime caused by injuries and illnesses. Each occupational safety and health program should include an evaluation plan. The plan should identify effective measures based on the goals and objectives of the program. Several considerations to include in the evaluation plan are:

1. The frequency of evaluation
2. The use of internal and external evaluators
3. The methodology used to conduct the evaluation
4. The final evaluation report

Specific components and referenced NFPA objectives

The question "what does an occupational safety and health program look like?" is not as important as the question "what should be included in the program?" In some cases, no one folder, book, or file contains all the various aspects of the program. Rather, specific elements of the program are contained within numerous forms, reports, documents, policies, and procedures. That notwithstanding, it is helpful to assemble a document that ties all these elements together for ease of reference, review, and evaluation. The following represents minimum components of an occupational safety and health program as required by NFPA Standard 1500.

- Organization components (NFPA 1500, Chapter 2)
  - Organizational statement (NFPA 1500, 2-1)
  - Risk management (NFPA 1250)
  - Fire department safety officer (NFPA 1521)
  - Occupational safety and health committee (NFPA 1500, 2-6)
- Training and education (NFPA 1500, Chapter 3)
  - NFPA professional qualification standards:
    - Instructor (1041), Driver pump operator (1002)
    - Fire fighter (1001), Airport fire fighter (1003)
    - Officer (1021), Wildland fire fighter (1051)
    - HazMat response (472)
    - Land-based fire fighters responding to marine vessel fires (1405)
  - Infection control program training requirements (NFPA 1581)
  - Live fire training evolutions (NFPA 1403)
- Vehicles, equipment, and drivers (NFPA 1500, Chapter 4)
  - Automotive fire apparatus (NFPA 1901)
  - Wildland fire apparatus (NFPA 1906)
  - Service tests of pumps on fire department apparatus (NFPA 1911)
  - Testing fire department aerial devices (NFPA 1914)
  - Ground ladders (NFPA 1931 and 1932))
  - Fire hose, couplings, and nozzles (NFPA 1961, 1962, and 1964)
  - Fire extinguishers (NFPA 10)
- Protective clothing and protective equipment (NFPA 1500, Chapter 5)
  - Station uniforms (NFPA 1975)
  - Protective clothing (NFPA 1971, 1976, 1977, 1999, 1991, 1992)
  - Self contained breathing apparatus (NFPA 1404 and 1981)
  - Life safety rope and systems (NFPA 1983)
- Emergency operations (NFPA 1500 Chapter 6)
  - Rapid intervention for rescue (NFPA 1500, 6-5)
  - Rehab (NFPA 1500, 6-6)
  - Civil unrest/terrorism (NFPA 1500, 6-7)
  - IMS (NFPA 1561)
  - EMS personnel responding to HazMat incidents (NFPA 473)

Facility safety (NFPA 1500, Chapter 7)
 Life safety code (NFPA 101)

Medical and physical (NFPA 1500, Chapter 8)
 Medical requirements for fire fighters (NFPA 1582)
 Infection control program (NFPA 1581)

Member assistance and wellness program (NFPA 1500, Chapter 9)

Critical incident stress program (NFPA 1500, Chapter 10)

Risk Management

Risk management (RM) plans are perhaps one of the more important, yet least understood, components of an occupational safety and health program. It is important to realize that the risk management plan should address all activities, both emergency and non-emergency, within a fire department. In some cases, city or county government risk management plans are developed. These plans usually include input and participation from all departments and divisions from within the local government. Although integrated into local government plans, NFPA Standard 1500 requires the development of a separate risk management plan for a fire department.

Several resources are available to assist with the development of a risk management plan. First, resources may be available at the local or state level. For example, many cities, towns, and counties have a risk management department. Second, the NFPA has several documents that can be helpful in developing a risk management plan. One such document is NFPA 1250, Recommended Practices in Emergency Services Organization Risk Management. This standard provides minimum criteria for developing, implementing, and evaluating a department's risk management plan. Another NFPA document is the Fire Department Health and Safety Standards Handbook. This document contains the complete text of NFPA 1500, 1521, 1561, 1581, and 1582. In addition, the document contains explanatory commentary intended to provide insight and understanding of the standard and to serve as a resource for implementing the provisions within the standard. Finally, a document called Risk Management Practices in the Fire Service is available free of charge from the USFA web site. Other risk management documents are also available on this site as a result of research conducted by students in the Executive Fire Officer program. (The Executive Fire Officer Program [EFOP] is an initiative of the United States Fire Administration/National Fire Academy designed to provide senior officers and others in key leadership roles.)

Risk management techniques have been used effectively in business and industry for many years. The classic risk management model involves five basic steps. These basic steps will be discussed briefly here.

Step 1: Identify the risk

The first step in developing a risk management plan is to identify potential risks. A risk can be defined as any chance of loss occurring. Risk identification is almost like trying to forecast the future. It involves looking at a variety of information sources and asking the simple question "what could go wrong?" This is vital to risk management; it is the very foundation upon which the plan will be developed. Because of this, risk identification should be conducted in a systematic process. In addition, the risks identified should be reasonable. Several sources of information to help with risk identification include:

- Past loss data (for example, look for injury and fatalities trends at the local, state, and national level)
- Members of the organization (for example, they may be able to tell you about near misses)
- Membership organizations (for example, the International Association of Fire Chiefs [IAFC] and NFPA)
- Consultants
- Professional journals and publications

Step 2: Evaluate the risk (frequency and severity)

The next step is to evaluate the risks using the concepts of frequency and severity. Frequency deals with how often a risk is likely to occur. Use of historical data, professional judgment, and tolerance to risk assist with determining the estimated frequency of a risk. Severity deals with the extent of loss if the risk occurs. Several factors used to estimate severity include direct and indirect costs associated with the loss, the impact to the organization, and the time and resources required to rectify the loss to its former status.

Step 3: Establish priorities

It would be nice to address all potential risks within a risk management plan. However, limited time and resources require that risks be prioritized. Some of the risks deserve more immediate attention than others. The process for establishing risk priorities is as much an art as it is a science. Consequently, the third step in the classic risk management model helps determine which of the identified risks should receive immediate action and which risks should wait.

The first consideration when establishing priorities is to look at the risk's frequency/severity ratings. A simple rating system of "low" and "high" can be used to provide some immediate information for establishing priorities. For example, most would readily agree that those risks with a high/high (high frequency and high severity) rating should receive immediate attention in a risk management plan. Most would also agree that those risks with a low/low rating would be located low on the priority list. The difficult risks to prioritize are those with low/high or high/low ratings. Several considerations used to evaluate these risks include the following:

- Cost vs. the benefits
- Cost of insurance
- Overall cost to implement
- How easy is the control measure to implementation
- How much time is required to implement
- How long it will take to realize the results
- Control measure estimated effectiveness

These factors should be considered together when establishing priorities for the low/high and high/low risks.

Step 4: Determine risk control techniques

The next step is to determine how to control the risks. In many cases, several risk control techniques are used for a potential risk. Risk control is the identification of techniques used to control the frequency and severity of potential risks. Risk control techniques can be grouped into the following categories: risk avoidance (simply avoid the risk all together), risk control (develop and implement measures that focus on controlling the frequency and severity of losses), and risk transfer (shifting the risk to someone else). In some cases, techniques from each of these groups will be used to address individual risks.

Step 5: Monitor the program

Program evaluation is a vital component of a risk management program. Without it, we would never really know if the program was meeting its goals and objectives or what part of the program is or is not working. Important elements in a program evaluation include:

- Determination of effectiveness indicators
- Frequency of monitoring
- Who conducts the evaluation
- Evaluation methodology
- Evaluation report

Risk management plan components

- Introduction
- RM program outline
- Purpose, scope, goals, and/or objectives of the program. Give a clear picture of the depth and breath of the program as well as projected outcomes.
- Authority/responsibility
- Monitoring/evaluating (who, what, where, when, how, etc.)
- Methodology (How does it work? How was the program developed? How will it be maintained?)
- Identified risk (a comprehensive list of "emergency" and "non-emergency" risks complete with several control methods for each risk)

Conclusion

Changes in role and scope of the fire service appear to have increased the risk to emergency service first responders. However, the development of a comprehensive occupational safety and health program can be used by departments to help manage and control occupational risk to their members. Prior to developing a program, the fire chief must identify and become familiar with the specific regulations and standards the department must comply with as promulgated or adopted by their department or local government. Preparation and planning are vital to the development of a comprehensive plan. Equally as important is appropriate administration and monitoring of the program.

Technical Resources

American National Standards Institute
http://www.ansi.org/

Center for Disease Control and Prevention (CDC)
http://www.cdc.gov/

Emergency Services Training Institute (ESTI)
http://www.estinstitue.com

Federal Emergency Management Agency (FEMA)
http://www.fema.gov

Fire Department Safety Officers Association (FDSOA)
http://www.fdsoa.org

International Association of Fire Chiefs (IAFC)
http://www.iafc.org

International Association of Fire Fighters (IAFF)
http://www.iaff.org

National Fire Protection Association (NFPA)
http://www.nfpa.org

National Institute for Occupational Safety and Health (NIOSH)
http://www.cdc.gov/niosh/homepage.html

Occupational Safety and Health Administration (OSHA)
http://www.osha.gov

OSHA's Fire Fighters Home Page
http://www.cdc.gov/niosh/firehome.html

US Fire Administration
http://www.usfa fema.gov/

9

FIRE SERVICE TRAINING AND EDUCATION

Russell J. Strickland

CHAPTER HIGHLIGHTS

- **Addresses what the chief needs to know about the training program and its management.**
- **Reviews certification based training and evaluations.**
- **Reviews types of training methodologies and resources.**

INTRODUCTION

The fire chief needs to know that the department will train together on the skills and performances necessary to accomplish the functions outlined within the organization's plan of operation. The personnel of the organization must be trained appropriately to do the things that are expected of them. This requires a clear identification of what the organization's mission and how the organization anticipates meeting its goals.

Changing times call for clear identification of where the "business" is today in respect to training, education, and experience coupled with technology available to weave it all together.

The challenge of explaining "everything you ever needed to know as a fire chief about training" in one chapter of a book that actually encompasses everything you ever needed to know as a fire chief, is that it can not be adequately accomplished. One great challenge is you the reader, and a second is the writer. You, the reader are anything from an 18-year-old active military person who is the chief of a small operation some place in our world, or even a local young individual who is chief of a small operation right here at home. The other extreme is that you are the new Fire Commissioner of the largest city in the world. Two different extremes, and certainly two different needs assessments—different, yes, but not completely separate. You will still direct forces, or have forces to "place the wet stuff on the red stuff." Your challenges will be different, if only at times in magnitude, but still different.

As the writer, I hope to offer my knowledge and opinion based upon 30 years of experience, most of which has been in the management and direction of a statewide training and education program. Thus, at times my perspective is very slanted toward that training focus and function—slanted to my environment and blind to the specifics of your environment that you have yet to share with me.

I present all this because a year ago our world of and within the emergency services changed. On September 11, 2001, our thought processes and planning strategies were formally introduced to the challenge of terrorism and what that will mean for us and for future leaders. Yes, for the last several years we talked about it, even trained a little, but we didn't prepare for what we have now experienced. This experience has and must refocus the basic core of our day-to-day operations, including our training and education. We must now be conscious and aware of the new bad word, "terrorism."

Everett E. Hudiburg, an individual who wrote and shared training experiences and opportunities with many said:

> *"A fire department in those days was simply a group of citizens, usually men, who had organized themselves to operate some sort of fire fighting equipment. Even then, little thought was given to the training of firefighters. Some of the larger cities used horses to pull equipment, and emphasis on all training was focused toward the horses. No organized attempt was made to provide training for men until about the middle 1920s."*

He relates his statement to the people of a local church who used lanterns to light their way to the church:

> *"So much like the little lanterns trying to light the way were a few individuals of those early years who attempted to let their lights shine into the gloom surrounding the vocation which sorely needed brightening. Now in our life we stand at the crest of a hill, look down the long slope into the years past, and recall of those noble men whose flickering lights attempted to pierce the darkness. We witness those lights here and there, and watch them become massed into a more brilliant day, our day. During our time we have witnessed the massing of not only minds, but hearts and hands."*

He goes onto to examine the past in order to move into the future through understanding the road education and training has traveled to reach today. That historical perspective sets the stage for our journey.

Challenge for the fire chief

Couple our historical past with our most recent events, throw in consensus standards, laws, regulations, rules, and procedures as promulgated locally, regionally, state-wide, and nationally, and you arrive to a point where the fire chief can't begin to "know everything"—even about training. We will hit the highlights, explore the necessary, plan for the absolute, and if time permits, pursue your interests.

It is fascinating to read the titles of the chapters of the first *Fire Chief's Handbook* and those of the most recent. "Training and Education" has evolved from the earlier chapter title of "The Fire Service Instructor." This also hints at where we are today and where we have been. Years ago, it was expected that the fire chief would lead the fire department in every function or activity. In today's changing world that is very difficult. And, in consideration of the size of the organization and the concept of span and control or management it is simply impossible. The fire chief has the responsibility to insure that what must be done, and what needs to be done, is done. However, he no longer accomplishes those tasks single-handedly.

Mark Sprenger, the CEO of *FireRescue World,* in a presentation about the training job, graphically and clearly states the "dreaded laundry list" of training topics that can be covered for firefighters. He clearly demonstrates that "the list" can be everything necessary for firefighters of the smallest organization in a local area to the necessities of the fire chief in the largest organization in the universe. It clearly places into perspective that you must first, and foremost, plan, and, that planning must be based upon an assessment of what you are going to do as an organization and why you do it.

The fifth edition of the *Fire Chief's Handbook* presented a chapter addressing training and education while another chapter addressed professional certification. We now combine professional certification with this chapter. This ties the certification to what may be law, or could be a voluntary compliance consideration in the scope of training. The previous edition of this chapter, written by William M. Kramer, made an excellent presentation of where we were at the time with training and education as well as predicting where we were going. Today, ten years later, it is still very accurate, with resources and systems in place for the fire chief to consider. Some of those same thoughts and words are expressed again.

So, after all that, you are still reading and know that you, as a fire chief, need to read some "wisdom" about training and education. Bring your lantern and let's examine training, education, and certification.

Training and Education

Training and education are two terms that must be clearly defined. Training in our business has clearly become the "skills and performances" that are taught to an individual. Webster defines *training* as "the education, instruction, or discipline of a person or thing that is being trained," and *train* as "to make proficient by instruction and practice, as in some art, profession, or work: to train soldiers," also, "to give the discipline and instruction, drill, practice, etc., designed to impart proficiency or efficiency."

Education has become more the "knowledge" or non-skill (theory) necessary for our profession. Webster defines *education* as "the act or process of imparting or acquiring general knowledge, developing the powers of reasoning and judgment, and generally of preparing oneself or others intellectually for mature life," and "the result produced by instruction, training, or study."

One word that appeared in both definitions is "instruction." Training and education both occur because of some type of instruction, which today can be very traditional to very, very, non-traditional.

A third definition that is essential to our profession is experience. Experience is clearly "learning" and thus can be considered as obtained through a non-traditional sense of instruction. Webster's definition is "knowledge or practical wisdom gained from what one has observed, encountered, or undergone."

It is essential that the fire service professional be well rounded and guided through the learning triad, which is a balance of all three—training, education and experience through their career. When individuals begin their careers as entry-level firefighters, clearly they will need more "training" or skills necessary to perform their jobs. As they advance or become more technically oriented, proper education with its theory and best practices is important. And finally, experience, which is continual learning, will provide real life "best practices."

Additionally, whatever you do and however you do it, today's fire service is and must be "professional." Not only is it essential that you train, educate, and provide experience, but also you should credential yourself and your organization the same as any other "profession."

All of these words and their definitions indicate to us that we are learners. We will learn knowledge, gain skills, and gather experience that will place us on a continuum of learning—a lifetime of learning. Thus, as professionals, we seek to maintain that which we have and strive to improve, enhance, and add to our knowledge so that we can perform in an excellent fashion.

The fire service has evolved from an organization whose single responsibility was fire suppression to an emergency services organization that provides fire suppression, fire prevention, fire code enforcement, fire investigation, fire inspection, plan review, emergency medical services (basic and advanced life support), hazardous materials mitigation, and specialized rescue operations (urban search and rescue, wilderness search and rescue, high-angle rescue, confined space rescue, trench collapse rescue, and water rescue). With these increased responsibilities come some of the greatest response challenges in our history, particularly that of terrorism events. Professionalism through learning is the key to our present and to our future.

THE TRAINING PROGRAM AND ITS MANAGEMENT

The fire service must train

That is a given. It is only logical, and simply makes sense in the technological environment. How training takes place is another discussion, but the fact of the matter is that you must train.

In order to have a training and education program, you will need to perform four things:

1. *Identify what your organization is going to do.* A major part of this process is to also identify or research what rules, regulations, laws, and standards may and do apply to what you have assessed that you are going to do. There are some excellent models for you to review and adopt as appropriate.
2. *Develop your training and education plan.* This plan should focus upon the results of your needs assessment, your organizations strategic planning and goals, and should simply and clearly address the "must" to know or do, the "need" to know or do, and finally the "nice" to know or do.
3. *Implement your plan.* This will take a plan in itself. You will need to identify all of the resources necessary to carry out your training and education plan. Additionally, as a chief officer you will need to lead and manage this implementation.
4. *Maintenance.* As with anything of value, your training and education plan will need to be maintained. You must continuously evaluate it and improve based upon changes in the "business," technology, methodology, and customer response.

Assessing your needs

In order to implement a logical training program you really need to start with the basis of your organization. A simple question is, "what does your organization do, what basic services does it provide, what is your plan of operations?" So, first thing you must do is assess your organization, or conduct a *needs assessment* for training that will tell you what you need to train for.

At this point, you will need to clearly define and delineate exactly what your organization does. What services does it provide, what degree or level with those services, and all functions that you perform as an organization. Your existing mission or vision statement is a logical starting point. If your organization has not defined its mission, then you should consider that process as the first step in defining your training agenda.

An organization might exist in the emergency services community today that will only respond to and function at wildland fires. If this is the case, then that organization does not really need to know about structural fire fighting. Another organization might respond to every call for help that is outside the scope of any other organization or department within that community or jurisdiction. This relates to the "call the fire department, they'll come to anything." And what does that mean? You need to clearly define that which you intend to respond to and mitigate or at least maintain and support until it is mitigated.

All of this focuses on the fact that you need to do and complete some organizational strategic planning, prior to identifying what you need to do with and about training. You need to have a plan to train and plan training. An excellent resource to start with for the strategic planning process is *Fire Department Strategic Planning: Creating Future Excellence* by Mark Wallace. He clearly demonstrates for the reader (chief) a 12-step planning process.

Knowing exactly what services your organization will provide (or nearly so), you can move to the next step.

A model to review for this is the National Fire Protection Association's (NFPA) 1201 Standard for Developing Fire Protection Services for the Public. This standard contains a chapter that addresses Training. This chapter will pose to you, the fire chief, major considerations for training in a fire department. The strongest point of this chapter is that it too starts with a needs assessment based upon the responsibilities consistent with what the organization does. You must identify what it is your organization does. Chapter 2 of this standard clearly identifies that the purpose of the fire department is a good model with which to start.

Develop your plan

Based upon the identification and assessment of your organization's functions and coupled with your research, you are ready to create a training plan and establish a training program. This is the point in the process that will be most involved, because it is at this point that you will create the detail that will guide your plan and program.

You have completed your assessment and identification of the absolute. Next address the planning—for the process to work you must plan for the people and resources that will make it work. Planning will be a daily event and strategic planning is imperative. This will focus on preparing the individuals at one level for the next step ("Rome wasn't built in a day"), or career advancement (upward mobility). So we'll need to focus on the training and education necessary to prepare individuals to perform additional or different functions.

We can identify training within three categories; *must* know, *need* to know, and *nice* to know. The first is that information or learning that an individual *must* know. Without any exceptions, this information is absolutely vital to the safety and successful conclusion of the operation in which the individual will be involved. Also in this category is all the information you identified through the research process to which the organization will be mandated.

Implementing your plan

The identification of the training needs is a point we have mentioned previously and is very clearly that which will drive and state what and why we need to do something. This step will define where the organization is at a designated time and where it will need to be in terms of knowledge, skills, and abilities or performances. It will also determine exactly what the need is and indicate whether it can be addressed by training, equipment, change in procedure, or any combination of these. It will identify who needs the training, equipment, or procedural change. Finally it will identify a method to address this need or achieve the challenge as established by the needs assessment.

Considerations within selection or identification of the performances are to clearly identify the learner, state exactly what is required of the learner, and specify the degree or how it will be measured by the learner. A consideration at this point is to identify already established performances that are available within the national professional qualifications standards and use those as the outcome performances you expect of and for certain classifications and levels of personnel.

The implementation process will require training facilities in order to support both didactic and skill training.

In planning for training, the fire department must arrange for classroom space that is comfortable, sufficiently large, and flexible in terms of seating configurations. Classroom facilities must be capable of accommodating training classes in basic fire fighting, emergency medical techniques, hazardous materials, response to terrorism, pre-incident planning, video conferencing, familiarization with SCBA (self-contained breathing apparatus), vehicle operations, the use of smaller tools and equipment, and a growing list of other disciplines. A modern fire training academy should include state-of-the-art classroom facilities as well as structural fire training buildings, SCBA training facilities, autos, tank trucks, aircraft, confined space, structures, flammable liquid and gas fires, and other simulated occupancies similar to those in the jurisdiction.

There is a constant challenge to avoid acquiring outdated equipment such as tank trucks that have obsolete value, but which an oil company may be happy to discard. The increasing environmental concern and regulations have also presented a challenge to fire training with air, water, and ground pollution. Many agencies have moved to flammable liquid fire simulators that burn flammable gas bubbled through water to simulate a liquid fire. The process reduces the air, water, and ground pollution.

Training facilities are complex and often appear out of reach of many budgets. There has been a growing use of consolidated training facilities that allow several jurisdictions to pool their resources to jointly construct and operate training facilities. A basic training facility with an administration/classroom building (90-student capacity), structural burn building, fire extinguisher training prop, forcible entry prop, ventilation prop, confined space prop, flammable liquids and gas props, vehicle extrication pad, and support building will cost in excess of three million dollars.

However, an ongoing level of maintenance, repair, and improvement in the buildings, vehicles, props, and equipment must support the initial investment. The changing environmental requirements and technology level of equipment and props will make the training academy a significant budget function.

Part of the training academy operation will be a constant vigil for safety. NFPA 1403 makes an important distinction between acquired structures and training center burn buildings. The standard recognizes that a structural burn building is much safer to use and is easier to manage safely than are acquired structures. Quite simply, at the training center burn building, it is easier to control the work site. The contents of formerly occupied buildings present a considerable hazard during live fire training, and often must be removed. There are also problems with exposure buildings, water supply, parking, security, and environment when using acquired structures. However the safety requirements of NFPA 1403 make an excellent guide for structural fire fighting training.

Additionally there are several NFPA standards you will need to consider with the operation and maintenance of a training program, particularly if you maintain a facility and conduct any type of practical skills training:

- NFPA 1401 Recommended Practice for Fire Service Training Reports and Records
- NFPA 1402 Guide to Building Fire Service Training Centers
- NFPA 1403 Standard on Live Fire Training Evolutions
- NFPA 1404 Standard for Fire Service Respiratory Protection Training
- NFPA 1405 Guide for Land-Based Fire fighters Who Respond to Marine Vessel Fires
- NFPA 1410 Standard on Training for Initial Emergency Scene Operations
- NFPA 1451 Standard for a Fire Service Vehicle Operations Training Program
- NFPA 1452 Guide for Training Fire Service Personnel to Conduct Dwelling Fire Safety

Evaluation and maintenance

Consider how you will maintain what you have just developed and implemented. What I mean here is "refresher training," "in-service," or whatever you want to call it such that everyone in your organization can do what it is you say they can do, naturally based upon your assessment, research, and development. You must also continue to reassess what you are doing as well as stay current on your research so that you will maintain your developed training plan.

MANDATORY TRAINING

The *must* know

Within the must know category of information or learning is the federally mandated requirements. Currently there are three basic federal regulations that all firefighters are subject to. These regulations address hazardous materials, respiratory protection, and bloodborne pathogens.

1. Occupational Safety and Health Administration (OSHA) Hazardous Waste Operations and Emergency Response (HAZWOPER) Training (29 CFR 1910.120)—this is the federal OSHA regulation that requires training for emergency response personnel who respond at identified levels of response to emergencies involving hazardous materials.
2. Occupational Safety and Health Administration (OSHA) Respiratory Protection Training (29 CFR 1910.134)—this is the federal OSHA regulation that requires the training, health monitoring, and fit testing for anyone who routinely utilized respiratory protection as part of their occupation.
3. Occupational Safety and Health Administration (OSHA) Occupational Exposure to Bloodborne Pathogens (29 CFR 1910.1030)—this is the federal OSHA regulation that addresses the need for an exposure control plan, education and training of employees, personal protective equipment, control, and vaccinations for individuals who are likely to be exposed to blood and other body fluids.

These federal regulations clearly identify areas that you must have your personnel trained and qualified for. Additionally, these laws require that the certifications must be maintained on an annual basis. Dependent upon your situation and your state's position, you may be required to address these.

Another consideration in the development of your plan is to research and review what similar and not so similar organizations have developed. It is an excellent opportunity to see what others have experienced and how they might have addressed and handled it.

There are several organizations that you might contact that will help in the research and review of other organizations:

- International Association of Fire Chiefs
- International Association of Firefighters
- National Fire Protection Association
- National Volunteer Fire Council
- United States Fire Administration

All of these organizations have information on fire organizations in this country. How do they do what you want to do and what do you think about how they do it? These can be excellent models as well as helping establish benchmarks with which you can strive to improve your program.

CERTIFICATION BASED TRAINING

The *need* to know

One method of addressing the *need* to know information and learning for your organization is to align your functions, based upon your needs assessment of your organization to the professional qualifications standards. This will create or utilize "certification based training and education." Certification based training prepares an individual to perform the competencies of the professional qualifications standards. This training and education focuses on the knowledge and skills necessary for the performances within a given standard, and can test and evaluate them to the standard. It is a "checks and balances" for your training program. It will produce emergency services personnel who are qualified, competent, and professional.

The result of this process is that your organization will produce individuals who are qualified for testing and evaluation to a standard. They will be certified either locally or at the state level as competent to a standard and the certification may be accredited by a national certification accreditation organization. This is the basis for the professional certification systems.

Professional qualifications of emergency services providers is necessary in today's ever changing and especially litigious society. The fire chief must not only assure that the highest degree of professionalism exists; but must now also assure that the professional is certified. This certification can be in many forms and by many methods. As with any certification system it must be realistic, credible, and must have passed some degree of validity measurement. These systems, once established, will impart upon the professional a readily identifiable sense of credibility and validity. These systems will establish the measures by which the professional will be judged and so certified to perform. These systems are, in general, a peer evaluation of minimum standards measurement.

The previous paragraph mentioned the evolution of our litigious society. That is one very clear reason for certification; just a few of the more obvious reasons for certification, and in particular national certification, are:

- *Recognition.* National, state, local, departmental, and peer recognition for demonstrating proficiency of the requirements for a national professional qualifications standard.
- *Credibility.* Certification is based on a national professional qualification standard that is constructed from the required performance proficiency for the fire service. This performance standard establishes a nationally recognized base for the measurement of fire service personnel abilities. Achievement of certification will establish credibility legally, functionally, and socially within the fire service community.
- *Professionalism.* Certification will provide that badge or symbol to identify a professional. It does not matter whether the individual is career or volunteer, from a small or large department, certification is the identifier of professionalism. It provides a statement of accomplishment, more than just mere words can provide.
- *Budget justification.* The fire service has become more complicated and we, like all service organizations must provide justification of and accountability for our existence. Certification provides a nationally recognized base to rationalize our profession as well as rewards for achievement.
- *Transferability.* National certification provides for improved mobility of fire service personnel. A certified fire service person has the ability to easily and quickly transfer from one department to another.

A collateral benefit of these certification systems is the standardization of our education and training programs. The standards we recognize today are performance standards, although many have become the core of our education and training programs. The intent of performance standards is to evaluate an individual regardless of their education, training, and experience. Through time, these standards have become the standard by which we evaluate an individual upon completion of an education and training program.

In the summer of 1993, the NFPA released the first professional qualifications standard revised utilizing a new format. This format states the performance objective and is followed by subsections that state the prerequisite skills and/or knowledge. The standard in this format addresses the education and training necessary (prerequisites) with the performance objective. Education and training programs can be developed to encompass the requirements of the standard by simply covering the prerequisites.

The early Wingspread conferences of 1966 and 1976 called for professionalism of the fire service. The first conference was held in February of 1966 when a group of fire service people gathered at the Wingspread conference center. This center is part of the Johnson Foundation in Racine, Wisconsin. The Johnson Foundation funded the conference and the publications resulting from the conferences. The people attending were from the fire service who felt the need for further study. They saw the fire service as thousands of individual fire department organizations attempting to cope with their vast numbers of responsibilities. These conferences centered on fire service administration, education, and research, examining avenues to enhance and develop all aspects of the fire service. One of these was certification. In 1970, as a result of the first Wingspread conference, a meeting in Williamsburg, Virginia was held. This meeting was the idea of Charles S. Morgan, then President of NFPA. He called together chief executives of principle fire service organizations for the purpose of establishing a line of communications to discuss issues of interest. The Williamsburg Conference resulted in the creation of the Joint Council of Fire Service Organizations. In 1971, this body appointed a committee to develop a national firefighter certification system. From this committee came the creation of the National Professional Qualification Board (NPQB or "Pro-board") in 1972 and the first technical committees for each type and level of fire service professional qualifications. These efforts resulted in the creation of the National Professional Qualifications System (NPQS) and in the establishment of the first state certification systems accredited by the NPQB. The first three states accredited were Iowa, Oregon, and Oklahoma. From there the rest is history.

Variety of certification systems

From a survey conducted in 1992 by the University of Illinois (with additional information compiled in 1993 by the University of Maryland), we learn that all of our states have some type of certification system in place (50). It is important to note that a certification system can be created on a local, county, regional, and state level as well as nationally. This means that a credible system based upon an acceptable standard can be utilized and recognized in different formats. The state may develop its own set of standards and criteria or the state may adopt any or all of the nationally recognized standards as published by the National Fire Protection Association (NFPA) or the American National Standards Institute (ANSI). These systems may be administered in a variety of ways. They currently exist from very bureaucratic styles of boards with mandatory standards to volunteer boards with voluntary standards. The important point is that all participants agree to the system and have reasonable expectations from the system.

Professional qualifications standards

NFPA 1000 *Standard on Fire Service Professional Qualifications Accreditation and Certification Systems.* This standard provides to the certification systems a guideline for the operation of the system. The minimum criteria for certification bodies as well as for the assessment and validation of the process to certify individuals is established within this standard. Any system interested in national accreditation of their certification system should follow and operate according to NFPA 1000.

NFPA 1001 *Standard for Professional Fire Fighter.* One of the four original professional qualification system standards, NFPA 1001 is perhaps the most widely utilized and definitive standard in the fire service. It is, in a sense, the starting point for all the standards in the professional series. At the entry level, the firefighter will be able to define, describe, and demonstrate a range of topics relating to all aspects of suppression activity, including the basics of fire behavior, fire fighting strategies and equipment, the fundamentals of emergency rescue, and methods of identifying a hazardous materials incident. As progression occurs through this standard, each of these topics will be investigated in greater depth—the expectation being that the more mature firefighter will have a better understanding of all aspects of his mission. At every level of this standard, firefighters will be expected to ably demonstrate a thorough knowledge of safety equipment including, but not limited to, self-contained breathing apparatus. It presents the job performance requirements Firefighter I and Firefighter II.

NFPA 1002 *Standard for Fire Apparatus Driver/Operator.* This standard applies to all legally licensed fire department apparatus drivers/operators, whether they be designated apparatus drivers or all firefighters who drive fire department vehicles under emergency response conditions. All firefighters must meet the requirements of Firefighter I in NFPA 1001 Standard for Fire Fighter Professional Qualifications, before being certified as a Fire Apparatus Driver/Operator. The driver/operator standard requires the candidate to perform simulated operations and responses on apparatus equipped with a fire pump, apparatus equipped with an aerial device, apparatus equipped with a tiller, apparatus designed for wildland fire suppression, as well as apparatus utilized for airport crash and rescue operations—although local demand will determine emphasis and order of priority. The candidate will also be responsible for knowledge of preventive maintenance and emergency response safety. In addition, all tests and routines necessary to maintain the safety and integrity of fire department vehicles will be examined. It presents the job performances for Driver/Operator—Pump, Aerial, Tiller, Airport Crash and Rescue, Wildland, and Mobile Water Supply.

NFPA 1003 *Standard for Airport Fire Fighters.* There is only one level of Airport Fire fighter in NFPA 1003, and to achieve it is indeed a commitment to professional growth. Airport firefighters are multifaceted public safety employees who are responsible for both the apparatus and human safety from fire and other hazards while on airport property. Under the mandates of 1003, they must be expert on airport procedures and layout, the details of both military and civilian aircraft, fire prevention, as well as the techniques of emergency rescue and emergency medical care. In addition, airport firefighters must possess all the knowledge of non-airport firefighters in terms of general fire suppression techniques, pre-planning, and record keeping.

NFPA 1006 *Standard for Rescue Technician Professional Qualifications.* NFPA 1006 is the standard that prepares the firefighter in several major technical rescue areas. It presents the job performance requirements for general site operations during a rescue incident and specific job performance requirements for rope rescue, surface water rescue, vehicle and machinery rescue, confined space rescue, structural collapse rescue, and trench rescue. The next edition will present the job performance requirements for sub terrain rescue, dive rescue, and wilderness rescue.

NFPA 1021 *Standard for Fire Officer.* NFPA 1021 is the standard that prepares a firefighter for taking a leadership position within the service. It recognizes four levels of progressive achievement, from the junior officer level to chief fire officer. It is in every aspect a comprehensive standard, asking candidates to perform to the highest levels of both fire ground command as well as fire department administrative management. As officers move through the fire officer series, they will be expected to augment fire training with knowledge of other disciplines including psychology, public administration, local, state, and national government, the law, as well as demonstrate an ability to communicate within and outside the service. Today's fire officers, at all levels, must take command not only of demanding multi-cultural suppression forces, but should also be prepared to assume leadership roles within the community. This standard presents the job performance requirements for Fire Officer I, Fire Officer II, Fire Officer III, and Fire Officer IV.

NFPA 1031 *Standard for Fire Inspector and Plan Examiner.* A fire inspector, at all levels of NFPA 1031, must be foremost an excellent communicator. The scope of their job is enormous considering that fire inspectors have the power to declare a building reasonably risk-free from all predictable fire hazards. In this capacity, the fire inspector works closely with community members and other professionals outside the fire department, including engineers, architects, and builders. Using locally-mandated codes and regulations, the fire inspector must be thoroughly expert on these codes in order to monitor building practices and materials. In order to achieve certification as a fire inspector, the candidate will be asked to describe aspects of fire behavior, to detail specific hazards and risks, and to identify all types of fire suppression systems, as well as emergency evacuation plans. This standard presents the job performance requirements for Fire Inspector I, Fire Inspector II, Fire Inspector III, Plan Examiner I, and Plan Examiner II.

NFPA 1033 *Standard for Fire Investigator.* Fire investigators are primarily responsible for the investigation of fires, explosions, and other property-destroying events of suspicious or accidental origin. The standard encompasses professional growth in both fire and legal disciplines; in fact many jurisdictions mandate that fire investigators be trained to meet the requirements of law enforcement officers. In this sense, fire investigators must be trained in both worlds, and must be able to move easily between investigatory and scientific modes. They will collect evidence and help prepare documents for courtroom presentations. Their area of expertise is vast, ranging from the procedures of salvage, overhaul, and rescue, to specific knowledge of hazardous materials, to the storage, handling, and use of flammable and combustible materials. To achieve certification as a fire investigator signifies the mastery of at least two challenging professions.

NFPA 1035 *Standard for Public Fire and Life Safety Educator.* This standard describes the three progressive levels of competence required for the public fire educator—that person or persons within a fire department or public safety agency who interacts regularly within communities, schools, and civic groups dispensing information and material demonstrating fire safety. NFPA 1035 mandates that the public fire educator be thoroughly familiar with the fire department servicing the community, although the public fire educator may be a civilian member of the fire service. Knowledge required for achieving certification in this standard ranges from techniques of building construction, general fire hazards, suppression systems, and a familiarity with all applicable community codes and standards. In addition, the public fire educator (all levels) will be able to demonstrate excellent written and oral communication skills, as well as methods of classroom and informal group instruction. This standard presents the job performance requirements for Public Fire and Life Safety Educator I, Public Fire and Life Safety Educator II, Public Fire and Life Safety Educator III, Public Information Officer, Juvenile Firesetter Intervention Specialist I, and Juvenile Firesetter Intervention Specialist II.

NFPA 1041 *Standard for Fire Service Instructor.* This standard establishes the minimum professional levels of competence required of fire service instructors. Three levels of instructorship are examined, each representing a particular phase in professional development. For instance, a Fire Service Instructor I will be asked to define terms and concepts relating to elementary levels of educational theory and to demonstrate several components of instructional techniques; while a Fire Service Instructor III will be examined on their capacity to administer and manage a fire service-training program. At all levels, candidates will be asked to demonstrate a range of skills, including goal-setting, record-keeping, and budget preparation. All candidates seeking certification as Fire Service Instructors should expect to be examined closely on their communication skills. This standard presents the job performance requirements for Instructor I, Instructor II, and Instructor III.

NFPA 1051 *Wildland Fire Fighter.* This standard presents the job performance requirements for Wildland Firefighter I, Wildland Firefighter II, Wildland Fire Officer I, Wildland Fire Officer II, Wildland/Urban Interface Protection Specialist, and Wildland/Urban Interface Coordinator. It addresses the necessities of wildland fire fighting at all levels of emergency services local involvement, including fireline safety, personal protective equipment, map reading, basic understanding of structural fire fighting strategy, and tactics in a wildland/urban interface.

NFPA 1061 *Public Safety Telecommunicator.* This standard presents the job performances requirements for Public Safety Telecommunicator I and Public Safety Telecommunicator II. It addresses performance of basic call talking, operating of the phone and radio systems, computer operations, controlling the conversation, listening techniques, and handling of special requests. These are addressed in several main headings of *receive information, process information, and disseminate information.*

NFPA 1071 *Emergency Vehicle Technician.* This standard presents the job performance requirements for Emergency Vehicle Technician I and Emergency Vehicle Technician II. It establishes knowledge and skill considerations addressing the organization of the maintenance facility, role of the technician, rules and regulations, tool safety, and vehicle operation.

NFPA 472 *Standard for Responders to Hazardous Materials Incidents.* An increased recognition of the importance of emergency responders to hazardous materials incidents has resulted in the creation of NFPA 472. This standard provides professional guidelines for those who encounter hazardous materials situations in the course of their normal duties, and of those for whom hazardous materials mitigation is their sole responsibility. At the awareness level, a candidate will be expected to recognize a hazardous materials emergency, be able to identify visible hazards, and to initiate the notification process. At the responder level, the candidate must be able to describe and demonstrate all methods of hazardous materials data collection and analysis, estimate potential risk, begin a response, and initiate the incident management system—among many other skills. At the level of incident commander, the candidate will be asked to meet all the competencies of both the awareness and operations responder levels, as well as all other training mandated by state and federal agencies. They will be involved with community-wide master planning for hazardous materials incidents, including risk-assessment and plans for the safety of citizens and their property during a hazardous materials incident.

NFPA 1521 *Fire Department Safety Officer.* This standard presents the objectives for the Fire Department Health and Safety Officer and the Fire Department Incident Safety Officer. It further addresses the requirements of the incident safety officer as they have application to fire suppression, hazardous materials, emergency medical services, and special operations.

NFPA 1600 *Disaster/Emergency Management and Business Continuity Programs.* Although this standard clearly is not a professional qualifications standard, it is being utilized to measure the basic ability of individuals to perform the program elements of disaster/emergency management. I mention it here because it is increasingly becoming a common practice that emergency management is part of the local fire service, and because of our changing environment—particularly as it relates to terrorism response—fire service leaders must be knowledgeable in this area. It is extremely important in the practice of emergency operations centers and unified command.

Major systems for accreditation of certification systems (nationally)

National Board on Fire Service Professional Qualifications (NBFSPQ). The National Board on Fire Service Professional Qualifications (NBFSPQ) was created after the dissolution of the Joint Council of National Fire Service Organizations in 1990. With the dissolution of the joint council, the former National Professional Qualifications Board had no "home" or sponsoring body for the National Professional Qualifications System (NPQS). The NBFSPQ was established to preserve the activities of the NPQS. Membership of the NBFSPQ consists of the International Association of Fire Chiefs (IAFC), the International Association of Arson Investigators (IAAI), and the National Fire Protection Association (NFPA).

When the NBFSPQ was originally formed in 1990, the International Association of Firefighters (IAFF) and the International Society of Fire Service Instructors (ISFSI) were members. They have since left the NBFSPQ. The NBFSPQ has a committee on accreditation (COA), serving the functions of the former NPQB. This committee continues the accreditation of state and local certification systems. As of September, 1993, there are 16 accredited certification systems, either state or Canadian province. These individual systems will issue voluntary national certificates with their certification in one or more levels of the appropriate standards. The NBFSPQ with its predecessor has issued over 15,000 national certificates.

International Fire Service Accreditation Congress (IFSAC). The International Fire Service Accreditation Congress (IFSAC) was established through the efforts of the Oklahoma State University in 1991. The Congress membership consists of all certification systems desirous of participation in the system. Any jurisdiction, state, or province that has a certification system may participate in the Congress. It was the result of a national meeting sponsored by the National Association of State Directors of Fire Training and Education. This meeting focused on the review and discussion of fire service certification and accreditation issues. The participants of this meeting were exploring an avenue of accreditation to meet the needs of their particular certification systems. Additionally, at the time, the future of the NPQB was uncertain with the dissolution of the Joint Council. The Congress has accredited approximately 15 state, local, or province certification systems. IFSAC has issued over 15,000 certifications through their accredited entities.

Chief Fire Officer Designation (CFOD). This is a relatively new program for the recognition of chief officers' achievements. It is an independent commission created under a trust agreement between the IAFC and the International City/County Management Association. The purpose of this program is to assist in the professional development of the fire and emergency service personnel by providing guidance for career planning through participation in the Professional Designation Program and to ensure continuous quality and improvement. This is a voluntary program designed to recognize individuals who can show their excellence in seven measured components including: Experience, Education, Professional Development, Professional Contributions, Association Membership, Community Involvement, and Technical Competencies. This is accomplished through an application process that is overseen by a board of directors.

To qualify for this certification requires an individual to create a portfolio that addresses each of the seven measurement components. Each component is graded independently with a minimum score necessary for success. A board of review with members of different backgrounds—including fire, emergency services, academia, and municipal groups—will score the components. The individual will receive a summary of the score and what might be necessary if minimum scores were not awarded. This is an excellent opportunity for the chief officer to benchmark their profession within a peer system.

For more information on this excellent chief officer career planning and tracking process, contact the Commission on Chief Fire Officer Designation.

Professional Development

The *nice* to know

The third category of training consideration is that which is *nice* to know. It may be excellent information, but dependent upon the operation and functioning assignment of the individual, it may not be necessary to know. An example of this is the firefighter who responds as part of an engine crew to a train derailment. Upon arrival, and from a very safe distance, the officer identifies that the cars visible from this distance contain dangerous chemical or commodities. There are several cars involved and all appear to be containing different, but dangerous commodities. The officer orders an immediate evacuation of the area and retreats the crew and apparatus to a safer area. The firefighter's job description and training level have brought firefighters to a level of response such that they can perform certain functions under the direct supervision and control of an officer. This scenario is beyond the firefighter's training. The fact that commodity A and commodity B, when mixed, will destroy all life within 500 feet would be "nice to know" information, but not necessary. How to obtain the information that the mixture will destroy all life is *must* know information, but the actual chemical or commodity reaction and the "how it will destroy all life" is really *nice* to know.

A little more to know

A fourth category of training that is seldom considered, but is included just as a thought here so that you see the entire picture, is the category of "don't really need to know."

An example of this is a firefighter who is functioning as a firefighter in the Florida Keys. That firefighter will never need to know about rescue situations involving snow avalanches. But, they might want to learn about those situations because they are considering moving to a department that conducts mitigation of those types of situations. The deciding factor is based upon your organization's needs assessment and it clearly indicates that your personnel will not need to know this type of information.

All of this is a constant reminder to check to see "why" you are doing something. The most precious commodities of any organization are the people and the organization's time. Waste of either of these is unacceptable.

Summary

We just reviewed a fairly simple three-point consideration of why we train. We need to consider in more depth the *must* and *need* to know information in order to complete your plan.

As mentioned previously, the *must* know information will be addressed and readdressed as necessary for all of your personnel. The *need* to know information can be approached from three points:

1. Initial training—the baseline training to perform at the identified level (Firefighter I, Fire Officer I, etc.), as well as the training to perform an identified function, event, or activity.
2. Refresher training—all training that reviews, and competencies as developed in the initial training, as well as any updates, changes to processes, policies, or performances that were a part of the initial training.
3. Enhancement training—all training that is beyond the first two points. Dependent upon the organization, this could be promotional training necessary for advancement within the organization, although some organizations may elect to provide this type of training as initial. Education provided through an institution of higher learning or education could be considered as enhancement for the individual.

Training Methods—Traditional/Non Traditional

Training is conducted in an increasing number of methods. We can approach them from the traditional classroom or academy setting to the non-traditional approach of distance learning with blended learning used to insure skill practice and teamwork, as well as performance evaluation.

Today there are a vast number of resources available for you to "develop" this training. One major consideration is to not waste time, effort, finances, and other resources in conducting any training. You need to research and review what else is on the market in our business. Why decide to develop from scratch a Firefighter I curriculum that is designed to clearly prepare the individual to perform the competencies of the national standard when there are literally hundreds of those curriculums available? Many of those curriculums are available from other organizations for little or no cost. The greatest challenge is to insure that whatever you are planning, it is sound, based upon your needs, and every step of the plan or training delivery has a reason why (with appropriate documentation).

Methodology

The specific methodology utilized for training and education is a study of and in itself, and clearly an art. We continue to study the "psychology of education" and add to the research and development of this study each and every time we develop a new curriculum, course, and/or program. The most important consideration is that your methodology addresses the learning style of your learners. Although this is difficult to accomplish, it will render the greatest success.

The fire service has been a very tradition-based operation. As such, the training methods have been based in a paramilitary style of delivery, similar to our style of emergency response delivery. We continue those methods today and attempt to blend them with other methods that will avail better utilization of time and financial resources. Distance learning is one of the greatest opportunities we have today. This opportunity can be anything from a self-study book to an online course. This online course can be coupled with practical skill and exercise sessions that will insure, not only the knowledge necessary, but the practical skill and performances as well.

To truly take advantage of all of these opportunities, the fire chief must be ready to explore each and every method available for program delivery and utilize that which will serve the organization the best. There is no right or wrong way, just different ways that will work in different situations.

One final thought with the methodology: don't forget to consider how the training you are conducting might be blended into education and recognized for higher education credit. The American Council on Education offers an evaluation program called the College Credit Recommendation Service. They will evaluate your training and education program against known standards within the academic arena and offer college credit recommendations. These recommendations can be taken to colleges and universities for possible acceptance. This adds additional value to your training program. It is also an excellent method to evaluate what you are doing and how well you are doing it.

Educational methodologies

I. Traditional
   A. Lecture
   B. Discussion
      1. Guided
      2. Conference
      3. Case Study
      4. Role-Playing
      5. Brainstorming
   C. Practical Skills / Demonstration
   D. Illustration
   E. Team Teaching
   F. Mentoring
II. Non-Traditional
   A. Individualized Instruction
   B. Self-Directed or Independent Learning
   C. Distance Learning
   D. Open Learning
   E. Computer-Assisted Instruction
      1. Blended Learning

Training Objectives—Educational Objectives

The training needs assessment identified the job performance requirements in the form of knowledge, skills, and abilities. Each performance requirement will necessitate the development of an educational objective. The instructional design individual—who may be the chief, the training section chief, or an instructor—must identify what is to be accomplished and how.

The result of the process is a behavioral objective that identifies a student behavior pattern that is to be modified. The construction of a behavior objective requires the identification of the action (such as define, discuss, apply, analyze, assemble, or interpret), the content or subject of the action, the minimum acceptable performance, and any special conditions.

NFPA 1041 Level II Instructor training will prepare the fire service instructor to identify needs and produce quality objectives. The evaluation of course objectives should be a continuing responsibility on the basis of the department's needs, student performance, and instructor ability.

Current educational methodology calls for the creation of a student performance objective that is a behavior objective. Then a series of enabling objectives are developed to establish a step-by-step process for learning.

The behavior objectives also are the basis for evaluation of both student performance and program performance. The objectives are the plans for education and training, the building blocks of learning, and the measuring device utilized to judge performance.

Training Records and Reports

Proper record keeping is a requirement for the emergency services organization. Training records are not only essential for the operation of the department in the form of internal documentation, but are usually also mandated by laws and regulations that drive the "must know" disciplines. While record keeping can be done manually, this approach makes little sense in today's technological world. If your department is one of the few still waiting to be computerized, training records might be a logical place to start. The types of documentation that need to be maintained include:

- Participants' names
- Lesson plan
- Instructor qualifications
- Competency and performance testing
- Number of hours of training
- Certification dates and levels

Your record keeping system should monitor required certification and refresher training needs to ensure adequate training and testing to maintain current certifications and skill competency. In addition, you should track total time and costs for department training. This process will not only assist you in budgeting and resource allocation, but will also give you a historical prospective of the training activity level in the department.

Your plan and implementation should also provide quality control on your curriculum, instructors, and student proficiencies. The utilization of a nationally recognized certification system will assist you in the quality control process.

Finally you need to compare the product to the plan to see if you are obtaining your objectives in the plan. You should also be able to measure training not only from the knowledge and skills testing of each student, but also in the performance level of your individual personnel, your companies, the department operations, and your service to the citizens of the community. The evaluation of your training program should be part of your department's strategic planning on an ongoing basis.

EVALUATING STUDENT PERFORMANCE

Student performance evaluations are generally conducted by two basic methods—either a written test or a practical skills test. With today's technology, these two methods can combine and expand to many different ways of conducting a test (such as Internet based written examinations and virtual reality performance simulators).

For our purposes, we will remain reasonably traditional and simple. Test questions or scenarios must be:

- Fair
- Based upon the training objectives and real world actions
- Valid and reliable

Fair in that the requested action as well as those involved with conducting evaluations should not discriminate or be biased towards any identifiable individuals or groups. The requested action should be based upon an occupational analysis (*i.e.*, evaluating what the individual will do in the real world). Validity and reliability are measurements through which the evaluations need to be vetted. Validity considers the review by subject matter experts (content validity) and review by test or evaluation experts (context validity) to ensure that the evaluation measures what it is suppose to measure. Reliability is the process of reviewing for consistency and accuracy of the evaluation. This is generally based upon some quantitative criteria.

The fire chief has several resources available for assistance in these areas. For example, with the consideration of pre-hire physical agility evaluation, the United States Department of Labor, in cooperation with the IAFF and the IAFC, has developed the Certified Physical Agility Training (CPAT). This program, with established criteria, is a consideration for the local department to measure an individual's ability to perform certain accepted firefighter functions. This program also develops the individual through a progressive physical training program and mentoring program to successfully pass the CPAT evaluation.

State and local fire service certification systems are available to assist the fire chief with the evaluation of personnel against the national professional qualifications. These systems are mentioned under the certification based training section of this chapter. They have established all of the necessary considerations for their evaluation processes, generally to nationally-accepted criteria as established by nationally accreditation organizations.

The main considerations for any evaluation can be based upon several questions:

- Does the evaluation evaluate what you intended it to?
- Is the evaluation a real and accurate representation of the real world?
- Does the evaluation discriminate against any identifiable group or groups of peoples?

- Is the evaluation valid and reliable?
- Are there established criteria by which the evaluation can be conducted?

These are some basic considerations in the evaluation of student performances. Based upon your needs assessment as well as established evaluation systems (certification systems) available to your department, you will need to decide to what degree you will evaluate your personnel.

An innovative way to both train and evaluate an individual is through "training in context." This method can be explained simply as a process whereby the individual is trained and coached to perform the skill and is then evaluated. This can be done through one process with four basic steps:

1. The performance expected is explained to the individual.
2. The final performance expected is illustrated or clearly demonstrated to the individual.
3. The individual practices the performance with coaching as necessary, striving for excellence in the performance.
4. The individual's practice is evaluated against the criteria established for performance. Information pertaining to the "knowledge" associated with the performance can be evaluated with the individual verbally as they perform. For example, an entire Firefighter I certification evaluation can be administered through practical "evolutions" with verbal responses.

EVALUATING THE TRAINING PROGRAM

Evaluating the training program can take different forms by different people with different levels of involvement. Traditionally we have evaluated training programs with great technical scrutiny during and after the first delivery. We should evaluate a training program during and after each delivery, periodically, and upon any significant change to the standards (if applicable, as the program is based upon defined standards). When we evaluate a training program there are several major items we want to involve:

- The instructor guide
- The student manual and student text
- The visuals
- The overall program delivery method

As we review, we are specifically looking for:

- Learners or students ability to perform the objectives

- Resources that we utilized as appropriately for the delivery
- Learners scores on evaluation instruments
- Completion of learners and instructors course evaluation documents (feedback)

A consideration of long term evaluation of the student and what the student learned and currently performs because of their completion of training is an area the National Fire Academy is involved with. They are using Kirkpatrick's four levels of evaluation. This is a sequential evaluation method that measures effectiveness of the training program over time.

- Level 1 Evaluation—Reactions
  - Immediate evaluation of the participant's reaction to the training program.
- Level 2 Evaluation—Learning
  - Moves beyond participant's reaction to assessment of participant's advancement in skills, knowledge, or attitude, requires use of tests conducted before the training and after the training.
- Level 3 Evaluation—Transfer
  - This evaluation looks at the participants' use of the knowledge obtained in their everyday activities.
- Level 4 Evaluation—Results
  - This evaluation requires the feedback from the participants' department or supervisor as to how well they are performing based upon the knowledge learned in the program.

This process is extensive but does cover the majority of the aspects of both the training program and the individual's application of learning how to do the job.

Training Resources

Local

Interact with other organizations within your immediate area. Find out what do they do, how to they do it, and what you can use of theirs. Another consideration is what the two organizations can do together that will be of mutual benefit. Development of a training program to meet a specific need will be better enhanced by involving people who will not only use the program, but bring new and fresh perspectives to the process. This is an excellent opportunity to consider regional opportunities within your locale.

State

Utilize your state training agency. Dependent upon the state you are located in, this agency may be located generally within one of several departments or areas of your state government or system. State training agencies are higher education based (such as located within institutions of higher learning), department of public safety based (to include state fire marshal's offices), independent based (generally with a commission or oversight board), or other state agency based (such as vocational education, insurance, emergency services, emergency management, etc.). To learn more about the state training agency or for information and assistance, contact the North American Fire Training Directors Association. They have an online site at *www.naftd.org*.

The state training agency may in some instances not only be able to provide you with the specific training program you are looking for, but they may have access to other state and federal resources that will assist you.

Federal

The National Fire Academy (NFA). In the recent past we have observed a tremendous growth in the federal fire focus. From earlier national reports such as *America Burning* in 1973 to the fire act grants of today, federal involvement and support has been greatly enhanced. Our primary focus today is with the Federal Emergency Management Agency (FEMA), which is the home agency for the United States Fire Administration (USFA). Within the USFA are the National Fire Academy and the Emergency Management Institute, both located at the National Emergency Training Center in Emmitsburg, Maryland.

The NFA offers a tremendous number of programs as well as direct training and education. This is offered through resident and direct academy deliveries (on campus in Emmitsburg as well as field sites), regional academy deliveries (off campus deliveries of courses in the FEMA regions), enfranchisement and endorsed course deliveries (opportunities for individual states to deliver academy hand-off courses, identified resident courses, and courses from their state or other states that have been evaluated and approved), distance learning, and hand-off courses. The NFA has developed numerous courses that are available for purchase by local departments.

Additionally, the NFA is the focal point for the Open Learning Fire Science Program conducted through institutions of higher learning. This Degree at a Distance program is a standardized list of courses with curriculums these institutions may utilize.

There is a growing number of chief fire officers who pursue and complete academic degrees in management, leadership, public administration, business administration, information technology, and many other disciplines. You must evaluate your own needs, as you did for the department, and plan for your future. It is a tribute to the increasing levels of training, education, and experience that we see fire chiefs progressing to city/county manager positions, becoming elected officials, and joining the ranks of senior managers in state and federal government. This trend is a clear result of the increasing professionalism demonstrated by the emergency service community and personnel.

The trend for higher education in the fire service is supported by the American Council on Education (ACE) process for college credit recommendations for vocational training. The National Fire Academy and several of the state training agencies have presented ACE with courses to evaluate. These credit recommendations provide many emergency services personnel with their start in pursuing an academic degree.

The Emergency Management Institute (EMI). This Institute is the training branch of FEMA for the study and training program development and delivery as it applies to preparedness, response, mitigation, and recovery from natural, human, or technological disasters. It is responsible to and a part of the USFA. They have developed excellent programs that are available only through your state and local emergency management. They offer generally one-week resident programs that address all activities and levels of operations within emergency management. Similar to the NFA, they have available college level curriculum for use by institutions of higher learning.

Proprietary programs. Too many to identify—simply perform a web search at *www.google.com* and see what you find. These are private vendors, for profit and not-for-profit organizations, and institutions that have training available for sale. Dependent upon your immediate need, mission, and budget, these organizations can offer you other methods for obtaining your training. They can provide training on location, at specified sites, and through a vast area of different highly technical delivery methods. It could be an excellent way to maintain certifications (blood borne pathogen training certification, for example).

Conclusion

Over the years, fire service work has become less physically demanding and more mentally challenging. Lighter weight hose, ladders, and breathing apparatus, combined with incident management procedures that rely on team actions, have reduced the need for individual firefighters to depend solely on strenuous physical effort. A properly functioning incident management system emphasizes that companies work as teams, reducing the frequency of one-on-one labor-intensive rescues. Meanwhile, as more firefighters have had to become trained as emergency medical technicians, hazardous materials technicians, public education specialists, and many other roles, the job has become more mentally challenging. Truly proficient firefighters must learn about hazardous materials, modern building construction techniques, response to terrorism, and a host of other technical subjects in greater detail than ever before.

This increased challenge of technology has brought with it the need to provide a professional firefighters with certification to prove their skill and performance levels. The public not only expects this professionalism and technological competency, but the courts have looked to certification and training records as significant elements of protection from the ever growing number of law suits.

The face or look of today's and tomorrow's firefighter is changing. In the past we concentrated on safely and effectively "putting out the fire." Through the years this has evolved into other areas such as emergency medical services and hazardous materials incident mitigation. Thus we concentrated on our firefighters being good, standard-based individuals (Firefighter II, EMTB, and Hazardous Materials Operations). However, after the events of September 11, 2001, the face is changing again—and quickly. Those events have taught us that we will respond, and be the first responders to incidents we have not planned for nor trained for. Our Firefighter II of today will need to be trained, not only as before, but will need to address the immediate and support operations of technical rescue (such as rope rescue, structural collapse, confined space, trench rescue, explosive mitigation, chemical, and biological considerations). These individuals will need to support specialty teams of individuals who will respond (such as local, state, and federal urban search and rescue teams). Thus, we will need to advance our training, based upon this needs assessment, to include those specialty topics, if only at the operations level. Additionally, the firefighter will need to work closely with other responders such as law enforcement and public health.

Today's fire department, regardless of size or configuration, is essentially incomplete and virtually nonfunctional without adequate education and training. If the leaders of the organization have attained an educational level that allows them to supervise and manage the organization effectively, the department's ability to achieve its complex mission is enhanced. Today's chief is forced to be both a manager of resources and a leader of people. While the fire service leaders of today are faced with many complex problems, current needs must also be balanced against future needs and a sense of direction. The difference between good fire departments and great fire departments, and the difference between capable leaders and great leaders, is a matter of knowledge—knowledge that is gained through training and education.

Section III

EQUIPMENT

10

FIRE DEPARTMENT APPARATUS

William C. Peters
Gary R. Pope

CHAPTER HIGHLIGHTS

- **Overview of needs assessment factors**
- **Review of NFPA 1901 and Annex Material**
- **Discussion of different types of apparatus and equipped features with advantages and disadvantages of each**

INTRODUCTION

The need for effective firefighting apparatus has been an issue since the very existence of this country. In early colonial times, some of the first settled communities were almost totally devastated to extinction by the ravages of fire.

Political leaders of the colonial communities attempted to meet these needs by purchasing imported hand pumpers to protect their communities. These hand pumpers were very expensive, primitive in design, and produced minimal output, which made them effective only on small fires. The more major incidents overran these apparatus very quickly, indicating a need for larger volume and less manpower-oriented apparatus. As the nation grew and technology expanded, apparatus to meet these characteristics was developed and has grown to what the fire apparatus industry reflects today.

The process of specifying and purchasing fire apparatus has been and remains a challenge to fire chiefs and political leaders throughout the country. Acquiring needed fire apparatus is a major endeavor for both small departments as well as large departments. As it was in the early days of this nation, fire apparatus represents a major investment to the communities. For this reason, fire apparatus must be dependable and remain in use for many years. This was not necessarily the case for some of the early rigs, particularly some of the steam-powered rigs.

The process to procure new apparatus is a long and cumbersome process and is very technically oriented. In many cases, the process can also get distorted because to many departments and fire chiefs, it is as much an emotional issue as it is a technical issue. Hopefully, the information contained within this chapter will help fire chiefs overcome obstacles related to the technical end of the process.

Conducting a needs assessment of the operations of the department, understanding the current NFPA Standards related to fire apparatus, and being familiar with the latest apparatus available in the industry should aid those persons responsible for the procurement of new fire apparatus. However, the emotional aspects need to be addressed from within the individual departments. Just keep in mind that good sense and sound technical knowledge should help overcome poor judgment based on emotional issues.

Needs Assessment

The first and most important part of the process in acquiring new fire apparatus for a department is understanding the operations of the department and the needs of the department as associated to the operations. This is not always as simple and as "cut and dry" as it would seem. The leadership of the department and their interpretation of the community's needs normally determine the way in which a department operates. The method for leadership to meet these needs can be accomplished in many different ways. The way they determine to meet the needs establishes the direction and design for the department.

For example, a rural community with limited volunteer manpower, long response distances, and good accessibility for apparatus may choose to prescribe to a "Quint Concept" for their operations. Whereas another department in a more urban community with traffic congestion, excessive EMS activity, and tight access to apartment buildings and townhouses, may prescribe to a "Paramedic Engine Concept" for the operations of the department.

In both cases, the primary concept for the operations of the department dictates some of the design of the fire apparatus to meet the needs. There are a variety of methods the leadership could have chosen to meet the needs of the community, but they chose the direction they did for one reason or another. There are many variations to these methods, and the department is not right or wrong for the methods chosen. However, a critical first step in procuring a piece of fire apparatus is the department's "Needs Assessment."

When conducting a needs assessment for purchasing new apparatus for a department, the following questions need to be addressed before a committee or group starts the investigative part of the process:

- What will be the primary function of the new vehicle?
- What other functions will the apparatus need to accomplish?
- What physical characteristics or restrictions are critical for the new vehicle?
- What features are preferred (meaning, they would be "nice to have") but are not necessarily required on the new vehicle?
- How many persons will the apparatus normally (comfortably) need to carry and how many persons will it need to carry as a maximum?
- Are there special operating conditions that the vehicle must meet?
- Are there special requirements for carrying equipment or powering equipment?
- What is the projected activity level of the vehicle?
- What persons will be operating the apparatus?
- How much funding is available for the purchase of the apparatus?

In many cases, the way these 10 questions are answered will very quickly narrow down the design of the apparatus. It is also important to remember that purchasing and designing fire apparatus is a series of decisions and compromises. In most cases, all of the preferred requirements cannot be purchased on a single piece of apparatus without adversely affecting the primary function of the apparatus. Many of these questions will also be germane to how the NFPA Standards apply to the apparatus.

It is important that these questions be answered up front so the committee or group can constantly focus on meeting these needs. Committees have to be careful that they do not purchase apparatus that does not meet the objectives of what they started out to accomplish. Some committees fall into the trap of purchasing apparatus based on what they have seen or on a convincing salesperson instead of basing the decision on meeting the fundamental needs of the departments operation. Remember that in most cases the department is purchasing apparatus with either public funds or funds raised by volunteer organizations from the public support. These funds should be spent wisely and not capriciously.

NFPA Standards

When purchasing new fire apparatus, departments rely on the most recognized standard, the NFPA 1901 Standard for Automotive Fire Apparatus. This standard serves as a guide to both the manufacturers that build new fire apparatus and the departments who are purchasing new apparatus. However, the document is primarily designed to be a user's document for the department purchasing the new apparatus.

The NFPA 1901 Standard for Automotive Fire Apparatus is developed as a *consensus* standard. A committee of approximately 30 voting members along with approximately 12 alternate or non-voting members and many task group members help develop and improve the document. The document is updated every three to five years, using input from the public through official public comments. The committee membership is made up of representatives from the fire service, manufacturers, consultants, and special interest groups. The committee monitors various issues and problems that occur with fire apparatus and attempt to develop standards that address those issues. A primary interest of the committee over the past years has been improving firefighter safety and reducing fire apparatus crashes.

The current edition of the NFPA 1901 encompasses various types of fire apparatus, which was a major change made in the 1996 edition. Prior to that edition, each designated type of fire apparatus was covered in a separate document. It was decided to combine all of the individual documents into one standard since there was a lot of duplication in the standards. Many of the vehicles shared commonality, and there was much movement towards combination type vehicles. (Note: NFPA 1901 does not include Wildland Apparatus; this is covered in NFPA 1906 Standard for Wildland Fire Apparatus.)

With this in mind, one of the primary things a user must do when beginning the process to design and purchase a new vehicle is to determine the primary function of the apparatus and what functions are secondary. This will show how to utilize the NFPA 1901 Standard. For example, would a pumper with a large tank have to meet different requirements than a tanker with a large pump? The requirements are based on primary use of the apparatus. Keep in mind that this standard is a minimum standard and that departments may want to exceed the requirements to meet the needs of the various localities.

Significant points of NFPA 1901

The NFPA 1901 Standard for Automotive Fire Apparatus, is a constantly changing and developing document. The original document originated in the early 1900s. The document has progressed considerably since that time to continually meet the current needs of the users. The following are some of the more significant points contained in the current standard and some that have changed in the latest revisions:

- Maximum number of personnel to ride on apparatus must be identified on the apparatus
- Driver and crew members must have a seat with seatbelts in a fully enclosed area
- Driver and crew areas must have noise reduction to protect occupants' hearing
- Maximum step heights are established
- Stepping and standing surfaces, interior and exterior, have to meet slip-resistant requirements

- Access handrails are required at cab entrances and locations where steps or ladders are present
- Tools and equipment stored in the cab and crew areas must be secured to prevent occupant injuries in the case of accidents or abrupt changes in direction or speed
- Minimum output levels and locations of work and step lighting are required
- Minimum levels of reflective stripping is required on all four sides of the apparatus
- Emergency warning lighting must meet minimum photometric output levels around the apparatus in responding and blocking the right-away modes
- Apparatus electrical system must meet performance requirements and may require load management under specified conditions
- ABS brake systems are required on all vehicles
- Auxiliary braking systems (transmission, driveline, or engine retarders) are required on apparatus 36,000 GVWR or greater
- Actual weight of the fully loaded apparatus must meet GVWR and GAWR
- Data plates are to be placed on the apparatus to indicate critical information related to operation and maintenance of the apparatus
- Minimum fire pump rating for a pumper is 750 gpm
- Minimum water tank size for a pumper was reduced to 300 gallons (to accommodate Quint fire apparatus)
- There are different ground ladder compliments required on aerial fire apparatus, pumper fire apparatus, and Quint fire apparatus
- New requirements for foam system performance were added (Mobile Foam Apparatus standard was incorporated into the NFPA 1901 Standard)
- Aerial ladder tip loads are rated in 250-pound increments with a 250-pound minimum tip load in all operating conditions
- Aerial ladder apparatus have required structural and stability levels and require third-party certification
- Interlocks are required for various pumping and other operating conditions to ensure safety of operators and other personnel
- Various performance tests are required for foam systems, line-voltage electrical systems, breathing air systems, and other operating systems on the apparatus
- Fully enclosed filling enclosures are required on apparatus equipped with SCBA filling capability
- Pump panel organization and color coding is required on apparatus equipped with fire pumps

- Intake relief valves and air bleeders are required on apparatus equipped with large-diameter gated intakes
- Additional documentation required of manufacturers for different performance test and manufacturing certifications

In addition to the NFPA Standards, there are other required standards and regulations applying to fire apparatus. Fire apparatus must comply with all applicable DOT, FMVSS, EPA, and individual state regulations. It is particularly important to know and understand the weight standards that will apply to the fire apparatus in the jurisdiction in which the vehicle will be operated. Generally, the manufacturers are cognizant of the applicable federal regulations, but they may not be fully aware of any state peculiarities.

Purchasers should also be aware of component limitations or restrictions that may apply. Some tires have speed and range limitations on use. Axles are rated differently for various applications. Engine and transmission ratings and certifications vary upon application. There are many other component limitations or application issues of which the purchaser must be aware.

NFPA Annex Material

The Annex Material (previously Appendix Material) of the NFPA 1901 Standard for Automotive Fire Apparatus is a combination of recommendations (Annex A) and worksheets (Annex B and C) that are designed to help users develop specifications and inspect the apparatus for compliance for new fire apparatus. Significant modifications and additions have been completed in these areas. The Apparatus Purchasing Specification Form in Annex B is a particularly useful form for the user to make sure all issues are clearly identified in the specifications.

It is also a very good format for the manufacturer to understand and make sure all needed information is obtained to ensure that their bid or proposal is accurate and complete. Most controversies occur because of the information not included in bid proposals. Trying to identify what is standard with the various manufacturers is confusing, difficult, and not necessarily consistent between the different manufacturers. In order to alleviate any misconceptions, as much detail as possible should be included in the specifications.

The Delivery Inspection Form and the As Delivered Weight Analysis Calculation Worksheet found in Annex B are organized worksheets designed to help the purchaser determine if the delivered apparatus meets the NFPA 1901 Standard. All fire apparatus are custom built to the users' requirements and do not always meet the minimum requirements of the NFPA 1901 Standard. There is no organization or oversight group that automatically checks or inspects new fire apparatus for compliance to the NFPA 1901 Standard. It is normally up to the purchaser to inspect for compliance or an independent third-party group can be contracted to conduct the compliance inspections.

The Worksheet for Determining Equipment Weight on Fire Apparatus found in Annex C of the latest revision of the NFPA 1901 Standard is a worksheet designed to help purchasers determine vehicle loading and can also be used for determining required compartment volume. Vehicle weight requirements have traditionally created some of the greatest safety problems with new fire apparatus. Great emphasis has been placed in this area of the new standards to help alleviate these problems in the future. Purchasers should pay close attention to issues related to weight and weight distribution of new apparatus.

The various charts, tables, and worksheets found in the NFPA 1901 Standard for Automotive Fire Apparatus can be very useful, particularly for new purchasers of fire apparatus. Purchasers should take full advantage of the many years of experience and development represented by this standard.

Apparatus Types and Characteristics

The following segments of this chapter are devoted to discussion of the various types of apparatus available and used by departments throughout this country today. The first seven types of apparatus discussed relate to the basic types as identified by NFPA 1901. The next segment discusses the combination type apparatus that has become prominent in fire departments today. The remaining segment addresses apparatus that does not necessarily conform or relate to the NFPA 1901 standard.

Pumper fire apparatus

Pumpers are the most common or baseline style apparatus built and used by the fire service. Pumpers, which may be also known as wagons or engines, are defined in NFPA 1901 Standard for Automotive Fire Apparatus as: fire apparatus with a permanently mounted fire pump of at least 750 gpm capacity, water tank, and hose body whose primary purpose is to combat structural and associated fires. These apparatus were previously called "triple-combination pumpers" because they incorporated three distinct components, namely pump, tank, and hose body.

In most departments, the pumper is the primary apparatus from which most operations are based. The primary purpose of a pumper is to provide personnel with sufficient apparatus and equipment to sustain an initial attack on structural fires. However, in many departments, this apparatus is used for many other purposes. Departments use pumpers to transport personnel and equipment for fire suppression, hazardous materials mitigation, technical rescues, EMS support and service, building inspections, preplanning, public education, and an endless number of other activities.

Normally pumpers are built to a fairly reasonable size since they are such a multi-use type apparatus. Usually the water tank size, hose carrying capacity, and compartment space are balanced to keep the apparatus maneuverable throughout the area in which the vehicle responds.

Cab configurations and chassis designs

When specifying a new pumper or any other fire apparatus, the first decisions that must be made refer to the number of personnel to be carried and the seating configuration. These decisions affect everything from the size of the cab to the location of the engine. Other issues that will come into play are overall size of the apparatus, activity level, and apparatus cost. Low-activity level departments with minimal funds may prefer commercial chassis. However, high-activity departments may find that custom chassis apparatus provide longer life expectancy, increased comfort levels, smaller overall vehicle size, greater adaptability, and increased ease of egress for firefighters making multiple runs each day. It has also been indicated that custom cabs may provide increased structural safety to crews when the apparatus are involved in accidents, particularly rollover situations.

The NFPA 1901 committee has been focusing much attention to crew safety in the last several revisions of the standard. New requirements have been placed in the standards that require equipment in the crew areas be extensively secured to protect firefighters in the event of an accident. Fire chiefs must monitor their personnel to make sure equipment that may endanger the personnel is not added to the crew areas after apparatus is placed in service. It is also imperative that officers make sure personnel are seated and belted when apparatus is in motion.

Other areas that have gained attention in recent years are cab noise levels and crew comfort. Long-term exposure to siren and engine noise has shown to have adverse effects on personnel health. Also, excessive heat levels generated by the apparatus' engine can have adverse effects on personnel's ability to perform at incidents, particularly in areas of that have high-ambient temperatures.

One of the factors that dramatically affects heat and noise levels experienced in the cabs is engine location. Engines that are totally out of the cab areas generate less obtrusive conditions than engines surrounded by the cab. Engines located at the forward portion of the cab, as with tilt-cab models, tend to produce less cab heat and noise than engines in the rear section of the cab as with cab-ahead designs.

Other chassis designs with rear engines and mid-mounted engines are available. However, they are normally more difficult to design around since the engine may be embedded in the body area that normally contains the water tank and equipment compartments. There are some jurisdictions that feel the advantage of crew comfort and noise reduction is worth the design and maintenance issues associated with the use of mid- and rear-engine chassis.

Fire pumps

Fire pumps range in size from 750 gpm to 2,000 gpm and even greater, depending on the water supply capacities and target hazard fire flow requirements. NFPA requires the fire pumps be rated in 250 gpm increments for consistency of rating and testing. Generally most pumper fire pumps are 1,000 gpm, 1,250 gpm, or 1,500 gpm.

The standard design fire pump found in United States is the centrifugal one- or two-stage pump. The pumps generally have cast-iron casings, bronze or brass impellers, on stainless steel shafts. Many departments prefer the single-stage pumps because of the simplicity of the design and since diesel engines operate well through the entire range of the pump. Two-stage pumps have a transfer valve, which requires more maintenance and operator training.

Departments that serve many high-rise building may want to consider two-stage pumps since normal operating pressures to supply systems may be better aligned to two-stage pumps. Two-stage pumps have two impellers mounted on a single shaft, which gives them the capability of operating in a series (pressure) or parallel (volume) mode through the use of the transfer valve. Departments whose operations require lower flows and even higher pressures may want to consider an added third-stage pump for this type operation. Generally these type pumps are set up for high-pressure booster line operations.

The most common design for powering fire pumps on pumpers is through the use of a transfer case (split driveline design). However, where pump-and-roll capability is needed or in cases where a transfer case-driven pump is not feasible, PTO-driven pumps can be used. Normally PTO-driven pumps are used with apparatus that either have smaller style pumps or have special requirements. There are a variety of pump and transfer case designs available where special applications may be needed. If straying from the normal design, it is imperative that the pump manufacturer is consulted to ensure the proper design for the application. Pump speeds and gear ratios are critical for efficient operations.

NFPA 1901 Standard for Automotive Fire Apparatus requires that fire pumps be tested at the manufacturer's facility and be certified by a third-party testing organization. Testing is also required for the fire pump and associated plumbing hydrostatically. Documentation of this test should accompany the new apparatus when delivered. If the purchaser requires special tests, it should be clearly noted in the specifications.

Discharges, intakes, and plumbing

Adequate numbers of discharges and intakes must be provided on the pumper to obtain rated pump capacity. The number and size of discharges vary on pumpers according to the department's requirements based on hose and appliance sizes used. Many discharges are placed and sized on pumpers to accommodate pre-connected attack lines while other discharges are sized to accommodate flow requirements, to match the size and type supply lines used, and to provide convenience for connecting lines in various operations.

NFPA 1901 recommends that the location of discharges and intakes, particularly large-diameter discharges and intakes, be carefully placed to provide safety to the pump operator. When possible, these discharges and intakes should be located in areas remote from positions where personnel would normally be located. Discharges and intakes 3 in. or larger require "slow-operating valves" to prevent the effects of water hammer. Many large-diameter discharges are gear-operated, electric, or air-operated to provide the slow-operating requirement.

Pump panel controls for discharges and intakes are required to be grouped with corresponding gauges and color-coded for ease of identification. Purchasers should require pump panel drawings in the design process to ensure the panels are easy to understand and that they comply with purchaser and NFPA requirements.

Purchasers may want to consider specifying stainless steel plumbing where the normal mineral content in the water promotes severe rusting or corroding of the plumbing. There are various grades and schedules of stainless steel that can be used. In cases where the weight associated with heavy gauges or schedules of stainless steel plumbing may adversely affect the apparatus, lighter gauges or schedules can be used. However, the purchaser must make sure the plumbing meets the required hydrostatic and operating pressures prescribed for the application.

Pressure regulation and intake relief valves

The NFPA 1901 Standard for Automotive Fire Apparatus requires a means for setting the discharge pressure of the fire pump. The device must maintain the pressure with no more than a 30 psi pressure rise with the fire pump operating in a range from 100 psi to 300 psi. There are normally two methods used to accomplish the required pressure control, either with relief valve systems or pressure governor systems. Both systems have proven to be very accurate and reliable means of accomplishing the task. Which system is chosen is more an operator preference than any other issue.

Relief valve systems operate on the principle of pilot-operated spring set valves opening and closing to accommodate changes in pump pressure. The excess pressure dumps back into the suction side of the pump. On some of the newer relief valve systems, a second intake relief valve is incorporated into the system for situations where the incoming pressure will not allow for the absorption of the excess discharge pressure. With these systems, the excess press will then be dumped in the atmosphere. It must also be kept in mind that on pumpers equipped with around-the-pump foam systems, these relief valve systems may circulate foam solution back into the booster tank and possibly onto the ground.

Pressure governor systems have shown a renewed surge in the industry. The modern electronic pressure governor systems interface very simply and efficiently with the electronically controlled fuel-injected engines. Most engine computers are equipped with cruise control circuitry that adapts with electronic pressure governor logic. Pressure governors are designed to control engine speed automatically to regulate discharge pressure. They allow engines to operate more fuel efficiently, more quietly, and with reduced wear by slowing the engine speed down when minimal flow or pressure is needed. The systems are very simple to operate and respond quickly to changes in pressure and flow.

The current NFPA 1901 Standard requires that any pump with more than one 3½" or larger intakes (not valved) must have an intake relief valve system adjustable from 90 psi to 185 psi. In addition, any 3½" or larger valved intake shall be equipped with an adjustable intake relief valve installed on the supply side of the valve. These intake relief valves will discharge to the atmosphere when the incoming pressure exceeds the set adjustment. Some departments require male hose threads to be installed on the relief valve discharges so that the water from the discharge side of the relief valve can be directed away from the pumper.

It should be noted that any time this is done, the hose must be secured so that the butt end of the hose does not whip around when a discharge occurs. Also, personnel must be cautioned not to place caps on the discharges since they will defeat the operations of the safety device. With the wide use of large diameter hose in today's fireground operations, it is imperative for the safety of the firefighters operating around fire apparatus that these intake relief valves are operable.

Water tanks

Water tank sizes generally range from 500-gallon capacity to 1,250-gallon capacities on standard single axle style pumpers. Water tank size has to be closely scrutinized to make sure the apparatus chassis does not become overloaded. Such items as axle ratings and engine location will dictate the water tank size. Normally, 500-gallon capacity tanks are more commonly found in urban areas where 750-gallon capacity and 1,000-gallon capacity tanks are found in the more rural areas.

Most manufacturers have gone to the plastic or fiberglass style water tanks to alleviate corroding and rusting problems associated with metal tanks. These tanks are durable, easy to repair, and lighter in weight. They are also very flexible in design parameters and can be fitted into various pumper configurations. Smaller water tanks are more commonly found on combination style apparatus such as Quints, where both weight and space are a prime issue.

The size, shape, and baffling of water tanks are important parameters to the safety of a pumper. Baffling, overflow requirements, and venting are issues addressed in the NFPA

1901 standard. Reputable manufacturers are normally well aware of these issues. The size and shape of the water tank have major effects on the stability and center of gravity of the loaded apparatus.

When specifying larger capacity water tanks such as 1,000-gallon and larger, axle capacities, weight distribution, frame strength, and applicable weight regulations should be examined closely. Normally, a tilt-cab style pumper with a 750-gallon capacity water tank will load normal weight capacity axles to their maximum limits. Remember that designing a pumper is a series of decisions and compromises. A gallon of water occupies 7.5 ft^3 of space. If you want a larger capacity water tank, it either has to go out or up.

Body construction and compartmentation

Each manufacturer has specific design methods for how they construct their bodies. The manufacturing construction design of a body has as much bearing on the strength of the body as the thickness, alloy, and tensile strength of the materials used. Construction designs should be evaluated closely to ensure the body would accommodate the needs of the department.

In the last 20 years, pumper bodies have transitioned more towards the use of aluminum and stainless steel. The resistance to rust, ease in manufacturing, and the lighter weight has made aluminum the most popular choice in materials. Aluminum is subject to corrosion and stress cracking but can be avoided with sound engineering designs and good manufacturing techniques. With aluminum, it is important to maintain dielectric barriers between dissimilar metals and use good priming and painting processes.

Purchasers should allow some flexibility in the construction designs of the pumper bodies to allow for the variations in construction methods of reputable manufacturers. It is also important that purchasers closely examine the construction techniques and quality control of the body builder to make sure the body will withstand the many years of use expected of the apparatus. The simple things like using dielectric tape or coatings and using stainless steel screws and fasteners can have major bearing on the future conditions of the body.

Compartment designs vary as much on pumpers as the shapes of snowflakes vary. Each department has a special design or a different need that requires an individual design of the compartments.

Generally there are two styles of compartment doors available today. The hinged door is the more traditional design but many departments have chosen to use roll-up or shutter style compartment doors. There are advantages to either type, depending on their application and location. Many pumpers are built using a combination of both types.

Roll-up doors have the advantage of not protruding out into the way of firefighters or into lanes of traffic when the doors are open. However, the drum roll occupies space at the front top opening of the compartment. In some cases, the loss of approximately 8–9 in. in door opening height may make a compartment not accommodate equipment to be carried on the apparatus.

On rescue style trucks where the compartments are especially tall, the loss of the compartment opening height may not be a factor. Also, roll-up doors do not tend to seal compartments as tight as hinged doors. Road dirt has a tendency to infiltrate roll-up doors more than with hinged doors. Roll-up doors are also subject to being jammed closed when loose equipment falls against the interior of the door. Heavy appliances can also damage the shutter tracks found on both sides of the door openings.

Hinged doors, on the other hand, have their own distinct problems. Lift-up style hinged doors are very difficult to see in rearview mirrors and are habitually known for being ripped off after being left open when apparatus leave the station. An open lift-up door can do substantial damage to the vehicle and to the fire station. Hinged doors have a tendency to get out of adjustment and leak. If the doors are large, the door weight will cause the door to come loose from the hinges. Many specifications call for hinged doors larger than 32–36 in. wide to be made as double doors to eliminate the excessive weight on the hinge.

Compartments on pumpers are required to be weather resistant, ventilated, and have provisions for drainage of moisture. All compartments greater that 4 $ft.^3$ of capacity are required to have compartment lighting. Many different methods of compartment lighting are available, including rope lights, LED lights, and standard incandescent lighting.

Whatever type of lighting is chosen, mounting locations should take in consideration protection for the lights and the ability to light the entire compartment. Some manufacturers choose to mount the compartment lights in the hinged doors to protect the lights and to give full lighting to the full face of the compartment. In some cases, multiple lights may be required to get full lighting throughout the compartment.

Compartment light door switches are known to be continual problems on apparatus. Many manufacturers have gone to sealed magnetic switches to alleviate some of the reliability issues. When designing a pumper, the department should address the location and design of door switches. On pumpers with horizontally hinged compartments, it is not uncommon to find door switches recessed into the upper extrusions or channels that may retain moisture and cause premature failing of the switch. The switches should be installed in a manner and location where they are protected and accessible for replacement.

Today's pumpers are much like high-rise office buildings where every inch of space is crucial. For this reason, some manufacturers have perfected storage methods that take advantage of every usable square inch of the body. The NFPA 1901 standard requires a minimum of 40 $ft.^3$ of storage space on a pumper. Most departments elect to require even larger capacity of space to accommodate the various tools and equipment used by the department.

The use of roll-out trays, roll-out and drop-down trays, slide-out tool boards, wheel well compartments, under body trays, drop-down ladder racks, through-the-tank storage compartments, and many other creative methods have been developed to take advantage of all usable space. There are also several tool bracket manufacturers that have developed a multitude of creative tool mounts for every type of tool and equipment. Committees and purchasers should take advantage of the annual trade shows to see the creativity in body compartment designs and equipment mounting.

Hose storage

Hose storage on pumpers has as many variations as compartment designs on pumpers. Each department must determine what methods of hose deployment work best for their operations and design their pumpers hose storage accordingly. The NFPA 1901 Standard requires 30 $ft.^3$ of storage for 2½ in. and larger hose and 7 $ft.^3$ of to accommodate pre-connected hose lines. Most pumpers will well exceed these capacities.

Generally, hose storage is divided into two types, supply hose storage and attack-line hose storage. Normally, supply hose storage is located so hose will deploy out the rear of the pumper. The supply hose storage is designed to accommodate long lays of supply-line. Many pumpers carry in excess of 2,000 ft. of supply line, and they may divide the storage area so dual lines can be deployed at one time. Many departments also carry multiple sizes of supply line.

When specifying a new pumper, the size hose and capacity should be specified so the hose storage area can be accurately calculated. Extra capacity should be included particularly if hose bed covers are not used. Hose piled up above the hose storage area can be blown out of the apparatus when traveling at road speeds, especially when lighter weight hose is used. Wind deflectors placed on the front of the supply hose bed will sometimes prevent air from getting under the top layers of hose.

Attack-line hose storage is accomplished in many different ways. Traditional departments have used rear hose beds with good success for many years. Other departments have chosen to use cross-mounted hose beds (crosslays) to better accomplish their operations. As pumpers have grown in height, departments have been faced with cross-mounted hose beds becoming higher above the ground since most are mounted above the pump panel area. To get the hose beds lower, the use of speed lays has also become popular. When using speed lays, keep in mind that they generally add length to the pumper's overall length and wheelbase.

Areas that are subject to long periods of bad weather conditions and areas that deal with flying brand problems and may choose to use metal hose bed covers. If considering a metal hose bed cover, overhead clearance in the fire station must be considered. Hose bed covers on both the supply hose and pre-connected hose will extend the life of the hose and keep the hose in a cleaner condition. Diesel exhaust emissions and other environmental factions attack the rubber and other synthetic compounds used in many typical fire hoses.

Initial Attack Fire Apparatus

Initial Attack Fire Apparatus as defined in the NFPA 1901 are: fire apparatus with a permanently mounted fire pump of at least 250 gpm capacity, water tank, and hose body whose primary purpose is to initiate a fire suppression attack on structural, vehicular, or vegetation fires, and to support associated fire department operations. As can be seen from the definition, these types of apparatus can vary dramatically according to the department's intended use.

Initial Attack Fire Apparatus used for structural operations was a hot item in the mid- and late 1970s. The use of Initial Attack Fire Apparatus for structural operations does not appear to be as prevalent today. However, many departments use these types of apparatus for all of the other functions defined. Generally these type apparatus are used in areas where there are access problems.

Cab and chassis designs

Normally most Initial Attack Fire Apparatus are constructed on commercial style chassis. In the past, the major problem associated with this style apparatus was designing a unit and not exceeding the limited GVWR of the suitable commercial chassis available. However today's choices of commercial chassis have improved substantially. The major commercial chassis manufacturers now offer chassis with both two-door and four-door cabs in the GVWR range that more closely meet the needs of initial attack apparatus.

In the past there was a void in suitable chassis in the GVWR capacity between 15,000 pounds and 26,000 pounds. There were also problems with availability of heavy-duty automatic transmissions with PTO outputs to drive fire pumps and other accessories. It is still very important for the purchaser of Initial Attack Fire Apparatus to match the chassis payload capacity to the design of the apparatus.

Location and drive designs of the fire pump and the location and capacity of the water tank have major effects on the finished apparatus. It should be remembered that raising spring capacities does not necessarily increase the GVWR or axle capacities. GVWR and axle ratings are based on a variety of factors including braking capacity, axle strength, usable horsepower transmitted to the ground, transmission capacity, frame strength, steering component capacities, and many others.

Large majorities of Initial Attack Fire Apparatus are built on four-wheel-drive chassis. There are large numbers of the commercial chassis available from the manufacturer as four-wheel drive. There are some situations that will require the chassis to be sent to an after market builder to have four-wheel-drive capability added. There are several reputable after market builders that can accomplish the conversion and re-rate the chassis accordingly.

Pumps

Initial Attack Fire Apparatus can have fire pumps with as little as 250 gpm capacities but may be found with much larger capacity pumps. Some rural areas use this style vehicle as a water supply apparatus since this style apparatus may have greater accessibility to rural water supplies than full-sized pumpers. Areas that use these for pump-and-roll type operations such as brush fires may want to consider separate engine-driven pumps. This allows constant pumping operations independent of the vehicle ground speed.

One of the more prolific enhancements of recent years is the addition of compressed air foam systems. These are used as initial attack on structure fires or brush fires. Several manufacturers are building packaged compressed air foam systems matched to fire pumps that work well for initial attack apparatus. Compressed air foam gives the limited amount of water normally carried on initial attack apparatus much greater fire suppression potential for both structural and brush fire operations. Some recognized authorities of foam systems claim that compressed air foam will increase efficiency of water as much as four times. This will allow a rig with a 300-gallon water tank to have the suppression potential of 1,200 gallons of plain water.

In applications where the apparatus is to be used in severe off-road applications, the pump drive casings and gearboxes should be protected and located where they do not encroach on the break-over angle. In some applications, the use of hydraulic drive systems allows greater flexibility in the mounting of the fire pump.

Discharges, intakes, and plumbing

The number of discharges and intakes on initial attack fire apparatus is normally limited in order to keep the pump panel simple and small. If the apparatus is to be used for accessing rural water supplies, front or rear suction intakes may be useful for drafting operations. When Class A or Class B foam systems are installed on the apparatus, the purchaser should consider stainless steel plumbing.

Pressure regulation and intake relief valves

On Initial Attack Fire Apparatus, the pumping systems are generally fairly simplistic. In most cases, pilot-operated discharge relief valve systems are used in conjunction with suction side relief valves. It should be remembered that when large diameter intakes are provided with shutoff valves, they must be provided with adjustable intake relief valves and air bleeders. Supplying an initial attack apparatus with large flows of water without adequate intake relief valves and air bleeders can create a severe safety issue.

Water tanks

The NFPA 1901 standard requires a minimum capacity water tank of 200 gallons on an initial attack apparatus. In many cases, larger capacity tanks are provided when the chassis GVWR accommodates the additional weight. The capacity and location of the water tank is a critical factor in the balance and axle loading of the apparatus. Overloaded axles will affect both on-road and off-road performance and safety.

Body construction and compartmentation

Weight is a major issue on this style apparatus, which promotes use of aluminum material bodies. Some departments have used lightweight constructed service industry style bodies to assemble initial attack apparatus. It should be noted that service industry bodies are constructed with an 8- to 10-year projected life span. Departments that plan on keeping a unit for longer periods should consider more durably built bodies.

Compartmentation is generally limited on initial attack apparatus due to the reduced size of the bodies. If there are special requirements for specific sized compartments for larger equipment and appliances, it should be noted early in the design of the apparatus. When the apparatus is to be used in severe off-road applications, the rear of the body may need to be tapered to accommodate larger angles of departure.

Hose storage

The hose use on Initial Attack Fire Apparatus is similar in function to the use found with pumpers. However, the required hose capacity for supply hose, 10 $ft.^3$, is one-third of the requirement is for pumpers. The allotted space is reduced, which in turn reduces the hoseload capacities. Initial attack apparatus usually work in conjunction with other full-sized pumpers and rely on their ability to carry additional quantities of hose and equipment. Initial Attack Fire Apparatus is still required to have a minimum of two pre-connected hose beds with each having a minimum of 3.5 $ft.^3$ of capacity. As with all apparatus, minimum requirements must be established up front when designing initial attack apparatus since options and capabilities will be limited in their construction.

Mobile Water Supply Apparatus

Mobile Water Supply Apparatus, commonly known as tankers, are normally a less complicated, more basic designed apparatus than most of the other types. Unfortunately it is also one of the more "under spec'd" apparatus found in use. In many cases tankers are "home-built" or converted vehicles. The NFPA 1901 committee monitors various fire apparatus accidents, and it has been noted that tankers are involved in a large percentage of fire apparatus accidents, particularly those involving rollovers and fatalities.

A Mobile Water Supply Apparatus as defined by NFPA 1901 is a vehicle designed primarily for transporting water to fire emergency scenes to be applied by other vehicles or pumping equipment. Since the primary purpose of the vehicle is to transport water, the majority of the requirements in NFPA 1901 relate to the water tank. The apparatus is required to have a pump but not necessarily a fire pump as defined on other vehicles.

Cab and chassis designs

A large percentage of Mobile Water Supply Apparatus are built on commercial-style chassis but there are a substantial number built on custom chassis as well. In many rural areas, mobile water supply apparatus serve as water and personnel transporters. Tankers have a tendency to be heavy, large vehicles with a high center of gravity, which makes them prone for rollovers when involved in accidents. The large mass pushing the vehicle can do considerable damage to the cab or passenger compartment when a rollover occurs.

A few key items should be considered when specifying chassis for use as tankers. Following the federal regulations for determining the number and size of axles should be considered. When axle loading becomes questionable, either consider adding additional axles or reducing water tank capacity. Fully lined frames are generally recommended for large tankers.

As much consideration or even more consideration should be given to braking and retarding the speed of the vehicle as is given to power or horsepower for acceleration. Top-end speed limitations should be considered, particularly on the larger style vehicles. Departments should also provide adequate training to personnel driving tankers, especially to those personnel who do not routinely drive large vehicles. Emphasis should be placed on deceleration, braking, and procedures associated with maintaining control of a large vehicle when a wheel drops off of the pavement.

Fire pump

A fully-rated fire pump is not required by the NFPA 1901 standard on Mobile Water Supply Fire Apparatus, but many departments do choose to have a fully capable fire pump. Since a major investment has already been committed to purchase the chassis, body, and water tank, the additional cost associated with a fully rated fire pump is not that significant. Adding the fully rated fire pump gives the capability of having both a tanker and a water supply pumper in a single vehicle. The pump manufacturers have developed some high-capacity fire pumps that are packaged very well and do not require the amount of space normally associated with fire pumps. Many traditional style tankers have pumps strictly designed to load and unload water. The ability to pump fire streams will be left up to other apparatus.

Departments that specify Mobile Water Supply Apparatus on single rear axle chassis have to watch axle weights very closely. The pump and transfer case weight of fully rated fire pumps have significant bearing on the weight of the finished vehicle. The more weight consumed by the fire pump, the less weight capacity is available for the water and water tank. Some manufacturers recommend not exceeding the 1,250-gallon capacity water tank on a chassis with a fully rated fire pump. Even at that capacity, it will still require a heavier than normal rated rear axle. It has also been found that using front-mount style fire pumps may create a better weight-balanced vehicle in some of the smaller capacity tankers. The front-mount pump also lends itself to efficient drafting operations.

Discharges, intakes, and plumbing

On tankers, the amount of discharges and plumbing associated to the pump is normally fairly minimal. NFPA 1901 requires a 2-in. minimum tank fill line, which is larger than the requirement for pumpers with smaller water tanks. Two preconnected hose line discharges are also required on Mobile Water Supply Apparatus.

The most significant requirements relate to filling the water tank and dumping water from the water tank. An external fill connection capable of filling the water tank at 1,000 gpm minimum is required along with the ability to dump water to the left, right, and rear of the apparatus. The dump valves must be capable of dumping 90% of the tank at 1,000 gpm or greater. Most modern tankers are provided with air or electrically actuated dump valves on the right, left, and rear of the vehicle. Controls for the valves are usually located in the cab and adjacent to the discharges. This allows for quick dumping where the operator does not need to leave the cab of the vehicle.

The side and rear dump valves vary in sizes but are normally in the 10- and 12-in. diameter range and are commonly butterfly style valves. A convenient option to have on these dump valves is the automatic extensions that activate when the valve is opened. Quick turnaround times are critical in efficient tanker operations. Anything that saves on operator time in loading and unloading influences turnaround time. When tankers are equipped with large diameter gated intakes, slow operating valves are required along with air bleeders and intake relief valves. Promoting safe operations when using large diameter hose cannot be over emphasized.

Pressure regulation and intake relief valves

Even though pumping systems on tankers are normally simplistic, they still require adequate pressure regulation and intake relief systems. Large quantities of water must be capable of being loaded and unloaded by requirement. It is critical that the pressure regulating system, particularly the intake relief systems, be sized to handle the volumes of water flow that will be encountered.

Water tanks

The water tank is the heart of the Mobile Water Supply Apparatus. Adhering to NFPA requirements on tank baffling and venting are paramount to ensure safe operations. The venting capacity should be matched to the filling and dumping capacities. If adequate venting is not provided, severe damage can occur with large flow filling and dumping.

Generally there are two style tanks found on Mobile Water Supply Fire Apparatus, either elliptical or square-cornered styles. Many rural areas prefer elliptical tanks because of the price and reduced forces created perpendicular to the chassis from the movement of the tank water. However, elliptical tanks do not lend themselves to carrying large quantities of hose. In situations where the apparatus will be required to carry large quantities of hose, most departments choose to purchase square-cornered tanks so hosebeds can be constructed on the top of the water tank.

The majority of water tanks today are constructed of manmade synthetic materials such as polyethylene or fiberglass. Even the elliptical style tanks can be constructed with the same materials. Some elliptical tanks are covered with bright finished aluminum to give the appearance of stainless steel tanks. If preferred, steel or stainless steel tanks are available. If it is anticipated that the tank will be used to contain potable water, a stainless steel tank should be considered. To do so, the tank must be sanitized according to local Health Department regulations.

When choosing the tank capacity, style or design of tank, and tank material, the chassis design and weight capacity must be closely monitored. Again, when in doubt about capacity, add an axle. Also, the design of the tank should be such that as low a center of gravity as possible can be achieved. Keep in mind that the higher the center of gravity of the apparatus, the higher the potential for a rollover.

Body construction and compartmentation

Many of the same considerations for pumper style apparatus should be considered with Mobile Water Supply Fire Apparatus. One common item found on tankers normally not found on pumpers is folding tanks. Many different styles of folding tank compartments are used today. One of the major considerations is the amount of manpower available to offload the portable tanks. Some very well-designed portable tank racks are available where the tanks are lowered down to a convenient level for minimal manpower to offload the tanks.

For efficient high flow capacity tanker operations, usually multiple portable tanks are required. As a general rule of thumb, the portable tank capacity should be double the size of the apparatus water tank. It generally is better to carry two portable tanks, each with enough capacity to contain the entire contents of the apparatus water tank. This will keep the portable tanks lighter and more manageable than one very large portable tank.

Hose storage

The NFPA 1901 standard requires 6 $ft.^3$ minimum of hose storage area for 2½ in. or larger hose and two areas each with 3½ $ft.^3$ minimum of space for 1½ in. or larger preconnected hose lines. This minimal area required by NFPA 1901 allows for the minimum amount of hose to load and unload water, to extinguish an unexpected small fire, and to protect the apparatus in the situation where the apparatus may become exposed to heat or fire. If the apparatus is going to be used other than as a water shuttle apparatus, more hose storage area may be necessary. When large amounts of hose are to be carried, a square-cornered tank rather than an elliptical tank may have to be considered.

Aerial Fire Apparatus

Aerial Fire Apparatus as defined by NFPA 1901 is a vehicle equipped with an aerial ladder, elevating platform, aerial ladder platform, or water tower that is designed and equipped to support firefighting and rescue operations by positioning personnel, handling materials, providing egress, or discharging water at positions elevated from the ground. As can be interpreted from this definition, a wide variety of apparatus are covered by this definition.

Aerial Fire Apparatus as we know them today are a relative newcomer to fire department use as compared to pumper fire apparatus. Peter Pirsch Company built the first all-metal aluminum aerial mounted on a fire apparatus chassis in 1936. Prior to the Pirsch metal ladder, aerials were constructed from wood. Early models were pulled by horses and raised manually by mechanical worm gear devices that took as many as six burly men to raise.

The first major advancement in aerial devices was the development of the manual *spring-assist* device. Several of the early fire apparatus builders, including Seagrave, Pirsch, and American LaFrance, used these style aerial devices with relatively good success. As cities have grown more "up" rather than "out," the need for more advanced and reliable aerial apparatus have fueled the development of the aerial apparatus found in use today. The improvement of gasoline and diesel powered apparatus and the development of hydraulic powered aerials have advanced the industry to where we are today.

There are Aerial Fire Apparatus available today that are more than double the height of the manual spring-assist type aerials. Modern Aerial Fire Apparatus are designed with much more operating capability and more stringent safety requirements. Some aerial apparatus have integrated the use of computers to further enhance the operations and safety of modern Aerial Fire Apparatus.

Aerial Fire Apparatus can be classified into three basic groups: straight aerials, platform aerials, and water towers. Each of the different types has different capabilities with distinct advantages and disadvantages. The different types can also be mounted on chassis in different configurations such as rear-mounted, mid-ship mounted, or tiller-mounted. Each of the mounting configurations also has their own distinct advantages and disadvantages. Choosing the right aerial for a locality is like choosing the proper dinner wine. The aerial apparatus has to be right for the application or situation. Where a red wine may be preferred with beef, a platform style aerial may be preferred for rescue of non-ambulatory victims.

When a locality is purchasing an aerial apparatus, it is imperative that all aspects of the locality are considered in the evaluation. Everything from available funding to apparatus bay space and target hazards to be protected will affect the process and outcome. The following is a discussion of the advantages and disadvantages of the various types and configurations of aerial apparatus. As with Ford and Chevrolet owners, you will find that users and operators of the various types of aerial apparatus are very opinionated as to which type is best.

Straight aerials

Straight style aerial devices are probably the most prevalent style in use today. The primary reasons are cost and versatility. A straight style aerial today will cost approximately $500,000 or more. To go to a platform style aerial will generally increase the cost by approximately 20 percent. The following are advantages/disadvantages of straight aerial apparatus:

Advantages:

- Lower purchase cost than platform style apparatus
- Can generally be kept smaller and more maneuverable than platform devices
- Provides a continuous means of egress for firefighters operating on upper floors or the roof of fire buildings

- Versatile; can be used as ladder, elevated stream device, or as a crane style device for rescues
- Simple to set up and operate
- Available in a multitude of operating lengths (75 ft. to 135 ft. in the United States and operating load capabilities (NFPA requires 250 pounds minimal in 250-pound increments)
- Lighter in weight than platform style apparatus
- Allow for flexibility in body and compartment designs
- Can be mounted on chassis in various configurations: rear-mounted, mid-mounted, or tiller-mounted

Disadvantages:

- Limited in elevated stream capacity compared to platform apparatus and water tower apparatus
- Not as adaptable for rescue of non-ambulatory victims as platform apparatus
- Does not provide protection for personnel operating at the tip of the aerial
- Limited space for equipment mounted at the tip of the aerial
- Limited load capacity of aerial devices
- Minimal operating room at the tip, no work platform
- No flexibility in reach such as up and over capability

Platform aerial apparatus

Platform aerial apparatus have grown in popularity since their inception in the early 1950s. Most major cities incorporate some mixture of platform style apparatus in their fire department fleets. There are four basic types of platform aerial apparatus in use today; aerial platforms (ladder-towers), telescopic boom platforms, articulating boom platforms, and articulating/telescopic boom platforms. Each of these apparatus has advantages and disadvantages.

Platform aerial apparatus also have various mounting configurations on chassis available such as rear-mount and mid-mount. Supthen did develop a tiller-mounted platform but there has been minimal industry interest to this configuration. The NFPA requires that platform aerial apparatus have a minimum load capacity of 750 pounds (dry) and a 1,000 gpm flow capacity with 500 pounds of weight in the platform. Rating is again in 250-pound increments. The platform is also required to have a minimum of 14 sq. ft. of floor space.

Aerial platform

The aerial platform, or what is sometimes known as the ladder-tower, is the most popular style platform aerial apparatus. The aerial platform is essentially an aerial ladder with a platform mounted on the end. It does incorporate some of the advantages of both a platform and a straight aerial. Most of the major manufacturers offer an aerial platform apparatus. These apparatus range from 85 ft. to 105 ft. in elevation with weight capacity ratings from 750 pounds to 1,000 pounds and water flow capacities up to 2,000 gpm. The most common *workhorse* versions of the aerial platforms are the 85 ft. and 95 ft. versions on rear-mounted chassis configurations. Aerial platforms are also available on mid-mounted chassis configurations where travel height or bay door height is limited.

Operational Attributes:

- Provide a stable work platform for firefighting and rescues
- Generally have higher water flow capability than straight aerials
- Provide continuous means of egress similar to straight aerial
- Have high equipment and personnel load capability at platform
- Require more area for setup than standard aerials due to increased outrigger footprint
- Rear-mount platforms have increased frontal overhang and blocked frontal visibility as result of platform over cab
- Aerial platform apparatus are dimensionally larger (length and height) than straight aerial ladder apparatus
- Aerial platform apparatus are generally heavier than straight aerial ladder apparatus (front axle weights should be monitored on large rear-mount apparatus)
- Hydraulic and electronic systems are more complex than straight aerial ladder apparatus
- Aerial platform apparatus are easier and smoother to operate than articulating platform apparatus

Telescopic boom platforms

There are currently two U.S. manufacturers, Aerialscope and Supthen, that offer a telescopic boom platform. These apparatus have a platform mounted on the end of a telescopic boom. The Aerialscope device is a full box-beam construction with the main boom constructed of steel and the extension booms constructed of aluminum. The apparatus incorporates a ladder mounted on top of the boom for escape purposes not necessarily for normal operations. The Supthen device is an open aircraft design box-beam boom constructed of riveted aluminum. Both of these apparatus are especially well designed for high-capacity water flow operations. A full box-beam style boom is very resilient against torsional loads associated with high capacity water flow operations. Aerialscope offers a 75-ft. and a 95-ft. version, both of which are mid-mounted style chassis configurations; and Supthen offers 70-ft. to 100-ft.versions that are also mid-mounted style configurations.

Operational attributes:

- Provide stable work platform for firefighting and rescue operations
- Generally provide higher flow elevated water stream capability than other types of aerial apparatus
- Escape ladders are not generally designed for normal egress and are not optimal for evacuating building occupants
- Box-beam construction of aerial make units more amenable to mid-mounted apparatus to keep overall travel height limited
- Higher water flow capabilities require larger outrigger footprint to maintain stability

Articulating boom platforms

The articulating boom platform was the first of the platform devices used in the fire service. These platforms consist of two booms that connect and articulate at a knuckle, allowing a different operating range than any of the other style apparatus. The first units were introduced in the 1950s. The "up and over" capability that is characteristic of the articulating boom platform made them popular, particularly in areas that have a number of buildings with parapet walls.

Original designs were based on equipment previously in use by the utility companies. The primary change made to the rigs was to incorporate the high flow elevated water stream capability. Standard articulating boom platforms range from 65 ft. to 85 ft. of elevation. These units are commonly known as Snorkel units, which comes from the name of the original company that produced these devices for the fire service. Snorkel units are still available today, even though their popularity has decreased where aerial and telescopic boom platforms have increased in popularity.

Operational attributes:

- Articulating boom platforms have an "up and over" capability not present in straight aerial platforms and telescopic platforms
- Operating height is limited on two-section booms because overall length of apparatus becomes excessive

- Operators have to be cognizant of location of the knuckle (pivot point) as well as the platform when operating the device
- Apparatus does not have continual means of egress
- Operating range is generally much different from aerial platforms and telescopic platforms (low angle reach is limited)
- Platform has more movement (bounce) than straight aerial platforms

Articulating/telescopic boom platforms

The articulating/telescopic boom platform is basically an enhancement of the articulating boom device. By making the booms of the device telescopic in conjunction with providing multiple articulating sections, it allows the device to be smaller in size, increase the operating height, and increase the operating range and flexibility. Some of these apparatus have operating ranges up to 230 ft. The combination of the telescopic booms and multiple articulating sections extremely alter and enhance the operating range of the device. It also makes the unit much more complicated and expensive compared to traditional devices.

Operational attributes:

- Units offered with extended height capability
- Combination of articulating booms and telescopic booms adds unique operating range capabilities
- Units are exceptionally complex for repairs and operation
- Generally have greater upward reach than outward reach
- Extended height units have large footprint outrigger system
- Operators have to be cognizant of location for the platform and the articulating knuckles
- Extended height units become very large and very heavy vehicles (may require additional axles for road travel weight limits)

Water towers

Water tower apparatus have been in use by fire departments prior to the existence of aerial ladders and even prior to the development of motorized fire apparatus. Early water tower apparatus were horse-drawn and constructed of wood. Today's water towers range from simple units in the 50 ft. range on up to complex units with three or more booms that extend up to 130 ft. in height.

Water towers are not a major market in the municipal fire service today as compared to the platform style apparatus with water flow capabilities. Water tower apparatus, particularly with high flow capability, are more common to the industrial fire service. It would be more likely to find a 130-ft. unit with 5,000 gpm foam capability at an industrial chemical plant complex than in a municipal setting.

Operational attributes:

- Limited elevated streams can be placed on relatively small pumper chassis without altering pumper design dramatically
- Smaller devices do not add excessive weight to the vehicle
- Lower costs are associated to basic water tower than aerial ladder or aerial platform apparatus
- Extremely high water flows can be accomplished with some water towers
- Units with three and more articulating booms offer unique operating ranges

Rear-mounted, mid-mounted, and tiller-mounted aerial apparatus

The terms *rear-mount, mid-mount,* and *tiller-mount,* when related to aerial apparatus, refer to the mounting location of the aerial turntable. Many of the various types of aerial apparatus can be mounted in any position. The following are some of the advantages and disadvantages of each mounting location.

Rear-mount apparatus advantages:

- Reduces overall length of apparatus
- Reduces rear overhang behind the rear axle
- Generally have increased body compartment area than mid-mounts
- Has greater area for operating at low angles, off each side and rear
- Weight balance on vehicle axles is tolerable particularly when no pump or water is on the vehicle

Rear-mount apparatus disadvantages:

- Rear-mount platforms have excessive frontal overhang
- Platform blocks upward frontal visibility
- Rear-mount platforms with pumps and water tanks can have weight problems particularly on the front axles
- Rear-mount units are more difficult to “spot” the turntable particularly when other apparatus take prime locations at the front of fire buildings
- Rear-mount apparatus usually have a higher travel height

Mid-mount apparatus advantages:

- Lower travel height
- Good frontal visibility
- Weight balance on axles is generally good when a pump and water tank are on the vehicle
- Easier to "spot" the turntable than rear-mounts

Mid-mount apparatus disadvantages:

- Excessive overhangs behind the rear axle
- Longer overall vehicle length
- May have to add counterbalance weight if a pump and water tank are not on the vehicle
- Generally results in less compartment space than rear-mount
- Low angle operating area is reduced to each side of vehicle
- Vehicle cab blocks low angle operations in the front of the apparatus
- Ground ladder storage is sometimes a problem

Tiller-mount apparatus advantages:

- Increases maneuverability of the apparatus
- Has increased compartment space on trailer
- Lowers the travel height of the vehicle

Tiller-mount apparatus disadvantages:

- Increased overall length of the apparatus (beyond 50 ft.)
- Requires two well trained operators
- Increased cost
- Not well suited for large fire pumps and water tanks

Quint Fire Apparatus

One of the areas that was included in the last revision of the NFPA 1901 Standard was related to Quint Fire Apparatus. The term *Quint* refers to having five distinct features; fire pump, water tank, hose bed, ground ladders, and aerial ladder. It was noticed by the NFPA 1901 committee that an increasing number of combination function apparatus, particularly Quint Fire Apparatus, were being used by fire departments. With this in mind, several significant changes were made to the standard.

The required water tank capacity was lowered to 300 gallons to help accommodate single-axle Quint Fire Apparatus, and the ground ladder compliment was reduced. The most popular Quint Fire Apparatus is built with a 75-ft. aerial ladder. Many departments choose the 75-ft. aerial ladder with a 300-gallon water tank so they can maintain reasonable axle weights on a two-axle chassis.

Special Service Fire Apparatus

The term *Special Service Apparatus* was incorporated in the NFPA 1901 Standard to cover the many different vehicles that do not fit into the other definitions. Special Service Apparatus generally refer to medium- and heavy-rescue squads, communications apparatus, command apparatus, air units, light units, hazardous materials units, and any other apparatus not fitting the previously mentioned classes. Some of the key areas in the current NFPA 1901 Standard that have been improved and upgraded to accommodate these apparatus are the sections on breathing air systems, winches, high-voltage electrical systems, and the Annex C, which relates to weights and dimensions on commonly-carried equipment.

One of the primary concerns with Special Service Fire Apparatus is the relationship of the weight and dimensions of equipment typically carried on these style apparatus compared to the compartment space and chassis weight carrying capacity. The following are some of the key points required in these sections of the NFPA 1901 Standard:

- Breathing air systems must be installed with hardware that maintains a 4–1 safety factor on projected pressure.
- Breathing air systems have to be purged and tested prior to delivery by the manufacturer.
- SCBA filling has to be in fully enclosed fill stations that have been type tested.
- Training and labeling has to be provided to the purchaser.
- High-voltage electrical systems have to conform to the applicable sections of the National Electric Code.
- Electrical components are to be listed for the type of application (wet or dry locations).
- Electrical wiring and components have to be labeled according to function.
- Labels are to be installed providing system rating information.
- Electrical system testing needs to be completed during pump testing to ensure adequate vehicle power is available for combined operations and to ensure generators do not contribute to overheating or other problems.
- Winches have to be installed to meet the winch manufacturer's installation requirements.

- Winch controls need to be located in safe operating locations.
- The winch cable and accessories have to be rated equal or greater than the winch.
- Command areas have to have two means of egress in case of an emergency.
- Noise levels in command areas can not exceed 80 db with the major operation components in operation.
- If the seating in the command areas are used during road travel, the seats are required to have seatbelts meeting the same requirements as in the cab.
- Floor surfaces in command areas are required to meet the slip resistant requirements for interior walking surfaces.

Mobile Foam Fire Apparatus

One of the other major changes in the NFPA 1901 Standard was the abolishment of the NFPA 11C Mobile Foam Apparatus document and merging the requirements into NFPA 1901. The old standard was primarily designed for industrial foam apparatus. The new requirements blended some of the industrial requirements with municipal fire apparatus requirements.

One of the most significant differences in the Mobile Foam Fire Apparatus from other fire apparatus is the difference in pump ratings for industrial style pumps. Industrial style pumps are pumps with capacities of 3,000 gpm and greater, which apply to a different testing standard than the standard for firefighting pumps. Other refinements have been made in the foam chapters to accommodate better testing requirements and to more closely relate to class A and B systems used on fire apparatus today.

Fire Apparatus Preventative Maintenance

A key component to an efficient fire department operation is dependable well-maintained fire apparatus. NFPA 1500 Standard on Fire Department Occupational Safety and Health Program requires that fire departments establish a preventative maintenance program for their fire apparatus. Some of the major requirements in establishing a program are the following:

- Must have an established program for checking and inspecting fire apparatus
- Must have a system for reporting deficiencies and problems found during the inspections of the apparatus
- Must have a responsible person to manage and oversee the program

- Must establish a list of "out of service" criteria for the apparatus
- Must establish a schedule for preventative maintenance
- Must have qualified persons conducting preventative maintenance and repairs
- Must have a system to maintain maintenance and repair records for the life of the apparatus

In order to help departments establish preventative maintenance programs, the NFPA developed a new standard for that purpose, NFPA 1915 Standard for Fire Apparatus Preventative Maintenance Program.

Establishing an aggressive preventative maintenance program and conducting the required annual testing on apparatus are essential to providing reliable apparatus.

11

FIRE DEPARTMENT APPARATUS SPECIFICATION AND PURCHASING

William C. Peters
Gary R. Pope

CHAPTER HIGHLIGHTS

- **Incorporates the needs analysis, replacement considerations, and financing information discussed in chapter 10, with a general understanding of organizing specifications, to produce a functional purchasing document.**
- **Guides the chief through the bidding, inspection, acceptance, and training phases of fire apparatus purchasing.**

INTRODUCTION

Specifying and purchasing fire apparatus can be a very difficult exercise for the fire chief who is unfamiliar with the procedures involved. Due to the infrequency of this process, few chiefs have the opportunity to become thoroughly familiar with the intricacies of apparatus purchasing. The common method utilized by many purchasers is to rely solely on the advice and guidance of an apparatus salesperson. This can add to the anxiety and apprehension of the chief who is concerned with obtaining suitable apparatus at a reasonable price for his community.

This chapter will help guide the chief through the purchasing maze in an orderly fashion by conducting proper research and evaluation and justification for the purchase.

The Process

The purchasing process is dependent upon many variables. Time, money, and the physical and tactical requirements of the apparatus are some of the considerations. The following outline covers the numerous steps that may be necessary to turn the initial proposal into a functioning piece of fire apparatus.

The Purchasing Outline

1. Determine who will research and formulate the specifications
    a. Fire chief
    b. Staff member(s)
    c. A committee
2. Establish and define the amount of time that is available from the beginning of the project to the delivery
3. Consider the financial implications and replacement options available
    a. Purchase
    b. Lease
    c. Refurbishment
4. Conduct research
    a. Basic types of apparatus
    b. Features and options available
    c. Manufacturer's reputation
5. Secure funding needed for the purchase
6. Outline preliminary requirements using NFPA Standards (Appendix C)
7. Determine acceptable manufacturers and request sample specifications and representative drawings
8. Using information provided, assemble a preliminary specification
9. Hold a pre-bid conference if required or desired
10. Distribute the final specifications for a public bid
11. Evaluate the bids and make a recommendation of award of contract based on the best value (not necessarily the lowest bid)

12. Attend a pre-construction conference with the manufacturer to discuss each detail of construction
13. Conduct other inspections as specified
    a. Completed chassis or pre-paint inspection
    b. Final inspection
14. Receive delivery and perform acceptance testing
15. Schedule manufacturer's training classes, if included in the specifications
16. When all members are proficient, the apparatus is placed in service

Who Should Formulate the Specifications?

In larger metropolitan fire departments, apparatus specifications and purchasing are usually addressed by an apparatus or maintenance officer, planning department, or support-services staff, with final specifications approval by the fire chief.

In smaller career departments without extensive staff personnel, the duties will usually be handled directly by the chief of department. In this case, the chief would be wise to receive input from deputies or possibly the company officers who will ultimately utilize the apparatus. Some seek the advice of larger departments who are more familiar with the process or the assistance of fire protection consultants. Considering the financial implications of the purchase, utilizing outside help is usually a wise decision.

Volunteer fire companies will usually form a committee to deal with the purchase. The makeup of this committee can be critical to the successful outcome of the project. While all members of the company should be able to voice their opinions, a select group of knowledgeable individuals should be charged with researching and evaluating the features and options that will ultimately become the finished apparatus.

Regardless of the size or the type of department, the following personnel should be considered for their input and expertise:

Company officers will be able to make suggestions relative to the tactical and operational objectives that they expect from the apparatus.

Drivers/operators/engineers should provide input into the performance expected of the vehicle. Items such as driving and operational controls, as well as pump panel layout should receive their attention.

Maintenance personnel should focus on the mechanical and maintenance portion of the vehicle. Concerns such as the electrical system, component size and performance expectations, and ease of operator preventative maintenance should be addressed.

Training officer will most likely be well informed about current practices and procedures that can be applied to enhance the operational efficiency of the apparatus. Innovations such as Class A foam, high output pre-connected attack lines, and large diameter hose (LDH) evolutions should be researched and considered for inclusion in the specifications.

Safety officer should focus on enhancing the general overall safety of the apparatus. Items such as ensuring a sufficient number of seats for responding personnel, adequate visual and audio warning devices, and evaluating steps and handrails should be the safety officer's area of responsibility.

All of these committee members will have to coordinate and interact with each other to facilitate the purchase. For instance, the safety officer might consider a siren or lighting configuration that would place an undue strain on the electrical system. The maintenance person should evaluate this condition and work with the safety officer to arrive at a compromise that is both adequate and practical.

Prior to the beginning of the committee process, ground rules should be established to determine how the final decisions will be made. Will it be by democratic vote of the entire fire company, or the apparatus committee, or will the chief make the ultimate decision after considering the other members' input? While this is a local decision, it is most important to define the process before embarking on the project as personalities can easily get involved and burden the entire process.

Timeframe

One of the first issues that must be addressed is the amount of time that is available until the delivery of the new apparatus is needed. Product research, sales presentations, preparation of specifications, holding a public bid, evaluating the results, and awarding a contract will usually take from 6 to 12 months.

The period of time from the signing of the contract to delivery will vary according to the manufacturer's backlog and the intricacies of the specifications. Taking 6 to 12 additional months to complete the unit is about average. Usually aerial units, special customization, or a fire apparatus chassis that is not manufactured by the apparatus builder will require additional time until final delivery.

Realistically speaking, 18 to 24 months should normally be allocated to complete the purchasing process.

If the apparatus being replaced is out of service due to a mechanical failure or accident, the time frame requirement may have to be significantly reduced. In this case, it might be wise to contact several manufacturers to see if stock or demonstrator units are available. Most manufacturers have demonstrator units, sometimes at reduced prices, that have been driven around the country to apparatus shows and displayed for potential customers. These units are usually equipped with the latest features and could be a viable option if time is of the essence.

Before rushing into the purchase of one of these units, the fire chief must be certain that it has the proper design and performance features to satisfy the department's requirements. It should be remembered that this apparatus will probably last 20 years and it should be fully capable of fulfilling the present mission of the department as well as carrying it into the future. If special requirements cannot be met, perhaps the community should consider borrowing a unit from a neighboring department or purchasing a used piece of equipment to fill the void until the new unit can be properly specified and built.

Purchase, Lease, or Refurbishment?

Just as the timeframe will have a direct impact on the replacement of the apparatus, so will the financial arrangements. In most cases where the municipal government is responsible for maintaining fire protection, outright purchasing using the capital improvements budget is normal. Funds for these major projects are borrowed and paid back over a period of time by selling municipal bonds. Bond purchasers receive a favorable return on their investment, and because in many cases it is tax free, the municipality enjoys a lower interest rate than the business community.

The premature replacement of apparatus due to accident, fire, or building collapse may significantly reduce the purchasing time frame. (Photo by Richard Sikora)

In many locations, the amount of bonded indebtedness that a municipality can carry at any one time is limited. Therefore, if the city can only borrow a certain amount, the fire department will be competing with other city agencies for their piece of the "capital budget pie." As well as providing fire protection, the governing body must consider roads, schools, libraries, parks, recreational facilities, and public buildings that require improvement. Obviously, each agency considers its projects to be of top priority. This is why considering the needs analysis and justification for replacement, discussed in chapter 12, is critically important.

When capital funding is unavailable, another viable alternative might be a lease/ purchase plan. Most manufacturers either provide leasing or can make the necessary arraignments for the purchaser. Similar to financing the family automobile, the apparatus is delivered and annual payments, for over 7 to 10 years are made to pay for the vehicle. When the last payment is made, title to the apparatus is turned over to the municipality. Obviously an interest charge is included so the total cost of the apparatus will be more than a straight purchase; but when funding is deficient, this is one method of maintaining proper fire protection at an affordable rate.

Many independent volunteer fire companies have found that while attempting to save the necessary funds for a purchase, inflation and routine price increases keep moving the project further into the future. With proper documentation, they too can qualify for a lease plan and enjoy the benefits of a reliable replacement while making the annual payments.

Insurance and maintenance during the lease period is usually the responsibility of the purchaser.

When discussing the funding for replacement, inevitably the question of refurbishing the present unit will be raised. It is wise for the fire department to possess a thorough understanding of the refurbishment process and the costs involved in order to intelligently answer questions that will arise.

Refurbishment is considerably more involved than just surface body work and a coat of new paint. With the proper research, there might be more reasons why refurbishment is not a viable alternative as opposed to why it is.

The following points should be explored before considering refurbishment:

- What is the condition of the vehicle's engine, transmission, driveline, and differential? Does it need to be replaced, rebuilt, upgraded, or converted from gas to diesel or standard transmission to automatic?
- What is the condition of the functional units of the apparatus? Is the pump of adequate capacity to meet your present and future operational objectives? Will it easily pass the annual pump test? Can the aerial device meet the extensive requirements of the annual inspection and load test as well as the five-year nondestructive test?

- Is the manufacturer still in business? Can repair parts be easily obtained now and will they be available in the future?
- In addition to obvious body work and painting, what unseen conditions might be encountered in the substructure or chassis that would require additional work?
- Is the apparatus able to conform to modern, more efficient firefighting procedures, such as pre-connected attack lines, large diameter hose equipment, larger water tanks, pre-piped deck gun, or class A and B foam systems?
- Is the aerial device long and strong enough to accomplish the tasks that it encounters? Are taller buildings out of reach of the aerial or are housing "setbacks" causing older light-duty aerials to be used at dangerously low angles?
- The final and most important question that must be asked: Is the fire department willing to delay the extensive safety upgrades of the current standards for another five to seven years?

All of these questions must be seriously evaluated before a decision to refurbish can be made. If a negative answer is obtained to one or more of them, it strengthens the justification for replacement rather than refurbishment.

Conducting Research

Whether a decision to purchase or lease was made, the chief or committee should plan on conducting extensive research into the basic types of apparatus, the features and options available, and the many manufacturers' reputations for quality, service, and warranty. There are several ways that this can be accomplished.

Advertisements

One of the first is a method that the manufacturers of all products use: Advertising. There are several national and many local fire service publications that rely on advertising for their revenue. In most cases, apparatus manufacturers will use this avenue to unveil new products, innovations, and special promotions. The reader's service card that usually accompanies the magazine will surely bring a rapid response from the advertisers selected.

These same publications often contain interesting articles about the latest apparatus and equipment trends, maintenance procedures, and problem-solving ideas. Much can be learned from the research that goes into preparing these essays.

Trade Shows

Another excellent way to conduct research is to attend the many trade shows that are presented around the country. National organizations such as the International Association of Fire Chiefs (IAFC) and the International Society of Fire Service Instructors (ISFSI) have shows in conjunction with their conferences that are held in different parts of the country. Acres of fire apparatus and equipment are on display for the customers' perusal. Many regional conferences are also held by local chiefs' organizations, apparatus maintenance groups, and instructor associations.

Apparatus manufacturers are well represented at these events and display their latest designs, innovations, and features. This provides the committee with the convenience of access to many manufacturers all in one location and knowledgeable factory personnel, who can answer most questions about their products. It also provides an opportunity to closely examine the workmanship and quality of the apparatus under consideration.

With the vast array of vehicles on hand, it is wise to take notes, photograph, or videotape the units and collect manufacturers' brochures and business cards. There is such an overwhelming amount of information available that it would be foolish to trust it all to memory.

Apparatus and equipment shows provide the best opportunity to inspect, evaluate, and ask questions about the units on display. (Photo by William Peters)

Apparatus Associations

Joining and participating in the activities of the many apparatus associations are other ways to gather valuable information. Many of these groups offer training programs on apparatus maintenance as well as information on the latest NFPA Standards and specification preparation. Presentations from the experts in the apparatus field are provided, and participants can benefit from the all-important opportunity for networking with others who share a common interest in apparatus. Many times valuable information can be obtained during friendly conversation with a counterpart from another part of the country.

Some of the national organizations include the IAFC apparatus maintenance section, the National Association of Emergency Vehicle Technicians (NAEVT), and the Fire Department Safety Officers Association (FDSOA). There are also numerous, well-organized, regional apparatus associations. An additional benefit of membership in these organizations is the newsletter that most publish on a regular basis. It provides an exchange of information on current apparatus trends, problems, solutions, and maintenance tips.

In most cases, you need not be an apparatus mechanic or technician to join these organizations; you need only a desire to advance your knowledge of apparatus technology.

Visiting apparatus manufacturer's facility provides the opportunity to examine apparatus in various stages of the building process. (Photo courtesy Saulsbury Fire Equipment)

Visit Manufacturing Facilities

Another way to gain insight into how a particular piece of fire apparatus is built is to visit the manufacturer's facility. Many routinely offer guided tours and most welcome an inspection of their facility. Examining apparatus that are in various stages of the building process is a very enlightening experience. It is not often that you can see the important subassemblies that make up the apparatus before they are hidden under sheet metal and numerous coats of shiny paint.

Inspecting apparatus that are finished and awaiting delivery can often stimulate new ideas or highlight different methods of accomplishing certain goals. As with the apparatus shows, record items of interest for future reference.

Visit Other Fire Departments

Visiting neighboring departments that have had a recent delivery of a certain piece of apparatus is another way of networking information. Most manufacturers would be happy to supply the names of the customers who have placed their apparatus in service. You could call for an appointment yourself or ask the apparatus dealer to set up the visit.

Each method has its advantages and disadvantages. In some cases the customer might be more candid if the dealer's representative is not present. The advantage to having him present is that he should be able to address questions about variations to the basic apparatus that the customer might not be able to accurately answer. A compromise of both methods might be the best way to get the complete picture. When you make the appointment for the visit, request that the apparatus salesperson join you an hour or more after you arrive to discuss options and variations to the apparatus that is being displayed.

On this type of visit, it is also wise to talk to both the officials who were responsible for the purchase and to the members who drive and operate the unit. This provides information about the manufacturer's reliability, conformance to specifications, and general disposition. This will hopefully provide a well-rounded opinion of the product and the manufacturer.

When calling for the appointment, ask if a copy of the bidding specifications and possibly a shop drawing are available for you to have. They will be valuable when researching products and making comparisons.

Decisions

Before moving on to the final research phase—sales presentations—several decisions will have to be made regarding the basic requirements of the unit. Developing an outline will provide the manufacturer with the necessary information about the department's wishes and will help expedite the presentation and estimated pricing.

This will also serve as a foundation on which to build and develop a complete set of bidding specifications.

Some suggestions of what to include in your basic requirements are:

- *Basic type of unit*—pumper, aerial, platform, rescue, mobile water supply, initial attack, or combination vehicle.
- *Chassis*—custom or commercial, maximum number of firefighters to be seated; diesel or gasoline engine; standard or automatic transmission; cab construction material (if there is a preference); cab type, split tilt, full tilt, fixed; special chassis requirements such as tight turning radius or short wheelbase.
- *Body*—standard or extra-large compartmentation; high side compartments; type of compartment doors, hinged or roll-up, body material (if there is a preference), hoseload.
- *Aerial Device*—aerial ladder, platform, telescoping or articulating boom, length, waterway output, outrigger spread, tip load requirements if they exceed the standard.
- *Pump*—rated gpm output, single or multistage, pressure control by relief valve or pressure governor, gauges or flowmeters, special requirements for suctions or discharges.
- *Tank*—size, construction material (if known), special requirements such as tank to pump flow rate, direct filling capabilities, tank dump systems, or foam capabilities.
- *Special options and features*—scene lighting, generator, booster reel, trash line, ladder rack, or pre-piped deckgun.

Sales Presentations

Based on your previous research, several acceptable manufacturers should be contacted for a sales presentation. The basic outline that was developed should be provided in advance and ballpark pricing requested. Most dealers would be happy to provide brochures, sales information and possibly recent shop drawings for your consideration. This material is a good starting point for developing your own specifications.

It should be remembered that the estimated price of the apparatus might not include all of your requirements such as bonding, inspection trips, delivery expense, equipment, and training. These will all have to be addressed before arriving at a specific dollar amount of funding to be requested.

It is also important that one member of the committee be designated to record the details of the meeting and to assemble the material provided in a neat orderly fashion for future reference.

Funding Commitment

After establishing approximate prices and delivery times, the committee should report back to the governing body for the approval of necessary funding. When seeking funding, provide an honest estimate but be careful not to cut yourself short. In the apparatus industry, there are frequent price fluctuations and several variables that have to be considered when the final bid is prepared. Also, ideas that were not originally addressed might be included in the final specifications. With the amount of work that is required, and the life expectancy of the apparatus, it is better to err on the high side rather than having to settle for less than desired or having to go back to the city officials to ask for more money, which could seriously jeopardize funding for the entire procurement project.

Approaching City Officials

Approaching city officials to request a large sum of money for the purchase of a piece of apparatus can be intimidating to say the least! This is especially true when the spokesperson does not have a background in public speaking.

There are several important items that should be understood before entering this arena:

- ***Be prepared!*** All areas of justification must be fully explored and documented. Anticipate all possible questions and have answers prepared that are rehearsed and flow naturally. (This is the type of preparation that the U.S. president goes through before a debate or news conference where important questions will be asked.) Being properly prepared will also help reduce the nervousness and anticipation that you might be experiencing. Never *wing it*!
- ***Stress escalating maintenance costs and inefficiency.*** Elected officials must be convinced that in the long run, this is a wise economic decision. Be prepared to answer why refurbishment is not economically feasible (if this is the case).

- ***Explore the issue of public safety.*** Are the citizens at risk because of an unreliable piece of apparatus? Don't make statements that can't be proven. Produce maintenance records highlighting the number of times the apparatus was out of service at the shop.
- ***Don't get emotional; stay calm.*** Remember, ultimately this is a business transaction, and it is always easier to do business in a rational manner.
- ***Don't threaten...no one likes to be "put in a corner."*** Threats in the form of recall elections or adverse publicity are counterproductive.
- ***Mention the liability factor.*** Subtly point out the liability involved if the apparatus fails to respond or if members are still riding without seats or proper restraints. The award of one lawsuit would easily cover the cost of the replacement apparatus.
- ***Don't count on major public support.*** Most taxpayers are more interested in how the purchase will impact the tax rate rather than how much more reliable their fire protection will be. (Nobody one ever visualizes a fire in *their* home.)
- ***Understand budget limitations.*** Emphasize your complete commitment to work with the administration by expressing an understanding of budget limitations and keeping the specifications basically "frill-free." Purchasing a combination apparatus and offering to take on additional duties such as EMS or HazMat could be proposed if it will help save money.
- ***Stress time and money.*** Discuss the time frame involved (18 to 24 months) and the anticipated price increases associated with delaying the order.
- ***Always be diplomatic!***

A recent magazine article by the chief of a volunteer department indicated that he very successfully educated his public officials by inviting them to tour the firehouse one evening. Prior to the date of their visit, a basic outline of the fire department's history, apparatus, training subjects, and number of man-hours committed, as well the type and number of calls the department responded to the previous year was delivered to each member of the city council.

During the visit, the governing body was shown how well the apparatus and equipment were being maintained. A demonstration of rescue techniques was given and problems (*i.e.*, the need for a diesel exhaust system) were pointed out.

At the conclusion of the visit, refreshments were served and the officials were treated to a video and slide presentation of recent local fires. After this enlightening experience, the governing body possessed a much better understanding of fire department operations and the sincerity of the personnel, as well as their needs.

Apparatus Requirements

Once funding is committed, the fire department must begin the process of developing the bidding specifications. The first step in this process is to determine which sections of the NFPA 1901 Standard applies to the purchase. This should be determined by the primary function of the apparatus. For instance, a 1250 gpm pumper equipped with a water tower should comply with the provisions of NFPA 1901 Chapter 5, "Pumper Fire Apparatus." An aerial apparatus, equipped with the same 1250 gpm pump, should be constructed in accordance with the provisions in NFPA 1901 Chapter 8, "Aerial Fire Apparatus."

The general requirements section of each apparatus standard contains the following statement:

> *Responsibility of Purchaser: It shall be the responsibility of the purchaser to specify the details of the apparatus, its required performance, the maximum number of firefighters to ride on the apparatus, and any hose, ground ladders, or equipment it will be required to carry that exceeds the minimum requirements of this standard.*

This effectively places the burden of specifying everything that is needed to perform the duties that are defined on the purchaser. In order to assist in this awesome task, the NFPA 1901 Standard contains a section entitled *Annex B* and *Annex C.* Utilizing the annex is not a requirement of the standard; it is included for informational purposes only. Annex B is divided into subject headings that follow the order of the standard. It is written in a question and answer type format, with an easy fill-in-the-blanks style. The section of the standard that deals with each question is listed in a column to the right of the page, making research of the requirements easier for the specifier.

Properly utilizing Annex B will help reduce the apprehension that can be present from concerns about missing critical specification requirements for the apparatus.

The completed Annex B along with all options and features that are required should then be given to several qualified apparatus manufacturers, requesting sample specifications that comply with your requirements.

Qualifying Manufacturers

What procedure should the fire department employ to determine the qualifications of a manufacturer?

From the previous research, the committee should evaluate each interested manufacturer on the following points:

- ***A commitment to fulfilling the requirements of the fire department with the least number of exceptions.*** Many times, if a particular feature or option is unavailable or difficult to provide, apparatus sales representatives might attempt to talk you out of it, rather than trying to comply with the requirement.
- ***The manufacturer's reputation.*** This is based on past performance and input about customer satisfaction obtained from other departments that you should have contacted as part of your research.
- ***Design and reliability factors.*** Is the apparatus produced by this manufacturer designed well?
- ***Ease of service and maintenance.*** Preventive maintenance is the key to maintaining apparatus in a state of readiness. If these routine chores are difficult, there is more of likelihood that they will be overlooked. The ease of maintenance correlates directly to the amount of downtime that will be required to repair and replacement parts.
- ***Manufacturer's delivery schedule.*** This is particularly important if the current apparatus is out of service. In some cases, it could be a major concern of the fire department.
- ***Service availability.*** A local dealer or private-affiliated service facility qualified to properly maintain the apparatus and provide reliable warranty service is required and should be carefully evaluated by the purchaser.
- ***Parts availability.*** The ready availability of replacement parts also correlates directly to the amount of downtime the unit will experience. If the manufacturer either has parts available at a local dealer or a good network of supply by express shipping, out-of-service time will be reduced considerably. This requirement is difficult to measure, except by interviewing present owners.
- ***Estimated price.*** While the price should be the least consideration in a critical purchase that might have a 20-year life span, we all live in the real world of budget constraints. The best apparatus could conceivably be priced considerably above the budget limitations.
- ***The stability of the firm.*** Fire apparatus represents a sizable investment, and it is critical that the manufacturer be available to provide service and parts into the future.

After considering all of the above points, certain manufacturers will be able to meet your requirements and should be deemed "qualified" to provide information for the development of your specifications.

The requirement of a local facility to provide service and warranty work for the apparatus is extremely important. (Photo by William Peters)

Using Sample Specifications

It is safe saying that most apparatus specifiers are not automotive engineers, nor do they possess the qualifications to tell a manufacturer what dimensions the frame rails must be, or how many cross-members are needed to provide the strength and stability necessary for the apparatus. For this reason, we must rely on sample specifications developed by people who do have the necessary knowledge.

By utilizing an established specification as a guide, we are provided with quantitative values for comparison with the other bidders. For instance, if the specifications indicate that the front bumper will be a certain size and thickness, it can easily be determined if the other bidders meet or exceed the requirements. If the specifications only state that a bumper will be provided, and no minimum standards are set, essentially any bumper that is installed will meet the specifications.

It is extremely important that the method of taking exceptions or explaining clarifications be provided. This fairly allows other bidders to present their version of compliance with the *intent* of the specifications. A bidder's exception will often exceed the minimum requirements of the specifications.

While using the information contained in a manufacturer's specification as a sample for your specs, there are many reasons why you should not use it as the bid document.

Certain "proprietary" items will undoubtedly be contained in the language that other bidders will either have a difficult or impossible time meeting. It is unlikely that a manufacturer providing specifications would include anything that they couldn't meet!

Because of computer generation of the specifications, there are likely to be confusing or conflicting requirements. For instance, in one specification, the cab portion was described with all of the standard features explained (*i.e.*, fixed side windows). If the customer desired side windows that slide open, the manufacturer would not alter the original wording of the specification, but list it as an option stating, "The fixed side windows will be replaced by sliders." This could be confusing for other bidders who are not familiar with the format of the specifications. It is better to describe the features correctly the first time in your own language.

Most manufacturers' specifications contain large areas of language (from several paragraphs to several pages) describing sections of the apparatus. This makes bid evaluation extremely difficult as the other bidders will probably take an exception to the entire section and address the components of their assembly in the same manner. If the spec writer takes the information provided in the sample, and converts it into smaller, more manageable units, the specification will be easier to understand, and bidders taking exception will have to address each element individually.

Another reason to rewrite the specification is that items you thought were included might have inadvertently been left out, or features that are standard with one manufacturer might be optional with other bidders. For instance if the supplier of the "sample" specs routinely provides a certain style of padding in the cab, it might not be sufficiently explained in the specifications for the other bidders to match.

Finally it could raise ethical questions by the other bidders. While perhaps there is nothing wrong with using manufacturer's specifications as a sample or guide to the way you expect the apparatus to be constructed, imagine the embarrassment at a bid hearing if you had to admit that the successful bidder actually *provided the bid document*!

Remember the NFPA Standards place the burden of specifying all of the details of the apparatus and its performance on the purchaser. Take the information contained in the sample specifications provided by reputable builders and modify it to meet the needs of your department.

Types of Specifications

Specifications can generally be put into three categories: design, performance, and a combination of the two, design-performance.

Design specifications contain a preconceived arrangement of detail and form, in which all of the details of construction are spelled out. Most specifications obtained from the manufacturers could be categorized as being heavily design-oriented. Often the design features of components are described down to the last nut and bolt.

There are certain advantages and disadvantages of each type of specification. The advantages of a design type specification include:

- Identifiable, everything is clearly defined.
- Uniformity of the fleet, which can benefit the maintenance, parts inventory, and training functions.
- Possibly easier to evaluate the bids. Other manufacturers will have to compare every item of their bid to the design features specified.
- More accurate estimations of the cost of the apparatus can be obtained for budgeting if specific components are identified.

The disadvantages of using a design specification are:

- Restrictive. Depending on the degree of design criteria used, the specification could be overly restrictive to other responsible bidders.
- Restricts innovation that could result in reduced effectiveness. By identifying every component, progressive new ideas will not be incorporated into the apparatus.
- Favoritism. As stated earlier, most specification writers are not automotive engineers and therefore incapable of designing a vehicle from the ground up. Therefore, many of the design features of the apparatus are obtained from using a manufacturer's specification. Realistically, incorporating an over-abundance of nonessential design features from one manufacturer, without liberally interpreting exceptions from the other bidders, could be construed as favoritism.
- Cost. When using a tight design specification from one manufacturer, there is a possibility that the cost of the apparatus will be higher, due to reduced competition. Other manufacturer's bids will be higher when trying to meet the design criteria specified if it differs from their normal method of construction.

Performance specifications are written around the required functional criteria of the apparatus. In this type of specification, all of the details of performance are outlined along with the associated testing necessary to quantify the results. Basically it tells the bidder what the apparatus must do but not necessarily how to accomplish it.

The advantages of a performance oriented specification are:

- Competitive. Allowing a manufacturer to build in its own style in order to meet your performance criteria will result in competitive pricing.

- Allows innovation. All new innovations in fire apparatus were ideas to meet certain performance objectives. Without restricting the design, better use can be made of innovative features.
- Better effectiveness. Allowing the manufacturer to use innovative ideas could result in a more effective apparatus.

Some of the disadvantages of the performance specification are:

- Interpretation of the specifications could be flawed due to the reduced number of design definitions.
- Measurement of performance could become a problem unless the testing criteria is clearly defined in the specifications.
- Uniformity of the fleet is lost. Maintaining uniformity for training, maintenance, and parts stock can be beneficial to the fire department. If a performance specification is used, it is unlikely that the fleet will remain uniform.

The third type is a combination design-performance specification. This type couples both the necessary conceptual details (design features) with the preconceived functional criteria (performance) to arrive at a specification that will provide an apparatus that is designed to fit your needs and will as perform as desired. The design elements of this type of specification should detail the important features while allowing each manufacturer to meet the nonessential requirements in their own manner.

For instance, some specifications are so descriptive about the components that make up the cab that they actually specify the diameter of the steering wheel and the number of spokes it will have! This unnecessarily causes other bidders to take an exception and explain their steering wheel. If the steering wheel has a specific performance requirement, such as the ability to adjust (tilt), that should be included, however the rest of its design should not be unnecessarily restrictive.

There are many advantages to using this type of combination specification. Because it is adaptable, there are actually no disadvantages.

Some of the advantages are:

- Practical. By combining the correct amount of important design features with the necessary performance, a practical specification is developed.
- Competitive. This type of specification allows for more competition between the bidders.
- Cost. Increased competition will help reduce the cost of the apparatus.
- Identifiable. Key components that are design in nature are identifiable and comparable for bid evaluation.

- Uniformity of certain components. For example, the fire pump can be maintained by outlining it as design rather than performance features.
- Adds credibility. Rather than using a manufacturer's specification word for word, the combination specification is really your specification.

Each type of specification can be modified as to the degree of difficulty that will be encountered meeting them. The term *tight* is often applied to a highly restrictive, design type specification and *loose* to one that is more relaxed. Often the degree of this variable will be controlled by the purchasing department or the community's administration. Some have very specific rules that must be followed about how restrictive a specification can be.

Open specifications are relatively free of specific components or manufacturing styles and do not restrict properly documented exceptions or clarifications. This type of specification is more competitive and will usually result in more bidders and sometimes a better price. Unfortunately, it might also attract bidders of lesser quality who can meet the relaxed specifications.

Tight bidding specifications are highly restrictive in nature and provide a detailed description of all components and construction designs. Often there is less tolerance for exceptions and deviations and will state "no exception" to certain requirements. A tight spec will usually limit the number of bidders, which will in turn increase costs, due to reduced competition. This type of specification is sometimes necessary to ensure delivery of a specific piece of apparatus.

General Requirements

As can be seen, specifications can be written in many shapes and forms. One of the biggest pitfalls in specification writing is a weak or poorly written set of general requirements. Sometimes referred to as the *boilerplate,* this section outlines the bidding instructions and defines the "ground rules" that apply to using the specifications. The general requirements are sometimes more important than the construction specifications in relation to determining the validity and evaluation of a manufacturer's bid.

Following are some of the important issues that should be completely explained in the general portion of the specifications. Remember, a manufacturer's sample specifications might not contain all of the items that the fire department considers important and certainly will not contain anything that would be difficult or impossible to meet.

- *Intent statement*—a general statement that describes the apparatus and the requirement to comply with the appropriate NFPA Standard as well as federal, state, and local motor vehicle laws.

- *Bid submission requirements*—the method and form of submitting a bid including the acceptable way of outlining deviations and exceptions (in bid order), and if contractor specifications and drawings are required. It is also wise to include, in very direct language, that the fire department's specifications will prevail over any proposal submitted, unless a properly defined exception was granted. This will help avoid a problem that sometimes surfaces when a bidder does not take exception but proposes to build according to his own standards regardless of the specifications.
- *Performance requirements*—performance requirements, including road and operational functions. It can simply indicate the appropriate NFPA paragraphs or might outline your own special requirements.
- *Delivery and payment terms*—the location of acceptance and location of delivery, whether at the factory or in your community. Payment terms should also be clearly defined. Some manufacturers offer a discount for pre-payment of components or progressive payments as the apparatus is being built. As a word of caution, require that the component being paid for, such as the chassis, be invoiced and a certificate of ownership be issued. Some fire departments have lost chassis that were pre-paid when the manufacturer suddenly went into bankruptcy.
- *Special construction requirements*—items such as whether a custom or commercial chassis is acceptable, the type of material to be used in the construction of the cab and body, and if one manufacturer is to build the entire apparatus. The latter is a very restrictive requirement that is sometimes used to help prevent the possibility of divided responsibility for warranty work. You must also realize that it will restrict the number of bidders on the project.
- *Approval drawings*—if approval drawings are required, the time when they will be delivered, as well as whether they or the written specification take precedence. It is wise to state that the fire department will make every effort to correct the approval drawings, but the written specifications, along with any corrections, will prevail.
- *Manufacturer's experience and reliability*—requirements that outline the criteria to establish a manufacturer's reliability and reputation. Items can include the number of years in business, a list of customers who can be contacted, and possibly a financial statement from a nationally recognized financial rating service. It might also be wise to include a statement if you are not willing to accept a "prototype" or first of a kind, apparatus built by the manufacturer. This will establish whether they have built apparatus similar to the one described in your specifications.
- *Bonding and insurance*—bonds are a form of insurance that bidder will comply with certain requirements of the bid. A bid bond insures that the bidder is responsible and will execute a contract if awarded the bid. This provides recourse

if for instance, the bidder submitted an inaccurate price or later decided that he did not wish to proceed with the contract. A performance bond is issued after the contract is awarded and indicates that the bidder will perform according to the provisions of the contract. If the contractor does not perform up to requirements, the bonding company is only responsible to provide a suitable replacement apparatus at the bid price. This does not mean that the department will receive a free fire truck, only that the bonding company will pay any difference between the bid price and the actual cost of the apparatus provided. It should be understood that while bonding provides an element of protection, it also increases the price of the apparatus.

Liability insurance is another item that is sometimes required. Multi-million dollar policies are maintained by the larger manufacturers; however, this might be difficult for some of the smaller ones. Some departments only include a phrase, such as: "The manufacturer will be responsible to defend any and all lawsuits resulting from the use of the apparatus."

- *Factory inspection trips*—factory trips are an essential part of the apparatus construction and inspection process. The number of trips, number of participants, and what associated costs will be borne by the bidder should all be included. Some departments require a set dollar amount for the trips to be included by all bidders and others pay for the trips themselves in order to eliminate any unfair advantage that a bidder who is geographically closer to the department might have.
- *Warranty and follow-up service*—a requirement for a factory authorized service center within easy travel distance of the purchaser is sometimes specified. Requiring the bidders to supply the size of the facility, number of employees, number of mobile units, and capabilities of the facility, can help determine if the service center meets the requirements of the purchaser. The location of where warranty service will be performed (at the fire station or at the repair facility) can also be specified.
- *Build time*—especially important if time is of the essence. The specifications should require the number of calendar days from the signing of the contract until delivery. Bidders should be warned that stating a time from the receipt of the chassis or other major component until delivery is unacceptable. Some purchasers include a penalty clause or liquidated damages for late delivery. If this is included, it is more likely that you will get a fair appraisal of the actual anticipated delivery date. The penalty should also cover apparatus that are delivered incomplete or not up to specifications.
- *Special requirements*—such as specific size, weight, or turning radius should be outlined.
- *Training*—if training by a representative of the manufacturer is desired or expected, it should be clearly outlined in the specifications. The number of days or

special hours including evenings and weekends for volunteer departments, should all be considered. Some departments require a program curriculum to be submitted in advance for the approval of the training officer. This could eliminate any conflict that might occur when the manufacturer's training program is not consistent with fire department standard operating procedures. Videotapes or handouts should also be included if desired. Some departments also include a training program for the maintenance mechanics who will maintain the apparatus. This could be particularly important especially if it is the department's first piece of apparatus produced by a particular manufacturer.

- *Warranty requirements*—the requirements as to the length of warranty and method of requesting warranty service on the apparatus. Individual components such as the engine, transmission, and pump will be covered by their respective manufacturer's warranties. Some purchaser's specifications require that the dealer act as the warranty agent, coordinating claims with the component builders. When the purchaser has specific warranty requirements, it is advisable that the bidders be warned that any difference between the warranty requirements stated in the specifications and the warranty offered by the bidder will be taken as an exception. Many times a bid will contain a page of warranty coverage and a phrase that the bidder's "stated" warranty applies. It is important that they cite the differences as exceptions.

CONTROLLING "FRILLS"

With the high cost of apparatus, and cities and towns struggling with a limited tax base, the specifier should limit the number of nonessential options that might be considered "frills." There is nothing that will halt a purchasing proposal faster than the governing body deciding that the fire department is oblivious to their serious budget concerns. Unnecessary options such as chrome wheels, a bell, fancy murals and logos painted on the rig, or an excessive number of inspection trips, all give the appearance of extravagance. (I once reviewed a specification that outlined four inspection trips, of five department members each! Apparently the whole fire company was going to the factory to supervise the building at one time or another!)

These unnecessary items should not be confused with legitimate options and features. Accessories such as anti-lock brakes, a pressure governor, or an auxiliary braking device can be fully justified in the name of safety or efficiency. By comparing the cost of the feature to potential maintenance costs or a liability claim that might be paid if a preventable accident occurs, the accessory can often be justified. For example, a reduction in brake maintenance costs by the use of a retarder could demonstrate the efficiency of the investment in the option.

Dividing the cost of a safety-related option by the life expectancy of the apparatus points out that the feature adds minimally to the overall averaged annual cost of the apparatus and can help justify the purchase.

However, it is difficult to justify the cost of options that do not enhance efficiency or operational safety in the same manner. For instance, it is doubtful that expensive gold-leaf can be justified when reflective lettering that adds to night visibility will not only identify the fire company but do it more safely.

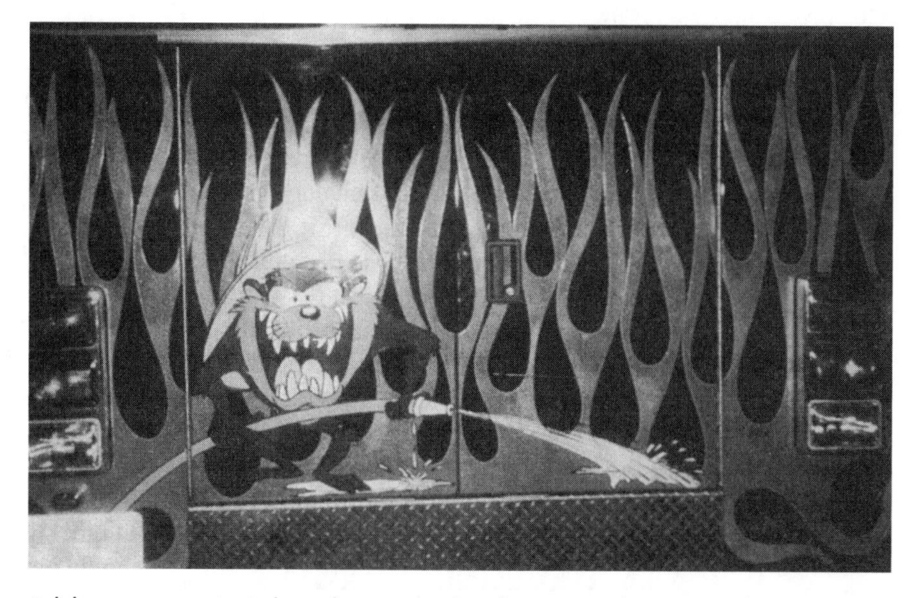

Including unnecessary options in the specifications such as fancy decorations give the appearance of extravagance and can jeopardize the funding for the project. (Photo by William Peters)

The Bid Process

When the fire department's committee is satisfied that their specifications have been refined and are complete, they are ready to be advertised for public bid. Some jurisdictions require an intermediate step known as a pre-bid conference. The conference is announced along with the availability of the preliminary bidding specifications. Some purchasing departments require that a manufacturer who intends to bid on the project be present at the pre-bid meeting to avoid contradictions or misunderstandings that might surface later.

At the meeting, each item of the specification is read and discussed by the participants. Minor changes and adjustments might be made at the conference; however, if a major change is proposed, the bid might have to be delayed while the fire department investigates the viability of allowing a change or substitution. An amendment containing all changes, clarifications, and corrections must be sent to all vendors who received the bid specifications.

While some consider this step an unnecessary inconvenience, at times it might save the fire department embarrassment, or help to expedite the bid process by uncovering an error, inconsistency, or misunderstanding.

When the final bidding specifications are distributed, the date, time, and location of the bid opening, as well as the method of submitting a bid, must be clearly stated. In most cases, the community's purchasing authority or governing body will handle the bid-opening process.

The fire department however, will most likely participate in the bid evaluation and recommendation phase of the procedure. Bid evaluation can sometimes be a very difficult task. By closely evaluating the proposals, the lowest *responsible* bidder must be determined. This should not be confused with a low bidder who does not meet the specifications or proposes a sub-standard product, hoping to get the award of contract based on price alone.

The first thing that should be examined is that the bid contains the required documentation for bonding, insurance, financial report, customer list, construction specifications, example drawings, and the general form of the bid proposal. Lacking important items such as a bid bond or insurance coverage should immediately disqualify the bid.

When evaluating apparatus bids, each item in the bidder's proposal should be compared to the fire department's specifications for compliance. If the bidder submitted his exceptions and corrections as instructed (in bid order), the process will go much more smoothly. A bid evaluation report should be organized, and each exception or deviation should be listed. In some cases, the bidder's exception might be equal or actually exceed the requirements of the fire department's specifications. In this case, it should be noted that the exception is granted. Be cautious of bidders who take exception to whole areas of the specification (*i.e.*, the body) and make a blanket statement such as, "The body will be constructed in accordance with our standard manufacturing methods." The competing bidders know how their products differ from the specifications, and it is their responsibility to outline the deviations if they want to be considered for the award.

Another important part of the bid evaluation process is to use the customer list that was provided by the bidder. Take the time to make some random telephone calls and network among other fire departments using the same apparatus. If an air of discontentment is present, you will usually find out quickly.

When all of the information is evaluated, the committee should prepare a written recommendation for the award of the contract. It should be based on the bidder who best meets the intent of the specifications, with the least number of exceptions and deviations. The price should be a secondary consideration. It is interesting to note that in some jurisdictions, the prices are submitted in a separate sealed envelope where they remain until after the bids are evaluated. The bids are evaluated strictly on their quality and ability to meet the specifications. After the evaluation is complete, only the prices of the manufacturers who met the specifications are revealed and the others are returned unopened.

Some jurisdictions conduct a post-bid hearing. Bidders are allowed to comment on the evaluation procedure and ask questions of the committee. If involved with a bid hearing, systematic preparation and justification for your decisions are essential. If the committee did its homework, it is unlikely that their decisions will be questioned.

If the governing body agrees with the fire department's recommendations, a contract will be awarded to the successful bidder and the apparatus construction process will begin.

Unfortunately at times, a difference of opinion will develop, usually based on a low bid that does not meet the specifications. If this occurs, the apparatus committee should detail each and every discrepancy between the requirements of the specifications and the manufacturer's bid. If the purchasing authority decides that it wants to accept less than was specified, the specifications should be rewritten and a new bid held.

Factory Inspections

Traveling to a manufacturer's facility for the purpose of inspecting the apparatus that is being built is an integral and necessary part of the purchasing process. Sometimes justifying these trips to the city's governing body might be harder than doing the inspections! When discussing trips with officials, it should be explained that they are not new or unusual. The NFPA Standards indicate that interim trips to the manufacturer's facility might be necessary to ensure compliance with the specifications, and the Appendix C questionnaire provides an area to indicate the number of trips and participants.

Another point is that some manufacturers require an inspection of the finished apparatus before shipping so that corrections or adjustments can easily be made. With the large investment that is being made in the apparatus, and the enormous number of variables involved, it is foolish to not comply with this requirement.

Often the first trip will be the preconstruction or pre-build conference. The purpose of this session is to meet with the manufacturer's product specialist or engineer to discuss variations and to establish your needs. Many times there are several ways to meet the intent of the specifications and often their suggestions make for a better overall finished product.

Before embarking on this journey, it is most important that the fire chief or designated representative have a full understanding of the purchasing rules and regulations regarding change orders. Change orders are official changes or clarifications to the specifications, and some might require an adjustment in price. Extensive changes that have a major impact on the bid price might not be allowed by law, as it could be construed as tainting the bid process.

It is equally important to document all clarifications and changes, whether there is a financial impact or not, and make them an addendum to the specifications. One member of the committee should be charged with maintaining accurate notes on the agreed changes so that a change letter can be composed after the meeting.

At the pre-build conference, there will be many decisions to be made, from where certain equipment will be mounted to the shape and design of the lettering. If a committee is handling the purchase and inspections, they should determine in advance how these decisions will be made. Some rely on the ranking officer's opinion while others use the democratic method of majority rule. In either case, it should be clearly understood *before* you are at the manufacturer's facility. Internal conflicts should also be resolved before leaving on the trip. It is totally unprofessional to be arguing and bickering at the meeting or trying to have company officials side with one opinion or the other. Approach the meeting as one customer purchasing one piece of apparatus.

If after the meeting an adjustment or clarification is requested, don't rely on verbal communications to accomplish the change. Follow up any telephone discussions with a letter or fax that should also become part of the specifications addendum.

Often there are two other inspection trips specified, a chassis or pre-paint inspection and the final inspection. The intermediate inspection, depending upon the fire department's

A preconstruction conference affords the fire department personnel the opportunity to discuss the various methods to meet the intent of the specifications with the manufacturer's representative. (Photo by Robert Milnes)

wishes, can be held either when the chassis is complete and before the body is installed, or after the body is installed but the apparatus is not yet painted. Each has its own advantages.

If you specify the chassis inspection, it is easier to examine the components of the chassis such as the pump, transmission, driveline, hoses, and wiring before they are covered by the body. The main advantage to the pre-paint inspection is that any corrections to the body that might be necessary are more easily accomplished since it will not involve refinishing the apparatus.

The final inspection is the most important trip of the three. While every manufacturer maintains a degree of quality control, it is *your* inspection that will determine the condition of the apparatus when it is delivered. It is extremely important that every item of the specifications be thoroughly and systematically checked for compliance, operation, and finish.

If you attempt to follow the order of the specifications to accomplish this, you'll quickly get exhausted jumping from one location to another. A better way is to prepare a checklist before departing on the inspection. Each item in the specifications is grouped by its physical location on the apparatus. For instance, the dashboard, cab controls, seats, interior finish, etc. are grouped in the cab section. The inspecting party can then check all items pertaining to the cab before moving on to another part of the apparatus. Utilizing this system will save a great deal of time, while ensuring that important items in the specifications are not overlooked.

The final portion of this inspection should include road and operational tests. While the apparatus is being driven on the road, observe if any unusual noise or vibration exist. This

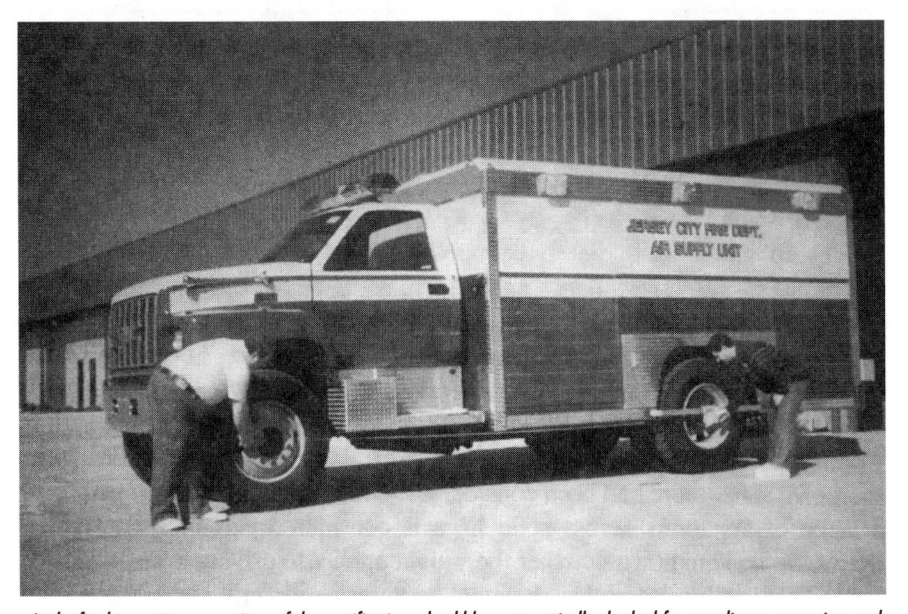

At the final inspection, every item of the specifications should be systematically checked for compliance, operation, and finish. (Photo by Robert Milnes)

should not be confused with the road test for acceptance, as outlined in the NFPA Standards. That should be conducted after the delivery, with a full tank of water and all equipment installed. Operating the aerial device or witnessing the pump being operated should also be part of the final inspection.

DELIVERY

The method of delivery and location of acceptance of the completed apparatus will also have to be specified. Delivery by road, "under its own power," is the preferred method since it allows a break-in period for the vehicle. Many specify a pre-delivery service at the apparatus dealer prior to delivery. This is similar to a new car "dealer-prep" where the oil is changed, the chassis lubricated, all systems are checked, and the unit is cleaned. Any difficulties that the transport driver had can be addressed at this time.

Some purchasers opt to take delivery at the factory and drive the apparatus to their community. If this is being considered, prior arrangements for payment, licensing, and proper insurance documentation will all be necessary.

Most specify that the delivery will take place at the purchaser's location. While this will add somewhat to the cost of the apparatus, there are several good reasons why this is the better method.

When the delivery takes place at the purchaser's location, the manufacturer maintains responsibility for the apparatus until it safely arrives at your fire station. Incidents such as an accident, mechanical malfunction, or unforeseen occurrences such as tire damage or broken glass will not be the responsibility of the purchaser.

In addition, most manufacturers utilize the services of a "drive-away" company that employs drivers with the proper licenses and maintains insurance on the vehicle until it is delivered. Most states have waived the requirement for a commercial drivers license (CDL) for firefighters in the performance of their duties within their state; however, there has been much debate as to whether transporting a piece of apparatus across state lines is considered part of a firefighter's duties. It is unknown if the waiver applies to drivers from outside the state, as this is a local decision. It would be very unfortunate if an apparatus being transported from the factory was involved in an accident and it was determined that the driver was not properly licensed to operate that class of vehicle in the state.

Acceptance

The term acceptance is defined as when the purchasing authority agrees with the contractor that the terms and conditions of the contract have been met. Acceptance tests are those tests that are performed at the time of delivery to determine compliance with the specifications. In most cases, the apparatus is conditionally accepted at the factory, pending the results of the acceptance tests at the purchaser's location.

The degree of acceptance testing is up to the purchaser and must be included in the specifications. At a minimum, it should include the road performance tests outlined in the NFPA Standards and the operational and capacity testing of the various apparatus systems. The road tests should be performed with a full water tank and full complement of equipment and personnel.

Tests for acceptance will vary according to the purchaser's specifications. Often an abbreviated version of the annual pump test will be required to be performed upon delivery.

At other times, extensive testing, going far beyond the normal operations will be required. The City of New York conducted a pumper acceptance test on a delivery of 1,000 gpm pumpers that involved pumping 800 gpm at 200 psi for 15 minutes, 500 gpm at 320 psi for 15 minutes, and pumping 250 gpm at 600 psi for 6 hours! When a purchaser does not have drafting facilities available to test a pumper, they will often rely on the third party certification tests performed at the factory to ensure compliance. Once again, the number and degree of acceptance testing is up to the purchaser and must be part of the specifications.

Acceptance tests are performed at the time of delivery to determine that the apparatus meets the performance criteria outlined in the specifications. (Photo by Greg Cariddi)

Tests for acceptance will vary according to the purchaser's requirements. The pumpers shown here are required to pump at 600 psi for 6 hours. That is more than twice the pressure required by the standard. (Photo by Greg Cariddi)

Training

Prior to being placed on duty, the firefighters who will use the new apparatus must be thoroughly trained. If the bidder is expected to provide this training, it too must be included in the specifications. Scheduling, number of days, and course content should all be outlined.

The scheduling and number of days will usually differ between career and volunteer departments. Career departments will most likely have to schedule the instruction to cover each shift. The number of days that the delivery engineer is required to spend with the department will vary depending on the work schedule. Volunteer departments might prefer to have the instructions in the evening or on the weekend when most of their members are available. Determining the number of days will depend on the class size and how much instruction is necessary to make them proficient in the use of the apparatus.

In most cases, the training period for aerial apparatus will be somewhat longer since much of the instruction is centered around the individual operators mastering the techniques of the operation. Pumpers are usually easier to learn since most of the functions of the operating controls are the similar to the previous apparatus. Previously trained pump operators can usually grasp the differences easily.

Sometimes driver training is also included in the specifications. This is an expensive, time-consuming process, especially when a fire company consists of numerous members. As a suggestion, two or three members could receive a "train the trainer" course of instruction and then go on to practice and qualify the other members of the fire company.

The degree of training to be provided by the apparatus manufacture after delivery is dependent upon the terms of the specifications and the complexity of the apparatus. (Photo by William Peters)

The course outline should be submitted to the fire chief or training officer prior to the beginning of the training. It should be reviewed for potential conflicts with the department's standard operating procedures (SOPs) and to confirm that all areas of training concern are covered. In addition, following the instructional outline will ensure that all members receive the same training and that nothing is inadvertently overlooked.

When considering the course outline material, operator preventative maintenance is a very important subject. All members should have a full understanding of how to properly perform daily or post-run apparatus maintenance. Some departments also include a block of training in shop preventative maintenance for their municipal or fire mechanics. This too could be a worthwhile investment, especially if the apparatus is different from the department's current inventory.

Providing course material such as a commercially produced videotape of the apparatus operations or student handouts should also be specified. Some departments wish to videotape the delivery engineer's presentation for future reference. If this is your intention, it should be stated in the specifications, as some instructors are more comfortable with being recorded than others.

Once the initial instructional period is over, the members must be encouraged to continuously practice and drill with the apparatus to hone their driving and operating skills. Training in apparatus operations must be an ongoing process.

SUMMARY

Preparing specifications and purchasing fire apparatus are major responsibilities that represent a sizable, long-term investment of community funds.

To approach this complex project in an orderly fashion, an outline of tasks, arranged in sequential order, should be established and followed. Allowing sufficient time to thoroughly complete all phases is also an important consideration in satisfactorily reaching the goal.

Intelligent decisions based on investigation, research and evaluation of available products will result in the needed purchasing justification. When undertaking this serious work, the fire chief must determine whether to be a follower, specifying a minimally compliant unit, or an innovator, taking advantage of the latest options and features.

References

Peters, W., Fire *Apparatus Purchasing Handbook,* Saddle Brook, New Jersey: Fire Engineering, 1994.

NFPA, *Fire Protection Handbook,* 17th edition, 1991.

Dolezal, F., *Repowering, Rehabilitating and Reconditioning,* International Association of Fire Chiefs Foundation, Fairfax, VA.

Eisner, H., "Show and Tell," *Firehouse Magazine,* December 1993, pp 72–73.

Peters, W., "Apparatus Bid Evaluation: Simplifying a Difficult Process," *Fire Engineering,* October 1993, pp 31–39.

Peters, W., "Is Refurbishment Right for You?" *Fire Engineering,* October 1992, pp 47–54.

Peters, W., "Inspection Trips, Your Vacation or Vacation?" *Fire Engineering,* May 1991, pp 16–23.

FEMA, *A Guide to Funding Alternatives for Fire and Emergency Medical Services Departments (FA141),* Emmitsburg, Maryland: Federal Emergency Management Agency, 1993.

12

PROTECTIVE CLOTHING AND EQUIPMENT

Robert Tutterow

CHAPTER HIGHLIGHTS

- **Provide the fire chief with the information needed to make informed choices on which personal protective equipment (PPE) is appropriate for use within their own department.**
- **Explain the applicability of National Fire Protection Association (NFPA) standards and the requirements of those standards.**
- **Provide other management information on the selection, care, maintenance, and use of PPE.**

INTRODUCTION

Personal protective equipment, including clothing, are items that have been used by members of the fire service since the service began. The type and complexity of this clothing and equipment has changed dramatically in over the past four decades and the pace of change has not slowed. Currently, there is a focus on how to maintain and manage this most valuable asset.

This chapter will provide the fire chief with the information needed to make informed choices on which PPE is appropriate for use within their own department. The chapter will explain the applicability of NFPA standards, the requirements of those standards, and provide other management information on the selection, care, maintenance, and use of PPE.

The original function of protective clothing was to shed water and to provide the firefighter with some minimal protection from falling debris. Today, this equipment protects firefighters from the extremes of heat and cold, supplies them with clean breathing air to use as they work, protects them from the hazards of falling debris and liquids, protects them from hazardous materials, and prevents the transmission of disease. PPE also provides alerts when a firefighter is in trouble as well as provides equipment for use in life safety situations.

NFPA Standards

One of the most comprehensive and exhaustive bases of information about firefighter protective clothing and equipment is contained in NFPA standards. A review of these standards is included later in the chapter. Every fire chief should have a basic knowledge of the NFPA process as well as the intent and content of the applicable NFPA standards for their department. The following is an overview of the NFPA standards process and the requirements of specific NFPA protective clothing and equipment standards.

Chiefs and other readers should be aware that requirements contained in any NFPA standard may be modified between editions by TIAs (Tentative Interim Amendments) issued by the NFPA and by the release of subsequent editions. Before you stake your career on a requirement of any NFPA standard, it is best to make sure that the requirement has not been modified by a TIA and to make sure that you have the most current edition of the standard.

NFPA standards are developed by committees consisting of users such as members of fire departments, manufacturers of protective clothing and equipment, outside organizations such as testing laboratories, and other interested parties. There are nine classifications of committee members. No one classification can hold more than one-third of the maximum 30 seats on a committee. The classifications are:

- Consumer
- Enforcing authority
- Insurance
- Labor
- Manufacturer
- Installer/maintainer
- User
- Special expert
- Applied research/testing lab

NFPA regulations prevent any one interest group from holding more than one-third of the membership of any committee. As a standard is created or revised, there are many ways a member of the fire service can have an impact on the contents. Prior to the development or revision of a standard, a call for proposals is made. Proposals can be submitted to the NFPA at any time and must be considered by the appropriate committee in the next revision cycle after the submission of the proposal. A proposal form is included with every NFPA standard or can be obtained by calling the NFPA; it is available on line at *www.nfpa.org*. Public comments may be mailed, faxed, or emailed to the NFPA. The proposal process allows you to provide direct input into the standards-making system. Anyone can make a public proposal. You do not have to be a member of the NFPA.

After a committee has completed a draft of a new or revised NFPA standard, the document is published as a Report On Proposals (ROP). All of the standards to be voted on by NFPA members at a fall or annual meeting are published together. For example, all standards to be considered at the NFPA's 2002 Annual Meeting are contained in one book, all those to be considered at the Fall 2002 meeting are contained in another and so on. Copies of ROPs are sent to many NFPA members and to anyone who requests them from the NFPA. Anyone may submit a public comment on any requirement contained in the ROP. This is your second opportunity to have direct input into the standards-making process. Like public proposals, public comments may be mailed, faxed, or emailed to NFPA. A form and time deadline are contained in every ROP to allow individuals or organizations to change, add to, or delete from a proposed standard or revised standard. The committee responsible for the standard must respond to every public comment submitted before the deadline.

Once the deadline for comments on the ROP has expired, committees meet to review and consider all public comments. If the comment is accepted in full or in part, the standard must be modified to reflect that action. If the comment is rejected in full or in part, the committee must give a reason for its rejection. A comment may also be held for further study if the committee needs more time to consider or research the information contained in the comment. Any comment held for further study automatically becomes a public proposal for the next edition of the standard. The result of all committee consideration of public comments is then published as Report On Comments (ROC). The ROC is sent to many NFPA members and anyone who requests it from the NFPA.

After the ROC has been published, members of the NFPA who attend the annual or fall meeting vote on the standard as proposed in the ROP and modified by the ROC. Discussion of specific requirements can occur and changes can be made on the floor of the meeting. After the standard is approved by the membership at a meeting, the standard goes through several other steps at the NFPA prior to being printed and issued to the public. Participation in the NFPA process is important for members of the fire service, especially fire chiefs. There is a misconception by some in the fire service that manufacturers dominate NFPA committees. This is simply not true. The fire service plays a major role in determining the output of NFPA committees that write standards for the fire service.

Nonetheless, there is still a need for more fire service participation. These standards impact every facet of our operations and we should have a hand in their development and processing.

The entire process for revising a standard takes 104 weeks. Here is a brief outline of the timeline:

Step 1

- Call for proposals is issued to amend existing document or for recommendations on a new document.
- The committee meets to act on proposals, to develop its own proposals, and to prepare its report.
- Committee votes by letter ballot on proposals. If two-thirds approve, the report goes forward. Lacking two-thirds approval, the report returns to the committee.
- Report is published for public review and comment (ROP).

Step 2

- Committee meets to act on public comments to develop its own comments and to prepare its report.
- Committee votes by letter ballot on comments. If two-thirds approve, supplementary reports go forward. Lacking two-thirds vote approval, supplementary reports return to the committee.
- Supplementary report is published for public review (ROC).

Step 3

- NFPA membership meets (May or November meeting) and acts on the committee report (ROP and ROC).
- Committee votes on any amendments to the report approved at NFPA May or November membership meeting.

Step 4

- Notification of intent to file an appeal to the standards council on association action must be filed within 20 days of the NFPA May or November membership meeting.
- The standards council decides, based on all evidence, whether or not to issue code or standard or to take other action, including hearing any appeals.

The term "NFPA Approved" is heard often. NFPA does not approve any process, design, installation, or procedure. Some NFPA standards, such as some NFPA protective clothing standards, require third party certification of a piece of clothing or a clothing design. The third party organization must not be affiliated with the manufacturer and must be an independent organization such as Underwriter's Laboratories (UL). This certification is intended to confirm that a piece of clothing will provide the level of protection required by a standard. Items of clothing or equipment that have received third party certification are referred to as "NFPA compliant" (they have been tested and found to comply with an NFPA standard). This is similar to a pump test for a new pumper performed by an organization such as UL.

It is hard to imagine any fire chief in North America who has not heard and read volumes of information about NFPA 1500 Standard On Fire Department Occupational Safety and Health Program. This safety standard has perhaps the highest impact on the operation of a fire department of any standard issued recently. The standard provides basic requirements for the provision of a safe working environment for firefighters. NFPA 1500 requires that an implementation plan be developed by any fire department that adopts the standard. In some cases the standard has been made statutory by the state and adoption is not at the option of the local fire department. Court cases have been influenced by the requirements of NFPA 1500 as a nationally recognized standard. If you go to court, the requirements of NFPA 1500 will be the subject of discussion, *whether or not it is adopted as law in your state or province.* The wise fire chief will remember that though NFPA standards can be used against a fire chief and his/her department, they can also be used to defend policy and actions.

There has been much apprehension and much misunderstanding about NFPA 1500. Many requirements attributed to the standard and many misunderstandings of the standard can be overcome by the basic step of carefully and thoughtfully reading the standard.

As of this writing, the most current edition of NFPA 1500 is the 2002 edition. Chapter 7 of the 2002 edition of NFPA 1500 covers requirements dealing with protective clothing and protective equipment. The following is a summary of the general requirements of that edition that relate to PPE:

The fire department is responsible for providing each member with protective clothing and protective equipment that will provide protection from the hazards to which the firefighter is likely to be exposed. (Firefighters who fight wildland fires should have wildland protective clothing and equipment, inner city structural firefighters should have structural protective clothing and equipment, firefighters who fight both types of fires should have both types of equipment.) The use of protective clothing is required when the member is exposed or potentially exposed to the hazards for which the clothing is provided.

PPE must be cleaned at least every six months as outlined in NFPA 1851 Standard on Selection, Care, and Maintenance of Structural Fire Fighting Protective Ensembles. This standard describes how fire chiefs should handle the process of selecting and caring for their PPE. Details of this standard will be described later.

When station/work uniforms are worn, they must comply with NFPA 1975 Standard on Station/Work Uniforms for Fire and Emergency Services. Firefighters must avoid wearing any clothing that may injure the firefighter if he or she is exposed to high levels of heat. Clothing made of fibers such as polyester or polypropylene may melt when exposed to heat and injure the firefighter worse than he or she would have been injured without the polyester or polypropylene clothing. This applies to uniform items as well as underwear, socks, tee shirts, and any other clothing items.

The fire department must provide for the cleaning of protective clothing and station/work uniforms. Cleaning can be done by an outside agency or company or may be done by the fire department at a central facility or with equipment provided in each fire station. If fire stations are equipped with washers, they must be used for the washing of protective and work clothing only. Washing will remove hazardous substances such as dirt, soot, products of combustion, and blood-borne pathogens from the clothing and lengthen the life of the clothing. The use of washing machines for protective and work clothing will *only* prevent cross-contamination of other items such as bedding, towels, and firefighter's personal clothing. Cross contamination could lead to the exposure of other fire department members or it could lead to the exposure of the firefighter's family.

Firefighters who engage in or are exposed to the hazards of structural fire fighting must be provided with structural fire fighting PPE. This includes protective coats and trousers or a protective coverall, helmets, hoods, gloves and footwear that complies with NFPA 1971 Standard Protective Ensemble for Structural Fire Fighting.

If protective coats and trousers are used as opposed to a protective coverall, a two-inch overlap between the two must be present. If protective coats are provided with extended wristlets that cover the hand, gloves may be provided which do not have extended wristlets. Coats with extended wristlets have been proven to dramatically reduce the incidence of burns to the hands and wrists. It is very difficult to don gloves with knit extended wristlets when wearing a coat with an extended wristlet, thus the allowance for gauntlet type gloves.

Firefighters who participate in fire fighting activities presenting high radiant heat exposure must be provided with protective clothing and equipment designed for those types of exposures. Aircraft Rescue Fire Fighting (ARFF) and flammable liquids fire fighting are examples of situations where high radiant heat levels are present. Protective coats, protective trousers, and protective coveralls must meet the requirements of NFPA 1976 Standard on Protective Ensemble for Proximity Fire Fighting. Protective covers, most often with an aluminized outer layer, must be placed over SCBA to provide additional radiant heat protection. Some departments choose to provide such coverage for helmets, gloves, and footwear.

Firefighters or other fire department members who perform emergency medical activities must be provided with protective equipment that meets the requirements of NFPA 1999 Standard on Protective Clothing for Emergency Medical Operations. These clothing items range from gloves to overalls and are designed to protect the caregiver from exposure to diseases such as hepatitis and other biological hazards. Medical gloves must be worn when providing emergency medical care and patient care cannot begin until the gloves are donned. If there is an incident where there are large splashes of body fluids, spurting of blood, or childbirth, then emergency medical garments and emergency medical face protection devices must be worn in addition to gloves.

Firefighters who engage in activities which may place them in contact with hazardous materials are required to have and use proper PPE, including SCBA. The type of clothing depends on the hazard presented by the chemical or chemicals, the form of the chemical or chemicals (solid, liquid, or gas) and the degree to which the firefighter will be exposed. Chemical protective clothing must be inspected and maintained according to the manufacturer's instructions. Following use, chemical protective clothing must be disposed of or decontaminated. If decontamination will not stop any deterioration or will compromise the integrity of the clothing, then it must be disposed. Disposal must be in accordance with the appropriate state and legal authorities.

Vapor-protective suits must meet the requirements of NFPA 1991 Standard on Vapor-Protective Ensembles for Hazardous Materials Emergencies. These suits must not be used alone for protection in any fire fighting or for protection from radiological, biological, or cryogenic agents, or in flammable or explosive atmospheres. It is permissible to use vapor-protective suits for protection against liquid splashes or solid chemicals and particulates.

Liquid splash-protective suits must meet the requirements of NFPA 1992 Standard on Liquid Splash-Protective Ensembles and Clothing for Hazardous Materials Emergencies. The specific suit selected must be chosen based on the suit manufacturer's penetration information, the chemicals known or suspected to be present, and several reference publications. Liquid splash suits are not designed for protective use against vapor hazards or unknown chemicals or chemical mixtures. In addition, liquid splash suits are not designed to protect against any chemicals known or suspected to be carcinogens, or that are otherwise toxic to skin. (The standard provides references for the list of these chemicals.) Liquid splash suits can be used for protection against solid chemicals and particulates.

Firefighters who are involved in wildland fire fighting must be provided and use PPE that is compliant with NFPA 1977 Standard on Protective Clothing and Equipment for Wildland Fire Fighting. This includes garments (shirts, jackets, trousers, or coveralls), helmets, gloves, footwear, and fire shelters.

Self Contained Breathing Apparatus (SCBA) must be provided for and used by members working in areas where the atmosphere is hazardous, suspected of being hazardous, or may rapidly become hazardous. Open circuit SCBA must be positive pressure and must meet the requirements of NFPA 1981 Standard on Open-Circuit Self-Contained Breathing Apparatus for the Fire Service. Closed circuit SCBA must be National Institute for Occupational Safety and Health/Mine Safety and Health Administration (NIOSH/MSHA) certified with a minimum service life of thirty minutes and must only be capable of operating in the positive pressure mode. Open circuit SCBA exhausts exhaled air to the outside atmosphere. Closed circuit SCBA retains exhaled air, removes some components such as carbon monoxide, adds oxygen, and return the air to the user to be rebreathed. Currently, there is very little, if any, use of closed circuit SCBA in the fire service.

Although not specifically protective clothing or equipment requirements, this chapter of NFPA 1500 also requires that: (1) members using SCBA must work in teams of two, and (2) the atmosphere of confined spaces must be monitored or firefighters operating in a confined space must be provided with and use SCBA. Members are prohibited from removing the SCBA in a hazardous atmosphere, such as in a rescue situation. (The thinking here is that firefighters will be safer working in teams, that many confined spaces contain hazardous atmospheres that may injure or kill a firefighter, and that a firefighter can serve a fire victim best by the expedient removal of the victim from the hazardous atmosphere, not by sharing their breathing air with the victim.)

The fire department must develop and implement a respiratory protection program that includes selection, use, inspection, maintenance, training, and air quality testing. Firefighters who use SCBA are required to be certified in the safe and proper use of the equipment at least once a year.

Full facepiece air-purifying respirators may be used only in non-IDLH atmospheres. These respirators must be NIOSH certified and meet all other OSHA requirements. Canisters and cartridges must be changed before the end of their service life.

Compressed air used in SCBA must meet an American National Standards Institute/Compressed Gas Association (ANSI/CGA) specification. If the fire department makes its own air, the air must be tested at least every three months by an accredited lab. If the fire department purchases air from a vendor, the vendor must provide certification that an accredited lab has tested the air and that it meets the requirements of the ANSI/CGA specification. According to the ANSI/CGA G7.1 Commodity Specification for Air, air used by firefighters must be minimum Grade D. NFPA 1500 also requires a dew point level of -65 °F or drier (224 ppm v/v or less), and a maximum particulate level of 5 mg/m^3 of air. The water vapor level (dew point) is particularly important in colder climates. Water may condense in the regulator when air with high levels of water vapor is used. This condensed water may freeze and lead to a malfunction of the regulator.

SCBA cylinders are required to be hydrostatically tested at intervals specified by the manufacturer of the cylinder or by governmental agencies. (Generally, steel air cylinders are required to be tested every five years and composite cylinders are required to be tested every three years. The best source of information on testing frequency is the SCBA or cylinder manufacturer.)

Qualitative or quantitative facepiece fit testing is required for all new firefighters and on an annual basis for all firefighters. This fit testing must also be performed if a new type of SCBA or facepiece is used during the year. Firefighters must have a properly fitted facepiece to operate in a hazardous atmosphere.

Qualitative fit testing is often performed by having the firefighter don a facepiece fitted with a filter that restricts airflow into the facepiece. The firefighter can still breathe but the restriction provided by the filter creates a negative pressure in the facepiece as the firefighter breathes. The filter is also designed to remove the smell of the pungent substance being used to perform the test. The firefighter is placed into a tent or enclosed space where a pungent substance such as banana oil is introduced. If the firefighter cannot smell the pungent substance after it is introduced into the area, the fit is considered acceptable. Although qualitative fit testing is an approved method, it has its drawbacks.

Quantitative fit testing is a preferred method of facepiece fit testing. This test is performed by having the firefighter don a facepiece that has an appropriate adapter so that the firefighter's seal is monitored by a machine. The machine provides a numerical value of each test exercise and then a computed fit factor that can be used as a benchmark for future fit testing the following year. The protection factor must be at least 500 for negative pressure facepieces. The machine produces records that become the historical record and tracking of the firefighters. Quantitative testing is not nearly as subjective as qualitative fit testing. In qualitative testing, it is quite possible for a firefighter to smell the irritant and report otherwise.

Nothing is allowed to interfere with the face to facepiece seal. (This includes facial hair such as beards, temple straps from sport eyeglasses, or anything else.) This requirement is effective in all situations, even if the firefighter can get a seal that passes fit testing with facial hair or other seal penetrations in place. Contact lenses and eyeglasses mounted inside of the facepiece are permitted for firefighters who need them to see. If contact lenses are worn, the firefighter must have demonstrated long-term successful use in wearing them.

Records of facepiece fit testing must be kept. These records have to include the firefighter's name, type of test performed, make and model of facepiece tested, and the pass/fail results.

Hoods are to be donned after the facepiece is in place. Donning the facepiece over the hood makes it almost impossible to produce and maintain an effective seal. In addition, the SCBA facepiece straps will be provided with thermal protection from the hood for an added measure of safety.

Personal Alert Safety System (PASS) devices are required to be worn and used by firefighters involved in fire suppression and other hazardous duties. The PASS device itself must meet the requirements of NFPA 1982 Standard on Personal Alert Safety Systems (PASS). The PASS device must be tested weekly and before each use. PASS devices are recommended for installation on the firefighter's protective clothing or they can be integrated into the SCBA. If an integrated SCBA PASS device is not detachable, then separate PASS devices should be provided for incidents where SCBA are not required. For example, a firefighter death in the Midwest at a wildland fire might have been averted if the firefighter had worn a PASS device. The firefighter became separated from his crew, became incapacitated, and died before his disappearance was discovered. A proper location for the PASS device that provides for use in all hazardous situations may be difficult to obtain. For example, placing the PASS device on the structural protective clothing does not guarantee that it will be worn at wildland fires. (In the case of the Midwestern firefighter, placement of the PASS device on the firefighter's structural fire protective coat would probably not have helped.) This is another validation of the need to work in teams and to look out for each other.

Protective clothing and equipment that is in use or in the possession of the fire department that met the edition of the appropriate NFPA standard when manufactured may continue to be used. All new clothing purchased by the fire department must meet the newest revision of the standard.

All life safety rope, harnesses, and hardware used by firefighters for supporting the weight of a human must meet the requirements of NFPA 1983 Standard on Fire Service Life Safety Rope and System Components. Life safety rope used to support the weight of fire department members or other persons during rescue may not be used for any other purpose. The type of device used depends upon the type of rope operation. Previous standards required that all life safety rope be destroyed after one use. This is no longer true. Rope may be reused if it is inspected before and after each use in accordance with manufacturer's instructions, the rope has not been damaged by heat, abrasion, or chemical exposure, has not been subjected to an impact load, and has not been exposed to anything known to deteriorate rope. If there is any question about the history of life-safety rope, it must be destroyed. The required inspection must be performed by a qualified person, be governed by a local procedure, and records must be kept.

Eye and face protection must be provided when called for. The helmet face shield provides limited protection for the eyes and face. Goggles provide eye protection, but minimal face protection. When engaging in operations such as automobile extrications where particulates and other flying objects may come up under the face shield, goggles must be worn. All eye and face protection must meet the requirements of ANZI Z87.1. When SCBAs are used the facepiece provides primary eye and face protection. Hearing protection must be provided for members riding or operating fire apparatus, operating tools and equipment, or in other situations where noise levels exceed 90 dBA. The fire department is required to engage in a hearing conservation program where harmful sources of noise are identified and controlled and audiometric testing of members is conducted. Hearing conservation is not just a protective equipment issue. It is an operational issue that deserves the attention of fire chiefs. Fire service job-related hearing loss is a problem that has been around for years and it is a compensable injury.

As discussed earlier in this chapter, the original intent of firefighter protective clothing was to keep the firefighter dry and to provide some limited form of protection from falling debris. At the time rubberized coats, pull up boots, and leather helmets were the fashion and provided reasonable protection for firefighters engaged in the fire fight from the outside of the building.

As members of the fire service we honor our traditions. Sometimes this honor is at the expense of firefighter safety. There are still fire departments today that fight interior structural fires dressed in rubberized pull-up boots, gloves, hoods, and helmets, none of which meet NFPA standards. A more prudent approach to firefighter safety is to provide the best of new technology to firefighters with equipment which meets or exceeds NFPA standards and allows them to perform their duties while benefiting from the extra margin of protection provided by their PPE. The emergency scene should not be a showcase of PPE firefighting traditions. The emergency scene should be a showcase of highly advanced, well-maintained PPE. The fire service's external customers and internal customers (firefighters) deserve no less. Fire service museums, musters, etc., provide a much better forum for displaying the older styles of PPE.

Structural Fire Fighting Ensemble

The primary NFPA standard for firefighter protective clothing is NFPA 1971 Standard on Protective Ensemble for Structural Fire Fighting. At the time of this writing the most current edition of this standard is dated 2000.

Like all of the NFPA protective clothing standards, NFPA 1971 is a performance document, not a technical specification. Fire departments should base their protective clothing purchasing specifications on the NFPA standard and add information on items such as garment design that best suit local conditions.

NFPA 1971 is not intended to specify protective clothing for any operation other than structural fire fighting. The standard does not specify requirements for proximity or approach fire fighting, or protection from other hazards such as chemical, radiological, or biological agents. Any jurisdiction may, at its option, exceed the requirements of the standard.

PPE is required to be certified as being in compliance with the standard by an approved certification organization. Labeling that indicates the certification is required. In addition, requirements for a certification program are outlined. A certification organization cannot be owned or controlled by a manufacturer. There can be no temporary or partial certifications.

A certification agency must have the laboratory equipment to conduct the test and must do so in accordance with the standard. Manufacturers are held to stringent dates of production to assure that the correct revision of the standard is applied to PPE. The certification agency must make at least two random and unannounced inspections of a manufacturer during a twelve-month period. In addition, certification agencies must have a program to investigate field reports of PPE failure and determine that the manufacturer has a recall system in place.

Additional labeling is required that provides use, maintenance, and cleaning advice to the user as well as providing identification of the materials used and information on the manufacturer. The elements of the structural fire fighting ensemble are:

- Coats and trousers (or coveralls)
- Helmets
- Hoods
- Gloves
- Footwear

Coats and trousers

NFPA 1971 requires that structural fire fighting protective coats, trousers, and coveralls be constructed of at least three layers. The outer shell is highly resistive to flame and heat, the moisture barrier provides protection from hot water and other fluids, and the thermal barrier provides additional insulation and protection from heat. The standard provides thermal protection for firefighters who wear the clothing by requiring a minimum Thermal Protective Performance (TPP). This value is obtained by exposing a composite of the outer shell, the moisture barrier, and the thermal liner to a test that is outlined in NFPA 1971. The test and TPP values in general are explained in more detail later in this section. The standard does not specify what materials are to be used in each layer, it only requires that the three layers be present and that they provide the stated minimum level of protection.

The standard provides that materials used to make up the composite have some minimum flame resistance, regardless of their location in the garment. A test is used which exposes a preconditioned sample of every material to flame. The flame is held in place for a specified period of time. The char length is measured after the flame is removed and the amount of time is measured from when the flame is removed until the material has stopped burning or glowing. This time is called *after flame* and it is measured in seconds. The procedure for conducting the test and the apparatus used to perform the test are specified in NFPA 1971.

Requirements for closure of the garment are outlined as are requirements for the visibility of retroreflective and fluorescent trim material. The retro-reflectivity of trim ensures firefighters can be seen at night and florescence of trim allows firefighters to be seen during the day. Both properties are effective at dawn and dusk. Protective coats must have at least one 360° circumferential band or staggered pattern that equates to a 360° circumferential band at least one inch from the bottom of the coat. The front of the coat must have at least one horizontal band at the chest area. No vertical trim is allowed on the front of the coat. The back of the coat shall have at least two vertical stripes perpendicular to the bottom trim on the coat on the left and right side of the coat back or one horizontal stripe at the shoulder blade area. Each coat sleeve shall have at least one 360° stripe. This stripe may be staggered to pattern with 360° visibility. Trousers must have at least one stripe on each leg between the knee and bottom hem.

Hardware is not permitted to penetrate from the outer shell through to the firefighter's body. Hardware items such as rivets, zippers, and tacks can conduct heat and cause burns to the body. The standard does allow hardware to touch the firefighter's body if it is covered on its external surface. There can be no continuous path of hardware from the outer shell to the firefighter's skin. For example, a rivet that goes from the exterior of the garment to the firefighter's skin would not be allowed.

A minimum four-inch collar is required. The collar must have an outer shell, moisture barrier, and a thermal barrier.

A minimum TPP of 35 is required. TPP is measured by exposing all layers of a protective garment to a standard test. The test uses radiant heat and direct flame contact to simulate a flashover. TPP roughly equates to seconds of protection from a flashover. TPP divided by two equals seconds of protection from second-degree burns. A TPP of 35 is designed to provide protection from a flashover for 17.5 seconds; after 17.5 seconds of exposure to a flashover, a firefighter could be expected to receive burns worse than second degree. It should be stressed that the TPP test does not guarantee a firefighter any fixed amount of time that they would be protected from a burn. The TPP test is a lab test performed in a standard manner; however, fires do not always perform in a standard manner. A firefighter exposed to a flashover should not stay in the flashover (no hurry, I've got 17.5 seconds) for any longer than necessary! Seam breaking strength is specified.

The materials used for the outer shell, moisture barrier, thermal barrier, collar liners, trim, and winter liner (if used), must be individually tested for flame resistance. Average char length and after flame characteristics are specified as well as a requirement that no material melts or drips when tested according to procedures defined in the standard.

Shrinkage resistance of materials used in the garment is specified and thermal resistance of all materials in the garment, not limited to the composite materials, is specified.

The tear resistance, char resistance, and water absorption properties of the outer shell and collar lining materials are specified. Additional requirements for moisture barriers, thermal barriers, winter liners, thread, trim, and hardware are also specified.

Moisture barriers are tested for water penetration, liquid or blood-borne pathogen penetration, and viral penetration. The seam seals of moisture barriers are also subjected to the same tests.

Many fire departments add extra layers of protective fabric to areas that may be compressed during fire fighting, such as knees and shoulders. Compression (such as from kneeling and SCBA straps) may reduce the protective qualities of protective clothing since air is squeezed from between the layers. Air is one of the best insulators known and when it is squeezed out, heat can pass through more freely. The shoulder and knee areas must be tested against heat transfer when the garments are compressed. The test is known as the Conductive and Compressive Heat Resistance (CCHR) test and must have a minimum value of 13.5 as determined in the test method of the standard.

All thread used in the assembly of garments must be inherently flame resistant and melting resistant. The same applies to zippers, hook and pile closures, hanger loops, emblems, and patches. Corrosion resistance is required for any metal components such as zippers, hook and "D" closures, and rivets.

The entire ensemble of coats and trousers (or coveralls) must be tested for leakage. This is a garment integrity test commonly referred to as the shower test. The ensemble is placed on a mannequin with the neck, wrist, and ankle areas sealed. The garment is subjected to shower of water from several nozzles at various angles for 20 minutes. The garment is then removed from the mannequin to determine if there are any leaks in the garment. The test is required for styles and composite materials offered by manufacturers.

The entire ensemble is also tested for total heat loss (THL). This is a test designed to test the breathability of an ensemble. Garments must have a THL of no less than 130 W/m^2 when tested using a Guarded Sweating Hot Plate. Typically, higher TPP values will cause lower THL values. Fire chiefs must compare the benefits of these values when making their decisions.

There are several other tests required in the standard to assure that firefighters are receiving a minimal level of protection. The tests described above are the more common tests in the standard and have the most impact. Below is a list of all the tests conducted for garments:

- Flame resistance
- Heat/thermal resistance
- TPP

- Thread melting
- Tear resistance
- Seam strength
- Cleaning/shrinkage resistance
- Water absorption resistance
- Water penetration resistance
- Liquid penetration resistance
- Viral penetration resistance
- Corrosion resistance
- Total heat loss
- Label durability
- Retro-reflectivity & fluorescence
- Overall liquid penetration
- Breaking strength
- Conductive & compressive heat resistance

In 2001, a companion document to NFPA 1971 was developed. NFPA 1851 Standard on Selection, Care, and Maintenance of Structural Fire Fighting Protective Ensembles provides fire chiefs with a user standard to assist in managing a PPE program. Appropriate portions of the standard will be described in conjunction with the description of NFPA 1971. The portions are:

- Selection
- Repair
- Inspection
- Storage
- Cleaning
- Retirement

Fire chiefs should take an encompassing approach to selecting protective coats and trousers or coveralls. The basis of this approach should be a risk assessment that includes the types of duties performed, frequency of use, past experience, incident operations, geographic location, and climate. Included in the risk assessment should be a determination if the item of PPE will interface with other items of PPE. For chiefs of smaller departments, benchmarking with similar departments can be very beneficial. For larger departments, it is often feasible to conduct field tests and evaluations. If evaluations are conducted, they must be conducted in a manner that is as equitable as possible. For example, every participant must have an opportunity to try all of the garments being evaluated.

When written specifications are used, the following minimum criteria must be included:

- Assure the garments are compliant with the latest revision of the standard.
- Stipulate areas where the specifications exceed the NFPA standard.
- Require manufacturers to include third-party certification documentation of the garments.
- Define the process for determining proper fit.
- Compare the bid submittals against the purchase specifications.

Many configurations of protective coats, trousers, and coveralls are available. Some use short coats and high pants, others use pants with near normal fit and coats that overlap to provide additional coverage. A minimum overlap between the coat and the trousers must be two inches. This coverage must be an overlap of all layers, not just the outer shell. When selecting coats and trousers, it is important to select garments that provide protection with the arms stretched over the head and when in a bending or crawling position. The selection should be made while wearing SCBA to make sure that there is a good interface among the components and that there is freedom of movement. One-piece protective coveralls are available that provide no gap between the coat and trouser. All of these configurations are aimed at providing the firefighter with the lightest weight, most effective garment possible. While the standard provides a minimum level of thermal protection, in general, the lighter the weight of the garment, the less physical stress will be placed on the firefighter.

Inspection of coats and trousers

There are two levels of inspection: routine and advanced. Any garments contaminated with hazardous materials or biological agents must be decontaminated before any inspection can occur. Fire chiefs should be sure their departments have a program in place to determine levels of inspection. The inspection process drives the need for cleaning, repair, and retirement.

Routine inspections are to be performed by the firefighter after each use. This inspection should include at least the following:

- Soiling
- Contamination from hazardous materials or biological agents
- Damaged or missing reflective/fluorescent trim

- Physical damage
  - Rips, tears, and cuts
  - Damaged or missing hardware and closure systems
  - Thermal damage such as charring, burn holes, and melting

Advanced inspections are conducted at least every twelve months and must be conducted by someone in the department who is knowledgeable about the garments. Records must be kept for advanced inspection. This includes all the items in the routine inspection in addition to the following:

- Loss of moisture barrier integrity as indicated by
  - Rips, tears, cuts, and abrasions
  - Discoloration
  - Thermal damage
- Evaluation of system fit and coat/trouser overlap
- Loss of seam integrity; broken or missing stitches
- Material integrity: UV or chemical degradation and loss of or a shift in liner material
- Wristlets: loss of elasticity, stretching, cuts, holes, and burns
- Reflective trim: reflectivity, breaking seams, and overall damage
- Label: legibility
- Closure functionality
- Liner attachment

Cleaning of coats and trousers

Dirty coats and trousers should never be washed in equipment that is not dedicated to the task of cleaning these items. This is most important to prevent cross contamination of cleaning equipment and to minimize exposure to others. Commercial dry-cleaning is not an accepted method of cleaning garments. If contract cleaning is used, the contractor must demonstrate knowledge of the fabrics used in coats and trousers and the manufacturers recommended methods of cleaning. NFPA 1851 outlines three methods of cleaning: routine, advanced, and specialized.

Routine cleaning is performed after every use that soils the garments. It is the responsibility of the firefighter to do routine cleaning. The process includes the following:

- Initiate cleaning at the scene if at all possible.
- Brush off dry debris.

- Gently rinse with a water hose.
- Scrub with a soft brush if necessary.
- Spot clean in a utility sink if necessary.
- Inspect for soiling and contamination and repeat if necessary or perform advanced cleaning.

Advanced cleaning must be performed when needed and at least every six months for garments that are in service. Advanced cleaning must be performed or managed by a person who is knowledgeable in cleaning protective garments. The basic process is:

- Brush off any dry debris.
- Clean in a utility sink or machine, or through a contract cleaner.
- Inspect for soiling and repeat if necessary.

If a fire chief decides to purchase equipment for cleaning garments, there are considerations to factor in the decision. A front-loading washer/extractor is highly preferred over top-loading home machines. The reasons include:

- No agitator that can cause mechanical damage to garments
- Multiple sizes to accommodate several sets of garments
- Less water consumption per pound of garments versus top-loading machines
- Typically are much heavier duty to accommodate the heavy garments

Drying of garments also deserves special consideration. The use of a tumbler dryer is not recommended unless no other means is available. The tumbling action will cause mechanical damage to the garments. The best process for drying is simply to air-dry. Air-drying can be accelerated by a forced air arrangement with a temperature not to exceed 130 °F. There are several manufacturers that offer garment-drying equipment. Garments should never be dried in direct sunlight.

Specialized cleaning is for garments that are contaminated with hazardous materials or biological agents. Garments that contain such contamination or are suspected to have this contamination should be isolated, bagged, tagged, and removed from service. Universal precautions should be used in handling contaminated garments. If the fire department has someone knowledgeable with protective garments and the contaminant, then the department can perform the decontamination. Otherwise, a contractor with knowledge and experience in this area must perform the decontamination. Contaminated elements must be shipped in accordance with federal, state and local regulations.

Repair of coats and trousers

An effective repair program will extend the life of coats and trousers. Garments should always be cleaned prior to repair. Coats and trousers should not be repaired by anyone who has not been trained to do so. A fire department can perform minor repairs. *All* repairs must be done in accordance with the manufacturer's recommendations. The manufacturer or the manufacturer's authorized repair facility must perform major repairs to the outer shell. Holes, tears, rips, and burned areas can be repaired by the fire department if the repair area does not exceed fifty square inches with a minimum one inch overlay over the damaged area. These types of repairs must use a patch rather than reattaching existing material. Any repair to a major A-seam that is over an inch long is considered a major repair. A major A-seam is a seam that is crucial to the integrity of the garment (much like a load-bearing structural member in building construction). Major B-seams can be repaired if there is no penetration of the moisture barrier and if the seam failure is less than one inch. A major B-seam is a seam in the moisture barrier or thermal barrier. If there is a question on whether a repair is major or minor, then the manufacturer should be consulted to make the determination.

Repairs to moisture barriers are not to be performed by fire departments. Only the manufacturer or a repair facility authorized by the manufacturer and the moisture barrier manufacturer can perform these repairs.

Common types of repairs include replacing of: retroreflective/fluorescent trim, replacement of hardware, zippers, hook and loop material, reinforcement material, and wristlets. As stated earlier, all repairs must be done in accordance with the manufacturer's instructions, whether performed by the fire department or a contractor.

Storage of coats and trousers

Proper storage of coats and trousers can also extend their life. For example, garments that are stored in sunlight will rapidly deteriorate. Also, UV degradation from artificial light will deteriorate garments, but at a much slower rate. It is important to store garments in a dry, well-ventilated area and never put garments that are still wet in storage. Storage areas should be free of any type contamination including hydrocarbons, hydraulic fluid, solvents, and hydrocarbon vapors.

Retirement of coats and trousers

Coats and trousers must be retired if it is not possible or cost effective to repair. This can be because of damage or soiling or decontamination. Any garments that were not compliant to the revision of the standard that was in effect when they were manufactured must be retired.

Any coat or trouser that has been retired shall be destroyed or conspicuously marked so that it will never be used again for live fire activity (including training). It is permissible to use retired garments in training evolutions that do not include live fire activities if the garments are marked as such.

HELMETS

Helmets are used to protect the wearer from falling debris and water. The components of a helmet are:

- Shell
- Energy absorbing system
- Retention system
- Fluorescent and retroreflective trim
- Ear covers
- Either a faceshield, goggles, or both

The helmet provides limited face protection through the use of a face shield. In the past, helmets were made of a leather or aluminum shell with no internal parts to protect the wearer's head. Today's helmets may have shells made of leather, fiberglass, Kevlar, or plastic (among other materials). Internal linings provide the wearer with additional protection from penetration and impact.

The field of vision allowed by the helmet when worn is specified. Four square inches of retroreflective material are required to be visible when the helmet is viewed from any angle.

The protection provided by the helmet from impacts coming from all sides and the top is specified. The resistance of the helmet to penetration, heat, and flame are specified as well as the electrical insulation provided. Performance of the retention system, ear covers, face shield and/or goggles, labels, and retroreflective markings are specified. Test methods to assess performance of each are specified in the standard. Below is a list of test methods conducted to assure the performance of helmets:

- Flame resistance
- Heat resistance
- Retention
- Shell retention
- Faceshield/goggle light transmittance
- Faceshield/goggle impact

- Top impact
- Label durability
- Impact acceleration
- Penetration
- Trim retro-reflectivity
- Faceshield/goggle scratch resistance
- Thread
- Hardware corrosion resistance
- Ear covers
- Label heat/flame resistance

Selection of helmets

There are three styles of helmets on the market today: traditional, conventional, and European. The style chosen is usually a reflection of the organizational culture of the fire department. There are choices available in shell material. These include: Kevlar, composite, fiberglass, and leather. Fire chiefs should study the characteristics of these materials (especially weight) when making a decision or approving a recommendation for their department. A good risk assessment is helpful in determining if the helmet should have faceshields, goggles, or both. At least one is required.

Inspection of helmets

When conducting an inspection of helmets, the following criteria must be followed:

- Soiling
- Contamination from hazardous materials or biological agents
- Physical damage including
  - cracks, crazing, dents, and abrasions
  - thermal damage such as bubbling, soft spots, warping, or discoloration
- Damage to ear covers such as cuts, rips, burn spots, charring, or melting
- Damaged or missing components to the suspension and retention system
- Damage to the faceshield and/or goggles including thermal, abrasive, scratches, or any properties that impair the ability to see through them

Cleaning of helmets

Helmets should never be machine cleaned. The manufacturer's instructions will provide direction about any special needs depending on the type of shell material of the helmet. The headbands, crown straps, earflaps, and suspension systems should be cleaned in a utility sink by hand. Solvents should never be used to clean faceshields or goggles.

Repair of helmets

Only personnel who are trained in helmet repair should repair helmets. Helmets should be cleaned before repair is performed. All repairs should be done in accordance with the manufacturer's recommendations and only parts from the original manufacturer or the manufacturer's authorized source can be used. Small scratches in helmet shells can be removed with a mild abrasive. If there is an indication of a crack, dent, abrasions, bubbling, soft spot, discoloration, or warping of the helmet, then the manufacturer must be consulted before repairs are made. Faceshields and/or goggles must be replaced if they are cracked, broken, warped, partially melted, and if cleaning does not provide adequate visible clarity. Helmet storage and retirement requirements are the same as those for coats and trousers.

Hoods

Fire fighting hoods are designed to provide limited protection to the head, face, and neck areas that are not protected by the helmet or the SCBA facepiece. Hoods must provide a minimum TPP (Thermal Protective Performance) of 20. This compares to a minimum level of 35 for coats and trousers. Hoods and thread used to sew them are subjected to flame and heat tests. There are also tests for burst strength, seam strength, and shrinkage.

Selection of hoods

Hoods are available in several flame retardant materials and thickness. There are also choices in the overall lengths of hoods. One of the primary considerations in hood selection is the interface area with the SCBA facepiece.

Inspection of hoods

Hoods must be inspected for the following:

- Soiling
- Contamination from hazardous materials or biological agents

- Physical damage such as rips, cuts, tears, charring, burn holes, and melting
- Loss of elasticity in face opening

Repair of hoods

Most departments do not attempt to repair hoods as they are considered disposable items. However, if repair is performed the hood must be cleaned first and the manufacturer's recommendations must be followed. Someone who has had training to do so must do the repair work.

The criterion for cleaning, retiring, and storing hoods is the same as for coats and trousers.

GLOVES

Protection of firefighters' hands from hazards on the fireground makes common sense. Gloves provide protection from:

- Cuts
- Abrasion
- Punctures
- Water penetration
- Liquid penetration
- Blood-borne pathogen penetration

Gloves must be designed to provide secure thermal protection, minimal interference with the use of firefighting tools, protection of the wrist, and non-irritating surfaces in contact with the wearer's skin. The minimum TPP (Thermal Protective Performance) is 35. This is the same as coats and trousers.

Firefighting gloves are subjected to the following tests:

- Flame resistance
- Heat resistance
- TPP
- Thread melting
- Tear resistance
- Burst strength

- Seam strength
- Puncture resistance
- Cut resistance
- Liquid penetration
- Viral penetration
- Corrosion resistance
- Overall liquid integrity
- Liner retention
- Dexterity
- Grip
- Label durability

Selection of gloves

There are many different types of gloves available in the market. Unfortunately, gloves that do not comply with NFPA standards can be found, often at a much lower cost than gloves that do meet the standard. Fire chiefs are cautioned to avoid the temptation to save money and increase the risk of injury and liability. Hand injuries are among the most expensive injuries to rehabilitate. Even a minor hand injury may prevent a firefighter from performing his or her duty for an extended period of time.

Firefighting glove shells are most often made of leather with flame resistant linings. Materials used for the outer shell include cowhide, pigskin, elkskin, other animal skins and Kevlar. Attachment of the lining material to the shell has been problematic through the years. Fire chiefs are encouraged to check references of other users and discuss liner retention with manufacturers. The better quality gloves have fewer problems. Gloves must have a moisture barrier.

Gloves are available with wristlets or with gauntlets. The decision between the two is best determined by examining the interface area between coat sleeve at the wrist area and the glove. If the coat has an extended wristlet, then a gauntlet type glove is appropriate. If the coat does not have an extended wristlet, then the wristlet style glove must be used.

The two most difficult problems for firefighting gloves are fit and dexterity. Gloves must be available in sizes XS, S, M, L, and XL. Each fire department should review the range of sizes provided by each manufacturer to determine if sizes are available that meet the needs of its members, especially female firefighters.

Inspection of gloves

Gloves must be inspected for the following:

- Soiling
- Flexibility
- Physical damage
- Rips, tears and cuts
- Shrinkage
- Inverted liner
- Thermal damage: charring, burn holes, and melting
- Contamination from hazardous materials or biological agents
- Loss of wristlet elasticity

Cleaning of gloves

Gloves must be cleaned in accordance with the manufacturer's instructions. Hand cleaning of gloves in a utility sink is appropriate. It is not recommended to machine-dry gloves with heat.

Repair of gloves

Generally gloves are not repaired but replaced. This is because most fire departments consider them to be of a disposable nature. However, if repairs are needed, a trained person must do them in accordance with the manufacturer's recommendations. Gloves should always be cleaned prior to performing repairs.

The storage and retirement criterion for gloves is the same as for coats and trousers.

Footwear

Firefighting footwear consists of a sole with heel, upper with lining, and an insole with puncture resistance and a crush resistant toecap (often known as a steel toe). Footwear is primarily designed to provide protection from physical and thermal elements. Metal parts are prohibited from penetrating from the outside to the inside of the boot and no metal parts (such as tacks) may be used to attach the sole to the boot.

Firefighting boots are subjected to the following tests:

- Flame resistance
- Heat resistance
- Liquid integrity
- Abrasion resistance
- Radiant heat resistance
- Electrical insulation
- Thread melting
- Bend resistance
- Label durability
- Puncture resistance
- Cut resistance
- Slip resistance
- Conductive heat resistance
- Liquid penetration resistance
- Impact compression tests
- Viral penetration resistance
- Corrosion resistance
- Eyelet and studpost attachment

Selection of footwear (boots)

Firefighter footwear is available in rubber, leather, and combination rubber/fabric materials. In the past, most firefighter's boots were made of rubber and a variety of rubber compounds. Many of us remember when boots came in two sizes—"too small and too large." Improved construction methods and sizing requirements in the standard have resulted in considerable improvement. Footwear must be at least eight inches in height. The size ranges are: 5–13 in men's and 5–10 in women's. Both ranges must include half sizes and a minimum of two widths.

Leather boots have become quite popular. Leather boots may provide a better fit and better ankle support than rubber boots but are often more expensive than rubber boots. In addition, leather boots are generally lighter in weight. The lighter weight has a multiplier effect on reducing the stress on the body. The expense of leather boots may be offset to an extent by the longer service life experienced in comparison to rubber boots.

Inspection of footwear

- Footwear must be inspected for the following:
- Soiling
- Contamination from hazardous material or biological agents
- Physical damage
  - Cuts, tears, punctures, cracking, or splitting
  - Thermal damage: charring, burn holes, and melting
  - Exposed/deformed steel toe, steel midsole and shank
  - Loss of seam integrity; delamination, broken or missing stitches
- Water resistance
- Closure system
- Excessive treadwear
- Lining condition: tears, excessive wear, and separation from outer layer
- Heel counter failure

Cleaning of footwear

Footwear should never be cleaned in a machine. The use of a utility sink is preferred. All cleaning must be done in accordance with the manufacturer's recommendations. Air-drying is the best method to dry boots.

Repair of footwear

With the exception of lace-up boots, when replacing of laces is possible, the manufacturer or the manufacturer's authorized repair center must do all boot repairs.

The storage and retirement of footwear are the same as those for trousers and coats.

Urban Search And Rescue (USAR)

NFPA 1951 Standard on Protective Ensemble for USAR Operations was released for the first time in 2001. USAR incidents are not directly related to other incidents such as structural firefighting, EMS, hazmat, wildland, etc. Examples of USAR operations include:

- Structural collapse
- Vehicle/person extrication
- Confined space entry
- Trench/cave-in rescue
- Rope rescue
- Other similar incidents

It had become obvious that a standard of protection was needed for these incidents as the common use of structural firefighting ensembles was not suited for such activities. As of this writing, few departments have USAR garments. However, this is expected to change soon. As fire chiefs lead their departments in reviewing their personal protective needs for the incidents they encounter, it usually reveals a need for this type of PPE.

USAR personal protective equipment is designed to provide limited protection for the following hazards:

- Physical
- Environmental
- Thermal
- Chemical
- Splash
- Bloodborne pathogens

The elements of the USAR ensemble include:

- Garments
- Helmets
- Gloves
- Footwear
- Eye and face protection

NFPA 1951 allows for both single-layer and multiple-layer garments. The total heat loss (THL) must equal or exceed 450 W/m^2 as compared to 130 for structural firefighting garments. This is the most defining difference between the two ensembles (*i.e.*, USAR garments are lighter weight and much more breathable). Otherwise, the performance tests for the two standards are similar.

Helmets can be either hat-style or cap-style. The eye and face protection can be either attached to the helmet or be separate. Eye and face protection must meet ANZI Z87.1 Practice for Occupational and Educational Eye and Face Protection requirements.

USAR gloves are not required to have a Thermal Protective Performance Test (TPP) like structural gloves. However, they are required to pass flame resistance testing, conductive heat testing, and heat shrinkage testing. Cut resistance and puncture resistance requirements are similar to that of structural firefighting gloves.

USAR footwear requirements very closely resemble the requirements for structural footwear.

Station/Work Uniforms for Firefighters

NFPA 1975 Standard on Station/Work Uniforms for Fire and Emergency Services specifies minimum properties for materials used to make firefighter station/work uniforms. It applies to fire departments who issue station/work uniforms. NFPA 1975 is not a standard on uniform design. The style and cut of the final uniform is not even discussed in the standard.

In 1999, the standard was revised to allow for cotton or wool uniforms. Prior to this, all fabrics used in making station/work uniforms must have been inherently flame resistant. The revised standard still has performance requirements that forbid the use of polyester. Uniforms should be provided that will not contribute to a firefighter injury or cause any reduction in the protection afforded by a firefighter's structural protective clothing.

Not all firefighters wear uniforms. Volunteer and paid-call firefighters, for example, respond in whatever clothing they are wearing when they are called to an emergency. Civilian clothing does not need to have the flame resistance that should be present in firefighter uniforms but is often exposed to the same hazards through this type of use. Firefighters who respond to emergencies in other than uniforms that comply with NFPA 1975 should be strongly encouraged to choose clothing that does not contain fabrics that can melt and cause more severe injuries if exposed to heat.

Station/work uniforms are required to be certified as being in compliance with the standard by an approved certification organization. In addition, requirements for a certification program are outlined. Additional labeling on or with the station/work uniform that provides use and maintenance advice to the user as well as providing identification of the manufacturer is required.

Protective Clothing for Proximity Fire Fighting

NFPA 1976 Standard on Protective Ensemble for Proximity Fire Fighting is designed to protect firefighters from fires where high levels of radiant heat are present. Examples of these types of fires are:

- Bulk flammable liquids
- Bulk flammable gases
- Bulk flammable material
- Aircraft rescue and fire fighting

In order to effectively attack the fire, firefighters must often approach the fire at a distance that exposes the firefighter and their protective clothing to levels of radiant heat that are beyond the protective capability of regular structural fire fighting protective clothing.

NFPA 1976 is intended to provide equivalent levels of protection to protective coats, trousers, and coveralls that meet the requirements of NFPA 1971 with added protection from radiant heat exposure.

Proximity protective clothing is required to be certified as being in compliance with the standard by an approved certification organization. Labeling that indicates the certification is required. In addition, requirements for a certification program are outlined. Additional labeling on or with the proximity protective clothing that provides use and maintenance advice to the user as well as providing identification of the manufacturer is required. Many of the requirements of these standards echo requirements of NFPA 1971.

Non-radiant reflective trim, leather wear-pads, and lettering are specifically prohibited by the standard. These items, if present, could limit the ability of the garment to reflect heat. In fact, these items could absorb heat and increase the chances of burning the firefighter. The outer shell material must have a radiant reflective capability as specified in the standard.

Protective Clothing and Equipment for Wildland Fire Fighting

NFPA 1977 Standard on Protective Clothing and Equipment for Wildland Fire Fighting is designed to protect wildland firefighters from external heat sources while not causing an extraordinary internal heat stress load. The standard includes requirements for protective clothing, helmets, gloves, footwear, and fire shelters to protect firefighters from the hazards of wildland fire fighting.

Wildland protective clothing, helmets, gloves, footwear, and fire shelters are required to be certified as being in compliance with the standard by an approved certification organization. Labeling that indicates the certification is required. In addition, requirements for a certification program are outlined. Requirements for wildland protective clothing include performance requirements such as fabric radiant heat protection, flame resistance, thermal shrinkage, tear resistance, seam strength, and other requirements. Testing procedures for each performance requirement are specified. Wildland protective footwear requirements are similar to the requirements for structural fire fighting. Protective toecaps, however, are not required by NFPA 1977. Wildland protective glove requirements are also similar to the requirements for structural fire fighting. Wildland protective helmet requirements include a weight limit, two design types, a requirement for a sweatband, and a requirement for a chinstrap. Face and neck shrouds, winter liners, and lamp brackets are specifically permitted, as well as other accessories. Performance requirements for wildland helmets include electrical resistance, penetration resistance, impact resistance, heat resistance, and requirements for separation of the suspension system from the helmet upon impact. Testing procedures for each requirement are included in the standard.

Protective shelters are required to meet a U.S. Forest Service specification and labeling must be provided.

Open-Circuit SCBA

NFPA 1981 Standard on Open-Circuit Self-Contained Breathing Apparatus for the Fire Service provides minimum requirements for respiratory protection when firefighters are in contaminated atmospheres. It is perhaps the single most important piece of protective equipment provided to firefighters. Fire departments with enforced mandatory SCBA use regulations, combined with effective search and operational tactics, often never experience another respiratory injury. The hazards of breathing any smoke are well known and SCBA units are designed to prevent this needless exposure.

Open-circuit SCBA used by the fire service are usually rated for 30-minute, 45-minute, or 60-minute service time. As anyone who has ever worn one can tell, the rating time almost never equals the actual service time experienced by a firefighter. The real life average duration of use on a 30-minute cylinder is 11 to 17 minutes. Programs such as "Smoke Divers" and general physical fitness programs for firefighters can extend this time but 30 minutes of working time is rarely, if ever, achieved.

SCBA are required to be certified as being in compliance with the standard by an approved certification organization. Labeling that indicates the certification is required. In addition, requirements for a certification program are outlined. Additional labeling on or with the SCBA is required which provides use and maintenance advice to the user as well as providing identification of the manufacturer.

The most recent revision of NFPA 1981 released in 2002 contains four major changes from previous revisions. The changes are:

1. All new SCBA must have two independent EOSTI (End-Of-Service-Time-Indicator). Each EOSTI must have a sensing mechanism and a signaling device and at least two human senses must be stimulated.
2. All SCBA must have a HUD (Heads Up Display) to alert wearers of their air supply.
3. All SCBA must be equipped with a RIC/UAC (Rapid Intervention Crew/ Universal Air Connection System). This consists of a male and female fitting (coupling) to allow replenishment of air to the breathing cylinder. The coupling must be compatible among all SCBA manufacturers.
4. All SCBA must be tested on a standardized breathing machine.

NFPA 1981 requires that all new SCBA units maintain a positive pressure in the facepiece during a specified test routine that simulates breathing under various physiological loads. The purpose of this test is to assure that firefighters working at an incident cannot overcome the ability of the SCBA to provide air and draw contaminated air into the facepiece from the outside environment.

The standard also requires that the SCBA withstand and operate in hot and cold environments, resist damage due to vibration, resist damage due to flame contact and heat exposure, resist corrosion, resist malfunction due to exposure to particulates, that the lens resists abrasion, and that communications through the facepiece meet a minimum standard.

Personal Alert Safety Systems (PASS)

NFPA 1982 Standard on Personal Alert Safety System*s* provides requirements for an automatic signaling alert if a firefighter becomes motionless or can be manually activated by the firefighter as a distress signal. PASS devices can be independent devices or they can be integrated into the SCBA. If an integrated PASS device cannot be disconnected and used as a stand-alone unit, fire chiefs should consider a second stand-alone PASS for incidents where SCBA are not required.

PASS devices were created in the late 1970s and early 1980s in response to a series of firefighter deaths that occurred in situations where other firefighters could have rescued the downed firefighter if they had known where the person was. California, through the Cal/OSHA standards, led the nation in developing the first standard for PASS devices. The first round of PASS devices developed for the fire service was saddled with sensitivity problems, cases and electronics that failed to withstand fire fighting exposure, and alarm sounds that were too close to the sound of a smoke detector and other fire alarm annunciators. Many of these problems have been solved through the development of new generations of PASS devices.

The requirements of NFPA 1982

The PASS device must be tested by the manufacturer and certified to be in compliance with the standard. The PASS must be labeled as compliant with the NFPA 1982 standard if testing proves it to be so. Manufacturer's use and maintenance instructions must also be provided with each unit.

All PASS devices shall have at least three modes: off, alarm, and sensing. The PASS actuation switch must be operable by a gloved hand. Two separate actions must be taken to change the PASS device from automatic mode to off in order to prevent accidental deactivation. A visual and audible indication must be present to indicate that the PASS is operational.

The PASS device must sound a pre-alert signal if it remains motionless for 20 seconds. The alarm signal must be activated after the device has been motionless for 30 seconds plus or minus 5 seconds. The alarm may be preempted if the unit is moved while in pre-alert.

The alarm signal tone must conform to frequency specifications and must produce a sound level of 95 dBA measured at a distance of 9.9 feet. The signal must be capable of continuous operation for at least one hour. A low battery signal is required.

The PASS device must be equipped with a retention system that allows it to be attached to the wearer, it cannot weigh more than 16 ounces, it must be intrinsically safe for use in hazardous atmospheres, it must have provision for water drainage, and it must resist corrosion. Test methods to evaluate the PASS device's compliance with the standard are detailed in the standard.

Rope and System Components

NFPA 1983 Standard on Fire Service Life Safety Rope and System Components provides the minimum requirements for the following life safety equipment:

- Life safety rope
- Escape rope
- Water rescue throw lines
- Life safety harnesses
- Belts
- Auxiliary equipment

The standard was developed partially in response to the deaths of two firefighters during an attempted rope rescue when a rope parted under the weight of a New York City firefighter picking up the weight of a trapped firefighter at a sixth floor window. Until the development of NFPA 1983, there were no standards for the design, performance, or testing of fire service rope, harnesses, or hardware. The standard does not apply to utility rope.

The requirements of the standard are summarized in the following paragraphs. Life safety rope, harnesses, and hardware are required to be certified as being in compliance with the standard by an approved certification organization. Labeling that indicates the certification is required. In addition, requirements for a certification program are outlined.

The standard specifies the load capacity, classification, fiber source, and construction type for life safety rope. Labeling is required within the rope and on a tag attached to the rope. The information on the labels includes the manufacturer's name, lot number, and the name of the certification organization. The tag must contain warnings about the use of the rope and the rope's working characteristics. The standard specifies strength characteristics for the rope, a method of classification, and test methods to assure the compliance of the rope.

Life safety harnesses are classified into three groups. The sizing, labeling, and performance requirements for harnesses are included in the standard. Hardware design, construction, labeling, and performance requirements are also detailed. Testing methods to assure compliance are also presented.

Hazmat Vapor and Liquid Splash Protection

NFPA 1991 Standard on Vapor-Protective Ensembles for Hazardous Materials Emergencies and NFPA 1992 Standard on Liquid Splash-Protective Ensembles and Clothing for Hazardous Materials Emergencies provide performance requirements for protective clothing for firefighters engaged in activities that bring them in contact with hazardous chemicals.

Each standard requires the compliance of the suit or garment to be certified by an independent certification agency. Labeling that indicates the certification is required. In addition, requirements for a certification program are outlined.

Additional labeling on the suit or garment is required. The labeling details the chemicals that the suit or garment has been designed and certified to protect against. The standard requires that each suit or garment be capable of withstanding exposure to a specific list of chemicals depending upon the garment. The suit or garment manufacturer in the technical data package may list additional chemical resistance.

The manufacturer with each suit or garment must provide a technical data package. The data package must include information on the suit or garment components, including type and material, chemical penetration resistance documentation, suit or garment component documentation, and physical property documentation—depending upon the suit or garment type.

Each standard also details design and performance requirements as well as methods to test or verify the performance requirements.

Chemical/Biological Protection (Terrorism Incidents)

In response to recent terrorism events, NFPA released a new standard in 2001 to provide emergency responders and fire chiefs with a minimum standard for protection in response to these incidents. The new standard is NFPA 1994 Standard on Protective Ensembles for Chemical/Biological Terrorism Incidents.

The standard is a bit unique in that it provides three classes of ensembles to choose among. The choice is determined by the perceived threat of the incident. The basic differences in the ensembles is their level of:

- Resistance to leakage of chemical and biological contaminants
- Resistance of the materials used in the ensembles to chemical warfare agents and toxic industrial chemicals
- Durability of the materials used in construction of the ensembles.

An ensemble consists of garments, gloves, and footwear. All three levels of the ensembles are restricted to one-time use. A review of each classification follows.

Class 1 ensembles

This is the highest level and intended for worst-case scenarios where the substance is unknown and is causing an immediate threat. The release is possibly still occurring. The duration of the ensemble is limited by the duration of the air supply from the SCBA. Most of the victims will appear to be dead or unconscious. These ensembles are required to:

- Be gas-tight by passing an inflation test with exhaust valves plugged.
- Show very low levels (less than 0.02%) for penetration of surrogate gas during an inward leakage test.
- Use materials that have the highest levels of permeation resistance to chemical warfare agents and toxic industrial chemical liquids and gases.

Class 2 ensembles

This is an intermediate level of protection. The substance has generally been identified and the release has subsided. Exposure is anticipated to be residual contaminated air from the release as well as contaminated surfaces at the scene. The duration of the ensemble is limited by the duration of the air supply in the SCBA (like Class 1). These ensembles are required to:

- Show no more than 2.0% leakage using the same surrogate gas as a Class 1 ensemble (not gas-tight).
- Pass a shower test to indicate resistance to liquid penetration (shower consists of spray nozzles from different directions).
- Use materials that are resistant to chemical warfare agents and toxic chemical liquids and gases as with Class 1, except at lower levels of concentration.

Class 3 ensembles

These ensembles offer the lowest levels of concentration. They are designed for use in the peripheral zone of the release or well after the release has occurred. There should be essentially no (but not necessarily none) exposure. The ensembles are intended for longer duration periods and it is expected that air-purifying respirators will be used in lieu of SCBA. These ensembles are required to:

- Be liquid tight by passing a short duration shower test. They do not have to be gas-tight.
- Provide permeation resistance to low concentrations of liquid chemical warfare agents and liquid toxic industrial chemicals where the liquid is permitted to evaporate during the test.

EMS

Since emergency medical operations have become an important part of the services delivered by fire departments, the NFPA developed NFPA 1999 Standard for Protective Clothing for Emergency Medical Operations.

The protection firefighters need from hazards posed by contact with liquid-borne pathogens is well documented. Liquid-borne pathogens are defined as an infectious bacteria or virus carried in human, animal, or clinical body fluids, organs, or tissues. The most commonly known liquid-borne pathogens are hepatitis and the AIDS virus.

There have been cases where a caregiver has been infected after contact with a person who has contracted one of these diseases. The caregiver can contract the disease if an open wound or cut, or any other mucous membrane such as the eyes or mouth, comes into contact with infected fluids. Extreme care should be taken to avoid contact with infected fluid. Since firefighters cannot know if any particular patient is infected, caution when in contact with any patient is a must. The standard covers requirements for emergency medical garments, gloves, and face protection devices.

Emergency medical protective clothing is required to be certified as being in compliance with the standard by an approved certification organization. Labeling that indicates the certification is required. In addition, requirements for a certification program are outlined. Additional labeling on or with the emergency medical protective clothing that provides use and maintenance advice to the user as well as providing identification of the manufacturer is required.

Upon request of the user or the purchaser, the manufacturer of the emergency medical protective clothing must provide a documentation package. The package must contain information on the materials used to make the item, construction methods, flame resistance characteristics, penetration resistance to liquid-borne pathogens (both after abrasion and after flexing), among other information. Each piece of emergency medical protective clothing must be watertight, and resist the penetration of a specific biological agent for one hour. Test methods are specified in the standard.

Emergency medical garments and gloves must be tested for tensile strength, puncture resistance, among other specific tests (such as a dexterity test for the gloves). Test methods are specified in the standard.

Managing the Process of Procuring PPE

One of the most technical and fast-changing areas of our service is the area of protective clothing. The large numbers of standards are largely unfamiliar to most fire service members. Hopefully, the summaries provided in this chapter will add to the understanding of these important standards. To add to the confusion, the technology of the fabrics, construction methods, and hardware that goes into this equipment is in a constant state of change.

While chief officers are not often directly responsible for the development of protective clothing specifications, they do have a responsibility to their firefighters and to those who ultimately pay for the equipment to make sure proper levels of protection are provided. Chief officers are also responsible to be sure that funds used to purchase protective clothing and equipment are spent wisely.

Begin slowly and incrementally

The specification and purchase of protective clothing and equipment is an extremely complicated and time-consuming process. Start out with an item of clothing that is easier to specify. Gloves and helmets are good first tries since the options associated with both of these items are limited and the number of manufacturers is relatively small (don't try and replace your entire SCBA inventory the first time you sit down to write a spec).

Use a team approach

Other members of your department will be interested in participating in the process—form a small team to help. Ask that a member of your department's occupational safety and health committee be appointed to help you. If your department has a member organization such as a labor union, ask one of the union officers to be involved. If you are a civilian or a uniformed member of the department assigned to staff, ask a member of your department that is more directly involved in service delivery to help. Get representation from all interested areas of your department, but resist the temptation to form a large committee. Try and keep your working team to no more than six members. Different teams can work on different pieces of protective clothing and equipment.

Talk to distributors and manufacturers

All of the major manufacturers of protective clothing and equipment operate a dealer or distributor network. Look in the yellow pages under fire or safety equipment and see what you find. Each year the NFPA publishes the *Buyer's Guide—Fire Protection and Fire Service Reference Directory* through the *NFPA Journal.* The guide lists the manufacturers of all types of fire equipment and provides names, addresses, and phone numbers for each. *Fire Engineering* magazine also publishes a very useful buyers guide edition each January. Write or call distributors and manufacturers, tell them you are in the process of writing a specification, and ask them to send you information on their products. Local distributors may be willing to visit your department, explain their products, and offer assistance in specification development. Advertisements in fire service publications such as *Fire Engineering* can be an excellent source of information.

Read the NFPA standard

Many of the misunderstandings of the requirements of any NFPA standard can be resolved by simply reading the document. Explanatory material for sections that have an asterisk (*) next to the paragraph number is contained in the back of the standard. The explanatory language is called the "Annex." It was formerly called the "Appendix." NFPA standards are available from the NFPA. Most fire prevention divisions maintain copies of all NFPA standards. If you do not have a copy, your county or state training organization may have a complete set. Other sources of NFPA documents are fire protection contractors, architects, consultants, county or city building departments, and other professional organizations.

Talk to your peers

Ask friends in other departments in your area about their experience. Ask them for copies of their specifications. There is no sense in reinventing the wheel—if a neighboring department has an established specification that incorporates all of the features that you want in your protective clothing, copy it. In cases where more than one department buys clothing and equipment off the same specification, cost savings may be realized by pooling your orders together. Also, check with your state agency about state contracts.

Understand impact of patents

Manufacturers of protective clothing and equipment are in business to make money. In order to protect themselves and assure that the money they spend on research and development is recovered, they may patent a process or feature that is offered by them. If you decide that the feature is something you want, specify it. Be aware that other manufacturers may not be able to provide the feature and that they may choose not to bid on your specification. You may find as you do your research that there are features that you do not want to do without. There is nothing wrong with specifying a feature that can only be provided by one manufacturer as long as you are aware of what you are doing.

Seek demo PPE for evaluation

Ask manufacturers and/or dealers to send you samples of their equipment or ask that they come to your department and demonstrate their clothing or equipment. For accounting reasons they may ask you to send a letter requesting the loan or may issue an invoice with the loan that allows you to return the item at no cost within a specified time period such as 60 days. If you plan on using the loaned clothing or equipment in a way that may get it dirty or destroyed, let the person loaning it to you know that. In most cases you or your department are responsible for anything that happens to loaned equipment when it is in your possession.

Attend seminars and trade shows

Each of the major fire service organizations holds meetings and shows around the country throughout the year. Many offer low cost one day passes. Trade shows provide an excellent opportunity to view products, interact with the manufacturers, and obtain answers to your questions. All of the major fire service manufacturers are in attendance at these shows. They are very willing to speak with you and provide information about their product line. These conferences also provide a tremendous opportunity to establish additional networking contacts with your peers. If a show is in your area or within driving distance, it is time well spent. Your department may also, if funding is available, be able to send you to a show.

Join local safety networking groups

In many parts of the country, safety officers and others interested in firefighter safety get together on a regular basis to discuss safety issues. Many times these discussions include protective clothing and equipment. Since Southern Area Fire Equipment Research (SAFER) was founded in Southern California in 1976, two other sister organizations have been established. Northern Area Fire Equipment Research (NAFER) exists in the Bay Area of California and a group called FIERO (Fire Industry Equipment Research Organization) exists in the southeastern part of the country. These organizations are an excellent example of information sharing, networking support, and vendor/user interaction. Many manufacturers, suppliers, and vendors are members and active participants in these organizations. Even if you are not able to attend the meetings in person, all four groups publish newsletters that carry the details of items discussed in their meetings.

Implementing new brands and models

Be cautious of replacing critical safety items such as SCBA and PASS devices in a piecemeal fashion. In times of extreme stress, firefighters will fall back on their training. A department operating with more than one SCBA model, and thus more than one mode of operation, may be placing firefighters at risk. In a time of extreme stress, such as when a firefighter is lost, will the firefighter have time or the presence of mind to remember that they are wearing SCBA model A or SCBA model B?

Seek comments on your draft specification

Send copies of your draft specifications to different vendors or manufacturers and ask for their comments. Ask them to provide suggested improvements and estimated costs. It is a sales representative's job to sell the product. Often, suggestions from vendors for specification modifications may, if adopted, give one vendor an advantage over the other. Be aware of this possibility and evaluate suggestions thoroughly.

Bid invitations

If you are purchasing a medium to large quantity of clothing or equipment, it may be to your advantage to advertise for and accept bids for the order. In many jurisdictions, bidding for any purchase over a certain dollar amount is required. Small or single item orders may not be worth the trouble and expense of a bid unless you are buying a stock item. In some cases, a local vendor may have exclusive distribution rights for a product. In cases like these, working directly with the vendor is encouraged. Always be aware of local and state laws on competitive bidding.

CONCLUSION

Protective clothing and equipment are the first line of protection for firefighters. Along with proper training and proper care, PPE can allow the firefighter to perform their job safely and effectively.

Technical Resources

NFPA address:

National Fire Protection Association
1 Batterymarch Park
P. O. Box 9101
Quincy, MA 02269
(617)-770-3000—Headquarters
(800)-344-3555—Customer Service
www.nfpa.org

Fire Engineering magazine address:

Fire Engineering
PennWell
21–00 Route 208 South
Fair Lawn, NJ 07410-2602
Tel (973) 251-5040
Fax (973) 251-5065
www.fireengineering.com

ANSI address:

American National Standards Institute
1430 Broadway
New York City, NY 10018
www.ansi.org

13

FIRE STATION AND FACILITY DESIGN

Daniel Redstone, FAIA, NCARB

CHAPTER HIGHLIGHTS

- Demonstrates and examines the processes required to develop a successful station.
- Reviews the need for the fire chief or another appointed officer to be involved throughout the design and construction process.
- Reviews the need to bring professionals on board your team that are knowledgeable of the needs of this special building type.

Identification of Needs

Public participation

In most communities, fire departments are user agencies of a city, village, or a township. In some parts of the country there are established fire districts that may include a number of political jurisdictions, and which can be as large as an entire county. Most facilities are funded through some form of bond issue, requiring a vote of the taxpayers.

Citizens committee

Many communities utilize an appointed building authority to guide capital improvement projects, both as a planning and as a funding agency. As the user agency, you will also need to solicit and obtain the political support of local elected officials, neighborhood groups, community businesses—including the local chambers of commerce, the insurance industry, and the support of firefighters and their neighbors and relatives within the community.

With or without a building authority, if you need to pass a bond issue, it always helps to have a "citizens committee" from the very beginning of the idea through the actual bond issue campaign. Like any political campaign, in order to be successful you must be able to sell your story to the public. Ideally, the citizens committee will be made up of a cross-section of the community, and will include citizens and municipal staff. For example, one jurisdiction put together a committee of eight, including four citizens (including two seniors and two pastors), the chief building official, the public works director, a representative of the business community, and the chief. The committee worked together throughout the needs assessment process, and gave credibility to the entire process. If your department is large enough, consider including a senior officer of the department. Including a member of the firefighters union or association is also recommended. Use this committee to confirm the department's mission statement and goals, and use them to guide the planning process.

Your first step is to identify what information you will need to sell the project and begin to think of what professional team members you will need to put together the necessary information that will go to your political subdivision, and eventually to the voters. Most departments need to request funding for initial needs assessments from their local governmental agency. Unless your operating budget can absorb these outside consultants, you will have to sell you plan to your political subdivision to obtain funding for even the beginning study.

Fire services consultant

If you are a growing community, an independent fire services consultant to conduct a fire protection study for your community can be very worthwhile. These consultants may be independent companies or part of the services of specialized insurance companies.

This type of consultant is experienced in analyzing service calls, response times, geographical patterns for responses and anticipated growth, as well as operational issues. If you are in a mature community with older existing stations, such a consultant may be of use in looking at a long-range plan of replacement facilities in locations that may better serve your community. Items that may be considered by this consultant include: population growth, location and condition of existing stations, utility infrastructure, personnel resources management, customer expectations, perceived strengths and weaknesses, and possible regional cooperation. You can find fire services consultants through your state or national fire chief organizations, and insurance bureaus.

The architect

The second type of consultant you will need on your team is an architect. Just what does an architect do? How do you find the right architect? How does your community select architects?

Architects are much like physicians. Architects listen to your issues and problems, and suggest solutions. Once a solution has been identified through a needs assessment, architects develop design and construction documents that permit contractors to build or renovate a facility. When construction commences, an architect will normally perform some oversight services during that phase of the project as well. The extent of the architect's duties during the construction phase is dependent on the client's specific needs, as well as the construction delivery method chosen.

Selecting an architect

Finding the right architect should be a simple process, and in most cases it is. Every state has at least one component of the American Institute of Architects (AIA). Some architects specialize in fire station design. As the chief of your department, you are looking for an architect with experience in fire station design and who comes with good recommendations from fellow chiefs.

Sometimes "your" architect may not be the same kind of architect that the political jurisdiction is looking for. You may be looking for experience; your community may be looking to engage the local architect or the low-bid architect. Some specialists team up with local architects to provide the local architect with expertise that might otherwise be lacking in your community.

Regardless of the methods described below, you will be selecting an architect based on a number of criteria. It is important to qualify such criteria into objective and subjective categories, so that the evaluation criteria can be properly weighted. Objective criteria generally have a stronger weighting than subjective criteria.

Objective criteria include:

- Recent programming experience
- Recent similar project related experience
- Examples of successful project budgeting
- Examples of successfully delivering a project within a previously established budget
- Previous involvement in different construction delivery methods
- Experience and knowledge with municipal projects

Subjective criteria include:

- Leadership capabilities; ability to think outside the box
- Being a team player; being able to move with consensus
- Ability to work with elected officials and governmental agencies
- Workload
- Professional involvement in the community and with professional organizations

References are important, but only if you actually call the reference and ask good questions. If a project is listed but there is no related reference, call that jurisdiction anyway. Make sure that you get the reference not only from the chief, but also from the mayor, city manager, or supervisor of the jurisdiction. What you really need to find out are the reference's opinions of the criteria listed above (objective and subjective). Call the appropriate building official and ask for an opinion as to the quality of the documents.

Resist the temptation of choosing an architect or other consultant because of fee. The *needs assessment process* is the foundation of the entire project, and while it is the least expensive phase of the process it will have the most significant impact on the subsequent design and construction phases. A good rule of thumb is that a needs assessment will cost about .5% of the eventual project cost. If the project is a $3,000,000 facility, an assessment may cost between $10,000 and $15,000, depending on your requirements. A good assessment will save tens of thousands of dollars for your jurisdiction.

More importantly, the needs assessment will be your selling tool to approving agencies and citizens alike. Combined with the recommendations of the fire service consultant and the space and site requirements of the needs assessment, you will have a complete picture of the project, including a good projection the funding requirements.

The needs assessment (the least expensive but most crucial step in the design process)

In order to ensure that you get an architect experienced in fire station design, be involved in the selection process, especially at the beginning of the process when you are hiring a firm to do the initial architectural needs assessment. Assist in writing the Request for Qualifications (RFQ) or a Request for Proposal (RFP). In a RFQ, the solicitation is to identify qualified architects (without consideration of a fee), while a RFP will normally identify a scope of services and request a fee for those services. A RFQ process is much preferred, as all architects are not equally qualified to be your architect. Ask agencies that have recently had projects contracted for examples of their RFQ. If you know an architect, ask for help or guidance.

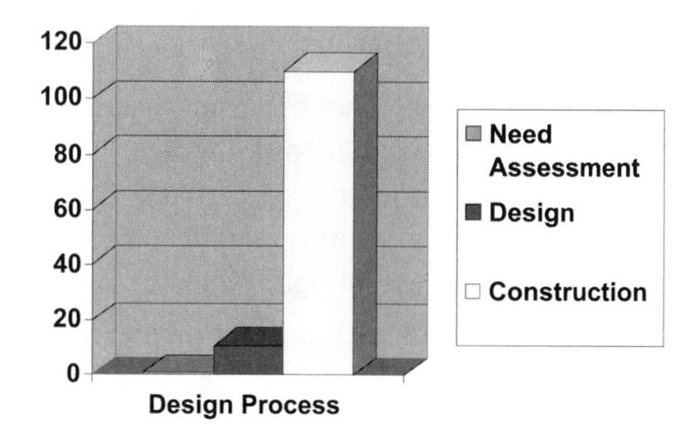

Qualifications Based Selection (QBS)

Many states use a process called Qualifications Based Selection (QBS). The QBS is modeled after the federal process of procurement for professional services. In this process, design professionals would normally submit their qualifications first. After review of the submission and interviews with the two or three top-ranked firms, negotiations with the top-ranked firm begin. It is during this negotiation that both sides can discuss what services are needed and strike a fair bargain can be negotiated. The new AIA documents have a checklist of services for the architect and client to review and discuss in detail. Services will also be dependent on what type of construction delivery process your community will use on this project.

If your jurisdiction has a bid requirement, check with the municipal attorney as to whether this requirement applies to professional services in many jurisdictions and states professional services are exempt from the bidding process. Enlist the municipal attorney to recommend the QBS process as a permissible method of selecting a design professional.

Requiring a professional to bid for services that are still to be defined can be very dangerous to the future well being of the project. The reason for potential danger is that unless the services are defined as to what the architect is to bid on, proposals will be received that reflect each architects' respective assumptions of what you had in mind, or need. As mentioned earlier, the assumption in writing this chapter is that you do not have significant experience in the design or construction process of a fire station, and therefore may not know specifically what you need from an architect.

If your jurisdiction has a building authority for its capital improvement needs, your position on the project team may change from captain to tailback. But even with a building authority (as the name implies, it has the authority over the money and the process), you need to maintain a very strong role in the process, as members of the authority are generally part time members. You need to be responsible for attending to the details of the building project from the initial stages of design to the completion of the "punch list."

The actual needs assessment

An Architectural Needs Assessment is the most important part of the design process. A successful Needs Assessment will form the basis for all of the efforts that follow, from establishing budgets and selling the community to actual design. Its importance cannot be overstated. Even if you are only doing an initial study, you can also use the QBS process to select an architect for this initial stage of your project.

The process of a Needs Assessment includes discussions between the design professional and the department, primarily the chief and command staff. Departmental objectives are identified and space requirements are developed. Space adjacency requirements are identified. A minimum site size is identified, to enable the community to identify potential site locations, or to confirm the adequacy of an existing parcel already identified as a potential site.

Through interviews with community stakeholders, community facilities that will be incorporated into the facility will be identified. This issue has implications for the budget, design, and site size. If the community attempts to justify the need for a large training room because of anticipated frequent community use, you need to be satisfied that your department will have sufficient availability of that space for departmental training. As training requirements become more extensive, public availability for shared training spaces may be reduced.

The final needs assessment must have buy-in by users and an endorsement by elected officials to be supported and campaigned for in a subsequent bond issue campaign. Properly done, the needs assessment process will achieve this objective.

Developing a checklist of spaces is a helpful exercise that will permit the thought process to identify needs and wants. It is important to prioritize the items identified, because when the project budget is first the total will often exceed what the community might be able to afford for the project. Prioritization will allow you to delete or reduce space requirements in order to be within funding limits.

Existing building condition survey

If you are dealing with an older facility and are considering a renovation or a replacement, an Existing Conditions Assessment is highly recommended. Even when everyone knows that the building is old, there are still many people in the community who do not understand that fire operations have changed over the years and that your existing facility is obsolete. Training, accommodating female firefighters, the issue of blood-borne pathogens, exposure to environmental issues, and OSHA requirements are but a few items that buildings as young as 10 years do not address. A formal assessment by an outside consultant will provide the objective opinion you need to confirm the building's condition.

In most cases, the architect engaged for the needs assessment will be able to perform this survey. Make sure that if renovation is seriously being considered as an option, that engineers review the structural, mechanical and electrical systems.

Project Budget and Funding

How much money do you need to build a new or renovate a fire station? Establishing the project budget is a critical step towards a successful project. In addition to the actual construction cost, a project budget will include such items as: site acquisition, site development, furniture, fixtures, and equipment (FF&E), security and communications, professional fees, bond costs, and miscellaneous owner costs.

Building costs

The completed needs assessment will provide the jurisdiction with a good, initial estimate of probable square footage for the project. The architect will assign a unit-price, dollars per square foot value to determine a bricks and mortar construction cost projection. While a more complete set of design documents would provide a more definitive estimate, your dollars at this stage of the project may not be able to afford more than professional fees for the needs assessment.

Site costs

Look at the site that you will use to build your station on. Does your jurisdiction own it? Do you need to acquire a parcel? Does the parcel have on it existing buildings that may have to be demolished? Whether you have identified the site through your fire operations consultant or with your own staff, the needs assessment will confirm the adequacy of that site or challenge it. If the site is too small, a one-story building might become a two-story building.

Site utilities must also be considered in developing a project budget. Are there utilities to the site, or must water and sewer be brought in from a distance? Are there topographical issues that need to be resolved?

If the jurisdiction has a long-range strategic plan to build more stations in the future, a funding proposal might include the acquisition of more than one site. Land never gets cheaper, and fire station locations tend to be situated in areas higher in density, therefore more valuable, property.

Furniture, fixtures, and equipment; security

Furniture, fixtures, and equipment (FF&E) as well as communications and security systems are two other categories that need to be budgeted for. Specialized equipment for fire station operations, such as vehicle exhaust systems, SCBA air systems, hose dryers, and high temperature washing machines, must be identified and budgets established for their purchase.

When building a new facility it is recommended that you budget and acquire new furniture as well. Older furniture is normally as obsolete as the buildings it sits in, and was not designed for today's technological and information age.

Communications systems are real necessities, especially since the events of September 11, 2001. How your fire station communicates with the rest of your jurisdiction's governmental and public safety operations will determine how much you need to budget for communications systems in a new building.

With today's computer technology, it is a relatively simple effort to combine your entire operation within one network. Remote locations, such as a station house or the chief's home, can be connected through Virtual Private Networks (VPNs), or through the Internet using a terminal server.

If the building will sit on a municipal campus, the building will likely be connected with a hard-wired cable connection. If the new building is remote from a civic campus, the jurisdiction's IT consultant will recommend communication solutions.

Bond and finance costs

Bond and finance costs must be budgeted. These costs normally are deducted from the proceeds of any bond issue, and will reduce the amount available for construction, FF&E, professional fees, and other associated costs of the project; included will be the professional fees for bond attorneys, commission fees paid, and the costs of financial advisors.

Operational costs

Operational costs have been forgotten on numerous occasions. Will your current operating millage be adequate to operate the new building? Remember, a new building, while much more efficient, will have significantly more square footage than the cramped and obsolete facilities you currently occupy. If you need additional operating funds for a new or expanded facility, you need to address the issue with the voters at the same time you are trying to sell a millage for a new building.

Professional fees

A project budget needs to include the fees to be paid to the architect and the other professionals that the project may need. Architects ordinarily include in their basic services normal structural, mechanical, and electrical engineering for the building itself. Other engineering disciplines, such as civil engineering (site design) may or may not be part of the architect's services unless negotiated. There will likely be additional services needed from the architect that are not normally included in basic services, and a budget should include a provision for them.

Other services that will be needed for the project include:

- Landscape architecture
- Interior design services (other than basic surface materials and colors)
- Environmental engineers
- Testing engineers
- Communications consultant
- Security consultant
- IT consultant

In successful projects, the architect and the jurisdiction will identify exactly what is needed for the project during the needs assessment.

Owner costs

In most owner-architect agreements, the owner is responsible for a number of items, such as:

- A complete topographical survey
- Soil borings (geo-technical investigations)
- Environmental testing and abatement costs, if any
- Reimbursable expenses of the design professional for travel, reproduction of documents, etc.
- Legal and accounting services

Millage campaign

Successful bond issues just don't happen; it takes a lot of hard work. Your participation in a millage campaign is critical. Your firefighters' involvement is also crucial.

A well thought out campaign is essential. For example, one might think that senior citizens usually will vote no on millage issues. As a fire chief, you know that 80% of your runs are for EMS services, and of that 80% there is a significant number that involve seniors. Rather than keeping a millage campaign quiet and hoping that the opposition will not vote (they always do), develop a strategy that makes senior citizens, who directly benefit from your services, your ally.

Remember also that most seniors vote by absentee ballot. On election day, it is likely that over 80% of them have already voted. Spend your efforts by meeting with this group well in advance of voting day.

The business and industrial community benefits directly from better fire protection in their insurance premiums. Even a one-point change in your jurisdiction's ISO ratings may reduce premiums significantly. One-half of the ISO rating reflects your physical and operational evaluation, and half is determined by your water service rating.

There are bond issue consultants who can assist your jurisdiction in developing a winning millage campaign. Find them and use them.

Site selection. It is important that the proposed site meets the response time criteria established by your department and that recommended by national standards. Direct access to main roads is a must. How will the driving patterns of a drive-through configuration impact the neighborhood is a question that will have a bearing on both the building design and the millage campaign.

In larger metropolitan areas, timely acquisition of a specific parcel can be a real problem. If you need to pay for a new parcel out of the proceeds of a bond issue, the identity of a parcel will no doubt be leaked out. Will you be able to tie-up or option the parcel before your interest in it becomes known? If your jurisdiction has funds to acquire the parcel before a bond issue is passed, does it have the strength to acquire it or condemn it prior to the bond issue?

In growing areas, identifying and acquiring parcels for future stations is an excellent strategy. Land acquisition requires reasonable investigation (due diligence) of the parcel, which may include wetland, woodland, and environmental issues, as well as availability of utilities. All of this takes time, and timing is everything. The involvement of an architect and a local real estate broker can be very helpful in identifying sites in a timely manner for further investigation. Remember to sit down with your planning consultant and building official as well. These officials can provide specific information regarding plans for extending sewer and water service, both prerequisites for significant population growth.

Site design. Site design and building design go hand in hand, each being an integral element of the other. It is important to look at the design of a site as part of the workflow of a fire station.

Fire apparatus must arrive and leave the fire station. While leaving quickly and safely is the most important element of a site design, how apparatus vehicles return to the station can be an important political issue for the station's neighbors. For this reason, a land parcel located at a corner of two streets, if it has the proper dimensions to permit a flow through design, may be considered an ideal site. When dealing with an internal site, the ability to design a flow-through station is dependent on a second means of egress or a parcel wide enough to accommodate the turning radius of the apparatus vehicles at the rear of the parcel.

Where land is expensive or in more dense urban areas, the site will dictate a "backing-in" solution. Here apparatus vehicles will back in from the same street that they enter onto when responding to a call for service.

Staff parking must be sufficient to your department's unique needs. In the case of full time departments, enough parking for two shifts is necessary to accommodate any overlap at shift change.

If the station will have a major training facility, parking needs to be provided for additional staff or for trainees from other departments. If you are pulling out staff from other stations for on-shift training, they may arrive on apparatus vehicles, in which case you will need a site design that will permit these vehicles to respond to their coverage area from the training site.

If the station is a central station, staff parking for the department's administrative personnel, such as fire marshal, arson investigators, training officer, GIS staff, etc. must be provided as well. Depending on the size of the site, and your own workflow, the on-duty area and the administrative area may not be in the same area of the new building.

Public (visitor) parking will vary in amount, based on the building program established by the community. Many elected officials look at a training room in a fire station as always available for citizens to use. They are unaware of the significant training that all firefighters must undertake on an ongoing basis. The bad news is that if your have a training room in your facility, you will need to provide parking for it, regardless of who uses it. The good news is also that a training room that can be used by the public may be a positive selling point in your bond issue.

Most zoning ordinances require parking to be screened from residential neighborhoods. You should anticipate that your community requires that your project comply with all zoning ordinances. While there are always exceptions to the rule, most communities believe it is good practice to set the example of compliance, rather than waiving requirements required of everyone else.

Building design

The spaces and equipment identified in the needs assessment program will become the basis for the design of your new facility. The architect will use the needs assessment as their checklist to help insure that the project's needs are met. Chances are that you have not been involved in the design of a fire station before. How will you work with the architect? Just what does an architect do? How involved should you become in the detail of the building?

If you do not have a working knowledge of the construction process, don't be bashful. Your level of involvement needs to be a combination of 1) your level of comfort, and 2) the involvement necessary to understand that what the architect is putting into the facility is what you specified. In other words, if you don't understand something, or you think the architect doesn't understand what you are trying to tell them, communicate!

An efficient layout of the spaces is the most critical element of your building. Time is of the essence in responding to a call for service, and a good layout will help minimize response time.

In order to understand layout, think of your building as a flow or process. Develop a thought process that tracks each task or duty one step at a time. This will become the natural flow for that task. For example, how do the firefighters maintain the equipment on the apparatus vehicles? Where would be an efficient location for a repair room to house tools and make repairs?

Department types

Paid departments have full time staff residing in the station. This means that there needs to be a close proximity to the living areas of the firefighters and the apparatus bay. The day room and the sleeping quarters need to have a clear path to the apparatus bay.

Building codes require special separation between the living and working areas of the station, and the apparatus bay or garage. The separation is for both fumes from apparatus and for the necessary fire separation.

Volunteer departments have a slightly different arrangement for responding to an alarm. Some departments respond to the station house; others have the first responder's to the station drive the apparatus, with the other volunteers responding directly to the location of the alarm.

Combination departments need both living quarters for the full time personnel who generally drive the apparatus.

Station types

All stations include apparatus bays. There are three types of stations that correspond in many cases to whether the department is a full-time or volunteer department, and to whether the fire jurisdiction has one or more stations.

Headquarters station or central station. This facility will normally house the fire department's central offices, training facilities, and a dispatch function, if required. Exceptions to housing the central offices at the station will occur in communities with a public safety department whose offices are combined in a police or city hall building. Outdoor training may occur at any location.

In larger communities, the central office will include provision for fire prevention, fire marshal, arson investigation, training officers, as well as for future GIS system development.

A full time station. This facility will have provision for living quarters, food preparation, day room, as well as a training area. Newer stations are also including quiet areas such as a library, where additional studying and training can occur. Internet connections should be provided in all new stations.

Non-staffed stations. This facility may by large or small, depending on how the department is organized. Troy, Michigan's five stations are organized as separate squads, each station having one squad of over 20 volunteers. Each station has its own training room, so that the squad trains together.

All departments need to resolve the storage and cleaning of gear, how hose drying operations are conducted after a fire, and what demands will be placed on the building throughout a complete firefighting and response cycle.

Essential building type

The 2000 International Building Code and previous codes consider a fire station to be an essential building type. This means that the building must be designed to withstand more significant wind and roof loads than a normal building type. The current code requires a factor of 1.50 be applied to the seismic factor (I_e), 1.2 for the snow factor (I_s), and 1.15 for the wind factor (I_w).

Make sure your architect's structural engineer is aware of these factors, which can be found in Table 1604.5 of the 2000 International Building Code.

Your building is in use in a round-the-clock environment. It is not just another building that shuts down at 5:00 p.m. Both the level of quality of the materials in the building and the systems that support your occupancy must be of significantly higher quality than in a typical building. The products for floors, walls, mechanical and electrical systems, and HVAC systems must all be chosen for their long-term value.

In evaluating a material or system, consider the following:

- Initial cost of material or system
- Normal operational cost
- Expected life of product in a round-the-clock operation (when will you have to replace this item?)
- Maintenance costs

This evaluation is sometimes called "life-cycle" costing, but what it really means is that your building needs to be durable over the long run. The cheapest system is normally not the best choice.

Typical spaces

The following are a list of typical spaces that will be found in many fire stations. Not all stations will have every function. It is the intent of this list to provide a checklist for discussion during the needs assessment as well as during the design process. Each of these spaces should be thought of for current needs and future growth.

Administrative spaces (headquarters)

Deputy chief's office

Reception/lobby/vestibule

Chief's office

Training officer room

Arson investigation office

Pre-function area for training room or breaks

Training room (possible community meeting room)

Storage room(s)

Conference room

Front counter

Lunch room

Fire marshal office

Communications room

Public rest rooms (if training room is used by the community)

Emergency operations center (could be same as training room)

File room

Apparatus space

Turnout gear storage

EMS storage

Apparatus maintenance shop area

Hose dryer

Washer/dryer

Apparatus bays

Hose storage

Equipment repair

EMS clean up room approved for blood born pathogen clean up

SCBA—cascade room with repair and clean up space

Spare SCBA and oxygen tank storage area

Mezzanine for use as a large storage area (Optional)

Small receiving area with telephones and portable radios with charges

Decontamination station (requires separate holding tank)

Full time station

Kitchen

Library/study/quiet area

Day room

Locker rooms (men and women)

Bathrooms (men and women)

Firefighter storage

Training/fitness room

Janitorial closets

Communications room

Sleeping room (with individual cubicles)

Support spaces

Mechanical rooms

Electrical/telephone closet

Janitorial closets

Hallways and vestibules

Emergency generator (normally outside)

Site requirements

Visitor parking

Staff parking

On-site circulation for apparatus

Trash receptacle

Outdoor training facilities

Equipment needs

Before designing a new station, you should consider your existing specialized equipment to determine what might be usable in the new station. Considerations include how it would be installed, who would install it, and how reusing equipment would impact ongoing operations during the switchover. Whatever you decide, whether purchasing new or transferring your existing equipment, you and the architect must consider the placement, size, and accessibility during the design process.

If you are currently in an older station, chances are that you do not have all of the equipment being installed in new stations. In addition to old-fashioned fire fighting, many of today's departments include EMS and handle hazardous materials

Do you need the equipment you currently have? A survey of how frequently you use the equipment may provide an answer to this question. Make sure to provide a large enough room size to operate, repair, and replace the equipment at a later time.

Some of the safety equipment being offered today includes:

- Vehicle exhaust extraction systems
- Carbon monoxide detection systems
- Heated air exchange for the apparatus bays
- Water-fill station inside the bays
- Hose washer and dryer systems with de-humidifiers
- Commercial washer and dryer for fire gear (high temperature)
- Washer and dryer for towels for towels and work clothes
- Decontamination shower with access to normal showers and locker rooms
- SCBA compressor with explosion-proof fill stations (SCBA compressor should have fresh air supplied from the outside)
- Oxygen bank with a separated, explosion-proof fill station
- Small compressor for equipment clean up stations

Remember to plan for a complete information technology network, which will be able to tie into the jurisdiction's computer network. With the growing use of GIS, and the integration of dispatch and emergency operations communication systems, your new facility needs to be able to accommodate these electronic requirements.

Some departments already have computers in their vehicles that not only provide full data on fire hydrant water pressure nearest to the call for service, but also are able to pull up diagrams of the home or building from the assessment records. The data may also identify any hazardous material that has been registered as required by law. Other jurisdictions have full access to building department submissions (electronic files) to enable fire marshal review without physically having a set of plans.

NFPA and OSHA requirements and recommendations

There are a number of NFPA publications that make reference to fire station facilities. You should be aware of them in your normal course of operation, but some of them are listed here for reference:

- NFPA 1402 Guide to Building Fire Services Training Centers, 2002 edition. This booklet provides an informative and illustrated guide to training facilities, including chapters on administration and classroom buildings, drill towers, and outside simulation areas.

- NFPA 1581 Standard on Fire Department Infection Control Program, 2000 edition. Chapter 3 of this publication addresses fire department facilities.
- NFPA 1500 Standard on Fire Department Occupational Safety and Health Program, 2002 edition. Chapter 9 of this publication addresses facility safety, including requirements for smoke and carbon monoxide detectors, infection control, and regular inspections.

Each of these publications references many additional NFPA publications on more specific topics. Each state also has its own OSHA requirements, which will impact the design of your facility. You should provide your architect with a copy of the OSHA requirements you must comply with.

Building aesthetics and style

Every single public building reflects on the community in which it is located. Public buildings reflect on a community's pride, its needs, and what its leadership wishes to leave for future generations to be proud of.

Every public building will have a normal life expectancy of at least 30 years and, in many cases, for much longer. The majestic city halls or capitol buildings and other great public buildings of the past 150 years that we visit when on vacation remind us of our heritage. (There are many examples of public buildings in Europe that are hundreds of years old—and still serving the public!)

The station that you will build in your community will be something that future generations will look back on as an example of their heritage. This undertaking will be your legacy for future generations.

Beauty is in the eyes of the beholder, and your community has many architectural critics. Since we are now in the 21st century, a new fire station or headquarters building should reflect today's technology, not only in its look, but also in its use of materials. Even a pre-engineered building type can be designed to be aesthetically pleasing, if you make the effort.

That is not to say that if your community is a historic New England village you should not maintain the character of the community, but in most instances that is simply not the case. However, if you plan to build in an historic district or section of town, be advised that there will be a significant amount of additional effort to design a building that satisfies the members of the historical commission or society in your community.

Construction delivery methods

Historically, governmental projects followed a traditional design-bid-build approach. In this approach, the architects, based on an approved needs assessment and budget, prepared the design and bidding documents, the project was bid to anyone who could post a bid-bond, and a contract for construction was awarded to the lowest (qualified) bidder. In this process, the architect typically assists the owner during the bidding phase, and once a construction contract is awarded, provides for the administration of the contract on behalf of the owner.

Most jurisdictions still follow this approach, which is required by procurement requirements of awarding a contract to the "low bidder." Everyone has heard horror stories about how terrible the low bidder performed, or that the low bidder went bankrupt, or failed to pay its subcontractors, etc. In reality these incidents are generally rare, because it not easy to be bonded by an insurance company.

One approach to help ensure a better level of contractors is to design a process requiring the submission of contractor qualifications as a precondition of bidding. The owner and architect then review the qualifications, including financial statements if appropriate. If a contractor appears to be too small, or for example a residential contractor wants to bid your $5,000,000 project, you would have a legitimate reason to deny that contractor the privilege of bidding the project.

In this traditional design-bid-build process, the contractor does not see the drawings until it is time to bid. Up to the time of award, the only members on the team are the owner (governmental agency), the user (fire department) and the architect. The architect will have done all of the estimating

It is recommended that if your jurisdiction is using this traditional approach, it reimburse its architect for the costs of an independent outside estimator to be a part of the team from the initial schematic design phase of the project. Controlling construction costs is vital to a successful project, and it is much easier to modify the program or quality if the budget cannot be adjusted.

The diagram on the following page illustrates the three elements of the building project and the need to monitor all three elements throughout the design process.

One major positive to using a general contractor is the single source of responsibility, as the owner is entering into only one construction contract. The general contractor is in charge of the coordination between trades, scheduling, and is responsible for the satisfactory completion of the work of all trades.

Another is the level of comfort afforded by knowing all of the costs at once, through the lump sum bid for all of the work.

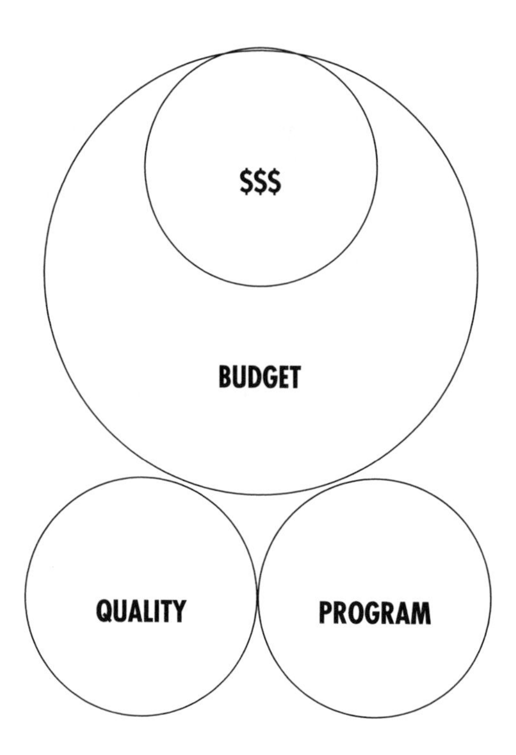

Construction management

A growing number of communities are using construction managers to help deliver the project. In this approach, the services of a construction entity are brought on board the team during the design phase. The construction manager assumes the responsibility for cost estimating, bidding, scheduling, and for offering alternatives for a more cost effective project.

The construction manager receives a fee for this work, plus is reimbursed for the "general conditions" of the project. This process is used more frequently where the project needs to be built quickly, and the project is fast-tracked by bidding the work out in separate bid packages, starting with a site and foundation package, and possibly issuing additional packages such as structure, enclosure, interiors, elevator, HVAC, electrical, and long-lead items.

Construction management is not recommended for a more simple facility such as a two- or three-bay, non-headquarters building.

Design-build

Over the past 10 years, the design-build approach has moved from primarily being a private sector delivery process to a more accepted public delivery process. Both the American Institute of Architects (AIA) and the Associated General Contractors have specific contract documents for this delivery method. This approach takes the concept of "single-source" responsibility to cover the entire design and construction process.

The AIA Owner-Design Builder Agreement has two parts, A and B. Part A covers the Schematic Design Phase and the establishment of a construction budget. If Part A is accepted by the owner, Part B is used to contract for the remaining design and construction phases, using the approved costs from A.

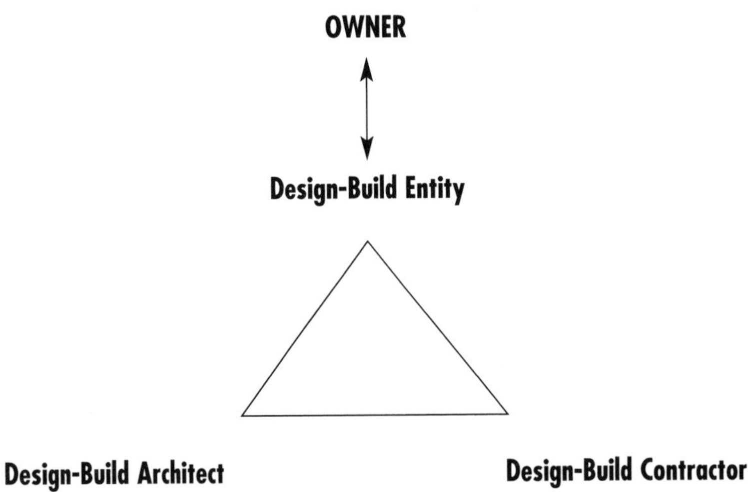

What are some of the issues that must be analyzed using a design-build process? First, the architect is now working for the design-build entity (DBE), not the owner (the DBE can be an architect, and sometimes is). The allegiance of the architect needs to be with the owner, in order to help maintain the level of quality for this essential building type.

Second, because voter-approved bonds finance most fire facilities, all of the money must be spent on the project approved by the voters. In most cases, leftover money cannot go into the municipality's general fund. As much of a challenge saving money is, it is even more challenging to design a facility to the limit without going over a set limit. A number of alternates can be provided to add if funds become available, including procuring additional equipment, if the bond issue so provides.

Third, there has to be a little more trust. If the design-build contractor is not also the design-build entity, the owner will be writing checks to the entity, which then has to pay the contractor. Issues of payment, waivers, and insurance must all be discussed and agreed to.

Thoughts for the future

The events of September 11, 2001, will forever change our lives. Fire departments are currently ill-equipped to handle many, if not all, of the possible acts of terrorism that may be attempted on our population. Chemical agents, nuclear attacks and germ warfare, or simple car bombs have all been mentioned as possible methods of terrorists.

What is clear is that the federal government is determined to develop strategies to develop strong, first responder strategies. Regions will no doubt share these facilities, as special equipment and technology will be costly. Funding sources for equipment and facilities to support anti-terrorism strategies will become available.

If you believe that your municipality may be or might want to become a candidate house for a future terror-disaster response team, you should begin to develop a strategy with your city commission or council to be awarded this facility. Unlike a typical fire station, we do not yet know what will be housed in such a facility. Equipment, special gear, and portable decontamination stations are all possible components. The easiest thing to do is to identify a parcel of land that could immediately be offered to situate this facility. Beyond identifying a site, everything else is speculation.

SUMMARY

The design and construction of a fire station is a lengthy and detailed process. A representative of the fire department needs to be actively involved throughout the process. Do not believe otherwise! In order to ensure acceptance, involve others in the process, from the citizens committee to your own firefighters association.

Through your involvement you will acquire knowledge of architecture, construction, finance, and be more politically involved that you thought possible. Remember to ask questions if you don't understand something.

A new building needs to house fire operations and offices in an efficient manner. The building must be built with a level of quality to maintain a round-the-clock operation for the next few decades. Its design must permit it to be maintained economically. It must be a comfortable place in which firefighters can live, train and work.

Finally, your new station must be a positive addition to the community's spirit, and a building that will be respected by future generations.

APPENDIX

Sample Fire Station Designs

Courtesy of Redstone Architects, Inc.

Southfield Fire Station No. 2

Detroit, Michigan, Engine Co. No. 5

Waterford Township, Michigan, Headquarters

Courtesy of Bob Luke, Luke & Kaye Architects Meridian, Mississippi

City of Philadelphia Fire Station No. 3

Courtesy of Steve Knarr, Horner & Shifrin Architects St. Louis, Missouri

Springdale FPD Headquarters Station

Photo Courtesty Lazlo Regos

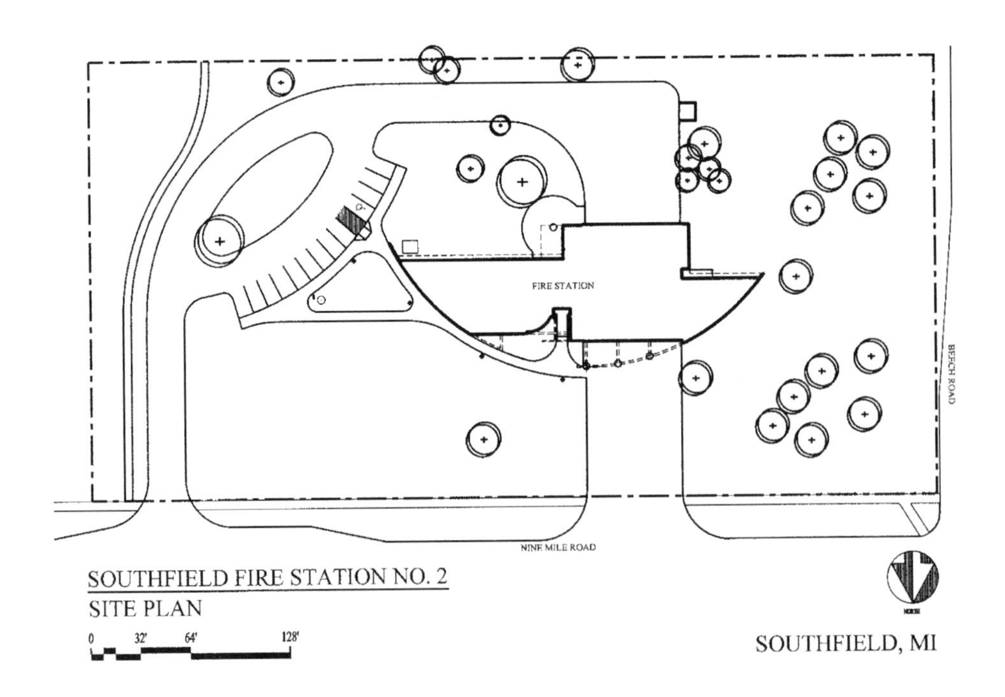

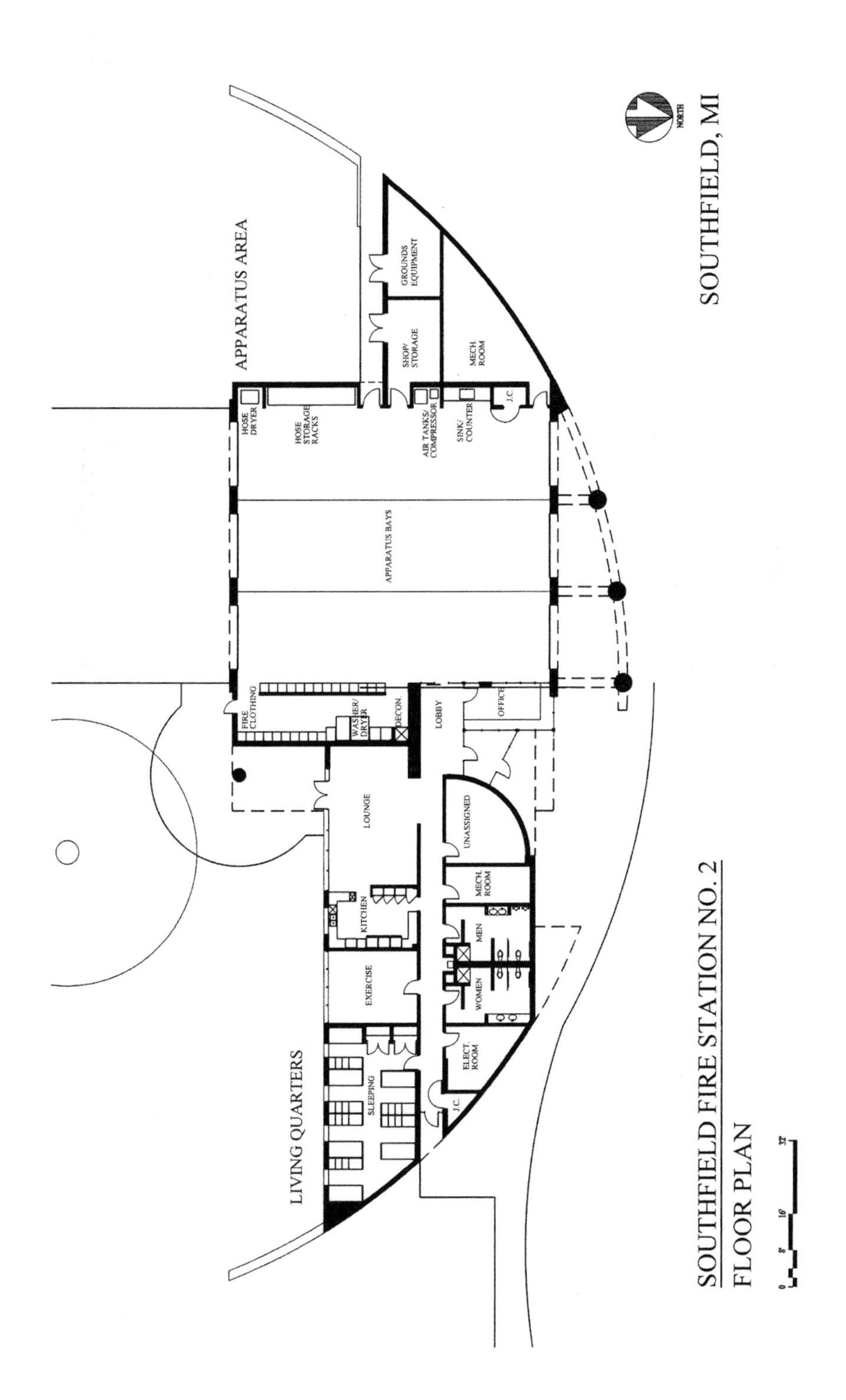
SOUTHFIELD FIRE STATION NO. 2
FLOOR PLAN
SOUTHFIELD, MI
NORTH
APPARATUS AREA
LIVING QUARTERS
GROUNDS EQUIPMENT
SHOP/ STORAGE
MECH. ROOM
J.C.
HOSE DRYER
HOSE STORAGE RACKS
AIR TANKS/ COMPRESSOR
SINK/ COUNTER
APPARATUS BAYS
FIRE CLOTHING
WASHER/ DRYER
DECON
LOBBY
OFFICE
LOUNGE
UNASSIGNED
MECH. ROOM
KITCHEN
MEN
EXERCISE
WOMEN
ELECT. ROOM
SLEEPING
J.C.

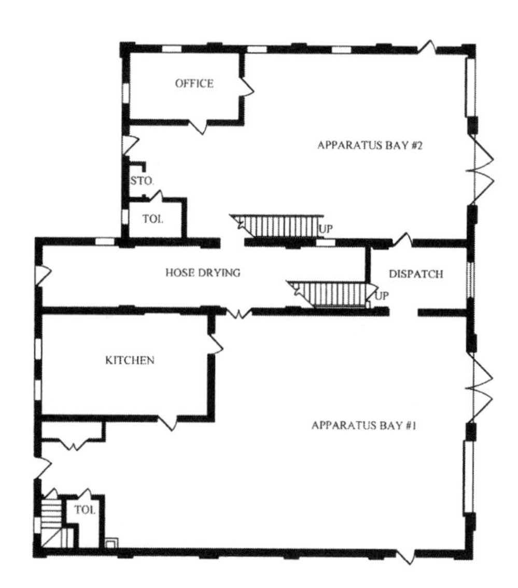

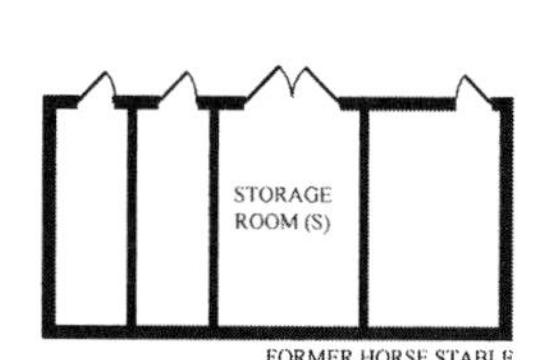

FIRST FLOOR PLAN (BEFORE)

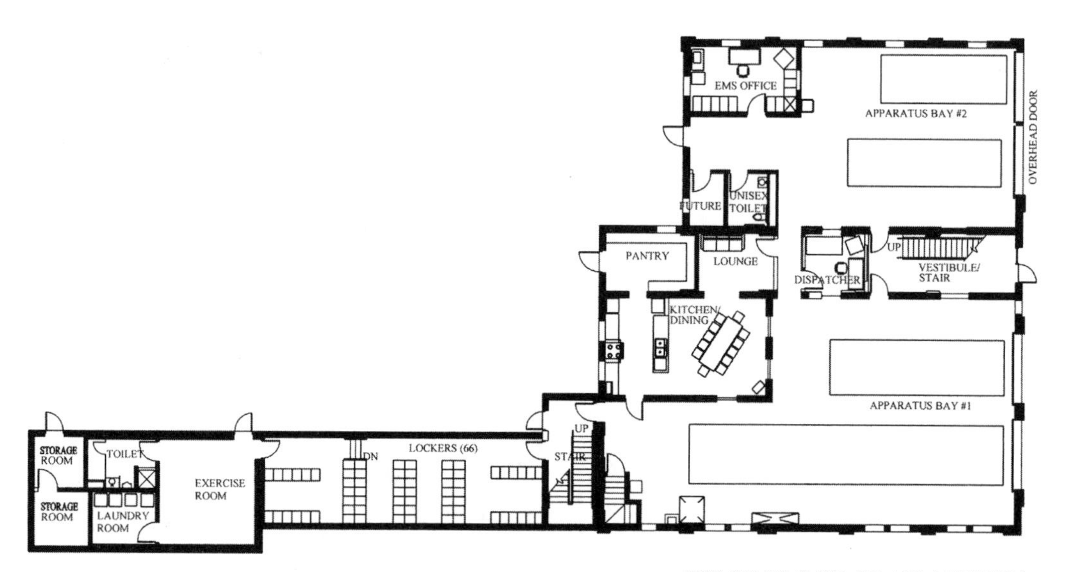

FIRST FLOOR PLAN (AFTER)

DETROIT ENGINE CO. NO. 5
RENOVATION AND ADDITION

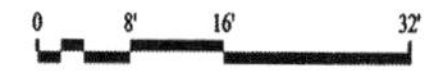

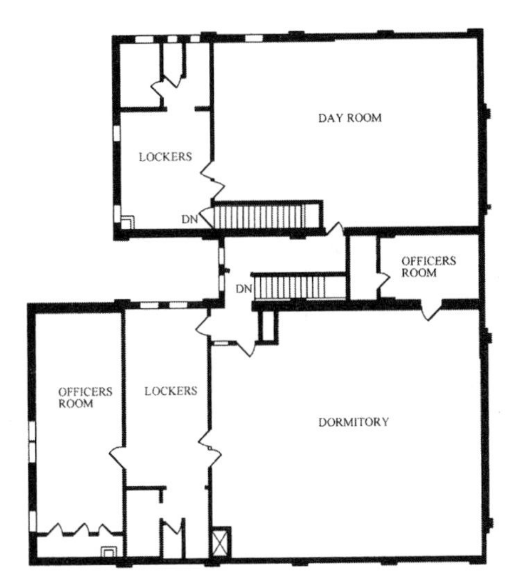

SECOND FLOOR PLAN (BEFORE)

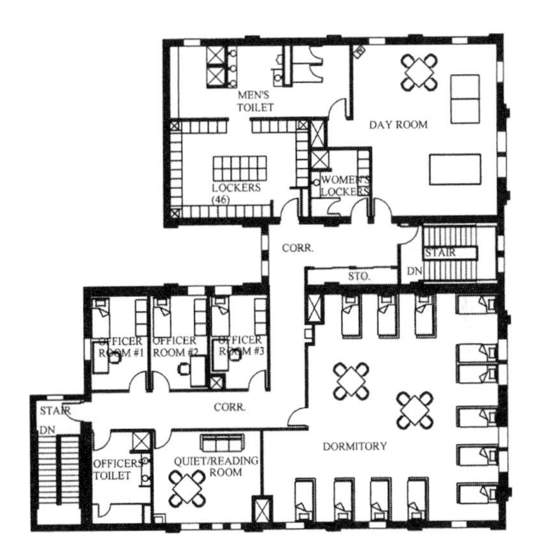

SECOND FLOOR PLAN (AFTER)

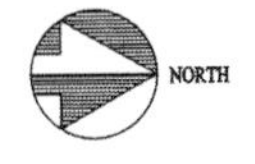

DETROIT ENGINE CO. NO. 5
RENOVATION AND ADDITION

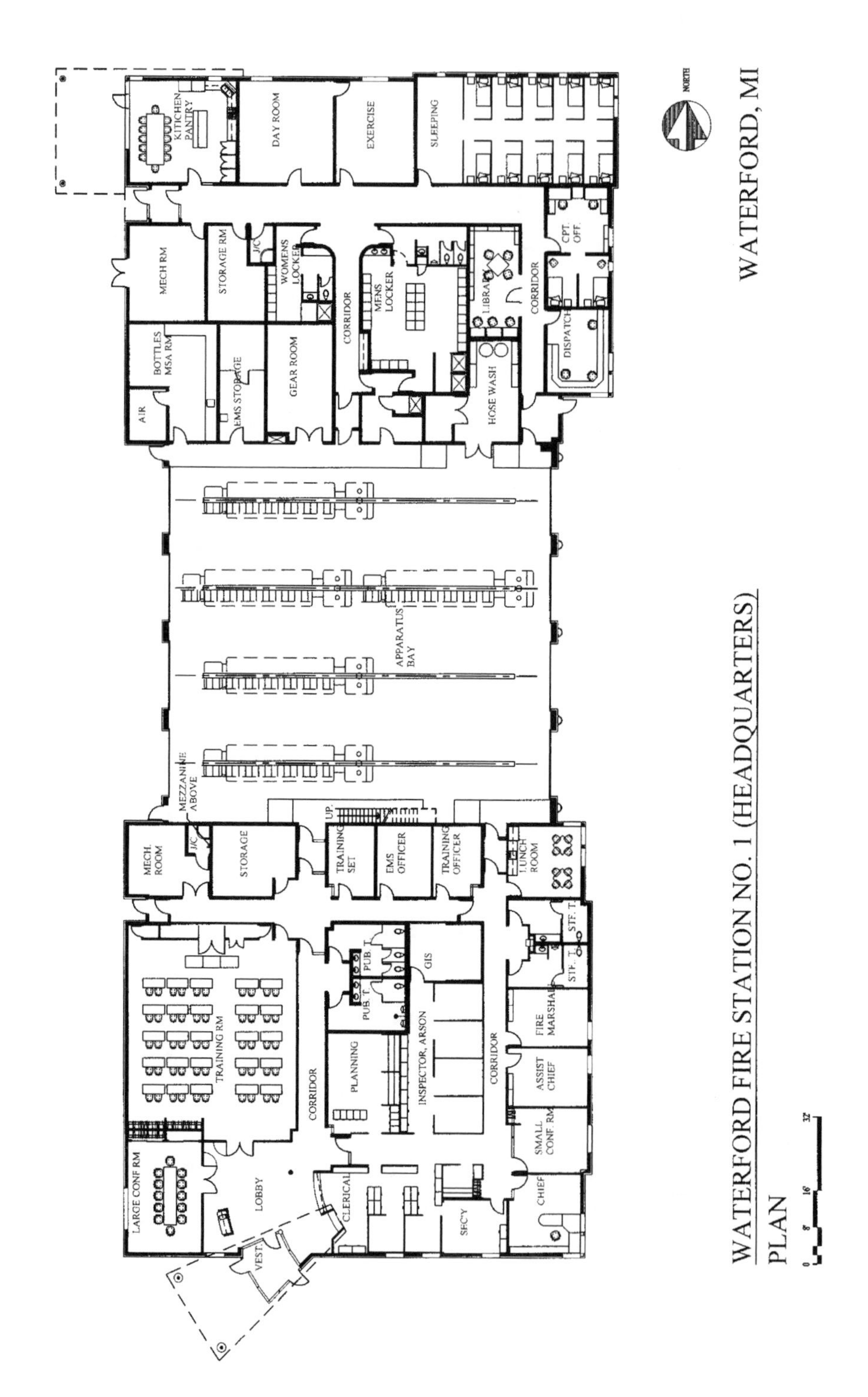

WATERFORD FIRE STATION NO. 1 (HEADQUARTERS)
PLAN
WATERFORD, MI
NORTH
KITCHEN PANTRY
DAY ROOM
EXERCISE
SLEEPING
MECH RM
STORAGE RM
WOMENS LOCKER
CORRIDOR
MENS LOCKER
LIBRARY
CPT. OFF.
DISPATCH
BOTTLES MSA RM
AIR
EMS STORAGE
GEAR ROOM
HOSE WASH
APPARATUS BAY
MEZZANINE ABOVE
MECH. ROOM
STORAGE
TRAINING SET
EMS OFFICER
TRAINING OFFICER
LUNCH ROOM
TRAINING RM
PUB. T.
GIS
PLANNING
INSPECTOR, ARSON
FIRE MARSHAL
ASSIST CHIEF
SMALL CONF. RM
LARGE CONF RM
LOBBY
CLERICAL
SEC'Y
CHIEF
VEST.

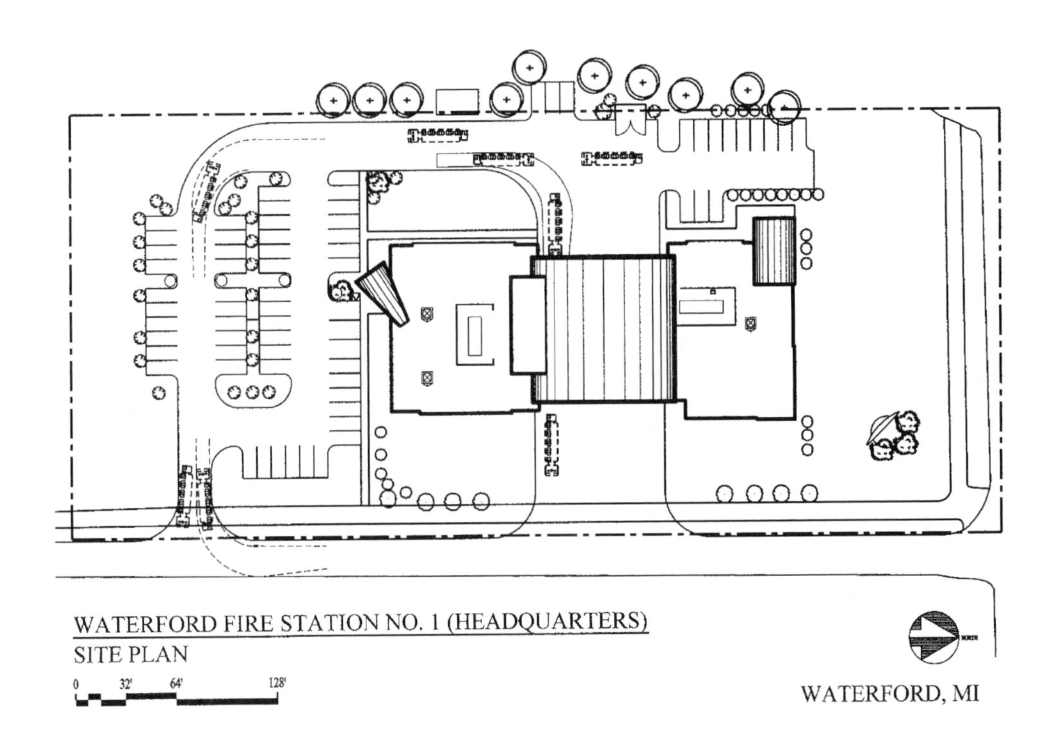

WATERFORD FIRE STATION NO. 1 (HEADQUARTERS)
SITE PLAN

WATERFORD, MI

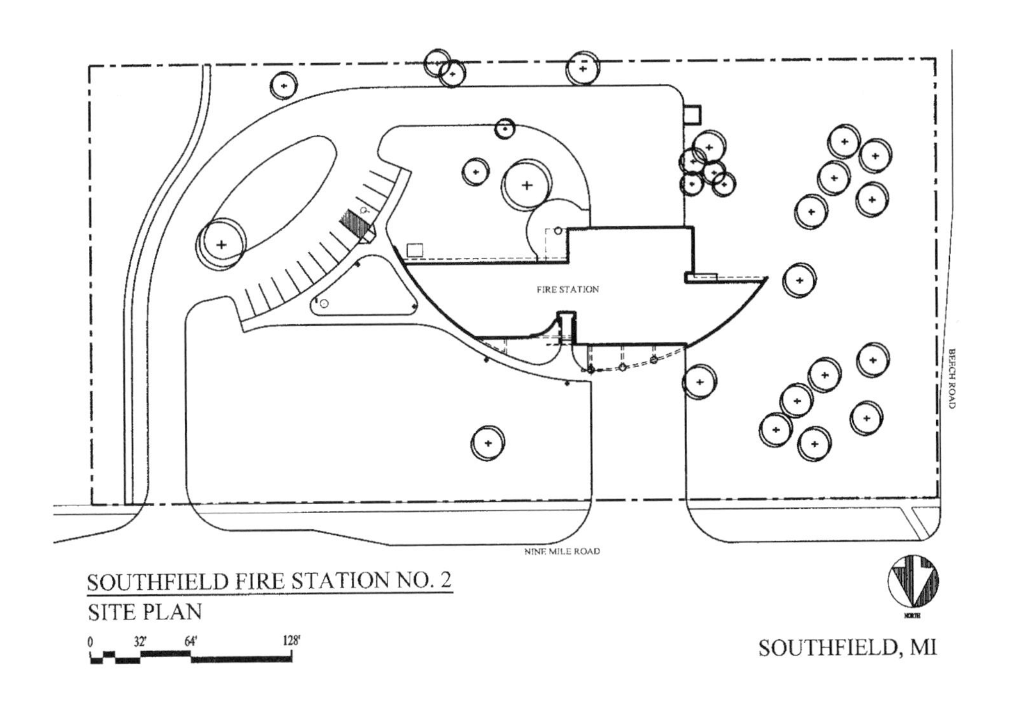

SOUTHFIELD FIRE STATION NO. 2
SITE PLAN

SOUTHFIELD, MI

Photo Courtesy Luke & Kaye Architects

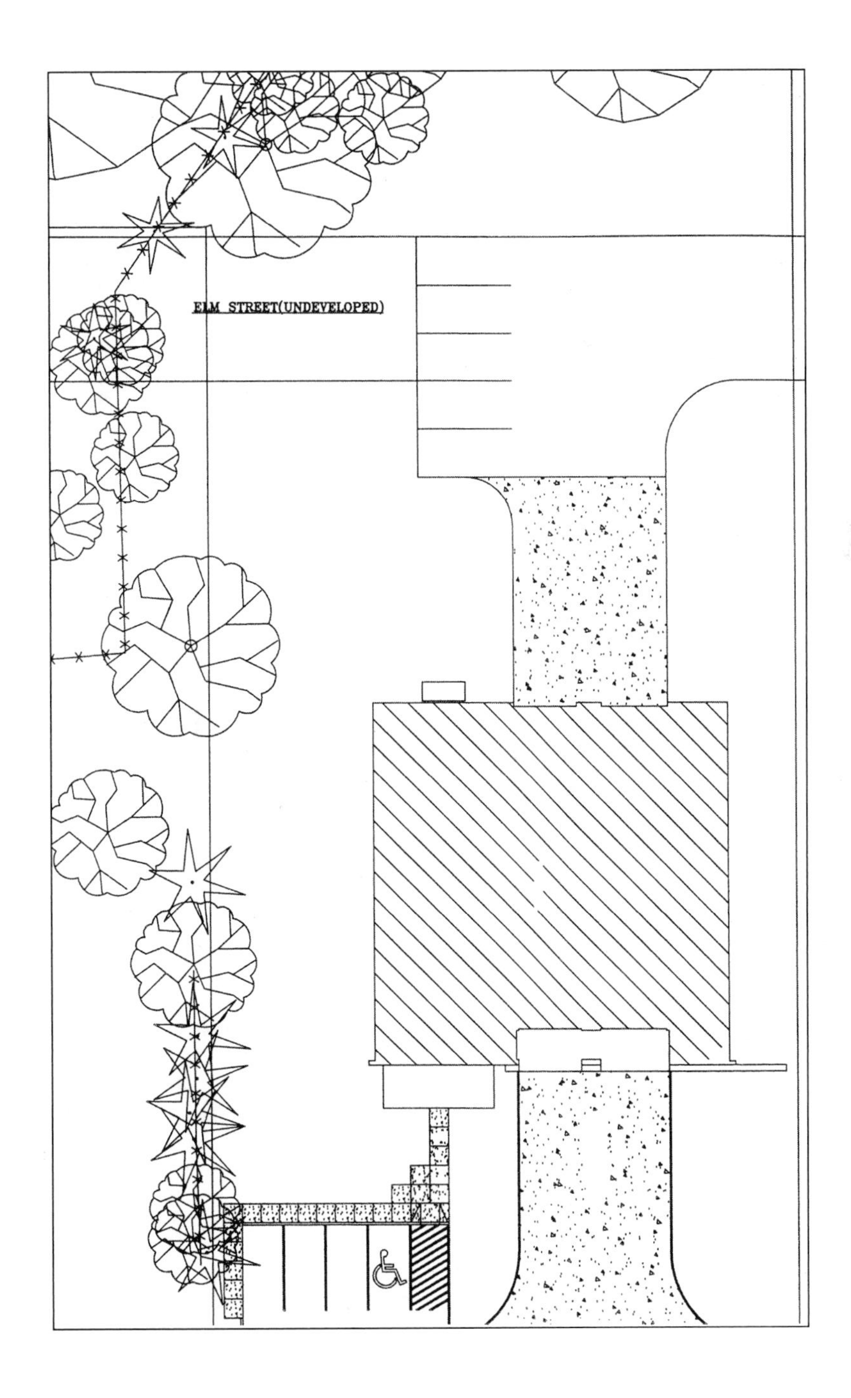
ELM STREET(UNDEVELOPED)

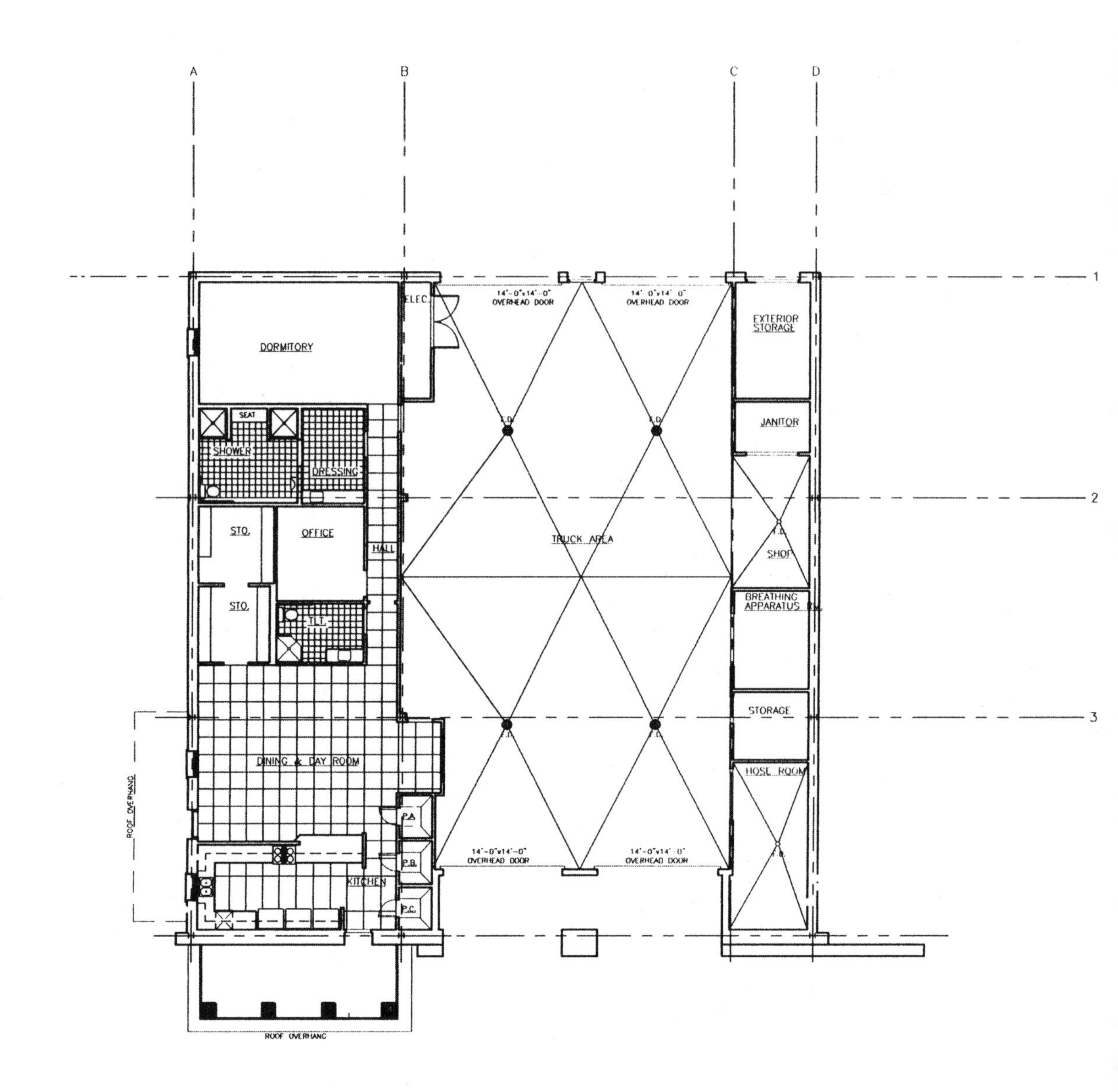
A
B
C
D
1
2
3
ELEC.
14'-0"x14'-0" OVERHEAD DOOR
14'-0"x14'-0" OVERHEAD DOOR
EXTERIOR STORAGE
DORMITORY
SEAT
SHOWER
DRESSING
JANITOR
STO.
OFFICE
HALL
TRUCK AREA
SHOP
STO.
T.T.
BREATHING APPARATUS
STORAGE
DINING & DAY ROOM
HOSE ROOM
ROOF OVERHANG
P.A.
P.B.
KITCHEN
P.C.
ROOF OVERHANG

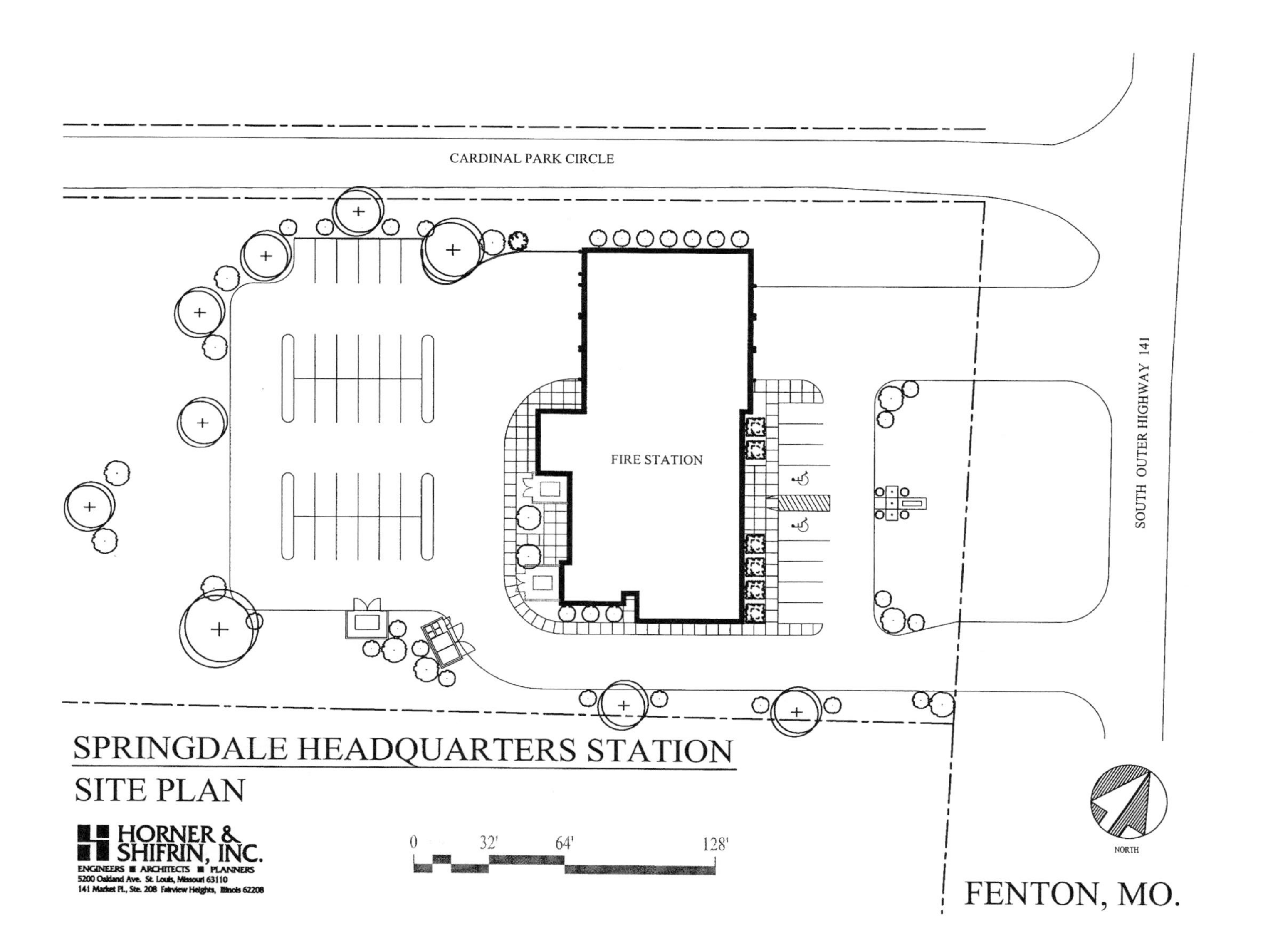
CARDINAL PARK CIRCLE
FIRE STATION
SOUTH OUTER HIGHWAY 141
SPRINGDALE HEADQUARTERS STATION
SITE PLAN
HORNER & SHIFRIN, INC.
ENGINEERS ■ ARCHITECTS ■ PLANNERS
5200 Oakland Ave. St. Louis, Missouri 63110
141 Market Pl., Ste. 208 Fairview Heights, Illinois 62208
0
32'
64'
128'
NORTH
FENTON, MO.

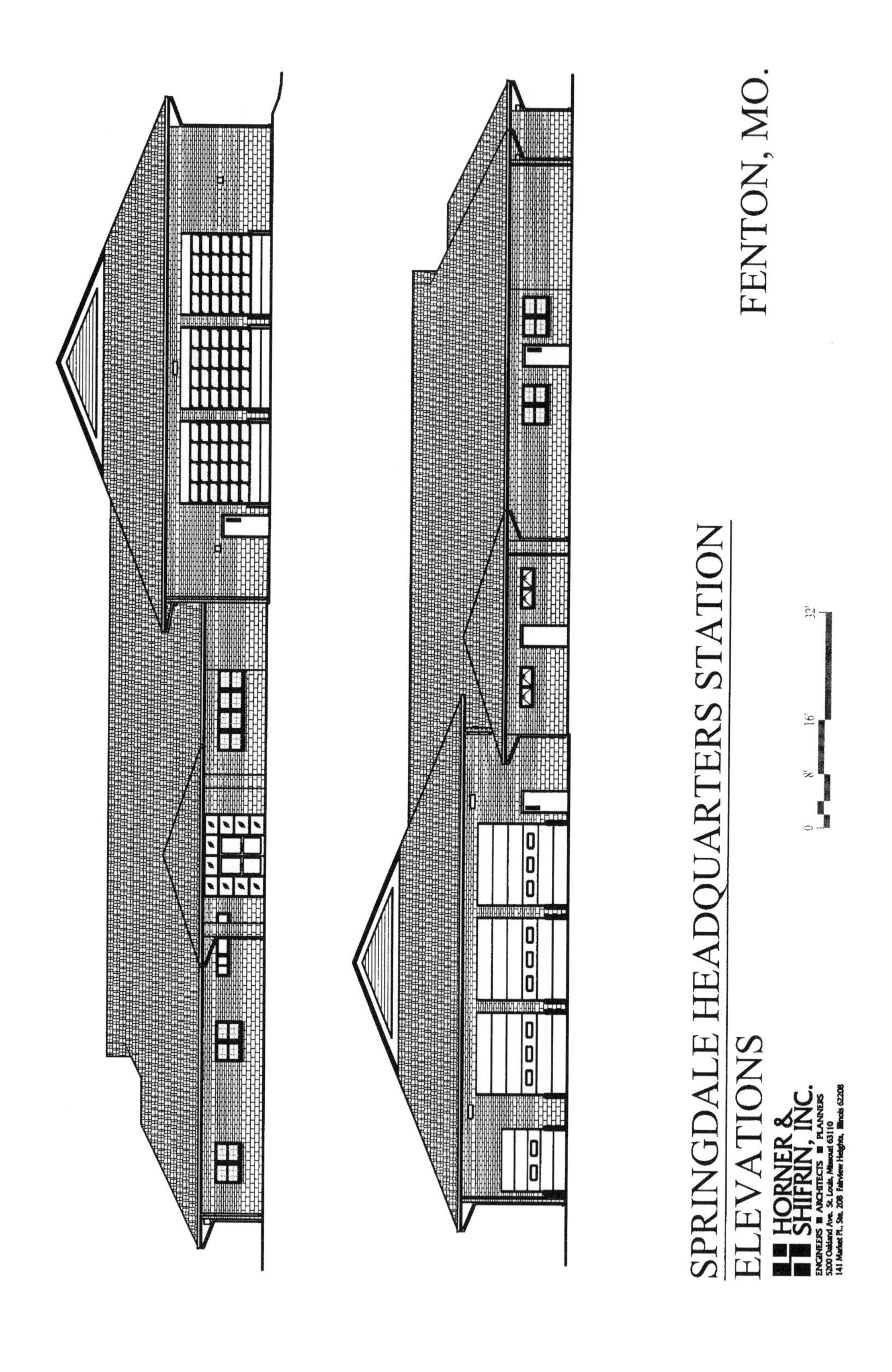
SPRINGDALE HEADQUARTERS STATION
FENTON, MO.
ELEVATIONS
0 8' 16' 32'
HORNER & SHIFRIN, INC.
ENGINEERS ■ ARCHITECTS ■ PLANNERS
5200 Oakland Ave. St. Louis, Missouri 63110
141 Market Pl., Ste. 208 Fairview Heights, Illinois 62208

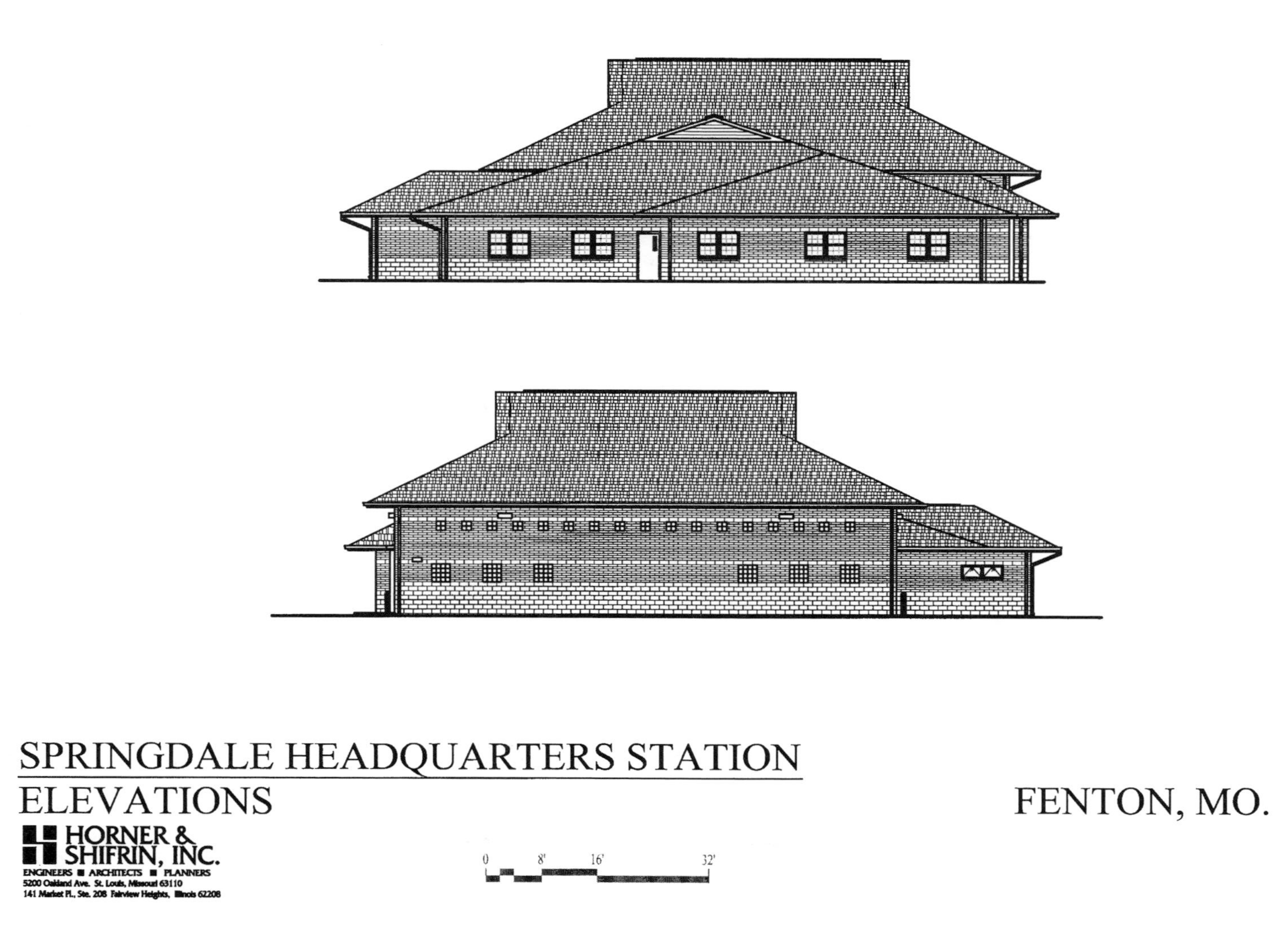
SPRINGDALE HEADQUARTERS STATION
ELEVATIONS
FENTON, MO.
HORNER & SHIFRIN, INC.
ENGINEERS ■ ARCHITECTS ■ PLANNERS
5200 Oakland Ave. St. Louis, Missouri 63110
141 Market Pl., Ste. 208 Fairview Heights, Illinois 62208
0
8'
16'
32'

Section IV

OPERATIONS

14

FIRE DEPARTMENT COMMUNICATIONS

Barry Furey

CHAPTER HIGHLIGHTS

- Common rules and regulations
- Computer Aided Dispatch and Computer-Telephone Integration
- Call taking and dispatching protocols
- Radio system terminology and configuration
- Recruiting, staffing, and training
- Security and safety issues

INTRODUCTION

Perhaps no subject is utilized more but less understood less than communication. After all, communication is utilized throughout the entire realm of fire service activities, and on every run. It begins with the initial report and receipt of the alarm, and ends with units signing back in service at quarters. In between, communications plays a role in a myriad of activities. Dispatch, response, size-up, and coordination between fireground teams are all dependent upon accurate and reliable communications. Without communications, the ladder company would not know that the engine crew has a charged line and is waiting for them to open up. Nor would the interior attack team be aware that the safety officer has spotted some cracks in an exterior wall and is ordering everyone out of the building.

The addition of the first responder role to many fire departments has also added the requirement to communicate patient information such as vital signs and telemetry. It has also led to many agencies requiring that dispatchers receive formal certification in EMD (Emergency Medical Dispatching) or carry EMT or paramedic status in order to properly handle these calls. While transmissions concerning fire operations may often deal with property such as exposure information and standpipe location, Emergency Medical Service (EMS) messages primarily involve people. That is not to say that life is not the first priority for both, but rather to note that there are slight differences in the way each must be handled. As the challenges faced by the fire service drastically change, communications also plays a vital role in the summoning of additional assistance needed to address them. Where once the dispatch of an extra alarm or mutual aid assignment was the most interoperability required, today's emergency communications plan must address the linking of a variety of agencies on the local, state, and federal level, some of which may not normally be considered as being part of the public safety family. The harsh reality of the 21st century is that the fire service—and especially fire service communications systems—must be prepared to deal with the myriad of conventional assignments as well as acts of deliberate terror.

Although communications is ever present, many fire service personnel only have a cursory knowledge of system design and operation. After all, the workings of a pump seem fairly straightforward and simple. The water enters the suction side and exits the discharge side. Many classrooms even have cut-away models or other visual aids to make this process even more apparent. It is a different thing, however, to understand how a 9-1-1 system, for example, can deliver data about the caller as the phone is being answered. And, even if the water source is several hundred feet away, it's easy to trace the supply line from the pond or hydrant to the pumper. It is quite another to follow the path of invisible electrons as they carry voice and data from the dispatch center to the incident scene several miles away. It can even be more baffling to know that this journey may not have been a direct one, but instead, a slightly circuitous route that utilized an array of hardware and software to complete.

Added to this mystique is the fact that communication is probably the least practiced of all subjects. During any rookie school, considerable time is spent raising ladders, stretching lines, and becoming proficient in SCBA. Tasks needed to function in the real world are learned through theory and practice—through textbooks and trial and error. However, perhaps no aspect of the fire service has changed more significantly or more rapidly than communications. Granted, there have been vast improvements in apparatus, turnout gear, and many other tools, but a careful look at equipment of a century and a half ago reveals that helmets, and axes, for example, have changed more in the materials used in their manufacture than in overall design. And, although we have gone from real to mechanical horsepower, pumps and ladders continue to utilize much of the same technology.

Much of what we take for granted in the way of fire service communications has been implemented during the past 40 years. During the 1960s, the home alerting receiver was introduced. While this still restricted the movement of the on-call individual, it did, for the first time, provide valuable and immediate voice instructions regarding the nature and location of the call. The same decade also saw the creation of the 9-1-1 nationwide emergency number, although deployment has been historically left up to local and state resources. While the majority of the U.S. population is protected, even at the start of the 21st century some communities were still contemplating implementation. One side effect of having a universal emergency number was the move toward the consolidated dispatch center. Here various arms of public safety as well as multiple municipalities could pool their resources in a centralized facility. This trend began in earnest during the 1980s, and continues today. Also in the 1980s came the advent of trunking technology, which utilized computers to effectively manage radio channel assignments. The 1980s, too, saw the introduction of the cell phone. Despite the tremendous impact already being felt by cellular devices, an even greater effect can be expected in the future. Telematics, based upon the delivery of data through wireless telephony is already routinely used to deliver real-time accident data, including the number of occupants involved.

The use of data also implies the use of computers, and the silicon chip has reshaped fire communications like every aspect of our everyday life. The CAD (Computer Aided Dispatch) system is a mainstay of modern departments. Replacing the flip charts, maps, and manual methods that had been used for years, CAD can provide instantaneous information on a variety of subjects. When this data needs to be passed to units in the field, human intervention is no longer required. The use of MDTs—Mobile Data Terminals—allows responders direct access to on-line databases through a keyboard and monitor in the vehicle. And, as computers become smaller, so-called PDAs (Personal Digital Assistants) can be used to secure the same information.

While inventions may not have come as far as science fiction writers may have guessed, technology such as Automatic Vehicle Location actually relies on satellites to accurately track apparatus, and satellites are also used to relay telephone conversations from previously inaccessible locations. The 1990s ushered in the era of the Internet, and many agencies now routinely use this tool as a means of public education, recruiting, and even disaster management. By using standard web browsers, incident commanders can utilize the World Wide Web to access resource data. To many in the fire service, the term "on line" has become as familiar as the term "hose line."

Where fire service communications goes from here is limited only by the imagination. While many of the improvements may come as a result of spin-offs from other commercial development, there is no reason to doubt that things once considered impossible will become commonplace. The electronic revolution remains in full swing, and it is highly likely that every firefighter will benefit from several significant upgrades in the area of communications during his or her career.

Call Handling and Dispatching

Types of service

The first, and often the most critical, aspect of emergency communications is the receipt of the initial call. The management of an incident prior to the delivery to first responders can be broken down into two phases, call handling and dispatch. This can be accomplished by a variety of methods, and although some specific requirements may be established by state legislation, is largely a matter of local choice.

The *Public Safety Answering Point* or PSAP is the place at which an emergency call is received. PSAPs may serve a joint function, in that the incident is both received and dispatched at the same facility. This is referred to as the *direct dispatch* method. However, some centers are designed to handle the telephone call only, with the actual assignment of apparatus occurring elsewhere. In such circumstances, a mechanism is required to assure that this dispatch occurs in a reasonable period of time. Using the *call relay* system, PSAP personnel obtain all the necessary information, then advise the fire department by phone or radio. Another possible option allows for the entry of alarm information into a CAD (Computer Aided Dispatch) system, which automatically sends and displays this data to the remotely located fire dispatcher. If the *call transfer* method is preferred, call takers immediately transfer all fire calls to the fire department. This occurs as soon as the nature of the emergency has been identified, and is typically done by dedicated telephone line. Address and number information can also de transferred along with voice if enhanced 9-1-1 service is utilized.

The simplest form of incident processing is known as *single stage dispatching.* Here the individual answering the telephone call is also responsible for making the radio assignment. This works well in smaller communities with lower call volumes, as it allows continuity of the call and is cost effective. However, in larger, high-volume environments, *two stage* or *dual stage* dispatching is preferred. Specialized call takers and dispatchers each play a part on every assignment. Call takers receive the initial information, then relay it to dispatchers by computer, card, or form. Dispatchers check for alarm assignments and alert the corresponding companies. While this does lessen the continuity of involvement gained when only a single individual handles the call, it has several significant advantages. The division of duties allows the call taker to remain on the line to provide instructions or to receive additional information while the dispatcher begins alerting first due units. This helps to speed response and improves the quality of information obtained. Also, since dispatchers must be constantly alert to on-the-air emergencies, the segregation of work responsibilities eliminates competition between ringing phones and radio traffic by allowing dispatchers to focus their attention on monitoring the assigned frequencies.

Where call taking and dispatching occur on the same premises, the facility is generally described as being either a *consolidated* or a *co-located* communications center. In truly consolidated centers, the entire operation, including staffing and supervision, is under a single common point of control. Co-located facilities consist of a number of agencies, each having responsibility for their own personnel, occupying the same physical plant.

Receiving alarms

Alarms may be received by a variety of methods and from a selection sources. The telephone is the most common means of reporting emergencies, with 9-1-1 being designated as the universal number in the United States. Agencies may also maintain published seven-digit telephone numbers. These may replace 9-1-1 as a means of contact where 9-1-1 is not in service, or be designed primarily for administrative matters. Regardless of the configuration, departments should anticipate the potential of occasionally receiving calls on non-emergency lines that require emergency dispatch. This holds true for 3-1-1 as well. Communications centers may also support dedicated telephone circuits or *ring down lines* that are directly connected to other facilities. No dialing is required. There are many applications for this type of service. One is to receive alarms from special facilities such as airport control towers. Another is to provide dispatchers with direct access to support services such as the local utility company.

Response may also be triggered by automated systems as well. These can take the form of municipally operated systems or private monitoring concerns. Some centers maintain on site registers to receive alarms while others are notified via telephone by alarm company *central stations*. Here, private concerns maintain watch over their subscribers, notifying appropriate jurisdictions when an actual alarm is received. The decision as how to address automatic alarms can be a matter of much discussion. Certain localities have taken to charging a fee for allowing alarm companies to connect receiving equipment in fire communications facilities. This provides a source of income and direct control over the entire alarm process. It also saves time, especially since many firms provide protection nationwide, complicating notification. Those favoring total privatization argue that monitoring alarms is both labor intensive and a potential source of liability. None of this applies to municipal alarm systems that serve street boxes and selected occupancies. However, it should be noted that the number of communities supporting the once familiar red pedestals has waned with the advent of other means of quickly reporting a fire, such as cellular phones. Community and state standards often mandate that certain classes of occupancy be equipped with automatic detection and notification systems, and specify the method of connection. Other regulations may also apply, such as prohibiting alarms from using automatic telephone dialers to notify the fire department. Permits for installers may also be required.

Caller interrogation

Immediately after answering the telephone, the call taker begins the most important part of incident processing—caller interrogation and information gathering. Lacking the proper address and specifics concerning the emergency prevents the dispatch of appropriate help. That is why these first few seconds are so important. There is an old anecdote concerning the car fire—that turns out to be in the garage—a garage that turns out to be in the basement—of what turns out to be an occupied home. While this is a flight of fancy, it serves to point out how every piece of information is important.

However, none is more important than a good address.

- The location of the call determines what stations will respond, and in multi-jurisdictional facilities, what department.
- The address may also indicate a special hazard occupancy, or receive special handling because it is located in a high value district.
- Extreme care must be taken where similar street names exist in a community to verify that units are being assigned to the proper address.
- Under no circumstances should the addresses automatically provided by E 9-1-1 be assumed to be correct. Callers may be reporting an incident at another location, or there may be errors in the database. There have been several high-profile cases caused by failure to verify addresses, none of which turned out favorably for the municipality involved.

The nature of the emergency must also be clearly defined.

- What's on fire?
- Where is it located in relationship to the caller?
- Is anyone in danger?
- Are hazardous materials involved? Even the type of vehicle involved is important. Obviously, there is quite a difference between a family sedan and gasoline tanker.

Dispatch centers may wish to establish their own set of guidelines regarding questions to be asked and advice or instructions to be given for a variety of common event types. Where an agency utilizes Emergency Medical Dispatch (EMD), however, a recognized authority should establish these protocols.

The receipt of alarms is a delicate balance between obtaining sufficient information to ensure an appropriate dispatch and protecting the safety of a caller. The call taker should be constantly vigilant for indicators that the reporting party may be in danger. Where escape is practical, many agencies directly advise the public to leave the affected area and call from a safer location. Where the victim is trapped, telephone contact should be maintained as long as possible, with updates being given to both the caller and rescue crews.

Methods of Alarm Receipt

9-1-1

In 1968 the Federal Government recognized that a universal emergency number would be instrumental in assuring timely access to public safety. Previously, there could be dozens of telephone numbers in service in a single county. It was not uncommon for the police, fire, and ambulance services in the same city to maintain independent seven digit listings. The advent of 9-1-1 provided an easily remembered and easy-to-dial number, which could be accessed for free, even from coin operated telephones. Additional features such as *line capture*, *forced disconnect*, and *automatic ringback* gave dispatchers the tools to seize a call to institute a trace, free up an emergency call, or automatically dial back a caller in case of hang up or accidental disconnect.

More important capabilities were included in the transition to *enhanced 9-1-1*. E 9-1-1, as it is commonly called, utilizes Automatic Number Identification (ANI) and Automatic Location Identification (ALI) to present telecommunicators with the calling party number and address. Unlike conventional caller ID, ANI/ALI cannot be blocked by the caller. E 9-1-1 also utilizes a function called *selective routing* that directs emergency calls to the proper PSAP. A database known as Master Street Address Guide , or MSAG, is created to correlate telephone number and address data with the appropriate public safety response. This is required because conventional telephone exchange boundaries do not correspond to government jurisdictional borders. From its humble beginnings in Haleyville, Alabama, 9-1-1 came to be accepted and acknowledged as *the* method of obtaining help. In 1999 Congress passed legislation that officially recognized that fact. However, few citizens needed an “act of Congress” to inform them. The National Emergency Number Association, in its 2001 Report Card to the Nation estimated that both the majority of the nation’s population and landmass were protected, and that 190,000,000 calls are made to 9-1-1 each year. It was further estimated that about one-fourth of these calls were made from wireless telephones, although many agencies report considerably higher percentages.

Wireless E 9-1-1

Providing enhanced wireless 9-1-1 has proven to be a challenge to technology. Because of this, the Federal Communications Commission promulgated rules that established deadlines and accuracy requirements.

- *Phase I* wireless E 9-1-1 provides the PSAP with a callback number.
- In addition, *Phase II* provides the approximate location of the call.
- Wireless location is provided by a *handset*, *network*, or *hybrid solution.*

- Handset solutions use a Global Positioning System chip in the telephone itself.
- Network solutions use a form of direction finding enabled in the cell tower.
- Hybrid solutions rely upon both.

At the time of publication, a few communities had instituted enhanced wireless 9-1-1, but it was by no means widespread. Two additional problems associated with cell phones are legacy telephones and automatic dialing. So-called legacy phones are devices handed down to family members or donated to charities for the sole purpose of calling 9-1-1. Unfortunately, most of these telephones do not completely support enhanced service, and cannot be called back by the dispatcher. This can be critical if the caller is disconnected or more information is needed. In technical terms, the FCC labels these units as "non subscriber activated handsets." The second issue—that of misdialed calls—is caused by the programming of a speed dial key on wireless telephones. Activating this key automatically dials 9-1-1. Unfortunately, this often occurs when the telephone has been placed in a pocket or purse. Many consumers do not even know that this feature is present until they receive a callback from the 9-1-1 center inquiring as to their welfare.

N-1-1 numbers

While 9-1-1 is a number familiar to the public and public safety alike, during the late 1990s, the use of other "1-1" numbers began to flourish. Some, like 4-1-1 for telephone company information had existed for decades. Others, such as 6-1-1 were commonly used, especially by wireless carriers, for repair service. The number 5-1-1 has been officially assigned to traffic and transportation, while 7-1-1 is designated as a nationwide relay service for the deaf. Social service agencies laid claim to 2-1-1, and in 1997 the FCC, in an effort to divert traffic from overburdened 9-1-1, created 3-1-1 for non-emergency requests for law-enforcement. The city of Dallas, Texas, has established 3-1-1 as a general request line for city services, with call takers for both this service and 9-1-1 under the auspices of the fire department. This is a departure from the norm where the police or an independent agency typically employs the answering personnel.

Telematics

Telematics is the name given to devices that automatically report incidents based upon monitored conditions. The most commonly encountered is *ACN* or *automatic crash notification.* For a fee, a motorist subscribes to a service that can provide a variety of functions such as roadside service, trip routing, and emergency reporting. When a signal indicating trouble, such as airbag deployment or the activation of an emergency button is received, this central station queries the wireless telephone of the subscriber, then notifies the authorities of the condition and location based upon coordinates provided by satellite.

There is almost an unlimited supply of information that can eventually be monitored such as vehicle speed, number of occupants, and whether or not there has been a rollover or ejection. Future upgrades might even relay real time video from an accident scene directly to responding units. While still in its infancy, telematics is a rapidly growing means of reporting emergencies, and available as an option on a variety of new automobiles. However, conventional telephone calls instituted by a human being are still the primary means of requesting help.

Dispatching

Regardless of the means by which a dispatcher receives the call, they are under obligation to assign the appropriate resources as quickly and as accurately as possible. This means utilizing the available resources to verify the proper assignment for the address, building type, and nature of the call. Time of day may also be a factor. Assignments may range from a single company, to multiple departments. Inter-jurisdictional responses are often referred to as *mutual aid.* No matter the number of units needed, telecommunicators must have the ability to deploy all resources under their control, and to account for the location and availability of these resources.

Notification can be accomplished in a number of ways. In-service apparatus may be dispatched by voice over radio, or by MDT. Dedicated phones may be installed in all stations to receive alarms, or a remote printer, connected to the CAD system may be installed on the apparatus floor. *Selective signaling* is a means whereby companies, departments, teams, or individuals may be dispatched without disturbing others. Tone alert receivers can be assigned to all fire halls with each having a unique set of frequencies. In career departments, "toning out" a station may open the public address system, turn on the bunkroom lights, and even open the bay door. In rural areas such actions may serve to retransmit information, or activate an audible device such as a siren or air horn. A common form of selective signaling is paging. *Paging* may be tone only, tone and voice, alphanumeric, and two-way. Although they may offer a range of useful features, and provide a large coverage area, commercial paging services are often not a good choice for emergency dispatching due to potential delays in activation. Agencies may therefore opt to maintain their own paging devices, although paging is not compatible will all frequencies and radio systems.

The initial dispatch is commonly called the *first alarm,* with additional pre-defined assignments increasing numerically. Incident commanders may opt to augment these responses with individual units, which are termed to be *special calls.* Career departments tend to dispatch by apparatus or companies, whereas volunteer organizations often assign calls to departments or stations. There are, of course, exceptions, and telecommunicators working in consolidated centers must be familiar with the methods used in all communities served.

The dispatcher must assure that the correct information is received and acknowledged by *all* units involved. In voice systems, this means that the location and call type should be announced at least twice. In all systems it means that verification must be made to assure that all assigned resources are en route to the proper location. Information relayed during the initial broadcast should include:

- Exact address, including apartment number, or building number if the location is part of a larger complex
- Business or commonplace name, if applicable
- Companies responding
- Classification of call, such as structure fire or chest pains
- Victim or patient information and location
- Known hazards
- Changes to normal response, such as substituted companies, road closures, etc.
- Method of receipt

The latter can be critical, even within the same classification of alarm. This should be specified on automatic fire alarms, as well, with companies being advised as to whether the call is for a water flow, detector, or manual pull station activation. Some departments may choose to include map reference numbers, water supply locations, or provide directions from the station to the emergency. The time and call sign should also be broadcast, as required.

Once the first alarm assignment is underway, the dispatcher must be alert to updates from callers or from field units that dictate additional actions. Any information that might impact response should be relayed immediately. Status changes such as arrival, first water, etc., should be logged, and the frequency carefully monitored for additional requests or distress calls. It is not uncommon for dispatchers to be responsible for multiple concurrent incidents on more than one channel. It is critical that the location of all companies be known at all times, from the initial notification until return to quarters.

Record keeping

Another critical aspect of fire department communications is record keeping. Although dispatch should not be thought of as a records department, it is the first step in collecting data that will generate all incident reports. Accurate logging of information is crucial to efficient operations, and plays a key role in firefighter safety. Record keeping can be divided into two areas, run information and references.

Data gathered during the initial report must be accurately maintained and a permanent log initiated. The minimum information captured should be:

- Address, including apartment or office number
- Exact nature of the emergency
- Call source (9-1-1, automatic alarm, by radio from field)
- Time and date of receipt
- Person receiving the call

Caller's name and telephone number, premise hazards, potential injuries or victims and specifics regarding the incident should also be tracked. This may be done manually using forms or cards, or electronically through a CAD. Another form of incident record is maintained in the form of an audio log. Master loggers record all information from telephone and radio channels, and save it on a variety of media. Reel-to-reel tape has given way to digital audiotape, VHS tape, CD, and DVD. Instant playback devices have also migrated from traditional tape to digital memory. Rather than providing permanent accounts, they are designed to give dispatchers the ability to recall selected telephone calls or radio traffic. Audio messages associated with emergencies can be reviewed word by word at the time they happen, or be stored indefinitely in electronic form.

Reference information can be defined as any data required to assist in the dispatch and support of response units.

- Some references consist of standard publications such as the local telephone directory and hazardous materials guides.
- Cross-reference or city directories are also helpful in finding specific addresses.
- Up-to-date lists should be maintained of all commonly called telephone numbers such as stations and medical facilities, with critical listings entered into speed dial devices.
- In addition to those resources used on a daily basis, it is important to maintain current information on what to do and who to call during disasters. Included here might be the home telephone numbers of key officials, federal agencies, and sources of specialized equipment.

It is imperative that all reference aids are checked for accuracy on a regular and reoccurring schedule.

Common Equipment and Communications Systems

CAD systems

Computers are a part of our everyday life, and the dispatch center is no exception. CAD systems serve as an effective means of entering and tracking calls for service. Early systems used mainframe computers and dumb terminals. A single machine did all of the computing. Recent trends have turned toward distributed processing, meaning that a certain amount of work is accomplished at each workstation. This assists in providing redundancy, or emergency backup. Some systems take this to another level known as fault tolerance, allowing for self-correction of the primary device rather than switching to a secondary. Whatever means used, including the temporary use of manual forms, the need to have an alternate plan should be clear. However, past this point, much of the hardware used in public safety networks can be found in most modern offices. It is the software that distinguishes how CAD systems function.

Keystones of this software are tables known as *geofiles,* which is short for *geographic files.* These form the basis for address verification that checks the street name and number entered for validity and responsible jurisdiction. Geofiles also support functions such as commonplace name, allowing lists of all schools or the location of all McDonald's restaurants to be recalled. Street name aliases give the ability to enter calls on roadways where segments or block ranges may be known by different or multiple names. Premise information links items such as hazardous materials storage or key holder information to a specific address. Geofiles further affect unit recommendation, or the computerized assignment of response. This is an automated version of the manual run card system that was the staple of fire dispatching for decades. Run cards are still used in many communities today, both as a primary means of dispatch and as a backup for CAD systems.

CAD software also serves to manage resources. Personnel and apparatus location and readiness can be constantly monitored and updated, as can skills and capacities. When event or incident records are created, specific information concerning the alarm is used as a base, and data from both the geographic and resource sides of the system are added and modified. CAD systems may also utilize interfaces that connect them to other computers. RMS (records management systems) may capture information directly from CAD in order to build run reports. This automation can go so far as to generate required state and NFIRS reports. Connection to MDTs (mobile data terminals) enables dispatchers to send messages and alarm information to apparatus, where it can be viewed on a screen. MDTs can also be configured to allow units to update their status, such as "responding," "on scene" and "available," or to access pre-plan diagrams or external databases. The addition of AVL (automatic vehicle location) allow for the tracking of apparatus by satellite using the GPS (Global Positioning System). GPS gained public exposure during the Gulf War, but has since been adopted by many transportation and public safety users. AVL may be tied to intelligent mapping that allows personnel to query on roadway attributes, and to provide accurate response routing based on a number of common factors.

While not primarily designed for information management, most CAD systems contain modules that allow for the generation of commonly required reports. Larger agencies may hire information system staff to develop custom routines, while smaller municipalities may opt to purchase third party software packages to accomplish greater levels of details. Because of the tremendous amount of variables involved in fire service dispatching, programs tend to be complex, and systems designed primarily for law enforcement use may not adequately support many common requirements such as move ups and reassignments. Regardless of the software used, the maintenance of accurate files is critical to the reliability of any CAD system and requires an appropriate commitment of resources.

CAD has also become married to the telephone through the introduction of CTI (Computer Telephony Integration). CTI replaces standard telephone sets with workstations, using the keyboard and screen to perform the answering, dialing, and recording functions.

Types of radios

Just as there are several types of apparatus, so too are there a variety of radios and radio systems. Portable radios are just that—portable. Designed to be held in the hand, they are extremely valuable to officers and fire ground crews as a means of maintaining contact and sharing information. Because of their size, they are typically less powerful than larger units, such as mobiles, that are permanently mounted in vehicles. Portable radios selected for fire service use should meet several criteria. Obviously, they must be rugged to withstand the rigors of search, rescue, suppression, and overhaul operations. It should go without saying that water resistance, as well as the ability to withstand being dropped are key components to meeting this need.

Another consideration is the ability to operate easily under extreme conditions. Knobs and controls should be large, and displays functional. The unit should be able to be turned on and off, channels changed, and transmissions easily made with a gloved hand. Physical size and available accessories should also be taken into account. Portable radios should be able to fit securely in designated pockets in turnout coats, or come equipped with a leather or composite holster. A collar microphone and rapid charge batteries are also useful options. Special adapters, throat or "bone" microphones allow for communications while wearing SCBA. Hazardous materials teams and others who routinely work in combustible atmospheres may wish to consider explosion proof models that are specially designed to reduce the potential for ignition of flammable vapors.

Mobile radios are higher in output, but rely on the electrical system of the apparatus to supply power. Vehicle mounted units should be installed so that operating their controls is not a hazard, and must be properly secured to prevent them from being damaged by normal in-cab activities. Correct mounting will also reduce the potential for the radio coming loose during collision and creating an additional hazard. While the antenna of a portable radio is an integral piece of the singular unit, mobile radio antennae are affixed to a vehicle surface. This mounting should be as high as possible, on a flat surface, free from obstructions in order to maximize transmission and reception. Mobile radios may also consist of two distinct pieces—the transceiver, containing the working portion of the radio, and the control head that provides the ability to adjust the volume and change channels.

It is important that mobile radios have sufficient audio power to allow operators to clearly hear transmissions despite the background distractions caused by engine noise, sirens, and air horns. This can be accomplished through amplified speakers or headsets. Apparatus may require multiple positions from which the radio can be operated. The pump panel, turntable, and tiller are all common choices. This typically results in the installation of additional control heads, or remote speakers and microphones. Weatherproof speakers and speaker enclosures and noise canceling microphones are commonly specified options for these applications. Departments that provide an emergency medical function may desire the ability for radio communications in the patient compartment of ambulances, as well.

Fixed site transmitters are often called base stations. These typically operate on commercial AC power, and critical transmitters should be connected to emergency generator circuits or UPS (Uninterruptible Power Supplies) to assure operation under unusual circumstances. It is also possible to convert some 12-volt mobile radios for fixed location use. These can be backed up with a standard automotive battery.

Radio systems

Some radio communications are carried out in the *simplex* mode. Here, transmission and reception takes place on a single frequency. The benefits of this arrangement are that no intermediary devices are required. Units talk directly to other units without the need for a system infrastructure. Typically, radios used in a simplex system are of relatively higher power than those used elsewhere. However, these benefits can also serve as limitations. Simplex systems may experience dead spots or gaps in coverage due to their reliance on individual radios to intercommunicate. Also, simplex communications in the lower frequency ranges are subject to a natural phenomenon called *skip* whereby the signals of departments hundreds of miles away bounce off the atmosphere and interfere with local operations. This can be minimized, to some degree, by the use of *tone-coded squelch,* a feature that is manufactured under names such as "Private Line" and "Channel Guard." Here, a non-audible tone is sent with each transmission in order to open up the squelch on receiving units. By assigning different tones, departments sharing the same frequency will not hear each other's messages.

Duplex systems do not depend solely on mobile and portable units to communicate, but instead rely on devices known as *repeaters* to relay conversations. Repeaters are generally higher power transmitters mounted on hilltops or in tall buildings, although specialized mobile repeaters do exist. The characteristic of duplex systems is, as the name suggests, the use of two frequencies instead of one. Transmissions are made on an input frequency that is received, boosted, and retransmitted by the repeater on the output portion of the pair. This arrangement is transparent to the user. Duplex systems can provide better coverage under a wider variety of conditions than can simplex; however, there is the drawback of requiring additional equipment to communicate. There may be some slight delays in retransmission; however, these should not be noticeable unless a user begins to talk prematurely. In these cases the first portion of the message may be lost or clipped. Many duplex systems provide for the use of a talk-around channel for limed range communications, or for use in case of repeater failure. This is so named because it operates in the simplex mode by bypassing, or "talking around" the repeater.

Satellite receivers are another commonly encountered piece of hardware. Lacking the capability to transmit, these devices are designed to boost reception by allowing weaker field units to gain access to the network. Satellite receivers improve communications in areas where system transmissions can be heard by mobile and portable radios, but terrain or building construction make it impossible for their signals to reach out. Where multiple receivers are used a *voting comparator* may be put into place. Since a signal may be received by more than one site, this provides a means of comparing these multiple inputs, and selecting or "voting" on the best audio to retransmit.

One specialized type of radio system is known as trunking. Conventional radio networks assign a channel or groups of channels to a particular service. The fire department, police department, emergency medical service and other agencies each have their unique frequencies. In many instances, they are unable to communicate with one another. In a trunked system, all frequencies are pooled, and services are programmed into *talk groups* that are part of an overall *fleet map.* A central computer known as the controller makes channel assignments based upon requests to transmit. The technology involved utilizes intelligent radios in the field, and allows different priorities to be assigned to different units. For example, where a trunked system serves an entire community, public safety radios can be given priority access over non-emergency departments. Capacity for inter-agency communications can be built into the fleet map, or can be accomplished by as-needed programming. In a sense, trunking operates similar to a telephone system. All telephone customers can talk to each other even though they do not have a dedicated pair of wires from their home or business to every other telephone in the network. They all share a common pool of wires, with their conversations being connected by electronic switches. Trunked radio users can similarly communicate over one of many shared channels assigned by the controller. The benefits of trunked systems are their efficient use of spectrum and their flexibility. By being part of a large channel pool, agencies expand their ability to handle major emergencies that demand increased use of airtime. The downside of trunking is the significant dependence upon hardware and software, being by far the most complex means of radio communication. Significant redundancy is normally provided, however, to minimize the potential for total failure. In early versions, users of trunking radios manufactured by different vendors could not talk to each other. Proprietary technology made this impossible. However, the Association of Public-Safety Communications Officials (APCO) recently adopted what are known as Project 25 Standards for interoperability. These establish a Common Air Interface (CAI) that permits such communications.

Specialized devices can be found inside the communications center. Communications consoles can refer to the electronics used to control radio networks, as well as to the furniture that houses them. Formerly a series of analog switches, many consoles now feature microprocessor-based solutions. Besides the facility to select or mute specific channels, consoles may contain features such as a telephone interface to allow connection between a field unit, and a telephone and radio cross patch to support conversation between two distinct channels. Encoders may be an integral part of the console or may be added as adjunct items. These are used to selectively alert stations or to page personnel.

Fire service radio bandwidth

Radios operate on different frequencies and in different bandwidths. Low band units utilize spectrum between 33.40 and 45.90 MHz. MHz is the abbreviation for *megahertz,* or million cycles per second. This refers to the *frequency* of the radio wave, or how many times it occurs in that specified period of time. The more complete cycles that occur, the higher the frequency. Oftentimes, the terms *channel* and *frequency* are used interchangeably. However, in the truest sense, a channel is nothing more than a label or identifier placed on a frequency or frequencies. For example, a frequency of 46.18 MHz may be known as the "dispatch channel" while the "mutual aid channel" may use a frequency of 45.88 MHz.

- *Low band* communications utilize the simplex mode.
- *High band* devices transmit and receive in the 154.65–159.21 MHz range, and may operate in either the simplex or duplex configurations.

Both of these bands actually contain additional frequencies, but the fire service may only secure licenses within those ranges specified.

The spectrum may be further refined into:

- *VHF* or very high frequency
- *UHF* or ultra high frequency

VHF contains both the high and low bands. Included in the UHF category are the 400, 800, 900, and 700 megahertz bands. The most recent of these to open to public safety are the 700 megahertz channels, which were reassigned from the television broadcast industry in the late 1990s. Also found here are microwave radios. Named literally for their small radio waves, microwave transmitters are used in fixed locations to provide point-to-point relay of information and data. Systems and transmitters can also be either analog or digital. Digital radio breaks both voice and data down into bursts of electronic codes during transmission and reassembles it into an intelligible message during the receipt. While there may be some performance differences noted between these two technologies in marginal coverage areas, most users will not know (or care) which type of system they are using during normal operations.

Radio coverage

Coverage, or the ability to communicate over a specified area, is obviously of concern to the fire service. Whether the primary function is to extinguish wild land blazes or provide emergency medical service to high-rise buildings, the need for the accurate and reliable relay of information is mission critical. Because of the importance of adequate communications to firefighter safety, specific performance guarantees should be stated at the time of system design, and these criteria should be periodically verified thereafter. Improper maintenance, deterioration of antennae and cabling, aggressive building construction in the community, and shifting populations can all have and adverse effect on previously adequate systems. Coverage can be measured in terms of signal strength utilizing a meter to electronically record levels of reception, and circuit merit to define the clarity and intelligibility of voice messages. In the latter, a series of predetermined phrases is broadcast to determine how well they can be received. Although slightly subjective, circuit merit does provide an important real world measure of effectiveness. This can be especially beneficial in simulcast systems that transmit from multiple locations simultaneously. These rely heavily on the proper timing of signals. Should this timing get out of phase, a significant difficulty can arise in understanding transmissions in areas where coverage overlaps.

A great number of factors influence the effectiveness of radio communications. Low band systems may be subject to skip, or interference from out of state users on the same frequencies, while UHF systems can be adversely affected by heavy foliage. Microwave channels may be subject to deterioration of signal from heavy precipitation, and sunspots and other atmospheric anomalies can impact almost any bandwidth. Terrain also plays a large role in the ability to communicate. Tall hills and deep valleys can present formidable challenges to radio waves, as can large deposits of iron ore. However, all hills and valleys are not necessarily natural. High rise buildings with reflectorized window coatings and large quantities of steel and concrete can be difficult to penetrate, as can sub-basement areas and buildings classified as "hardened" construction. Products such as interior wall panels embedded with metallic particles will also have a negative impact. Designed for use in areas where cellular telephone calls are discouraged, these panels have the unwanted side effect of limiting public safety radio reception, as well.

Engineers specify coverage based upon percentages; the higher the better. Because of the variables involved, it is almost impossible to achieve a guarantee of 100% reliability, but for the fire service, 95% or greater coverage should be specified. However, a number without further refinement is insufficient to insure an adequate system. The following all serve as additional qualifiers:

- Is coverage to be measured in buildings or outdoors?
- If inside measurements are to be utilized, will these be assumed to involve light construction such as houses, medium construction such as small apartment and commercial complexes, or heavy construction such as large manufacturing or specialized facilities?
- Will exterior measurements be taken with portable or mobile radios, and will the user be stationary or in motion?
- Are the coverage requirements measured throughout the service areas or only on the fringe?
- Will the measurements apply to the ability of field units to hear dispatch, dispatch to hear field units, or both?
- How will coverage tests be performed, at how many sites, and what remedies are available should a system initially not meet specifications?

There are several things that can be done to improve reception in a public safety radio network. Surprisingly, the first does not require an equipment upgrade. Providing initial and follow-up training to all end users can increase understanding and operational efficiency. Another low cost solution may be as simple as replacing antennae on portable radios with more efficient models. However, nothing can replace a properly designed system that has an appropriate number of sites and sufficient power to blanket the service area. Unfortunately, some municipalities, when faced with budgetary concerns, may reduce the number of locations specified in order to lower costs. After all, a remote radio facility and equipment can carry a price tag well in excess of a million dollars, making for a tempting target. However, such economics can prove foolish long term, and can even foster dangerous conditions by creating additional marginal areas of operation.

While area wide deficiencies normally require major upgrades to correct, problems with specific buildings can often be addressed through the use of passive antenna systems or bi-directional amplifiers. These are designed to boost the signal level inside the target structures in which they are installed. Just as sprinklers and specialized fire protection systems may be mandated by code for certain occupancies, some communities are now requiring that amplifiers be installed in structures that would normally impede adequate communications. Boston, Massachusetts; Grapevine, Texas; Scottsdale, Arizona; and Burbank, California, among others, have enacted applicable legislation. Broward County,

Florida, has gone even further by requiring that new construction not impede emergency communications in any fashion. This includes the erection of large buildings that could cast shadows on or interfere with radio coverage overall, rather than just in the interior of the structure itself.

Communications system design

The overall design of an emergency services communications system must take into account the purpose for which it is being used. It must be reliable, and have at least one level of backup to account for unforeseen circumstances. Just as a backup hose line is stretched at a working fire, so, too must a level of redundancy exist in the radio network. Local conditions play a large role in determining the specifications and design. The following all exert influence:

- How many radios must be supported?
- How many alarms are answered annually?
- How large an area must be covered?
- What is the terrain?
- Will voice, data, or both be sent over the system?
- How many dispatch points will be utilized?
- What natural hazards must be considered?
- Do any local ordinances restrict tower location or height?

A properly designed system will in many cases be transparent to the end user, as it will operate flawlessly under a wide variety of conditions. Poor planning and design, however, can create an impediment to daily operations and compromise the safety of citizens and first responders alike.

Reverse 9-1-1

Reverse 9-1-1 is the name generically applied to the ability to selectively call, by telephone, a specific group of people or residents of a geographic area. This is accomplished by computer in the PSAP, based upon a listing of telephone numbers maintained at that location. A prepackaged or customized message can be broadcast based upon an area selected from a map. From a fire service perspective, this concept lends itself to issuing shelter or evacuation orders to populations effected by hazardous materials releases.

Fire Service Communications Facilities

Communications centers serve a specialized need, and, as such, have specialized design requirements. Just as fire stations and fire apparatus must be laid out in an efficient fashion and possess the components necessary for optimum performance, so, too, must fire department dispatch facilities be up to the task assigned.

The configuration of communications centers varies widely from community to community, and is dependent upon several factors. In a small town, it may be a single desk or position, perhaps in the local sheriff's office or city hall. This individual may be responsible not only for fire, police, and emergency medical dispatching, but for a variety of other duties as well. It is not uncommon for this lone dispatcher to be responsible for the highway department radio, for serving as the switchboard operator for all of local government, for handling all walk in requests for permits, reports, and records, and potentially serving as a jailer, as well, should the holding cell be in use.

On the opposite end of the spectrum is the large metropolis having dozens of telecommunicators on duty, each with a particular job or function. Specialized call takers manage all contact with the public, and enter required information into a computer system for action by the dispatchers who are responsible for assigning the appropriate units. On-floor supervision assists on special incidents and helps to maintain the flow of exceptionally high call volumes. The "super centers" may actually be consolidated undertakings serving multiple communities, or may be segregated as to function, with local fire, law enforcement, and EMS agencies each operating a separate facility. Regardless of the size and makeup, there are several key ingredients required to create a fire dispatch center.

The first is a telephone system. Since the telephone is the primary means of receiving alarms, this should go without saying. However, even in the smallest community emergency lines should appear on at least two telephone instruments to provide a level of redundancy. In addition to the emergency numbers, dispatchers should have access to the required administrative lines as well, including at least one unlisted number. Maintaining this private number assures that a line will remain free for making outgoing telephone calls even during the most critical incidents. Some agencies also utilize wireless telephones as a means of providing another level of backup for their dispatchers. These devices are generally kept on charge until needed, but can readily be put to use in case of a major system failure or if additional calling capacity is quickly needed.

Obviously, there is also a great variety of features and sophistication available when it comes to telephone systems. The most basic technology is known as 1A2, which represents the standard home telephone. While a facility may have multiple instruments, and these instruments may share multiple lines, there is little intelligence past the ability to place a call on hold. Some form of electronic switch is required to allow for more advanced functions such as transfer and conference calling. Of course, with this advancement comes the potential for failure. Whereas 1A2 telephones operate from voltage on the telephone lines, electronic devices depend on a microprocessor and commercial electric power. Therefore, it becomes critical from the fire service perspective to provide both redundant processing and backup electrical sources for these more advanced electronics.

If a community has the 9-1-1 emergency number, equipment will have to be present to support the specialized features. The 9-1-1 can be broken down into two broad categories—basic and enhanced. Basic 9-1-1 is just that; providing the public with three-digit emergency access. However, additional capabilities exist with the enhanced version. The two primary benefits are ANI (Automatic Number Identification) and ALI (Automatic Location Identification) that provide the telecommunicator with instantaneous information about the caller's number and location. Receipt of this data requires an ANI/ALI display. This can be a stand-alone monitor, an interface that allows for import of the number and address directly to a CAD system, or both.

The equipment utilized in the receipt and dispatch of emergency calls requires space to operate and support. Again, in small communities, standard phone sets and desktop consoles may be all that is required, whereas centers that serve major populations will no doubt need multiple free standing consoles with integrated electronics for the dispatchers and large climate controlled facilities for the racks and switches needed to make them all work. Additionally, space will be required for administration and support.

The actual operations area of a modern fire dispatch center must be well lit, and free from glare. Since much dependence is put upon computerized records and aids, the ability to easily see and read display screens is a must. This can be accomplished through the use of indirect lighting whereby fixtures are mounted with the bulbs facing the ceiling, allowing only reflected light to hit work surfaces. Specialty task lighting may also be used, enabling dispatchers to individually aim and focus illumination as required. The ceiling, walls, and floors should be designed to reduce excessive noise, which can be distracting and interfere with the clear transmission of alarms. The use of fabric, carpet, and other acoustically rated materials is often specified. Ceiling tile and wall finishes that have a high NRC (Noise Reduction Coefficient) are the most efficient. While carpet helps to control sound, it must be remembered that a dispatch center operates around the clock, and therefore receives almost five times the normal traffic of a 9 to 5 office. Because of this, only high-grade stain resistant carpet suited for commercial use should be considered. Special anti-static weaves are also available to help protect sensitive electronics. Regardless of the interior finishes chosen, they must all be durable and conform to the appropriate building codes.

Carpet tiles may be used to cover a raised floor, which is used to run cables and electrical power into the center. The area below the floor must be kept neat and clean, and appropriate wiring used and fire detection and protection provided, where applicable. The heating, ventilating, and air conditioning must be designed and sized to account for 24-hour occupancy and the presence of heat generating equipment. Even in the northern climes, cooling can be a larger issue than heating. It is important to provide a continuous air exchange in order to keep personnel comfortable and alert, with HEPA or electrostatic filtration added to remove dust particles from the air. There should also be a means of quickly closing the air intake in case of a nearby hazardous materials release, or biological or chemical attack.

Smoke and/or heat detectors must be provided as per code, and especially in plenums, and fire stopping installed in cable chases. Portable fire extinguishers compatible with the class of hazard should be provided, and staff trained in their use. Fixed, inert gas systems are commonly specified. Until the 1980s, halogenated hydrocarbon (Halon) was utilized due to its suitability in protecting delicate electronic devices, such as computers. However, as fluorocarbons became implicated in Ozone depletion, a search was made for alternate solutions. Today, agents tend to be a mix of inert gasses such as Argon and Nitrogen. Two of the more popular are manufactured under the names *Intergen* and *FM200.*

In addition to the safety and security of the communications center, attention is now also being paid to the welfare and comfort of the dispatchers. An increasing number of agencies now routinely work 10- and 12-hour shifts, and many facilities are short staffed, making it appropriate to focus on this issue. Regulations such as the Americans with Disabilities Act (ADA) and agencies such as OSHA all have an impact on the work environment. However, more frequently, ergonomic issues are being considered as part of equipment and facility design. This relationship between man and machine has had significant recent influence. While many dispatch centers had formerly been housed in hardened basements, more and more are being built or remodeled at or above ground level in order to benefit from natural light, and to lessen the feelings of isolation and confinement. Additional windows may also be added to allow for public viewing. For example, the fire dispatch center in Caddo Parish, Louisiana, contains a wall of specially bonded glass that can be electrically switched from a mirror finish to clear to accommodate scheduled tours while providing privacy at other times.

Heavy-duty chairs, designed for 24/7 use are also now readily available. These offer radical improvements over standard office models including large, stable bases, increased lumbar support, and replaceable parts. Dispatch consoles have also undergone change. One of the biggest challenges faced in the past was the integration of a variety of non-standardized components and controls. This typically resulted in a hodge-podge of devices that were mounted and stacked in a variety of configurations. Obviously, this was ineffective, as it took up a significant amount of space and was difficult for the dispatcher to monitor and operate. An increasing number of consoles now utilize CRT (Cathode Ray Tube) control. Computer monitors are used to replace analog switches and dials. Improved flexibility allows for the addition of new systems to the screens through software, rather than through physical expansion of the workspace. A beneficial byproduct is that all controls remain within sight and within easy reach of the dispatcher. In addition to controlling the radio and telephone, many consoles now allow telecommunicators to control their environment, as well. Built in units provide personalized heating and cooling and augment the main building systems. These may also include heated footrests and radiant panels.

Areas that support the dispatch operation are also critical, and so is their design. Equipment rooms must be spacious enough to accommodate computer, radio, alarm, and telephone equipment and to provide sufficient power and cooling. As is the case with all critical components, an alternate source of electricity must be supplied. This can be accomplished through a generator, UPS, or both. Capacities must be such as to carry at least the critical portions of the facility for several hours. Configuration should allow for automatic switching to alternate sources in case of commercial power outages. A manual override feature is also essential in case the mechanical transfer method fails. Electric generating plants may be powered by a variety of sources, however, a completely independent fuel supply such as an on-site diesel tank is usually recommended. UPS devices also serve to regulate voltage on a daily basis to protect equipment from spikes and surges. Dispatchers require the ability to remotely monitor the status of the building's electrical system, as well as all critical functions such as fire and intrusion alarms.

For centers having a uniformed staff, locker rooms provide a convenient changing area. When equipped with showers, they facilitate extended stays, as may be required during a disaster or severe weather. Some means of accommodating overnight stays will also come in handy. This need not be a dedicated area, but rather a suitable location that can be converted into a dormitory by means of folding cots or inflatable beds. A well-appointed kitchen is also welcomed during such protracted operations, but is also a must for daily routines. Since many dispatchers may not be able to leave the area for meal breaks, the kitchen should allow for the preparation and storage of food for an entire shift. If access is provided to a main fire station kitchen area, a small kitchenette that can double as a break room may be located in close proximity to dispatch. Newly constructed facilities often have areas set aside for relaxation such as exercise rooms and quiet zones in an effort to reduce stress.

Mechanical rooms may actually be larger than expected for the relative size of the complex due to the need for redundancy. Housing dual sets of air conditioners and heaters, they help to assure comfort even in the most severe circumstances. Administrative offices for communications centers will be little different than those found elsewhere in the fire service. Consideration should be given to providing a robust LAN (local area network) to assist in the delivery and sharing of data. A large conference room is also a plus, especially if there is not a recognized EOC (Emergency Operations Center). It will provide a meeting place for command officers to consult on issues without gathering directly in the dispatch center itself.

Although complex in nature, the dispatch center is largely designed through simple necessities. Additional modifications may be added by the locality. The San Francisco 9-1-1 facility is mounted on a giant pendulum in order to withstand earthquakes. The Madison County, Alabama, facility has only one exposed wall—it is otherwise covered in earth as a precaution against tornadoes. The roof and three exterior walls are effectively below grade. Regardless of the locale, the facility should be able to withstand predictably severe weather, and be located away from known hazardous materials locations and above the flood plain. For more information on universal requirements, refer to the National Fire Protection Association 1221 Standard for the Installation Maintenance and Use of Emergency Services Communications Systems.

Security

The communications center should be a safe and secure environment, ready to operate under almost any condition. Since communications is the heart of any emergency service, security is a must. A minimum of two solid locked doors must stand between the public and dispatchers, with any street level glass being bullet proof. Electronic surveillance systems that include alarms and cameras should be installed at key points, and an intercom and electric door strikers put in place to provide dispatchers with control over all entrances.

In some circumstances a completely secured grounds will also be provided trough the use of vehicle access passes, fences, and exterior cameras and sensors. Thought must be given to maintaining an adequate distance between vehicular traffic and the center, and parking garages underneath the facility are ill advised. If they are present, then they must be extremely secure, with their use restricted to authorized individuals only. Regardless of how it is arranged, protection from unauthorized entry is a must.

There are a variety of access control systems available. In the simplest sense a standard key may be issued. However, where keys are utilized, it is almost impossible to prevent unauthorized duplication. Even requiring that employees turn in keys upon termination is no guarantee that integrity is maintained. Combination locks are another option. Here again, combinations must be periodically changed, and employees must refrain from providing unauthorized personnel with the door code or the sense of security is truly false.

Card readers are a significant step up in that they can be utilized as part of a full facility solution that defines various levels of access. Cards can also double as photo ids, and can be decommissioned upon the telecommunicator leaving service. Retinal and fingerprint scans use optical devices to measure individual characteristics, and are used in many high security operations. Unlike cards, fingerprints and retinas cannot be "borrowed" by other users. Reliability of these devices classified as biometrics is improving, but cost is still relatively high in comparison to other options.

Cameras should be positioned to provide a full view of critical interior and exterior areas, especially near points of entry or sensitive equipment. The installation of a time-lapse video recorder serves to maintain a permanent record. Exterior cameras should be chosen with weather and lighting conditions in mind, and can be configured to scan, or to activate monitors only when motion is detected. Monitors should be of sufficient size and clarity to allow for easy viewing, and should be located in plain sight of the dispatcher. Consideration should be given to the protection of selected areas with intrusion alarms. This is especially applicable to equipment rooms and exterior doors and windows that are removed from the communications center. A means and procedure for quickly notifying law enforcement should also be put in place. Some facilities install a *panic alarm* at the receptionist desk or supervisor's console. Arrangements should also be made for periodic checks of the facility by regular patrols. In extreme cases, law enforcement personnel or private security may be assigned full time.

When the dispatch room shares space in a building with another operation, it must be surrounded by fire rated walls. If it is located in a fire station, consideration must also be given to assure that apparatus exhaust cannot penetrate the space or enter the air handling system.

Dedicated telephone and electrical circuits should serve dispatch, and not be shared with other uses. The outside of the building and the grounds should be well lighted, fenced, and absent of hiding places. Large containers that could conceal explosives should be removed from public areas, and access to the facility limited. All unknown parties should be challenged and held in a secure area until such time as their need to enter has been verified. Deliveries should be scheduled, and all service, other than emergency repairs, should be accomplished during regular weekday hours. Even food delivery persons should be instructed to wait outside the secure zone. A log of all visitors must be maintained, and certain areas of the building may require an escort. Mail should be carefully screened, and proper notifications made regarding suspicious letters and packages. At no time should parcels be accepted directly into dispatch. The need for security also extends to remote locations, as well. Hilltop radio sites need to be protected from unauthorized intrusion by many of the same means described above. Dispatchers must have the capacity to monitor security and operational parameters.

Redundancy and backups are other significant concerns. No single failure, whether accidental or deliberate, should be capable of compromising the ability to receive and respond to emergency calls. Telephone and alarm circuits entering the facility should be duplicated, and a backup facility identified. This PSAP need not have the full capacity of the primary location, but should be available at a moment's notice should an evacuation be required. Plans for temporary diversion of traffic to a neighboring dispatch center should also be formalized.

Although security and safety considerations may often be more closely associated with law enforcement, the harsh reality of our modern world dictates that fire departments turn appropriate attention toward maintaining secure communications. No factor affects security more than individual responsibility. Telecommunicators must clearly understand the importance of policies and procedures, and management must be diligent in enforcement. The entire communications team must be vigilant for physical signs of threats in the workplace and verbal indications contained in incoming calls. Training should be provided in weapons of mass destruction awareness, and all information received concerning potential threats or danger should be immediately relayed to the appropriate units. Responders, too, share in this responsibility, and need to report all suspicious activities observed in the field. Care must be taken when relaying sensitive information to do so, when possible, by secure means. Scanners can easily monitor most radio systems. This includes 800 MHz trunked networks that previously were difficult to scan. The same applies to cellular telephones. Public telephone networks, dedicated phone circuits, and encrypted radio devices are advisable. Mobile data networks also reduce the potential for unauthorized reception.

Safety should be a primary concern behind all communications, but takes an even more elevated role in the 21st century. New challenges created by domestic terrorism demand an even higher level of awareness and concern. Despite this, any number of common occurrences can quickly and seriously degrade crew safety. While it would be impractical to list every concern, severe weather rates high on the list. In addition to the use of public warning systems, dispatchers should provide units in the field with specific updates on approaching storms and deteriorating conditions. This is also a consideration when alerting crews who may have been asleep during the onset of a storm, and who may not be prepared for the current weather.

Communications Organization and Management

Organization

The dispatch function may be organized in a variety of ways. Communications can be a uniformed division of the fire department, staffed and managed solely by firefighters and fire officers. Employees may or may not be represented by a bargaining unit. Dispatch can also be accomplished on a municipal, county, or regional level. In these cases, control may be placed in the hands of a law enforcement agency, such as the sheriff, or a joint-powers agreement may be drawn up that creates a management board. Here, a committee is appointed to oversee a manager or director who is responsible for the day-to-day operation. Similar arrangements can be seen with special communications districts that are political subdivisions created solely to provide 9-1-1 service. Just as fire protection districts serve a defined area, communications districts have established boundaries and legally mandated powers. Through privatization, communities may contract with a commercial service for their emergency communications needs. In rural areas, this may be no more than the fire department entering into an agreement with a local business to answer their emergency telephone number. In more complex scenarios, this can entail a multi-year contract with a specialized firm that provides both employees and equipment. Where control is not directly vested in the fire department, avenues for input, change, and corrective action must be defined. Because fire service call volume tends to be lesser than that of law enforcement in most communities, there is often a fear that the fire department needs will become subservient to those of the police in consolidated facilities. This need not be the case, and there are numerous examples of successfully managed consolidated centers that well serve the needs of every agency and community involved.

Effective management

Just as each incident and station must have an identified chain of command, so too must a dispatch center have leadership. There are numerous functions required of management. Among them are:

- Long range planning
- Establishing goals and objectives
- Personnel management
- Staffing
- Budgeting
- Systems design, acquisition, and maintenance
- Public representation of the facility
- Liaison with user agencies
- Assuring legal compliance

The skills required to perform these tasks are no different than those demanded for any fire service leadership position. However, as is the case with any division, a successful manager or supervisor must have a knowledge base sufficient to enable good decision making in the particular specialty. Practical experience is especially desirable for supervisors who directly control dispatch personnel, and who may be required to pitch in during emergencies. It is understandable that an engine company crew might more likely follow a leader that they have seen at the end of a line. Similarly, telecommunicators may show more loyalty and respect to someone whom they have observed at the end of a headset. A clearly defined chain-of-command must also be in place, so that all personnel understand the responsibilities and roles of every job category. This is important not only during daily operations, but especially so in cases where the chief operating officer is not present and critical decisions must be made.

Funding

Where dispatch exists as a function of the fire department, funding will be an issue addressed through the annual municipal budget. Communications will exist as a specific budget activity, or as a subset of a larger function such as "support activities" that may include the Records Division. It is likely that the bulk of revenue will come from local sources such as property and sales taxes, augmented by state and federal grants, as applicable. The fire department will compete with all other government agencies for a share of the pot, and communications will vie internally with suppression, inspection, and administration for dollars.

In consolidated or multi-jurisdictional centers, funding will come from all of the communities involved. Most likely, some sort of funding formula will be developed to adequately assess the individual cost of operation. This normally takes into account the number of calls answered, services provided, and personnel assigned. Money may also be received through 9-1-1 service fees placed on telephone bills. The majority of states have such legislation, with some or all of the collections returning to the local level. An increasing number support these fees on cellular telephones, as well. Generally, these moneys may be used for emergency communications purposes only, and cannot be spent on apparatus or suppression personnel. For major projects such as the construction of a new facility, many communities may turn to the issuance of bonds. This provides a targeted source of income outside the normal taxation structure. Where fire protection is provided through subscription or financed by donations, dispatch may also be dependent upon these same avenues. Special tariffs or fees may also come into play. Some may involve automatic fire alarms. Dispatch centers may choose to monitor private alarm systems for a fee, or the community may enact alarm ordinances that fine subscribers for repeated false alarms. Both can be a source of revenue, however local statutes will dictate whether the proceeds will go directly to the fire service or whether they must be deposited in a general municipal fund.

Supply and inventory

Maintaining an adequate stock of supplies is another often overlooked management responsibility. Large fire departments may have a supply officer or an entire division responsible for the procurement and stocking of goods. Running out of critical items in a dispatch center is akin to running out of water on the fireground. Therefore, it is imperative that a system for monitoring the level of critical supplies and providing replacements exists. Included among the items needed to support typical communications centers are:

- General office supplies
- Department specific forms
- Reference aids
- Computer expendables such as paper, ribbons, and media
- Spare computer parts, especially keyboards and monitors
- Spare radio parts
- Spare telephone parts, including headsets
- Extra tapes, discs, and parts for logging recorders
- Housekeeping, cleaning, and paper products
- Building system supplies like HVAC filters and spare light bulbs
- Special system supplies, such as oil and anti-freeze for generators, and batteries
- Extra dispatch chairs and replacement parts
- Food, blankets, and cots to support staff during major emergencies

The size and function of the communications center will determine the amount required. However, sufficient stock of critical material should be maintained in house to permit operation during a disaster when regular delivery of supplies may be interrupted. Agencies must periodically assess their needs, and assure that a procedure is in place that tracks both usage and stock on hand. Inventory control may require that request forms are submitted for the issuance of certain items, or may be as simple as regular restocking based upon past depletion experience. Systems can be as basic as a clipboard and paper hung by a supply locker, and as complex as barcodes and scanners that track supplies by means of a centralized computer.

Maintenance and repair

Although only occasionally responsible for performing repairs, managers must establish a mechanism for the maintenance and repair of all communications equipment. This includes, but is not limited to radios, telephones, computers, and the physical plant. A schedule for regular preventive maintenance should be created, with special attention given to mechanical systems such as generators. Regular testing accompanied by documentation of such testing should be required. Computer systems must be regularly backed-up and data stored in suitable locations. For security reasons, these may include safes, fireproof vaults, or off-site facilities. On-line files may also demand periodic purging in order to assure optimum performance.

Emergency repairs must also be given adequate consideration. Dispatchers require around the clock access to service in order to address critical breakdowns. It is imperative that these contact lists be kept up to date. The same holds true for emergency action plans that deal with temporary steps to be taken in the result of equipment failure.

Maintenance can be provided by in-house sources through the employment of technicians, or through contract with private concerns. Where contracts exist, care should be taken to ensure that critical components are covered outside of the normal 9:00 a.m. to 5:00 p.m. workday, and that a reasonable maximum response time is guaranteed. Penalty clauses may be inserted, and proof of insurance, equal opportunity compliance, and other local legal requirements addressed.

Staffing and personnel

Adequate trained staffing of fire department communications centers is a must. While individual companies respond to only a fraction of the incidents that occur in any community, dispatchers and call takers essentially "touch" every call. They are the first point of contact from the citizen, and especially to the citizen in crisis. To the caller, they are the fire department, and their action or inaction can arguably have as much impact on the successful outcome of an emergency as the conduct of the first responders. Because of this, telecommunicators must be chosen with great care, provided proper guidance, training, and tools, and a reasonable number of personnel must be assigned to duty. Without any one of these elements present, difficulties can arise.

Staffing

The exact number of personnel required to staff an emergency dispatch facility is subject to much discussion, as it is dependent upon a number of variables. Obviously, the number of alarms handled is a prime factor. A busy department will no doubt need more staffing than a less active agency. This is largely affected by the size of the community. The Insurance Services Offices, as part of their Public Protection Classification program establishes basic guidelines in this area, requiring that agencies handling a minimum number of calls have at least two persons on duty at all times. In addition to having a direct impact upon operations, the communications criteria established by the Fire Suppression Rating Schedule counts for a full 10% of a community's insurance rating.

However, call volume by itself is just one factor. For example, what duties do communications personnel perform? If they are involved in both call taking and dispatch, it will obviously require more resources than if they just receive screened and prioritized calls that transferred from another agency. The same is true for the addition of clerical or record keeping duties that may conflict with emergency call handling. The type of service involved will also help to establish requirements. If Emergency Medical Dispatch is utilized, and pre-arrival instructions given, it is reasonable to assume that call-takers will spend a protracted time on the telephone. Additional personnel may therefore be required to meet call standard benchmarks and to assure that incoming emergencies are addressed in a timely fashion. While there is no universally accepted "law" concerning the maximum amount of time that an emergency line should ring, or what constitutes an excessive delay in call handling within the 9-1-1 center, there are some reasonable assumptions. It is quite clear that a timely response is critical to both fire and emergency medical incidents, and that the timeline for this response actually begins when the citizen discovers the threat. Anything that adversely affects the prompt delivery of service is unacceptable. It is not only telephone traffic that will help to determine the required staff. Dispatchers can only be expected to monitor a limited number of frequencies, and to be responsible for a finite number of units. Regardless of improved technology, it is unreasonable to ignore the danger created by assigning an excessive workload.

The number of agencies involved also has an obvious bearing. Facilities that provide dispatching for fire, EMS, sheriff, and police will no doubt need to assign personnel to each specialized task, except in the smallest communities. Even in rural areas, there will be a minimum number of telecommunicators needed in order to assure around the clock staffing. Work schedules are another prime consideration. As a rule of thumb, it takes at least 4.5 employees to provide a single on-duty person 24 hours per day. The number of break periods assigned and benefits such as sick leave and vacations can alter this figure. Further adjustments come from the exact schedule utilized. Several formulas are available that allow for further refinement and more exact calculation. There may also be differences between average staffing designed to address "normal" periods of operation, and staffing aimed at dealing with major emergencies and special events. For example, departments may choose to increase the number of on-duty dispatchers for Halloween and the Fourth of July based upon past experience, and may institute a recall of personnel during severe weather conditions. Perhaps the most difficult situation to address is the occasional spike in call volume caused by heavily reported incidents. These large concentrations of incoming calls are occurring more frequently because of the popularity of wireless telephones. It is not unusual to receive dozens of reports for a car fire at a busy intersection. In the past, only one or two calls may have been generated by this relatively minor alarm. Unfortunately, most governments are not in the position to base their staffing for any service based upon worst-case scenarios, and during these periods citizens may experience slight delays.

The Fair Labor Standards Act controls the numbers of hours that can be worked in any given week. While special considerations are given to uniformed personnel, dispatch center regulations are the same as those for any standard office. Still, the use of 8-, 10-, or 12-hour days coupled with a variety of shift assignments and days off can produce a variety of results. Some agencies utilize assigned shifts whereby the same squads are always assigned to days, afternoons, and evenings. Rotating shifts require that all squads spend a certain amount of time working each shift. The direction in which this rotation is made, and the number of days assigned to each shift can have a significant influence on the Circadian rhythm, or body clock. Studies have shown that improper scheduling can increase stress, reduce performance, disrupt sleep patterns, and increase fatigue and illness. Without a labor contract, shift assignments and vacations may be at the discretion of the employer, or may be tied to seniority. Although some agencies still utilize dispatch as a light duty assignment, the increased technical nature of the job has transformed it into a highly specialized position requiring a well-trained and dedicated staff.

Recruiting and testing

During the late 1990s and early 2000s, a staffing crisis existed in public safety communications. The combination of low pay, high stress, poor chances for advancement, unattractive work schedules, and competition from the private sector during periods of economic growth seriously reduced the applicant pool and increased the rate of turnover. Since many agencies do not offer a preferential retirement plan to support personnel, there can be a lack of long-term commitment to communications as a career.

That notwithstanding, fire departments must focus on obtaining the best-qualified people to fill these critical positions. This may be a function of a government-wide human resources department, or may fall upon the division commander. Attracting qualified candidates is crucial, and can be accomplished through regular job postings, advertising in the local media, and by word of mouth. Experienced dispatchers are among the most effective tools for recruiting. The use of photographs in newspaper solicitations will help to set you apart from the rest, and mentioning the terms "9-1-1" and "emergency" in the heading can help gain additional interest.

Regardless of the method used to recruit, there must obviously be established minimum standards and a mechanism in place to verify individual qualifications. The NFPA as well as several states have developed formal requirements for dispatch personnel. Many communities use these as a base upon which to build a comprehensive checklist. Among the most commonly desired attributes:

- Minimum age of 18, or 21 depending upon local standards
- High school graduate or GED
- Basic typing skills and keyboard knowledge
- Lack of criminal record
- Good moral character
- Ability to multi-task
- Ability to work under pressure

Candidates may be subject to a battery of formal tests and personal interviews, some of which may be given only after a conditional offer of employment. Psychological examinations and drug screenings are common examples. Although clear speech, good hearing, and visual acuity may be required to perform the basic functions of a fire department dispatcher, all qualifications and hiring practices must comply with the Americans with Disabilities Act and other appropriate regulations. Written exams to test knowledge of communications practices and local geography may be developed in house, or prepackaged tests that focus on universal skill sets can be used. Where permitted, departments may wish to maintain open postings that allow for continuous receipt of applications. One low cost means of judging the aptitude of a potential hire is to invite that person to observe operations for a shift. It is not unusual for an otherwise qualified candidate to excuse themselves from further consideration after seeing the rigors of the job first hand.

Training

As is the case with suppression personnel, training is critical. Nothing can take the place of an organized course of instruction that provides an opportunity to learn and practice required skills. Here, too, training can be custom designed by the employing agency, or one of many nationally recognized classes can be used. APCO, the Association of Public-safety Communications Officials presents a Basic Telecommunicator Course in person or over the Web. A self-paced study guide is also available. NENA, the National Emergency Number Association offers registration as an ENP or emergency number professional, and dispatcher certification is also available from the International Municipal Signal Association. There are also several approved courses produced by private sector sources. Regardless of the instructional methods used, basic information—including local SOPs—must be included. A typical curriculum should address:

- Telephone procedures
- Equipment/system operation
- FCC rules and regulations
- Alarm/response assignments
- Department organization
- Local SOPs
- Disaster/emergency procedures

Even the smallest of agencies must have some sort of formalized training program to ensure uniformity of operations. On-the-job training, or learning by the seat of the pants is no longer an acceptable alternative. The use of modern educational techniques such as CAI, (computer assisted instruction) and simulation are finding their way into telecommunicator training. Obviously, suppression personnel benefit from experiencing fire conditions under the controlled atmosphere of the training grounds. So, too, will the performance of dispatchers improve through hands-on practice. To this end, special databases can be established in dispatch computers that allow for the receipt and assignment of "calls" without interfering with live dispatch. Hardware and software that approximate telephone and radio traffic help to make the experience even more lifelike. These simulations can be reinforced with tape-recorded copies of actual incidents.

Quality assurance

One of the more important managerial functions is that of quality assurance. Quality assurance serves as a means of measuring performance, and is critical in determining the effectiveness of training and in targeting future areas of educational reinforcement and revision of policy.

It is essential that service levels are both defined and consistently achieved. Fire department communications is an active and complex entity, however administrators typically have at their disposal tools that can greatly assist in providing an accurate picture of employee and unit efficiency:

- Tape-recorded logs of call handling and dispatching may be randomly reviewed and compared to established criteria. Similar records of critical or unusual incidents can serve as training aids, pointing out lessons learned from these emergencies.
- CAD records can be used to determine the accuracy of entries, as well as the elapsed time between call receipt and dispatch.
- Telephone Management Information Systems (MIS) are valuable in determining the number of calls handled by each employee, as well as answer and process times. These can help to identify both exceptional and problem employees.

Quality assurance is especially critical where EMD is used. It is impossible to overemphasize the need to constantly validate that the appropriate protocols are being used and the approved instructions being given. Independent review or assistance from a designated physician is highly recommended. Private sector solutions such as agent monitoring software and customer follow up programs may also prove beneficial in creating QA programs for communications operations.

Liability

Training and quality assurance can go a long way to reduce liability incurred by fire department communications centers. Despite the large numbers of calls received during emergencies, an overwhelming majority of these facilities have exemplary track records. Still, managers must understand their exposure to liability, and must take action to lessen the risks.

In order for liability to exist, several conditions must be present. These are:

- A *duty* to perform a service must exist. This duty may be established by law, or charter, or may be created through the agency's own standard operating procedures. Duty may also be inferred from a reasonable standard of care. In other words, how do other communities handle similar situations, or how would a reasonable person respond if faced with the same set of circumstances. Even if no lawful responsibility exists, an employee may overstep his or her responsibility and assume a duty. This can create a situation known as *vicarious liability,* or assuming the liability of another.
- There must be a *breach* of that duty. This can be through an *error of omission* (not performing a required task or service) or an *error of commission* (not performing that task properly).
- There must be *damages.* In other words, something bad has happened.
- The breach of duty must be the *proximate cause* of these damages.

If one or more of these elements are missing, then liability does not exist. However, it is easy to see how less than careful handling of critical information can lead to problems. Every state has its own set of regulations covering government liability standards, typically known as *municipal torts*. These may place a lesser liability on cities and counties than on the public, and may limit awards. Certain legal actions such as civil suits, ADA complaints, and Civil Rights violations fall outside these boundaries. Regardless of legislation, fire departments and individual employees can suffer stiff penalties for failure to comply in these areas.

While professional liability insurance can be purchased to afford some protection from suit, the best protection is a comprehensive program of training and quality assurance. Time stamped recording devices such as CAD systems and voice loggers also serve to provide protection against unwarranted claims by maintaining an accurate representation of events.

Procedures, Regulations, and Regulatory Agencies

Standard operating procedures

Standardized practices are required in every area of the fire service, from the specification of hose threads to communications procedures. A well-defined set of SOPs is therefore essential in establishing and ensuring the exchange of information along with improving firefighter safety. There are many components to consider in this area:

- Calling protocol—will messages take the form of "to-from" or "from-to," as in "Engine 4 from Chief 1" or "Chief 1 to Engine 4"?
- Unit Identification—what manner of numbering system will be assigned to departments, stations, companies, and individuals?
- Phonetic alphabet—which one will be used?
- Time—military (24 hour) or conventional clock?
- Plain speech or ten-codes?

Regardless of the choices made, it is critical that every firefighter utilizes the same set of protocols. This utilization should also extend to all agencies that may be expected to participate in disaster or special operations. Plans must also be made that allow these departments to communicate, as neighboring communities may be on different frequencies.

- The adoption of the ICS (incident command system) will provide structure to both on-scene operations and communications. By providing a single point of fireground contact to the dispatcher, and by clearly identifying the function of all crews, ICS simplifies what can otherwise be a source of confusion.

- Maintaining radio discipline at emergency scenes is critical to firefighter safety. Transmissions should be kept to a minimum in order to keep the channel as clear as possible for emergency messages.
- In ideal situations, every member assigned to the incident will have a personal portable radio. If this is not possible, all crews should have at least one radio, which can be used to coordinate their actions and call for help, if needed. This is particularly critical for those assigned to search and rescue or interior attack.
- All personnel should immediately report any hazardous condition observed, and a pre-determined evacuation signal should be established.

There are numerous documented cases of communications playing a role in firefighter fatalities, including the inability to notify interior crews of changing conditions. Among these is the World Trade Center collapse of September 11, 2001, the largest loss of responder life in history. Firefighters must understand that no communications system is foolproof, and that large steel framed structures and reinforced below-grade areas may severely hamper radio communications.

Some radio networks support push-to-talk transmitter identification and emergency buttons. When these devices are activated, an alert notification is sent electronically to the dispatcher. Where emergency buttons are provided, a written procedure must be established concerning their use and outlining the appropriate response to activation.

An understanding of the responsibility of company officers, command officers, and dispatch concerning functions and duties must exist. Although some of this authority will be defined through job descriptions and organization charts, there needs to be clear-cut responsibility concerning notifications, the call back of off duty personnel, and a wide range of incident related functions. Some departments may choose to conduct certain activities in the field, while others assume them to be part of the dispatcher's duties. Some actions may be carried out only after a direct order by the incident commander, while some may be triggered by simply meeting a criteria such as the dispatch of a third alarm. Regardless of how and when such activities occur, it is critical to identify who is responsible. In special cases mobile command posts or communications vans may be dispatched. This is normally done during protracted operations, where these vehicles may serve as both a remote dispatch and emergency operations center, coordinating efforts directly from the scene.

Plans may also exist that identify specific purposes for radio channels. One may be assigned to dispatch, for example, and another to fireground communications. Systems that cover large areas may also designate regional or sector channels. Where emergency medical service is provided, additional frequencies may be utilized for this purpose, with hospital communications and patient telemetry utilizing yet another set of resources.

Regulatory and Advisory Agencies

NFPA

The fire service, like all branches of public safety, is subject to a variety of regulations and governed by a number of regulatory agencies. Many of these rules and rule making bodies have a direct effect on communications. The National Fire Protection Association has developed several applicable standards.

Two of these specifically address issues surrounding dispatch. Standard 1221 defines the Installation, Maintenance, and Use of Emergency Services Communications Systems, while 1061 sets the Standard for Professional Qualifications for Public Safety Telecommunicators. Other documents that may influence communications operations or communications center construction and maintenance include: Standard 262, Method of Test for Flame Travel and Smoke of Wires and Cables for Use in Air-Handling Spaces; Standard 110, for Emergency and Standby Power Systems; Standard 76, for Emergency and Standby Power Systems; Standard 75, for the Protection of Electronic Computer/Data Processing Equipment; and Standard 72, the National Fire Alarm Code.

NIOSH/OSHA

NIOSH, the National Institute of Occupational Safety and Health has developed guidelines that are applicable to dispatch centers. Musculoskeletal injuries such as Carpal Tunnel Syndrome have increased rapidly in the workplace, and telecommunicators, faced with repetitive tasks, certainly are not immune. (See Appendix)

OSHA, the Occupational Safety and Health Administration is charged with creating a safe workplace. OSHA regulations cover telecommunicators and service personnel as well as suppression crews. Tower repair crews, for example, must follow safety precautions for climbing and for exposure to radio waves. Even bodies such as the Federal Aviation Administration play a role through placing limitations on tower heights near active runways.

FCC and FCC regulations

No agency has more impact upon fire service communications than the FCC. The Federal Communications Commission is an independent government agency reporting directly to Congress. Created in 1934, it consists of five commissioners appointed by the President. The FCC promulgates regulations concerning radio and telephone that are applicable in the United States and all U.S. territories. The following paragraphs outline the most important of these.

Licensing—the FCC establishes licensing requirements and procedures which assure that only those eligible receive authority to operate in a particular service. This includes establishing frequency coordination guidelines designed to reduce interference. Generally, this means that agencies assigned to the same frequency (who are not part of the same system or who do not use the channel for mutual aid purposes) are physically separated by a minimum number of miles. The licensing process also may place restrictions on output power and antenna heights in order to provide additional protection. Licensing requirements are periodically reviewed and modified. The Universal Licensing System, or ULS, allows applications to be submitted electronically.

Spectrum management—Since radio frequencies are a limited natural resource, an organized plan is required to assure that they are efficiently used. The FCC apportions sections of bandwidth according to function. For example, the fire service is allocated a portion of spectrum assigned to public safety, which is, in turn, a segment of a larger group classified as *land mobile* service. Broadcast television, AM/FM radio stations, military, aircraft, maritime, personal communications services, business, petroleum, and forestry make up just a few of the many competing uses.

Because of the great demand for bandwidth, the commission has taken to auctioning off certain channels to the highest bidder. Much of this has been driven by the explosive growth of the cellular and wireless telephony industry, and has provided significant cash flow to the federal government. Some of the auctions have, however, required some public safety services to vacate previously used microwave channels, however these agencies were compensated for this relocation. It is anticipated that as competition increases, the practice of frequency auctions will continue.

Refarming is another means of improving spectrum efficiency. One part of refarming is the reapportionment of spectrum allocated to services; switching blocks of frequencies formerly assigned to one use to another. The most visible aspect of refarming is the narrowing of frequencies. As part of a multi-year plan, the FCC is assigning new channels that will eventually be only one-fourth as wide as their predecessors. Imagine that a parking garage represents the available radio spectrum. By dividing the parking places in half—and then in half again—you will eventually wind up with four times as many places as you started with. Obviously, you will also need narrower cars or, in this case, radios which operate on a more precise bandwidth. These devices have been available for some time now, and current licensing reflects these new frequency requirements.

Rules and regulations—Some Federal Communications Commission rules are universal, while others pertain to specific services. Part 90 of the FCC rules governs fire department communications. Included here are:

- Eligibility requirements for licensees
- Licensing and frequency coordination procedures
- Specific frequencies assigned to the fire service

- Power and use limitations for all channels
- Paging operations
- Transmitter technical requirements
- Operating guidelines
- Antenna height limitations and tower marking
- Prohibited uses

While there are literally hundreds of rules outlined on this document, among the most important are:

- Radios may only be operated by agencies having appropriate licenses and by individuals having the authority to do so.
- Priority is given at all times to transmissions regarding safety of life or property.
- Users shall not create harmful or intentional interference.
- Communications must pertain to the class of service. (In other words, daily fire department operations.) Public safety users cannot, by law, transmit, for example, what would be considered "programming."
- Stations must identify themselves every 30 minutes or after each group of transmissions. Similar requirements exist for commercial broadcast stations, as evidenced by the frequent displays of network logos and local affiliate call signs. Fire departments may opt to meet this requirement by either having dispatchers make verbal announcements, or through the use of an automatic Morse code identifier.
- Profanity is prohibited.
- Licenses must be posted and maintenance records retained.

It's easy to see that most of the FCC rules and regulations fall under the category of common sense. However, failure to comply with or flagrant violation of these rules can result in substantial fines and, in severe cases, revocation of a license.

The World Wide Web

The Internet has become a major part of our daily life, impacting education, commerce, and communications. It should come as no surprise, therefore, that the web has also made itself known to the fire service. An increasing number of departments have a web presence, a portion of which may be dedicated to public education and requests for information. However, it is the technology used to manage the Internet that has shown considerable promise. *Voice over IP* utilizes routing and network protocols typically associated with data to carry voice, and holds promise for the future delivery of 9-1-1. Voice over IP is already in place internally in many telecommunications application. HTML and XTML, languages used in Web creation, can also be used to author online reference files for the dispatch center. Software has also been developed that allows Incident Commanders to utilize standard web browsers to remotely access CAD systems and other centralized data resources. In this manner, individuals having the appropriate password can use almost any laptop to manage emergencies. This is extremely cost effective as it eliminates the need for specialized software for remote workstations. The World Wide Web is also an excellent research tool.

Summary

In order to adequately address demands for service, fire departments must embrace new technologies. Nowhere is technology expanding at a faster rate than in the area of communications. The ability to share data and information has never been greater, but based upon past experience, the cutting edge technology of today will become the quaint antiquities of tomorrow. With two to three generations of computers being produced each decade, quantum leaps in our ability to store and exchange messages will likely continue to grow at an explosive rate. Of course, with this growth will come challenges. The advent of wireless telephony has greatly reduced emergency reporting times in rural areas and on limited access highways. However, the widespread use of cell phones has also had negative impacts on public safety answering points. Public wireless communications networks may also become jammed with calls following a major event. Without a priority access plan for emergency service users they may be of little value as incident management tools. The fire service must be positioned to fully utilize the benefits of new technology while finding ways to minimize the associated problems.

Reliable communications is an absolutely necessity. The safety of the public and first responders depends upon it. As communications is used in one form or another on every alarm, it is accurate to say that no other aspect of fire department operations has a greater impact. At times both simple yet complex, we rely on communications to report emergencies, receive alarms, dispatch equipment, coordinate on-scene operations, and summon additional assistance. Communications advises us of the type and location of the call and the vital signs of the patient. We use it to warn others when a fire is spreading, and to inform them when it is under control. In short, as fire service professionals we rely on communications to help us carry out the basic functions of our daily assignments. Yet, for all the focus on technology, successful fire department communications require an adequate and well-trained staff. Therefore, future focus needs to be not only on the acquisition of new technology, but on the recruitment and retention of professional personnel.

Internet Resources

The following sites provide on line information concerning fire department communications:

Association of Public-safety Communications Officials International (APCO)
www.apco911.org

Federal Communications Commission (FCC)
www.fcc.gov

Federal Emergency Management Agency (FEMA)
www.fema.gov

International Municipal Signal Association (IMSA)
www.imsasafety.org

National Emergency Number Association (NENA)
www.nena9-1-1.org

National Institute of Occupational Safety and Health (NIOSH)
www.cdc.gov

Occupational Safety and Health Administration (OSHA)
www.osha.gov

Public Safety Wireless Network (PSWN)
www.pswn.gov

Glossary of Acronyms

AAT. Above Average Terrain. The height of an antenna in relationship to ground level.

ACD. Automatic Call Distributor. A device that automatically routes calls to an available attendant, typically found in a larger communications center having multiple call takers.

ACN. Automatic Crash Notification. A system that monitors vehicular condition and automatically notifies a central station should an accident occur, providing the location and particulars about the crash.

ADA. Americans with Disabilities Act. Federal legislation requiring equal access to services for all Americans.

ALI. Automatic Location Information. A function of enhanced 9-1-1 that automatically provides the dispatcher with the address of the calling party.

ANI. Automatic Number Information. A function of enhanced 9-1-1 that automatically provides the dispatcher with the telephone number of the calling party. Also refers to the same feature of a radio system that automatically displays the unit number of an apparatus when transmitting.

APCO. Association of Public Safety Communications Officials. The oldest and largest organization of public safety communications personnel. Active in education and training, as well as FCC matters.

ARRL. Amateur Radio Relay League. Ham radio operators.

AVL. Automatic Vehicle Location. The ability to track the location, position, and travel of apparatus, and to accurately display same on an electronic map.

CAD. Computer Aided Dispatch. A computer system designed to track incidents and apparatus, and to make appropriate unit recommendations based upon location and type of call.

CAI. Common Air Interface. Technology associated with trunked radio systems that allows radios manufactured by different vendors to intercommunicate.

CAS. Call Associated Signalling. A method of delivering ANI/ALI information on wireless E 9-1-1 calls.

CPE. Customer Premise Equipment. The telephone devices and equipment located in a facility such as a fire dispatch center.

CPS. Cycles Per Second. An older reference to frequency, derived upon how many times a radio wave occurred in one second of time. Term has been replaced in most circles by MHz or GHz. (See listings.)

CPU. Central Processing Unit. The portion of a computer that performs operational routines. May also be used to refer to a computer in general.

DTMF. Dual Tone Multi Frequency. A signaling pattern used for touch-tone telephone dialing and in certain alerting systems.

E 9-1-1. Enhanced 9-1-1. The provision of location and number information automatically on incoming emergency calls.

EMD. Emergency Medical Dispatch. A formalized means of establishing protocols for emergency medical calls and providing pre-arrival instructions.

EOC. Emergency Operations Center. A facility at which department heads and government officials meet to manage a disaster.

ESN. Emergency Service Number. A response zone used in the creation of geographic databases. All addresses having the same ESN are assigned to the same public safety response agencies.

FCC. Federal Communications Commission. National regulatory body that establishes rules and regulations for radio and telephone services.

GHZ. Gigahertz. One billion hertz, or one billion cycles per second. May be used as a reference to frequency, particularly of a microwave radio, or to the processor speed of a computer.

GPS. Global Positioning System. A means of pinpointing the location of an earthbound object by use of satellites.

IMSA. International Municipal Signaling Association. An association having a long time interest in fire alarm and fire communications operation and licensing.

ISP. Internet Service Provider. Any company that supplies access to the Worldwide Web.

ITS. Intelligent Transportation System. A "smart highway" containing the technology to monitor and manage traffic flow. May be accomplished through sensors, electronic signs, cameras, and a variety of tools.

IVR. Interactive Voice Response or Interactive Voice Recording. An Automated means of answering telephone line that provides callers with a choice of options. Example: Press 1 for Fire Prevention, Press 2 for Training Division, etc.

LEC. Local Exchange Carrier. A conventional telephone company.

LED. Light Emitting Diode. An electronic component commonly used for indicator lamps and displays in communications equipment.

LEOS. Low Earth Orbit Satellite. A satellite in close proximity to the earth (in space terms), typically used for communications or location technology.

MDT. Mobile Data Terminal. An in-vehicle computer, often linked to the dispatch system and external databases.

MHz. Megahertz. One million hertz or one million cycles per second. Used in reference to a radio frequency, as in 46.18 MHz, or to the speed of a computer processor.

MSAG. Master Street Address Guide. A telephone data table that is used to match telephone numbers and addresses with the appropriate response agencies. Supports enhanced 9-1-1.

MVT. Mobile Video Terminal. An MDT possessing video capacities.

NASNA. National Association of State Nine-one-one Administrators. Professional group comprised of chief executive officers responsible for 9-1-1 implementation and operations.

NENA. National Emergency Number Association. A group heavily involved in 9-1-1 issues.

NCAS. Non Call Associated Signaling. A method of transferring ANI/ALI data on wireless E 9-1-1 calls.

NOC. Network Operations Center. A telephone company facility designed to monitor and control the telephone network.

PDA. Personal Digital Assistant. A miniature hand-held computer.

PSAP. Public Safety Answering Point. Any facility that answers emergency telephone calls from the public.

PSTN. Public Switched Telephone Network. The equipment and cabling owned and operated by the telephone company.

PSWAC. Public Safety Wireless Advisory Committee.

PSWN. Public Safety Wireless Network. A joint venture of the Treasury Department and Department of Justice designed to foster seamless interoperability in public safety communications.

PTT. Push To Talk. The feature, or the switch involved, that enables the ability to push a button and transmit on a radio.

RAN. Recorded Announcement. An automatic message placed on an incoming line, especially an emergency number, to provide instructions to callers.

TDD. Telecommunications Device for the Deaf. A keyboard and modem device specifically designed to the deaf and hearing-impaired to communicate by phone.

TTY. Teletypewriter. Any device using a keyboard to send messages over a telephone line. May be used to communicate with the deaf.

UHF. Ultra High Frequency. A portion of the radio spectrum that contains the 400, 700, and 800 MHz bands. The fire service is allocated channels in each of these bands.

VHF. Very High Frequency. A portion of the radio spectrum that contains the low and high public service bands.

APPENDIX

The following questions come from the NIOSH computer workstation checklist:

1. Does the workstation ensure proper worker posture, such as: horizontal thighs? vertical lower legs? feet flat on floor or footrest? neutral wrists?
2. Does the chair adjust easily? have a padded seat with a rounded front? have an adjustable backrest? provide lumbar support? have casters?
3. Are the height and tilt of the work surface on which the keyboard is located adjustable?
4. Is the keyboard detachable?
5. Do keying actions require minimal force?
6. Is there an adjustable document holder?
7. Are arm rests provided where needed?
8. Are glare and reflections avoided?
9. Does the monitor have brightness and contrast controls?
10. Do the operators judge the distance between eyes and work to be satisfactory for their viewing needs?
11. Is there sufficient space for knees and feet?
12. Can the workstation be used for either right- or left-handed activity?
13. Are adequate rest breaks provided for task demands?
14. Are high stroke rates avoided by: job rotation? self-pacing? adjusting the job to the skill of the worker?
15. Are employees trained in: proper postures? proper work methods? when and how to adjust their workstations? how to seek assistance for their concerns?

15

FIRE DEPARTMENT WATER SUPPLY

Jack L. Cottet

CHAPTER HIGHLIGHTS

- Formulas for required fire flow (theoretical and/or fireground)
- Definitions of needed fire flow and basic fire flow
- Five duties of a water supply officer
- Two solutions to water supply problems in the rural/urban interface

INTRODUCTION

Water remains the most often used fire suppression agent in structural and wildland fire fighting. Most fire apparatus is designed with the primary purpose of supplying water to a fire. Firefighting tactics center on exposure protection, confinement, and extinguishment with water. Whether in a large city, a rural area, or a forest, water is the extinguishing agent the fire service uses most often. This chapter will focus on fire suppression water supply and examine the chief officer's responsibilities for assuring adequate water supply in municipal and rural areas.

Fire Flow Requirements

Generally, fire flow requirements (the amount of water necessary to control a fire incident) are determined for specific buildings—especially large buildings, structural complexes, or target hazards. Fire flow requirements can be set for areas that contain similar construction features and occupancies. An example would be an apartment complex or residential neighborhood. Necessary fire flows also should be determined for buildings and complexes in areas not served by municipal water systems so that adequate alternative means of supplying water can be developed. In setting fire flow requirements you must consider building height and area, type of construction, contents or fire loading, and the proximity of adjacent buildings.

The most basic question with respect to water for fire fighting is how much is needed. Among the many approaches for determining fire flows is the much-respected Insurance Services Office's (ISO) formula for Needed Fire Flow, which was published in 1980. Several other methods that have more direct application to fireground use are those from Iowa State University, the National Fire Academy, and NFPA Standard 1231 Water Supplies for Suburban and Rural Fire Fighting.

Needed Fire Flow

The ISO Needed Fire Flow (NFF) is defined as the rate of flow expressed in gallons per minute at 20 psi residual pressure necessary to control a fire incident within a specific structure. Needed Fire Flow is used to determine the adequacy of a water distribution system during the classification of a city using the *Fire Suppression Rating Schedule.* NFF is determined based on the construction, occupancy (building use), exposures, and the protection provided openings in party walls or passageways. NFF cannot be greater than 12,000 gpm (8,000 for frame and ordinary construction), nor less than 500 gpm. It is rounded to the nearest 250 gpm up to 2,500 gpm and the nearest 500 gpm for flows that exceed 2,500 gpm. The recommended fire flow for residential properties (such as single family dwellings that do not exceed two stories in height) depends on the distance between structures.

Distance	Fire flow
>100 ft.	500 gpm
31 to 100 ft.	750 gpm
11 to 30 ft.	1,000 gpm
<10 ft.	1,500 gpm

Basic Fire Flow

Another term used in municipal public protection classification is Basic Fire Flow (BFF). The BFF is used to determine the number of needed engine companies. The fifth highest Needed Fire Flow in the city is selected as the Basic Fire Flow and is considered a representative flow for the community. The maximum BFF in the ISO process is 3,500 gpm.

Fire flow duration can vary from two to ten hours depending on the quantity required. The duration for the maximum Basic Fire Flow requirement of 3,500 gpm is three hours. For requirements of 2,500 gpm or less the required duration is two hours and increases by one hour for each 1,000 gpm required.

Gallons per Minute Required	Duration in Hours
< 2,000	2
3,000	3
4,000	4
5,000	5
6,000	6
7,000	7
8,000	8
9,000	9
>10,000	10

The ISO uses the formula: $F = 18\ C\ (A)\ 0.5$

Where:

F = required fire flow in gallons per minute,
C = coefficient related to the construction type, and
A = total floor area of all floors except the basement in the building.

A is modified for fire-resistive construction. The fire flow is increased or decreased by 25% depending on the occupancy. The figure can then be reduced for automatic sprinkler protection. Finally, a percentage ranging from 5% to 25% is added for exposed structures within 150 ft. Anyone who wishes to use this guide should get a copy and become familiar with the details of its application.

Regardless of what formula or method is used to arrive at a required fire flow estimation, the primary thing to keep in mind is that, in all cases, this is merely an estimate of the amount of water that might be needed for fire control. It is best to apply it conservatively, that is to assume that additional water could well be needed depending upon conditions at the time. Fire flow estimates, like so many items in the fire protection field, should be considered minimums that must be achieved, as opposed to the approach that "we can probably make do with less."

Fire flow testing

Once a theoretical fire flow requirement has been determined for a building or an area, the fire department should test the water supply provided from the available source(s) (water system, static source, or fire department operations [relays or shuttles]) to determine if it can meet the need adequately.

There are two fallacies associated with the testing of pressurized water distribution systems. First, a high-pressure reading on a hydrant (static pressure) does not necessarily equate to a large, or adequate, amount of water being available for fire fighting. Second, the amount of water measured flowing from a single hydrant is not necessarily the total amount of water available in the area for fire or emergency operations. Proper fire flow tests have to be conducted according to standardized procedures to determine the amount of water actually available in an area. Some of the texts that outline the procedures for conducting fire flow tests include the *Fire Protection Handbook* published by the NFPA, the IFSTA-validated *Water Supplies for Fire Protection,* and *Fire Protection Hydraulics and Water Supply Analysis* by Pat Brock. Water system testing should be coordinated with the water department and it is the fire chief's responsibility to seek the cooperation of the water department. For excellent information on rural water supplies (static and flowing water, hose relays, and apparatus shuttles) consult *The Fire Department Water Supply Handbook* by William Eckman.

FIREGROUND FORMULAS

Iowa formula

The "Iowa Formula," developed by the Iowa State University, is based on the quantity of water in gallons per minute that when changed to steam will displace the oxygen in an enclosed space. This is basically a field method that can be applied on the fireground. The formula is:

$$\text{Required Flow in gpm} = \frac{L \times W \times H}{100}$$

Where:

L = Length,
W = Width, and
H = Height of the enclosed area (all measured in feet)

An example in a one-story building $30W$ by $80L$ by $10H$ would be:

$$\text{Required Fire} = \frac{(30)(80)(10)}{100}$$

Required Fire Flow = 240 gallons per minute

Essentially, the formula requires one gallon per 100 cu. ft. of space. Since this formula considers only the size of the building and not the occupancy or fire loading, the dimensions of the entire building generally are used. Various authors and some texts recommend that the quantity determined using this formula should be multiplied by a factor of 2, 3, or 4 to compensate for inefficient water application and occupancy hazards.

It should also be recognized that the answer determined by using this method will invariably be lower by a substantial amount than that obtained either with the ISO or NFA formulas. This fact further supports the idea that fire flow determination is not a precise science and should be regarded as an estimate only. It seems wisest to use the method that results in the highest gallons per minute value to achieve the most conservative answer to the question.

In considering the Iowa method, it should be recognized that the original testing and experimentation involved did not include the high fire loads and extensive amounts of plastic materials prevalent in modern buildings. Many people are of the opinion that the water flows calculated using this method are too conservative and that efficient fire control is less likely using these smaller amounts of water.

This method also considers that the building will be relatively intact and that a great deal of the extinguishment activity will occur as the result of steam generated by the water application being trapped in the fire area and assisting with extinguishment. Where the fire has self vented or fire department ventilation has been accomplished, the effect of steam is greatly reduced.

National Fire Academy formula

The fireground formula developed by the National Fire Academy for Required Fire Flow is:

$$\text{Required Flow in gpm} = \frac{L \times W \times \textit{Number of Floors Involved}}{3}$$

Where:

L = Length, and
W = Width of the structure (both in feet)

The Required Fire Flow is for a fully-involved structure. If only a portion of the building is involved, the amount should be reduced proportionally, generally to 50% or 25% of the calculated required flow. In multistory buildings, floors below and those not involved in the fire are not considered in the calculation. Add an additional one-fourth of the Required Fire Flow for each exposed building or exposed floor in a multistoried building.

Using this formula for the same building as used in the Iowa formula (an 80 ft. x 30 ft. one-story structure), the flow rate needed is 2,400 divided by 3 = 800 gpm. This amount is nearly four times that of the Iowa method. In considering this substantial difference many people conclude that the NFA formula is probably a more realistic estimation of the quantity for water and that it is more likely to result in successful fire control than the Iowa method.

Planning to meet determined fire flows

Once the tests have been run and the results computed, the amount of water available can be compared to the estimated water requirement determined earlier. This will quickly show the adequacy or inadequacy of the system, static source, or fire department operations. This computation is used by ISO in their community grading process by comparing the available flow rate versus the required and calculating the percentage of deficiency. This calculation is done using the simple formula: Available divided by Required, with the answer multiplied by 100. Thus, if a building needs 2,500 gpm, but the flow tests show only 1,500 gpm is available, the percentage of deficiency is 40%.

If adequate water is not available, the fire department must develop two action plans. First, the staff must devise operations to immediately provide additional water to the site. This could involve relay operations, the use of large diameter hose, water shuttles, auxiliary equipment, or the initiation of automatic/mutual aid.

Second, the fire chief should work with the water department and municipal officials to plan infrastructure improvements on a long range basis that will increase the available water supply for the area. In rural areas this could mean working with township supervisors or county commissioners to develop water supply points. Often, simple suggestions by the fire department can make substantial improvements in water delivery. For example, pointing out where water mains can be cross-connected to form a grid or looped with large feeder mains can be very helpful. This is an opportunity for the fire department to state its case for direct involvement in future water system planning and overall long-range planning of the entire infrastructure. Another avenue to pursue may be the installation of dry hydrants or drafting sites to help supplement the water available from a water system. Care must be taken in such arrangements to avoid any possibility of water supply contamination by cross connection—in other words, pumpers being supplied by a municipal water system should not have their water supply supplemented by connections to pumpers pumping from non-potable supplies such as dry hydrants or drafting sites.

Fire Department Needs

The fire department should be directly involved in community planning, especially in the area of water supply for proposed important new housing tracts, shopping or apartment complexes, or industrial parks. Plans should be reviewed before construction begins. Close cooperation with water department officials ensures that they will understand fire protection needs, which will exceed those for domestic consumption. Plan for adequate protection during the building construction phases of new developments. Water supply demands must include domestic consumption, fire protection systems, and additional water for fire department suppression activities. The municipal water system must be designed to provide the needed water for each of these requirements. Areas of concern include water storage capacity, main sizes, establishing grids with numerous cross connections so there are parallel mains to provide water flow from multiple directions (reducing friction loss), and large feeder mains looping the grid. Additional items are valves, hydrant branch sizes and valves, and an adequate number of hydrants suitably located with proper spacing. The location of fire hydrants in relation to normal approach routes of fire apparatus is an important consideration. Where large fire flows are required, several hydrants will be needed relatively close to the building in order to allow the larger amount of water to be obtained by several pumpers. Be careful to locate hydrants far enough away from the structure so that they will be usable during a fire, generally a distance of at least 50 feet is used as the minimal separation distance. Hydrants are limited to a maximum flow of approximately 3,000 gallons per minute. Although dual pumping operations are possible on hydrants, it is a good idea to avoid this need if possible and provide additional hydrants where large flows are available and needed. Planned review of new buildings is important in water supply analysis because hydrant and placement accessibility for fire department operations are vital. Do not overlook hydrant location with respect to automatic sprinkler or standpipe connections—hydrants should not be more than 100 ft. from these fire department connections.

Water System Maps

Another aspect inherent in developing a good working relationship with the water department is to obtain maps of the water system. Such maps are invaluable in pre-incident planning and for the incident command staff during emergencies. Adequate water system maps are needed in the command post at major fires. All operational chief officers' vehicles should have water maps, at least for the sections of the city to which they regularly respond. With today's technology, maps can be scanned into a computer and brought up on a terminal while a unit is responding. Include information on hydrant locations, main sizes, and available fire flow. Make sure that dispatchers have access to water maps on their computer. Institute procedures for continuously updating the computer-based maps.

In your work with the water department, suggest system updates to improve firefighting capability, including recommendations to connect dead-end mains and create loops. Replace older, undersized mains and/or improve primary feeder mains. Replace, relocate, or add fire hydrants. Seek the installation of new mains where necessary as replacements or for expansion. Also, seek regular examinations of the condition and actual diameter of mains affected by encrustation or tuberculation. Inspection, testing, and servicing of fire hydrants is required twice a year and is an important element in the ISO evaluation of a community's water system.

Adequacy and reliability

Officials of the water utility, government administrators, and elected boards or councils often do not understand the water system concerns relating to fire protection. It then becomes the fire chief's responsibility to explain to those in authority what these requirements are and why they are necessary. Although water department personnel understand sources, storage, and distribution (at least for domestic consumption), they often do not realize the fire service need for immediate large water flows to quickly control a spreading fire, nor the concepts of adequacy and reliability.

Adequacy is defined as the water system's ability to deliver maximum daily domestic consumption plus the required fire flows at various locations. This includes having an adequate number of fire hydrants within a reasonable distance to deliver the required flow. In most situations the maximum a fire department pumper can deliver from a single hydrant is 2,000 gallons per minute. This obviously depends upon pump size, large diameter hose, and the ability to apply the water using multiple master stream devices.

Reliability is defined as the ability to continuously supply water even if a part of the system is out of service or other factors are present resulting in difficulty in supplying the amount of water needed. In other words, the system is redundant if there is more than one way to provide an ample water supply to meet fire flow requirements. A big part of the reliability component lies in storage of water within the system. Most water systems have tanks or reservoirs containing many hundreds of thousands of gallons which can allow the maintenance of a continuous flow rate for fire fighting. It should be recognized that large flows applied for long durations consume significant quantities of water. For example, a flow rate of 2,500 gpm for a duration of two hours is 300,000 gallons of water. The principle of reliability addresses whether the system can supply that flow rate with one or more pumps out of service, etc.

Sources

Water systems obtain water from a variety of sources, including reservoirs that collect runoff, rivers, lakes, underground aquifers, and wells. Most static sources are adequate unless there are extended periods of drought. Wells can go dry or the water table can drop, but most problems are associated with insufficient capacity or the number of sources. The major difficulties with sources are the adequacy of the transmission lines bringing water to the system and their reliability. In simple terms, do they carry enough water, or, if a transmission line fails, is there a second line or alternative method of using the source of water? Again, the principle of adequacy and reliability can be seen at work in this discussion. A prolonged fire during the peak domestic consumption phase can cause severe problems for the fire department in many cases. A water system that has more than one source of water increases the reliability factor, but is a relatively rare situation.

Storage

Most water systems maintain several hours of treated water storage to allow for peaks in consumption, breakdowns in the system, a treatment facility that cannot keep up, and, in many areas, to provide pressure to the system. Water can be stored in underground reservoirs, in ground storage tanks (usually placed on high ground), and in elevated storage tanks. In some cases, it is necessary to pump from the storage into the system; in others, gravity is used to provide the pressure source, which is a more reliable method. A pump malfunction or power outage can nullify the storage capacity of a pump-reliant system. The adequacy of the storage is usually the greater problem; it must be sufficient for maximum daily consumption and the required fire flow duration measured during a period of the highest possible consumption, coupled with the worst case scenario for replenishment from the source of supply.

Distribution

Distribution systems are considered reliable if there is a good cross-connected grid that provides water from multiple directions. Reliability also is enhanced if the system has adequate valving so that only short sections of pipe must be closed for repairs. Of course, adequacy depends on pipes sized to meet both the domestic consumption and fire protection requirements. Small mains in older sections of a city or in rural water districts, long dead ends, the lack of a grid system, or buildup of foreign materials in the mains can decrease distribution system adequacy. Comparing current fire flow test results with past tests will indicate changes in adequacy. The minimum main size recommended for fire protection is six inches, which should be used only in low-density residential neighborhoods and should be cross-connected frequently. The American Water Works Association has developed standards for the pipes and hydrants used in distribution systems.

Residential areas should have 6-in. mains interconnected at 600-ft. intervals with 8-in. mains. In business and industrial areas recommended minimum sizes are 8-in. mains with cross connections to 12-in. mains every 600 feet. On principal streets and in long mains that are not cross-connected at frequent intervals, 12-in. mains should be used.

It should be noted that in older water systems, large mains with a closely spaced grid are rare and the reverse is more typical. By bolstering the distribution grid, an older system can be updated to provide increased fire flows needed by modern buildings.

Hydrant use

Fire departments often fail to make good use of the system capacity and adjacent hydrants, especially at large fires and where hydrants are spaced relatively far apart. Departments have a tendency to select hydrants which are conveniently close to the fire location as opposed to selecting those where the flow rates may be greater but the distances longer. With the prevalence of large diameter hose today, this tendency has been somewhat reduced but it is still a major concern.

Hydrants on large diameter water mains with good pressure generally will supply two large pumpers without difficulty, depending upon main sizes and pressures. A large-volume, low-pressure system also will commonly support two pumpers, with the pumps developing the necessary fire stream pressures, provided that one or more of the pumpers is placed directly at the hydrant. In sections of a distribution system where large mains are cross-connected at relatively short intervals, a single hydrant may even supply three or more pumpers. A guide for hooking up additional pumpers is the residual pressure as read on the pump's intake gauge as each pumper is connected to the hydrant and discharges its streams.

If a substantial drop in the residual pressure occurs after charging the initial line, this indicates that a great deal of the potential water flow is being used and is a warning sign for the pump operator. Keeping an adequate residual pressure is an important part of safe pump operation. A figure of 20 psi is considered a safe minimum to maintain. No additional pumpers should be placed at the hydrant and no increase in the draw on the system should be permitted once this figure is reached, and once it is, this fact should be reported to the water supply officer or to command. If additional water supply will be needed and the system has already been reduced to a residual of 20 psi, alternative sources, such as a tanker shuttle or drafting sites, will be needed.

Where hydrants do provide large flows of water, a technique called dual pumping can be used. Dual pumping is sometimes mistakenly called tandem pumping. The differences between tandem and dual pumping should be well understood. With tandem pumping, the pumps are connected to a single hydrant, but in such a way that the first pumper discharges into the intake of the second unit, which then increases the pressure even further while the volume remains limited to that provided by the original pumper. In dual pumping operations, the first pumper is connected to a strong hydrant and discharges water to the fire. The second pumper is then connected intake to intake or to available additional discharge outlets on the hydrant, uses the surplus water not being pumped by the first pumper, and discharges directly to the fire.

Placing two (or three) pumpers on a single hydrant offers the following operational advantages:

- The speed with which the second or third engine company can stretch handlines to supply heavy stream appliances.
- The reduction in the amount of hose used at large fires. The same amount of water can be directed on the fire with fewer lengths of hose. Remaining hose can be used for additional lines to increase water delivery on the fire. When less hose is used, the time to place companies back in service can be reduced, a plus when staffing is low and/or rehabilitation is required.
- Grouping apparatus close together improves the efficiency of the operation. Fewer blocked streets and less stopped traffic mean less private and public transportation is inconvenienced.
- There is more effective use of pump pressure that is normally wasted in friction loss when hose stretches are long.
- Shorter hoselines means that water from deck guns, portable monitors, or elevated master streams reaches the fire faster and increases the potential discharge capacity since pump pressures can be used more efficiently. This can be the difference between an offensive attack with early extinguishment and retreating to a defensive operation.

Water System Maintenance

Water system maintenance with respect to available fire flow is a major concern. If annual flow tests show progressive reductions in flows, the chief needs to communicate this information to water department officials, ask whether they can explain the lower flow, and cooperate with them to remedy the problem. The fire chief may have to take the added step of helping the water department administration justify improvements or major renovations to clean and line older water mains.

Hydrant maintenance

In some cases the fire department is responsible for hydrant maintenance. However, no matter who is resonsible, it is up to the chief to see that hydrant maintenance is done properly and on schedule, that repairs are completed quickly, and that adequate records are maintained. Hydrant inspections should be conducted at least twice yearly. Inspections conducted in the spring can detect damage done during the freezing winter months and inspections in the fall help to assure that hydrants are ready for service before cold weather approaches. For dry barrel hydrants, the inspection process must assure that hydrants properly drain or are pumped out to prevent freezing in cold weather. A good maintenance inspection should include:

- Checking visually for hydrant damage
  - Struck and out of line
  - Cracks in barrel, bonnet, or caps
  - Damaged operating nut
  - Missing caps or chains
  - Foreign objects or debris in the barrel
  - Obstructions that have been added around the hydrant
- Performing a pressure test
  - Checking for leaks in bonnet, stem packing, and nozzle caulkings
  - Checking for a defective drain valve
  - Checking for underground leaks
- Flushing the hydrant
  - Checking its operation and stem stiffness
  - Flowing until discharge is reasonably clear
- Checking the drain operation
- Checking the condition of outlet threads with a female coupling
- Ensuring free movement of hydrant cap chains; removal of excess paint from chains and swivels
- Checking cap gaskets
- Lubricating cap and outlet threads
- Lubricating hydrant if required
- Painting if necessary and color coding according to national standards

Hydrants located in areas of cold climates must be checked frequently. If necessary, thaw with steam, pump on a regular schedule, or treat against freezing.

Some common hydrant installation problems are:

- Hydrant facing the wrong direction
- Outlets located too high
- Outlets located too low, especially the pumper connection
- Obstructions to outlets or hydrant wrench use
- Inaccessible location or too far from street
- Hidden locations

Any hydrant found to be unusable for fire fighting must be noticeably marked so that a responding fire company does not attempt to use it in an emergency. A large prominent sign should be used for this purpose and has the side benefit of encouraging the water department to hasten repairs, as it often leads to public concern about the reduction in fire protection that it constitutes. Any fire companies that respond to the area should be notified that the hydrant is out of service and re-notified when it is returned to service.

Hydrant branch valves also may need to be operated and maintained. If individual hydrant flows are found to have decreased, the water department should be asked to check valves on the lines to make sure that all are fully open.

Records should be kept of hydrant inspections, maintenance, and repairs; individual hydrant flows and area fire flow availability; and recommendations for water system improvements. These records should be stored in a computerized database where they can be easily accessed and updated.

HYDRANTS

The fire chief or the department's appointed water supply officer should further assist the water department in selecting locations for new or replacement hydrants. In addition to choosing appropriate locations, selecting the type of hydrant also is important. Hydrants vary from one to six outlets of various sizes. The most common type is a three-way hydrant with two 2½-in. outlets and one pumper (usually 4½-in.) connection. However, the department may prefer a hydrant that has two pumper connections installed on large mains at target hazards. These hydrants need a large branch main (8-in. minimum) and barrel size to accommodate large flows. Hydrant barrels range in size up to 16 inches. Be sure to make suitable arrangements for protecting operating nuts, caps, and valves on hydrants from unauthorized use. If special wrenches or tools are required, be sure responding units have them.

The department should use a hydrant color coding or marking system meeting national standards. Hydrants should be marked to indicate individual flows by painting the bonnet and caps according to NFPA 291 Recommended Practice for Fire Flow Testing and Marking of Hydrants.

Capacity of 1,500 gpm or over = Light blue

1,000 gpm to 1,499 gpm = Green

500 gpm to 999 gpm = Orange

less than 500 gpm = Red

The standard also requires hydrants to be painted chrome yellow in color and the color code applied to the bonnet and caps.

An additional reason for using the nationally recognized color coding system as opposed to a local system is that mutual aid apparatus may be called upon to use the hydrants and may not be familiar with a local system. For the same reason, outlet connections on hydrants should be standardized as 4½-in. for steamers and 2½-in. for hose outlets.

An alternative used in some communities is to code the hydrants based on water main size or stencil the main size on hydrants. Fire flow results also can be stenciled on a hydrant for the use of pump operators. Some fire departments have a specific paint scheme to indicate the limitations of hydrants on dead-end mains. The visibility of hydrants should be improved by using a paint for the barrel that reflects headlights of responding units. This type of paint usually has a reflective glass bead to improve visibility. Other departments increase hydrant visibility by using distinctive reflective markers, usually blue, in the center of the street. Reflective paint can be used on the street (to paint an arrow, for example) to show hydrant locations. However, in areas where snow or ice would cover the street, the hydrants should be marked with a flag or post visible above snow accumulations. Fire departments in areas where the amount of snow accumulation is a problem must have plans in place to get snow removed from around hydrants to allow for visibility and unimpeded use. A good public relations program is to have local residents adopt a hydrant and keep it visible and accessible during the winter.

Water Supply Officer

Every department, no matter what size, should consider having a Water Supply Officer (WSO) who assists the incident commander by monitoring water for fire suppression use and fireground operations on a day-to-day basis, both for large and normal fires. Since the WSO's primary objective is to ensure an adequate water supply during emergencies, there must be close liaison with water department officials. The WSO's responsibilities include acquiring and keeping up-to-date water system maps; developing procedures for notification of out-of-service hydrants; scheduling and/or monitoring inspections, ensuring that hydrants are maintained and repaired; recommending system improvements; and planning fire suppression needs for water system expansion.

The WSO needs to work with the fire prevention bureau to review and plan the number, location, and size of new hydrants needed for new properties. Another function is to assist operational officers to develop Standard Operation Procedures (SOPs) necessary for adequate water supply. In many cases, special alternate operations will be needed to produce the required fire flow.

The WSO needs to be included in writing specifications for new fire apparatus. The WSO will ensure versatility in design for water delivery adequate hose, fittings, adapters, intakes and outlets to achieve the best performance from the water supply and should provide input on pump and water tank capacities to be purchased.

On the fireground, the Water Supply Officer must determine the amount of water being delivered and consult with command to determine if additional water will be needed, then calculate how much more water is available, or whether special water supply operations need to be initiated. Pump operators should be taught to record static pressures before pumping. The WSO can collect the static pressures, check the residual pressures, and determine the amount of additional water available. The WSO should always respond to multiple-alarm fires. If a WSO is not available, a temporary WSO should be appointed. Sometimes the water department may need to respond to the scene or change normal operating procedures to increase the flow. The WSO may recommend that additional pumpers be dispatched as water supply units or that special pumping operations are established.

Standard Operating Procedures

A recommended procedure is to have each pumper operator record, by mechanical or electronic means, the location of the pumper and hydrants, type and size of hydrant connection, static pressure, number and size of discharge lines for estimating flow (using flow meters to obtain total flow is more accurate, especially if automatic nozzles are used), and residual pressure. Then the additional capacity of the water system can be calculated. In some departments, a Water Supply Company is assigned on an extra alarm to provide additional pumping capacity, hose for water supply use, and staff. To limit radio traffic, assign firefighters to collect the water supply information from the pumping companies and bring it to the WSO. Then they can assist in analysis, and be used to correct or improve water supply operations. Some solutions are as simple as adding another intake line between the hydrant and the pumper. Review department SOPs to make sure they provide adequate water supplies. Depending upon the water system capabilities, forward or reverse lays may be required in areas where hydrants are widely spaced; reverse or split lays may also be needed depending upon conditions. SOPs should also be written to maximize pumper capability. For example, a 1,500 gpm pumper with two 1½-in. lines operating at a major working fire is *not* being used to its capability. Similarly a pumper being supplied by a single four-inch supply line several hundred feet from a hydrant is probably not receiving an adequate supply based upon the size of the hose and the distance involved. Departments that use 2½-in. and 3-in. hose must be trained in the use of parallel and siamese lines to increase fireground water supply. Other pumping operations that increase fireground flow include relays; tandem, dual, and supplemental pumping; and the use of large diameter hose or tanker shuttle operations.

Placement of Pumpers at Hydrants

There are several ways to connect more than one pumper to the same hydrant; the number of outlets on the hydrant and their direction and other factors all affect the choice of method.

Two pumpers can take their water from a single hydrant if the first pumper is connected to the 4½-in. connection and the second pumper to the other connection or connections. A hydrant gate valve(s) or gated wye should be placed on the other connections(s) by the first pumper. An alternative is to connect the first pumper to the 4½-in. connection of the hydrant and the second pumper to the unused intake connection of the first pumper. Having a large valve on the intake of either or both pumpers expedites this operation. This is known as dual pumping.

Combining the two methods will place three pumpers on one hydrant. The first pumper connects to the pumper outlet of the hydrant, the second pumper attaches to the unused intake of the first pumper (dual), and the third pumper to the other outlet(s). The third pumper can be connected by dual lines from the gated outlets or a gated wye, or by using an increaser from a 2½-in. outlet to a larger threaded hoseline.

If the large hydrant connection is not used when hooking up a pumper, much of the capacity of the system is wasted, as the pressure to move a high volume is expended overcoming the friction loss in the small diameter supply lines. Reduce this friction loss by using short, large diameter intake lines or soft sleeves.

Other Pumping Operations

Relay pumping uses large diameter or multiple hose lines to pump water from a source pumper to an attack pumper located more than 1,000 feet away. Depending on distance, hose size, and required flow, additional intermediate pumpers may be needed. In the northeast, relays that use large diameter hose for distances of one and a half miles are not uncommon. Tandem pumping is a short relay between two pumpers, often referred to as a wagon and a pumper. The attack pumper (wagon) makes a straight lay and the pumper connects to the source and pumps through the line. An alternative is to have a second pumper lay hose from the discharge of a unit already connected to a source (preferably with large diameter hose) and connect the line to its intake. In this operation, the advantage is that the wagon or second pumper receives water under pressure and can use it to build its own discharge pressure. To remember tandem, think of one pumper behind the other as with the wheels of a tandem axle. Dual pumping is two pumpers connected to the same hydrant. Although they can connect to different outlets, normally this is done by connecting the intake side of the two pumps together. The objective is to get full use of the hydrant with minimum hose lays. Think of dual pumping as one pumper next to the other, like dual wheels.

Supplemental pumping is a pumper at a secondary source supplying additional water to one or more pumpers at the scene that already are working from a water supply, but in need of more water to provide effective streams. Of the several ways to accomplish this, the most common is to have an engine company connect to a large main on an adjacent street and discharge into the unused intake of the pumper(s) as the fire scene.

SUPPLYING FIRE SYSTEMS

The fire department often will find it necessary to provide support to private fire protection systems such as automatic sprinklers or standpipes. Private yard hydrants on the lines supplying these systems should not be used by the fire department, because they can rob the system of needed supply. Similarly, hydrants on the public water supply to these systems may be incapable of supplying the fire protection system and additional handlines for the fire department. These hydrants should be tested before an incident and the results recorded in the prefire plan. The key here is that the sprinkler system should receive priority in terms of water supply above other fire department water needs.

Buildings equipped with standpipe systems must be supported by the first-arriving engine company. This is essential for a dry standpipe if it is to be used. Special supply operations are required for standpipe systems equipped with pressure-reducing valves. Large diameter hose is recommended for supporting both types of systems. Consideration must be given to how standpipe and sprinkler systems can be supported if the fire department connection has been damaged.

SUBURBAN AND RURAL WATER SUPPLY

Water supply requirements for fire suppression in rural or suburban areas actually do not differ from those in urban areas served by a municipal water system. In terms of water supply, the differences are where the supply is found and how it is delivered from the source to the fireground. Essentially, determining fire flow requirements, prefire planning with emphasis on water supply, SOPs for delivering water in rural districts, assigning a water supply officer, apparatus considerations, and developing rural water sources are very similar in scope to the operations necessary where a water system is in place. What does differ is that in rural or suburban areas not provided with hydrants the need to develop an adequate water supply can easily consume as much or more in terms of resources than the attack on fire itself.

Minimum Water Supply formula

The methods for estimating required fire flow described earlier also can be used in rural situations. Another method often used in areas that are beyond a water system is described in NFPA Standard 1142 Water Supplies for Suburban and Rural Fire Fighting. This system bases the Minimum Water Supply (MWS) on the total amount of water that should be necessary for a fire in the structure, given in gallons, on the size of the structure, an occupancy classification which relates to the fire load, and a construction class. When there are exposures, the MWS is multiplied by a factor of 1.5.

The formula is:

$$\text{MWS} = \frac{\textit{Volume of the Structure} \times \textit{Construction Classification} \times \textit{Exposure Factor}}{\textit{Occupancy Hazard}}$$

For a two-story frame dwelling of 40 x 50, the volume is the length multiplied by the width multiplied by the height. For the area below a peaked roof, use one-half of the height of that section. For a dwelling with a peaked roof, the figures would be 40 x 50 x (20 + ½ x 8) = 48,000 cu. ft. The Occupancy Hazard (OH) figure ranges from 3 for severe hazards to 7 for light hazards. This dwelling would be a light hazard with an OH of 7. The Construction Classifications range from 0.5 for fire resistive to 1.5 for wood frame; however, frame dwellings have the classification of 1.0. Thus, the MWS in this example would be:

$$\frac{48{,}000 \times 1}{7} = 6{,}857 \text{ gallons}$$

The fire department should have the capability of developing the water needed using one or more methods such as dry hydrants, tanker shuttles, or large diameter hose relays, depending on the total water supply required. Of particular importance in rural areas are the larger target hazards such as schools, day care centers, and homes with multiple elderly, or physically or mentally challenged occupants. Large potential loss structures, including manufacturing, processing, storage, and entertainment facilities, frequently have water for domestic use only and pose a major fire challenge for fire departments in rural areas.

Once the required fire flow or MWS based on Standard 1142 is determined, chief officers must establish a plan to provide the needed water supply in a timely and efficient manner. They must develop pre-incident plans that will determine both necessary resources and how these resources will be brought together on the fireground. A vital portion of the plan is to identify water sources available for the rural structure and how to move the water from the source to the fire in quantities adequate to meet the determined fire flow requirement. Calculations should be worked out to determine the size and amount of hose and the number of pumpers, or the number of mobile water supply apparatus required. It may be possible to control fires in smaller buildings with the water carried on responding units. Prefire plans should include the location, distance, and amount of flow available at the water source and its reliability and accessibility. If special operations might be needed (for example, making a hole with an augur or other device through the ice at a drafting site) this too should be indicated on pre-incident plans.

ISO Requirements

The Insurance Services Office will credit the water supply provided by fire departments through a relay or tanker shuttle operation within five miles of the fire station if 250 gpm can be established in five minutes after arrival and sustained for two hours. ISO will credit increased amounts that can be developed within 15 minutes and sustained. Mutual aid can be used to achieve the maximum flow. The relay or shuttle must be demonstrated to receive the credit and a lower insurance classification.

Departments that elect to meet these requirements will need to do considerable planning before attempting the demonstration. Develop fill sites or sources that offer a minimum of 1,000 gpm. Determine an adequate first and second alarm response. One potential solution maybe to create a Strike Team that consists of five tankers and a fill site unit for shuttles, or five pumpers with adequate hose for relay operations. If a shuttle is used, there must be large storage capacity at the unloading site and proper scene organization to accomplish this goal. Water supply shuttle operations require frequent practice to maintain individual and department skill levels.

Water Sources

Fire departments often overlook the numerous water sources in rural areas, some of which can be developed with minimal expenditures and work. Frequently a source which seems to be of minimal capacity will be ignored without having been tested to determine its actual capacity. The appendix of Standard 1142 gives valuable guidance on how to evaluate the capacity of such water sources as a flowing stream. Others, especially around target hazards, may have to be constructed. Survey the static sources in your district: lakes, ponds, rivers, streams, irrigation canals, swimming pools (above- and in-ground, inside and out), and gravel pits, and record the sources on water supply maps. Give the maps wide distribution. Many departments mark static sources with signs for easy recognition. Calculate the capacities of these sources and record them, with any seasonal changes that may occur noted. Formulae for calculating the water in streams, swimming pools, and ponds can be found in *Planning for Water Supply and Distribution in the Wildland/Urban Interface* (an NFPA publication). This text also provides good information on how to install dry or drafting hydrants and construct cisterns.

Considerations at static sources

Are the sources directly accessible to a pumper or will hoselines have to be stretched from multiple portable pumps? Has the owner given permission to use the source in emergency situations? What preparatory work is necessary (*e.g.*, putting a gate in a fence, dumping gravel or stones for access, or installing a dry hydrant)? Can stream sites be improved by installing a manhole in the bridge pavement for inserting the hard suction, alleviating the need to place the hose over the bridge rail, building a weir to enhance the volume available, or sinking barrels in the stream bed to provide ample depth for drafting? Are there problems of silt and debris that must be overcome to avoid pump damage? If multiple portable pumps are to be used, arrange them to discharge into portable tanks and then draft from the portable tank with a standard pumper. Manifolding portable pumps to discharge into common headers or into several intakes of a pumper generally has the result of decreasing the capabilities of the portable pumps and should be avoided.

Do not overlook the issue of seasonal accessibility. Is the static source reliable during all seasons of the year? Consider heavy human or vehicle traffic that could cause time delays and safety problems; soft ground in the spring, thin ice, thick ice that can be walked on but must be pierced, and lower water levels during the summer that require long horizontal or vertical lays of suction hose are all factors to consider. Information on static sources is available in IFSTA's *Water Supplies For Fire Protection* and in NFPA Standard 1142, which also details constructing water sources such as cisterns.

SOPs for non-hydrant areas

Standard Operating Procedures are important for non-hydrant areas, where an adequate response on the first alarm often means bringing in additional water supply units. For large structures, target hazards, and where distances are great, it may be necessary to establish mobile water supply strike teams of five mobile water supply apparatus plus a fill site pumper for loading operations. As a rough rule of thumb, you can assume that each tanker involved in a water shuttle operation will contribute between 150 gpm and 200 gpm to the overall water supply. Thus if the needed fire flow is 1,250 gpm, assume at least six to eight tankers will be needed. The best way to assure a good water supply in these situations is to use an automatic aid system. In other words if the fire could require the use of six tankers, have them dispatched automatically on first alarm along with adequate fill site personnel and pumpers so that an immediate water shuttle can be established. If the water supply is to be provided through a relay operation, enough pumpers with adequate hose to transport the fire flow must respond. Large diameter hose greatly enhances relay operations; however, in some cases the hose resources may not be available locally for a lengthy relay. Again automatic mutual aid is the solution. Relays of multiple 2½-in. and/or 3-in. hoselines usually should be limited to 2,500 feet. Large diameter hose relays using 5-in. hose of 1½ miles and more have been used with good success. In one LDH relay the responding fire departments were able to lay 7,000 feet of hose and supply 1,200 gpm in about 15 minutes. An important part of your SOP is to have an engine company respond to the nearest water source on the first alarm to set up a tanker filling site or act as the source pumper for a relay. Use the largest capacity pumper available for this evaluation and equip all pumpers with the necessary fittings and adapters to accommodate local apparatus, hose sizes, and hydrants.

Suburban/Rural Water Supply Officer

Water supply officers in suburban and rural areas have to perform a variety of duties, both fireground and non-fire. During an incident the WSO should confer with the first-arriving units immediately to be sure they are aware of the available water sources and the SOPs to use them, and to initiate operations if they are needed. The WSO then assumes responsibility for overall water supply by knowing how much is being used, maintaining an adequate supply, and planning for additional needs. Meeting additional needs means initiating shuttles or pumper relays, requesting apparatus as needed for these operations, and assigning personnel to leading and unloading sites for shuttles. The WSO maintains apparatus in reserve, at least one pumper and a tanker, for breakdowns or additional supplies. The water supply officer must coordinate and communicate water supply orders and brief the fireground commander on water supply conditions. Before an incident, the rural water supply officer needs to:

- Determine water supply needs (initial attack, fire flow, total Minimum Water Supply, duration of need).
- Determine deficient areas and make recommendations to improve supply.
- Maintain maps of water supply sources.
- Survey and test water supply sources.
- Maintain records of inspections and tests of the sources.
- Determine locations for constructing static sources and dry hydrants.
- Ask property owners for permission to use their water supplies as static sources.
- Help to write rural water supply SOPs based on available resources.
- Train officers and firefighters in rural water supply operations.
- Research new equipment and methods to improve water supply operations.
- Communicate with adjacent WSOs, share resource data and operational improvements, and conduct joint training exercises.

Apparatus

Two types of apparatus are often used in rural water supply operations: those used for attacking the fire, relaying water, or tanker filling operations, and those used for transporting water. The first type can be standard structural pumpers or pumper-tankers with water tanks of 1,000 gallons or more. These apparatus must meet the design and performance specifications of NFPA Standard 1901 Pumper Fire Apparatus. Their designs should be versatile, with sufficient hose loads and equipment to meet the variable needs of rural firefighting. Apparatus used primarily to transport water must meet the design and performance specifications of NFPA 1901 also and may have pumps; a pump is not a requirement but it does increase the versatility of such units greatly.

Mobile water supply apparatus must be constructed for safety and efficiency. They should have tank capacities that will not overload the chassis, or cause weight to be poorly distributed. Often a small unit with a low center of gravity has better transport capability than a larger vehicle. Loading and unloading rates must be at least 1,000 gallons per minute. Many properly designed units far exceed this minimum. An adequate chassis, power, and brakes; proper tank baffling; and proper tank security will ensure good road handling. Units should be capable of unloading to either side or the rear of the vehicle; this can be accomplished simply with a flexible hose or with gravity dump valves located on each side in addition to the rear. Gravity unloading can be enhanced by water-jet-assisted dumps which create a venturi action. If unloading is accomplished by pumping there must be adequate tank-to-pump plumbing for the required flows. No matter how the tank is unloaded there must be adequate venting and air movement across the top of the tank. NFPA 1142, 1901, and *The Fire Department Water Supply Handbook* all provide details on the mobile water supply apparatus construction.

Auxiliary tank vehicles for major fires or natural disasters can include bulk milk trucks, commercial tank vehicles, street flushers, and concrete mixers. Such units may be able to be pressed into service to augment normal fire department apparatus. Establish procedures for their response, loading, unloading, and subsequent cleaning, when required. They can be extremely useful if plans have been made in advance and SOPs developed.

Apparatus flow capability

The flow capability of fire department mobile water supply apparatus can be estimated by the formula:

$$Q = \frac{V}{A + B + C}$$

Where:

Q = Delivered gallons per minute
V = Vehicle tank capacity, gallons
A = Unloading time, minutes
B = Roundtrip travel time, minutes and
C = Loading time, minutes.

If a mobile water supply unit with a 2,000-gallon tank is in a shuttle that has a loading and unloading time of 2 minutes, and a travel time of 6 minutes round trip, the flow capability can be calculated as:

$$Q = \frac{2{,}000}{2 + 6 + 2}$$

Q = 200 gallons per minute

The Insurance Services Office calculates the travel time for apparatus with the formula:

$T = 0.65 + 1.7\,D$

Where:

T = Time in minutes
D = Distance in miles

It is very possible that a tanker with a relatively small capacity tank can provide as much water if not more than a larger unit when you consider that fill and unloading times will be increased with a larger tank and travel times may also be more. Numerous test have been conducted which show that the actual gpm delivery from tankers greater than 1,500 gallons varies little. At the same time the cost involved in larger tankers will increase substantially once they exceed 2,000 gallons.

Shuttle operations

The key to shuttle operations is to *keep all units moving!* Any time a tanker is stopped, it is not transporting the needed water supply. The best way to minimize downtime is to improve loading and unloading times, and then to make sure that units are not obstructed after they are loaded or unloaded. Numerous items of assistance in this area include quick couplings on hose and tank inlets, sufficient large fill openings, filling stations and apparatus, properly placed large vents, stream shapers for pumping directly from outlets into portable tanks, large unloading valves, jet dumps, sufficient air movements across the top of the tank (baffle openings), and remotely controlled unloading valves and vents.

An important component of water movement is operational organization; plan for water movement in advance for both relays and shuttles. For relays, this includes determining whether the constant pressure method will be used, selecting source pumper, and maintaining the relay by using a dump line when attack lines are temporarily shut down. The constant pressure relay is designed so that all units discharge at the same pressure and maintain a constant flow volume. The distance and hoselines between units are matched. If more water is needed, additional hoselines are laid between units. The relay is maintained by a dump line that is opened to discharge water when firefighting lines are shut down. That way the other pumpers in the relay do not have to constantly decrease and increase pump pressure to maintain the relay. For a relay operation a pumper should be dispatched immediately to the source and set up. As stated earlier, this should be the largest capacity pumper available.

The two major concerns in shuttle operations are the distance from the source to the fire and the number and carrying capacity of the tank. To keep the unit transporting water, consider the number and placement of portable tanks for unloading, assignment of officers and personnel to the loading and unloading site(s), supplying attack pumpers with a short relay from water supply pumpers at the unloading site where fireground space is limited, and how units are going to load and unload most effectively. To maintain flow rates on the fireground of 500 gallons per minute or more, three or more portable tanks are required. The attack pumper should have a corner of the center tank placed at its intake, with additional tanks placed adjacent to it for easy access by shuttle units. Water then is transferred from the side tanks into the middle one for use by providing water powered siphons connected to hard suction hose used to link the tanks together.

Both loading and unloading sites need to be managed; someone must organize the site and maintain communications (aided by staff to direct traffic), make and break connections, open vents, and perform loading or unloading operations. Sometimes it is necessary or desirable to unload a tanker at some distance from the fireground. In those cases a pump on the tanker can relay water to the portable tanks or nurse tanker or to the attack pumper.

Units can unload through a gravity dump, a jet-assisted dump, by pumping the water off, or with a combination of dumping and pumping. Vehicles should be checked for the quickest method and this used when conditions permit. The most efficient methods of filling are to use multiple hose lines to direct tank inlets, large diameter hose inlets, or large flow top-filling devices. Familiarity obtained by frequent training sessions with the units in the shuttle will allow loading site personnel to be prepared to fill them quickly.

Whenever possible use a circular traffic flow pattern so that vehicles in the shuttle do not have to pass each other on narrow roads. In other words, have all of them move in the same circular direction when both full and empty. This may increase travel distance, but the increased safety and reduced time make up for it. In training tanker drivers, instruct them to remain in their trucks at all times during the shuttle. Outside duties such as opening or closing tank unloading or loading valves should be handled by the ground crew at the site, not the tanker personnel. The purpose for this rule is to help avoid time delays when it is time to move the tanker, which occur when the driver has left the truck for any reason.

Another factor to bear in mind is to attempt to use tankers of approximately the same size in a shuttle if possible. The existence of one or two very large tankers in a shuttle can often create a bottleneck at the filling site and the unloading area.

Coordinating operations into an organized water supply plan can be enhanced with SOPs for non-hydrant areas. SOPs for positioning and operations of attack pumpers, water supply pumpers, and tankers should be included. With good operations, large delivery handlines and master stream attacks can be easily achieved, resulting in quick knockdown and early extinguishment. Train all members in the evolutions and drill everyone in the SOPs for relays, large diameter hose, and tanker shuttles.

SUMMARY

This chapter has discussed water supply concerns for chief officers in urban, suburban, and rural areas. We have looked at how to determine needed water and several methods for approaching water supply problems. In essence, using water supply officers, planning water supply needs, and working with water department officials at all levels leads to better water availability for fire protection. It is up to the chief officer to see that good water supply operating procedures are developed and used in the department, and to provide the leadership and encouragement to help your department overcome any water supply problems.

In the past it was common to use the excuse that "the water supply was inadequate" and therefore the building was destroyed. That excuse is no longer valid nor was it ever really valid. Today there is little reason for an inadequate supply of water in either urban or rural areas and chief officers must devote increased time to the skills needed to develop water supplies to the level they are needed.

Firefighters in rural areas particularly should adopt a "can do" attitude when it comes to water supply—there is no reason to assume that in rural areas water supply is by necessity always going to be a problem. Firefighters all over the United States have shown that with training, practice, and ingenuity the drawbacks of a rural environment can be overcome and high volume water supplies can be generated.

16

INCIDENT COMMAND

Nick Brunacini

CHAPTER HIGHLIGHTS

- The incident commander's role in incident operations
- Incident operations are conducted around completing the tactical priorities of rescue, fire control, property conservation, and customer stabilization
- Discussion of various incident command systems and their deployment

PURPOSE AND SCOPE

Our customers call us for help when they are having a really bad day. Most often these calls for help will require the efforts of multiple fire units (and in many instances units and resources from other agencies) to stabilize and control. There is a standard list of activities (tactical priorities) that we are responsible for when we respond to the emergency scene. They are to:

- Protect life
- Stabilize the incident
- Stop the loss

For structural firefighting these tactical priorities are:

- Rescue
- Fire control

- Property conservation
- Customer stabilization

All of our incident activities must be directed toward completing these tactical priorities. The person responsible for making sure that incident scene resources effectively and safely complete the tactical priorities is the incident commander (IC).

The IC's role in incident operations is to manage the collective efforts of all the incident scene responders and resources under a single incident action plan. The incident action plan defines how the tactical priorities will be accomplished and should also describe the safety plan to ensure that everyone goes home after the event is over.

Well-managed incidents produce better outcomes than poorly-managed incidents. Well-managed incidents are safer and incur less damage (from both the incident problem and our actions required to solve the problem). The incident command system must:

- Be used at every incident
- Expand to meet the needs of each incident
- Ensure that each member understands their role and responsibilities
- Be designed to solve the incident problem

Incident operations are conducted around completing the tactical priorities of rescue, fire control, property conservation, and customer stabilization. These priorities are listed in the order of importance. It doesn't make much sense to assign firefighters to do salvage activities before the fire area has been searched. After each tactical priority is completed, it is announced over the tactical radio channel. These are the tactical completion announcements:

Rescue	"All clear"
Fire control	"Under control"
Property conservation	"Loss stopped"

Firefighter safety is one of the main reasons we use an incident management system. Structural firefighting is very hazardous work. Fire operations must be conducted within a well-managed risk management plan. The risk management plan defines how much risk firefighters can take. Here is an example of a simple, straightforward risk management plan:

We will begin our response with the assumption that we can protect both life and property. Based on this assumption we will (1) take great risk to protect savable lives; (2) take a small risk, in a highly calculated manner, to protect savable property; and (3) not take any risk to protect what is already lost (lives or property). This risk management plan is aligned with the tactical priorities. The tactical priorities provide us with measurable and obtainable goals to conduct the different phases of incident operations. The risk management plan defines the level of risk firefighters can expose themselves to while completing the tactical priorities.

The incident commander is the overall scene boss. The IC maintains control of all incident scene activities by performing the functions of command. The functions of command define and describe what it means to be in command.

Assume, confirm, and position command

If incident operations are to be well managed, they must begin that way. It is very difficult, and sometimes almost impossible, to get operations under control if the first three or four companies on the scene have all taken action independent from one another. It is impossible to execute a well thought out plan of action when all the incident players on a different page and doing their own thing. The solution to this problem is to have a single IC in place from the very beginning of incident operations. This is accomplished when the initial arriving person (usually a company officer) assumes command. Just as bad as having no one in charge is having more than one person who claims to be the IC. Only one person can be in charge of managing all the incident scene resources. It is important to note that while only one person can be the IC, the IC should be supported to whatever degree is required in order to effectively manage the incident.

An IC is established when the initial arriving member/officer arrives on the scene and gives an initial on-scene report, including the assumption of command, over the tactical radio channel. An on-scene report should contain the following information:

1. Unit ID and their location – *"Engine 1 is on the scene..."*
2. Description of the incident site – *"...of a medium-sized two-story house..."*
3. Description of the incident problem – *"...with a working fire on the second floor."*
4. Action taken – *"Engine 1 is laying a supply and advancing a 1½" attack line to the second floor for search, rescue, and fire control."*
5. Declaration of strategy (for structure fires) – *"We will be operating with an offensive strategy."*
6. Assumption of command – *"Engine 1 will establish the Main Street incident command post."*

When the initial IC is on a "working" piece of equipment (engine or ladder company, etc.) they have an option of where the IC will be located (physical location). There are three different command modes/positions:

1. *Investigative mode.* The investigative mode is used for incidents that do not have an obvious problem (smells and bells). The initial IC goes with their crew and investigates further.
2. *Fast attack mode.* The fast attack mode is used when the initial IC is faced with a working situation and his/her immediate, physical involvement in solving the

incident problem will make a difference. The IC assists/supervises/manages with control efforts and runs the incident over the portable radio.

3. *Command mode.* The command position is utilized for incidents that require a strong command presence from the onset. The IC runs the incident while staying on the apparatus (establishes a command post). The IC may not always stay with the original apparatus, but will move to the most advantageous position, depending on department policy. A defensive fire is an example of an incident where the IC (company officer) may elect to operate in the command mode. Chief officers assuming command should always operate in the command mode. The actions will greatly depend on the available resources and the policies and procedures of the department.

After the initial arriving responder assumes command and becomes the IC, a quick size-up should be conducted.

Situation evaluation

Before the IC can take effective action, a size-up of incident conditions and critical factors must be conducted. In most cases, the initial IC has to take action once assuming command. Situation evaluation begins with the dispatch. Dispatch information should include the location of the incident, the reported problem (*i.e.*, fire, explosion, etc.) along with type of structure (*i.e.*, house, commercial, etc.), which units are assigned to the incident, and the tactical radio channel the incident will be using. This information gives responding crews a general idea of what to expect upon arrival. These critical factors include (but are not limited to):

Customer profile (elderly, family, etc.)
Fire stage
Time
Type of occupancy
Apparatus staging/access
Resources present
Personnel safety
Life hazard
Exposures
Building concerns
Fuel (what is burning and potential exposures)
Non-fire hazards/problems
Action being taken
Special circumstances

The IC uses these different critical factors for the size-up process. This information is provided to the IC in three different ways.

1. *Visual*—the informational items that the IC can see. This is limited by the IC's position.
2. *Reconnaissance*—this information is gathered from reports that the IC receives from people that are operating in those areas.

3. *Pre-plan*—this is information that is gathered ahead of the event. The advantage of pre-plan information is that it is available before the incident occurs. This information consists of structural concerns, layout of the building/site, occupancy/life hazard concerns, etc. Pre-plans need to be easy to access and have only the info needed to carry out incident operations (using a set of blueprints to run the fire generally doesn't work very well).

After the IC conducts a quick size-up, decides on the overall strategy, and formulates an incident action plan, units are assigned and the incident problem is attacked.

Communications

The incident commander uses communications to link together the three different levels of the incident scene organization. These three levels are strategic, tactical, and task.

1. *Strategic*—The strategic level is what the IC and command staff operate on. The strategic level is responsible for managing the strategy and the overall management of the tactical priorities.
2. *Tactical*—The tactical level is managed by sector officers. This level is responsible for managing the different geographical and functional pieces of the incident site.
3. *Task*—The task level is composed of the companies and personnel that are actually doing the work (searching, fire control, treating patients, etc.).

Each of these levels plays their role in completing the incident goals. The three different levels serve as the template for the IC to build an effective organization. The type of organization used to manage the incident will mirror the communications model the IC uses to manage all incident activities.

The IC uses the radio to manage the incident. The radio is the tool the IC uses to identify the strategy and communicate the incident action plan, make assignments, and coordinate incident activities. If the IC is knocked off the air for any reason, their ability to manage the incident is lost, and they become little more than a frustrated spectator. As command is transferred and upgraded (moved from a fast-attacking company officer to an IC in a command post, operating in the command mode) the communications position is improved. The most notable change in this transfer is the new IC's ability to monitor and control the communications process.

The communications system used at the incident scene is built around hardware (the radio system) and standard operating procedures.

Hardware. This is the equipment we use to communicate. Radios (mobile, portable, etc.), mobile digital terminals (MDTs), mobile computers, cell phones, fax machines, etc. are all different pieces of communications equipment that fire departments use to conduct incident operations. The communications system varies from organization to organization, but the goal is the same—to quickly connect the customer that needs our service to our dispatch center. In many large urban systems this is done with a 911 reporting system, satellite-tracking system for apparatus, and other high-tech goodies that facilitate the process. When the call comes in, the dispatch center confirms the address and the nature of the emergency. This information is entered into the computer-aided dispatch system and the closest appropriate resource(s) is selected for dispatch. In smaller systems, this may be accomplished by calling the local sheriff's office and paging out the local volunteer fire department. Regardless of what system is used the goal is the same—to get the customer whatever help they need.

When units arrive on the scene, the communications system has to operate in a way that keeps all of the incident players connected. Since the workers that depend on that system often operate in very hazardous environments, the hardware must be rugged and dependable. If companies operating inside burning structures get into trouble they need to be able to immediately get hold of the IC to communicate their situation and get whatever assistance they need. It is for this reason that many fire departments equip every member that operates within the hazard zone with their own portable radio. At the very minimum each team that enters the hazard zone must have a portable radio.

Standard operating procedures. The communications process must be written down (SOPs), understood, and practiced by all the incident participants. An effective communications system allows all the participants to exchange important information while maximizing the available airtime—leaving it available for urgent communications. Most of the communication that takes place at the incident is done face-to-face (on the tactical and task level). While this is the best way to communicate, it is impossible for the IC to have a face-to-face dialogue with everyone operating at the incident site. The IC would quickly become overwhelmed with the long line of people at the command post, all waiting for their turn to speak to the incident boss. Most of the IC's communications take place over the radio. The IC uses the radio to request resources, assign and track resources, receive and provide progress reports, and to periodically check that all assigned resources are okay. This is the reason the IC operates at a command post (it is impossible to be in an effective communications position while sprinting around the incident site).

Order model. The IC gets units into action by identifying a tactical need, securing an available unit that is ready to go to work, and then ordering them into action. These communications should follow an order model, such as:

"Command to Engine 1."

"This is Engine 1 – go ahead Command."

"I want you to lay a 2½" line to the north side and advance a 2½" attack line into the east exposure for search, rescue, and to keep the fire from extending. You will be East Sector."

"Engine 1 copy – lay a 2½" line to the north and a 2½" attack line to the east exposure for search, rescue, and to keep fire from extending. We will be East Sector."

Having the receiver "parrot" the order back ensures that they completely understood the order and eliminates confusion.

Clear text. Personnel should use clear text when talking over the radio. Simple, understandable text is preferable to the use of codes (systems that use "10" codes can sound like a foreign language—and many times it takes as long to learn all the codes as it does to learn a foreign language).

Deployment

This function of command describes how resources are assigned to an incident (the dispatch process), how they get cycled through incident operations (assignment by the IC), how they are tracked (accountability), and ultimately how resources are rehabilitated and put back into service.

Deployment begins when the dispatch center turns the customer's service request into a dispatch. Units are normally assigned to incidents in "alarms." These range from single unit request (single company fire response, EMS calls, special duty assignments, etc.) escalating up into major alarms for large-scale incidents. Most departments bring more resources to the scene by requesting additional alarms. An alarm can be four engines, two ladders, and the associated support equipment. Regardless of how many and what kind of resources make up an alarm, it needs to be a standard response for the participants of the local system. When the IC calls for a second alarm (or the local equivalent), four engines, two ladders, and a battalion chief (or the local version) are on their way. An aspect of size-up is the IC making the determination of whether or not there are enough resources ordered to handle the situation.

Staging. The next aspect of deployment is staging procedures. There are two types of staging: Level 1 and Level 2. Level 1 staging is in effect for the units on the initial alarm (in many systems this is for the first four engines and two ladders). When the initial unit arrives to the scene and assumes command it automatically puts the remaining responders into Level 1 staging. For engine companies, this means that they don't pass their last water source (*i.e.*, hydrant). For ladder companies they do not pass their last access point to the incident site. As units stop short of the scene they announce their location and that they are staged, for example: "Engine 1 staged west."

This alerts the IC that uncommitted resources that are ready to go to work have arrived on the scene. Just as importantly, it gives the IC the capability to assign resources according to plan. Staging procedures eliminate resources from coming directly into the scene and taking independent action.

Level 2 staging is used to group units into resource pools. Level 2 staging is used when calling for greater alarms, placing resources closer to where the work is being done or anywhere else it is called for. Level 2 staging is managed by the staging sector officer. This person is normally the first officer assigned to the staging area. Level 2 staging should be located close enough to the scene to provide quick and easy access, but not so close that they are in the way.

Accountability. The IC must keep track of all of the incident resources—particularly those operating in the hazard zone. This is accomplished on the strategic level by using a tactical worksheet. The IC uses the tactical worksheet to record units assigned to the incident and their status (staged, assigned, etc.). As the IC assigns units to the incident, their approximate position is indicated and their assignment is noted. Filling out and maintaining the tactical worksheet provides the IC a quick and simple system to know who is on the scene, a general idea of where they are, and what they are doing.

Accountability on the tactical level is accomplished by making sector assignments (subdividing the incident scene into more manageable pieces). Sector officers are responsible for keeping track of the personnel assigned to their sectors. Sector officers must be aware of which units and resources are assigned to their sectors along with where those resources are (physical location).

Task level accountability is managed within the crew. This is the most important part of the accountability system. Crews must remain intact—particularly when operating in the hazard zone. This means that crews come in together, stay together, and leave together. The hazard zone is full of things that can kill and injure us. The incident management system is designed with a set of checks and balances that allow us the flexibility to provide service (complete the tactical priorities) and survive the event. No combination of system elements can outperform this ancient tactical/task rule—stay together and always maintain an exit path (don't leave your hose lines) out of the structure.

Many fire departments utilize some type of hardware-based (passports, cow tags, etc) system to facilitate tracking resources. These systems must be customized to fit the local responders that use the system and they must be designed to track the users as they enter and exit the hazard zone.

One key component of the accountability system is a personnel accountability report (PAR). The IC receives PARs from companies/sectors assigned to the incident scene. A PAR indicates that the crew, company or sector is intact and okay. The IC receives PARs when units exit the hazard zone, when a sudden hazardous event occurs, at predetermined time intervals, when switching strategies (offensive to defensive), or any other time the IC has need to check on the status of personnel.

De-committing resources. As assigned units exit working sectors they will cycle through the rehabilitation sector (rehab) for fluids, food, air, rest, and medical evaluation (if required). Most, if not all, assigned resources end up going through the rehab sector.

At large-scale working incidents, rehab can end up looking like a "day after Christmas sale" at the local outlet mall. Many times half of the resources assigned to the incident may be located in rehab. The rehabilitation sector must be staffed with enough people to manage the rehab needs of companies and personnel cycling through. The management of rehab must include keeping track of assigned units. After rehabilitated units are ready for reassignment, they should be moved out of the rehabilitation sector into a forward staging area. This area can be adjacent to the rehabilitation sector, but should be managed by a different person if at all possible.

Once the incident is brought under control the IC may begin to place units back into service. Before the incident is terminated and units are made available, the IC must make sure that units are fit for service. For most routine incidents this may be as simple as Engine 1 reporting that they are "assembled and ready to go." For more significant, emotionally-charged incidents, the IC might need to provide some type of debriefing prior to putting units back into service.

Identify strategy and formulate an incident action plan (IAP)

When the IC arrives to the scene, a quick size-up of the incident critical factors, decision on the appropriate strategy, and formulation of an IAP that begins to stabilize the incident problem is necessary. Effective and safe incident operations are the product of a well thought out and conceived IAP (after determining the correct strategy). The strategy identifies where workers will operate. The offensive strategy defines inside operations—attacking the fire inside the structure. Offensive operations are centered on achieving an "all clear" in the fire area and adjoining exposures, controlling the fire, and conserving property. The defensive strategy is used when the fire has burned beyond the control capabilities of an interior attack. Defensive operations are centered on achieving an "all clear" for the fire building and the exposures, keeping the fire from extending into uninvolved property, and knocking down the main body of fire.

As we stated in our risk management plan, we will always begin incident operations under the assumption that we can save lives and property. Whenever possible we will operate in the offensive strategy for this very reason. Occasionally the IC may be faced with a tactical situation where conditions are turning defensive, but the incident organization has not achieved an "all clear" on the fire area. Operating in offensive positions in defensive conditions is a marginal situation. The only reason we would ever expose ourselves to this level of risk is if we were attempting to rescue savable lives. An example of an acceptable marginal operation would be pulling up to a house with deteriorating conditions and being met by a person that just exited the structure telling you that "someone is still trapped in the back bedroom." We will do everything in our power to save that trapped person. Marginal operations are not reserved as an alibi to conduct interior operations any time the urge strikes us. Pulling up to a well-involved fire (defensive conditions) in a boarded-up vacant building at 3:00 a.m. indicates using the defensive strategy—there is nothing for us to save (unless information to the contrary is obtained).

The goal of fire attack is to put the fire out. When the fire grows past a certain point (for whatever reason) it cannot be controlled from interior positions with small diameter hand lines. If the fire is generating an amount of heat that requires 1200gpm of water to overcome, it will continue to burn until we can match the heat production with direct water application (or until the fire runs out of fuel—whichever comes first). Most interior attack hand lines flow between 125–250 gpm. To make matters worse, fires begin to adversely affect the components that hold the building up within a very short period of time (minutes). We don't have all day to assemble the required number of firefighters needed to drag four or five heavy-hitting attack lines inside to overwhelm the fire with water. A very large and personal piece of this time equation is not only how long it takes to get firefighters in place, but also how long it takes to get them out of those positions when things get out of hand (*i.e.*, the fire gets bigger, the structure begins to show signs of deterioration, etc.). When firefighters continue to operate in the interior or within the collapse zones in deteriorating conditions, they are operating in "no man's land." These early-stage defensive conditions can be (and sadly are) lethal. This represents the primary reason we have, and use, two different strategies—firefighter safety and firefighter survival. These two strategies align with our risk management plan. If the fire is advanced, it has consumed any savable lives and property. We will not risk our lives at all for what is already lost. We respond to the emergency, and base our actions (company/tactical operations) on a size-up and well thought out attack plan. If we can control the fire using the offensive strategy, that's what we do. If the conditions have deteriorated past the point of offensive actions, then we use the defensive strategy. Personalizing the required strategy (*i.e.*, offensive tactics are good; defensive tactics are a personal failure) is not warranted and is potentially very dangerous. In situations where the fire is too big to control utilizing the offensive strategy, switching to the defensive strategy and flowing "big water" may be the only chance we have to extinguish the fire and the save what is left of the building, protecting the community in the process.

Incident action plan. After the IC determines the proper strategy, an IAP must be formulated. The IAP must define how the tactical priorities are going to be achieved. The IAP for offensive fires should describe which areas get searched first, what direction to attack the fire from, how many attack lines will be needed to control the fire, which sides of the fire to check for extension and what order they will be addressed in, what type of ventilation will be needed to support the attack, etc. An example IAP for a room and contents fire in a single-story, 2,000 ft^2 house follows.

Offensive operation – *First arriving engine company to secure water supply and advance an attack line to the seat of the fire for search, rescue, and fire control. Support initial attack with positive pressure ventilation and utility control. Second attack line to finish search of remainder of the structure. After fire knockdown open ceiling in room of origin to check for fire extension. Early salvage and customer support...*

IAPs for defensive fires should identify the most dangerous avenue of firespread, the most effective use of water application, how many resources will be needed to control the fire, which exposures and what order they will be searched, etc. An example IAP for a defensive fire in a medium-sized, single-story commercial structure follows.

Defensive operation – *Secure water supplies to both the front and rear of the structure. Engine companies to use master streams for fire knockdown. Set up ladder pipes on side A and C for elevated master streams onto the main body of fire. No one is to go inside the structure and all personnel and equipment are to be kept out of possible collapse zones. Provide exposure protection if needed. Provide any needed customer support, food, and bottled water for the troops.*

The IAP needs to brief and describe the current phase of incident operations. If the IC is in the middle of managing what is forecasted to be a prolonged search, it is premature to begin planning for salvage activities. The IAP must be updated, revised, or completely changed when conditions change. It is part of normal operations to have used two or three different IAPs during the course of an incident. The key to incident action planning is that the IC is in charge of the process. Everyone operating at the incident must be part of the IC's overall plan (all the different incident players need to have their own plan for whatever their objective is, but this must fit into the overall IAP that the IC has for the incident).

The advantage of the initial radio report is that it wraps the first five functions of command together with a brief radio transmission that is shared with all the incident players. The initial on-scene report identifies that someone has arrived to the scene, describes the incident problem along with the strategy and action being taken to solve the problem (the initial IAP), and confirms that someone has taken command. This initial radio report takes less than a minute to voice over the radio, but it sets up the next (and most important) ten minutes of the incident. We solve most of our customer's problems (particularly for offensive fires) during this initial 10-minute window.

Organization

The incident management system must expand to allow the IC to manage the eventual number of personnel and resources that are required to solve the incident problem. Most management authorities put the effective span of control somewhere between four to six. How many units you can manage depends on several things: where you are (physical location), what you are doing, and what you're trying to manage. A fast-attacking IC isn't going to be able to manage 10 different units while leading the attack inside a burning building. This same IC should be able to manage getting the initial attack in place by taking the appropriate action with their own crew and assigning the next few units (this level of command solves the incident problem most of the time; when it doesn't, command needs to be transferred to an outside position to someone who will operate in the command mode). If the incident is spread over large or separate areas (commercial buildings, exposures, etc.) or multiple units will be operating in a particular area, the officer of the first company assigned to these areas should be assigned sector officer responsibilities.

The IC uses sectors (many systems refer to this management subunit as a "sector"; in some other systems a sector is called either a "division" [a geographical subdivision] or a "group" [a function subdivision]) to decentralize managing incident activities around the incident site. Sector officers manage different areas or functions around the incident scene. They provide management and supervision directly where the work is taking place, managing the details of that work. This takes a huge load off of the IC, allowing more focus on the strategic level. When assigning sector responsibilities the IC needs to give the new sector officer the objectives for the sector and the resources assigned to the sector. As work progresses in the sector, the sector officer should report back to the IC. These progress reports should include conditions in the sector, actions being taken within the sector and any additional resource needs within the sector.

Sector officers manage the activities within their assigned area on the tactical level. The sectors officer's main focus is to complete the tactical priorities within their sector for geographic sectors (*i.e.*, North Sector, Sector A, Sector 2, etc.) or complete their assigned function for functional sectors (*i.e.*, Vent Sector, Water Resource Sector, etc.). As the IC continues to assign sector responsibilities around the incident site, managing too many sectors may become overwhelming. Before this happens the IC needs to expand the command organization in order to maintain the ability to effectively manage the incident.

Command teams. Some fire departments are adopting a command team approach for incident management. Command teams are designed to put a reinforced command component in place quickly during the beginning of incident operations. One example of a command team is grouping together an IC, support officer, and senior advisor. The support officer does just that—supports the IC. Upon arrival to the command post, the support officer challenges and verifies the IC's strategy and IAP. This causes both the IC and support officer to take a minute to look at the incident critical factors and make sure that the current operation matches those conditions. It also helps to identify any uncovered areas or needs. The support officer then takes over the tactical worksheet and the tracking of the incident resources, allowing the IC to focus on what is going on in the operational sectors. The senior advisor is the third member of the command team and is responsible for managing the command post and helping to connect the different pieces of the incident organization. The senior advisor verifies that the strategy and IAP match the incident conditions. The senior advisor then makes sure that the incident organization has the right resources in place to effectively and safely manage the incident. This team approach is very similar to a pilot, co-pilot, and navigator working together to fly an airplane. The principle is the same—to land safely and make sure everyone goes home okay after it is over.

The expanded organization. One of the things the command team (or IC) must maintain is the constant ability to control the position and function of all assigned resources. A big part of this capability is achieved by creating a properly sized incident organization to manage the personnel operating at the incident site. For most of our day-to-day operations, the IC uses sectors to accomplish this. Some incidents, because of their size or complexity, will require a larger incident organization. At these types of incidents, the IC may quickly become overloaded with trying to manage too many sectors. Before this becomes a problem the IC can implement branch officers. Branch officers provide coordination between the strategic (IC) and tactical (sectors) levels of the organization. An example of a branch level incident would be a fire that requires a large evacuation effort. These two distinctly different operations can be split; the IC could assign a branch officer to manage the evacuation while the IC and command team manages the fire attack. In most cases where the IC implements branch officers, they will be located outside the command post and will operate on a separate radio channel, if possible.

The IC and command team can further expand the incident organization by implementing section positions. Section chiefs can be assigned to manage the following:

- Safety
- Logistics
- Planning
- Administration
- Operations

Section chiefs are responsible for managing the incident activities/support in their assigned area. Section chiefs use the same organization (sectors) to divide the work and responsibility in their assigned areas. An example of this is for the logistics sector to manage the rehab and staging sectors.

The IC uses an incident organization to manage the incident. The size of the organization is driven by incident requirements and needs. The IC only implements the parts of the organization that are required to safely and effectively handle the situation.

Review, evaluate, and revise

The IC performed the prior six functions of command to be placed in a position to perform this function. Once the IC gets the initial attack in place an evaluation of the effectiveness of incident operations must be undertaken. This evaluation should be based on completing the tactical priorities and the overall safety picture for the workers (particularly those operating in the hazard zone). The longer a structure burns, the more unstable it becomes. Offensive fire operations are designed to quickly control the fire. If the initial attack isn't controlling the fire, the IC has to figure out why and adjust the attack accordingly.

The evaluation process should start with a review of the critical incident factors. The IC needs to fill in any unknown critical information. This information should begin with the factors that impact the safety of operating personnel (hazard zone workers). The IC will begin incident operations based on a certain number of assumptions (we rarely, if ever, begin incident operations with complete information). The IC verifies that these assumptions are correct by assigning units to those positions in order to complete the current tactical priority and find out what conditions are present. This information is used to keep the strategy correct and the IAP current.

When the IC performs an initial attack on the incident problem it will react in one of two ways; getting better or getting worse. If the control efforts are working, the IC adds any other pieces needed to complete the tactical priorities. If conditions have deteriorated to the point where interior offensive operations are no longer effective or too dangerous to keep in place, the IC must keep the troops from becoming casualties by switching strategies and moving everyone out of the hazard zone.

The ultimate test of the incident management system is if the IC can control the position and function of all assigned resources. The IC uses this capability to put the right amount of resources in the right place in order to solve the incident problem. The IC must also evaluate if enough uncommitted resources (responding/staged) are available to get the job done. Always "holding a little back" keeps the IC ahead of the deployment curve. This uncommitted resource can be plugged into any uncovered tactical holes, and are also available for any urgent need (*i.e.*, using them as rapid intervention teams).

Another evaluation item that affects the IC's ability to manage is if a properly sized organizational structure (sectors, command teams, branches, etc.) has been implemented. Not using a large enough organization can lead to companies taking independent action (free enterprise). Making the organization too large leads to micromanagement (never have more people commanding than working). The properly sized organization coordinates getting productive work done, eliminating situations where companies search the same area five times while other areas don't get searched at all. It also places enough supervision in work areas that the IC can effectively manage the entire operation (completing the tactical priorities) while keeping the workers safe.

Continue, transfer, and terminate command

Incident operations begin with an IC. This initial IC is usually an engine company officer. For offensive situations, this IC will normally use the fast attack command mode. Operating in this command position ends in one of two ways—the incident is quickly brought under control, or command is transferred to someone who will be operating in the command mode (outside the hazard zone, at a command post). This transfer drastically improves the IC's ability to command and manage the incident. Being stuffed into a protective envelope that is designed to allow its wearer to survive an 1800° flashover while assisting with task level activities inside a building that is on fire is a pretty tough position to run an incident from. There is a better way to make sure the incident starts off and stays under control; the incident management system must be designed to quickly and seamlessly transfer and upgrade the position of the IC (when the current IC is operating in the fast attack mode), thereby improving the level of command.

There must be an SOP for the transfer of command. This eliminates any confusion regarding who is in command. The following procedure outlines a simple and easy process for transferring command.

1. The person who is going to assume command arrives on the scene (it isn't a very good idea to transfer command of an incident if you are not there). If there is an in-place IC operating at the incident, it is generally preferable to have them remain in charge until the "next in" IC actually arrives to the scene.
2. The arrival of ranking officers to the incident scene does not automatically mean that they are in command. The person in command is transferred to (IC #2) is usually a ranking officer (*i.e.*, chief). We transfer command in order to improve the quality of command. The first transfer of command will normally be to a person whose job duties include commanding incidents (in mid-sized to large departments this person is normally a battalion chief). For most of the local incidents our departments respond to, this level of command gets the job done. As the incident escalates beyond this level it is ideal to support the IC and assemble a command team to manage the incident. As more command players arrive on the scene they report to the command post (or Level 2 staging) and get plugged into the incident management system wherever they are needed.
3. Command is transferred when the person transferring command notifies the current IC (over the radio or face-to-face) and formally announces "I will assume command." Part of this dialogue should include an update on the incident conditions, action taken, assigned resources, and any other needs or significant information. If the transfer is done face-to-face, the new IC needs to update the dispatch center over the tactical radio channel that he/she has assumed command of the incident.

The new IC should be operating in a command position. The command post offers more powerful radios, lighting, protection from the elements, a place to write and record assignments and other important information, space for command staff and helpers, reference materials, etc., as someone would expect to find in an office. The IC is in effect setting up an office when operating in the command mode.

Operating out of a fixed stationary command post puts the IC in a position where the incident operation can be managed on the strategic level. The command post becomes the place where the IC can build a command staff (command team and sections). Engine and ladder companies carry the tools and equipment they need to be effective (hose, nozzles, PPE, saws, etc); the command post offers the IC the set of tools needed to effectively and safely manage the incident.

The incident scene is the best place to learn, but the worse place to teach. Taking time at the end of the incident to critique the operation allows us to reinforce what went right and make any needed corrections. A critique should be a review of the incident designed to improve future performance—not a public bloodletting. Post-incident critiques can be conducted in several different ways. Most of our day-to-day operations can be critiqued right after the event in a very informal way in a short period of time. These sessions are always best facilitated by the IC. Larger events, where there were lots of lessons learned, may be more suited to a department-wide critique that is held several days (or weeks) after the event.

The IC needs to make sure that the victims have been taken care of before terminating the incident (customer stabilization). The IC must ensure that the fire is completely out (no rekindles), the property is secure, hazards are eliminated, and that the customer has the post-incident support they are going to require. This can be as simple as calling the customer's family to the scene, contacting the customer's insurance company (some jurisdictions may have grave legal concerns with what is said to an insurance company and this should be dictated by department policy), or requesting the Red Cross to provide support. This very important final step of incident operations is far superior to the old system we used to use. We would load our hose, tell the customer how sorry we were for their loss, and drive back to the station—leaving the customer standing in the front yard of their burned out home (or business) with a dazed look in their eyes. Adding another few minutes to the end of the incident and connecting the customer to the people/organizations that are going to assist them with the next phase of recovery is excellent customer service. It is also the way we would want our family and loved ones treated in the same situation.

Before the incident is terminated and companies are put back into service, the IC should ensure that workers have been properly rehabilitated. If the incident was emotionally charged (serious injury or death), the IC should take a moment to meet with the troops and make sure they are okay. Firefighters seem to do the best when they are talking with one another, making sense of what just happened. When their boss takes the time at the end of the event to make sure that everyone is all right, it sends a strong message that the system cares about them. If someone is having trouble dealing with what happened at an incident, critical incident debriefing should be provided.

17

COMPANY OPERATIONS

FEBV Staff

CHAPTER HIGHLIGHTS

- The principles of department organization
- Incident Command Systems
- Companies as the keystone to departmental operations
- Training and standards

INTRODUCTION

Well-developed company operations are critical to the effectiveness and safety of any fire department. Sometimes a misconception exists that the term "company operations" refers to such issues as apparatus and equipment types, response scenarios, suppression tactics, tools, or new technologies. In reality, "operations" means much less—and much more. It is essentially *the process required to produce the desired effect.* Company operations can be defined as the most basic manpower, organization, and training principles necessary to produce the desired emergency response.

The details and procedures for developing effective fire companies are myriad, and many excellent resources are provided by the National Fire Academy, IFSTA, NFPA, FEMA, and other professional organizations. Our purpose in this chapter is to provide a broad overview of the principles and functions of company operations.

In order to get the full picture of how company operations fit into and affect total department organization and services, it is helpful to have background on the varied roles of fire departments and their organizational principles.

Fire Department Services

From times as early as the Roman Empire, fire departments were established to keep the village from burning down. As the village grew into a town, a city, and ultimately a metropolis, the fire department grew with it, developing new manpower methods and technologies for increasingly complex emergency situations. Fire suppression and rescue has been the traditional purpose of departments for centuries.

But the firefighter's job description has changed significantly, specifically since the 1960s. With the fire service capacity for rapid emergency response, functions have expanded to include

- Emergency medical service/life support services
- Emergency transportation
- Hazardous materials response
- Technical rescue–high angle, trench, structural collapse, dive teams/swift water rescue
- Search and rescue
- Social service assistance (health care access, counseling, chaplain services)
- Community education services

This broad variety of community services can only be carried out through a department of highly developed company teams, each trained for specific types of response. The fire companies are the department's frontline practitioners of fire suppression and emergency response, but their effectiveness is made possible by the department's overall structure.

Fire Department Organization

Fire departments vary greatly in size and scope due to the size, economics, and needs of the community served. They range from volunteer departments, serving small communities that could not otherwise afford service, to paid departments manned by thousands of firefighters and medical technicians for large cities and other heavy populations. Besides municipal and sometimes independent services, some departments are developed to serve state and federal government or the military for specialized protection areas.

No matter the department size or community served, all successful fire departments share certain principles of organization for effective and safe function, including

- *Chain of command:* an established command hierarchy from the lowest to highest departmental level, ensuring that each subordinate reports to one supervisor.

- *Supervisory limits:* the number of personnel one supervisor can effectively manage. While the number can vary according to circumstance, a general fire service guideline is up to five or six firefighters per one supervisor.
- *Division of labor:* to ensure that all responsibilities are assigned and to prevent duplication of effort.
- *Disciplines and regulations:* written procedures to set boundaries and enforcements for expected individual and departmental performance.

The basic fire department staffing consists of company units that allow a department to meet the demands of labor division, communication, and chain of command. A simple diagram of a basic department company and management structure is found in Figure 17–1. The company operations division is the backbone of any fire department. As company operations increase to satisfy increased service community, more divisions are added for finance and administration, training, public education, communications, and other areas that advance and support company operations.

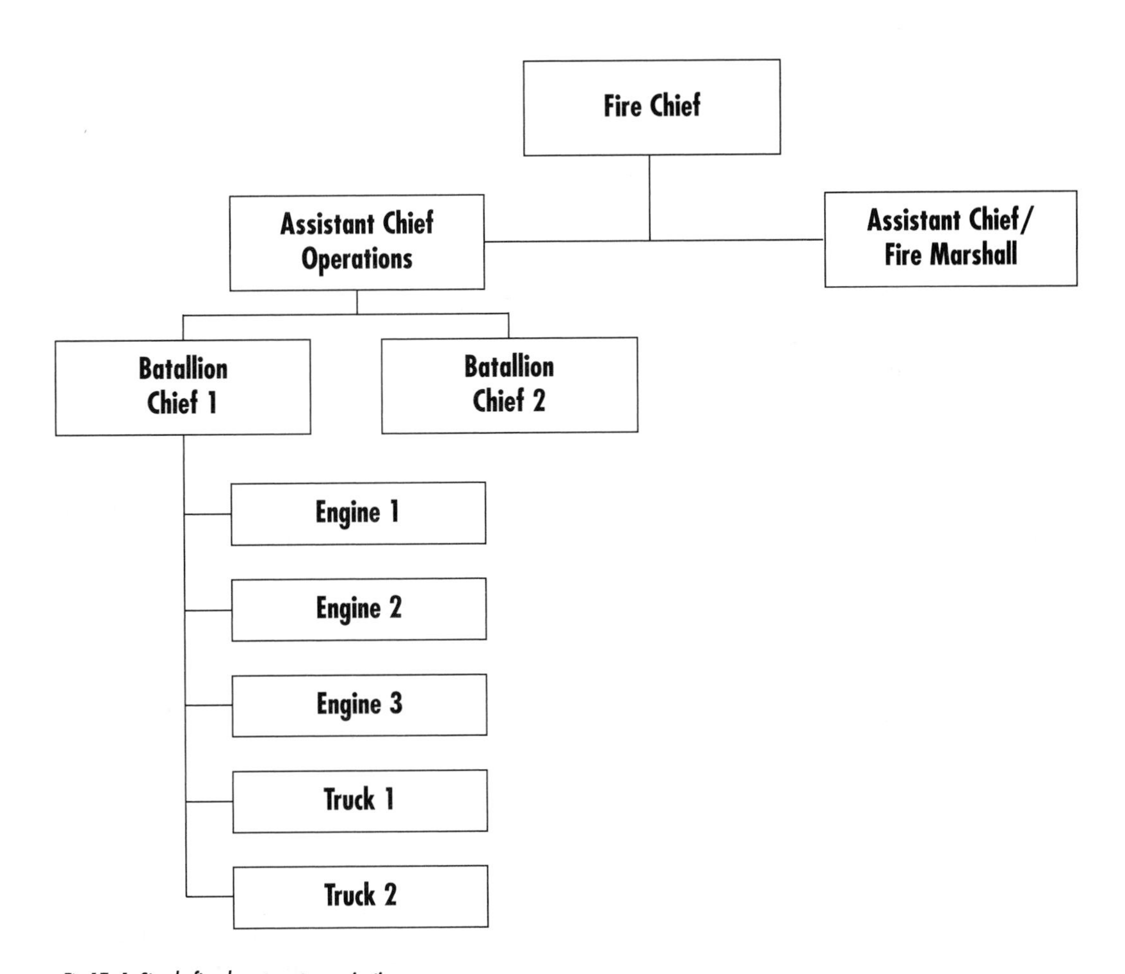

Fig 17–1 Simple fire department organization

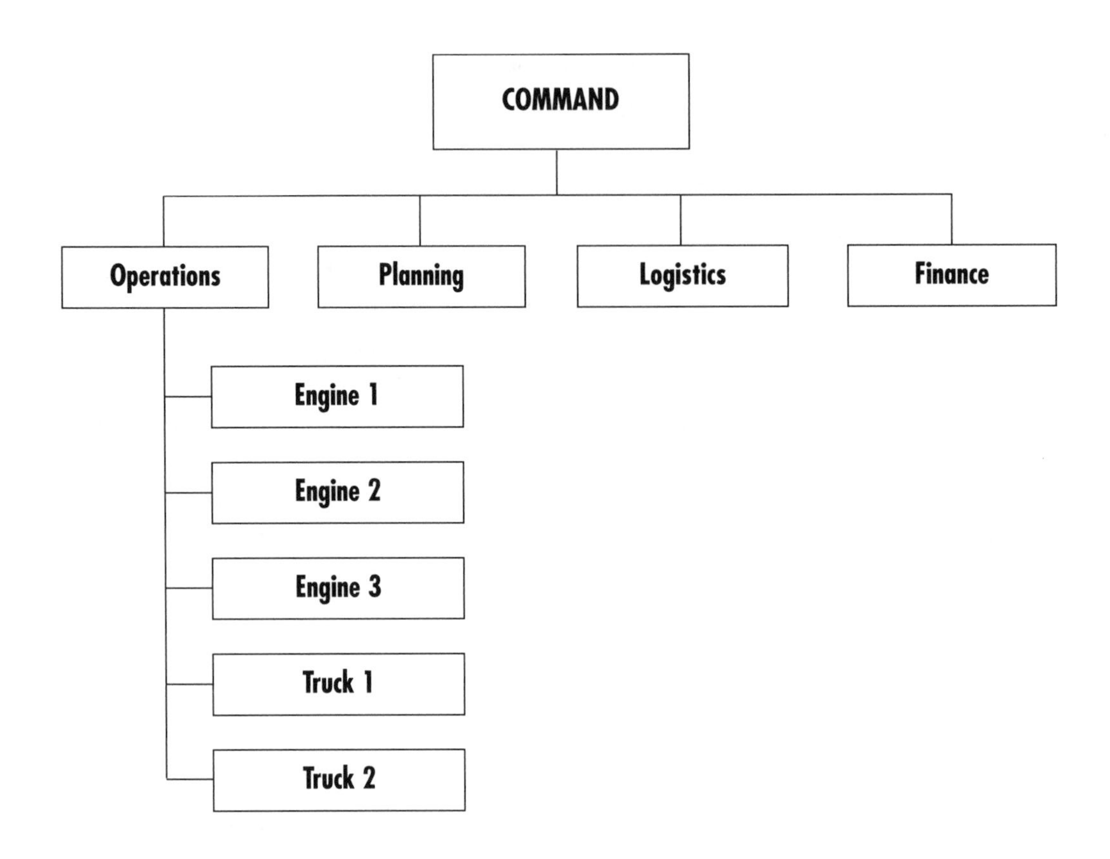

Fig 17–2 Simple ICS structure

Incident command

One of the primary tools to implement the four guiding principles for company performance is the *Incident Command System (ICS)* or *Incident Management System (IMS).* Originally developed in California during the 1960s as a means of managing large-scale disaster operations, the systems have been modified by the National Fire Academy and local jurisdictions to adapt to a variety of emergency operations. The system is flexible enough for any emergency incident type or size and contains procedures for controlling personnel, facilities, equipment, and communications.

Figure 17–2 shows a diagram of a simple ICS structure and the functions of companies within the system.

Types of Companies

The company is the standard fire department operating unit, consisting of firefighters, firefighter/emt, or firefighter/paramedics assigned to a specific fire apparatus.

Company members are trained to act as a team to control hostile fires and certain other emergencies. The company may work alone or in harmony with other companies to accomplish the same goal. The overall strategy may be well conceived, but company implementation determines how the job gets done, for better or worse.

The various types of companies found in fire departments of North America include:

Engine company: A standard fire department work unit consisting of a pumper and the staffing assigned to it. Staffing would include a first line supervisor, someone designated to drive the vehicle, and firefighters, all working at an assigned station to provide the basic response unit for a fire department.

Ladder/truck company: A fire department work unit consisting of a specialized vehicle and the staffing assigned to it. Staffing would include a first line supervisor, someone designated to drive the vehicle (2 people if the rear of the vehicle is steerable), and firefighters, all working at an assigned station. The ladder truck company has a special set of skills and duties at fire scenes that use equipment carried on these large vehicles. These duties may include use of the powerful hydraulic ladder or other extensive ground ladders carried on the truck.

Squad or rescue company: Duties vary from company to company, but the rescue company is primarily responsible for removal of victims from danger, including vehicle extrications and other types of structural entrapment. Rescue companies often provide assistance to engine companies.

Heavy rescue company: A rescue company that includes a heavy rescue vehicle equipped with specialized rescue equipment. Staffing would include a first line supervisor, someone designated to drive the vehicle and firefighters/rescue technicians, all working at an assigned station. The heavy rescue company is equipped to handle most rescue situations in a given community as well as perform a number of specialized tasks at fire scenes and emergency medical incidents.

Special companies: A fire department work unit consisting of a specialized vehicle and the staffing assigned to it. Staffing would include a first line supervisor, someone designated to drive the vehicle and assigned firefighters with special skills. Some special companies might include *Hazardous Materials*, *Wildland*, *Support*, *Special Operations*, *Marine*, and *Airport Rescue and Fire Fighting*, among others.

Company Operations

For the majority of fire departments in North America, company functions can include (but are not limited to)

- Fire suppression
- Forcible entry

- Ventilation
- Search and rescue
- Technical rescue
- EMS
- Hazardous materials
- Utility control
- Salvage work
- Fire inspections
- Community education

The two fire company types—engine companies and ladder/truck companies—are each organized and trained for specific functions, although these functions can vary according to the size of the department and needs or hazards of the service area. In a very small department, a single company will possibly be responsible for functions that would normally be handled by two or more companies.

Although further detail is provided below on engine and ladder company responsibilities, all fire company assignments are constantly subject to the complexity of emergency situations.

Engine companies. Water application is still the primary tool for fire suppression, and engine companies hold the critical responsibility of supplying water to the fire scene. Engine companies perform the basic tasks of providing and managing the water stream for extinguishment, as well as exposure protection, by deploying hoselines into and around structures and supplying water to aerial apparatus.

Basic engine company functions can include

- Use of initial attack and backup hose lines
- Operating master stream devices
- Tactical use of protective systems
- Rescue
- Forcible entry
- Overhaul

As needs vary from community to community, engine companies may at times also provide response to general assistance situations such as medical emergencies and accidents.

Ladder/truck companies. Working in conjunction with engine companies, ladder or truck companies perform most of the non-water related functions in fire suppression. The ladder truck contains the aerial devices necessary to upper level firefighting and rescue, as well as ground ladders and other rescue equipment. Company firefighters are responsible for providing structural access and exits, and working to control heat, smoke, and gases on the fire scene. Ladder company functions can include:

- Forcible entry
- Search and rescue

- Ventilation
- Reconnaissance
- Utility control
- Salvage
- Overhaul

It is often necessary that the ladder company perform two or more of these functions simultaneously, which emphasizes the need for highly-trained cooperation among crew members. Just as with engine companies, ladder companies sometime respond to medical emergencies, accidents, and other assistance needs.

Engine company

Training

Training in all aspects of fire department operations is essential. This includes both individual skills and knowledge as well as company operations. All companies should participate in regularly scheduled company and multiple company drills. One way to measure company performance is through the use of minimum company standards as outlined in NFPA 1410, Standard On Training For Initial Emergency Scene Operations.

> *This standard contains the minimum requirements for evaluating training for initial fire suppression and rescue procedures....—NFPA 1410 Scope*

Some of the areas covered by the standard are:

1. Methods of evaluating evolutions
2. Logistics of conducting evaluations
3. Performance for handlines
4. Performance for master streams
5. Performance for automatic sprinkler system support
6. Performance for truck company operations

The NFPA 1410 appendix contains illustrations of handline operations.

Ladder company

Staffing

Each fire company is staffed by a company officer, an apparatus driver, and one or more firefighters, all with multiple responsibilities. The company officer not only provides supervision at the scene but also holds responsibility for various equipment and station inspections, certain company training tasks, and company administrative and reporting tasks as needed. Apparatus driver/operators transport the crew safely to the scene, perform firefighting duties, keep trucks and apparatus maintained and serviced, and act as the company's second-in-command.

In addition to basic company staffing and leadership, departments with multiple companies and stations will also add shift captains and/or battalion chiefs to the line of command, for expanded company organization and supervision.

Staffing trends

In an independent agency survey conducted on 250 U.S. cities in 2001, the following trends were noted for engine company staffing (Table 6–1):

Table 6–1 Median Personnel Numbers per Company

City Population	Medium Average Personnel Per Company	
	Engine Co.	Ladder Co.
Under 100,000	3.09	2.4
100,000–500,000	3.50	3.4
500,000–1,000,00	3.60	3.7
1,000,000 and above	4.00	4.5

National standards

The 2001 edition of NFPA 1710 has established a minimum staffing level of four on-duty personnel per fire company. For companies serving high hazard or high incident frequency jurisdictions, the NFPA standard is a minimum five or six on-duty personnel. Fire suppression companies are to be staffed with numbers that take into consideration issues of populace safety, potential property loss, fireground hazards, and other tactical and safety issues.

Other special equipment or apparatus companies assisting engine and ladder companies shall be staffed according to the requirements of their hazard level or other contributing factors, as designated by their jurisdictional authority. This includes staffing of EMS units, which must take into consideration minimal levels needed to provide patient care and member safety.

REFERENCES

Clark, William E., *Firefighting Principles & Practices,* Fire Engineering Books and Videos (1991).

Mittendorf, John, *Truck Company Operations,* Fire Engineering Books and Videos (1998).

Fire Service Orientation and Terminology, 3rd Edition, IFSTA (1993).

NFPA 1410 Standard on Training for Initial Emergency Scene Operations, NFPA (2000).

NFPA 1710 Standard for the Organization and Deployment of Fire Suppression Operations, Emergency Medical Operations, and Special Operations to the Public by Career Fire Departments, NFPA (2001).

18

RESCUE OPERATIONS

Larry Collins

This chapter on fire service rescue operations is dedicated to
the firefighters and police officers in New York City
who willingly put themselves in harm's way to rescue others
on that dark day in September of 2001;
to the passengers of Flight 93
who prevented far worse losses
while giving their own lives in the service of others;
to those who risked their lives at the Pentagon;
and to all the other firefighters and rescuers
who were more than willing to do the same
if they had been confronted with similar attacks
on their cities and towns.

CHAPTER HIGHLIGHTS

- In-depth review of Urban Search and Rescue (US&R, a.k.a., USAR) operations on the state, federal, and international level of operational capability
- Sample equipment lists for rescue apparatus
- ICS checklists and procedures for various rescue scenarios
- Coverage of all related rescue operation NFPA Standards
- Review of all major rescue scenarios and how to prepare for and respond to them

INTRODUCTION

If the main mission of the typical modern fire department is indeed to protect lives, the environment, and property (in that order), it follows then that rescue operations are a primary role of firefighters. The public assumes the fire department is prepared to extract trapped victims from practically any situation, and there is an assumption that firefighters are prepared to identify, request, and coordinate special resources needed to get the job done. Most progressive firefighters and officers understand this and, consequently, the concept of saving lives through improved rescue capabilities has become part of the fire service mantra.

Some "traditionalists" in the fire service have decried the expansion of fire department responsibility into fields like rescue, hazardous material response, swiftwater rescue, regional and national disaster management, and now terrorism response. In their view, the demands of these ancillary responsibilities dilute the effectiveness of firefighters on the fireground. In the view of some traditionalists, the fire service should limit the scope of its responsibilities in order to concentrate on fireground operations. In other words, tasks like technical rescue should be the responsibility of some other agency—but this ignores two important facts.

First, it is the public that determines the scope of fire department responsibilities. Every time a citizen dials 9-1-1 the public is defining the mission of the fire department. And it is an indisputable fact that the public expects the fire department to conduct effective, professional, and timely rescue operations, as well as handling other emergencies like hazardous materials releases, emergency medical emergencies, floods, earthquakes, terrorist attacks, and other life and death situations.

Second, training, equipping, and preparing firefighters to conduct more effective rescue operations has the inherent benefit of making them more "well rounded." This improves their problem-solving abilities, giving them new tools to conduct fireground operations such as forcible entry, ventilation, access, etc., and improving their ability to rescue themselves and their colleagues when fireground mishaps occur. Rescue-trained firefighters are, by definition, better firefighters.

In short, the fire service's considerable investment in improving local, regional, and national rescue capabilities is a win-win proposition.

Emerging Rescue Challenges and Hazards

Immediately following the September 11th terrorist attacks, one observer was heard to lament, "Everything has changed, and nothing will ever be the same." The level and nature of danger and challenge faced by the fire and rescue services of all Western nations, and the context with which we view them, have been inextricably altered by the emergence of increasingly frequent, lethal, and destructive terrorist attacks—and the likelihood that they will continue to occur in the coming years. In the post-September 11th world, firefighters, rescuers, officers, and other decision-makers must view everything we do in a new light.

In terms of rescue, the scope and severity of hazards facing modern firefighters are far different than those that confronted our predecessors. And not all of these dangers are related to terrorism. The ever-expanding concentration of the world's populations in regions prone to natural and manmade disasters ensures that rescue-related emergencies and disasters will be ever-more frequent. Aging cities and infrastructures increase the frequency with which people will become trapped in collapsed buildings and construction accidents, especially when earthquakes, floods, and other disasters occur.

The continued industrialization of society ensures that workers will continue to become trapped in machinery, in confined spaces, mines and tunnels, in refineries and industrial settings, as well as vehicle mishaps of all types. Although some "traditional" rescue hazards (like those related to deep coal mining and automobile manufacturing) have been somewhat reduced by machines that have replaced much of the human labor, and by safety regulations that have drastically slashed the number of miners killed every year, the July, 2002 entrapment of nine coal miners in Pennsylvania demonstrated that one mishap can have devastating consequences and create challenges that test the mettle of the world's most qualified rescue authorities. In Pennsylvania, the problems began when miners using old mining maps inadvertently breached an earthen wall separating the "working" mine from one that had been abandoned in 1959 (and which had become inundated with 60 million gallons of water that burst through the wall like a collapsing dam, flooding the new mine and trapping the workers). Fortunately, the world's best mine rescue authorities were among those at the scene, and after nearly 80 hours of round-the-clock operations, they managed to rescue all nine miners alive from the tunnel in which they were trapped in water, 240 feet below the surface. It was one of the most brilliant examples of how rescue training, experience, technology, and good fortune can somehow combine to create a positive outcome, even in the direst of circumstances. It demonstrated the effectiveness of multi-tiered response to technical rescue, and showed us why rescuers simply cannot (and should not) call off search and rescue operations until every potential option for survival has been exhausted.

The increasing mobility of our respective societies increases the likelihood of rescue-related mishaps and disasters like airline crashes, train derailments, ship collisions and sinkings, subway mishaps, and other transportation accidents. The popularity of outdoor sports and adventure activities guarantees that people will become trapped in increasing numbers on cliffs, in vehicles "over the side," in avalanches, on rivers and streams, in landslides, and other wilderness predicaments.

Simply put, people will become trapped in ever-more complex and dangerous predicaments, requiring the fire/rescue services to respond in kind. Naturally, the danger of entrapment is not restricted to citizens. Fireground operations continue to trap, injure, and kill firefighters, who must then rely on their own survival skills and the rescue skills of fellow firefighters (including organized approaches like rapid intervention crews, FAST teams, etc.).

Rapid intervention operations themselves are increasingly hazardous because of wide-ranging factors: lightweight construction collapses rapidly under fire attack; façades continue to peel off burning buildings; truss roof, unreinforced masonry construction, and other "traditional" fireground hazards are still found in many older or renovated buildings; fire loads include more plastics and highly flammable, toxic smoke-producing materials; tightly sealed mid-rise and high-rise buildings create different fire dynamics; inspectors, builders, and designers sometimes cut corners (and they themselves are sometimes surprised by the range of forces to which their structures may be subjected); and now because terrorists have found a potent weapon of terror in massive fires caused by explosions and the crashing of airliners and tanker trucks into large occupied buildings (complicated in some cases by secondary attacks), ultimately resulting in collapse and other catastrophes.

Two Rescue Revolutions

The first rescue revolution in North America occurred in 1915 when the Fire Department of New York (FDNY) implemented the first rescue company,[1] whose job it was to rescue firefighters lost, trapped, or injured on the fireground, to conduct specialized rescue operations, and to provide forcible entry capabilities at the scene of difficult fires. Staffed by a captain, a lieutenant, and up to eight firefighters, the FDNY's first rescue company set the standard for a new approach to rescue and fireground operations.

In 1917 the Boston Fire Department established its first rescue company, equipped and staffed to conduct the most difficult rescues and to give fireground commanders unprecedented options for forcible entry, fire attack, and the rescue of downed firefighters. In 1929 the Chicago Fire Department followed suit with the creation of three heavy rescue companies. These pioneering agencies quickly demonstrated efficacy of specialized rescue units whose primary job is to handle high-risk rescue operations, including what's now called rapid intervention (the rescue of fellow firefighters who become lost, trapped, or otherwise incapacitated on the fireground and other emergency scenes).

The 1980s saw what might be called the second revolution in rescue operations. It included an emerging recognition that fire departments in many regions have at least an implied (and in some cases legal) responsibility to conduct technical rescue operations in a professional manner, using sound tactics and strategies, the latest equipment, and ensuring a reasonable level of safety for firefighters and other safety personnel. The second "rescue revolution" included a recognition that fire/rescue agencies have a responsibility to ensure timely rescue (*e.g.*, rapid intervention) for firefighters and other safety personnel who themselves become lost, trapped, or injured during the course of fireground and rescue operations.

The second revolution—which continues to this day—is notable for a long string of successful rescues under seemingly non-survivable conditions that have repeatedly demonstrated how fire service-based rescue capabilities improve survivability for trapped victims. Victims are being located and rescued faster than ever, with fewer preventable complications. Firefighters, armed with newfound knowledge and experience about the rescue and disaster conditions that confront them (and with effective, time-proven solutions and improved rescue equipment) are becoming trapped, lost, or injured less often during the course of search and rescue operations.

The second rescue revolution has been actualized in the public mind by the increasing number of successful rescues and disaster operations. Citizens have come to expect that their fire department can fix practically any emergency problem, and that they can rely on timely, professional, and effective rescue response from the fire department and the supporting state and federal agencies.

With the spread of this revolution, North America is experiencing a dramatic proliferation of fire department-based rescue and US&R programs. Today it can be said that the typical municipal fire department provides firefighters, rescuers, fire chiefs, and emergency managers with the tools to effectively manage most rescue emergencies; and that they have access to specialized US&R/rescue units with the ability to handle highly complex search and rescue operations that once might have defied their efforts. During major disasters they are supported by a network of state and federal teams trained and equipped to manage the most daunting rescue-related disasters. In short, the fire/rescue agencies of North America maintain the highest level of preparedness for rescue.

A Global Approach to Rescue

Naturally, the trend toward improved rescue and disaster response isn't limited to the United States. Indeed, the fire services of Great Britain, Ireland, Australia, South Africa, Israel, Italy, France, Canada, Russia, Japan, Taiwan, New Zealand, Switzerland, Germany, Turkey, Greece, and many other nations have greatly advanced their capabilities to manage urban search and technical rescue operations, and a number of these countries now sponsor travel-ready urban search and rescue teams that can be deployed anywhere in the world to assist other nations when earthquakes and other US&R-related disasters strike.

This newfound global focus on rescue operations vastly improved the means by which firefighters and other rescuers in many nations are able to detect, locate, treat, and extract trapped victims in practically any situation. Mirroring this paradigm shift is an ever-expanding pattern of dramatic survival stories—successful rescues accomplished under conditions that might have resulted in failure before the advent of formal rescue programs. There's strong correlation between the two phenomena: At the same time firefighters and rescuers are experiencing a revolution in the availability of better tools, training, and systems for locating and rescuing trapped people, more trapped people are surviving complex entrapment under conditions that once would have meant almost certain death.

The First Line of Defense

It's well known that firefighters are the nation's first line of defense when fire occurs, when there's a medical or hazardous materials emergency, when terrorists strike, or when victims are trapped by some sort of mishap. And they are frequently the primary rescuers when technical rescues and disasters occur. Modern firefighters are adept at rescuing people from a wide variety of predicaments, including victims trapped on cliffs, in collapsed trenches and excavations, in machinery, within collapsed structures, in swift-moving water, beneath avalanches and land slides, within mud and debris flows, in transportation accidents, and innumerable other rescue emergencies.

Today the general public recognizes that firefighters are often responsible for managing the most critical moments of emergency incidents, when time-critical decisions and actions have the most lasting and profound effect on the lives of affected victims, setting the stage for all the "consequence management" operations that follow.

The new emphasis on rescue has several positive implications for the fire service: the expanding awareness that rescue is a primary role of modern firefighters, the understanding that rescue cannot be separated from the other fire service missions, and proof that improved personnel and citizen safety is a natural byproduct of more comprehensive rescue service delivery. It's been proven through experience that improving the ability of firefighters to manage difficult rescue emergencies has the side-effect of making them better-rounded and more accustomed to solving complex problems. Firefighters trained and experienced in rescue are more attuned to "thinking outside the box" when faced with unusual conditions. They are more practiced at complicated problem solving under time-critical conditions. In short, improving rescue has the proven effect of making better firefighters and (in turn) better departments.

The new emphasis on rescue was never more evident than in the aftermath of disasters like the 1993 World Trade Center bombing attack, the 1994 Northridge earthquake, dozens of hurricanes and tornadoes in the 1990s, the 1995 Oklahoma City terrorist bombing, and of course the terrorist attacks of September 11, 2001.

And never has the absence of rescue companies and other specially trained rescue units and officers been more devastating than the aftermath of the collapse of the World Trade Center towers, which took the lives of all the on-duty personnel assigned to all five FDNY Rescue companies (and those who were getting off duty at shift change), personnel from several squad companies, the fire chief, the deputy chief of the special operations command, more than half of New York City's FEMA US&R task force, along with all those engine and truck companies and so many key members who would normally be charged with managing the consequences of the largest and most deadly structural collapses in history.

Prior to the 9-11 attacks, some people had difficulty imagining a situation so catastrophic that most of the local rescue units would be decimated, that hundreds of firefighters might be lost in an instant, and that thousands of firefighters and rescuers from across the nation would be required. For these people it might be even more difficult to estimate how such an event could be managed. But it's a measure of the nation's progress in fire service-based rescue that thousands of New York firefighters and police officers, supported by other regional and state resources, and backed by twenty 68-person FEMA US&R task forces, were able to fill in behind the lost rescue companies and the many lost rescuers, working round the clock in a demonstration of diligence and efficiency for many weeks after the collapse of the World Trade Center towers, without the loss of a single additional rescuer.

It is also, of course, a testament to the fortitude of the FDNY and the City of New York that the disaster was managed as well as it was. It's a sad irony that US&R task forces converging on New York from around the nation were there helping to look for the very firefighters and rescuers whose job it would normally be to handle structure collapse rescue operations there. Today, under the direction and influence of surviving rescue company, squad company, and special operations command, the FDNY rescue program has regrouped and is carrying on where the members who were lost in the 9-11 attacks left off.

Since 1991, the U.S. federal government, through the Federal Emergency Management Agency (FEMA—now supported by the U.S. Fire Administration), has operated what has been described as the world's most formidable nationwide network of US&R task forces. The success of this system in dozens of major disasters is the result of a remarkable partnership between FEMA and the nation's fire services. It is a model of rescue efficiency and effectiveness that's emulated by many nations around the world. The FEMA National US&R Response System is—by any standard—among the most effective and imaginative programs ever launched by the United States government, and it continues to improve and adapt as the US&R task forces are called upon to help local and state fire/rescue agencies manage ever-larger, more complex, and more deadly disasters.

FEMA's national US&R system ensures a rapid response of multidiscipline teams of highly trained and experienced rescuers, with the right equipment, anywhere in the United States and all U.S. territories. Since 1991, the FEMA US&R task forces (and the US&R incident support teams that respond to support and coordinate their operations) have built a formidable base of experience in applying rescue tactics and strategies that have proven effective in a wide range of manmade and natural disasters, including some that lasted weeks. As an added bonus to the sponsoring agencies and the citizens they serve, the experience gained by personnel assigned to the US&R task forces is ultimately transferred back to the "home turf" as rescuers return to their own fire/rescue agencies from disasters around the nation.

It must be emphasized that state and FEMA US&R task forces are not intended to take over incident command; the local incident commander remains in charge throughout the operations. Rather, these task forces (and the FEMA IST [Incident Support Team]) are designed to *augment* local resources to conduct and manage (as necessary, depending on the state of local emergency management and fire department chief officer status in the aftermath of a disaster) search and rescue operations at the scene of disasters and major incidents; and the FEMA IST is specifically dispatched to manage the operations of the FEMA US&R task forces. This is an example of why firefighters and officers need to clearly understand the distinctions, capabilities, limitations, etc., of these resources.

On the international front, the U.S. State Department—through the Agency for International Aid, Office of Foreign Disaster Assistance (OFDA)—in cooperation with the United Nations dispatches three US&R task forces to rescue-related disasters around the world. Each of these teams is an existing FEMA US&R task force strategically located for international response and chosen after a competitive selection process with parameters specific to response to all manner of international disaster search and rescue operations, as well as standards related to maintaining international cooperation and relations and representing the United States Government in a variety of capacities related to disaster planning, preparedness, capacity building, and instruction on foreign soil.

This advanced rescue capability is part of the United States' growing effort to provide humanitarian aid to people trapped in disasters on foreign soil. It has become commonplace for OFDA US&R task forces to join US&R teams from other nations at international disasters, where they work together to conduct dramatic and complex search and rescue operations around the clock.

Beyond the basic humanitarian considerations and moral obligations that are the primary concern of fire and rescue professionals, there is a little-discussed advantage of maintaining US&R task forces capable of deploying to disasters around the world: The OFDA international US&R response system provides advanced urban search and rescue capabilities that can quickly respond to terrorist attacks and other disasters that trap people in U.S. facilities abroad. This is an important consideration at a time when international terrorism is on the rise, and with the ever-present potential for other disasters affecting U.S. facilities (or places where U.S. citizens live and work on foreign soil). Recent events clearly demonstrate that U.S. citizens and those of other western nations are at risk from the consequences of terrorist attacks on foreign soil. Therefore, it makes sense to maintain an effective international US&R response capability—and the United States has done just that.

Beyond those obvious needs, there is yet another advantage to the deployment of U.S. urban search and rescue task forces to international disasters: the transfer of experience and knowledge gained by US&R task force personnel deployed to international disasters. Firefighters and other rescuers assigned to these teams bring home invaluable lessons from dealing with the consequences of disasters around the world, and many of these lessons are gradually incorporated into local disaster planning, training, and response.

The experience gained by these rescuers at international disasters cannot be replicated in training and simulations. For the first time in history, a collective pool of knowledge and experience from disasters nationwide has been established in a systematic way that ensures it will be applied to future disasters. For the past two decades, U.S. citizens (and citizens of other nations that sponsor international US&R teams) have been the benefactors of this "trickle-down" effect, whereby information and experience gained by the international US&R task forces is gradually disseminated to agencies that deal with similar hazards. The ability to bring that kind of experience home before the next disaster is clearly an advantage for local fire/rescue agencies and nations that participate in the international response system.

TERRORISM

Since the last printing of *The Fire Chief's Handbook,* the greatest changes in the world are related to modern terrorism, which has introduced a new and dangerous dimension to the fire service. Today, there are too many individual terrorists and terrorist groups with the ability and the willingness to create "urban canyons," to trap, injure, or kill hundreds, thousands, and perhaps tens of thousands of innocent people. The evolving doctrine of terrorism virtually ensures that certain groups and individuals will employ methods scarcely conceivable just a decade ago.

The methods used by terrorists will range from hijacking fully fueled airliners and cargo planes and flying them into high-rise buildings, sports stadiums, centers of government, and nuclear power plants; to stealing gasoline tankers and driving them at high speed into government buildings (which already happened in Sacramento in April, 2001); from planting bombs in tanker trucks, boats, yachts, and cargo ships, and piloting them into densely populated coastal areas to be detonated; to derailing trains carrying chemicals and explosives. Terrorists will attempt to flood subways, train tunnels, and car tunnels; as well as cause immense fires that trap thousands of people and compel the fire department to commit its personnel to virtual suicide missions.

As security clamps down on the largest venues and modes of attacks, terrorists will resort to modes of operation that are more difficult to prevent. They will use car bombs, sabotage, and snipers to cause terror and panic. Suicide/homicide bombers will begin walking into crowded malls and public gatherings and schools wearing explosives and detonating themselves. They will begin targeting places that are less likely, with the intent of creating a sense of insecurity everywhere. This is already occurring in places like Israel and parts of the U.K. and other Western nations. Only the most naïve would be led to believe that it won't happen in the United States.

Furthermore, some terrorist groups and individuals will be tempted to initiate terrorist attacks that purposely target firefighters and rescuers responding to disasters, an especially hurtful (and, in the view of terrorists, effective) tactic to employ when fire/rescue resources are already stretched to the limit. They will be inclined to commit acts like exploding dirty bombs in populated areas, knowing that the explosive irradiation of cities and the natural environment will take a heavy and long-lasting toll; to cause dam failure that creates massive urban floods; or to use insidious biological and chemical weapons that exceed the personal protection levels of the public and firefighters.

Even worse, some of these groups appear willing and capable of obtaining and detonating full-fledged nuclear devices in one or more Western cities, possibly in simultaneous attacks. In short, there is growing potential for firefighters to be confronted by a number of "worst case scenarios" that could potentially dwarf the 9-11 attacks in terms of lethality and rescue problems. In light of these factors, it's becoming clear that the rescue challenges related to terrorism are unprecedented in history.

The new paradigm in terrorism is not restricted to the United States. Perhaps it seems like an American experience at the moment because most Americans are unaccustomed to dealing with the insidious terrorism that has gripped many other parts of the world for decades. Today and for the foreseeable future, the free nations of the world are going to be plagued by the consequences of terrorism. The terrorists (and the entities and ideals they claim to represent) have vowed to harm us and the people that we, as firefighters, are sworn to protect. Unfortunately, they have a record of success that invites other groups and individuals to attempt similar attacks, or to "up the ante" by causing ever-more lethal events, and "moments in time" that no one will forget.

The primary role of firefighters in terrorism response

It's a little-spoken truth that firefighters and other first responders have become participants in the domestic battle against the "asymmetrical" effects of terrorism. They are part of the "consequence management" aspect of the anti-terrorism efforts of their respective nations. Firefighters and other first responders are a bellwether, an indicator by which the public gauges the effectiveness of the ability of the government to confine and control the effects of a terrorist attack. In the 9-11 attacks, firefighters and other first responders were the primary combatants in the first volleys of this new war.

Consequently, firefighters and other first responders carry the burden of being the first wave of response to an attack on our societies by those who seek to destroy them. Some may bristle at the following association, but the case can be made that the proper response of fire/rescue resources to a terrorist attack should closely resemble (in a metaphoric way) that of ants swarming out to protect their colony when it comes under attack.

If firefighters, fire/rescue agencies, and law enforcement respond purposefully, rapidly, and effectively to acts of terror, if they quickly rescue, treat, and remove the injured and quickly care for the dead in a well-organized manner that assures the public the government is in control of the situation, they help limit the consequences of the attack, thereby reducing its "effectiveness" (from the terrorists' perspective). If the consequences of terrorist attacks are consistently held to a minimum through the rapid, massive, and effective response of fire/rescue agencies and law enforcement, it may have a dampening effect on the terrorist efforts because it blunts some of the impact of the attacks and helps thwart the attackers' motivation of causing panic and forcing the public to question the government's ability to protect them.

If, on the other hand, the emergency response is slow, timid, haphazard, poorly organized, and ineffective, then the public will be compelled toward panic and questioning the government's ability to remain in control. This would play right into the hands of the terrorists, who will have gained a double victory by conducting a successful attack and placing the government between the proverbial rock and hard spot.

Terrorism isn't always imported from other nations. There are any number of American-born terrorists and terrorist organizations operating inside the United States and other free nations, prepared to conduct asymmetrical operations using leaderless cells and other strategies and tactics gleaned from their study of international terrorist groups. The existence of a sort of "shadow world" of domestic terrorists was driven home by the Oklahoma City bombing and other recent attacks by American-born terrorists who populate various regions of the United States. In the aftermath of the 9-11 attacks some of these groups and individuals may have become less visible, but they are out there all the same, waiting for a shift in the public's sentiments—waiting for their opportunity to strike out.

In order for us to do our jobs properly and protect those we've sworn to serve, the strategies, tactics, equipment, and philosophy adopted by today's fire departments and rescue agencies must be equal to (or greater than) the challenges posed by the new paradigm of terrorism and the related hazards that will follow. Fire departments, their chief officers, and the line firefighters must have the freedom, imagination, and (most important) the motivation to think fast, to plan far ahead, to react quickly to fluid conditions, to ramp up their response capabilities in a pro-active manner, and to stay one step ahead of the terrorists (rather than being ten steps behind, which unfortunately has been our experience in past years).

This is going to require a major change in the way most fire/rescue agencies operate. It requires a change in the way fire/rescue decision-makers view and respond to terrorism and its inevitable side effects. It will require cultural changes within organizations that are prone to heavy reliance on traditional ways of doing things and resistant to rapid change. Not only must today's firefighters and rescuers (and their supervisors) be receptive to rapid adjustments to stay ahead of rapidly changing hazards; they must demand it in the interest of firefighter safety and because the public expects the fire department to be on top of the situation, even when it's rapidly evolving.

"We need to ensure that all firefighters and rescue personnel remember that the posture of an ostrich has never proven effective in solving dangerous and complex problems," says Geoff Williams, Deputy Fire Master of the Central Scotland Fire Brigades, a renowned agent of change in the U.K. fire and rescue services. "The future is like an enormous flood. When faced with it, there are two choices. You can divert the course of the flood—or you can change your own course. To commence the necessary changes of course in the fire/rescue services," continues Williams, "We must accept that we cannot afford to waste precious time trying to re-invent the wheel. If we needed convincing about the truth of this statement, the tragic events of 9-11 confirmed this with interest. We need fire service leaders who do not 'view the world through a straw.'"

As is often the case with adversity and challenge, this is a time pregnant with the potential for great innovation and improvement that will forever change the face of the fire and rescue services by consolidating our resolve, coordinating our efforts, compelling us to finally speak one "language" at emergencies, prodding us to speak with one voice with respect to funding and legislation, to adopt common terminology, to embrace the Incident Command System (ICS) and SEMS (Standard Emergency Management System) principles, to seek out new technology and concepts that will improve our ability to protect lives and property, and to improve the safety of firefighters and others whose job it is to locate and rescue missing and trapped people.

There is also an unspoken responsibility to honor the memory of the innocent people who lost their lives in the 9-11 attacks, including the firefighters and other rescuers who made the ultimate sacrifice trying to save lives when the World Trade Centers towers collapsed. Tragically, some of the world's most experienced and innovative firefighters and rescuers were lost in the WTC collapse. It is a sad irony that the very collapse search and rescue methods, tools, teams, and systems they helped develop to reduce suffering and save lives across the United States and around the world, were in turn required to help locate, rescue, and recover them and their colleagues beneath the fallen towers.

Buried within this story is a metaphor about a process of experience, research, development, implementation, and improvement coming full circle. You, the reader, are now part of this circle, and the lessons you take away from the experiences of those who passed before, documented here and elsewhere, have a material effect on the outcome of future rescue emergencies and disasters. It starts with acknowledging that rescue is a primary mission of the fire service, inseparable from fire fighting, emergency medical services, hazardous materials response, and other fire department functions.

Rescue-Related Disasters

In addition to "daily" rescue emergencies, fire departments are responsible for conducting search and rescue operations when disaster strikes. In disastrous events like earthquakes, floods, train crashes, airliner crashes, landslides, mud slides, tornadoes, hurricanes, avalanches, and terrorist attacks, the fire department is usually first on the scene and last to leave. Firefighters are expected to demonstrate their expertise in the process of quickly and accurately assessing the conditions, requesting the necessary resources to get the job done right, organizing and performing round-the-clock search and rescue operations, treating the injured (including those still trapped), and generally mitigating all threats to life. Firefighters are also expected to manage the process of recovering the dead in an organized and respectful manner, and even making the scene reasonably safe for the appropriate authorities to gather evidence and conduct post-disaster investigations.

Consistent with the newfound emphasis on rescue, and the constant raising of the bar on rescue standards, modern fire departments routinely conduct complex and extremely high-risk rescue operations with a level of timeliness, effectiveness, experience, rescuer safety, and personnel accountability unheard of just a decade ago.

Whereas rescue in past eras could sometimes be characterized as an *ad hoc* affair conducted by well-meaning personnel (who were often inadequately trained and ill-equipped, but who persevered through the use of common sense, their own emergency experiences, and the application of knowledge they gained while working in various trades), many of today's rescue operations can be described as efficient and well-choreographed operations evolutions, conducted by teams of firefighters who have been properly trained and equipped for the task, who have done it before (either in training, in simulation exercises, in preplanning sessions, or in actual rescue emergencies).

The sharing of information, training, techniques, and technology among fire departments in various nations is unprecedented, and the result is an increasing sum of lives saved in "daily" rescues and large disasters alike. Today it can truly be said that many fire departments in the U.S. and elsewhere are applying a more global approach to rescue than at any other time. It has led to many improvements in the timeliness, safety and effectiveness of rescue operations in the United States and abroad. The common threat of terrorism and disaster affecting many nations makes this spirit of international coordination and communication ever-more essential for the fire service.

Progressive fire departments have taken the lead in advanced local and regional rescue capabilities by establishing rescue companies, US&R units, swiftwater rescue teams, and technical rescue teams. Countless fire department-based technical rescue teams have sprung up in urban and non-urban areas across North America and other parts of the world. State and federal government agencies have become integrally involved in coordinating or providing advanced rescue capabilities in support of first responders.

The entire concept of rescue has indeed undergone a dramatic revolution in many places. And yet, some things always remain the same, like the need for well-trained, highly experienced firefightighters willing to take certain calculated risks in order to achieve the strategic and tactical rescue-related objectives of incident commanders, and for fire department administrations to support their efforts to maintain the highest state of preparedness for these operations.

Staffing and Personnel Selection for Rescue Units

In some regions and organizations, the labor-management relations (including local memorandums of understanding and labor contracts that specify how fire department units shall be staffed, trained, and organized) are a determining factor in who does rescue. In some departments the captain of the rescue company has the authority to interview and hand-select the personnel who will be assigned to the company.

Other fire/rescue agencies are guided by strict agreements between management and labor that specify documented qualifications (*e.g.*, completion of rescue-related training courses meeting NFPA 1670 Standards for selected rescue tracks, minimum levels of experience, participation in continuing education, maintenance of selected certifications, etc.) and seniority as the only criteria for being assigned to open positions on rescue or US&R units.

A typical example is this: If firefighters or officers of the appropriate rank have completed the requisite training and certifications, they may be eligible to submit requests for transfer to US&R or rescue companies; and if more than one qualified member bids for a position that opens up, the person with the most seniority automatically gets the position. Some agencies even apply these rules to the selection process for FEMA US&R task forces (with FEMA-mandated training, medical, and other requirements) and other special assignments.

Other agencies employ a time-honored method whereby the rescue company or US&R unit captain is allowed to hand-pick the most qualified personnel based on factors like the individual's interest in the position, their level of training (including, in some cases, meeting NFPA 1670 Standards and selected certifications), fire/rescue experience, current or previous employment in a trade that's deemed pertinent to rescue (*e.g.*, construction, commercial diving, tunneling, etc.), demonstrated rescue skills, and ability to work as part of a team under high-risk conditions. This approach allows the rescue or US&R company supervisor to tailor personnel selections to achieve a certain goal of teamwork, expertise, and daily operational effectiveness. It also places a great deal of responsibility on the supervisor to make the right choices. But it is a proven approach that yields very positive results for units expected to function at maximum capacity in high-risk environments.

Both of these approaches are common in the United States, depending largely on regional and organizational traditions, demonstrated effectiveness, and labor-management agreements and rules. Both have their potential merits and drawbacks.

Some agencies have even assigned performance or hazard bonuses to rescue or US&R unit positions. The issue of bonuses for assignment to a rescue or US&R unit is like the proverbial double-edged sword: on the positive side, it makes sense to reward dedicated and highly skilled employees who maintain a high level of performance and take extraordinary risks in the normal performance of their duties. The argument can be made that "rescue" bonuses will encourage highly experienced and trained employees to remain in their specialized positions for years to come. It can be seen as a reward for taking extraordinary risks, which is appropriate. Finally, if the member dies in the course of their extra-risk duties, the bonus may be applied to the benefits that will be paid to their dependents.

On the negative side, there is the potential for certain employees to complete the minimum level of training required to "bid" a rescue bonus position, and then (once they are in the position) to comply with only the minimum standards of performance in order to maintain the rescue bonus until the day they retire. There is the potential that some members who would normally recognize they may not be "cut out" for the rescue unit, may remain anyway in order to maintain the bonus, thereby taking away a position that might be filled by another member who's far better suited for a rescue or US&R unit. Finally, some members who might otherwise leave the unit as they approach the end of their careers because the rigors of the rescue unit are too demanding, may decide to "stick it out" until the day they retire, in order for the bonus to be figured into their retirements benefits.

Bonuses for rescue work are obviously a sensitive topic, and there are many ways to look at the issue. However, many fire/rescue agencies that attach bonuses to positions on rescue units can attest that both the positive and negative effects (as described above) can be detected by objective observers. This is not to say that bonuses should or should not be attached to rescue positions; only that the decision-makers should be aware of the potential for double-edged results and have a plan in place to deal with the negative ones.

Whatever the method for selecting rescue or US&R unit members, there should be some method to ascertain the level of performance and for making adjustments as necessary to maintain reliable performance of all members (including provisions for removing personnel who are not performing to the required level for reasonable operational effectiveness and personnel safety).

US&R vs Rescue

In order for firefighters and chief officers everywhere to establish a common ground for the discussion of rescue, it's important to dispel the oft-held myth that "urban search and rescue" is somehow different than rescue operations. The evolution, development, and operation of local, regional, and national urban search and rescue units and programs since the 1980s, have taught us that rescue operations and US&R are one and the same, with certain regional and organizational differences. The term "USAR" was first used around 1988 in Los Angeles County to differentiate fire department-based technical rescue programs from "traditional" volunteer-based or law enforcement-based Mountain Search and Rescue (SAR) teams. Eventually the acronym USAR stuck, even though it often also refers to municipal fire department companies that respond to rescue emergencies in both urban and non-urban settings. Even the spelling of the acronym for urban search and rescue differs between some agencies. Many fire/rescue agencies use the designator "USAR" to identify their technical rescue units. Other agencies, including FEMA, use the acronym "US&R" to identify urban search and rescue resources, even though they are, in some instances, identical. (In order to emphasize standardization, the FEMA acronym is used throughout this chapter.)

That concept does not seem to be well understood in regions where the term US&R is imprecisely used to refer exclusively to FEMA US&R task forces or other resources so specialized that they're called upon only in times of disaster or very unusual rescues. It's true that the nation's FEMA US&R task forces are intended to conduct urban search and rescue operations at the scene of federally-declared disasters (and that they are among the world's most expert and potent resources for managing large-scale disaster SAR operations). However, the term US&R is not exclusive to these teams. Many fire departments use the term "USAR" synonymously with "rescue" to refer to companies of firefighters who specialize in technical rescue and fireground operations every day in their local jurisdictions. It's an indication of the connectivity between rescue companies and US&R units that some firefighters and officers assigned to them are also selected for disaster-related rescue duty as members of FEMA US&R task forces and FEMA US&R ISTs. US&R units are synonymous with rescue companies in many fire departments in the United States and in other nations. Particularly in the western United States, US&R is a designation given to fire department units whose primary role is technical rescue and firefighting in urban, suburban, and even in wilderness areas, mountains, deserts, oceans, and forests. It can be said that rescue companies are to midwest and east coast fire departments as US&R units are to west coast fire departments. In many fire/rescue departments there is little between an "east coast style" rescue company and a so-called "west coast style" US&R unit—except perhaps that some west coast US&R units respond to technical rescues in the mountains, the desert, the ocean, and other wilderness areas in addition to major fires and technical rescue emergencies in the city.

Every day, fire department first responders, backed up by fire department-based rescue companies and US&R companies, locate and extract people from every imaginable predicament. Firefighters assigned to rescue companies and US&R companies even rescue animals (large, small, domesticated, and wild) from every imaginable predicament. It should come as no surprise that a vast cross-section of the public would classify the rescue of trapped animals as a reasonable priority of the fire department (particularly when it involves the rescue of their own pet or animal stuck in a tree, on a cliff, in the mud, in a ravine, in a hole in the ground, or some other predicament). Fortunately, many animal rescues are resolvable by adapting the very same tools and methods used to save humans—with obvious exceptions. In addition to the moral and humanitarian obligation to assist trapped animals (especially those who are led into sticky predicaments through the actions of their human friends), these rescues are yet another way for fire fighters and other rescuers to get valuable experience. The primary limiting factor is rescuer safety, which should always be a primary consideration of the incident commander in these cases. Obviously, the risk-vs-gain calculation for rescuing animals differs from the decision to place rescuers in harm's way to save people, and sometimes it's simply too dangerous for human rescuers to be placed in harm's way to attempt the extraction of an animal.

There are also differences in the manner of service delivery selected by individual agencies. Some fire/rescue departments deploy fully staffed rescue companies of six or seven personnel assigned to special divisions (for example, the FDNY's five 6-person rescue companies assigned to the Chief of FDNY's Special Operations Command), which is clearly the "Cadillac" of rescue system approaches.

Other agencies combine specially trained engine companies and dedicated rescue/US&R companies in the same fire station to create six-person US&R/rescue task forces (perhaps the next-best approach if the dual missions of jurisdictional fire/EMS/rescue responsibility can be balanced against the technical rescue role and the extensive training, planning, exercises, research, and development that goes along with it).

Common "Daily" Rescue Emergencies

- Submerged victims (dive emergencies)
- Passenger vehicle or bus extractions
- Structure collapses
- Marine rescues
- Overturned heavy equipment
- Mountain rescues
- Snow and ice rescues
- Rapid intervention operations on the fireground
- Vehicles "over the side"
- Cliff rescues
- Confined space rescues
- Swiftwater rescues
- Mine and tunnel rescues
- Coastal rescues (boats on the rocks, etc.)
- Other situations where victims are reported trapped
- Transportation accidents (plane crashes, train wrecks, ship collisions)
- Technical animal rescues requiring fire department assistance
- Machine/industrial entrapment rescues
- Trench/excavation collapse rescue
- Semi trucks or heavy equipment on top of occupied automobiles
- "Jumpers" (potential suicide jumpers or hostage situations on bridges, rooftops, cliffs, dams, and other physical rescue locations)

Next on the scale of rescue system approaches is that where a fire department trains selected engine or truck companies to "jump over" to the US&R or rescue unit in their fire station when technical rescues occur. This approach is employed by many fire/rescue agencies whose budgets (or budget priorities) do not support fully dedicated rescue/US&R company staffing or the US&R/rescue task force concept. It can work if the members are diligent and if their supervisors give them the leeway to accomplish their dual roles of "first-in" fire/EMS and rescue and multiple-jurisdiction rescue response. However, the "jump over" approach can be problematic because unless the company officer is absolutely diligent in drilling their crew (including after hours as necessary to keep up with the rescue duties after completing the normal engine company functions), the technical rescue role will likely be subjugated by the other normal requirements and demands of a standard engine company. It's worse if the rescue/US&R engine is not allowed to leave its first-in jurisdiction when necessary to conduct training with other units and agencies, to "pre-plan" target rescue hazards, to conduct planning and coordination functions, and to conduct post-incident reviews at the site of unusual rescues. The result, if not done right, may be a crew that's out of practice with certain rescue skills as well as equipment that does not react in a timely manner to changing rescue conditions.

Lower on the comparative scale of effectiveness and timely response are rescue systems whereby fire/rescue departments provide training for personnel who are dispersed among various engine/truck/squad company assignments throughout the department, relying on these individuals (sometimes including off-duty members) to assemble and respond in the event of a technical rescue or disaster. In a variation of this, some agencies band their trained manpower together to form regional US&R or technical rescue teams, consolidating their various resources into one cohesive team.

Still lower on the effectiveness scale are agencies that have no formal rescue or US&R units, whose trained personnel (if there are any) are scattered around the department with no way to assemble and use them, and no plan to utilize mutual aid rescue resources.

Lowest yet on the "relative rescue safety/effectiveness/preparedness scale" are departments that simply haven't addressed the issue of rescue at all. Fortunately, these are relatively few and far between in the United States and other industrialized nations.

Each system of personnel selection for rescue or US&R units, and the deployment of these specialized resources, has its own merits and disadvantages But one universal statement can be made about models of rescue service delivery: Regardless of the particular method or staffing or deployment, the use of multi-tiered response systems (*e.g.*, first responders augmented by well-trained rescue or US&R units, including those meeting the new NFPA 1670 Standard and pertinent certifications) and the employment of rapid intervention protocol for rescue incidents is a proven way to ensure timely and effective rescue service to trapped victims—with a reasonable assurance that any would-be rescuers who become lost, trapped, or injured will have a reasonable chance of themselves being rescued.

Clearly, the best rescue systems are those that are backed up by a wide swath of first responders (not assigned to rescue units) who have extensive rescue/US&R training through their fire department's recruit academy and the normal training and continuing education for all firefighters; and those who have (either on their own or through their department) attended specialized rescue courses. Fire/rescue agencies that encourage and support this level of training, knowledge, and skill among all their line firefighters and officers, are in the best position to manage the full range of "daily" rescue emergencies and rescue-related disasters.

WHAT RESCUE "RESOURCE TYPING" IS AND WHY IT IS NECESSARY

Even with good planning and preparation, it's simply not feasible for every fire/rescue agency to possess all the internal resources and specialized equipment and units necessary to properly and safely manage the full range of rescue emergencies that might occur. And when disaster strikes, the resources of any single fire department may be overtaxed (the very definition of disaster indicates an event that may outstrip local resources). The concept of mutual aid in the fire/rescue services was developed to address both of these situations.

In order to have an organized system of mutual aid resources that can be called upon to perform designated operations or respond to certain hazards during large and/or complex rescue emergencies and disasters, it is first necessary to identify the type and capabilities of the resources that are available. This can pose a problem because of the simple fact that there is such a wide variety of emergency resources available in the modern world, each with its own set of characteristics, capabilities, and limitations.

Resource "typing" is closely related to the incident command system in this way: for an incident commander to request additional resources (especially those that require mutual aid response), they must be able to quantify the type and number of resources they need. Looking at this from the fireground perspective, even the most common resource in the fire service world, the "standard" engine company, can come in many variations based on size, weight, pump capacity, water tank capacity, size and length of hose, two-wheel drive vs four-wheel drive, ladders, rescue equipment, and staffing.

Taking a non-rescue example for a moment, if the incident commander at the scene of a wildland/urban interface fire needs twenty mutual aid engine companies to protect structures in a residential neighborhood, and another ten mutual aid engine companies to drive rugged dirt motorways into the mountains above the same neighborhood to stretch thousands of feet of hose lays through the hills in a flanking action, they must understand the capabilities and limitations of different types of engine companies or risk losing the neighborhood because the wrong mutual aid resources were sent and they were incapable of performing the tasks they had in mind.

The same dynamic applies to rescue operations, with this caveat: The variations between different types of rescue resources are sometimes even more striking, and fewer fire/rescue officials are thoroughly familiar with the capabilities and limitations of each. If an incident commander at the scene of a structure collapse disaster needs ten mutual aid rescue companies with certain capabilities, and one or more FEMA US&R task forces to augment the rescue companies, and the other units to conduct round-the-clock operations, they need to understand the characteristics of each of these resources and how (and where) to use them to the best advantage, or they risk losing the lives of trapped victims and rescuers alike.

In order to ensure the right types of resources are being requested, dispatched, and used, there is a need to develop and use fairly precise definitions and standards when typing emergency resources for mutual aid, especially with regard to rescue. This process of characterization of resources is known as "resource typing," and it is already an ongoing activity in many parts of the U.S. and other nations.

Once the resources of a city, county, region, state, or nation are typed, they can then be inventoried with a reasonable degree of assurance that when an incident commander requests a certain type of resource, that is what they will get. Not only does this make life easier for the incident commander, but it's good for the people who are waiting to be rescued (or whose property is threatened).

How resource typing is conducted

Typing of fire, rescue, EMS, helicopter, hazmat, wildland, and other categories of resources is conducted at various levels, including within municipal fire departments, within county or regional mutual aid systems, within states, and within federal fire/rescue agencies like the U.S. Forest Service. In places like California, where mutual aid in some form is used repeatedly every day of the year (and where some disasters have literally required the response and assignment of more than one thousand mutual aid fire/rescue units to manage the consequences of a single event), the typing of every type of fire/rescue resource available to the county, regional, and state mutual aid systems is legally mandated.

Looking back to the fireground example, engine companies in California and other places are typed by their pump capacity, water tank capacity, length and size of hose, and the number of personnel. For example, a Type I Engine Company is essentially one that might be seen in any municipal fire department serving an urban or suburban area, which can be assigned to conduct interior structural firefighting operations during disasters like urban conflagrations—or to perform structural protection during urban/wildland interface fires. A Type II Engine Company is one that carries less water, has lesser pumping capacity, and is built to deal with more rugged conditions than the typical city street. A Type III Engine Company is built for even more austere conditions, including steep fire roads and off-road areas, with certain parameters for water tank size, pump capacity, etc. (See Appendix 1)

Example of mixed resource typing: strike teams and task forces

In accordance with the incident command system, similar resources can be grouped together in strike teams. For example, in California, a Type I Engine Company Strike

Team refers to five Type I Engines that are assembled as one unit and supervised by a strike team leader who is typically a battalion chief with his/her own command vehicle. In California, a Type I Rescue Company Strike Team consists of two Type I Rescue Companies with a battalion chief that respond as a single unit capable of performing the most advanced rescue operations.

Mixed resources can be grouped together as task forces to address particular problems requiring the use of mixed resources. Some task forces are permanent, with specific staffing and equipment requirements. Others can be assembled at the discretion of the incident commander to address specific problems or hazards.

For example, two Type I Engine Companies combined with a water tender can be dispatched as a Fire Attack task force. This type of resource is applicable in the aftermath of earthquakes, when water mains and other traditional sources of firefighting water may be disrupted. Similarly, two Type I Rescue Companies could be combined with a Type I Engine Company (to provide water for fire protection and manpower during rescue operations, and to supply water for concrete cutting tools that require water to keep blades cool), a Type I Bulldozer (for emergency road clearing), and a commander as a rescue task force, to respond to structure collapses in the aftermath of a damaging earthquake.

Another, more familiar, example of a task force is the FEMA US&R task force. This is a team of 68 specially-trained firefighters and rescuers, who are assembled in times of disaster into five major components (command, search, rescue, medical, and technical support), and equipped to respond anywhere in the United States by air, by ground, by rail, or by ship (as necessary) to complete a specific specialized task: locating and rescuing deeply entombed victims of structure collapses, and rescuing people in other forms of rescue-related disasters.

An incident commander faced with the effects of a disastrous earthquake might be compelled to request mutual aid in the form of individual components (*i.e.*, "Respond two 50-ton cranes"), or larger increments (*i.e.*, "Respond three Type I Engine Company strike teams, two Type I Fire Attack task forces, two rescue task forces, and Two FEMA US&R task forces."). (See Appendix 2 for equipment recommendations)

Rescue Successes Breed a New Paradigm

Clearly there is a close connection between today's highly advanced rescue capabilities and the higher survival rates for trapped victims. For those of us who have seen or experienced the success of complicated rescues that once would have defied us, no other explanation is necessary. These experiences leave us with an intrinsic understanding that a vigorous and pro-active approach to unusual rescue situations will result in the saving of lives that once would have been lost. We also understand that these new rescue capabilities might one day be called upon to save our own lives if things go wrong on the scene of an emergency and we ourselves become trapped. From the perspective of both the rescuers and the public we serve, the development of better rescue systems is truly a win-win situation.

Today, when a victim survives a particularly riveting entrapment, it's often called a miracle. But many firefighters understand that the real miracle may be the sea change in the way modern fire/rescue services (sometimes with the assistance of specialized state and federal teams) manage rescue operations.

The new approach to rescue emphasizes factors like innovative training, better equipment, and more intelligent planning by well-informed decision-makers who constantly ask the question "what if?" It's a philosophy that emphasizes adherence to accepted incident command system principles; increased reliance on fully staffed rescue companies and US&R units for daily fire/rescue operations; and better adherence to personnel safety principles, including the use of redundant safety systems, rapid intervention, and other related protocols. In combination with the time-honored fire/rescue service traditions of rapid response, hard work, sacrifice, and the application of common sense, these factors often lead to dramatic, well-planned rescues that look miraculous to the public.

As a result of the second rescue revolution, modern fire department rescue companies (and US&R companies, US&R task forces, swiftwater rescue teams, US&R truck companies, etc.) are commonplace. Fire service leaders understand that rescue and US&R units are vehicles in which highly trained personnel and specialized rescue equipment can be concentrated in one place, thereby creating a vast base of knowledge and experience.

Dedicated and fully staffed rescue units allow fire departments to experience, review, quantify, and implement the lessons from repeated rescue operations, thereby reducing the chance of repeating the same mistakes, reduce the loss of valuable rescue experience, and ensuring a force prepared to manage the rescue emergencies and disasters that affect them. And while the discussion thus far has centered on rescue companies, US&R companies, and other specialized rescue resources, the basic principles of rescue apply equally to the typical truck company, engine company, brush patrol unit, fire/rescue helicopter, chief officers, disaster planners, rescue instructors, and others with responsibility for conducting or managing rescue.

Improved Rescue Standards

Rescue developed in a rather fragmented pattern in the North American fire service and in most other nations. One reason was the absence of recognized national and international standards and laws related to rescue. It took decades for the fire service to develop the level of experience, supported by research, development, and testing, needed to develop standards for rescue. It took even longer for those standards to be adopted. Today there are a number of recognized standards for fire service rescue training, staffing, equipment, and systems. It's yet another part of the current rescue revolution. As standards continue to evolve, rescue will become more widely recognized as a professional discipline. For the purposes of training, these disciplinary standards can be used as guides to set the direction of rescue programs.

The following rescue-related regulations and standards provide the basis for developing an effective rescue training program, and for determining the competency of rescue firefighters:

- NFPA 1670 Standard on Technical Rescue
- NFPA 1006 Standard on Professional Competencies for Technical Rescue
- NFPA 1983 Standard on Life Safety Rope, Harnesses, and Hardware
- NFPA 1470 Standard on Search and Rescue Training for Structural Collapse
- NFPA 1500 Standard of Fire Department Occupational Safety and Health Programs
- NFPA 1561 Standard on Fire Department Incident Management Systems
- OSHA 1910.134 Respiratory Protection
- OSHA 1926.650 Trench and Excavations
- OSHA 1910.146 Permit Required Confined Spaces for General Industry

More on NFPA 1670

No discussion on urban search and technical rescue training would be complete without mention of NFPA 1670 Standard on Operations and Training for Technical Rescue Incidents, adopted by the National Fire Protection Association in 1999. This national consensus standard should be considered the baseline for establishing US&R and technical rescue systems. Its origins lie with NFPA 1470 Standard on Structural Collapse Training and Operations and NFPA 1983 and NFPA 472 Standard on Hazardous Materials Operations, which established benchmarks for fire/rescue services engaged in those operations.

The effect of NFPA 1670—the new standard for technical rescue—on the fire service is far reaching. The origin of NFPA 1670 is also worth noting. While NFPA Standards 1470, 1983, and 1470, developed in the 1980s and 1990s, covered specific aspects of certain types of technical search and rescue operations, there was no all-encompassing standard for the full range of disciplines that come under the heading "rescue." Consequently, fire departments and other public safety agencies ran into problems on several fronts.

First, since there was no nationally-recognized regulation guiding the establishment of formal rescue systems, well-meaning administrators who wanted to enhance the safety of their personnel were hamstrung by the lack of a specific standard they could cite as evidence of the need to enact meaningful requirements for rescue and US&R units.

Second, the lack of a national standard meant there was little impetus to prompt less-willing fire/rescue agencies to adopt formal rules about training, staffing and equipment for rescue.

Third, some labor unions protested when fire departments attempted to establish formal rules for training and staffing, and restricting personnel from working rescue units without first having completed certain technical rescue training courses or maintaining locally developed rescue certifications. Without a recognized national or international standard upon which fire departments could base requirements for staffing rescue units, some unions had a field day with fire administrations whose hands were figuratively tied behind their backs. Without a national standard there were no "teeth" to many local training standards for rescue. Conversely, the lack of a national standard that could be adopted locally had the effect of leaving many firefighters in the dark about exactly which rescue courses to attend and which rescue-related certifications to obtain and maintain.

Fourth, without a national standard, it was up to each individual city, county, state, and training organization to establish its own requirements (that often varied widely, if they bothered to establish standards at all). Naturally this led to significant differences in rescuer competence from one state or county to the next, between different fire/rescue agencies, and even from one unit to the next within the same agency.

This need to address the safety and operational needs of firefighters and fire/rescue agencies responsible for conducting rescue formed the genesis of the birth of NFPA 1670. Here, then, is a review of the main points of this trailblazing national standard for rescue.

One of the interesting features of NFPA 1670 is the career path, an algorithm of sorts that allows progress from awareness to first responder to operations to technician. It's a proven approach, one employed by a number of fire/rescue agencies across North America.

NFPA 1670 includes seven main disciplines that represent the bulk of fire department rescue responsibilities nationwide. They are:

- Collapse rescue
- High angle rescue
- Trench rescue
- Industrial/machinery rescue
- Water rescue
- Confined space rescue
- Vehicle extrication rescue

There are subsets of these specific disciplines, available to fire departments that have a need to train personnel to perform them based on local and regional conditions. NFPA 1670 provides the framework upon which any agency can base this training and competency testing. Within each rescue discipline, NFPA 1670 identifies three distinct operational levels (Awareness, Operational, and Technician) based on knowledge and competency.

The *Awareness Level* is essentially for first responders who must—to operate effectively and with a reasonable level of personal safety—understand how to size up, assess, and manage the rescue scene until the victim is rescued or until the arrival of Operational or Technician level units arrive to augment them. They should understand the basic dynamics and hazards of the rescue emergency, they should be prepared to evaluate the need for additional assistance (and call for it without delay), they should be prepared to safely commit themselves to the incident (or, if the danger exceeds the level of protection, to establish an exclusion zone and deny entry),[2] and they should be prepared to control and manage the rescue scene.

The *Operations Level* includes personnel who have met the Awareness Level requirements as a baseline, and who have gone on to received additional instruction and experience that allows them to review the size-up of the Awareness Level rescuers, to determine probable or confirmed victim location and survivability, to make the rescue scene reasonably safe for emergency operations to proceed, to protect victims and package them as necessary, to use (and supervise the use of) rescue equipment, and to safely and effectively apply appropriate rescue tactics and strategies. Essentially, these are the secondary responders to rescue emergencies.

The *Technician Level* represents the highest level of training, experience, and competency. Technician-Level rescuers should meet or exceed the Awareness and Operations Level requirements as a baseline. They should be prepared to assess and reassess the rescue problem and devise an appropriate strategy, supported by proper tactics, to solve it while maintaining a reasonable level of safety for rescuers. They should be able to provide personal protection for the victim and rescuers alike during the course of operations. They should be prepared to advise the incident commander about specialized resources and to supervise or conduct advanced methods in complex search and rescue situations. In short, NFPA 1670 Technician Level personnel should be the rescue experts for your agency.

As noted by Chase Sargent in a recent *Fire Engineering* magazine article, "The standard also allows different geographical areas of the country to compare programs and service levels based on specific knowledge, skills, and abilities (KSAs) rather than program names or descriptions. Finally, people from the east coast, west coast, and midwest can talk about levels of service based on KSAs instead of program names and descriptions. Organizations will be able to speak the same language and create a more efficient program evaluation nationally. This benefits all organizations when discussing reciprocity, cross-certification, and service levels, allowing all to use common terminology. For example, FEMA US&R documents are eliminating the reference to such things as 'Rescue Systems I or equivalent,' opting instead to provide specific KSAs as outlined in the NFPA 1670 standard."[3]

Like the other NFPA Standards, 1670 is not binding, and some fire/rescue agencies may choose not to adopt it. But because it is a recognized national standard, agencies that don't adopt it (or exceed its requirements) could find it being used against them in liability lawsuits and other legal cases. Not only that, but in the event of a "rescue gone bad" (especially one that's filmed by news cameras, aired live, and rebroadcast over and over) the news media and public officials could cite NFPA 1670 to argue that the agency was not adequately prepared to manage a local rescue hazard.

For those who ask "should my agency adopt NFPA 1670 as a minimum standard?" the answer might well be another question: Does your agency respond to rescue emergencies? If the answer is yes, it's probably a good idea to adopt NFPA 1670 and, as appropriate, other rescue-related standards.

Selected excerpts from 1670

Section 2-1.6: *The Agency Having Jurisdiction (AHJ) shall provide for training in the responsibilities that are commensurate with the identified operational capability of each member.*

Section 2-1.6.1: *The AHJ shall provide for the necessary continuing education to maintain all requirements of the organization's identified level of capability.*

Sections 2-5.1.1 and 2-5.1.2: *All personnel shall receive training related to the hazards and risks associated with technical rescue operations...and shall receive training for conducting rescue operations in a safe and effective manner....*

Section 4-1.2: *The AHJ shall evaluate the effects of severe weather, extreme heights, and other difficult conditions to determine whether the present training program has prepared the organization to operate safely.*

Section A-2-5.4: *The AHJ should address the possibility of members of the organization having physical and/or psychological disorders (*e.g., *fear of heights, fear of enclosed spaces) that can impair their ability to perform rescue in a specific environment.*

Section A-3-2.2: *The AHJ should train members to recognize the personal hazards they encounter and to use the methods needed to mitigate these hazards in order to help ensure their safety.*

More about NFPA 1006

Fire/rescue agencies should also be aware of NFPA 1006, Standard for Professional Competence for Responders to Technical Rescue Incidents. NFPA 1006 is a national consensus standard intended to augment NFPA 1670 by providing the framework by which fire/rescue agencies can ensure the readiness and competence of their personnel to conduct rescue operations. NFPA 1006 (Section 2.1) says, in part:

> *Because technical rescue is inherently dangerous and rescue technicians are frequently required to perform rigorous activities in adverse conditions, regional*

and national safety standards shall be included in agency policies and procedures. Rescue technicians shall complete all activities in the safest possible manner and shall follow national, federal, state, provincial, and local safety standards as they apply to the rescue situation.

More about NFPA 1500

The effects of NFPA 1500, the Health and Safety Standard, are equally far-reaching. This standard is representative of the new emphasis in reducing injuries and death among fire/rescue agency personnel. Gone are the days when firefighter deaths were accepted as par for the course; something to be expected, for us simply to bear. Instead, firefighter safety regulations have given us sophisticated methods to combat the tragic death tolls that plague our profession.

Those responsible for developing and managing fire department rescue programs should be familiar with the firefighter safety and survival requirements established by NFPA 1500.

Rescue Equipment

Most of us have heard the phrase "The right tool for the right job." Perhaps nowhere is this saying more relevant than in a discussion about rescue. By its very nature, rescue involves the use of specialized tools and/or the innovative use of conventional tools to free people from unusual situations.

And the range of predicaments in which people will find themselves has no boundaries. There is another saying that goes "If you build it, people will find it." In rescue, we can modify that phrase to say "If there is some new way for people to get trapped, they will figure out a way to get themselves stuck in it."

Consequently, the tools that modern fire departments need to have on hand is equally diverse, dependent in part on the range of local hazards, training, preferences, traditions, experience, and budget. Obviously it's impossible for the average engine company to carry every tool that they might need in the course of a technical rescue operation. Truck/ladder companies are more suitable for carrying certain tools that are more specific to truck company operations like forcible entry and vehicle extrication.

Rescue companies and US&R companies in particular have to be considered (in part) as big tool boxes that contain every necessary piece of rescue equipment not carried on the other units. While it's not possible within the space of this chapter to discuss the merits of every piece of equipment on a typical US&R or rescue company, it may be helpful to review the equipment lists of US&R and rescue companies of various fire/rescue agencies, in order to compare notes and determine whether equipment specified in some regions may also be applicable to other regions.

With that in mind, Appendix 3 is a sample list of equipment carried by US&R/rescue companies of one fire department.[4] There is no endorsement here of any particular brand or manufacturer of tools and equipment. The best advice is for fire/rescue agencies to explore the full range of options available to them.

The Question of Jurisdiction

When it comes to rescue, so many fire departments cover a wide variety of terrains and hazards that it's not practical or even advisable to separately discuss fire departments whose jurisdiction is primarily "urban" from those that are "rural" or wilderness. The fact is, some "urban" fire departments also have jurisdiction over "non-urban" terrain, and their personnel are dispatched to technical rescues in wilderness areas or the ocean, in addition to the typical rescues that occur in urban areas. Likewise, some "rural" fire/rescue agencies are called upon to conduct rescues that are usually seen in the city.

Clearly, the lines separating urban and rural fire/rescue departments are increasingly blurred by the combined effects of the suburbanization of America, the advance of better search and rescue-related technologies, new rescue capabilities, and ever-expanding fire department rescue responsibilities.

In fire departments where dedicated fire department rescue companies or urban search and technical rescue programs have been in place for years, technical search and rescue has long been considered a priority and a primary mission responsibility, and therefore the local rescue capabilities were maintained at a high level of performance and it's clear who is responsible for managing rescue operations.

In other regions, fire service rescue as a formal discipline has evolved only during the past two decades, radically changing the face of local fire departments and regional mutual aid capabilities. New rescue programs have evolved in fire departments whose members for decades operated without the benefit of formal rescue resources. And many existing rescue systems have been reinvigorated by the newfound emphasis on technical search and rescue and disaster rescue.

Considering these changes, the issue of jurisdictional responsibility is worthy of discussion. In some places the advent of fire department-based urban search and technical rescue capabilities is causing conflict between law enforcement teams and volunteer citizen groups who traditionally managed rescue, and firefighters who now are equally (if not better) trained and equipped to do the job, and who often arrive on the scene faster than the various search and rescue groups.

Conflict sometimes occurs where volunteer or law enforcement-based teams assume they hold jurisdiction over rescue operations, and who are sometimes surprised to arrive on the scene to find firefighters already engaged in difficult search and rescue operations. In some mountain communities where volunteer or law enforcement teams have for years managed mountain search and rescue without substantial fire department assistance, it is increasingly common for them to find rescue-trained firefighters committed to mountain rescue operations, and this creates conflict over the question: "Who should be performing rescue, and under whose guidelines and command?" This conflict also occurs in some cities where law enforcement has historically taken on some responsibility for technical and/or vehicle extrication rescue operations, but where the fire department has since become equally or better equipped and trained, and often arrives on the scene first.

The answer to the question "Who is responsible?" can be complicated, but it usually starts with strategic positioning of fire stations for timely response to daily fire, EMS, and other emergencies. In most cases, because of the location of fire stations and mobile units, local fire departments are first on the scene of rescue incidents with companies staffed, trained, and equipped to conduct these operations—regardless of whether they occur in the city, in the desert, on the ocean, in the mountains, or elsewhere. And the public has come to expect the local fire/rescue department to manage these emergencies.

Likewise, if the local fire department finds itself overwhelmed by the scale or complexity of a particular rescue or disaster, the public expects the chief officers to request appropriate assistance without delay, in order to mitigate the hazard with minimum loss of lives. Because of the responsibility to render aid without delay, and because of the need for firefighters and paramedics to reach the victim to start stabilization, medical treatment, and packaging, it's necessary for firefighters to make access to the victim and begin preparing their extraction whether or not a volunteer search and rescue team has arrived on the scene.

In short, few things are more fundamental for modern firefighters than maintaining constant readiness to properly manage the search for, and the rescue of, victims (including other firefighters in rapid intervention situations) during the course of fireground operations and rescue-related emergencies and disasters alike.

Evaluating Local Hazards Related to Rescue

An accurate hazard evaluation allows fire department leaders to develop rescue capabilities commensurate with the local and regional risks. Properly employed, accurate risk information can help guide fire/rescue officials in the plans, systems, and resources capable of properly managing the consequences of mishaps in industry, at home, on roads and highways, at sea, in the mountains, and practically anywhere else where "daily" rescues occur. By recognizing the benefits of properly assessing the potential for "daily" rescue emergencies, and by supporting efforts to quantify them, emergency officials will be ahead of the proverbial power curve in terms of preparing their respective agencies to manage rescue emergencies.

All firefighters and officers should be trained to recognize local hazards that result in rescue emergencies. In order to conduct a realistic evaluation of rescue hazards, it's necessary to have an understanding of local conditions; to recognize probable rescue emergencies; and to be alert for signs of trouble that aren't obvious to the untrained eye. A physical inspection of local terrain, buildings, infrastructure, public assemblies, transportation facilities, and other factors is an effective means of determining key rescue hazards. The presence of one or more of the following potential rescue hazards within a fire department's jurisdiction indicates a possible rescue problem that should be addressed by plans and resources to effectively manage them:

Potential Rescue Hazards

Natural rivers or streams, whether flowing year-round or seasonal only
Buildings undergoing renovation or demolition
Mountains
Industrial complexes
Flood control channels
Ports, harbors, marinas
Geographically unstable areas subject to landslides and mud slides
Airports
An ocean
Wildland fires in steep terrain (indicating the potential for massive post-fire flooding and mud and debris flows, as well as increased runoff in urban areas)
Public works construction (dams, tunnels, storm channels, etc.)
Regions subject to flash floods, rockslides, or mud and debris flows
Amusement parks
Nuclear power plants and nuclear waste repositories
Aging buildings, especially those of unreinforced masonry construction
New construction sites, including those with excavations, trenches, and caissons
Potential terrorism targets such as embassies, churches, mosques, and government buildings
Refineries
Water treatment plants
Chemical plants
Mines and tunnels
Rock quarries
Coastal cliffs and bluffs
Dams
"Arizona" crossings (where roads cross streams and river beds)
Railroads and rail yards
Subway or metro-rail systems
Any other condition that may lead to people becoming physically trapped

It's helpful to review the location and type of rescue incidents based on official records such as fire/rescue dispatch and response data, NFIRS (National Fire Incident Reporting System), special studies, and other official records. Unofficial sources of information about local rescue patterns and trends may include police and fire department dispatchers, the local coroner's office, newspapers and other news media outlets, and firefighters and officers (active and retired) with a broad sense of the jurisdiction's history. It may be helpful to check with outside agencies like the police department, public works, hospitals, and other agencies or facilities that deal with public safety data that may indicate telltale pattern of rescue emergencies.

MULTI-TIERED RESCUE/US&R SYSTEMS

A modern truth of rescue is this: Few fire/rescue agencies have the wherewithal to train, equip, exercise, and ensure the high level of experience to make every single first responder capable of managing every type of rescue emergency that might confront them in their career. Here is another modern truth: The age of specialization in the fire service is here: no single firefighter has enough time in the day to become a true expert in every discipline for which the modern fire service is responsible. Fortunately, there is an answer to the ever-increasing demands on the fire service to expand its scope of responsibility: Multi-tiered rescue response. Just as in the delivery of EMS (Emergency Medical Service) and hazardous materials response, where multi-tiered response has proven to be the best approach, so it is with rescue.

First responders should be trained and equipped to begin rescue operations at a basic level, before the arrival of specialized units. Since most rescues are conducted by first responders, this approach is useful for most rescue problems. Among the most effective fire/rescue departments are those that have adopted a system whereby the first responders are trained to NFPA 1670 awareness or operational standards, or other pertinent training levels, and equipped to support awareness or operational level rescue tasks. This allows the first responders to manage the consequences of minor to moderately-difficult rescues, and to establish the baseline for the next tier of response to assist with more complex rescues. Rescue companies, US&R companies, US&R task forces, swiftwater rescue teams, helicopter rescue teams, or other specialized rescue units are trained and equipped to back up the first responders. In these systems, such companies are dispatched on the first alarm and/or special-called by incident commanders to provide timely backup for the first-arriving engines, trucks, paramedic units, and other first-responder units.

Furthermore, the most effective rescue response systems include provisions for requesting outside resources when the situation demands even more specialized rescue service or when disasters overwhelm the local rescue resources. Fire/rescue departments in the United States always have the option of requesting regional or state rescue teams, including the formidable FEMA US&R task forces (68-person teams of rescue experts prepared to respond anywhere in the nation to provide expert search and rescue capabilities at the scene of disasters). This concept of multi-tiered rescue response, and the adoption of well-defined rescue standards such as NFPA 1673, is being implemented across North America and in an increasing number of nations around the world. It is quantitatively improving the survival of victims and the safety of firefighters and other rescuers.

FDNY Rescue System

All five FDNY Rescue companies report to the special operations command, which also supervises other specialized units. Rescue companies are assigned to their own fire stations.

Although all five FDNY Rescue companies are staffed with six personnel trained and equipped for the full range of rescue and fireground situations likely to be encountered in New York City, each rescue specializes in selected disciplines like structure and trench collapse, high angle rescue, dive rescue, etc.

Because the FDNY rescue companies are assigned to the special operations command, there is direct interaction and uninterrupted communication between the command and the rescue company personnel on a daily basis. The special operations command directs the daily training, equipment, planning, and other activities of the rescue companies, in order to ensure they are in a constant state of readiness to provide the best service to the FDNY's engines, trucks, and other "operations" units. At major rescue incidents an officer of the special operations command can be counted on to be there to help the incident commander manage specialized operations. It is a system that has worked well for years.

The FDNY squad companies are specially trained and equipped engine companies designated as the second tier of their three-tiered rescue program. These units carry much of the equipment assigned to rescue companies, and they also carry basic hazmat equipment. All squad company personnel are trained as rescue technicians as well as hazmat operations-level. They ride engine companies that are designed to carry all this extra equipment. In essence, they augment the first responders until the arrival of the rescue companies. Because there are more squad companies than rescue companies, they usually arrive on the scene to begin setting up rescue operations before the rescue companies arrive.

FEMA US&R Task Forces

Based on a mandate from Congress in 1990, and using the example of urban search and rescue task force models already in use by the Miami-Dade (Florida) Fire and Rescue Department, the Fairfax County (Virginia) Fire and Rescue Department, the California Office of Emergency Services, and other agencies (including those of other nations that field international disaster search and rescue teams), FEMA embarked on an ambitious program to develop 25 operationally functional, 56-person, multi-discipline Urban Search and Rescue (US&R) task forces based in major metropolitan fire departments around the U.S. and consisting of specially trained firefighters and civilians.

Multi-disciplinary working groups consisting of experienced urban search and rescue firefighters and other rescue practitioners were selected from around the United States and convened to develop the standards for equipment, training, experience, administration, and methodology. Many of these firefighters and rescuers came from fire departments in places where structure collapse and other disasters are a common element of the environment. Others were members of teams that had been deployed to disasters in the U.S. and abroad. It was a unique synthesis of expertise drawn from across the nation to tackle a problem that knows no borders. By 1992, FEMA US&R task forces had gone from concept to reality, with the first deployment taking place in October of that year when Hurricane Iniki struck Hawaii. Today the FEMA US&R task force system represents one of the most formidable programs of its type in history.

FEMA US&R task forces are part of the U.S. Federal Response Plan, which is divided into Emergency Support Functions that designate the role of all federal agencies in the case of various types of national disasters. Emergency Support Function 9 (ESF-9) is designed for management of urban search and rescue-related disasters. Under ESF-9, FEMA is the lead federal agency for coordination of US&R operations, and seven other federal agencies are listed as support agencies. When federally-declared disasters strike, an ESF-9 Emergency Response Team (ERT) is dispatched to assess the need for federal assets to assist local and state authorities. The ERT provides an immediate source of expertise and counsel and is prepared to expedite the process of supporting the local and state agencies through the deployment of federal resources. ESF-4 covers fire-related disasters; ESF-10 is for hazmat catastrophes, and ESF-6 is for mass care and includes shelter, food, and emergency first aid.

As of this writing, the staffing of each US&R task force has been upgraded to 68 persons who make up four six-person rescue squads, two two-person technical search squads, four canine search teams, two medical squads, and specialists in heavy equipment, logistics, communications, structural engineering, technical information, hazardous materials response, and weapons of mass destruction (WMD). These rescuers are organized under the following "Teams" within each US&R task force: search, rescue, medical, and technical. There are two leaders for each team, and each position on the US&R task force is duplicated, in order to allow the entire task force to be split in two for round-the-clock operations. All members of every FEMA US&R task force are being equipped and trained to operate in environments contaminated by nuclear, biological, and chemical agents. Additional hazmat specialists are being added to the FEMA US&R task force rosters in order to address the hazards presented by weapons of mass destruction attacks.

After Hurricane Iniki came a series of disasters that resulted in the deployment of US&R task forces, including a clutch of hurricanes, the 1994 Northridge earthquake, the 1995 Oklahoma City bombing, and of course the 9-11 attacks. In terms of improving the effectiveness and coordination of urban search and rescue operations (as well as improving safety for personnel operating in extremely hostile environments like collapsed structures), the success and advantages of this national disaster search and rescue program quickly became evident. The recognition of the success of this approach created a call for additional task forces to ensure faster disaster rescue service to other parts of the nation. Based on this understanding that more US&R task forces were needed to address the disaster potential faced by the U.S., the FEMA US&R system has since grown to 28 task forces, with the addition of new training and equipment that allows them to conduct search, rescue, and recovery operations following terrorist attacks involving Weapons of Mass Destruction (WMD).

In terms of equipment and mobility, US&R task force capabilities are significant. Each task force has a cache of 50 tons of equipment that's maintained in a constant state of readiness for immediate deployment by ground, by fixed wing transport, or by helicopter. Each task force responds to the disaster site with everything necessary to operate at a disaster site for days or weeks, including food, water, shelter, and medical supplies for 10 days. They arrive at the scene of disaster with a wide array of specialized search tools, rescue devices, medical equipment, logistical capabilities, and even specially trained canines that the local agencies affected by the disaster may lack.

The equipment caches are packed in special waterproof containers, arranged on military-grade pallets, generally loaded on 40-foot trailers, and stored at the home base for each task force. Each task force is located near a military base, from which deployment can be expedited by flying in military cargo planes. FEMA requires the task forces be ready to be in the air within six hours of activation to a disaster. This helps ensure the timeliest response possible for the nation's most advanced disaster rescue teams. In cases where distances are lesser (and air transportation is not necessary), US&R task forces can generally be on the road within a few hours of notification.

Most important, the FEMA US&R task forces bring a level of experience that is rarely matched. By virtue of their response to disasters like the Northridge quake, the Oklahoma City bombing, hurricanes and tornadoes, the 9-11 attacks, and whatever disasters may befall their home jurisdictions, members of FEMA US&R task forces are among the most experienced rescuers in the world. Through years of training, planning, exercises, and response to actual disasters, many personnel assigned to these task forces have "been there and done that." Anyone who recognizes the value of experience in the decision-making process on the emergency scene must also acknowledge that the FEMA US&R task forces are tapping into sources of training and experience that were unobtainable just two decades ago. For the local incident commander faced with the first actual structure collapse disaster they have ever experienced, or overwhelmed by the need for advanced rescue capabilities that aren't locally available, the arrival of FEMA US&R task forces should be a welcome sign—and should be a mandatory part of any disaster plan in places where US&R-related disasters are a potential hazard.

Even though the nation's FEMA US&R task forces are staffed mostly by rescue-trained firefighters who specialize in structure collapse rescue and other disaster search and rescue operations (many of whom are highly experienced rescue practitioners assigned full-time to rescue companies and US&R units in their home jurisdictions), this point is not universally understood in the fire/rescue services. Instead of bunker gear or turnouts (that are overly restrictive and hot for long-term technical rescue and collapse SAR operations), firefighters and civilians assigned to FEMA US&R task force members wear more appropriate Nomex jumpsuits, reinforced wildland firefighting gear, or other specialized garb that may be brown, black, dark blue, or yellow in color (depending on regional norms), with appropriate reflective striping, pockets, and markings.

Instead of wearing structural firefighting helmets, the firefighters and civilians assigned to FEMA US&R task forces are equipped with helmets designed for use in collapse rescue and other technical rescue operations. This "non traditional fire department" attire may lead some local chief officers and firefighters to believe that the FEMA US&R task forces are non-firefighters or (heaven forbid) even bureaucrats who have arrived in town to take over their emergency operations. This erroneous notion simply must be dispelled if local incident commanders expect to properly manage disaster rescue operations that exceed the local, state, or regional capacities and require activation of the federal response plan.

Misunderstandings about the makeup, capabilities, and mission, and expertise of FEMA US&R task forces can (and have) lead some local officers and firefighters to mistakenly characterize the members of US&R task forces as "non firefighters" or FEMA office workers, which in the eyes of the local incident commander may mean that these teams are not going to be assigned to collapse zones or other high-risk operations. Such fundamental misunderstandings have, at times, led to delays in assigning these formidable disaster rescue resources where they can do the most good in support of the local firefighters and rescuers. It has also led to delays in getting the most help to the most victims in the least amount of time with limited resources, which should be the main priority of every incident commander at the scene of any disaster.

It's understandable that some fire department officers will feel more comfortable working with (and giving assignments to) firefighter-based teams who are more familiar with fire department operating protocols, than with civilian teams whose backgrounds may be unknown, and who may not be accustomed to operating in the semi-military command systems used by fire departments. But it's also the responsibility of the incident commander and the other decision-makers to understand the nature, capabilities, limitations, and availability of resources available to them in times of disasters. Given the ever-present potential for natural and manmade disasters, and the probability of increased domestic and international terrorism in the coming years, it is incumbent on officers and firefighters alike to acquaint themselves with the FEMA US&R task force system, in order to better utilize these specialized resources in the event of a rescue-related disaster.[5]

Today, with the National Urban Search and Rescue System entering into its second decade of operation, there is little excuse for any chief officer or firefighter in the United States who does not understand that the FEMA urban search and rescue task forces are made up mostly of highly experienced, well trained, and well-equipped firefighters, supported by highly trained and experienced specially trained structural engineers, emergency room physicians, paramedics, and canine search teams who are accustomed to working with firefighters at the scene of major disasters.

Likewise, there is little excuse for local incident commanders who don't understand that the FEMA US&R task forces and other federal search and rescue resources are going to report to the local IC and work for them. Fire chiefs should ensure that their officers and firefighters are aware of the nation's disaster rescue resources, how and when to request them, and how to employ their specialized skills to achieve the best result for victims who may still be trapped. In addition, it's important for the incident commander and their officers to consider the use of US&R task forces to augment their existing resources for round-the-clock search and rescue operations, to relieve weary firefighters and rescuers, and to provide an extra layer of backup during long-term disaster operations.

As of this writing, the FEMA US&R task force system has grown to include 28 US&R task forces. FEMA was handed a mandate in 1990 to "morph" itself into an emergency response agency by developing a national urban search and rescue system for disasters, and the nation's fire/rescue services stepped up to help FEMA make it happen with such speed and at such a cost value to the federal government that it has rarely been equaled. Rarely has a government program achieved its stated goals with such clarity of vision and purpose—and achieved them in such short measure. And rarely has there been such a successful partnership between the federal and state governments and local fire/rescue agencies and civilian experts.

FEMA US&R task forces provide round-the-clock rescue operations

One advantage of FEMA US&R task forces is their ability to operate nonstop from the beginning of the disaster until the end. They are designed to ensure 24-hour operations, a critical requirement when people are trapped alive in collapsed buildings or other deadly predicaments where survival is often based on being located and extracted without delay.

For structure collapse operations, US&R task force operations are predicated on the five standard phases of structure collapse search and rescue:

1. Size-up and Recon (Stage 1)
2. Surface Rescue (Stage 2), which includes the primary search
3. Void Space Search (Stage 3), which includes the secondary search
4. Selected Debris Removal (Stage 4), which includes carefully dissecting the building with heavy equipment and rescue personnel and looking for live victims and the deceased
5. General Debris Removal (Stage 5), including the clearance of all material using heavy equipment and other methods

The US&R task forces are prepared to conduct advanced search and rescue operation in all manner of other disasters, including landslides, tornadoes, earthquakes, hurricanes, floods, mud and debris flows, avalanches, transportation disasters like plane crashes and train derailments, dam failure, explosions and, of course, terrorist attacks. The most likely scenario for deployment of FEMA US&R task forces is a large scale disaster with actual or potential structure collapse with multiple victims trapped in predicaments that require advanced search and rescue capabilities.

US&R operations may last days or even weeks, depending on the scope of destruction, the survivability factors, and the availability of resources. Recognizing this stark fact, all FEMA US&R task forces are designed to provide 24-hour operations without cessation for days and weeks—as long as it takes to find and rescue all victims in a given disaster. In the case of a disaster so large that victims are trapped for such lengthy periods, US&R task force leaders and the incident support team may assign entire task forces to work straight through for a period of time (usually 16 to 24 hours), at which point a 12-hour rotation will begin, with only half the task force working at any given time while the other half eats and rests.

There is another mode of operation known as a "blitz," whereby an entire US&R task force works straight through, at maximum capacity, until the operation is completed. Generally the blitz is reserved for individual incidents with limited victims, limited geographical and time parameters, and higher potential for live victims.

FEMA US&R incident support teams

In order to coordinate and supervise the efforts of federal US&R task forces, and to provide liaisons between the incident commander and the US&R task force leaders, FEMA fields three national US&R Incident Support Teams-Advanced (IST-A), identified by the designators Red, White, or Blue. The IST-As are specially trained teams, each comprised of 20 highly experienced firefighters, officers, and US&R authorities selected from FEMA US&R task forces around the nation.

While each IST-A member is an active member of a US&R task force, they have additional skills, training, and experience that make them a valuable ally to the incident commander. Since the US&R task forces are there to serve the needs of the local IC, and to provide specialized urban search and rescue capabilities that may not be locally available (especially in times of disaster), they are an extension of the local emergency response. If the incident commander understands the capabilities and limitations of the US&R task forces, and if the IST-A is allowed to work to organize and coordinate the task forces, the incident commander is assured of getting the most "bang for their buck." This may be a difficult proposition for agencies accustomed to managing emergencies without outside assistance, but in very large or complex disasters, the best strategy is sometimes to request outside assistance in the form of highly experienced IST-As and US&R task forces, and to let them work for you. The FEMA US&R task forces and the IST-A are not going to take over the incident; they are there to work for the incident commander to help achieve the goals of locating and rescuing victims, stabilizing the scene, providing advanced rapid intervention capabilities, recovering the dead, and other special functions.

Each US&R IST-A represents the advance elements of a complete incident support teams, which is generally staffed with 39 people for very large or complex disasters. The role of the IST-A is to get to the disaster site as soon as possible to begin establishing conditions that will allow for timely and effective use of the US&R task forces. If the incident proves to be so large or complex that a full IST is required, the remaining positions are then filled to augment the existing IST-A, bringing it to full IST strength. At least one IST-A is dispatched whenever one or more FEMA US&R task forces are activated, in order to assist the local incident commander with command and control/coordination of federal US&R resources.

During disaster operations, IST leaders provide input to the local incident commander from the IST and task force operations leaders. They help the local incident commander with tasks like identifying operational goals during each operational period, evaluating structure collapse situations, establishing the conditions under which search and rescue operations can proceed in a timely and reasonably safe manner, and they provide other strategic and tactical assistance.

Once agreement is achieved with the local commanders, the IST leaders convey the strategic and tactical goals to the IST operations officers, who are responsible for developing an operational plan to make it happen. The operations plan is distributed by the IST plans section to all assigned US&R task forces and other resources involved with the search and rescue operations. The IST leaders and operations officers conduct briefings with US&R task force leaders, local commanders, and other affected parties before each operational period, and they provide continuous coordination and supervision of US&R operations and support.

Based on the IST operational plan, each US&R task force planning section develops its own written US&R tactical plan specific to that task force's operations during every operational period. The plan is the guide for all the task force elements to complete their assignments every operational period.

This is a "thumbnail sketch" of the standard FEMA US&R system approach to ensure that FEMA US&R task forces operate in a manner consistent with the needs of the local incident commander, and to ensure that all resources at the disaster are "on the same page." This is particularly important at large-scale and/or complex disasters, where different resources can potentially find themselves employing opposing tactics if the action plan is not transparent (or if it is not effectively communicated).

In order to assure that these US&R strategic objectives are completely understood, FEMA protocol calls for the IST leaders to draft a written Memorandum of Understanding for US&R operations with the local incident commander. The agreement specifies the purpose of the presence of the US&R task forces, establishes performance bench marks, and spells out demobilization protocols. This is another example of the extent to which the FEMA US&R system is designed to support the needs of the local entities.

The ISTs operate in the same work rotation mode as the US&R task forces. Here is a typical daily routine for ISTs operating in the 12-hour rotation mode: It might include shift changes at 0700 and 1900 hours. For the day shift, work begins at 0500 with a shift change briefing from the "night shift" and a tour of the collapse area. The status of strategic and tactical objectives are reviewed and observed, complications noted, and operations plan changes discussed. This is followed by a formal IST operations briefing at 0700 hours, attended by all IST members, US&R task force leaders, and representatives of local and state agencies and any other entities with operational responsibilities.

Prior to each night shift, the same briefing process and operations briefing are held from 1700 to 2000 hours. Consequently, each 12-hour operational period actually consists of a 16-hour shift for IST members and others with operational command/control/coordination responsibilities. It's a working schedule that's sustainable for days, weeks, or even months if necessary, under the often-austere conditions of a disaster site.

FEMA US&R in Los Angeles County

To effectively manage the consequences of the wide range of US&R-related emergencies and disasters that affect Los Angeles County, the LACoFD utilizes a three-tiered US&R system consisting of first responders (all field personnel), secondary responders (the US&R technicians assigned to US&R task forces and other units, augmented when appropriate by air operations helicopters, hazardous materials squads, and swiftwater rescue teams), and state and federal US&R task forces. The third tier of LACoFD US&R response includes CATF-2 (California Task Force #2), the department's FEMA US&R task force (that also has a 15-person swiftwater rescue task force component). CATF-2 is augmented by the nation's 27 other FEMA US&R task forces, eight of which are located in California. As is the case for every US&R task force, CATF-2 is designated as a local resource for disasters within the LACoFD's jurisdiction, and as a regional resource for other local disasters. All eight California-based US&R task forces are considered state resources for disasters within California until a federal disaster is declared, at which time they become federal resources under FEMA's direction.

It should be noted that the department's US&R companies are patterned in certain respects after the FDNY rescue companies and other successful rescue models, with several differences owing to various staffing considerations and differences in local rescue like earthquakes, floods, and mountains. The point is that the LACoFD didn't set out to "reinvent the wheel." Rather, great care was taken to investigate other existing rescue systems and to heed lessons learned by other fire/rescue agencies across the United States and around the world.

All personnel assigned to US&R task forces are certified as LACoFD US&R Technicians (which exceeds NFPA 1670 standards). Each US&R task force consists of one three-person engine company (staffed with three US&R technicians) combined with one US&R company (also staffed with three US&R technicians). They are capable of providing technical assistance, advanced search and rescue capabilities, and rapid intervention support to the incident commander and first responders.

Each LACoFD US&R task force carries the following designations: California Type I Swiftwater Rescue Team, California Type I US&R Company, L.A. County Fire Department Confined Space Rescue Team, Helo/Swiftwater Rescue Team, Helo/High Rise Team, and Helo/Deployed Open Water Rescue Team. California Mine and Tunnel Rescue Team certification is pending.

Each US&R task force is capable of operating as an independent company when necessary to manage EMS incidents, vehicle fires, and other "single engine" responses. They may also operate independently to facilitate training, planning, preparedness, and public education activities.

The US&R companies provide EMS, fire fighting, and other emergency capabilities within their assigned fire station's jurisdictional response areas. They also reinforce battalion fire and EMS response capabilities. Not only does this allow faster compliance with OSHA-mandated "two-in, two-out" responsibilities at the scene of structure fires, but it provides extra manpower to pull hose lines, perform ventilation, search for victims, and other essential fireground tasks. The equipment carried on the US&R companies (including thermal imaging and other specialized tools) enhances the fireground search and rescue and fire fighting capabilities.

The LACoFD has developed a standard response matrix to help ensure that appropriate resources are dispatched to US&R-related incidents. The incident commander has the option of upgrading or adding units to the selected response matrix.

Each US&R task force is capable of being transported by one of the LACoFD's Bell 412 or Skorsky Firehawk via helicopters to reduce response time, to make quick access to remote areas, or to assist air operations crews during difficult technical rescues. For certain incidents (*e.g.*, minor and major marine disasters, some high-rise fires, some mountain rescues) some or all US&R task forces are transported by helicopter to facilitate unique deployment options. This includes open water rescue swimmer team insertion, helo/high-rise team deployment onto the roofs of burning high-rise buildings, hoist insertion for some mountain rescues, helo deployment for some dive first responder operations, swiftwater/helo operations, and other specialized tactical operations.

LACoFD Air Operations units have the option of requesting US&R task force personnel to assist with difficult cliff rescues, vehicles over the side, water rescues, and other high-risk operations. In these instances, US&R task force personnel may be assigned to be inserted into the scene via hoist, rappel, "one skid," or landing the copter. They are prepared to assist air operations personnel with packaging patients, moving them to a pickup location, and assisting with pickoff rescues, and managing the hoist cable connection to the pre-rig of the rescue litter.

Whenever air operations units are dispatched to assist outside agencies with technical rescues in mountainous areas, high-rise fires, open water rescues, swiftwater rescues, and other US&R-related emergencies, they shall be supported by the dispatch of at least one US&R task force. The intent is to ensure the presence of LACoFD US&R technicians trained to assist air operations personnel with hoist rescues, short hauls, and other helicopter rescue operations, and to ensure the presence of a US&R-trained rapid intervention crew.

FEMA US&R task forces for WMD disasters

FEMA US&R task forces are trained and equipped to conduct search and rescue operations in environments contaminated by WMD events, including those involving nuclear, chemical, biological, explosive, or incendiary devices and hazards. Consequently, the US&R task forces are quickly becoming a primary support resource for local fire/rescue departments in the event of WMD terrorist attacks. Because they are based in major metropolitan fire departments in some of the highest target areas, the US&R task forces are a local resource for those areas, ensuring a rapid response.

The WMD training and experience gained by the personnel assigned to FEMA US&R task forces will gradually be transferred to their home agencies and surrounding agencies. And when a WMD disaster occurs, the local incident commanders are more likely to be familiar with the resources that can be applied to the consequence management, by virtue of their exposure to the FEMA US&R WMD task forces. As a result of these efforts, people trapped in collapsed buildings after a WMD attack will be given the maximum chance of survival through the deployment of teams that are ready to monitor the disaster site for contamination, to determine where the most viable patients are located, and to help facilitate locating and extracting them with a reasonable degree of safety for rescuers.

These and other developments represent a sea change in disaster rescue theories and practices nationwide. Their origins, in part, lie with the fire service, which has helped lead the movement for better rescue capabilities. As a result of all these efforts, timely expert response to technical search and rescue emergencies and disasters occurs today on a scale scarcely dreamed of just ten or twenty years ago.

ICS/SEMS Command Issues

Effective command of rescue operations does not require one to be an expert in urban search and rescue or technical rescue. The same qualities that define a good fireground commander are often found in good rescue incident commanders.

However, regardless of the strength and capabilities of the rescue resources, they cannot be used to their maximum effectiveness unless all responding agencies, firefighters, rescuers, and law enforcement are using some form of the incident command system (ICS) or standard emergency management system (SEMS). This is especially true with regard to disasters, which by their very nature require extraordinarily strong command and control to ensure the most good is being done for the most people with resources that may be overtaxed or overwhelmed.

This author has had the honor of teaching urban search and rescue and disaster response across the United States and in other nations (including Turkey, the United Kingdom, Australia, and Mexico); and in every instance it has become apparent that if the incident command system or SEMS is not being used as a baseline for command and control of emergency operations, rescue operations are hampered. Some responders unfamiliar with ICS have mistakenly developed the impression that ICS/SEMS is some sort of an entity—an actual object or organization—rather than a tool.

We must be absolutely clear about this: ICS/SEMS is not an agency, a school of thought, or an organization. Rather, ICS/SEMS is a generic emergency management tool designed to provide for a single incident commander (or, in some cases, a unified command), who builds or reduces an incident command organization based on the needs of the particular emergency incident they are commanding.

ICS/SEMS allows the incident commander to delegate responsibility to designated officers the major functions of incident safety, operations, planning, logistics, and (in very large or long-term incidents) administration/finance. ICS/SEMS is consistent with the time-proven command-and-control concepts of manageable span of control (the optimum and maximum number of persons that one supervisor can command with maximum effectiveness), delegation of authority, company unity, and unity of command.

Within the major function of the operations section, the incident commander of a major rescue emergency (through the operations section chief) can manage significant functions like search and rescue branch, as well as a suppression branch or hazmat branch (if necessary). Within the scope of authority of the branch directors would be functions like search group, rescue group, medical group, hazmat group, air operations, staging areas, and geographical divisions. In disasters of sufficient size/magnitude to warrant the deployment of FEMA US&R task forces, the incident commander could integrate the FEMA incident support team into the command structure to help manage the search and rescue operations and/or the operations of the US&R task forces.

For smaller incidents, the functions of search, rescue, medical, hazmat, EMS, air operations, staging, and divisions could be managed by the operations section chief. In many cases, "daily" rescue incidents are resolved with just three ICS positions: incident commander, rescue group leader, and medical group leader. If the incident also involves fireground or hazmat operations, the incident commander may assign an operations section chief to manage the rescue, suppression, hazmat, and medical groups.

It should be evident based on these brief examples that ICS/SEMS is highly flexible and expandable (or shrinkable), depending on the needs of the incident. It's popular to describe ICS/SEMS as a sort of tool box that the incident commander opens to grab the tools they need to handle the incident.

The simple fact is, large and/or complex emergency operations cannot be conducted with maximum efficiency and a reasonable level of safety for trapped victims and rescuers unless there is an effective, modular, expandable, and highly flexible command and control system in place before the emergency occurs. Therefore, it is incumbent on all fire/rescue agencies (and assisting law enforcement and government agencies) to adopt ICS/SEMS and use it every day to ensure that it become ingrained in the local, regional, state, and national emergency response cultures.

It should be emphasized that ICS/SEMS is not intended to be placed on a shelf and used only during disaster operations. That is probably the biggest mistake that an agency can make. ICS/SEMS are tools that can (and should) be used to enhance the safety and effectiveness of "every day" fireground, EMS, hazmat, and rescue operations. Daily use of ICS/SEMS for smaller incidents will help ensure that all incident commanders, officers, firefighters, rescuers, and responding agencies are prepared to use these powerful tools when disaster strikes.

Responsibilities of the incident commander

Rescue commanders are responsible for the following tasks: initiating first responder operations within the parameters of their agency's rescue/US&R training and guidelines; recognizing if and when the safe working limits of first responders have been reached; ensuring effective actions are being taken by qualified personnel to secure and stabilize the scene until the arrival of US&R task forces, helicopters, and other secondary responders; and then utilizing these resources to their greatest advantage when they arrive.

The rescue incident commander is responsible for requesting additional resources based on the need for specialized equipment and trained personnel to establish rigging, operate rope systems, staff litter teams and (when applicable) rapid intervention crews, and to provide manpower for hauling systems and carrying patients out of back-country locations; for requesting or approving helicopter transportation of rescue resources when appropriate (to expedite resolution of the incident and to improve personnel safety). Decisions should be made that comply with applicable worker safety regulations for operations in environments characterized as immediately dangerous to life and health (IDLH), where "two-in, two-out," rapid intervention, and personnel accountability protocols are required.

The incident commander should refrain from canceling rescue resources until one or more of the following benchmarks are achieved:

- Rescue of the victims has been completed, and rescuers have exited any IDLH environment or other special-hazard situation.
- Successful and safe rescue of the victim has been positively assured (*i.e.*, beyond a reasonable doubt) based on the eyewitness assessment of a qualified rescuer; and rescuers are no longer located in an IDLH environment.
- A size-up indicates that the incident is strictly a US&R first responder operation, and does not require secondary responder capabilities, nor is there a need for secondary responders to perform the function of safety officer, structural safety assessment, rapid intervention crew, or another assignment based on their specialized capabilities.
- The incident has been confirmed to be a false alarm (*i.e.*, there is no victim, or the victim has been located and determined to be in a location of safety with no need for technical operations to relocate the victim to receive EMS assistance, etc.).

Responsibilities of rescue/US&R company officers

Rescue/US&R company officers are responsible for ensuring constant readiness of their companies to support the first-arriving firefighters in order to effectively and safely manage the consequences of rescue emergencies anywhere within a fire department's jurisdiction. This includes planning, conducting, and participating in mission-specific training exercises; preplanning "high probability" and "extreme risk" rescue sites within a department's jurisdiction; and coordinating with allied agencies to ensure maximum effectiveness. This also includes maintaining all pertinent rescue skills.

In agencies that have access to helicopters, rescue/US&R company officers may be responsible for assessing the transportation needs for incidents in remote or inaccessible areas, and requesting helicopter transportation for rescue company members and equipment when airborne response is likely to materially impact the outcome or personnel safety of the incident. They may also have the authority to order the rescue company to respond via helicopter for technical rescues where the rescue company's assistance may be required to conduct hoist rescues, short haul rescues, or ground-based rescue in conjunction with the helicopter crew.

Key Points on Managing Rescue Emergencies

The majority of US&R-related incidents will be resolved by first responders. Recognizing this, it becomes evident that all first responders should be prepared to size-up and continually evaluate US&R-related emergencies in order to:

- Recognize critical cues and life hazards for victims as well as rescuers, including those that indicate the need for secondary responders to help manage them.
- Initiate effective first responder operations within the scope of their training and equipment.
- Recognize the boundary between the safe limits of first responder and secondary responder operations, and to understand when it's best to concentrate the efforts of first responders on stabilizing the situation as best they can until the arrival of secondary responders.
- Request additional resources based on the need for specialized equipment and trained personnel to conduct stabilization, search, rescue, and rapid intervention operations.
- Recognize when it's appropriate to expedite the response of US&R/rescue units and other resources by requesting or approving helicopter transportation.

Response issues

It's been proven effective in reducing rescuer injuries and fatalities and improving the survivability for trapped victims if fire departments respond rescue and US&R units on the first alarm, whenever there's an indication that their specialized training and equipment may be needed. Generally, it's prudent to dispatch rescue/US&R units on the first alarm to the following types of emergencies, in order to expand and reinforce the capabilities and operational safety of the first responder firefighters. Such situations include the following:

- "Fire Fighter Down" (Mayday) situations on the fireground, including instances where personnel have become trapped by the collapse of burning structures or high-piled stock, fire fighters "through the roof" or "through the floor," or other fireground entrapments. This also includes fire fighters lost or missing during fireground operations, as well as personnel injured in locations (*i.e.*, upper floors, basements, roof tops, etc.) where technical rescue methods are required to remove them to a place of safety.
- Rapid intervention operations to locate and extract personnel who become trapped, lost, or injured during the course of non-fireground emergencies in mountainous terrain, in rivers and flood control channels, by mud and debris flows or landslides, and other non-fireground rapid intervention situations.

- Emergencies involving the collapse of buildings, trenches, excavations, tunnels, bridges, and other structures, with causes that include (but are not limited to) earthquakes, explosions, design problems, material defects, slope failure, floods, collisions (*i.e.*, vehicles in structures, train derailment, and aircraft crashes), sabotage or terrorist acts, or fire damage.
- Traffic collisions with vehicles "over-the-side" on mountain roads, bridges, and all other "high-angle" situations requiring technical rescue methods for insertion of personnel and extraction of patients.
- Victims trapped or stranded on cliffs, hillsides, ice chutes, rocks, waterfalls, trails, and other mountain rescue situations (regardless of whether there are any injuries), or any other "person trapped" situations requiring the use of ropes, cables, helicopters, or other technical means to remove victims to safety and (when necessary) proper medical treatment.
- Aircraft crashes, aircraft fires, and "alert" standbys for aircraft in trouble during flight.
- Train and subway derailments, collisions, and fires.
- Marine emergencies where people may require rescue from the open water (*i.e.*, those who abandon ship to escape fires and capsizings), including ship collisions, capsized boats, sinking ships or boats with people aboard, fires aboard occupied ships or boats, watercraft "on the rocks," aircraft crashes in the ocean, etc (assuming the rescue/US&R units are trained and equipped for marine emergencies).
- Traffic collisions and other vehicular mishaps involving reports of people trapped in (or beneath) tractor-trailers, bulldozers, earthmovers, cranes, motor homes, or other large vehicles.
- Victims trapped on bridges, window-washing platforms, cranes, high-rise buildings, and other "urban" high-angle predicaments.
- Industrial accidents where people are reported trapped, entrapment in machinery, beneath heavy objects or materials, and other workplace environments.
- All confined space rescues, including engulfment, electrocution, entrapment, injury, and unconscious or missing victims in confined space environments.
- All reports of people trapped or missing in mines and tunnels, including subterranean storm channels, waterworks, train tunnels, and subway tunnels (including fires, floods, collapse, explosions, and other mishaps within tunnels and mines).
- Victims trapped or swept away in rivers, streams, flood control channels, flash floods, fast-rise flooding, and mud and debris flows.
- Potential suicide and hostage situations in high-risk environments, where the safety of law enforcement crisis negotiators requires fire department support in

the form of belay lines (to deal with subjects in high-angle environments), observation equipment (thermal imaging systems, search camera, trapped person locator, fiber-optic systems, etc.), cutting and breaching (in order to obtain access to subjects and victims), and other unusual needs. Typical situations include potential "jumpers" or hostage incidents on bridges, buildings, and other high-angle situations, or similar situations in confined spaces and tunnels, and certain industrial settings.

- Landslides, avalanches, rockslides, and other emergencies caused by geological failure.
- When rescue helicopter personnel require the assistance of additional personnel who are trained to conduct helicopter hoist operations, short-haul rescues, open water deployment, helicopter high-rise operations, or victim packaging for emergency missions in the mountains, the ocean, cranes and towers, and other unusual search and/or rescue missions (assuming the rescue/US&R units are trained in helicopter rescue operations).
- Ground-based searches requiring thermal imaging, night vision, trapped person locators, fiber-optic systems, search cameras, void space search, and other scenarios where victims are missing.
- Helicopter-based searches requiring hand-held thermal imaging systems, hand-held night vision systems, or other equipment carried on US&R/rescue units.
- High-rise fires. (US&R/rescue units in some departments are designated as helo/high-rise teams, capable of being deployed onto the rooftop via helicopter, in order to conduct rooftop ventilation, helicopter evacuation of victims from the roof, search and rescue on the upper floors using Biopack SCBAs, and fire attack.)
- Refinery fires (where victims may become trapped on cracking towers and other high-angle predicaments, and where other US&R capabilities may be used).
- Train fires (for forcible entry, thermal imaging, etc.).
- Victims trapped in submerged vehicles.
- Any emergency where the incident commander determines that US&R task force thermal imaging, shoring, lifting, hauling, searching, cutting, breaching, or other capabilities may be needed.

It's also the practice of some fire departments to dispatch a rescue company or US&R unit (including US&R task force fire stations) to all multi-alarm structural fires. Some departments assign US&R/rescue units to augment the existing rapid intervention crew, to conduct structural safety evaluations, to be prepared to shore and stabilize structures in danger of collapsing after fire operations, to perform specialized tasks such as thermal imaging, and other assignments that make use of the specialized capabilities of these units.

En route to the rescue scene

Incident commanders should strongly consider continuing the response of US&R/ rescue resources until the victim has been extracted, and until all personnel are out of potential danger zones (*i.e.*, trenches, building collapses, over the side, confined spaces, high angle, swiftwater, etc.). As long as rescuers and victims are in the hazard zone, the IC should continue the US&R resources to provide technical advice, to act as trained safety officers, to provide backup rescue capabilities, and to assist with the rescue as needed.

Based on the incident type, the incident text, and other information, the incident commander should consider the potential need for additional resources. Typical "special resource requests" for US&R incidents include the following:

- Extra engine or truck companies (or camp crews in wildland areas) to provide extra manpower for manpower-intensive emergencies like high-angle rescue, trench rescues, structural collapse, confined space, etc.
- Shoring or collapse units for trench rescues, structural collapse, mud and debris flows, heavy extrications involving tractor-trailers, and other incidents requiring extensive shoring and cribbing.
- Hazardous materials units for confined space rescue, tunnel rescue, and other incidents that require atmospheric monitoring and other hazmat capabilities.
- Heavy equipment resources like bulldozers, loaders, track hoes, cranes, etc., to advise or assist with heavy lifting operations, structure collapse operations, trench/excavation collapse rescues, and other such incidents where heavy equipment may be required.
- Heavy wreckers to assist with heavy lifting operations, entrapments involving tractor-trailers, stabilization of vehicles in precarious positions, railway accidents, etc.
- Rescue companies or US&R units from other agencies to assist with large or complex rescue/US&R incidents.
- US&R canine search teams for structural collapse, mud and debris flows, avalanches, trench/excavation accidents, water rescues, and other situations involving missing victims.
- Structural engineers for technical support during collapse operations.
- Industrial-size dirt vacuums, air knives, and other mechanical devices to remove soil and other material from trapped victims.
- Crisis negotiation teams to deal with subjects threatening to jump from bridges, buildings, towers, or other high-angle situation; to deal with hostage situations; and to address other emergencies that fall under the purview of crisis negotiators.

As necessary, consult (via radio, mobile digital terminal (MDT), or cell phone) with US&R resources for recommendations on other special resources. Consider the response time of US&R/rescue resources and other special units to your incident. In some cases, it will help to expedite their response by using helicopters. Consider the need for helicopter hoist operations and other aerial capabilities, and request them early in the incident to reduce reflex time.

At the scene

An eight-sided size-up assesses the sides, top, and bottom of the involved structure(s), as well as a "rotary" assessment of a 360 degree sweep of the ground level surrounding the building(s), and a 360 degree "rotary" assessment of the air space above the building(s). The rotary assessment includes looking for fall hazards from adjacent buildings (after explosions or collapses, etc.), the potential of secondary collapse of surrounding buildings, and any hazards on the ground like ruptured gas and water mains, and any other hazards around the site that can affect emergency operations. Conduct an eight-sided size-up to assess (and then report) the conditions in the chart below.

Conditions to Assess and Report After Size-Up

- The exact location of the incident and the best access for ground units and helicopters.
- The nature of the victim's predicament.
- Is the victim injured?
- If the victim falls, is dislodged from the present position, or gets hit by falling or rolling debris, can injury or entrapment worsen?
- Is the victim able to assist in the rescue?
- If trapped on a cliff, in a trench, or a similar predicament, can the victim move to a safer location/position until help arrives?
- Is this a situation which is best handled with a helicopter hoist rescue, a ground-based rescue, or some combination thereof?
- Are any victims missing?
- What is the exact extrication problem?
- If a vehicle is involved, does it require stabilization (*i.e.*, cables, ropes, etc.) to prevent movement?
- If a damaged structure or trench is involved, what stabilization measures are required to prevent secondary collapse?

- Begin stabilizing the incident and/or victim as much as possible until victim can be extracted.
- Ensure, whenever possible, continued medical treatment of trapped victims throughout size-up, stabilization, and extrication operations.
- Consider the need for a helicopter to stand by for immediate aerial transportation of the victim following extrication.
- Recognize critical cues and life hazards for victims as well as rescuers, including those that indicate the need for secondary responders (US&R task forces) to help manage them.
- Initiate effective first responder operations within the scope of their training and equipment.
- Assume command and name the incident.
- Recognize the boundary between the safe limits of first responder and secondary responder operations, and to understand when it's best to concentrate the efforts of first responders on stabilizing the situation as best they can until the arrival of secondary responders.
- Request additional resources based on the need for specialized equipment and trained personnel to conduct stabilization, search, rescue, and rapid intervention operations.
- Develop a strategy to perform the rescue if on-scene personnel have the proper training and equipment to conduct the operation safely and effectively.
- If the incident exceeds the capabilities of first responders, develop a strategy to protect the patient, protect rescuers, and stabilize the incident until the arrival of US&R/rescue units. If necessary, isolate the danger zone and deny entry until US&R units arrive. Consider the need to provide indirect assistance to the patient (i.e., lowering an oxygen mask and helmet to the patient, providing ventilation, preventing secondary collapse, etc.).
- Consider the need for unified command on large or complex incidents.
- Be prepared for long term operations because of unforeseen complications. Many technical rescues take twice as long to safely complete than initially anticipated.
- Consider rotating crews and providing rehab to reduce fatigue and maintain good working strength until the rescue is accomplished.

Implement rescue plan (incident action plan)

Develop a strategy to perform the rescue if on-scene (first responder) personnel have the proper training and equipment to conduct the operation safely and effectively.

If the complexity and technical demands of the incident exceed the capabilities of first responders, develop a strategy to protect the patient, protect rescuers, and stabilize the incident until the arrival of US&R task forces. If necessary, isolate the danger zone or the IDLH Zone and deny entry until US&R units arrive. Consider the need to provide indirect assistance to the patient (*i.e.*, lowering an oxygen mask and helmet to the patient, providing ventilation, preventing secondary collapse, etc.).

If the response of US&R/rescue units is extended, consult with rescue officers via radio or MDT regarding special needs. Request additional resources early to reduce the reflex time (a measure of the time between calling for a resource and the arrival of the resource), thereby avoid critical delays. Consider the use of unified command on large or complex multi-agency incidents.

Be prepared for long-term operations, as unforeseen complications may necessitate that technical rescues take two or three times as long to complete than initially anticipated.

For manpower-intensive operations, consider rotating rescuers every 30 minutes (or another appropriate time frame) and providing rehab to reduce fatigue and maintain good working strength until the rescue is accomplished. Apply LCES (Lookout, Communication, Escape Route, and Safety) principles to the operation.

Applying LCES to rescue operations

LCES is the development of strategy and the employment of tactics for ensuring firefighter safety in rescue operations. The concept was originally developed by wildland firefighters as a guideline to avoid common mistakes that result in fireground fatalities. For years LCES has been a mainstay for wildland firefighters whose every move is influenced by it.

Progressive fire departments have adapted LCES for use in non-wildland situations like interior structure fire fighting, terrorism incidents, hazardous materials response, and rescue. LCES as a rescue concept became prominent when members of FEMA US&R task forces included it in their operational action plans at the Oklahoma City bombing in 1995. In places like Los Angeles County, LCES is commonly used in all forms of high-risk rescue emergencies.

Before committing personnel to the danger zone of a rescue scene, the incident commander should always ensure that LCES has somehow been addressed and that all members are aware of them. The following is a brief review of LCES as it applies to rescue.

Lookout. Some member of the team (or another reliable responder like a firefighter, police officer, structural engineer, construction worker, public works member, etc.) should be assigned to observe the rescue scene for signs of impending secondary collapse, secondary explosion, fire, frayed ropes, avalanches, rock slides, flash floods, mud and debris flow, or other immediate life hazards that can kill rescuers. It may be necessary to place the lookout in the basket of an aerial platform, on an aerial ladder, on an adjacent building, on a mountainside, or even in a helicopter to ensure they can view the entire rescue scene.

It may be necessary to post more than one lookout. It may be necessary to use devices like plumb bobs and other tools that can indicate movement of a building preceding secondary collapse. It may be necessary to mount P-S Wave detectors to identify earthquake aftershocks before their destructive waves arrive and cause secondary collapse, thereby giving rescuers a few seconds to get into Safe Refuge Areas or Safety Zones.

Communication. Every rescue operation should have a clear communication plan that includes designated radio channels for certain functions and teams. But the communication plan must extend beyond the use of radios, that are subject to failure (and that can be lost or damaged during the course of rescue operations). Firefighters engaged in rescue must be familiar with other forms of communication such as whistles, air horns, hand signals, and voice commands. All personnel operating in and around the rescue site should be familiar with the communication plan, and each officer should ensure their firefighters are using the components of the plan appropriately.

Clear position designations are also critical to communications. The use of identification vests, properly marked helmets, armbands, or other identifiers should be mandatory. In a disaster setting when these methods may not be available to everyone (including personnel who report to duty from home), the use of marker pens to hand-print designations on shirts, helmets, or even on arms, is preferable to the chaos that occurs when everyone looks the same and no one can identify who is responsible for what. In disaster-prone areas, pre-designated caches of armbands, helmets, and vests can assist in the process of communication.

Communication also includes the use of clear and concise incident action plans (IAP) that coincide with what's actually happening in the rescue zone. In other words, it's helpful if the IAP matches (to a reasonable degree) the actual conditions on the scene. Too many times firefighters will look to the IAP for guidance and discover that it's outdated, that it's not accurate, and that it doesn't begin to convey the actual operations that are occurring. IAPs are a form of communication that can be a great help if they are accurate and well thought out. Conversely, poorly developed IAPs that don't match the reality on the ground can sometimes make the situation worse by misdirecting tactics and strategy.

Escape route. Every firefighter engaged in high-risk rescue (and every supervisor and the incident commander, if possible) should have a clear idea of the primary and alternate escape routes. Every officer should brief their crew or team on the chosen escape routes during each entry into an IDLH environment. The escape routes should be reevaluated as conditions change. If necessary they should be revised.

Escape routes should be the fastest, safest way out of the danger zone, or the fastest and most direct way to a safe refuge. In the event of a secondary collapse, a fire, a secondary explosion, flooding, or another unexpected event, preplanned escape routes may save the lives of firefighters and other rescuers. If necessary, the escape route(s) should be identified by fluorescent spray paint markings, signs, fireline tape, lumber crayons, and/or other clearly identifiable methods.

Escape routes can take different and sometimes unique forms. The Dade County (Florida) and Fairfax County (Virginia) rescuers are members of two United States US&R task forces dispatched under the auspices of the OFDA, the Agency for International Development, and the U.S. State Department, which coordinate the operations of American urban search and rescue assets during disasters on foreign soil. During search and rescue operations following an earthquake that shook the Philippines in 1992, US&R specialists from Dade County and Fairfax County found that rapid escape through the corridors of an overturned hotel was unfeasible during aftershocks. To expedite egress from the collapsed building, they stacked mattresses outside windows. The agreed-upon escape route was this: rescuers would simply scramble to the designated windows and dive out, one at a time, each rescuer rolling off the mattresses just in time for the next team member to land safely.

Stacking mattresses as the escape route might seem comical to some who have never operated inside a collapsed building with aftershocks continuing to strike, but it was clearly a simple and workable plan—one that was successfully used to evacuate rescuers from the collapse zone numerous times over a period of several days. When faced with unusual conditions, it's important for team leaders and officers to "think outside the box" when addressing the safety needs of their fellow rescuers.

Safe zone. Team leaders and officers should identify at least one safe zone, an area safe from secondary collapse and other hazards, into which rescuers can retreat in the event of an aftershock, an explosion, a secondary collapse, or another unplanned event. The safe zone may be outside a building (and beyond the collapse zone, usually the same distance as the height of the building), or it may be beneath a rock ledge, or inside a building, or on a hillside.

If the rescue operations are occurring inside a structure and escape to the outside will take too much time or is otherwise unfeasible, the safe zone may be designated within a stairwell or another fortified area within a building.

In some cases, it may be necessary to construct a safe zone inside a damaged structure, fortifying it through the use of shoring, cribbing, or other methods. Everyone entering the danger zone should be clearly aware of the safe zone(s). In the case of an unplanned event, the team leaders or officers should conduct "head counts" at the safe zones in order to ensure that all rescuers made it to safety (and to determine if some are in need of assistance).

Rescuer safety

The safety of firefighters and other rescuers is a key factor in the success of a rescue operation. Not only is firefighter/rescuer entrapment, injury, or death during rescue operations a tragedy in itself, but it also causes a cascading effect that reduces the chance of survival for the original victim. On the fireground, many fire departments rely on rescue companies and US&R units to perform rapid intervention operations, or to augment existing rapid intervention capabilities.

Fire/rescue agencies across the U.S. have proven that fire companies that specialize in rescue are one of the best assurances that firefighters will be located, accessed, and extracted in a timely manner if they become lost, trapped, or seriously injured on the fireground. A number of fire departments have made it standard policy to dispatch US&R companies and rescue companies to all multiple-alarm fires for the sole purpose of upgrading the first alarm rapid intervention resources by providing units with specialized search and extrication equipment operated by personnel specially trained and highly experienced in rescuer people (including rescuing other firefighters in fireground situations).

This is part of the new revolution in rescue: deploying specially trained and equipped fire department rescue and US&R units to rescue firefighters who become lost, trapped, or injured while battling fires.

Every firefighter and rescuer should be considered as an unofficial safety officer, constantly alert to danger, always playing the "what if" game, prepared to alert other members to take protective action in the event of an unanticipated event like secondary collapse, secondary explosion, earthquake aftershocks, fire, etc. Even when there's a designated safety officer at the site of a rescue, every other firefighter/rescuer should consider himself a safety officer in their own right. If we have this mindset, then we are more vigilant for situations that can lead to the entrapment, injury, or death of ourselves and our colleagues.

In rescue there are no "acceptable casualties," even if we understand that casualties will occur among firefighters and rescuers from time to time. If a rescuer is lost, seriously injured, or killed, it often means something went wrong, and it often (but not always) might have been preventable. We all know that an aftershock or secondary explosion might bring down a damaged building, killing rescuers and trapped victims alike, regardless of how much shoring has been placed, and regardless of how many other precautions might have been taken.

Even a small and unexpected secondary collapse can kill rescuers within a damaged building. Yet we are compelled by our jobs to consider committing personnel into damaged buildings to conduct search and rescue following a collapse. We are compelled to commit firefighters and other rescuers into all manner of situations where—if something goes wrong—the result may be tragedy. Such decisions always involve some level of calculated risk. The key is to reduce the risks to a reasonable level whenever possible through effective planning, training, equipment procurement, and rational emergency response.

One of the advantages of establishing a formal rescue program is that a fire department (or a group of fire departments, in the case of regional systems) have resources whose job it is to ensure that rescue operations are conducted effectively and with a reasonable degree of safety based on solid planning, training, equipment, research and development, and implementation of time-proven tactics and the employment of good strategies.

Rescue systems can be clearing houses for rescue information, repositories of rescue experience, technical support for incident commanders who may not be familiar with the details of rescue, but who understand the effective use of specialized resources to get unusual jobs done. In other words, there are a number of systematic approaches that will assist fire department officers to effectively manage rescue, IF the proper preparatory work has been done.

Operational retreat

All rescuers and team members must be familiar with the signal to begin an operational retreat (immediate exit of the danger zone or building in cases where secondary collapse or some other secondary event is imminent, or when another immediate life hazard has been discovered). An operational retreat system is necessary for the incident commander to make sure all personnel are safe and accounted for, and to determine if any rescuers are missing, trapped, or injured.

Rapid Intervention for Rescue and Firefighting Operations

Rapid intervention is included in the rescue chapter for one important reason: nowhere else are the rescue skills of firefighters more important than when their own colleagues become lost, trapped, or injured at the scene of a rescue or on the fireground. Rapid intervention is the most basic and essential form of rescue, because it involves saving the lives of those who have voluntarily placed themselves in harm's way to help others—and from another perspective—because the effective and timely rescue of downed or lost firefighters allows the other fire/rescue personnel to resume their original mission of saving the lives of citizens and protecting property and the environment.

When a rapid intervention (or Mayday) situation occurs, the context and priorities of emergency operations immediately change. The first priorities are to identify what just happened, what secondary events are likely to happen unless stabilization precautions are taken, and to locate and extract any trapped firefighters and rescuers. Firefighters may be forced to halt their original assigned tasks in order to assist the trapped/injured/lost colleague(s). In some cases this has the unfortunate effect of diverting help away from the original victim, at least temporarily. Aside from the potential tragic consequences to the trapped/lost firefighter or rescuer, the effect on the original victim is yet another reason to avoid the need for rapid intervention operations to begin with by taking prudent precautions.

Unless the incident commander has wisely ensured the immediate availability of a rapid intervention crew (RIC) with a solid RIC plan (and additional resources at their disposal in case of an "adverse event") the situation can quickly spin out of control. Furthermore, entrapment or death of rescuers may cause hesitation on the part of others who may (wisely in some cases) choose not to place themselves at risk of the same fate. A rapid intervention situation may have the overall effect of reducing the effectiveness of rescue operations, not just by reducing the effectiveness of rescuers, but also by causing additional mental trauma to their teammates. All these factors add up to a reduced chance of survival for the original victim.

The best solution is to prevent rapid intervention situations from happening in the first place, when possible, through the use of effective strategies and tactics that recognize accepted risk-vs-gain equations and use reasonable precautions to ensure the safety of firefighters and other rescuers.

If you, as a rescuer, become lost, trapped, or injured under these conditions, you should expect that the incident commander will immediately mount a rapid intervention operation to locate and rescue you from harm, before "normal" search and rescue operations resume, or simultaneous to other search and rescue efforts. If you aren't sure if the incident commander has made provisions for this, you may want to ensure these measures are in place before your crew is committed so far into the IDLH zone that escape may be impossible if something goes wrong.

Conversely, all firefighters and rescuers should be prepared to conduct rapid intervention operations to rescue other team members who become lost, trapped, or injured during the course of collapse search and rescue operations. Rescue companies and fire department-based US&R units should be prepared to conduct timely rapid intervention operations on the fireground, or to augment existing RIC resources. All incident commanders should be prepared to manage RIC operations. It should be standard protocol for the incident commander to designate a RIC, to develop a rapid intervention plan, and to ensure sufficient resources are in place before committing crews to an IDLH situation.

Whether at the scene of a technical rescue or on the fireground, the designated RIC should be in a full state of readiness outside the IDLH zone, dressed in "full battle gear" to immediately launch a rapid intervention effort if something goes wrong. In a disaster setting we all understand that the incident commander or the officer in charge (OIC) may not have the luxury of designating RIC teams whose only role is RIC. All the same, the incident commander or OIC should still have a plan to rescue personnel who become lost, injured, or trapped, even in the middle of a disaster.

Generally, a RIC plan should include a radio channel for RIC operations, a RIC team officer, protocol for conducting search and rescue in the particular environment at hand, and provisions for augmenting the RIC with additional manpower and equipment as necessary. There should be a medical group ready to take over treatment of the RIC subject once they are removed from the IDLH zone. The medical group should include paramedics, and ambulance, and perhaps even a medevac helicopter on standby for high risk operations.

There should be a standard radio designation for RIC operations, as well as other notification and communication protocol. When a rescuer is found in need of assistance, a standard radio call like "Rescuer Down," "Firefighter Down," or "Mayday" (depending on the individual agency's protocols) should be issued. The IC or OIC should call for radio silence to ascertain the downed rescuer's location and predicament, and a RIC operation should be launched immediately, using the designated RIC channel.

In short, rapid intervention is not just for confined space rescue and interior firefighting. There should be a RIC capability for every type of rescue where firefighters and other rescuers are in danger of becoming lost, trapped, or badly injured (requiring physical extrication) if something goes wrong.

RIC concepts for rescue and fire emergencies

RIC operations are not limited only to fireground scenarios. Modern progressive fire departments understand the need for RIC capabilities to be applied to every emergency involving high life hazards, where a mistake or mishap may result in firefighters or other rescuers becoming lost, trapped, or injured.

Emergencies Typically Requiring RIC Capabilities

Fireground operations
Confined space rescue
High-angle rescue
Dive rescue
Collapse search and rescue
Landslides and mudslides
Marine disaster/marine rescue
Deep shaft rescue
Mud and debris flows
Flood rescue
Swiftwater rescue
Avalanche rescue
Other high-risk search and rescue operations
Helicopter-based technical rescue

RIC operations on arrival

- Determine apparatus placement for best access to RIC equipment (avoid collapse zone).
- While RIC leader reports to incident commander, RIC members establish equipment pool.

- RIC leader reports to incident command for briefing and review of:
  - Current situation/status (what is happening, where are victims, what is their predicament, where are units and personnel)
  - Operational mode (offensive, defensive, rescue, etc.)
  - Personnel accountability
  - RIC frequency and contingencies
- RIC leader briefs RIC and together they evaluate the incident site.
- Take a "hot lap" around the scene.
- Evaluate structural integrity.
- Access points; make additional access points as necessary. Make them large, because they are also exit points.
- Throw additional ladders (or request additional ladders to be placed).
- Evaluate incident-specific hazards.
- Observe effectiveness of current tactics and strategy to anticipate possible problems and RIC situations.
- If offensive fireground operations are occurring, evaluate interior and exterior.
- If the fireground situation is defensive, evaluate exterior.
- If a technical rescue, evaluate the actual rescue site and surrounding areas.
- Observe for unusual signs (*e.g.*, terrorism, secondary devices, secondary collapse potential, additional flood surges, aftershocks, etc.).
- Discuss contingencies.
- Remember: no side jobs or freelancing. If you're assigned as RIC, you and your crew should perform no other task that prevents them from mounting an immediate rapid intervention operation. The IC should request sufficient resources to address other fireground and/or rescue needs.
- Review RIC equipment pool to ensure it is appropriate for the hazards at hand. (See Appendix 4)

Rescue Training

Intelligent and realistic rescue training is one of the most critical factors in the ability of a fire department to locate and extract trapped or missing victims without serious injury or death to firefighters. There are innumerable options for training firefighters to conduct rescue operations in a professional manner with a reasonable level of safety. The number of formal training options is expanding every year, allowing fire departments to tailor their rescue training programs to the needs of the local agencies.

There is an old saying that "those who can, do; and those who can't...become trainers." That may have been true at one time, but it's time to lay that saying to rest and bury it. While it's true that there will always be a need for highly experienced and accomplished retirees, those injured in the line of duty, and those who otherwise no longer operate in the line of duty, to teach, it is more important than ever to ensure you have a cadre of instructors who are active duty members that actually conduct rescue operations.

It's important to ensure that the information being taught is up to date, that it's actually being used in the field, and that it actually works under emergency conditions. It's more important than ever to review case studies in order to highlight successes (what worked) as well as failures (what didn't work!). It's necessary to examine how things are done in different agencies across the nation (and, yes around the world) to select the most practical and successful training points about tactics, strategy, methods, etc. Sometimes the best way to do that is to use trainers who are actively engaged in managing the emergencies, as well as those with the wherewithal to conduct research and development.

Clearly, it's also important to select instructors who know how to teach. We all can remember instructors who might have been brilliant tacticians or those who clearly excelled in doing the job, but who simply were terrible instructors because they failed to understand the methodology of teaching to highly motivated adults, or they had not themselves been instructed in the basics of teaching.

This is one reason why some states require instructors to complete teaching courses like Fire Instructor 1-A and Fire Instructor 1-B prior to being certified to instruct these topics. It's one reason many fire departments require their fireground and US&R instructors to complete similar fire instructor courses.

Nationally noted emergency service risk management expert Gordon Graham illustrates the importance of good training by pointing out that there are essentially two types of emergencies: "Low Risk, High Frequency," and "High Risk, Low Frequency."

Whereas, Graham explains, the typical firefighter may be accustomed to fighting fires in single family residential buildings or handling EMS calls (considered "high frequency, low risk incidents" in many municipalities), the same firefighter may have no experience being lowered into a deep shaft confined space to rescue an unconscious victim who's physically trapped (a "low frequency, high risk" incident). This is where effective training is one of the only hedges we have against lack of experience. The good thing is that effective training also helps firefighters to operate safely and effectively in "high frequency, low risk" emergencies by correcting bad habits and introducing new information about familiar (high frequency) hazards.

Naturally, effective rescue training has some inherent risks that mirror those found in emergency operations. In fact, it can be stated that the best rescue training includes realistic simulations that mirror some of the real-life (potentially injurious or even lethal) conditions found on the rescue scene. The difference is this: whereas under actual rescue conditions there is always a level of uncertainty about changing conditions and hazards, in training we take whatever precautions are necessary (including measures like the use of belays, backup air supply, rapid intervention crews, etc.) to insure that personnel are not seriously injured or killed during training sessions.

How can training hazards be reduced to a reasonable level? One method increased use of didactic training in order to:

- Communicate policies and procedures
- Conduct preplanning and "blackboard simulations"
- Critique past incidents

But classroom sessions are no substitute for effective manipulative training under realistic rescue conditions. For first responders and rescue technicians alike, realistic manipulative training is a key to safe and effective operations. For that reason it's important for supervisors to embrace rescue training and encourage firefighters to practice as if their lives depend on it—because survival may depend on instant reactions to changing conditions, and the person you rescue may be your own partner or teammate.

NFPA 1670, OSHA regulations, and other recognized standards form the basis for deciding what training is required to accomplish the desired level of readiness to manage local and regional hazards. Each fire/rescue agency should correlate its training standards with the local and regional hazards it faces. State fire marshals, the National Fire Academy, and FEMA have taken lead roles in developing, adopting, and requiring ever-more stringent, quantifiable, and effective rescue training standards. These are positive trends that support the efforts of local fire/rescue agencies to adopt the required effective and appropriate training for the rescue hazards likely to confront their personnel.

Operating at Rescue-Related Disasters

General disaster rescue operational planning

Recent disasters have reinforced many lessons about emergency preparedness and tactics and strategies that were learned in other disasters. Unfortunately, devastating disasters have taught new lessons about managing incidents involving structural collapse, mass casualties, long-term search/rescue operations, and terrorism.

Effective disaster planning is based upon a realistic assessment of the types of disasters which are possible in a particular locale. It used to be that extensive disaster planning was conducted only in areas prone to earthquakes, hurricanes, floods, tornadoes, and other natural disasters. Clearly, every community needs to reevaluate its exposure to potential disasters, whether natural or man-made.

Fire departments should plan for worst-case scenarios. Disaster planning should include worst-case scenarios (*i.e.*, high-impact time of day, multiple simultaneous events, adverse weather, etc.) to avoid being caught flat-footed. Modern fire departments with comprehensive rescue/US&R programs and with plans for quickly requesting and utilizing mutual aid resources (including FEMA US&R task forces) are assured of the highest levels of effectiveness and safety when disaster strikes. Develop and maintain written disaster plans. These should be "living" documents (subject to periodic updates) which include the following items:

- A disaster communications plan including the use of radios, telephones, cellular phones, mobile data terminals, local television and radio stations (for messages to the public or to recall personnel to duty), and "runners" to physically carry written messages between commanders. (As much as possible, communications plans should remain simple to avoid unnecessary confusion.)
- Policy requiring fire units to conduct immediate district "windshield" damage surveys to determine the status of critical structures and facilities (*i.e.*, hospitals, dams, etc.) after earthquakes, hurricanes, tornadoes, explosions, and other events that may cause widespread damage.
- A policy regarding the decision of individual fire units to take emergency action before completing their district surveys. This is necessary to prevent units from committing themselves to the first fire or other serious situation they come across (and possibly missing worse situations down the street).
- Specific command modes for different levels of disaster (*i.e.*, battalion command, fire station command, area command, unified command, etc.).
- A plan for making rapid assessments of structural integrity of buildings.
- Standard marking to reflect damage to structures based on the international US&R building marking system.
- Planning for immediate access to lumber, tools, generators, lighting, and other equipment from local vendors during disaster rescue operations. This should be planned by every fire station and rescue team.
- Planning to feed and shelter firefighters and rescue personnel for extended periods.
- A plan to require off-duty firefighters and emergency personnel to report to work after assuring safety of their families when disasters occur.
- A plan to continuously update personnel with critical information during disaster operations (*i.e.*, where is the worst damage, how are other areas faring, etc.). This

may help relieve anxiety of firefighters and others on duty by informing them of conditions at home.

- A plan to relieve on-duty members by oncoming firefighters so they may assure the welfare of their families (and then return to disaster duty).
- A plan to request and use mutual aid resources and special assistance from your city, county, state, and the federal government (including FEMA US&R task forces).

Assume the infrastructure and utilities will be disrupted in the area of explosions, earthquakes, and other destructive events. Utility damage may create extra danger and require additional safety precautions. Check damaged buildings continually for structural stability (particularly when shifting loads, adverse weather, and aftershocks may further impair the building's ability to remain standing). Recent earthquakes and the World Trade Center collapse are examples of this. Local fire agencies affected by disasters should update pre-attack plans for fire fighting and rescue operations in damaged buildings. Some buildings may need major shoring operations to make them safe for entry, search, and rescue. Remember that fire alarm/fire extinguishing systems may be inoperable in buildings damaged by explosions and other destructive events. Structures with damaged fire systems may require 24-hour fire watch or evacuation, and some commercial processes and occupancies may need to be closed down until repairs can be made.

The use of aerial reconnaissance can expedite damage assessments and improve command and control capabilities on large-scale incidents. Local fire commanders from company officers to the chief should be trained to recognize situations that require special search and rescue resources (*i.e.*, FEMA US&R task forces). This is an ongoing process nationwide that needs to be expedited, particularly in areas with high disaster potential. Be prepared to request special resources (*i.e.*, FEMA US&R task forces, Hazmat Teams, etc.) for major rescue operations as soon as the need becomes evident.

Safety evaluations and consequent marking of buildings by city, county, state, and private structural engineers and building inspectors must be consistent with one another, using a standard format easily recognized by firefighters and other emergency responders from across the nation. This will help fire/rescue personnel to quickly determine whether a particular structure is safe to enter for rescue and fire fighting after a disaster. Unless fires are causing life-threatening conflagrations or threatening trapped victims, rescue operations should be established as the highest initial priority in order to save the most possible lives. Unless wind, wood shake roofs, and other exposure problems exist, many structural fires will burn themselves out eventually. Obviously this will be a judgment call based on the particular conditions found. The point is this: there is a natural tendency for fire units to amass at any visible fires. However, it is counterproductive to commit fire fighting forces to fires with low spread potential if there are live victims trapped and waiting for help in collapsed buildings elsewhere.

When water mains are ruptured, local commanders should consider requesting water tenders, engine companies, pool pumps, and other resources to supply firefighting water. Consider establishing fire attack task forces (*i.e.*, two engines paired with one water tender) for firefighting operations. It is especially important to maintain firefighting capabilities at major incidents where many rescuers might be subject to life threatening conditions if a fire broke out somewhere in the building.

During disasters, firefighters may suffer sensory overload which will limit their ability to operate effectively under rapidly changing conditions. This is sometimes exacerbated by disaster plans that include confusing communications plans and other factors that create extra confusion. To assure maximum effectiveness, local disaster plans should be as consistent with normal operations as possible, and periodic disaster drills should reinforce these plans. Fire departments that conduct periodic disaster exercises tend to manage disasters most effectively because their personnel have had the opportunity to operate within the disaster system during non-emergency conditions and the procedures are more ingrained in their response.

Structural collapses and other disasters generally require tremendous amounts of manpower and equipment for long periods of time. Be prepared for long-term operations when disaster strikes. This is especially true in cases of major structural collapse. Victims have survived in collapsed buildings for as long as 16 days. Do not call off search and rescue operations until all options have been exhausted. Rely on special knowledge of US&R-trained personnel and members of FEMA US&R task forces. It may take days or weeks to search every void space for trapped victims.

Patients trapped for extended periods of time sometimes require treatment during extrication. Good working knowledge of confined space EMS, crush syndrome, compartment syndrome, and other disaster medical problems should be mandatory for local firefighters, paramedics, physicians, and others. Trapped patients often require intravenous fluid replacement and other appropriate treatment if rescue will be lengthy (and assuming medical capabilities are available to render treatment). If not, you may spend valuable time rescuing a patient who later suffers a preventable death.

Crisis management and consequence management of terrorist attacks

The best solution to terrorism is preventing attacks (a step known as crisis management in law enforcement parlance). But when domestic or foreign terrorists succeed, the most important solution is a rapid, measured, appropriate, and effective emergency response (the phase known as consequence management) to ensure that trapped, injured, and missing people are given the best care available in the shortest amount of time. In the event of an actual terrorist attack, a rapid and effective response to the consequences is the best way to defeat the intentions of the terrorists. This is one reason why effective rescue operations are critical in the moments, hours, days, and weeks following a major terrorist attack that involves trapped and/or missing victims.

With modern terrorism taking its inevitable course toward ever-more spectacular (and increasingly lethal) attacks perpetrated by both foreign and domestic groups and individuals, fire/rescue professionals must consider the scope of their mission with a far wider perspective than ever before. A nagging concern for modern firefighters is the ever-increasing frequency with which they are confronted by explosive devices and other dangerous weapons during incidents which appear (on the surface) to be "routine." Well informed firefighters understand that concerns about terrorism are no longer limited to foreign places. Truly well-informed people understand that domestic terrorism is not new; that terrorism has always been a potential threat even in the United States.

As a consequence of the current trends, fire fighters can anticipate more bombs, booby-traps, and other dangerous weapons for which there may be no outward warning signs. Recent examples include fire fighters finding booby-trapped apartments during fireground operations, Los Angeles City firefighters who encountered a large fuel and ammonium nitrate bomb in the back of a burning truck they had just extinguished in the parking lot of an IRS building, Kansas City firefighters who were killed when an intentionally set fire caused the detonation of explosives at a construction site, and Bakersfield, California firefighters who narrowly escaped death when a bomb exploded inside a burning passenger van. In short, there is no end to the combination of otherwise "benign" circumstances in which firefighters may be confronted by explosive devices.

The modern fire service is now confronted with a new paradigm: The emergence of both domestic and international terrorism with the means for incredible lethality, and a "terrorist doctrine" that makes it acceptable and even inviting to these groups to attempt to inflict huge civilian and rescuer casualties. Many of these groups (both foreign and domestic) have adopted the "leaderless cell" and "sleeper cell" modes of operation, which makes it all the more difficult for law enforcement agencies to locate and disarm them.

Consequently, the result will be continued attacks with more lethal results for citizens and rescuers, for the foreseeable future. It didn't take the 9-11 disasters to alert observant fire/rescue professionals to this new fact of life. The signs have been there for all to see since events like the 1993 World Trade Center bombing, the Oklahoma City bombing, and the 1998 bombings of U.S. embassies in East Africa. The East Africa bombings, which required a massive international response that included both of the United States' current international US&R task forces from the Fairfax and Miami-Dade Fire and Rescue Departments, to locate and remove more than 200 victims from the collapsed buildings. This raises an interesting point. What will the United States' US&R response be to future terrorist attacks that require more than two US&R task forces?

As the first responders to terrorist attacks, firefighters must look far beyond the obvious in the process of planning for (and responding to) acts of terror committed by foreign and domestic groups and individuals. Today, being extra vigilant to the winds of change, and adopting a doctrine of "thinking outside the box" is a mere starting point for progressive fire/rescue departments.

From the events of September 11, 2001, and beyond, it becomes clear that the fire service must start to think very far "outside the box." Firefighters responding to airline crashes and other unusual emergencies must approach these events with a higher index of suspicion for terrorism (including the potential for secondary devices or attacks) than in the past; and that we must give serious consideration to scenarios that might have been dismissed as preposterous just a decade ago. In 1995, in a published case study of urban search and rescue operations at the Oklahoma City bombing, this author wrote:

> *The Oklahoma City bombing, the 1993 World Trade Center bombing and other recent terrorist incidents dramatically demonstrated that urban terrorism has arrived on the shores of the United States. Practically any city or town may be the target of terrorists with a wide variety of agendas. These developments should be of grave concern to the Fire Service, whose members are generally first on scene and, therefore, extremely vulnerable to secondary attack (a common terrorist tactic more commonly seen in other parts of the world), secondary collapse due to structural instability, and other life hazards. The potential for terrorism raises the specter of future incidents in which massive damage is inflicted upon large, multi-story buildings.*
>
> *This is not the first time United States governmental facilities have been targeted by terrorists. The Oklahoma disaster was preceded in the 1980's by equally horrendous bombings of the Marine compound and the U.S. Embassy in Beirut. Unfortunately, there were no FEMA US&R task forces available to help during those incidents. Today, United States embassies and other government buildings in other nations are under constant threat of terrorist bombings. The possibility of further domestic terrorism must be taken seriously by the Fire Service.*

Rescue and terrorism from domestic and international sources

Following the 9-11 attacks, many people seemed inclined to focus exclusively on the threat of terrorism emanating from other lands (particularly the Middle East). This is a deadly mistake. While foreign terrorists are a huge concern, it is a colossal error (and a sign of serious naïveté) for firefighters to assume the threat of domestic terrorism has somehow been eliminated or is no longer a concern. Today there are many domestic terrorist groups and individuals operating within the borders of the U.S. Many of these groups are fully intent on bringing down the government of the United States, some to establish a so-called "all-white" nation in the Northwest, others to bring in some other form of government more to their liking. And the members of some groups have already demonstrated their willingness to commit all manner of atrocities in order to achieve their goals.

Based on the current progression of terrorism, it's obvious that the future will bring additional paradigms in terrorism in North America, including suicide bombers, the use of dirty bombs and other nuclear devices, the expanded use of secondary and tertiary explosive devices to wipe out fire/rescue professionals, and attacks aimed at provoking the government to respond in such a way as to cause a backlash leading to martial law and a possible civil war between citizens and the government.

Consider, for example, the recent case of the Montana militia whose leaders were arrested in February 2002 and charged with planning to kill as many judges, prosecutors, police officers, and firefighters as possible in order to provoke the state to activate the National Guard. The group, billing itself as Project Seven, was found to have amassed 30,000 rounds of ammunition and hundreds of weapons, as well as intelligence on the personal lives of police officers, firefighters, prosecutors, and judges (and their families), who were on hit lists found in the leaders' homes. According to investigators, Project Seven members planned to attack and kill National Guardsmen and other law enforcement officials who would be dispatched to Montana in response to the murders. According to documents seized during raids, the group ultimately intended to ignite a domestic war that would topple the federal and state governments.

Groups like Project Seven can no longer be considered isolated cases or anomalies unlikely to replicate. The fact is, domestic terrorism groups have sprouted across many parts of the United States, and they are fully prepared to take many innocent lives. The Oklahoma City bombing proved that beyond any doubt. Some of the domestic terrorist groups may lie low for a period of time after events like the 9-11 attacks. Inevitably some of these groups will surface to strike out against the government when things quiet down. Or they may establish yet another new paradigm by conducting terrorism when the government appears most vulnerable, like in the immediate aftermath of major attacks by foreign groups.

Weapons of mass destruction and rescue operations

Many anti-terrorism programs have been narrowly focused on potential biological and chemical attacks, to the exclusion of the ever-present danger of (increasingly) likely scenarios like conventional explosives and nuclear attacks (including "dirty bombs" that use conventional explosives to contaminate victims, firefighters, EMS personnel, police officers, civilian rescuers, and property with radioactive materials).

It's interesting to note that many fire departments increased their readiness to manage the consequences of structural collapse just as they began turning in their civil defense radiological monitoring devices after the fall of the Berlin wall and collapse of the Soviet Union (and the supposed end of the threat of nuclear attack). The potential for terrorists to obtain and use nuclear materials (perhaps in combination with conventional explosives or dirty bombs) to attack U.S. targets apparently was not considered valid by emergency planners. Consequently, many of the nation's first responders are left without reliable radiological monitoring capabilities. This is a huge tactical and strategic error that almost certainly will come back to haunt us.

An emerging terrorist threat is the potential use of dirty bombs, which can cause massive structural collapse and radiological contamination at the same site. Without radiological monitoring equipment, the first responders have no way to determine if there is a nuclear component to a terrorist bombing, nor are they capable of designating proper exclusion zones, nor are they capable of determining how long they can safely operate in a contaminated area. Without dosimeters, first responders have no way of documenting their total exposure doses for a given emergency operation.

Urban canyons and WMD

The history of the United States is replete with attacks by various domestic groups and individuals who employed explosives, fire bombs, sabotage, firearms, mail bombs, and even chemical and biological agents. Despite the variety of weapons used, the use of explosives has always constituted the major threat. Terrorists often see explosions or other spectacular events like airplanes flying into buildings as dramatic statements of their power to disrupt our societies when they are aggrieved. Moreover, explosive attacks create moments in time and instants of terror that few witnesses will forget.

In terms of potential for structural collapse and other consequences that would require the use of urban search and technical rescue resources, bombings (including nuclear devices and dirty bombs) appear to be the most potent and prevalent threat. According to the State Department and the Justice Department, bombs have been—and remain—by far the most common weapons used in terrorist attacks in the United States. Explosive devices are favored by both domestic and international terrorists and groups. Bombs are used with ever-greater frequency in the violent settling of personal conflicts in the home and at work. In the United States, the most frequently used explosive devices are pipe bombs and other small bombs. In California alone, more than 400 pipe bomb incidents are reported each year. Since 1990, the incidence of pipe bombings has increased by more than 50%.

Explosive devices are of special concern because they can sometimes be built by people who lack a great deal of technical expertise or resources. The materials required to produce bombs are readily available to the public. And even crudely-made explosive devices may be used with deadly accuracy. Powerful bombs may be hidden in mail boxes, small packages, luggage, and other items for which the index of suspicion would normally be low. Larger ones may be hidden in cars, trucks, airplanes, trains, ships, and cargo containers. Powerful bomb blasts can bring down large buildings, bridges, dams, mountain sides, and other large structures whose collapse may trap many people and require round-the-clock operations of sophisticated urban search and rescue resources for many days.

Today the most destructive threat is probably a nuclear device, one of the few true WMDs. One would be naïve—perhaps even negligent, given the course of recent events—to discount the potential for one or more nuclear devices to be detonated in the West in the foreseeable future.

Geoff Williams, Fire Master of the Central Scotland Fire Brigade, has warned for years about the need for fire/rescue agencies in various nations to be prepared for catastrophic events like earthquakes and huge detonations by terrorist groups that create "urban canyons," events that leave entire sections of a city slashed to bedrock, with surrounding buildings as the canyon walls.

For years Williams was a voice in the wilderness on this particular issue because it was difficult for many fire/rescue officials to imagine an event with such force to create such an urban canyon. But now we have all seen an urban canyon in the form of the 9-11 attacks on New York, and it's a certainty that—even as you are reading this book—some terrorist or terrorist organization is plotting, discussing, or dreaming about an event that will make the World Trade Center attacks look like a foreshock preceding something even bigger, perhaps in one or more different cities in the U.S. or some ally nation. And certainly there is the ever-present potential for a catastrophic earthquake to cause similar—if not worse—destruction in some cities.

The question is, how does the fire service plan and prepare for such an event, and how will fire departments manage the consequences when it happens? Fortunately (if the word fortunate can be used here) there is an answer to this. We employ the same basic strategies and resources that we have used to manage the consequences of other terrorist attacks and earthquake disasters. Only we do it quicker, on a much larger scale, with many more resources, with much better coordination, and for much longer periods of time.

The downside is that we may suffer many more casualties among the first responders, and perhaps also among the secondary responders. But the point is this: we handle the consequences and do not allow the evil of men or the capriciousness of nature to overcome us. This means that fire chiefs, company officers, firefighters, and rescue, haz mat, and EMS specialists must be prepared to quickly expand the scope of their operation in the aftermath of such an event, even if we are employing the same basic strategies and tactics to handle the individual and composite problems that will confront us.

Rescuers will still do rescue. They will just do much more of it, for a much longer period of time. Urban search and rescue teams will still conduct US&R operations using the five stages of structure collapse search and rescue, they will just do it on a much larger scale. FEMA US&R task forces (trained and equipped to operate in WMD environments) will still be deployed, but they will probably be reinforced by US&R teams from other nations.

We will still employee the concepts of ICS/SEMS, LCES (Lookout, Communications, Escape Routes, Safe Zones), structural triage, medical triage, risk-vs-gain decision-making, personnel accountability, operational retreat, and rapid intervention—we will just do it on a scale that was not imagined just a decade ago. If the urban canyon and surrounding areas are contaminated by radiation or other poisons, there may be a need for wholesale evacuation. We will determine the perimeter, contain the area, do what we can to reduce casualties, and handle the problem. If command-level leaders are lost, there will be a succession plan based on ICS principles. In short, the problems associated with an urban canyon event are ultimately manageable.

Before the 9-11 attacks, the only terrorist attack on American soil comparable to the World Trade Center and Pentagon attacks was the Oklahoma City bombing in April 1995. Today we have been thrust into an entirely new paradigm of terrorism. The 9-11 attacks shocked the world and the fire/rescue services to recognize the growing dimensions of modern terrorism. Particularly in the United States, which for so long had felt somewhat immune to terrorism, there is a deeper understanding that the fire/rescue services must change the way they do business. And the U.S. fire/rescue services must start looking with greater interest at the lessons already learned by firefighters and rescuers in places like Israel, Ireland, Scotland, England, and South Africa.

Earthquake Rescue Operations

With the exception of catastrophic events like the collapse of the World Trade Center twin towers in the 9-11 terrorist attacks, the problems encountered in most single collapsed structures pale in comparison with the potential effects of large disasters like earthquakes, explosions, tornadoes, landslides, tsunamis, and other events capable of causing dozens or even thousands of structures to fail simultaneously.

Earthquakes are among the most common cause of multiple collapse situations. Since many of us live and work in regions where the threat of catastrophe from earthquakes is ever-present, earthquake search and rescue is an issue that demands the attention of fire/rescue professionals. In places like Istanbul (where 50,000 to 100,000 structures are feared vulnerable to collapse during a future earthquake), Los Angeles (where experts estimate up to 20,000 people may die in a future quake), and Tokyo (where more than 100,000 people died in a single earthquake in the 1900s), the dangers and problems presented by the sudden collapse of numerous structures are very real indeed. The number of large cities that share significant seismic risks is too long to mention here. Of course, the threat of earthquakes is most palpable in places that have already experienced great earthquake losses in recent memory.

In the aftermath of damaging earthquakes, firefighters are responsible for locating, treating, and rescuing trapped victims, extinguishing fires before they reach trapped people and/or cause conflagrations, and responding to the tsunamis, collapsing dams, and other quake-induced events.

In the United States, earthquakes have drastically affected the building (and destruction) of some cities. During the 1800s the flow of the Mississippi River was, for a time, reversed by a series of earthquakes whose strength has seldom been witnessed. In those days the New Madrid Fault zone was sparsely populated, so the human effect was not particularly evident. But today this region is densely clustered with apartments, high-rises, industrial complexes and chemical factories whose designers apparently forgot or ignored the lessons of the New Madrid earthquakes. A similar event today would cause a catastrophe unprecedented in U.S. history.

In the Pacific Northwest, Native American legends describe the earth shaking and floods that rival those of the Bible. Scientists recently discovered corroboration in the form of ancient inland forests drowned by salt water. It is stark evidence that earthquakes caused vast coastal areas to drop below sea level, allowing the ocean to sweep inland like massive flash floods from which escape would have been unlikely. An earthquake of that magnitude would cause damage and life loss that few people wish to consider. (See Appendix 5)

Case study: Northridge earthquake US&R operations

The 1994 the Northridge quake struck Los Angeles on a blind thrust fault whose presence wasn't even known to seismologists. The eastern end of the Santa Susana Mountains grew more than a foot in a matter of seconds and rose several more inches during ensuing aftershocks. Large chunks of the San Fernando Valley were raised or lowered by the thrusting fault, permanently changing the elevation of many neighborhoods by as much as nine inches. It was merely part of the process by which the mountains, valleys, and flood plains of Southern California are formed, destroyed, and reformed, shaping the settlement of Los Angeles and neighboring cities.

The largest life loss occurred at the Northridge Meadows Apartments, a large, U-shaped, three story apartment complex with 160 units of Type V construction, wood framing, with exterior stucco and interior sheet rock. The first floor was concrete slab on grade, and consisted of apartment units mixed with "tuck under" carports; the carports were supported by 3-in. diameter steel pipe columns. The "open" areas of the carports create a condition commonly known as a "soft" first floor.

The Northridge Meadows apartments collapsed when the building "racked" to the north as the east and west lateral supports disintegrated, and the walls running perpendicular to the east and west walls hinged at the base and top. The top two floors moved eight feet to the north, then crushed the first level. The second floor joists came to rest on beds, dressers, washing machines, parked cars, and other solid objects. In some places the second floor was flush with the ground; in others there were 1- to 2-ft. voids. Many victims were killed outright; others were trapped in various predicaments.

First responder firefighters arrive. First arriving Los Angeles City Fire Department (LAFD) units were confronted with darkness, aftershocks, gas leaks, electrical hazards, hundreds of frantic citizens, and dozens of trapped victims inside the complex. So complete was the collapse of the first floor that fire units actually drove past The Meadows while completing their district damage surveys; in the darkness, they did not notice that the building had been three stories before the quake.

Because the available resources were overtaxed by fires, explosions, multi-casualty situations, and trapped victims across the northern half of Los Angeles County, most firefighters conducted initial search and rescue operations with little assistance from US&R units. Using hand tools like saws, pry bars, and air bags, they pulled many victims from under collapses.

For several hours, LAFD firefighters conducted extensive extrication operations to free a dozen others at The Meadows. Their efforts focused on opening void spaces to find victims, and extricating live victims as they were discovered. Bodies were being found throughout the first floor of the complex as well. Extreme fire and rescue activity elsewhere left personnel at the Northridge Meadows short-handed under severely taxing conditions. Sharp aftershocks created an ever-present potential for trapping firefighters as they worked through the day.

FEMA US&R task force operations at the Northridge Meadows apartment collapse. During the first hours after the quake, the L.A. County Fire Department (LACoFD) was heavily involved with collapse rescue operations, fire fighting, hazmat incidents, medical incidents, and damage surveys. Aerial surveys with US&R-trained personnel in five paramedic/rescue helicopters allowed the damage surveys to be completed rapidly. Finally all the major collapse sites had been cleared except for The Meadows, Northridge Fashion Square parking structure, and a couple other sites.

The FEMA US&R task force from the LACoFD was activated to assist the L.A. City Fire Department at The Meadows collapse. Identified as CATF-2, the team met with LAFD commanders in front of The Meadows apartments in a 12-vehicle caravan (including two 40-ft. flat-bed tractor-trailers carrying the equipment cache; two heavy duty stake side trucks carrying extra generators, lighting, and shoring; two of the department's field division US&R trailer caches with shoring and tools; an RTD bus loaded with rescuers, and various command vehicles).

The CATF-2 base of operations occupied three lanes for a block in front of the apartment complex. The entire area had been cordoned off to traffic—except the large news media camp occupying the fourth lane and several vacant parking lots across the street. LAFD commanders were confident that most of the live victims had been rescued, with at least 13 confirmed fatalities discovered thus far. However, they knew the sheer size of the apartment complex left many void spaces to be searched. That assessment was confirmed by a CATF-2 Structures Specialist, who estimated that potentially hundreds of void spaces required a physical search to assure no live victims were left behind.

A US&R task force operational plan was quickly implemented to shore up the building as necessary, to systematically cut into it, and take any other actions necessary to search every void space which might contain live victims, as well as to recover the deceased. Two canine search teams were sent in showing interest in several locations that later became the site of intense search and recovery operations.

Since Collapse Rescue Stages 1 and 2 were completed, and Stage 3 operations had been completed in some parts of the building, it was decided to continue with Stage 3 (void space search) operations. Pairing CATF-2 search team personnel with the four six-person CATF-2 rescue squads worked well because searching void spaces required extensive use of "rescue" cutting and extrication equipment and techniques. Search cameras and fiber optic devices from the technical search teams would be used in tandem with the more "traditional" methods and tools. Meanwhile the CATF-2 canine search squads would roam and be available when needed to check specific sites within the complex.

A detailed map was obtained by the search team manager, and the entire apartment complex was re-drawn by hand on a larger scale to help the task force leaders and the rescue and search team managers to conduct better personnel accountability and real-time tracking of the task force's four rescue squads. The various CATF-2 rescue squads began a systematic search of every apartment, hallway, elevator shaft, stairwell, parking garage, and laundry area in the entire complex. In consideration of the time of the earthquake (0430 hours), bedrooms and bathrooms would be high priority for initial search efforts.

At approximately 1600 hours, LACoFD US&R-1 (the department's central US&R company at the time) arrived at The Meadows after assisting LAFD firefighters with the daring nine-hour rescue of a street sweeper truck driver who had been trapped on the bottom floor when the Northridge Fashion Center parking structure collapsed. Under the command of Captain Wayne Ibers, US&R-1 was assigned to assist with the search process while remaining available to respond to other collapse operations and technical rescues that might occur that night in Los Angeles County.

The search team members and the four rescue squads began at the front of the complex and methodically worked their way toward the rear. The basic strategy for getting into the void spaces between the first and second floors was this: firefighters assigned as technical search specialists would make a hole in the floors of the second level to make a preliminary search with the search camera.

If no victim was sighted with the search camera, a rescue squad would begin tunneling through the ceilings of the collapsed first floor areas. After cutting away carpeting, crews pounded the concrete with sledge hammers to break up the concrete. After the concrete was shattered and cleared, task force members used chain saws, circular saws, axes, and other tools to "strip cut" access holes in the floor, through which other firefighters would enter in teams of two to conduct physical searches. For the firefighters of CATF-2, this was a familiar process: many of the same tactics (with an obvious twist) are used to ventilate burning buildings.

It became immediately evident that there would be a problem correlating "like" spaces between the different floors: The floor plan was identical for all three floors of the apartment complex, and one could normally assume that cutting downward from a second-floor bedroom would land you in an identical bedroom on the first floor. However, because the building shifted laterally (greater than 10 feet in places) before the two top floors collapsed onto the first level during the earthquake, there was significant displacement between the floors. Cutting downward from the kitchen of the second floor might land you in a bathroom. And so on.

Another problem was the need for additional search cameras, Trapped Person Locators, and other electronic search equipment. At the time, only one of each of these items was included in the US&R equipment caches (additional devices have since been assigned to every FEMA US&R task force) to allow multiple simultaneous technical search operations.

"Blitz" search and rescue operation at The Meadows. By the evening of January 17, the Northridge Meadows Apartments were practically the only remaining location in the city with potential for trapped victims. Therefore, based on parameters established by FEMA, an all-night blitz was the agree-upon mode of operation for CATF-2. All task force rescue squads and search squads were committed simultaneously for a nonstop operation.

Void space search operations. Searching hundreds of void spaces was a time consuming and labor-intensive process exacerbated by the lateral shift of the building. After cutting each access hole, US&R task force members crawled into the collapsed first floor, sometimes 40 or 50 feet away from the escape hole, searching every potential void space in which a live victim might be found. The firefighters worked in pairs at all times, with an attendant positioned at the opening to track their progress. Fiber optics and search cameras were used when they were available.

Even with the availability of technical search equipment, firefighters were ultimately forced to crawl beneath the building—experiencing significant aftershocks from vulnerable positions—to physically check every remote void space to ensure that no victims were left behind. The aftershocks, which rattled the entire neighborhood, sending telephone poles swinging and jolting The Meadows apartments, impressed upon everyone the need to carefully evaluate the structural integrity and to place solid shoring wherever necessary. Of course it was not feasible to completely shore every single vulnerable point in this large, block-sized complex. So the rescue squad members—and their supervisors and the incident safety officer—continually made risk-vs-gain evaluations as they methodically moved through the building.

Every member entering void spaces was in constant visual and voice communication with a safety member positioned at the hole. Shoring was used where needed. Sometimes rope lines were used with tenders; however, the rope frequently tangled and created more problems when attempting to exit.

Void space entry was hazardous due to the continuing aftershocks, which caused substantial shifting of the building, and the decision to place rescuers "in the hole" was made cautiously. In some situations a physical search of the void was the only method to determine whether live victims might still be trapped. Occasionally, the search process was hastened by reports from relatives that certain residents were still missing. Search of these units was given a high priority.

US&R task force command. Search and rescue operations continued through the night into the next day. After several hours, the LACoFD US&R task force developed a regular routine and a pace that everyone felt comfortable with, and operations proceeded smoothly and efficiently. One aspect which was clearly helpful was the leadership, which set clear goals and adjusted the strategy as necessary throughout the operation. Captain Mike Minor, the Search Team Manager, had a vast background in technical rescue in excess of 25 years, including rescue operations in the Mexico City and Sylmar quakes. Throughout the night, Minor organized the search in a logical sequence and prevented squads from stepping all over each other, missing locations, or performing redundant tasks. He set up a personnel tracking system and, with others in the base of operations, maintained a real time record of the positions of all personnel. This was a key element to the operation.

After several hours it became apparent that, due to the size of The Meadows complex and the hundreds of void spaces, it would take all night and possibly another full day to complete the secondary search. After discussing this with other task force members, the task force leaders asked the LAFD incident commander to request an additional FEMA US&R task force. The IC forwarded the request to the Northridge Quake incident command post. Within minutes, the California Office of Emergency Services mobilized the Riverside US&R task force (CATF-6), which had been staged at a nearby location.

CATF-6 arrived at The Meadows at 2230 and began setting up its base of operations adjacent to the rear of the complex. The decision was made to split The Meadows into two divisions: The front half (East Division) to be worked by CATF-2; and the rear half (West Division) to be handled by CATF-6.

Aftershocks. Aftershocks were a constant hazard. It seemed that the areas where the first floors had already collapsed were fairly stable; the main shock seemed to have taken them down as far as they were likely to go. But there was sufficient damage and instability to cause further collapse in the event of a major aftershock. Each jolt rocked the entire complex sufficiently to warrant extreme caution; even minor building shifts might crush fire fighters crawling inside the collapsed first floor.

In some sections of The Meadows, all three floors were still intact (but badly buckled and tilted, looking as if they were on the verge of collapse). These areas were considered to be especially dangerous. The apartment complex directly north of The Meadows (called The Northridge Apartments) was also severely buckled and on the verge of collapse. It, too, had a "soft" first floor with carports, and had suffered extensive collapse. The

Northridge complex posed a danger to all rescuers operating in the 400-ft. alley between the two complexes. Later, this complex would become the site of an intensive search.

The CATF-2 structure specialists were invaluable to search operations at The Meadows. They spent the evening advising task force leaders and rescue squad leaders as they entered, searched and cut into the building. Their advice was sought as squads prepared to cut load-bearing structural members to access void spaces and victims they found. In an after-action report on the operation, one of them described the process (somewhat tongue-in-cheek) as follows:

> *The assessment team entered the front building from the north west corner where...buildings were still somewhat intact. In this scenario, 'somewhat intact' is a very subjective term. Essentially it means that we could climb up and around them and into the building corridors without major obstruction from twisted metal and debris. Under ordinary circumstances such wreckages would be avoided. One determines a sense of the stability of the collapsed structure more or less intuitively. Generally one weighs such factors as the weight of structure overhead...and the since the initial (quake) shock. In addition one is led on by the firefighters who have a 'better feel' for such things; generally they look at it for about ten seconds, see nothing falling and walk on in. Unless you can point out an obvious hazard, they will not stop.*

Their point was this: the firefighters are so focused on rescuing live victims that they sometimes enter extremely hazardous locations without a second thought. It is important for leaders and all members to evaluate the integrity of the building before entering. Shoring is sometimes required to allow safe entry and rescue operations. The constant advice of structural engineers throughout The Northridge Meadows operations was extremely helpful, and prevented several unsafe acts. This is one of the many proven advantages of the FEMA US&R task force concept.

At approximately 2245 hours, one of the rescue squads from CATF-2 working in the East Division discovered a body under one of the apartments. The squad leader radioed to the rescue team leader that removal of the body would require extensive shoring, demolition, and lifting. Meanwhile, the search for live victims proceeded elsewhere in the complex.

Multiple fires erupt as electricity is restored. Just after 2300 hours, firefighters noted that the street lights suddenly came on. Electric service to the grid had been restored. Everyone at the base of operations stopped what they were doing, because lights were also visible in The Meadows apartment complex. This wasn't right; the utility company had assured firefighters that they had cut electricity to the complex.

Knowing that any fire could quickly become an immediate life threat to the dozens of firefighters engaged in void space search, the CATF-2 and CATF-6 leaders radioed an alert to be vigilant for live wires and signs of fire. The CATF-2 assistant task force leader made a quick reconnaissance of the complex, finding several apartments inside The

Meadows lighted. At this point all personnel began an operational retreat from all void spaces until power to the complex could be severed.

Within moments several fires erupted in various parts of the multi-story complex. The LAFD immediately dispatched several fire units, which made initial attacks on several working fires. They discovered several more working fires in the west end of The Meadows, and requested additional units. A second alarm was transmitted to attack more than a dozen working fires.

This event demonstrated the potential dangers of operating inside damaged buildings. All personnel and leaders must remain vigilant for dangerous events and maintain escape routes from each location (particularly when rescuers are crawling in tight void spaces where aftershocks may trap them and cause fires to ignite). Firefighting resources, ambulances, or rescue helicopters should remain in rapid intervention or fire watch mode whenever possible.

Workers from the L.A. Department of Water and Power discovered that one of the main circuit breakers was located in a vault in an alley adjacent to The Meadows. Part of the complex had collapsed into the alley, and the second floor covered the vault. Firefighters using chain saws cut their way through to the entrance of the vault, allowing DWP workers to disconnect the power. Fire extinguishment and power cessation took nearly one hour. During this time, task force members rehabbed while leaders reviewed progress and plans.

Once electricity was confirmed to be off, both US&R task forces returned to the task of searching for remaining victims in The Meadows. The final victim was removed at 0900 hours on January 18, 28 hours after the quake had struck. By 0930, both US&R task forces were pulling their personnel and equipment from inside The Meadows. Both task forces conducted briefings, then began equipment rehabilitation and packing. CATF-2 and CATF-6 were en route to their home bases by 1130 hours.

The Northridge quake was, at the time, the most destructive disaster in the history of the United States. In terms of life loss, the quake was amazingly sparing. The same earthquake occurring in the middle of the afternoon on a typical workday would probably have killed hundreds or perhaps thousands, and the collapse rescue operations might have lasted days. A quake of that magnitude lasting just another 15 seconds may have doubled the toll. It was an example of how local, regional, state, and national fire/rescue resources can be mobilized rapidly and efficiently to provide timely rescue service to people trapped across a broad swath of devastation.

Unfortunately, the Northridge quake has since been eclipsed by more devastating disasters, including the Oklahoma City bombing and the 9-11 attacks. Improvements in the nation's urban search and rescue systems that resulted from the Northridge quake were implemented in these large life-loss terrorist-related disasters.

It is indeed a sad fact that this nation's rescue and disaster systems are becoming more practiced and experienced as a result of disasters on a scale few had imagined when the

FEMA US&R response system was first developed beginning in 1989. It's a sadder statement that the nation's US&R resources, and those of local fire/rescue agencies, are likely to be severely challenged by future manmade and natural disasters in the years to come. At the same time, it should provide some measure of comfort to people who find themselves trapped in these events that the most effective rescue systems known to mankind are being deployed rapidly and effectively in order to locate and extricate them while there is still time.

Structure Collapse Search and Rescue Operations

Newfound understanding about survivable void spaces in certain construction styles, combined with recent innovations in rescue operations and recognition and rapid treatment of trauma-related maladies such as *crush syndrome* and *compartment syndrome* while victims are still being extracted, have greatly improved the chances that trapped victims will be rescued alive and survive over the long term.

There are many examples of cases where building design appears to provide a significant survival link, even during failure, include the collapse of reinforced concrete parking structures, freeway overpasses, and double-deck freeways and bridges. When these structures fail, experienced fire fighters know to search for live victims next to (or inside of) automobiles and between beams within the collapse zone, because these are typically places where survivable spaces are found.

This knowledge, combined with experience gleaned from earthquakes and explosions, accounted for the decision to conduct round-the-clock "rescue mode" operations for nearly 16 days following the 1995 Oklahoma City bombing, and for weeks after the collapse of the World Trade Centers. It will have a profound effect on long-term search and rescue operations in future collapse search and rescue operations.

Structure Collapse Patterns

Lean-to floor collapse. Occurs when one of the supporting walls fails or when floor joists break at one end. Usually creates large void spaces.

Cantilever. Occurs when one end of the floor or roof section is still attached to portions of the wall. The other end hangs unsupported. This type of collapse is extremely dangerous because of unsupported sections.

V-shaped void. Occurs when heavy loads cause the floor to collapse at the center. Victims trapped above the V-collapse will usually be found in the bottom end of the collapse. Victims below the V-collapse will be found in the void spaces away from the "V."

Pancake collapse. Occurs when bearing walls or columns collapse, causing the upper floors to pancake down on the floors below. Victims may be found between floors or in voids created by furniture which supports floors.

A-frame collapse. When floors or roof sections collapse on either side of bearing wall. Victims may be found in void spaces on either side of wall, or at the downward-sloping end of collapsed floors/roof.

Combination collapse. Some buildings collapse in a way that leaves more than one of the above-mentioned characteristics. For example, when part of the Pentagon collapsed following the 9-11 attacks, it fell into a *pancake* configuration that was *leaning* against part of the structure that remained standing. Locating and extracting victims from within this *lean-to/pancake collapse* was an engineering problem that challenged some the nation's most experienced rescuers.

The World Trade Center disaster was another example of a combination collapse because there were so many variations of collapse within the 16-acre rescue site. Firefighters and US&R task forces operating at the WTC disaster encountered practically every configuration of collapse imaginable.

Mid-rise pancake collapse. Multi-story buildings experience mid-level, full-floor collapses. Many of these mid-level collapses occur in modern structures, including those of reinforced concrete construction, steel frame construction, and even earthquake-engineered buildings. In a mid-rise pancake collapse, one or more entire floors pancake down on top of one another. Furniture and structural components may prevent the floors from contacting one another, resulting in potential void spaces throughout the entire floor where live victims may be trapped. The floors immediately above and/or below the collapsed floors may hold the weight of the building. It's possible for small secondary collapses to quickly progress into catastrophic collapse of the entire structure (and possibly crash into adjacent buildings, many of which were already on the verge of collapse themselves).

These scenarios create nightmarish conditions for rescuers. At these sites, incident commanders and supervisors must decide whether it's safe (or even possible) to commit rescuers to the following tasks:

- Conduct thorough interior searches (which may require cutting and breaching floors and walls and sending rescuers crawling into void spaces to physically search for victims)
- Treatment of trapped victims
- "Surgical" removal of debris from between floors to reach trapped victims and clear egress paths
- Penetration of the building horizontally and/or vertically
- Large shoring and stabilization operations to prevent secondary collapse

- Lifting operations requiring "sectioning" of concrete slabs, rigging for crane work, use of air bags, and other technical work
- Victim removal from the collapse and from upper floors

With a mid-level pancake collapse in a large building of reinforced concrete, steel, or masonry construction, each of the above-mentioned operations would be exceedingly dangerous, exposing rescuers to tremendous hazards. As demonstrated in many recent disasters, secondary collapse is a major concern whenever rescuers are operating inside badly damaged buildings. Add to this the effect of external factors such as wind, rain, aftershocks, and rescuers removing various parts of the building in search of victims, and the danger rises exponentially. These decisions are further complicated in large disasters because there may be few backup rescue capabilities available to assist if something goes wrong.

The decision to commit rescuers to a mid-level collapse carries a grave responsibility. The importance of collapse rescue recognition and training for fire officers and other rescuers cannot be overemphasized. Neither can the value of technical assistance from qualified structural engineers and shoring experts. In the United States, the rapid response of FEMA US&R task forces may be a life-saving resource, because each US&R task force includes two specially-trained structural engineers, two specially-trained heavy equipment operators, many shoring experts, and other necessary resources for properly managing major collapse operations.

En Route to Collapse Incidents

Consider the possible cause of the reported collapse, such as a terrorist attack, a transportation accident (*e.g.*, plane crash into a building), an intense fire, a natural gas leak and explosion, a mud slide, a flood, or even an avalanche. Each of these causes is often associated with particular hazards. Always consider any indication of terrorist bombing, which *might* be accompanied by the release of nuclear, chemical, or biological agents.

The collapse search and rescue strategy used by many fire departments with experience in such matters is based, in part, on the standard Five Stages of collapse search and rescue as developed in the United States.

It's important to note that certain strategic objectives and tactics may be employed through the course of all Five Stages of collapse rescue, depending on the conditions that rescuers encounter. The Stages are not divided by empirical walls, and rescuers must us a certain element of finesse and judgment to ensure that the most is being done to locate and rescue (or recover) all victims trapped in the collapse. Therefore, the Five Stages of collapse rescue are simply a framework within which all tasks required to extract all the victims proceed during the course of a collapse search and rescue operation.

Stage 1: Size-up and recon

- Conduct a "six-sided" size-up of the involved building(s).
- What time of day is it?
- What day of the week?
- What is the building's occupancy type (*e.g.*, offices and commercial vs residential or schools and hospitals, etc.)?
- How many potential victims are missing or trapped?
- Are there fires, gas leaks, flooding, or other hazards that my take the lives of victims before they can be located and rescued?
- What is the situation in the immediate area of the collapse?
- What is the condition of the area surrounding the actual collapse?
- How is the building constructed?
- Which type of collapse (pancake, lean-to floor collapse, lean-to cantilever, V-shape void)?
- Are secondary devices present?
- Is there a hazmat problem?
- Is there a need for additional resources?
- Name the incident, request resources, establish command (be prepared to establish unified command).

Note: Size-up and reconnaissance is a process that may last for days (or even weeks) in large or complex disasters, through all five Stages.

Use standard US&R Search Markings to identify the results of your searches! Every building that has been searched should have the Search Marking spray-painted on the front of the building to signify that the search has been completed. Otherwise, other crews may waste valuable time searching the same buildings repeatedly. In a disaster situation, such redundant searching will drain valuable resources which could otherwise be used for other essential tasks. The result may be the loss of many lives, which might have been saved elsewhere.

Stage 2: Surface rescue

This includes the rescue of all victims who can be extracted without cutting, breaching, or heavy lifting operations. Once the victims who are obviously visible, lightly trapped, and readily reachable are rescued, the temptation to declare the start of the "recovery" phase should be avoided until all potential survivable void spaces are searched for live victims.

Stage 3: Void space search operations

Physically searching every void space that rescuers can find, looking for viable victims trapped in survivable void spaces. During the void space search phase, well-trained and equipped firefighters and US&R task forces tunnel their way through the building using special tools, rope rescue, mining and tunneling, and structural stabilization methods. They use fiber-optic and ground-penetrating radar technology, special search cameras, extremely sensitive acoustic- and vibration-sensing instruments, search dogs, and direct visual and voice contact to locate victims trapped within void spaces created when the structure collapsed. In many cases firefighters and other rescuers must squeeze through cracks and void spaces, crawling through the interior of collapsed buildings, in order to positively determine whether victims are trapped. This is extremely hazardous (but essential) duty.

Stage 4: Selective debris removal

After all known survivable void spaces are searched, selective debris removal begins. During this phase, rescuers work with heavy equipment operators, structural engineers, construction and demolition contractors, and others to "de-layer" the collapse zone from top to bottom. As upper layers of the buildings are peeled away, additional void space search operations are generally conducted in order to check newly-accessible parts of the building for potential survivors.

After all known survivable void spaces are searched, selective debris removal begins. During this phase, the firefighters and US&R task forces work closely with heavy equipment operators, structural engineers, construction and demolition contractors, and others to take the building apart piece by piece, usually from top to bottom. As upper layers of the buildings are selectively peeled away like an onion, additional void space search operations are conducted in order to check newly-accessible parts of the building for potential survivors. This process can take hours, days, or weeks. At the World Trade Center collapse, where more than 1.4 million tons of debris was removed in more than 98,000 truck loads, the selective removal of debris—in concert with general debris removal—lasted more than seven months.

In many cases (especially in large buildings), void space searches with selective debris removal (or combinations of both) should continue until the entire building has been dismantled and all possible survivors located and extracted. These operations are extremely dangerous because of the instability of damaged buildings, as well as the continuing aftershocks that accompany major earthquakes. Without proper training, equipment, and experience, personnel conducting these operations can cause the building to collapse, killing rescuers and victims alike.

It is during these stages of a disaster that some of the most difficult, complex, and time-consuming rescues are made—the so-called miracles of disaster response. These operations are the "bread and butter" of modern urban search and rescue teams (or task forces), both domestic and international. Until these two phases of rescue have been completed, officials should refrain from declaring that the "recovery phase" has begun.

Sometimes the strategy of void space searches (Stage 3) is alternated with selective debris removal (Stage 4) until the entire collapse zone is dismantled and all victims are located and extracted. These are high-risk operations because the stability of the building was compromised by the original event, and is often made far worse during Stage 4.

Structural stabilization, generally through employment of shoring, tie-backs, and other means, is something that may be required from the first arrival of units until the end of Stage 4 or even Stage 5. At the Pentagon, some sections of building on the perimeter of the actual collapse required stabilization to prevent the rest of the building from coming down, even as the last bits of collapse debris were being removed.

Void spaces need to be searched whenever they are encountered. Therefore, even after Stage 3 has been completed and Stage 4 (selected debris removal) has begun, additional voids may be uncovered by heavy equipment, and those voids must be searched for additional victims before debris removal continues.

During Stages 1 through 4, rescuers typically encounter the most difficult, complex, and time-consuming rescue problems (often resulting in the so-called "miracle rescues" that follow some disasters). Locating and rescuing deeply entombed victims from collapsed structures is "bread and butter" for modern fire department US&R programs. It's a specialty of FEMA's 28 US&R task forces, and a mainstay of the America's internationally-deployed US&R task forces. Until the first 4 Stages have been completed (leaving Stage 5—General Debris Removal—as the last option), officials should generally refrain from declaring that the "recovery phase" has begun.

Medical treatment, or "rubble pile medicine," may be conducted through the first four stages. Rescuers may find themselves treating one trapped victim for many hours in one section of the collapse, while Stages 1, 2, 3, and 4 proceed in other areas of the building.

Stage 5: General debris removal

This may include the use of heavy equipment to bulldoze, demolish, and remove large sections of building, as well as the debris piles left behind. It may involve hundreds of rescuers or laborers participating in the hand removal of tons of debris not accessible to heavy equipment. A search for victims in the rubble is sometimes continued through Stage 5.

Removal of overhead hazards is another potential multi-stage task. At many collapse operations, suspended debris must be removed throughout the operation to prevent death or injury to rescuers. Other examples of multi-stage tactics are canine search operations and technical search, which may be required to search different parts of a collapsed structure during the initial reconnaissance, then repeatedly as the collapse area is "de-layered."

Helicopter Rescue Operations

What tool is more useful than helicopters for plucking people from cliffs, canyons, mountain tops, floods, ice chutes, the roofs of burning high-rise buildings, and other "difficult access" predicaments? Considering the adverse conditions of weather, terrain, and unusual emergency situations under which public safety helicopters are called upon to rescue victims, few tools are more effective. Helicopters may be used to insert rescuers into a hostile or inaccessible rescue scene, and then to extract rescuers and victims in a timely manner.

And that is why many progressive fire/rescue agencies include helicopter rescue operations in their repertoire of rescue options. For agencies that have their own helicopters, this includes regular training of helicopter rescue crews (and training ground-based rescue or US&R companies to work with helicopter crews) to conduct the full range of rescues likely to occur within the jurisdiction and surrounding areas. For agencies that don't have their own helicopters, this includes identifying outside helicopter rescue resources and developing plans to request and utilize them when conditions call for helicopter rescue operations.

Helicopters are unsurpassed as aerial resources because of their multi-mission capabilities. They can quickly gain access to remote areas, cutting hours off the time that might normally be required for rescue teams to reach victims of backcountry accidents. They can shuttle manpower and equipment to the top of burning high-rise buildings, saving precious minutes off the time normally required for firefighters to climb the stairs. They can land right at the scene. They can perform "one-skid hot landings" under power when the terrain in uneven or unpredictable. They can hover over the scene, allowing firefighters and other rescuers to be hoisted into the incident on a cable, or lowered by rope or rappelling. They can provide overhead lighting for a rescue scene, and they can conduct searches with night vision and thermal imaging. In short, helicopters can get to places that are otherwise difficult or time-consuming to reach, and they can extract victims and rescuers with equal speed and utility.

With such versatility and success come certain expectations. Movies and the news tend to fuel a perception that helicopters can perform practically any maneuver, any time, under any conditions. The public has been conditioned to expect that the day is saved when helicopters show up. Increasingly, this expectation is turning to demand that helicopters get into the mix when things are at their worst.

Sometimes these demands lead to conflict. Practically every year in the United States there are disputes about whether helicopters should have been employed to rescue some stranded person or another. It's become rather commonplace for public officials and the news media to call into question the tactics employed by helicopter rescue crews and other rescuers, particularly after prominent events that are televised live.

The incident commander should understand basic hazards associated with helicopter rescue operations, and they should know when to use them. When a helicopter rescue is requested, they should take appropriate precautions. This is not to say that helicopter rescue is excessively dangerous. To the contrary, helicopters are frequently the safest method available to rescue some victims. But the risks associated with helicopters must be taken into consideration when developing emergency policies and procedures, when incident commanders are making decisions about methods of rescue, and during the training process for helicopter rescue teams.

The point is this: The incident commander has a responsibility to used helicopters with the understanding that it's not a risk-free operation. A risk-vs-gain evaluation should determine the best course of action, and when conditions call for the use of helicopters for rescue, the IC should make the request without delay. But the IC should also take appropriate precautions to safeguard the helicopter crews by ensuring rapid intervention capabilities are immediately ready in case of a mishap.

Canceling the assisting ground units before the helicopter has completed the rescue should be discouraged. In many cases the IC should consider having ground-based rescuers proceed to make plans for a ground-based operation in case the helicopter crew decides that a helicopter rescue is unfeasible, and in case there is a mishap. If the rescuer gets into trouble, or if the helicopter suffers a mishap, it's a matter of life and death to have personnel immediately available to conduct a rapid intervention.

In extreme cases it may be appropriate to have another helicopter "on station" to back up the crew of the first helicopter. This may seem to be overkill to some, but those who have seen or experienced a helicopter rescue mishap understand that seconds count, and that the incident commander has a responsibility to have a contingency plan in place.

Helicopter hoist rescue operations

Helicopter-mounted hoist rescue systems are often employed to conduct these operations. Some hoist rescue systems are hydraulically-operated, others electric. Some hoist systems extend and retract faster than others, and depending on the manufacturer they have a number of varying features.

Despite the obvious advantages of helicopters equipped with hoist capabilities, and despite the fact that they are among the most effective and versatile forms of technical rescue, rescue professionals must be aware of the hazards associated with any rescue option. And so it is with hoisting. Along those lines, this is a good place to discuss certain limitations and dangers associated with hoist rescue operations.

Typically a rescue hoist mounted on a helicopter is rated for a 600-lb. load (working strength) that will accommodate one rescuer and one victim in a direct vertical (not side-loading) lift. This is a far lower strength rating than the typical ½-in. kernmantle rescue rope employed by most fire/rescue services. And the cable is generally deployed by itself, without any sort of belay or "safety line." It's impractical to deploy two cables because it's likely the rescuer will spin several times during a typical hoist operation, leaving the cables twisted together and making it impossible for them to retract back onto the hoist spool. Consequently, if the cable separates during a hoist rescue operation, the rescuer (and perhaps the victim) will simply fall.

At the end of the typical modern rescue hoist cable is a swedge, a locking "hook," and a free-spinning hand-hold ring to prevent the victim from grabbing onto the cable itself and unwinding it as they begin to spin. The wound strands of a hoist cable are vulnerable to "unwinding" *if* the end of the cable is not attached to a free-spinning device and *if* the rescuer or victim spins while holding the cable itself, *or* if the free-spinning hand-hold freezes up.

If not properly tensioned—or if the cable begins to unwind—the cable can "spider" (fray in such a way as to resemble a rat's nest) as it is retracted into the housing and wraps around the barrel. This can put tension on the cable sufficient to make it separate or break. The result may be a fatal fall.

Steel cable used for rescue is like rope in that it stretches to a certain degree. The degree of stretch in a cable is proportional to the length that's deployed and the load (*i.e.*, the rescuer and/or victim). The longer the distance that cable is deployed below a helicopter, the more stretch there will be. More stretch in the cable equals more "forgiveness" to insults like shock-loading. A rescuer who falls and shock-loads a longer cable is less likely to experience a catastrophic failure, whereas a firefighter who shock loads a shorter section of cable is more prone to experiencing total cable failure and a devastating fall. Ironically, the most dangerous part of a hoist mission is often the last few feet before the rescuer and victim reach the safety of the helicopter cabin.

If, just inches from the safety of the cabin, the rescuer and/or victim falls from the skid, or otherwise shock-loads the cable with only a short section deployed, the result can be a fatal fall because the shorter cable is less forgiving of shock loading. It's a sad fact that a number of catastrophic hoist cable failures have occurred while rescuers/victims were stepping into or out of the cabin of the helicopter during hoist rescue operations.

Hoisting in swiftwater situations. Helicopters routinely conduct standard hoist rescue operations to pluck victims to safety from houses, rocks, cars, other predicaments where the victim is in a static condition (where the force of moving water is not pushing on them). Just like rescuing a victim from a cliff, plucking a person from a non-moving object in the middle of a stream or river is a reasonable use of the helicopter.

But hoist rescue operations are not always appropriate for extracting victims from swiftwater rescue situations. Excessive danger may result from several conditions common to hoist rescue operations where the victim is in the moving water. First is the limit of a particular hoist rescue mechanism. As previously mentioned, common helicopter hoist systems are rated to lift a maximum of 600 lbs., with the cable lifting vertically (no side loading). The problem with swiftwater rescue and helicopter hoist system is that victims being swept downstream are in the grip of the water, and this may create dramatic side-loading forces, raising the potential for immediate (and life threatening) damage to the hoist cable, the hoist motor, the hoist mechanism, or the support by which the hoist mechanism is attached to the airframe of the helicopter. Failure of any of these components may be catastrophic for the victim/rescuer and possibly to the crewman who is operating the hoist.

Another potential danger is side-loading of the hoist system, as a result of the victim or rescuer who is exposed to the power of fast-moving water, or who may be entrapped in strainers or heavy debris. This may result in a situation in which excessive side-loading is exerted on the hoist arm, which in turn may upset the delicate balance of the helicopter. Because the helicopter is generally hovering close to the ground when conducting swiftwater rescues (with a cable and victim/rescuer attached to it, no less) there is little or no room for recovery if something goes wrong. Yet another danger is related to the potential for the hoist cable to snag on heavy debris in the water, or on trees, bridges, or other objects.

Therefore, many progressive agencies refrain from using helicopter hoists when the victim is in moving water. The alternative is to use a "short haul" rescue system employing a standard rescue rope that's anchored in the copter.

Short haul systems for swiftwater and flood rescue operations

Helicopter short haul rescues in moving water generally work like this: A rope of varying length (depending on the task and the procedures established by each agency) is attached to a rescuer "belly band" that encircles the floor of the cabin and the belly of the copter, or within the cabin. A rescuer (and/or a rescue device for the victim) is attached to the end of the rope. The pilot flies the helicopter to a point at which the rescuer can snag the victim, or where the victim can climb into the rescue device. Then the pilot ferries the rescuer and/or victim (dangling beneath the copter) to a safe location, where the rescuer/victim can be lowered to the ground beneath the still-hovering helicopter.

Short hauling is a skill that generally takes a certain amount of experience and training to maintain a safe margin, and one requiring excellent depth perception. It can be difficult for the pilot to ensure adequate clearance when poor weather and darkness reduce visibility and depth perception, or when wind and rain are buffeting the copter. A crewman is usually assigned the job of acting as a second set of eyes to observe the rescuer/victim at the end of the rope, and to communicate with the pilot to ensure adequate clearance over obstacles. At night or in poor weather, even with night vision goggles, the crewman's depth perception may be impaired, adding to the potential dangers of short hauls.

Standard evolutions for helo/swiftwater rescue teams

Dynamic swimmer free evolution: The rescue swimmer enters the water independent of the helicopter. After they make contact and the rescuer secures the victim using a Cearly Strap, the helicopter (which has been tracking the rescuer) lowers the short haul rope. The rescuer clips himself and the victim into the rope, and the copter (still matching their speed in the current) lifts them from the water and deposits them on shore.

Dynamic tethered rescuer evolution: The rescuer is lowered from the copter on the short haul line. When the victim is dangling about 30 feet below the copter, the crewman locks the rope off at the anchor. The pilot, guided by the crewman, matches the speed of the victim and lowers the rescuer into the water. The rescuer contacts the victim, wraps a Cearly Strap around him, and signals to be raised. The copter, matching the water's speed, lifts the rescuer and victim to the shore.

Static swimmer free evolution: The rescue swimmer proceeds to a victim who is stranded on a stationary object or in a pond or lake (no current). The rescuer attaches a Cearly Strap to the victim and signals for the copter. The copter hovers overhead and the crewman lowers the short haul rope. The rescuer connects himself and the victim to the rope, and both are short-hauled to safety.

Static tethered rescuer evolution: The rescuer is lowered from the copter on the short haul rope to a victim who is stranded on a stationary object or who is in a pond or lake (no current). The rescuer places a Cearley Strap on the victim, attaches the victim to the rope, and signals to be lifted to safety.

Cinch harness rescue evolution: The Cearley Cinch Harness, a victim capturing device designed for static and dynamic helicopter rescues, including swiftwater rescue, is lowered from the copter at the end of the short haul line to a victim. The victim places the cinch harness over their head and slips it below their arm pits. The crewman, standing on the skid, "jerks" the rope to dislodge the Velcro on the cinch harness, which allows the harness to cinch up on the victim. The crewman, talking on a "hot mic" (voice-activated microphone), instructs the pilot to raise the victim from the water. The victim is short-hauled to shore (or, in some cases, to a boat). In some cases, a rescuer may be lowered on the same rope, so they can place the victim in the cinch harness. This may be necessary if the victim is injured, unconscious, or is a child.

One requirement for short haul operations is maintaining sufficient altitude to avoid dragging the rescuer/victim through tree-tops or striking other objects. This author remembers conducting numerous short haul operations, but one in particular stands out—the rescue of a one-year-old boy trapped underwater in a river. After plucking the child from where he was wedged between a tree and a rock, the author was short-hauled about 400 feet in the air to "overfly" a set of power lines that crossed 300 feet above the canyon as the pilot looked for a place to set him on the ground, still dangling beneath the copter...and all the while performing CPR on the baby. At times like that the rescuer has no control over fate insofar as lethal obstacles are concerned; it is strictly up to good training and experience of the helicopter pilot and crew to ensure sufficient clearance to avoid disasters.

More than once, short haul operations have ended in tragedy when things went wrong, so it's incumbent on everyone involved to understand that it's not a risk-free operation.

The bottom line is this: For any rescuer who agrees to be suspended beneath a helicopter in an emergency or in training, there is an assumption of risk that they may have to be punched free of the craft in order to save the copter, and the crew aboard it, as well as bystanders and fire/rescue personnel on the ground.

Helicopter-based high-rise operations

It's no revelation that it can be an advantage to use helicopters to place specially trained and equipped firefighters and equipment on the roof during some high-rise fires emergencies and disasters. This rapid deployment to the top allows firefighters to perform a variety of critical tasks like size-up, ventilation, search, rescue, firefighting, and rapid intervention—without the need for firefighters to climb many flights of stairs just to start the job.

Helicopters have already played a prominent role in past disasters that took place in high-rise buildings. Some examples include:

- The MGM Grand fire in Las Vegas, which killed more than 80 people. Helicopters from the U.S. Coast Guard, the Air National Guard, and other agencies rescued victims from the roof during the course of this fire.
- The One Meridian fire in Philadelphia. In a desperate attempt to locate and rescue three firefighters lost on upper floors, an ad-hoc helo/high-rise team was assembled using engine company personnel who were inserted on the roof. While these firefighters were unable to locate the original three missing firefighters, they did manage to direct other lost firefighters to safety.
- The First Interstate fire, where Los Angeles Fire Department personnel were confronted with 4 floors (levels 12 through 16) of a 63-story building fully involved. A pre-established LAFD Helo/High-Rise Team was deployed to the roof in early morning darkness in order to open the stairwells for improved ventilation, search the upper floors for a number of missing victims, and other tasks.
- Baltimore Fire Department's Rescue Company 1 was deployed by a Maryland State Trooper helicopter to the roof of a burning high-rise hotel at night, managing conduct topside size-up, open the stairwells, and conduct search and rescue operations on the upper floors until the fire was knocked down by companies working their way up from the ground floor.
- In the aftermath of the 1993 bombing of the World Trade Center, the NYPD deployed a team of Emergency Service Unit members to the roof of the North

Tower. The NYPD was roundly criticized for failing to participate in a unified command structure, for failing to coordinate the helicopter deployment with the FDNY fire command, for deploying personnel without proper personnel protective clothing, for deploying personnel who were not experienced fighting fires (and especially high-rise fires), and for its failure to establish and maintain communication with the incident commander and the many fire department units operating in the building.

- When terrorists set fire to a high-rise hotel in Puerto Rico in the 1980s, helicopters were used to deploy firefighters to the roof and to extract victims from atop the smoke-filled building.
- Helicopters attempted to rescue people trapped atop a flaming high-rise in San Paulo, Brazil. The intensity of the fire and the thermal columns and smoke prevented effective helicopter operations, and many people died.

The use of helicopters for high-rise emergencies is naturally a topic of some controversy. It's a somewhat risky tactic that's rarely employed for a number of reasons. Most fire/rescue agencies lack their own air force to develop, practice, and perfect these methods. Sometimes the coordination between fire departments and agencies that have their own helicopters could be improved. And some fire department leaders are not fully aware of the capabilities and limitations of helicopters and well-trained helicopter firefighting/rescue crews.

Some have argued that helicopters have limited application in high-rise fires; that the time-tested method of firefighters entering at the ground floor and working their way up the stairwells from below to attack the fire and conduct other tasks (sometimes using elevators to transport equipment and/or manpower to levels below the fire) remains the most effective approach. Some have expressed concern that rotor wash may fan flames or change the thermal balance in the building in a way that adversely affects firefighters. There are legitimate concerns that thermal columns and smoke may prevent helicopters from reaching the roof or cause them to crash; that there is no guarantee the copter can get back to the roof to retrieve the firefighters; and that conditions may be untenable for firefighters to operate above the fire.

On the other hand, others in the fire and rescue services argue that helicopters are an under-utilized tool that could—if properly used—expedite placing properly trained and equipped teams of firefighters on the roof to assess the conditions and report them to the incident commander (who, armed with more data, can make better decisions). They argue that helicoptering specially trained companies of firefighters to the roof will help accomplish the following goals (all in a more timely manner than may be possible by firefighters hiking up from the lower floors):

- Control of crowds escaping onto the roof from smoke-choked stairwells and fiery floors (and, potentially, from blast-damage related to terrorism), and preventing them, through their presence and their organized actions, from leaping from the roof
- Organizing and conducting rooftop helicopter evacuations
- Opening stairwell doors to help facilitate vertical ventilation to make the upper floors more tenable for trapped victims
- Searching for people missing and trapped in heavy smoke conditions (aided by thermal imaging systems, closed circuit SCBA, and other special equipment)
- Conducting interior fire attacks from above the fire to reduce vertical extension and protect routes of escape until other firefighters can reach the fire from below
- Protecting victims who are being "sheltered in place"
- Conducting rapid intervention operations
- Conducting other essential tasks

Intuition suggests that helicopter deployment of well trained companies of firefighters and officers has the potential for getting a faster assessment of conditions above the fire; for evaluating structural integrity, for getting fresh personnel in position to establish effective vertical ventilation faster; for safeguarding or removing trapped occupants; for getting water on the fire faster (thereby preventing fire spread and reducing the potential for structural collapse); and generally improve the situation for those unlucky to be trapped above the fire and for those attempting to command the incident and control the fire.

Until these concerns can be answered with certainty, fire departments with high-rise buildings in their jurisdictions are well advised to be prepared to try it when a "worst case scenario" makes it necessary, even if that means practicing with local outside agencies (including law enforcement and the military) that can provide helicopters for this purpose. Fire departments with their own helicopters and high-rise hazards should certainly develop the capability to deploy specially trained firefighters to the roof when conditions clearly call for this tactic.

The 9-11 World Trade Center attack naturally raised questions about helicopter high-rise operations. But the conditions that confronted the FDNY on September 11, 2001 were so extraordinary that it's difficult to draw definitive conclusions about the ultimate efficacy of using helicopters to high-rise fire/rescue operations. To some, the WTC attack raised more questions than conclusions.

The one thing we can say with authority about deploying personnel to the rooftops of burning high-rises is this: Police officers and other personnel without full turnouts and SCBA, without extensive high-rise firefighting experience, without fire department communications, and operating outside the fireground command system, are ill-prepared to operate effectively on the floors above a high-rise fire, and they are more likely to be a liability.

We can also definitively say that police and other "non-fire-department" agencies with helicopters on the scene of a burning high-rise have a responsibility to coordinate all their efforts and operations with the jurisdictional fire department incident commander, to establish open radio and other communications with the firefighting forces operating inside the building (and their commanders), to immediately share critical intelligence with the fire department, and to operate under the aegis of the fire department. To do anything less is to endanger the lives of citizens, firefighters, and other people affected by the event. Law enforcement agencies that refuse to operate by these rules simply confirm the argument that more fire departments should operate their own helicopter fleets in order to effectively execute the fire department mission without interference of the uncooperative law enforcement agencies.

Helo/high-rise teams in use. Some fire departments have moved forward on developing specially trained teams of firefighters and officer to the roof to improve their high-rise firefighting capabilities. Some of these fire departments have developed teams of firefighters from rescue companies, US&R task force fire stations, truck companies, or other designated units whose job it is to be inserted by helicopter on the roof of burning high-rise buildings. And the experience thus far indicates favorable results.

The firefighters assigned to a helo/high-rise team should be experienced in fighting difficult fires. High-rise firefighting experience is obviously an advantage. All helo/high-rise teams members should wear full turnouts and SCBA (preferably a "re-breather" closed circuit SCBA like a Biopack or Draeger, which allow at least four hours of rated working time, and up to eight hours if the wearer becomes trapped and conserves their air/oxygen; or a one-hour SCBA bottle); a minimum of a Class III rescue harness (either internally worn beneath the turnouts or donned over the turnouts); one or more thermal imaging systems; night vision devices where available; forcible entry tools (including tools to take down antennae and other flight hazards on the roof); an "officers" high-rise hose pack, a rope pack; a water extinguisher; victim harnesses; and other tools. (See Appendix 6 for recommended procedures.)

High-Angle Rescue

High-angle rescues are among the most common rescue emergencies to which fire departments respond. High-angle rescues include (but are not limited to) the following types of emergency situations:

- Traffic collisions with vehicles "over-the-side" on mountain roads
- People trapped or stranded on cliffs, hillsides, ice chutes, rocks, waterfalls, dams, or trails
- Aircraft crashes in the mountains on into high-rise buildings

- People stranded by high water, mud and debris flows, landslides, or rockslides
- Any other situations that require the use of ropes, cables, helicopters, or other technical means to remove victims to safety and (when necessary) to provide proper medical treatment

It should be noted that reports from hikers and other reporting parties are notoriously incorrect in terms of injuries to victims and other specifics about the victims' predicaments. In a significant portion of cases, victims originally reported as "not injured" are in fact injured and requiring BLS and ALS treatment. In a large proportion of instances the reporting party can only see or hear the victim from a long distance (*i.e.*, across a canyon, from the bottom or top of a cliff, etc.). In many other cases the informant is communicating third-party information from other hikers, cell phone callers, passersby, etc. Therefore, in cases where victims are reported trapped, stranded, lost, or otherwise in need of physical rescue, it should be assumed that the victims are injured, until departmental personnel can positively ascertain that they are *not* in need of medical assessment or medical care.

Confined Space Rescue

Confined space rescues are among the most hazardous incidents that firefighters are exposed to in the course of their duties. Nationwide, there is an inordinately high incidence of mortality among fire fighters and other would-be rescuers during confined space rescue operations. These operations need not result in preventable loss of rescuers if proper planning, training, procurement of equipment, and multi-agency cooperation are included as part of the approach to confined space rescue emergencies.

Properly trained high angle rescue resources are required to ensure safe entry and extraction from deep shafts. IDLH (immediately dangerous to life and health) atmospheres need not result in the death of victims if proper ventilation can be conducted and they can be rescued in a timely manner. Incident commanders should prepare for long-term operations, because many confined space rescues take hours to complete. Above all, the incident commander should strive to ensure that rescuers do not become victims unnecessarily.

As in other situations where extreme dangers are present, chief officers and incident commanders have a moral responsibility to ensure that firefighters are sent into these situations with adequate awareness, training, procedures, and equipment which will allow them to complete their jobs in a reasonably safe manner. There are also legal imperatives for maintaining a working knowledge about confined space rescue. Some worker safety laws place direct responsibility for compliance with safe confined space rescue practices on chief officers, Incident commanders, and other emergency supervisors. In some localities, failure to comply with worker safety laws—even under emergency conditions—can result in citations, fines, and even criminal prosecution for supervisors. Some fire departments have already been cited for failure to comply with legally mandated confined space safe practices.

At a minimum, firefighters and other rescuers should be trained to recognize the hazards of confined space rescue situations. Procedures should be implemented to assure safe management of these incidents. Finally, some type of local confined space rescue capability should be established for rapid response. In some cases, existing rescue squads and/or hazardous materials teams can be trained to properly handle confined space rescue operations.

Whenever rescuers are sent into a confined space, it is critically important to support them with atmospheric monitoring capabilities, appropriate respiratory protection, good communications, well trained rope rescue teams (to manage vertical entry operations), back-up rescue teams, a clear rescue plan, and other essential safety mechanisms. Without these components in place, firefighters are operating beyond the boundaries of reasonable safety, with lethal dangers waiting around every corner and little chance to recover from major mistakes or unexpected complications.

What is a confined space?

A confined space meets the following description:

- The space is large enough for a person to enter.
- The space has limited or restricted means of entry and egress.
- The space is not designed for continuous human occupancy.

These descriptions cover a broad range of situations, including deep shafts, wells, refinery facilities, water and gas piping and containers, and at the site of tunneling and mining operations, well casings, caissons, silos, piping, air vents, vaults (including below-grade electrical vaults), storage tanks, and others.

Many of the hazards associated with entering confined spaces are capable of causing rapid injury, illness or death, or long-term disease. Fatalities often occur when workers and would-be rescuers fail to recognize basic hazards and take appropriate precautions to protect themselves. Deaths also occur when adverse events (including mistakes made by workers or rescuers) occur in an environment in which recovery and rescue is impeded by restricted access, confined working spaces, and other factors.

Until proven otherwise, all confined space environments should be considered IDLH atmospheres. The biggest danger is likely to be atmospheric (*i.e.*, lack of oxygen, too much oxygen, an explosive atmosphere, or the presence of toxic substances in the air). In every case it should be assumed that IDLH hazards will exist throughout the duration of the entry.

It should also be assumed that the victim and rescuers may need physical assistance from the space in the event of an "adverse event" like the collapse of non-shored excavation walls, failure of a breathing air line, failure of the rope system, or other negative events. Backup rescue is a need which may be difficult to address, in part because there may only be room in the space for one entrant at a time. Backup rescue operations to assist the primary entry team may be dangerous, complicated, and time-consuming. The wise IC will recognized this fact early in the incident and take appropriate steps to ensure that an effective backup rescue plan is established before entry teams enter any confined space.

The IC should be cognizant of the potential for a dangerous situation to be created by the actual rescue work being performed in a confined space. Certain cutting, shoring, and lifting operations carry a certain level of inherent danger in any environment. The danger is multiplied if these tasks must be performed in a confined space. For example, the cutting of metal with torches is accompanied by toxic byproduct gases that might normally dissipate in open air. If this task is required to free a trapped worker in a confined space, it may lead to a concentration of gases which, combined with other atmospheric hazards (*i.e.*, oxygen deficiency, etc), may quickly reach IDLH levels. Accordingly, OSHA and Mine/Tunnel Safety regulations require all employees (including fire and rescue personnel) who are expected to assist in a confined space rescue to be protected from all existing and potential hazards through proper equipment, training and procedures.

Some concerns of the IC should be atmospheric and fall hazards. Whenever a victim has reportedly become incapacitated or is missing in a confined space, IDLH atmospheres should be suspected until proven otherwise by direct monitoring with appropriate instruments. Oxygen deficient atmospheres are a major hazard in many confined spaces. A 1985 OSHA study revealed that of 173 confined space fatalities, 67 were in non-tested, oxygen deficient atmospheres.

Even a small reduction of the oxygen percentage indicates that some process is actively reducing the level of oxygen in the space. It may be the result of consumption by fermentation, bacterial, or chemical reactions; absorption of hazardous substances into the lining of the shaft; displacement by other gases formed within the space or introduced from the outside; purging operations; oxidation from rusting steel casings or curing concrete shaft casings; or even as a result of respirations from a trapped victim, which can reduce the oxygen level to dangerous or lethal levels. This hazard may be especially pronounced at the bottom of a deep shaft or other confined space where air circulation is severely reduced.

Combustible atmospheres may ignite or explode if a source of ignition is present or is introduced. Combustible agents may include naturally occurring gases, vapors from liquids such as fuels or solvents, or dusts of combustible materials.

Combustibles are considered hazardous when they reach 10% of their Lower Explosive Limits (LEL). Some flammable gases may flow into deep shafts naturally or be introduced by workers into the space accidentally. An oxygen enriched atmosphere (23.5%+) increases the potential for ignition. Different gases, heavier or lighter than air, will seek lower or higher levels (stratification) in a deep confined space such as a vertical shaft. Desorption of chemicals from the walls of the confined space may cause a combustible atmosphere. Dusts may become combustible in certain concentrations. Generally, dusts are considered combustible when particulates reduce visibility to less than five feet, but some materials may reach dangerous concentrations long before this happens.

The atmosphere of a confined space might contain asphyxiants and irritants that may cause disease, illness, injury, or death. Effects can be immediate or delayed. Carbon monoxide (CO) may be found in some confined spaces, formed by incomplete combustion of fuels containing carbon and in decomposition of organic matter. CO is odorless and colorless and may quickly reach lethal levels in a confined space and give little to no warning before a victim or rescuer is overcome. Hydrogen sulfide gas is produced from the natural decomposition of sulfur-bearing organic matter. Raw sewage can produce extremely high concentrations of hydrogen sulfide. Exposure to low concentrations may cause pulmonary complications. Exposure to higher concentrations may cause unconsciousness and death in seconds. The rotten egg odor may not seem present at higher concentrations because of resulting paralysis of the olfactory nerve which controls the sense of smell.

Mechanical hazards are frequently present in industrial settings. Power machines must be isolated and locked out/tagged out and a guard posted, if necessary, to prevent reactivation. Restrictive entry and egress openings often contribute to the dangers of confined space rescues. The number, location, and size of openings affect the time required to enter and exit the space. Communication problems are a main consideration in carrying out a safe entry. Effective communication links (*i.e.*, hardwire or radio, sometimes with repeaters) between entrants and personnel outside the space must be established prior to entry and maintained throughout the operation. Line-of-sight visual monitoring of entrants should be maintained when possible. In potentially flammable or explosive atmospheres, radio communications must be intrinsically safe by law.

Naturally, the lack of natural light is a problem in most confined spaces. Few situations could be worse for a rescuer than becoming lost in a shaft or tunnel without light because batteries have run low. Accordingly, OSHA requires that three separate intrinsically safe light sources be carried by each entry team member. The light sources may be of any type as long as they are intrinsically safe. To meet this requirement and provide some level of flexibility to meet changing conditions in a confined space, some agencies provide one helmet-mounted headlamp, one hand-held flashlight, and cylume light sticks to each rescuer.

Temperature and humidity variations in a confined space may cause ill effects for rescuers and victims alike, and may hamper rescue efforts. Frequent rotation of rescue personnel may be necessary to stave off the effects of cold, heat, water, and other environmental conditions. Noise may be intensified because of reverberation in a confined space. Loud noises in this environment will not only hamper communication between the rescuer and the topside, but will simply annoy rescuers, limit their ability to concentrate, raise stress levels, and decrease their effectiveness.

Some subterranean confined spaces penetrate water-bearing layers of earth, causing a water hazard in addition to the other typical hazards. Confined spaces may also bisect existing utility lines, including sewers and water pipes, with the resulting potential for failure and inundation of the confined space, which is likely to be catastrophic for rescuer and victim alike. Rain or flooding will certainly create problems for rescuers working in and around a subterranean space. Failure to aggressively address any of these hazards at the scene of a shaft rescue may have catastrophic results. During several rescue operations in natural caves in England, emergency dams and dikes were built by entry teams to divert the flow of rain water away from the cave openings. Similar tactics may be required for some confined space rescues.

Engulfment and collapse hazards exist in many confined space situations. Even some above-ground confined spaces (*i.e.*, grain silos, borax silos, etc) may have engulfment/collapse hazards similar to those found in below-grade shafts. In some cases, a "crust" bridge may develop as material quantity decreases in the shaft or space, creating a thin surface above a void space that will not support the weight of rescuers but looks deceptively solid.

Arrival at confined space rescue sites

The first-arriving company should determine what has happened, the potential number of victims, the probability of victim survival, and potential hazards to rescuers. Interview witnesses, the reporting party, and the site supervisor to obtain this information. If work was being conducted in the confined space, request from the job supervisor the required Entry Permit which should contain information about the space, entrants, and potential hazards. Obtain Material Safety Data Sheets (MSDS) if available. Victim information should include the number of entrants, their location, age, sex, general health, special health conditions, what they were doing in the space, and what protective clothing and equipment was being used.

A full reconnaissance of the area may assist in obtaining information about the situation, including the number and types of entry points, potential or known hazards, etc. The information obtained by an accurate recon will assist in determining the type of operation (*i.e.*, search, rescue, or body recovery) and the need for special resources.

Determine if the situation is a potential rescue or a probable body recovery, based on the events which took place, the IDLH nature of the environment, the length of exposure, and other pertinent factors. Use a realistic risk-vs-benefit equation when developing the incident action plan and rescue plan.

Incident command

The organizational structure for confined space rescue incidents should be consistent with standard ICS principles. Proper management may require some level of unified command. Participants may include representatives from law enforcement, local utility companies, ambulance companies, OSHA inspectors, mine/entry teams, mine and tunnel safety authorities, private contractors, and others.

The command post should be located outside (but within line of site of) the immediate operational area. All personnel not immediately assigned to specific duties should be staged outside the operational area, but at an easily accessible location. The first-arriving unit should determine the appropriate command post and staging locations and have the dispatcher relay the locations to all responding fire department companies and other assisting agencies.

The IC should assign a trained safety officer, preferably someone who is trained in confined space rescue and understands what the hazards are and how best to address them. Assign a trained rescue group leader to develop a rescue plan and supervise overall rescue operations. Assign a medical group leader to establish treatment for the victim after removal from the confined space (and, as necessary, any rescuers who become trapped or exposed to IDLH substances). If victims are missing, consider the establishing a search group leader.

Establish an exclusion zone (the danger zone into which no personnel should enter without suitable safety equipment). Establish an operational zone (a 100-ft. perimeter around the exclusion zone), in which primary and backup entry team operations and support operations will be conducted. Outside the operational zone, establish an equipment pool, logistics, rehab, and entry control. Outside the perimeter of the incident, establish a zone for media, bystanders, and public officials who will often gravitate toward extended deep shaft operations. Use fire line tape and other barriers. It may be necessary to post police around the perimeter to keep bystanders at a safe distance.

Training and qualifications

Generally speaking, only personnel who have completed a recognized confined space rescue course, and who have maintained the required level of continuing education, should attempt entry to conduct confined space rescue operations. It would of course be beneficial if the incident commander has an equal level of training, but this is admittedly a rare situation and is not absolutely necessary to command the incident. However, it is prudent policy to ensure that chief officers receive confined space first responder training. This will improve their ability to perform an accurate size up, recognize hazards, and run a safe and effective rescue operation. Additional training and certifications (examples might include mine/tunnel rescue, Rescue Systems I, dive rescue, swiftwater rescue, etc.) may be necessary for the entry teams to safely deal with specific hazards which may be found in local confined space environments.

First responder actions

For first responders who do not have the advantage of formal confined space rescue training, yet who may be confronted with an immediate rescue situation prior to the arrival of a certified confined space entry team, initial efforts should generally concentrate on the following priorities: (1) preventing additional victims; (2) establishing an effective incident command structure; (3) considering ways in which rescuers may safely initiate indirect rescue operations without entering the confined space.

Indirect rescue methods may include such things as lowering harnesses and ropes to the victim for assisted self rescue, lowering SCBA to the victim to provide respiratory protection if they are conscious, lowering air hoses and blowers to ventilate the space and provide fresh air ventilation, and other tasks that will assist the victim without placing rescuers inside the danger zone.

Naturally it is difficult for firefighters and other rescuers to resist the first impulse to enter the danger zone to save a life. This is where good training, a thorough understanding of the risk-vs-gain equations, and strong command and personnel discipline are required. In these situations, the IC has to keep in mind that committing unprepared first responders into a confined space may result in a far worse rescue situation if they become trapped or overcome inside it.

Monitoring prior to entry

The atmosphere of all confined spaces should be tested prior to any entry operations, and prior to working near the entrance. Remember that toxic gases may emanate from the space and contaminate surrounding areas. Monitoring should continue throughout the incident, and a written record should be maintained until the incident is over. Self-contained breathing apparatus (SCBA) or supplied-air breathing apparatus (SABA) should always be worn by rescuers in the shaft and near the opening when the atmosphere is IDLH or is potentially IDLH.

In atmospheres below 19.5% oxygen, the use of self-contained breathing apparatus or supplied-air breathing apparatus is required by law for workers and rescuers in the United States.

Supervisors who allow it to occur may be subject to prosecution, and agencies may be subject to citation (this has already occurred as a result of previous confined space incidents).

Respiratory protection

It may be impossible for a rescuer to wear standard firefighting SCBA on their back during entry into a narrow space. Rescuers should not enter confined spaces if they have to remove their SCBA to fit through the entrance. Lowering the SCBA bottle/back pack after the rescuer is not acceptable and is strictly forbidden by some regulations. The justification is simple. There have been cases where the bottle/backpack was dropped, ripping the mask off the face of the rescuer, who was then exposed to IDLH atmosphere, sometimes with fatal results.

Although supplied air systems are the most effective respiratory protection for confined space rescues, even they may prove to be cumbersome for rescuers operating in some narrow confines. Consider all the objects which must be passed into the entrance of a confined space to support some rescue operations: Ropes, hardware, tag lines, air lines, air blower tubes, atmospheric monitors, hardwire communications lines, lighting, patient extraction litters and harnesses. These items will not only clog the entrance to the space and limit emergency egress for the rescuers, but they will reduce the amount of available light and present additional entanglement hazards for the rescuer and victim.

Vertical entry and egress

An approved rescue harness should be worn by all primary and backup rescuers working near the opening of any vertical-entry confined space. At a minimum, the harnesses should be NFPA-certified for high angle rescue operations and capable of suspending and maintaining a rescuer in an upright position

Rapid intervention

Back-up rescuers should remain at the ready outside confined spaces to provide immediate assistance in case of emergency. There should be at least one backup rescuer for each primary entry team member. Backup entry teams should wear appropriate protective equipment, including a separate breathing air source available and ready for immediate use. Essentially, they should be wearing the same equipment as the primary entry team.

Entry

Only approved (intrinsically safe) lighting and electronic equipment rated for hazardous atmospheres should be used in confined spaces. An effective means of communication between personnel inside the space and the back-up rescuers should be maintained throughout entry. If hazardous materials are involved, standard hazardous materials procedures should also be enforced.

Viability of the victim should be evaluated prior to committing personnel into a confined space situation. If conditions are such that survival of any victims is considered impossible, great care should be taken in deciding whether to risk the lives of rescuers to recover bodies and perform any other non-life saving tasks. The decision to commit entry teams will be even more difficult and important when the exact condition of the victim and other conditions within a confined space are not known. Rescuers have died under these conditions, in part because of the urgency that accompanied situations in which a victim was known to be trapped in a confined space, but the exact conditions in the space could not be determined from the surface. This is where experience, training, and good judgment on the part of the IC and technical specialists is of critical importance.

Atmospheric monitoring

Atmospheric monitoring of IDLH spaces is usually done with direct reading instruments before rescuers are allowed to enter the confined space. Remember that all confined spaces are considered IDLH until proven otherwise. Whenever a person crosses the plane of the entrance, they are considered to have entered the space, or the *exclusion zone.* To prevent accidental exposure to IDLH products, avoid placing any body part in or near the opening of the space to assess the situation or to communicate without proper protection.

Conduct atmospheric testing of the surrounding area outside and near the point of entry to provide a baseline for further testing. If the entrance is covered, consider taking the initial reading with the cover in place until IDLH protection and ventilation are ready. If it is necessary to remove the cover to test the interior, the cover can be opened enough to allow for a remote probe to be placed into the space. Test from the point of entry and preferably every four feet vertically until reaching the bottom of the space. This may require the use of ropes to lower and raise the testing instruments. Do this in accordance with the manufacturer's instructions for obtaining samples. Then proceed with ventilation operations.

Atmospheric testing should be conducted continuously through the rescue effort in all potential IDLH confined spaces. Monitoring by a three or four range monitor should minimally include testing for oxygen percentage, percentage of combustibles (% of LEL), and toxicity. The atmospheric monitor must be intrinsically safe to use inside the space and must be calibrated per manufacturer's specifications.

A minimum of three monitors should be used during the incident, and all should be calibrated in clean air prior to use. Monitors should be assigned to

1) The primary entry team
2) The back-up entry team
3) At the point of entry

Ventilation

Ventilation of confined spaces requires consideration of the following factors:

- Is the victim unconscious? Is their unconsciousness due to a physical injury or the atmosphere in the confined space? Is the proper equipment on scene to safely perform the operation?
- Ventilation, once started, should continue throughout the operation unless it creates additional hazards.
- When ventilating, consider where the exhausted atmosphere is going. Is it IDLH, and will its removal create a hazardous atmosphere somewhere else?
- Ventilate where it will do the most good for the operation.
- Will the introduction of air into a space that contains a flammable gas bring it to within its explosive limits, thereby increasing the hazard of ignition?
- Entry into a space should be considered only after complete ventilation and continued monitoring of the atmosphere has demonstrated that it is safe.
- Blow fresh air in instead of sucking toxic fumes out. Consider additional openings that may help with ventilation and that will not counteract positive pressure ventilation.
- Care should be taken to prevent recirculation of contaminated air or circulation of exhaust gases from generators or vehicle exhaust into the space.
- Consider the use of built in ventilation systems, or positive pressure ventilation.
- Ventilate at all levels due to possible stratification of gases, especially in deep shaft situations. Air streams need to create continuous turbulence throughout the space to achieve proper ventilation/air exchange.
- Water fog nozzles do not produce small enough water droplets to inert a deep confined space atmosphere. They may also increase flammability due to increase air flow into the space (and may even drown the victim if not used carefully).
- Remember, the only way to assess ventilation effectiveness is to continuously monitor the atmosphere!

Lock out, tag out

Lock-out, tag-out, blank-out procedures are required to prevent severe life hazard resulting from accidental reactivations of machinery or the introduction of toxic substances.

- All electrical, mechanical, or other forms of energy must be shut down and de-energized prior to entry.

- All valves, switches, gates, or other control devices must be locked out with a keyed padlock and tag that says "DO NOT REMOVE" or "DO NOT TOUCH."
- Hydraulic lines and pipelines must be blanked or blinded by disconnecting or using provided steel plate blank out system. The system shall then be bled to assure deactivation.
- The key should stay with the person who places the lock or given to a responsible person, such as the safety officer, attendant, or the incident commander.
- If the device cannot be secured and locked out it is necessary to tag the switch and station an entry team member at the device or switch to prevent activation.
- Locate a responsible party intimately familiar with the systems to assist.

Point of entry control

Point-of-entry control is an absolute must in confined space entry. The following points should be considered:

- Names of entry team members, their function, and time of entry should be logged on a written record.
- Maximum entry time for each individual entrant must be closely watched and strictly enforced.
- Written records of atmospheric monitoring must be maintained throughout the incident.
- A backup entry team equal in number and equal in protective clothing and equipment must be ready at the entrance to the space at all times. Their names should be recorded.
- Visual monitoring of entrants should be maintained at all times.
- The assigned safety officer must maintain oversight over these operations.

Standard positions

The following positions, with their attendant duties and responsibilities, have been found to be essential for successful management of many confined space rescues.

Rescue group leader or operations officer. Supervises the overall rescue operation, including all aspects of search and rescue that might be required in addition to the confined space entry.

Entry group supervisor. Has overall responsibility for safe confined space entry operations.

Entry team manager. Supervises the attendant, the rigging team, and the litter team, and other critical operations at the point of entry.

Confined space safety officer. Responsible for ensuring adequate safety procedures, for alerting the IC to any unsafe situations, and for halting any unsafe operations.

Air supply manager. Responsible for ensuring that a continuous supply of air is maintained to all rescue entrants and the victim (when umbilical air is supplied to the victim) throughout the incident. The air supply manager is also responsible for managing ventilation operations and breathing air resources (*e.g.*, mobile air units, breathing air caches, etc.).

Primary and backup entry teams. Each entry team may consist of one, two, or more rescuers, depending on the situation. Generally, for confined space rescues, two-person entry teams are desirable, but this number will obviously have to decreased to one if that is all that the space will allow. It is also acceptable to have more than two entry team members.

Attendant. Monitors the condition and progress of the rescuers in the space from just outside the point of entry.

Rigging team. Generally consists of the US&R company or rescue company, or another team qualified to supervise high angle rope systems and other methods of inserting and removing entry teams and the victim from the space.

Line tenders. Consists of a minimum of one tender per entrant, to tend the safety rope of entry teams as they make their way through the confined space.

Litter team. Consists of four personnel to transport the victim from the point of entry to the medical group.

Ventilation team. Works under supervision of the air supply manager to ensure continuous and effective fresh air exchange in the confined space.

Air source manager. Works under the supervision of the air supply officer to ensure a constant supply of breathing air from mobile air units, SCBA caches, etc.

Assistant to the attendant has been found to be a necessary position to allow the attendant to do his/her job. The assistant is responsible for documenting atmospheric monitoring, coordinating various essential tasks for which the attendant has no time, and for generally allowing the attendant to concentrate on their duties.

The proposition that 19 people are required to properly manage a confined space rescue may see outlandish. For some departments this is the equivalent of a second alarm structure fire assignment. And it is a significant commitment of resources for the largest of fire departments. Be that as it may, years of experience in actual confined space incidents (as well as review of disastrous rescue attempts which resulted in fatalities to would-be rescuers) have demonstrated that these positions are sometimes required to ensure reasonable safety for the entry team and effective rescue of the victim for certain "working" confined space rescues.

However, we live and work in the real world, where some fire departments may not have sufficient numbers of trained personnel to staff all the positions, and where a victim's dire situation may compel firefighters to take a higher risks by entering before a full-blown confined space entry system can be set up. This is why the LACoFD and other fire/rescue agencies have adopted a protocol known as "limited confined space entry."

Limited confined space entry

When the victim's predicament is such that they may suffer permanent damage or death if they remain in the confined space for another moment, when they are visible from the portal, if the portal is sufficiently large for firefighters to enter while wearing SCBA, when there simply is not time to establish the standard confined space entry system, and when the incident commander determines that the risk-vs-gain equation is such that a high-risk rescue must be attempted, firefighters (including first responders who are not formally trained in confined space rescue) may opt for a "limited" confined space entry.

The limited confined space entry requires a minimum of three persons, which will even allow a three-person first responder engine company to attempt the rescue—one attendant (the company officer), one primary rescuer, and one backup rescuer. This rescue procedure complies with OSHA standards (if barely), and it allows an immediate rescue attempt with just three members.

However, it must be emphasized that a limited entry is only to be attempted when all of the following parameters are met: the victim is visible from the outside of the confined space; the entrance will allow the primary and backup rescuers to wear SCBA on their back during the entire rescue (and they will wear the SCBA); the victim's predicament warrants the extra risk to firefighters; it will be assumed that there is an atmospheric hazard until proven otherwise through monitoring; and full turnouts (or a US&R/Rescue ensemble meeting NFPA 1781) must be worn in case of flash fire or the presence of contaminants.

If the victim is not visible, the space is too small to enter with SCBA on the backs of the rescuers, the victim is reasonably safe for the time being, or if no SCBA are available, the incident commander should wait until sufficient confined space-trained resources arrive to set up a full-blown confined space entry system.

The attendant (officer) has overall responsibility for coordinating a safe confined space entry operation and extraction of the victim. The officer must continue the response of other supporting units until the rescue is completed and all personnel have exited the high-hazard zone. He/she must assume that there is an atmospheric hazard and possibly other hazards present, and therefore mandate that the primary and secondary rescuers wear full turnouts (or a US&R/rescue ensemble meeting NFPA 1781) and SCBA during the entire rescue.

The primary rescuer must comply with the aforementioned PPE requirements and remain visible to the attendant during the entire rescue. The rescuer must be attached to a safety rope or a vertical extraction device that will all the backup rescue and attendant to pull them out of the confined space if anything goes wrong.

The backup rescuer must comply with all the above requirements, and must remain in a constant state of readiness to conduct a rapid intervention if the primary rescuer becomes trapped or disabled.

Deep shaft rescues

A particularly dangerous type of confined space rescue is deep shaft rescue. OSHA defines a deep shaft as any excavation more than 20 ft. in depth whose depth-to-width ratio is 5 to 1 or greater. The general term "deep shaft" includes all manner of vertical mines, ventilation holes for subway systems, bore holes for deep pilings at construction sites, old wells, other vertical below-ground passageways, and abandoned septic tanks, natural shafts, caves, sink holes, and any other spaces that meet the 5 to 1 depth to width ratio, even if they are less than 20 ft. in depth.

Safe and effective rescue in the deep shaft environment requires a mix of disciplines, including confined space rescue, high angle rescue, collapse rescue, mine and tunnel rescue, and others. If the shaft is filled with water, dive rescue or swiftwater rescue techniques may also be required.

Any rescuer sent into a confined space is operating in a foreign and often hostile environment. He is vulnerable to many lethal hazards, which can strike without warning. Deep shaft rescues are doubly dangerous because small openings and tight quarters may prevent others from helping if trouble occurs.

If the operation requires rescuers to be lowered into a shaft or other deep space, a trained high angle entry team, supervised by a qualified officer, should be assembled and assigned the task of establishing and operating a high angle system. A thorough review of the conditions and requirements should be made by an officer or other member trained in Rescue Systems I, Rigging for Rescue, or another recognized high angle rescue course. This individual may be assigned as extrication officer. The extrication officer should be responsible for developing and operating a safe and effective high angle entry/egress system, with appropriate safety redundancy (*i.e.*, dual rope systems, etc.) and backup rescue capability.

Trench and excavation rescues

Workers trapped in collapsed trenches and excavations are a typical rescue emergency, accounting for about 100 worker deaths each year and dozens of injuries (and an unknown number of "near misses"). These emergencies may be more prevalent in communities experiencing high levels of construction and renovation, and where dangerous soil conditions exist. (See Appendix 7)

About Trenches and Excavations

A trench is a narrow excavation that is deeper than it is wide, whereas an excavation is a man-made cavity or depression in the earth's surface. Each is subject to failure (and then a greater then 50% chance for secondary failure afterward!) due to a number of factors that may affect the rescue operations.

Sometimes the victim of a collapse is completely buried, and the first challenge is to stabilize the site and locate them before time runs out. In other cases the worker is partially buried and still visible at the surface, leaving firefighters with the problem of extracting them without causing a secondary collapse, further injury, or a fatal mishap that takes the lives of rescuers or the victim.

In responding to dozens of trench and excavation collapses through the years, this author has found one common trait: The need to move fast to keep the victim's airway clear, reduce the pressure on their chest (to allow them to continue inhaling), and treat them for other injuries, while moving with a sufficient amount of caution to prevent a secondary collapse while stabilizing the area, extracting soil and debris from around the victim, and removing them and any rescuers. In other words, trench and excavation rescues (much like structure collapse operations) are both a race against time and a sort of engineering problem where one wrong move might bring the whole thing in on the victim and rescuers, making the situation far worse.

As is often the case in technical rescue, the key to success begins with well-trained first responding firefighters prepared to assess the scene; request additional equipment and materials (*e.g.*, lumber for shoring, tools for digging, etc.), isolate and deny entry by unauthorized persons (including, in many cases, construction workers unfamiliar with the hazards of indiscriminate rescue attempts—like digging for victims with back-hoes and trenching machines); begin moving back the spoil pile (if it's too close to the trench lip); and begin stabilization operations while uncovering or protecting the trapped victim's airway and ventilating the site. Then, to reinforce the first responders, rescue or US&R units trained and equipped to advise the incident commander strategy and tactics; perform advanced shoring and digging operations (including the use of dirt vacuums and air knives); establish vertical lifting operations as necessary; monitor the site for atmospheric hazards, to request and use devices like Hydro-Vacs; and to provide rapid intervention capabilities.

Protection from collapse

The danger of trenches and excavations are typically reduced through one or more of the following methods:

Sloping. Cutting back the walls to a maximum allowable slope, generally the angle closest to perpendicular at which the particular soil type will remain at rest.

Shoring. Placing timber shores, screw jacks, air shores, or speed shores in the trench to support the walls and prevent movement. Shoring should be capable of handling existing and expected loads, including those imposed during a rescue operation. Whether in a routine construction situation or in a rescue, the shoring should be built beginning near the surface and working downward to more dangerous depths. Here are standard components of shoring:

Plywood sheeting. 1⅛" or ¾" thickness (obviously thicker is better for stabilization) to hold back running soil and loose material.

Uprights. These are vertical supports of 2" thick (or thicker) timber, installed parallel with each other, extending from the top of trench to within 2 ft. of the bottom (except in running soil, where uprights should touch the bottom).

Struts. Horizontal supports to spread the pressure exerted by the timber jacks, screw jacks, or hydraulic jacks that will be used to pressurize the walls.

Trench shield. Inserting steel plates welded to a heavy steel frame, from which the workers can do their job in relative safety.

In trenches greater than 4 ft. in depth, many states require escape paths (ladders, stairways, or ramps) to be located no further apart than 25 ft. along the trench, and at each end to allow rapid escape in case of a collapse, flood, or other mishap.

As a general rule, trenches and excavations of depths of 5 ft. or greater (and those less than 5 ft. deep in hazardous soils) should be sloped, shored, or otherwise protected. The spoil piles should not be closer than 2 ft. from the edge of the trench.

Arrival/size-up

Establish an exclusion zone 50 ft. around the collapse (into which no personnel should enter without suitable safety equipment, without a job to do, etc); an operational zone (a 150-ft. perimeter around the exclusion zone), where essential entry team activities, shoring, the command post, and other necessary operations will be conducted; and a support zone for 300 ft. around the site for the equipment pool, logistics, rehab, entry control, etc. Outside the perimeter of the incident, establish a zone for media, bystanders, and public officials who will often gravitate toward extended deep shaft operations. Use fire line tape and other barriers. It may be necessary to post police around the perimeter to keep bystanders at a safe distance.

What type of collapse occurred? Was it a *spoil in* (where the spoil pile falls into the trench, sometimes causing part of the lip to shear away as well); a *lip-in* (where too much weight at the edge causes the edge to fall away); a *shear in* (where the loss of moisture in a long-opened trench causes the wall to fail); or a *slough-in* (where a section of the wall fails, often with an undercut area left)?

What is the soil type? Is it *compact*, *disturbed*, *running*, *saturated*, or *virgin* soil? Each has its own characteristics, with a higher level of danger of collapse from *running*, *saturated*, and *disturbed* soils. It is Type A, B, or C...or rock? (See Appendix 7)

Are the stress cracks in the soil indicating an impending secondary collapse? How deep was the trench? Were there utilities that could rupture in a collapse? What equipment is available for shoring and digging?

Are victims are missing in the collapse area? If so, how many? Where were they last seen? What were they doing? Was there a pipe, an auger, shoring, or other items that may have provided a shield against the soil (*e.g.*, is it possible the victim is in an air pocket or a void?).

Always assume the victim may be alive until proven otherwise. In too many cases, the victim has been alive and awaiting help while completely buried, while rescuers at the surface were writing them off as obviously dead. As professional rescuers, it is our job to understand that there is a potential for live rescue even when the non-professionals are declaring the accident unsurvivable.

Incident command

The incident commander should identify the incident, request additional resources as needed, and establish the following positions:

- Operations
- Rescue group supervisor (rescue/US&R company captain?)
- Entry team manager (rescue/US&R officer)
- Attendant (rescue/US&R member)
- Safety officer
- Primary rescue team (rescue/US&R members)
- Rapid intervention team (rescue/US&R members)
- Backup rescue teams (for personnel rotations, rescue/US&R members)
- Shoring teams (rescue/US&R/truck company members)
- High-angle system manager (where appropriate, rescue/US&R truck company)
- Liaison officer (with construction companies, police, etc.)
- Medical group leader
- Air supply manager (if appropriate)
- Litter team
- Cutting team (to cut shoring dimensions)
- Public information officer

The IC should ensure that hazards are controlled, including removal of workers from the trench/excavation; elimination of vibrations from traffic, trains, heavy equipment; crowd control; keeping news helicopters a safe distance above the scene to eliminate noise; establishing lighting and ventilation. He should establish a staging area, an equipment pool, a shoring material pool, a cutting station, a rehab area, a media area (away from the collapse zone but where the media can still do their job of reporting), and ambulance staging.

Tactics

Approach from the ends of the trench to view the side walls. Place ladders at the ends and at 25-ft. increments. Place edge protection (4x8 sheets of plywood or 2x10 or 2x12 timbers placed along all edges of the trench/excavation to distribute the load of rescuers and equipment).

Create a safe zone for the victim, or at the *point last seen* (PLS). This can be done by placing one "quick shore" (sometimes with an upright to spread the weight) on either side of the victim or the PLS, within 4 ft. of each other. If time and material allows, slip a sheet of plywood between the upright (or quick shore) and the soil. But don't hesitate if there are no plywood sheets immediately available. What we want to do is pressurize the walls around the victim to prevent secondary collapse, giving us just enough time and space to unbury the victim's head or to keep their airway clear.

If necessary, a ladder can be placed here and a primary rescue team member can climb in (with a tag line attached to them as a tracer) and begin working to immediately uncover the head and chest. Or, if a vertical approach is necessary, the rescuer may be lowered into the safe zone via rope system. The rescuer should not leave the safe zone until the adjacent areas have been properly shored. In some cases the victim can be uncovered and raised vertically out of the safe zone without ever expanding the shored area beyond the safe zone.

If the victim is visible, provide SCBA mask (umbilical is best) or other airway protection and a "tracer line" to victim in case of secondary collapse, remove soil from the head and chest first, and work downward. Consider requesting a helicopter for medevac once the patient is rescued. Consider requesting an emergency room board-certified physician at the scene to advise and assist paramedics in evaluating and treating the victim for crush syndrome and other significant trauma issues.

If the victim is missing, consider using a *life detector, trapped person locator,* or another acoustic/seismic sensing device to detect signs of life. Consider requesting US&R search canines to detect the location of human scent. Consider inserting teams of rescuers with hand tools to begin digging for victims (after the walls are stabilized sufficiently). Do not assume the victim is dead until they are actually located and physically examined by a paramedic, EMT, or other qualified member.

Once the victim is located, consider establishing vertical lifting capabilities (build harnesses on victim as they are uncovered, starting with wrists, chest, pelvic, etc.), so they may be lifted vertically if that proves less time consuming and safer. This may require the placement of ladders (including heavy-duty-rated aerial ladders) or cranes as high points over the rescue site. Consider the use of hydro vac trucks (at least two, perhaps three to allow for rotation as they are filled up or in case of mechanical failure) to speed the process of uncovering the victim. No vibrations for 300 ft. in all directions!

Monitor the atmosphere and begin ventilation of the trench/excavation. Consider requesting a hazardous materials unit to conduct and supervise these tasks, as well as other related functions.

Develop a shoring plan and start putting shoring in place. Consider the following:

- Protection for the side walls. Some teams begin by inserting uprights, and then they install the sheeting by sliding the plywood over them from the opposite sides of trench. In theory, each upright provides 4 ft. of force outward from the pressurized strut, thus creating 4-ft. safe zones. In this case, the maximum travel distance from the center of any upright is 4 ft. in some states.
- Overlapping of zones increasing the safety factor for rescue operations.
- In hard compact soils, horizontal separation of shores and uprights can be a maximum of 8 ft. in some states.
- In loose running soils, the horizontal separation of shores should be no more than 4 ft. in some states.
- Place the horizontal struts, beginning 18 to 24 in. below the trench lip and progressing downward. Place the next strut 4 feet maximum below the top strut, and then place additional struts every four feet working downward. Place the last strut within 24 in. of the trench bottom.

Assign an appropriate number of primary rescuers (with a rapid intervention team in place) into the trench/excavation once it is shored, in order to conduct the uncovering operation and other essential tasks. In some cases it's best to work from the sides, digging away the dirt and placing it elsewhere in the trench (or in buckets attached to ropes, allowing firefighters working above the trench to haul out the soil by bucketful). Using a dirt vacuum or hydro vac truck often expedites the removal of dirt from around the victim and from within the trench or excavation. Using an air knife to loosen clumped or compacted soil may further reduce the time needed to uncover the victim.

In one recent case in Los Angeles, LAFD firefighters were able to work from an open ended trench cut into a hillside, using shovels to rapid and incessantly pull away soil from around a construction worker buried to their chest. Meanwhile a US&R company and truck company were creating a safe zone around the victim and further shoring up the trench. Practically an entire first alarm structure assignment was on the scene, and the firefighters took turns furiously digging away the dirt from around the victim, rotating into the digger positions about every 5 or 10 minutes (ensuring that fresh firefighters were always digging, and moving them out as soon as they started to slow down).

The worker was mostly unburied before all the shoring was installed. The firefighters were able to pull the victim out laterally (in fact, they practically walked out). The entire rescue was completed in about 40 minutes because the firefighters were able to get close to the victim from the open end of the trench; simultaneous shoring and safe zone operations were taking place; the incident commander organized a "round robin" rotation of firefighters who dug furiously and nonstop; and the IC also took care to continue the response of the US&R resources to provide reinforcement and rapid intervention capabilities.

We aren't always fortunate to find victims trapped under such favorable conditions; many trench and excavation rescues take hours to complete...sometimes half a day or more. Sometimes victims are pinned between a slough and an unstable building; sometimes they are buried in a collapse that ruptures a water or gas main; sometimes they are buried so deeply that it may take hours of dangerous digging just to get near them; sometimes they are suffering the effects of crush syndrome or mechanical asphyxiation; and sometimes we simply cannot find them beneath tons of soil and debris.

Just as in other technical rescue situations, the incident commander and the first responders need to be prepared for any eventuality and have a plan in place to address it. Rescue companies and US&R units should be well-practiced at the skills and tactics to make these rescues happen, and the US&R or rescue company officers need to be prepared to advise the IC on the best strategies based on the particular conditions.

Marine Rescue Operations

Marine disasters and marine emergencies include situations where victims are in need of rescue on the open sea (or, in some cases, lakes) resulting from airplane crashes, capsized boats, boat collisions, boat fires, etc. Many fire departments have some jurisdiction over marine rescue operations, and some departments participate in multi-agency marine disaster response systems. This helps ensure a coordinated response of marine-based, air-based, and land-based fire/rescue/EMS resources to effect timely rescue and recovery, provide rapid transportation to shore, and process (as necessary) through a "landside" multi-casualty incident system. In terms of rescue, this can mean several potential tasks for fire/rescue units.

Ultimate responsibility for off-shore search and rescue operations along most U.S. coasts is assigned to the Coast Guard, usually identified as the Federal Search and Rescue Coordinator. In some cases—including the coastal waters off places like Los Angeles County—there are agreements recognizing the first on-scene agency (including local fire/rescue agencies) to establish command. Upon arrival of the Coast Guard, unified command is sometimes utilized thereafter.

Regardless of the command issues, some fire department rescue companies and US&R units have responsibility for conducting the actual rescue operations in the water during marine disasters. Some of these units may be responding on boats, IRBs, or PWCs, while others may respond by helicopter. Some are dive rescue teams, others are surface rescue teams, or rescue swimmer teams. The first goal is to get to victims before they slip beneath the surface, and to provide them with flotation. The next goal is to get them out of the water so they may be removed from the hazards (*e.g.*, cold water and LPMAs [Large Predatory Marine Animals]), placed on boats, and taken to shore for triage and treatment.

For rescue and US&R companies, this may involve extensive training in marine rescue, in the safe methods of helicopter-based deployment into the ocean environment, in personal survival (including survival in the open sea and/or helicopter ditching procedures and HEEDS/dunker training) skills, and other skills unique to marine rescue (see Appendix 8). They also need the right equipment to get the job done safely.

Vehicle Extrication

Vehicle extrication is almost certainly the most common type of technical rescue performed by the fire service every day. The basic concepts of vehicle extrication are understood and practiced by most fire departments, to the point where it's practically a matter of routine. So it may be the unusual cases that capture the attention of the media and public. The following are some basic points of vehicle extrication.

En route

Based on the dispatch information, consider the need for additional resources. If hydraulic rescue tools may be required, insure they are en route with sufficient manpower to operate them and to perform other essential tasks. One engine company and a typical two-person paramedic unit are generally not sufficient to properly extricate a patient, provide a safety hose line, treat the patients, and provide effective command at the same time. Don't hesitate to request sufficient units to handle the manpower demands.

If the incident is reported to be something more extensive than a typical vehicle entrapment (*e.g.*, and overturned vehicle, a "vehicle over the side" on a mountain road, a semi-truck on top of a car, a person trapped in a semi-truck accident, a train-vs-auto collision, a vehicle into the water, etc.), request sufficient rescue or US&R resources to get the job done in a timely manner, with a minimum of reflex time. Be prepared to use rescue air bags for lifting, spreading, and stabilization; cribbing and shoring for vehicle stabilization; special cutting tools, fiber-optic viewing equipment to visualize entrapped extremities during extrication; and other special capabilities to expedite the rescue.

If the incident is a vehicle beneath a semi-truck or other large vehicle, request the response of at least one heavy wrecker (preferably two). It's important to specify (when possible) a wrecker equipped with a rotating, extending boom. Some wreckers have a radio control which allows very precise lifting maneuvers while the operators observe the movement inch-by-inch. The additional heavy wrecker may be required to provide a tandem lift or for additional stabilization and lifting. Some departments like the Los Angeles Fire Department have their own heavy wreckers that perform a dual role as rescue or US&R units. This helps ensure highly experienced operators who understand the concepts of rescue.

Arrival on scene

- Size-up for special hazards and needs
- Determine extent of entrapment and develop rescue strategy
- Stabilize vehicle (flatten tires, chock wheels, use cribbing and shoring to support overturned vehicles, remove fire hazards) to make the scene safe for the victim and rescuers
- Provide safety hose line and traffic control
- Begin simultaneous treatment and extrication
- For vehicles with air bags, take appropriate safety measures. Attempt to deactivate the air bag system. Before cutting or moving the steering column, wrap chain around the wheel to secure it from accidental deployment of the air bag. Consider the potential hazards from side air bags, etc.
- Establish a shoring/cribbing team if the vehicle must be lifted or rolled up to free a victim. Do not lift or move vehicles without cribbing and shoring (lift an inch, crib an inch!) to prevent catastrophic movement of the vehicle.
- Assign a safety officer
- Follow standard techniques and methods, but remain flexible to adapt to changing situations
- Have an alternate plan. It's customary during complicated rescues to develop "backup" rescue plans in case the initial tactics and strategy prove unsuccessful. In keeping with this principle, make alternate plans to be implemented if the situation suddenly deteriorates, or if unexpected complications arise.

River and Flood Rescue

River and flood rescue (or *swiftwater rescue*) represents one of the most dramatic examples of the new rescue revolution. During the past two decades, swiftwater rescue has become a standard operation for many fire departments across the nation and around the world. The public has come to expect firefighters to pluck people from raging rivers, flooding streams, flash floods, and concrete-lined flood control channels that flow faster than any natural river on Earth.

No longer are the citizens surprised to see firefighters stepping off engine companies and truck companies wearing running shoes and PFDs (personal flotation devices) over their lightweight Nomex wildland firefighting gear, backed up by other firefighters wearing dry suits and rescue helmets arriving in rescue companies or by helicopter. It's commonplace to see firefighters successfully plucking victims from rivers, streams, and flash floods under conditions that might have stymied them just a decade or two ago, before the availability of formal swiftwater rescue training, elaborate swiftwater rescue systems, and before many fire/rescue personnel had been made aware of the life-threatening dangers associated with moving water.

With the advent of the age of waterway rescue preplans, designed to intercept victims from flood control channels moving 20 mph to 30 mph before they're forever lost, it's not uncommon for the news helicopters to film fire department units strategically deployed at designated rescue points for miles ahead of a victim, all poised to attempt plucking the victim from the water as the victim is swept past each rescue point. And now that firefighters are being trained to conduct contact rescues in a reasonably safe manner (and with the proper personal protective equipment), it's not uncommon to find specially-equipped firefighters entering flood control channels and other treacherous waters to conduct contact rescues, which sometimes entail grabbing the victim and protecting them while being swept downstream to the next rescue point, where fighters on bridges and the shoreline will attempt to pluck them from the water.

It's increasingly common to see well-trained fire department helicopter rescue crews successfully snatching victims from flood waters under conditions that defy shore-based or boat-based rescue operations. Rescue companies and US&R units are more commonly equipped with motorized inflatable rescue boats that can be deployed to rescue victims from wide rivers or flooded neighborhoods, or which can be controlled by simple or elaborate rope systems in water too powerful for motorized boats.

Many fire departments have established swiftwater rescue teams equipped with the latest gear and trained to perform the most demanding and hazardous rescues. Sometimes these teams consist of personnel normally assigned to US&R companies or rescue companies, hired on their off-duty days to augment existing resources by staffing swiftwater rescue teams during major storm periods. In other cases, rescue company and US&R unit personnel are assigned to augment the existing staffing of helicopter rescue teams to create five-person or six-person helicopter-based swiftwater rescue teams that are prepared to handle the worst-case scenarios.

Some cities and counties have established multi-agency swiftwater rescue systems, whereby the multi-tiered swiftwater rescue capabilities of multiple agencies are coordinated to ensure seamless swiftwater search and rescue operations. In some of these systems, swiftwater rescue teams are strategically deployed to provide regional coverage, and these teams respond across city or county borders based on the need to get the closest, most well-trained rescuers to the scene when victims become lost or trapped in swiftwater and flood situations.

As a result of all these efforts—tailored to the conditions and needs and available resources of each region, county, or city—lives that otherwise would have been lost are now being saved practically on a daily basis, and certainly when major storms and flood events occur. (See Appendix 9)

Recognition and assessment of flood and swiftwater hazards

Swiftwater/flood hazard evaluation is sometimes a daunting task. To conduct a realistic evaluation, one needs to understand the dynamics of moving water, its expected behavior in natural and manmade surroundings, and its effects on victims and rescuers. It also helps to recognize signs of trouble that aren't obvious to the untrained eye.

Swiftwater hazard evaluation should include not only the physical environment such as terrain and drainages, but also a study of weather patterns, the local history of swiftwater and flood rescue emergencies, and other interpretive factors. The patterns that emerge may identify geographic locations where trouble can be expected under certain conditions.

If any of the following hazards are found, there is a potential swiftwater/flood rescue problem:

- Natural rivers or streams
- Flood control channels
- Above-grade levees
- Dams
- "Arizona" crossings
- Coastal subduction zones that may generate tsunamis and/or cause the land to subside
- Offshore thrust faults (tsunami hazard)
- Regions subject to flash floods, rockslides, or mud and debris flows

- Coastal areas subject to hurricanes
- Coastal zones subject to undersea landslides in deep offshore canyons
- Proximity to active volcanoes (mud and debris flow hazard)

Swiftwater rescue training

River and flood rescue training that meets NFPA 1670 or other recognized standards is usually sufficient to address most conventional and unconventional river and flood hazards. NFPA 1670 includes new standards for water rescue "career paths," including *Water Rescue Dive Technician, Ice Rescue Technician, Surf Rescue Technician,* and *Swiftwater Rescue Technician.* NFPA 670 also includes standards for awareness and operational levels in each of these water rescue-related disciplines. A number of states have extensive programs to train fire and rescue personnel to manage the consequences of floods, swiftwater situations, and even mud and debris flows.

Training sources include private companies that specialize in technical rescue or swiftwater rescue. A number of fire departments now conduct formal river and flood rescue courses as part of their standard curriculum for fire fighter training. Some fire departments include swiftwater rescue and US&R training in their recruit academies. And several community college fire science programs now offer river and flood rescue training.

"Jumpers" and Other Crisis Negotiation Rescue Operations

Fire departments, are called upon to assist at the scene of "jumpers" and other self-threat emergencies that Dr. (and Sergeant) Barry Perrou, head of the L.A. County Sheriff Department Crisis Negotiation program, calls "suicides in progress." In L.A. County and a growing number of other communities, law enforcement agencies are increasingly likely to call upon the local fire/rescue agency to provide backup and safety support for crisis negotiator working with subjects on top of buildings, bridges, cliffs, and other high angle situations; as well as in confined spaces and other locations that qualify as *immediately dangerous to life and health* (IDLH).

Tommy Langone, assigned to the NYPD's Emergency Services Unit (ESU) until his untimely death in the collapse of the World Trade Center towers after the 9-11 attacks, was an early practitioner of high angle rescue methods to protect crisis negotiators dealing with potential "jumpers."

Langone himself performed a daring "snag" of a man threatening to jump from the upper stories of a building, while other members of the ESU belayed him and two colleagues, and while other members of the ESU set up to catch the man from an apartment below. In a closely choreographed move after all other options had failed, and at the point it became apparent the man was indeed going to jump, Langone lunged for him and managed to grab him just as the man launched himself. Langone and their partners managed to hold the man while, emerging from the apartment below, other ESU officers reached up and pulled the man's legs, with Langone and their partners allowing just enough "play" for the officers below to pull the subject into the window, ending the incident. It was an example of how firefighters and police officers can use high angle rescue maneuvers to resolve "jumper" incidents.

This author has responded to many "suicide in progress" calls, including one where the subject attempted to take a firefighter off the bridge with him (long before there was any discussion of belaying firefighters and crisis negotiators before they approached subjects threatening to jump). In a number of instances, a coordinated effort by firefighters establishing high-angle rescue protection for negotiators, and establishing "fall bags" and other safety devices below (while the subject was purposely distracted), resolved the situation in minutes rather than hours.

In managing suicides in progress, Dr. Perrou refers to the so-called Five C's: *command*, *contain*, *communicate*, *coordinate*, and *control*. To this, firefighters and rescuers might add LCES. The point is not to allow rescuers to become victims. It's important, then, for fire departments to establish good working relationships with local crisis negotiators, and to work out plans to handle these incidents in a logical and effective manner. (See Appendix 10)

Crushing Problems and Recognition

Firefighters and paramedics know from experience and training that appearances can be deceiving when it comes to medical issues. Although classic Crush Syndrome normally doesn't occur until after about four hours of entrapment, there is always a danger of rapid hemorrhage or hypotension when the pressure of entrapment is released. This is one reason that the term "smiling death" has been coined for crushing injuries. The victim is frequently alert—and sometimes euphoric—as the time of their release nears, but may deteriorate rapidly or slowly on a seemingly unstoppable spiral toward death following their release. In the case of massive hemorrhage following the release of pressure, profound hypovolemia and hypotension may quickly bring about a patient's demise if precautions aren't taken *before* the pressure is released.

And if the crushing mechanism continued for much longer, the initial stages of true Crush Syndrome might not be far behind. Constant compression of a muscle mass in one or more extremities (resulting in continued arterial perfusion but poor venous return) robs the tissue of oxygen-carrying blood and raises the level of carbonic acid, followed by a shift to anaerobic metabolism. At a certain point, the pressurized muscle cells stretch to the point that myglobin begins leaking from their membranes. This, combined with heightening levels of potassium, calcium, albumin, and oxygen-free radicals, causes metabolic damage to the extremity. After several hours, the inside of the compressed extremity becomes a sort of high-pressure toxic soup trapped behind a dam.

As the pressure is released, the "dam" ruptures, spilling the toxic soup into the central circulatory system, back toward the heart, lungs, brain, and other vital organs. These chemicals can bring on a variety of secondary maladies, including ventricular fibrillation (in minutes), pulmonary embolism (minutes to hours later), adult respiratory distress syndrome (in hours), renal failure from solidified myoglobin that clogs kidney tubules (hours to days later), traumatic rhabdomyolysis or myoglobin and albumin in the blood serum (hours to days), and opportunistic infections (days to weeks later).

High pressure, brief duration crush problems

Examples include victims affected by explosive forces; those crushed by machinery, structural collapse, or heavy debris, who were quickly released; victims briefly trapped between the bumpers of two automobiles; and other high-pressure, short duration events. Higher survival rates may be seen when upper extremities and pelvic girdle are involved. Morbidity increases when head, abdomen, and chest are affected. These forces may result in traumatic amputation of extremities, in combination with traumatic crush affects to other parts of the body.

Low pressure, long duration crush problems

Examples include victims "lightly" entrapped in structural collapse; "lightly" entrapped by the partial collapse of trenches and excavations; certain vehicle extrication situations that meet this criteria; and some unconscious victims and other so-called "lack of spontaneous movement" situations. Survivability decreases when head, chest, and abdomen are involved.

High pressure, long duration crush problems

Examples include major, long-term extrication operations resulting from structural collapse, cave-ins, vehicle collisions, industrial accidents, etc. May be found in combination with high pressure, brief duration causes (*i.e.*, explosions followed by entrapment in the resulting collapse of structures). Post-rescue amputation is a common necessity when extremities are involved in high pressure, long duration crush incidents. Death prior to extrication is not uncommon when the head, chest, and abdomen are involved.

Medical considerations

Regardless whether the fire department is the provider of paramedic care of other advanced prehospital treatment, firefighters engaged in rescue should be aware of certain medical considerations that can affect the ultimate outcome of the rescue. Too many patients have survived hours without any form of medical intervention during difficult extrications, only to die hours or days later from complications that could have been avoided if the incident commander had recognized the need to treat certain conditions during the extrication.

In some cases the conditions at the scene (*e.g.*, confined areas, potential for secondary collapse, etc.) may prevent paramedics from establishing intravenous lines and performing other treatments during the extrication. This author (a paramedic himself for more than two decades) has been involved in a number of difficult and lengthy extrications where there simply was no room for paramedics to get into the space occupied by the victim; or where the conditions were such that medical treatment would have delayed the extraction and posed additional danger to the victim and rescuers alike. But, whenever, possible, it's an advantage to the patient if the incident commander and the rescue officers ensure the patient is being constantly evaluated from the medical standpoint, and is receiving appropriate treatment during the extraction process.

The principle medical concerns during rescue include crushing injuries, amputations, and others related to the trauma of entrapment. Traumatic crush injuries are associated with a variety of causes, including structural collapse (related to earthquakes, explosions, landslides, floods, construction accidents, and other causes); entrapment in traffic collisions (from passenger space intrusion or direct impact with moving vehicles); industrial accidents; entrapment in the collapse of trenches and excavations; and even situations in which the lack of spontaneous movement (*i.e.*, deep sleep, intoxication, or coma) results in long-duration pressure on tissues.

Treatment

Effective treatment for crush injury includes oxygen therapy, maintaining blood pressure and other vitals, and tactics like proper intravenous administration of crystalloid fluids while monitoring to maintain a urinary pH > 6.5 with timely infusions of sodium bicarbonate during the extrication process, however long it may take. Obviously the most effective treatment of all is to rescue trapped patients as quickly as possible within reasonably safe limits as determined for each situation.

The need for elaborate monitoring and medical judgment for crush injuries is one reason that paramedics across the nation are being trained to recognize and treat these maladies. It's been deemed sufficiently important that the LACoFD's medical director is requested by US&R captains and incident commanders to respond to many long-term extrication operations, to augment the paramedics who are already on the scene and to provide advanced treatment options that exceed that paramedic scope of practice. It's also why each of the nation's 27 FEMA US&R task forces includes two specially trained emergency room physicians and at least four crush syndrome-trained paramedics.

Aggressive medical intervention during extrication is a concept that should be considered by all fire/rescue agencies. Not only will it benefit citizens who are trapped, but it can also help firefighters and other rescuers who find themselves trapped by secondary collapse and other adverse events.

Conclusion

The new paradigms we have seen in rescue-related hazards in recent years, combined with the innovative solutions that have come about to meet those challenges, and the inevitable changes we face in the coming years, are all proof that the evolution of rescue is not a sprint, but a decathlon full of hurdles and challenges.

If the primary mission of the fire service is the protection of lives under virtual any emergency circumstance that the fire department is called upon to handle, then it follows that rescue is a primary role of most firefighters. Today the public assumes that local fire departments are prepared to extract trapped victims from practically any situation. Because fire departments have established a long record of success in practically every type of rescue, the public assumes that firefighters are prepared to accurately handle whatever may come. Through good rescue planning and preparation, the fire service will continue to deliver on its promise of providing expert rescue service in the future.

References

Downey, Raymond, *The Rescue Company.*

Czajkowski, John, "NFPA 1670 Hits the Streets," *FireRescue Magazine,* March 2001.

Sargent, Chase, NFPA 1670, "New Standards For Technical Rescue," *Fire Engineering* magazine, October 1999.

California Governors Office of Emergency Services FIRESCOPE ICS. US&R 120-1 (Appendix B); ICS 420-1; and Urban Search and Rescue O.E.S. Resource Evaluation Form. *www.firescope.oes.ca.us.gov* OES FIRESCOPE OCC Document Control Unit, 2524 Mulberry Street, Riverside, California, 92501-2200 (909) 782-4174 Fax (909) 782-4239.

"Incident Command System Swiftwater/Flood Search and Rescue Recommended Training, Skills, and Equipment List" ICS-SF-SAR 020-1, January 24, 2001, FIRESCOPE Document Control Unit.

"Report From CATF-2," Larry Collins, *NFPA Journal,* Fourth Third Quarter, 1995.

"In The Last Piles Of Rubble, Fresh Pangs Of Loss," *New York Times,* March 17, 2002.

LACoFD *Confined Space Rescue Technician Training Manual,* 1996.

County of Los Angeles Fire Department *Trench Rescue Manual.*

County of Los Angeles Fire Department (LACoFD) *Standard Equipment Inventory for USAR Companies.*

For more detailed information on the National Urban Search and Rescue Response System, including details on each FEMA US&R task force, see the information contained further in this chapter and refer to the FEMA web site: *www.FEMA.gov* and search the phrase urban search and rescue.

The *Firehawk* is the civilian fire/rescue version of the UH-60 Blackhawk helicopter used by the U.S. military.

"Report From CATF-2," Larry Collins, *NFPA Journal,* Fourth Third Quarter, 1995.

"In The Last Piles Of Rubble, Fresh Pangs Of Loss," *New York Times,* March 17, 2002.

APPENDIX 1

RESOURCE TYPING

For the purposes of this appendix, we'll examine the California Urban Search and Rescue resource typing system as an example of Rescue Company Resource Typing. The California Fire Service has multiple tiers of US&R capabilities and resources that are in addition to the state's eight FEMA Urban Search and Rescue Task Forces. These levels have been designated as *Basic, Light, Medium,* and *Heavy*—corresponding approximately to the type of structural collapse search and rescue and other technical rescue operations they are equipped, trained, and staffed to perform.

Because of the ever-present potential for catastrophic earthquakes in the state, and the significant demands of these disasters, California's Rescue/US&R typing system was historically focused on response to earthquake disasters. But this system is also intended for timely, coordinated, and large-scale response to other disasters and major rescue emergencies, including dam failure, landslides, floods, mud and debris flows, avalanches, mud slides, tsunamis, explosions, underground emergencies, and the emerging threat of disastrous terrorist attacks. Naturally, there is new emphasis on a standardized *typing* system for the effective management of disasters related to terrorism.

The California Governor's Office of Emergency Services (OES) is tasked with typing all fire/rescue resources and including them in the state's master mutual aid system, which is divided into six geographical regions. To ensure consistency and accountability statewide, the inspection and typing of US&R resources is conducted by OES staff members of the special operations unit and the local OES fire and rescue branch assistant chief. They make on-site inspections of the resources, including a physical inventory of apparatus and all the equipment and personnel records, in order to ascertain the appropriate category for mutual aid "typing."

California US&R/rescue resource typing levels[1]

Light operational level. The light level represents the minimum capability to conduct safe and effective search and rescue operations at structure collapse incidents involving the collapse or failure of light frame construction and basic rope rescue operations.

Medium operational level. The medium level represents the minimum capability to conduct safe and effective search and rescue operations at structure collapse incidents involving the collapse or failure of reinforced and unreinforced masonry (URM), concrete tilt-up, and heavy timber construction.

Heavy operational level. The heavy level represents the minimum capability to conduct safe and effective search and rescue operations at structure collapse incidents involving the collapse or failure of reinforced concrete or steel frame construction and confined space rescue operations.

Four levels of US&R operational capability and recommended minimum training requirements

Basic operational level. The basic operational level in California represents the minimum capability to operate safely and effectively at a structural collapse incident. Personnel at this level shall be competent at surface rescue and rescue involving minimal removal of debris and building contents to extricate easily accessible victims from non-collapsed structures. Rescue operations would include removal of victims from under furniture, appliances, and the surface of a debris pile. Training at the basic level should at a minimum include the following:

- Size-up of existing and potential conditions and the identification of the resources necessary to conduct safe and effective urban search and rescue operations
- The process for implementing the incident command system (ICS)
- The procedures for the acquisition, coordination, and utilization of resources
- The procedures for implementing site control and scene management
- The identification, utilization, and proper care of personal protective equipment required for operations at structural collapse incidents
- The identification of construction types, characteristics, and expected behavior of each type in a collapse incident
- The identification of four types of collapse patterns and potential victim locations
- The recognition of the potential for secondary collapse
- Recognition of the general hazards associated with a structural collapse and the actions necessary for the safe mitigation of those hazards
- The procedures for implementation of a structural identification marking system and a structural hazard marking system
- Procedures for conducting searches at structural collapse incidents using appropriate methods for the type of collapse
- The procedures for implementation of a search marking system

- Procedures for the extrication of victims from structural collapse incidents
- Procedures for providing initial medical care to victims

Light operational level. Personnel shall meet all basic level training requirements. In addition, personnel shall be trained in hazard recognition, equipment use and techniques required to operate safely and effectively at structural collapse incidents involving the collapse or failure of light frame construction and basic rope rescue as specified below:

- Personnel shall be trained to recognize the unique hazards associated with the collapse or failure of light frame construction. Training should include but not be limited to the following:
  - Recognition of the building materials and structural components associated with light frame construction
  - Recognition of unstable collapse and failure zones of light frame ordinary construction
- Recognition of collapse patterns and probable victim locations associated with light frame construction
- Personnel shall have a working knowledge of the resources and procedures for performing search operations intended to locate victims who are not readily visible and who are trapped inside and beneath debris of light frame construction.
  - Training should include but not be limited to the following:
  - Types of search resources: Urban Search and Rescue Dogs, Optical Instruments (Search Cameras), Seismic/Acoustic Instruments (Listening Devices)
  - Capabilities of search resources
  - Acquisition of search resources
- Personnel shall be trained in the procedures for performing access operations intended to reach victims trapped inside and beneath debris associated with light frame construction. Training should include but not be limited to the following:
  - Lifting techniques to safely and efficiently lift structural components of walls, floors or roofs
  - Shoring techniques to safely and efficiently construct temporary structures needed to stabilize and support structural components to prevent movement of walls, floors or roofs
  - Breaching techniques to safely and efficiently create openings in structural components of walls, floors or roofs
  - Operating appropriate tools and equipment to safely and efficiently accomplish the above tasks

- Personnel shall be trained in the procedures for performing extrication operations involving packaging, treating and removing victims trapped inside and beneath debris associated with light frame construction. Training should include but not be limited to the following:
  - Packaging victims within confined areas
  - Removing victims from elevated or below grade areas
  - Providing initial medical treatment to victims at a minimum to the BLS (basic life support) level
  - Operating appropriate tools and equipment to safely and efficiently accomplish the above tasks

Medium operational level. Personnel shall meet all light level training requirements. In addition, personnel shall be trained in hazard recognition, equipment use and techniques required to operate safely and effectively at structural collapse incidents involving the collapse or failure of reinforced and unreinforced masonry (URM), concrete tilt-up, and heavy timber construction.

Heavy operational level. Personnel shall meet all medium level training requirements. In addition, personnel shall be trained in hazard recognition, equipment use and techniques required to operate safely and effectively at structural collapse incidents involving the collapse or failure of reinforced concrete or steel frame construction and confined space rescue.

References

[1] Governors Office of Emergency Services FIRESCOPE ICS. US&R 120-1 (Appendix B); ICS 420-1; and Urban Search and Rescue O.E.S. Resource Evaluation Form. *www.firescope.oes.ca.us.gov* OES FIRESCOPE OCC Document Control Unit, 2524 Mulberry Street, Riverside, California, 92501-2200 (909) 782-4174 Fax (909) 782-4239

APPENDIX 2

FOUR LEVELS OF EQUIPMENT INVENTORIES FOR CALIFORNIA US&R TYPING

These lists identify the minimum amount of tools and equipment needed to provide a safe and acceptable level of service for each of the four levels of US&R operational capability. The amount, size, and type of equipment listed can be increased to provide a higher degree of safety and service in each level of US&R operational capability.

US&R basic level equipment inventory

2 8-10 LB Sledge Hammer

2 3-4 LB Sledge Hammer

2 Cold Chisel (1" x 7⅞")

4 Pinch Point Pry Bar (60")

2 Claw Wrecking Bar (3')

2 Hacksaw (Heavy Duty)

3 Carbide Hacksaw Blade 1 package

2 Crosscut Handsaw (26")

1 Cribbing & Wedge Kit (See Tool Info Sheet)

1 First Aid Kit (See Tool Info Sheet)

1 Trauma Kit (See Tool Info Sheet)

2 Blanket (Disposable)

1 Backboard w/ 2 Straps

1 Bolt Cutter (30")

1 Scoop Shovel "D" Handle

1 Building Marking Kit (See Tool Info Sheet)

1 Axe (Flat Head)

1 Axe (Pick Head)

US&R light level equipment inventory

1 US&R Basic Equipment Inventory

2 Friction Device (See Tool Info Sheet)

6 Camming Devices (See Tool Info Sheet)

1 Litter & Complete Pre-rig (See Tool Info Sheet)

2 Edge Protection (See Tool Info Sheet)

2 Commercial Harness (Class II or better)

2 3-4 LB Short Sledge Hammer

3 Tape Measure (25')

1 Shovel, Long Handle RD Pt.

2 Tri or Speed Square

1 Nails (See Tool Info Sheet)

2 Rolls Duct Tape

2 150' x ½" Kernmantle, Static, NFPA Approved

12 Carabiner (Locking "D," 11 mm)

3 Pulley, Rescue (2" or 4")

1 Webbing Kit (See Tool Info Sheet)

2 Pick Off Straps (See Tool Info Sheet)

6 Steel Pickets (1" x 4')

1 Chain saw (See Tool Info Sheet)

1 Shovel, Long Handle SQ Pt.

2 Framing Hammer (24 oz.)

2 Carpenter Belts

2 Hydraulic Jacks (minimum 5 Ton)

US&R medium level equipment inventory

1 US&R Basic Equipment Inventory & Light Equipment Inventory

1 Bolt Cutters (Heavy Duty, 42")

4 Floodlight (500 WT)

1 Junction Box (4 Outlet w/ GFI)

1 Circular Saw (12") w/ 2½ gal. fuel

12 Circular Saw Blades (12" Metal Cutting)

1 Pressurized Water Spray Can

1 Rotary Hammer Bit Kit (See Tool Info Sheet)

1 Saw, Electric (10¼")

12 Skill Saw Blade (10¼" Metal Cutting)

12 Sawzall Blades (Wood)

2 Rope (300' x ½") (Static Kernmantle NFPA Approved)

3 Pulley, Rescue (2" or 4")

12 Carabiner (Locking "D," 11 mm)

1 Etrier Set

2 Shovel, Folding, Short

8 Ellis Clamps

8 4' x 4' x 8' Lumber

1 Pipe Cutter, Multi-Wheel (1½")

2 Hi-Lift Jack w/ Extension Tube

1 Come Along (½ Ton)

1 Tool Kit (See Tool Info Sheet)

1 Demolition Hammer, Large (See Tool Info Sheet)

1 Ventilation Fan (See Tool Info Sheet)

1 Air Bag Set (3 Bag, 50 Ton w/ 3 spare air bottles)

1 Generator (5 KW)

6 Extension Cords (50')

1 Wye Electrical Adapter

2 Circular Saw Blades (12" Carbide Tip)

2 Circular Saw Blade (12" Diamond, Continuous Rim)

1 Rotary Hammer (1½")

1 Anchor Kit (See Tool Info Sheet)

2 Skill Saw Blade (10¼" Carbide Tip)

1 Sawzall

18 Sawzall Blades (Metal)

2 Rope (20' x ½") (Static Kernmantle NFPA Approved)

2 Friction Device (See Tool Info Sheet)

1 Webbing Kit (See Tool Info Sheet)

2 Commercial Harness (Class II or Better)

4 Haul Buckets (Metal or Canvas)

1 Ellis Jack

6 Screw Jacks, Pairs (1½")

6 Pipe (6' x 1½", Schedule 40)

1 Cribbing & Wedge Kit (See Tool Info Sheet)

1 Chain Set (See Tool Info Sheet)

1 Demolition Hammer, Small (See Tool Info Sheet)

1 Electrical Detection Device (See Tool Info Sheet)

1 3 Range Air Monitor

US&R heavy level equipment inventory

1 US&R Basic Equipment Inventory

1 US&R Light Equipment Inventory

1 US&R Medium Equipment Inventory

6 SCBA (with PAL & 1 Spare Bottle each)

3 Supplied Air Breathing Apparatus (SABA)

Umbilical System w/ Escape Bottles & 250' hose each

1 3 Range Air Monitor

1 Tripod (Human Rated, 7' – 9' w/ hauling system)

2 Full Body Harness (Class III or Better)

1 Ventilation Fan (See Tool Info Sheet)

1 Circular Saw (16") w/ 2½ gal. fuel

2 Circular Saw Blade (16" Diamond, Continuous Rim)

2 Circular Saw Blade (16" Carbide Tip)

1 Pressurized Water Spray Can

6 Canister Type Respirators

24 Replacement canisters for Respirators

1 Generator (5 KW)

4 Floodlight (500 WT)

6 Extension Cords (50')

1 Junction Box (4 Outlet w/ GFI)

1 Wye Electrical Adapter

1 Rotary Hammer (1½")

1 Rotary Hammer Bit Kit (See Tool Info Sheet)

1 Sawzall

12 Sawzall Blades (Wood)

18 Sawzall Blades (Metal)

1 Drill (½", variable speed)

1 Drill Bit Set (Steel, ⅛" – ⅝")

1 Drill Bit Set (Carbide Tip, ¼" – ⅝")

1 Chain Saw, 12" Electric w/ spare carbide tip chain

1 Rebar Cutter (1" capacity)

1 Cutting Torch (See Tool Info Sheet)

1 Come Along (½ ton)

1 Demolition Hammer, Small (See Tool Info Sheet)

1 Demolition Hammer, Large (See Tool Info Sheet)

1 Extrication Stretcher for Confined Areas

2 Shovel, Folding, Short

1 Mechanical Axe (High Voltage)

1 Mechanical Grabber (High Voltage)

2 Pair Lineman Gloves (High Voltage)

1 Upgrade High Pressure Air Bags to a Total of 245 Tons

1 Air Bag Regulator, Control Valve w/ 2 additional hose

2 Building Marking Kits (See Tool Info Sheet)

1 Cribbing & Wedge Kit (See Tool Info Sheet)

1 Ram Set Powder Actuated Nail Gun (w/ 150 red charges)

1 Box Ram Set Nails w/ Washers (2½")

1 Box Ram Set Nails w/ Washers (3½")

1 Green Stone Wheel (to sharpen carbide tips on tools)

1 Nails (See Tool Info Sheet)

2 Tri or Speed Squares

2 Framing Hammers (24 oz)

2 Carpenter Belts

1 Level (6")

1 Level (4')

1 Nail Gun, Pneumatic (Framing Type, 6p–16p)

1 Case Nail Gun Nails (8p)

1 Case Nail Gun Nails (16p)

32 Ellis Clamps

1 Ellis Jack

8 Post Screw Jacks

12 Screw Jacks, Pairs (1½")

12 Pipe (6' x 1½", Schedule 40)

12 Steel Pickets (1" x 4')

1 Case Orange Spray Paint (Line marking, downward application type)

1 Case Duct Tape

1 Technical Search Device (See Tool Info Sheet)

1 Hydraulic Rescue Tool (See Tool Info Sheet)

US&R tool information sheet

Anchor Kit 1 box 3/8" x 5"
Hilti Kwick Bolt concrete anchors

25 ea.3/8" SMC Stainless Steel
Anchor Plates

25 ea. 3/8" Drop Forged H/D eye nuts

Anchors & Plates are for rope system
anchor points

Building Marking Kit 2 ea.Orange Spray
Paint, Line Marking (downward)
application type

4 ea.Lumber Chalk

2 ea.Lumber Crayon (red)

2 ea.Lumber Crayon (yellow)

4 ea.Lumber Pencil

Camming Device Prusik Loop
(7 mm or 8 mm) or Gibb's Ascender
or combination of each

Chain Saw Gasoline or Electric w/ carbide
tip chain & one spare chain and bar oil

Gasoline: 2½ gal. spare fuel and
oil mixture

Electric: Need electric power source and
100' of extension cord

Chain Set 1 ea.1' w/ a grab hook
on each end

1 ea.5' w/ a grab hook & a slip hook

1 ea.10' w/ a grab hook & a slip hook

1 ea.20' w/ a grab hook & a slip hook

(All chain is 3/8", grade 7 or better)

Cribbing & Wedge Kit 24 ea.4" x 4" x 18"

24 ea.2" x 4" x 18"

12 pr 4" x 4" x 18" wedges

12 pr 2" x 4" x 12" wedges

Containers to store & carry

Cutting Torch One or more Plasma Cutter,
Exothermic Torch w/ 50 rods,
Heavy Duty Oxy/Acetylene Torch
w/ spare O2 cylinder or other similar device

Demolition Hammer, Large Electric,
pneumatic, or gasoline
(60 lbs. Minimum)

2 ea.Bull Point Bits

2 ea.Chisel Point Bits

Demolition Hammer, Small Electric,
pneumatic, or gasoline
(30–45 lbs. Minimum)

2 ea.Bull Point Bits

2 ea.Chisel Point Bits

Edge Protection Commercial Edge Rollers, canvas tarps, split fire hose or any combination of each

Electrical Detection Device Hot Stick Electrical Alert Device, Volt/Ohmmeter, or other device to alert crew members of electrical current

First Aid Kit Basic first aid supplies for minor injuries to six victims or crew members.

Bandages, elastic bandages, Band-aids, eye wash, 4" x 4" gauze pads, gauze dressings, triangular, etc.

Friction Device Figure 8 w/ ears or Brake Bar Rack or one of each

Hydraulic Rescue Tool Gasoline, electric or manual device w/ 10,000 lbs. minimum force. Able to cut, spread and pull.

Gasoline: $2\frac{1}{2}$ gal. spare fuel & oil.

Litter & Complete Pre-rig Litter capable & rated for horizontal & vertical lift & hoist. Pre-rig can be commercial or preassembled to include adjustment & attachment capability.

Nails 25 lbs. 16d vinyl coated (Green Sinkers)

25 lbs. 8d vinyl coated (Green Sinkers)

25 lbs. 16p Duplex *(Note: High humidity areas may require cadmium coated nails to prevent rust during long term storage)*

Pick Off Strap Webbing strap with one "D" ring at one end and one "V" ring adjuster on webbing strap.

(Webbing: $1\frac{3}{4}$" wide with 10,000 lbs. rating, minimum 42" long) (Hardware strength 5,000lbs. rating)

Rotary Hammer Bit Kit 1 ea.Carbide Tip Bits, $\frac{3}{8}$", $\frac{1}{2}$", $\frac{3}{4}$", 1", $1\frac{1}{2}$", 2"

2 ea.Bull Point Bits

Appropriate adapters for bits and depth range capability

Technical Search Device
One or more of the following:
Optical Instruments (Search Cameras),
Seismic/Acoustic Instruments
(Listening Devices).

Tool Kit 1 ea.12" Crescent Wrench

1 ea.8" Crescent Wrench

1 ea.Slip Joint Pliers

1 ea.Channel Lock Pliers

1 ea.Wire Side Cutters

1 ea.½" Socket Set w/ Ratchet
& 6" extension

1 ea.½" Breaker Bar

1 ea.Ball Peen Hammer

1 set Standard Head Screwdrivers

1 set Phillips Head Screwdrivers

Any other tools required for
maintenance and repair of
equipment in cache

Trauma Kit
Basic supplies to treat trauma
injuries to six victims or crew members.
ALS type equipment (i.e. IV solutions, drugs, etc.) is not listed but may be carried if authorized. Examples of items to carry include: Large trauma dressings, splints, airways, bag valve respirator w/ large and small masks, etc.

Ventilation Fan Electric or gasoline powered
w/ extension tube to direct air flow

Webbing Kit 6 ea.1" x 5'
6 ea.1" x 20 '
6 ea.1" x 12'
6 ea.1" x 15'

All webbing is spiral weave nylon,
4,000 lb. minimum tensile strength.
Each webbing length must be a
different color.

APPENDIX 3

Sample Inventory of US&R/Rescue Company Equipment List

Equipment name

Camera, Thermal Imaging System

Radio, Handheld, Intrinsically Safe, Command Frequency

Technical Search Camera, Snake-Eye

Axe

Air Bag, Rescue, 21.8 ton, 11.1" lift

Air Bag, Rescue, 12 ton, 8" lift

Drill, Wood ½" x 18"

Reducer, 2½" to ⅝"

Drill, Wood ¾" x 18"

Driver, Screw, Flat 10"

Chisel, Pneumatic

Air Bag, Rescue, 73.4 Ton, 20" Lift

Air Bag, Rescue, 3.2 Ton, 3.5" Lift, (Rectangle)

Drill, Pneumatic ⅜"

Chains, Tire, Snow

ArcAir Rods, Box, ⅜"

Saw, Rotary, Blade, 16", Concrete, Dry

Ram Set Nails, 3.5"

Rotary Hammer Bit Kit

Radio, Handheld, Intrinsically Safe, Tactical Frequencies

Wrench, Torque

Tote Set, 2 pc.

Air Bag, Rescue, 17 ton, 9.2" lift

Chains, Tire

Wrench, Impact ¾" Drive

Chain, cutting, diamond

Air Bag, Rescue, 43.8 ton, 15.5" lift

Wrench, Spark Plug

Air Bag, Rescue, 1.5 ton, 3" lift

Air Bag, Rescue, 4.8 ton, 5" lift

Air Bag, Rescue, 35 ton, 10" lift, (rectangle)

Rotary Hammer & Charger

Chain, Tensioner

Ram Set Bolts, 7" x ½"

Couplings, set

Adapter, Chuck SDS

Saw, Rotary, Blade, 14", Concrete, Wet

Ram Set Bolts, 9" x 1"	Ram Set Bolts, 7" x ¾"
Ram Set Bolts, 5.5" x ½"	Saw, Rotary, Blade, 16", Diamond
Saw, Rotary, Blade, 14", Concrete, Dry	Saw, Rotary, Blade, 16", Concrete, Wet
Saw, Rotary, Blade, 16", Carbide Tip (Wood)	Saw, Rotary, Blade, 16", Metal Cutting
Blade, Wood Cutting, 16"	Boat Motor, 25 hp Outboard w/ Fuel tank
SABA Air hose, 100' sec.	Saw, Rotary, 14", K-700
Fiber Optic Technical Search System	Boat, Rescue, Inflatable
Drill/Hammer Drill-24V	Pneumatic Rescue System (Airgun 40)
Bit, Pilot Point Set–19 piece	EMS Backpack For Mountain Rescue, Collapse, Confined Space, and Other Limited Access Patient Treatment
Chuck, ½" w/ key for DW5351	Map Books and Maps for applicable counties
Chairs, folding, for Confined Space Rescue Equipment Donning and Standby Team Readiness, and for Personnel Rehab	
Building Triage/Marking Kit	Rotary Hammer, 1½", Kit, electric
Stanley Rescue System Parts Kit	Bolt Cutter, 36"
Pump, Hand-Fuel	Bolt Cutter, 12"
Life Detector Technical Search System & Accessories	Hose, Garden
Power Unit Hydraulic, Stanley-HP1 Rescue System	Chainsaw, Diamond-Stanley Rescue System, Hydraulic, DS11
Saw, Skil, Worm Drive 7¼"	Cut-Off Saw, Stanley Rescue System, 14", Hydraulic, CO23
Technical Search Device, Olympus Fiber-Optic	SABA Air Cart Air Supply System
Breaker, 90#, Stanley Rescue System, BR89	Bit, Moil Point-Stanley Rescue System Breaker
Pump, Trench Shore,	Ducting, Ventilation, Accordian, 15'
Chainsaw, Gas	Half-Backboard, LSP (Miller Half Back), For Confined Space Vertical Extraction and Other Limited Access Vertical Rescues

Saw, Reciprocating, 18 v Rechargeable

Saw, Rotary, 16", Partner K-1200,

Chain Kit, for Combo Hydraulic Spreader

Air Horns, Hand-Held

Tool, Release, Trench Shore

Fluid, Hydraulic, Trench Shore

Hydraulic Rescue Tool System Power Unit

Kit, Air Bag, Rescue, Low Pressure

Hose, Airbag, Rescue, 16' ,Black

Wrench, Pipe, 14"

Air Bag, Rescue, Master Control Package

Resuscitator with Airways, Bag Valve Mask, etc., In Backpack

Atmospheric Monitor Kit, Hardened Case, w/ Organizer

Saw Reciprocating—24V

Drill, Hammer Rotary

Bit, Wood Eater Set—5 piece

Chainsaw, Electric, 16"

Rescue Frame

Kit, Trailer Hitch for Towing PWC Trailer

Air Knife System

Swiftwater Rescue Hose Inflator System

Chainsaw Bar, 20"

Swiftwater Rescue Board

Adapter, Rotary Hammer, Spline to Taper Shank

Kit, Pneumatic Shoring Feet, in Hardened Carrying Case

Heavy Duty Lockout Tags, 1 package

Cutting Torch, Oxy-Acetylene,

4:1 mech. Advantage Rope System, Pre-rigged—In High Angle Rescue Packs

Rotary Hammer Bit, Core Bit Adapter

Ice Chest for Biopack SCBA Canisters & Rehab

Ram, Hydraulic Rescue System, 20"

Ram, Hydraulic Rescue System, 30"

Ram, Hydraulic Rescue System, 60"

Breaker, Hammer, 60#

Saw Parts & Repair Kit, in Ammo Box

Confined Space Communications Kit, in Hardened Case

Lock-out Tag-out Kit, in Hardened Case

Extinguisher, Dry Chemical

Ear Plugs, 1 package

Spinner, Rivet

Burn Sheet, Set

Dirt Vacuum System

Clipboard, Technical Rescue, With Technical Rescue ICS Worksheets and Other Incident Documentation Items

Swiftwater Rescue Line Thrower System

Chainsaw Chain Blade, 16", carbide tip

MRE (Meals Rescue to Eat), Case, for Disaster Operations and Overnight "Hike In" Rescues in the Mountains

Confined Space Scene Management Kit, case

Night Vision Goggle Systems

Radio, Hand-Held, Mountain Rescue Frequencies

Rope, Rescue, 7/16", 200'

Water, Bottled, Case, for Disaster Operations, Long Term Rescues, and Rehab

Confined Space Communications—Operator ext. Mic/mute switch, 1'

Hook, Grappling

Tripod Rescue System

Rescue Straps, (Cearley Straps), for Helo/Swiftwater, Marine Rescue, and Cliff Rescue Operations

High-Rise "Officer's" Hose Pack

Litter, Rescue, Ferno (For PWC Rescues)

Confined Space Communications Command Module

Bolt Cutter, 42"

Pliers, Needle nose

Liter w/ Pre-rig

Ram Bar, Sliding

Shovel, Round Point, 3'

Night Vision Monocle System

Ram Set, Powder Actuated Nail Gun

Wrench, Crescent, 8"

Lifeguard Rescue Can

Blanket, Rescue, Reflective

SCBA w/ masks (1 per Post Position)

Hammer, Sledge 16#

Confined Space Communications Talk Box

Rope, Rescue, 1/2" 250'

Kit, IRB (Inflatable Rescue Boat)

RIC (Rapid Intervention) Pack, Air w/ Bottle

Rapid Intervention Straps—CMC—One Per Post Position.

Alarms, Personal Safety System (One Per Post Position)

Helicopter-High-Rise Rescue Pack

Litter, Rescue, Junkins

Breaker, Bosch, Hand held

Atmospheric Monitor Calibration Kit

Dikes, Crosscut

Pliers, Lineman's

Chain, 20'

Technical Rescue System-Search Camera

Shovel, Long Handle, SQ Pt.

Atmospheric Monitor, Sampling Pump

Kit, Tripod, Pneumatic

Pneumatic Shoring System

Wrench, Crescent, 12"

Confined Space Communications--Operator external Mic/Mute switch and cord, 20'

Set, Index, Drill

Victim Rescue Harnesses—4

Spray Paint, Fluorescent Orange, Case (For Structural and Victim Marking Systems)

Anchor Bolt-box—½" x 7"

Anchor Bolt-box—¾" x 7"

Rebar Cutter, Hand Pump

Saw, Beam, 16¼"

Bit, Concrete 1/2" L-10 Depth 10

Bolt Cutter, 14"

Electrical Adaptors—Various—In Kit

Electrical Adaptors

Rope Bags, For All Ropes Carried on USAR Company

Bio Pak Closed Circuit SCBAs—One Per Post Position

Ice and Snow Rescue Kit—One Per Post Position

Pump, Water, Stanley Rescue System Chain Saw

Wrench, Combination 1¼", Stanley Tool Tool

Plug, Spark, Stanley Rescue System Power Unit

Saw, Concrete, Circular, Stanley Rescue System

Filter, Air, Power Unit, Stanley Rescue System

Filter, Oil-Stanley Rescue System

Bar, Chain Saw, Stanley Rescue System

Rebar Cutter, Electric, 1" capacity

Litter, Confined Space Rescue (SKED Sled)

Probe, Rescue, Telescoping, 6', Stainless Steel

Combination Tool, Hydraulic Spreader/Cutter

Ram, Rescue, 40"

Anchor Bolt-Box–1" x 9"

Rope, Rescue, 600' (2 sets)

Bit, Concrete ⅜" L-10 Depth 10

Clipboard—Medical (For EMS Documentation)

Pneumatic Shore-Regulator

Set, Wrench, Ignition

Bit, Concrete 1" L-10 Depth 10

Sprocket, Drive, Chain Saw, Stanley Rescue System

Dive Rescue Kit—One Per Post Position

Avalanche Poles—Kit

Wrench, Crescent, 10"

Wrench, Combination 1 1/16, Stanley

Pre-rig Rings, Steel, "O"

Electrical Adapter

Rope, Utility, ½" 200' in Bag

Pulleys, Knot Pass 2"—In High Angle Rescue Packs

Ducting, Ventilation, Accordion, 25', 2 Sets, For Confined Space and Mine/Tunnel Rescue

Pre-Rig Systems for Litters

Rotary Hammer, battery powered.

Blade, Rotary, Metal 16"

Prussicks—Long—In High Angle Rescue Packs

Rope, Anchor, 25'—
In High Angle Rescue Packs

Edge Rollers—In High Angle Rescue Packs

Rescue Platforms—10 Person Marine
Rescue Platforms—For Helicopter
Deployment into Ocean During
Marine Disaster Operations.

Search and Recon Kit—In Backpacks

Rods, Cutting, 50/box x 4 Boxes—
For Arc-Air Rescue Metal Cutting System

Pressurized Water Spray Can, Hudson type

Air Bag, Rescue, Hose, 16', Yellow

Chainsaw Chain, 20", Carbide Tip, Spare

Chainsaw Chain, 16", standard, spare

Axes, Flathead

Kit, Chain Breaker Stanley

Helmet, Victim, Rescue

Filter, Hydraulic, Stanley Rescue System

Bit, Chisel Point–Stanley Rescue
System Breaker

Digital Camera—Still

Come-Alongs, ½ ton

Straps, 2" x 20', Tie down

Air Horn Canister, spare

Hydraulic Jacks, 12 ton, Bottle

Oil, Hydraulic 5 gals.,
Stanley Rescue System

Respirator, Dust and Mist

Edge Pad, canvas—In High Angle
Rescue Packs

High Angle Rescue Pack—Industrial
(Steel Hardware)

High Angle Rescue Pack—Pony pack—
Mountain (Aluminum Hardware)

Sawzalls, w/ Hardened Cases, &
blades, Electric

Plumb Bob

Brooms, Corn

Anchor Bolt—box—½" x 5½"

Air Bag, Rescue, Dual Controller

Funnel, 7½" diameter

Cutting Torch Cylinder, Oxygen

Cutting Torch Cylinder, Acetylene

Axes, Pickhead, w/ Belt—
One Per Post Position

Wrecking Bar, 3'

Air Bag, Rescue, Hose, 16', Black

Pliers, Vice Grips

Wrench, Allen 15 pc. Set

Kootenai Carriages—In High Angle
Rescue Packs

Hoses, Hydraulic, Stanley Rescue
System, Twin, 50'

Bars, Digging, 16 lb., 72"

Devices, Friction, 8 plate, Aluminum—
In High Angle Rescue Packs

Air Bag, Rescue, Hose, 16', Red

Pneumatic Shore, 36.3" to 58.0"

Rope, Rescue, 200'

Shovel, Long Handle, Round Point

Prussics—Short—In High Angle Rescue Packs

Collection Plate-Aluminum—In High Angle Rescue Packs

Radio, Hand-Held, USAR "Talk Around" Entry Frequency

Point, Bull 12" SDS Max Tool

Salvage Covers

Radio, Hand-Held, Aviation

Sigg Bottle, 1 qt.

Sprocket, Nose, Bar, Stanley Rescue System

Brady Single Pole Circuit Breaker Lockout, 1 package

Shores, Speed, Trench shores

Lights, Portable, Ground, 500wt

Atmospheric Monitors—One Per Post Position

Sawzall Blade, 1 Package, Wood

Screwdriver, Common

Radiation Monitoring Kit—Civil Defense

Pneumatic Shore, 55.5" to 87.3"

Swiftwater Strobe Lights, on Each PFD

Pneumatic Shore, Extension 36"

Device, Friction, 8 plate, Steel—In High Angle Rescue Packs

Confined Space Communications Cable, 200' w/ Connectors,

Pneumatic Adaptors—Kit

Hydraulic Adaptors—Kit

Rope, Rescue, 300'

Electrical Adapter Kit

Chisel, Cold 7/8" x 11" SDS Max Tool

Bank Chargers, Battery, Various

Tygon Tubing, (1 ea. 25', 50')

Cribbing, 4 x 4 x 24—Kits

Oil, Engine—2-Cycle 8 oz.

Pick-off Straps—In High Angle Rescue Packs

Swiftwater Rescue Equipment Packs

Bags, Drop—One Per Post Position

Brady Multi-Pole Lock-out

ArcAir Rods, Box, 3/8"

Charger, 1 hr. w/ 3-stage charging f/24v DeWalt Cordless

Sawzall Blades, 1 package, metal

Pneumatic Shore, 18.8" to 24.5"

Radio Batteries, VHF and UHF—Charged

Vests, Orange, Position Assignment—Kit

Underwater Radio Kit

Hydraulic Rescue System Hose, 20' sec.

Tape, Duct—Box

Carabiner-X large-steel—In High Angle Rescue Packs

Helmets, Helicopter Operations—One Per Post Position

Headlamps, Pelican Versa Bright II—Communications
One Per Post Position

Edge Protection, Ultra Pro—
In High Angle Rescue Packs

Bit, 1½" Wood Eater

2 Personal Watercraft, For River and Flood Rescue Operations, and for Submerged Victim Search Operations

Trailer, Personal Watercraft, Rescue

SABA Escape Bottles, Spare, 10 Minutes—One Per Post Position

Light, Hand, King Pelican

Fuel Can, 1 gal., Safety

Horse Rescue Harness—Large

Mask, Swim—One Per Post Position

Pulleys—4"—In High Angle Rescue Packs

Pneumatic Shore, Extension 12"

Swiftwater Throw Bags. w/ 70' Rope

Cordage, Utility 10M—In High Angle Rescue Packs

Horse Rescue Harness—Medium

Fins, swim, pair—One Per Post Position

Cyalume Light Sticks—
Boxes With Different Colors

Saw, Electric, 16" (beam)

Saw, Rotary, Blade, 14", Diamond

Air Bag Hose, Green, 32 ft.

Grip hoist, TU28, w/ 50' cable

Air Bag Hose, Blue, 32 ft.

Confined Space Rescue

Cable, 50', w/ connector

Confined Space Rescue Communications—
Face Mask Rescue Set, w /speaker

Confined Space Rescue Communication Cable, 100', w/ connector

Bit, 1/2" Straight Shank

Bit, ⅜" Straight Shank

Swiftwater Personal Flotation Device, (PFD)'s—One Per Post Position, One for Victim

Rotary Hammer, Hilti, Battery, 36 volt

Swiftwater Knives, One per PFD

SABA Mask Kit, Case

Harness, Full-Body, Class 3—
One Per Post Position

Blade, Rotary, Concrete 16"

Pneumatic Shore, 24.8" to 35.4"

Radiation Pager—One Per Post Position

Blade, Diamond continuous rim, 16"

Dog Rescue Harness

Whistles, On Each PFD

GPS (Global Positioning System), Hand-Held

Saw, Circular, Electric, 10¼"

Air Chisel/Impact Wrench, Pneumatic, Kit

Distance Measuring Device

Hand Truck

Mechanical Grabber (Electrical Line)

SABA Air Hose Reel, w/ 250' Air line, victim

Shovel, Scoop, "D" handle

Parts Cleaning Station Flow-Thru Brush

Confined Space Rescue Communications Boom Mic Wind Screen

Nut Driver Set, Standard

Screw Driver Bit Set, 60 pc.

Ratchet, 5 1/2" Stubby flex head, 3/8"

Pliers, Lineman's

Pliers Set, Reach Needle Nose

Pliers Set, Locking

Nut Driver Set, Metric

Handle, Spinner, 1/4"

Wrench Set, Adjustable, 5 pc.

Chainsaw Bar, 16"

Respirators—APR—One Per Post Position

Wrench, Combo, Metric, 23mm

Binoculars, 10 x 25

Slings, 2" x 20'—In High Angle Rescue Packs

Brady Plug Lockout

Wrench, Combo, Metric, 6mm

Wrench, Combo, Metric, 25mm

Wrench, Combo, Metric, 20mm

Wrench Set, Combo, 12 pt., 17 pc.

Socket Tray Set, 3 pc.

Socket Tray Set, 3 pc.

Socket Set, Bit, 19 pc.

Anchor Eye Nut, 1/4"

Level, 4'

Drill Bit Set, Steel, 1/8"–5/8"

Lumber Crayon, Red, Box

Parts Cleaning Station

Air Bag Hose, 32', Red

Lumber Crayon, Yellow, Box

Search Camera Reel System—300' With Standard Camera/ Two-Way Communication Head

Ratchet, Standard Tooth Flex Head, 3/8"

Punch & Chisel Set, 24 pc.

Pliers, Arc Joint

Short Arm Hex Set, Metric, 11 pc.

Short Arm Hex Set, Standard, 11 pc.

Multi-Meter, Digital LCD

Handle, Flex "T," 1/2"

Respirator Canister, Replacement, Case

Electric Detection Device—"Hot Stick"

Pliers Set, Pro

Anchor Plate, Hanger, stainless steel, box

Ram, Hydraulic Rescue, Accessory Kit

Label Machine

Wrench, Combo, Std., 1 5/16"

Screw Driver Set

Backboard, w/ Straps (Miller Board)

Wrench Set, Combo, Metric, 12pt., 17 pc.

Socket Wrench Set, 235 pc.

Socket Tray Set, 3 pc.

Socket Tray Set, 3 pc.

Wrench, Combo, Metric, 26mm

Saw, Rotary, Kit

Drill, Variable speed, 1/2", DeWalt

Drill Bit Set, Carbide, 1/4"–5/8"

Die Grinder, Pneumatic, Kit

LaserPointer

Anchor Eye Nut, 3/8"

Anchor Eye Nut, 1/2"

Video Camera, w/ Light, Spare Battery, In Case

Awning, EZ-UP—For Rehab and Equipment Pool

Metal Detector

Ladder, Little Giant

Search Cam Video Transmitter and Connector

Hall Runners

Nail Gun Nails, case 8p

Tow Straps—Nylon

Hose, 1 3/4", 50' sec.
Cutter)

Ducting, Lay-Flat L

Swiftwater Rescue Briefcase w/ log book, Waterway Rescue Preplans, Inundation Maps and Tsunami Inundation Maps

Nail Gun Nails, Case,16p

Breaker Bit, Chisel, 1", 1 1/8" Hex

Air Bag, Rescue, Hose, Black 32 ft.

Rope, Tagline 300" 8mm in Bag

Exothermic Cutting Torch,

Target Hazard Fire Attack Preplans

Funnel, 3 1/2" dia.

Ground-Penetrating Radar—Technical Search—Field Ready System

Ventilation Blower, Electric, Intrinsically Safe

Water, Drinking, case

Dosimeters—One Per Post Position

Nail Kit

Awning Anchor Kit, Heavy Duty

Awning Heavy Duty Leg Set, EZ-UP—For Rehab and Equipment Pool

Level A Entry Suits—One Per Post Position

Stanley Breaker, 45#, hyd., BR45

Anchor Eye Nut, 5/8"

Fuses, Hwy, Box

WMD Terrorism Antidote Kit

Nail Gun, Pneumatic, kit

Tool Chest

Mechanical Axe, (Electrical Line

Respirator parts kit

Pickett Anchor set—in Backpack for Mountain Rescues, Mine/Tunnel Rescues, Swiftwater Rescues, etc.

Breaker Bit, Bull Point, 1 1/8" Hex

Chainsaw Kit,

Air Bag, Rescue, Hose, 32', Yellow

Screwdriver, Straight

Pliers, Pr, Slip Joint, 10"

SABA air hose Kit, case

Hydraulic Rescue Tool Pump, Manual, w/ foot pedal

Rotary Hammer Bit, 1"

Ventilation Blower Conductive Duct, 8" x 4'

Tape, Electrical, Black, Roll

Ventilation Duct Coupler

Rotary Hammer Bit, 3/8"

Hacksaw, (Heavy Duty), with 10 blades

Camera, Digital

Electrical Junction Box, 4 outlet, w/ GFI

Electrical Adapter, Wye

Saw, Skill, Blade, 7 1/4" carbide tip

Halligan

Ventilation Duct Storage Bag

Bauman Bag Vertical Extrication C-Spine Immobilizer

Ram Set Studs, Box, 3/8"

Nail Puller, (Catpawels) Packs

Hammer, framing, 24 oz. straight claw

Ram Set Booster, Red, Hilti

Ram Set Booster, Yellow, Hilti

Camera, Digital, Motion Picture

Cutting Torch Cutting Tip #000

Compasses, Type 3, One Per Post Position

Cutting Torch Welding Tip #00

Cutting Torch Tip, Multiflame #6

Saw, Rotary, Blade, 14", Carbide Tip (Wood)

Carabiners, Large, locking, aluminum—In High Angle Rescue Packs

Carabiners, Lg., Locking, Steel—In High Angle Rescue Packs

Ram Set Fasteners, 1 3/4" x 7/16"

Wedges, Pr. 6 x 2 x 18—Case

Carabiner, Standard, locking, aluminum

Palm Pilot for IT Applications

Breaker, Chain

Rotary Hammer Bit, coring

Pipe Cutter, Multi-Wheel

Crosscut Hand saw

Ventilation Blower 12" Duct Adapter

Level, Line

Thermal Imaging System—Handheld—With Video Transmitter Link

P-S Wave Detectors

Smart Levels

Sledge Hammer, 10 lb.

Oxygen Bottle, 55 cu. ft.

Tape Measure, 25'

Brake Bars—In High Angle Rescue

Sledge Hammer, 3 lb.

Dewatering Pumps

Cutting Torch Cutting Tip #2

Carpenters pencil, box

Exothermic Torch 3/8" Conversion Kit

Cutting Torch Cutting Tip #1

Cutting Torch Tip, Multiflame #8

Cutting Torch Cutting Tip #00

Exothermic Torch Washers

Electrical Cords, 50' 12-3

Ram Set Fasteners, 2.25" x 3/8"

Wedges, Pr., 6 x 6 x 30—Case

Lumber, Plywood, 1" and 2"

Wedges, Pr. 2 x 4 x 18

Picket Anchors, 1" x 4', Cold Rolled Steel

Webbing, Green, 5'—
In High Angle Rescue Packs

Cribbing, 6 x 6 x 30—Cases

Air Bag, Rescue, Pneumatic Components Kit

Auxiliary Electrical Adaptors, Kit, Assorted

Chainsaw Air Filter

Cutting Torch Wrench, 8 MC

Saw, Rotary, Drive Belt, spare

Rotary Hammer Bit, $\frac{7}{8}$", spline shank

Level, 6"

Engraver, Electric

Pliers, Tongue & Groove, 14"

Screw Driver, Phillips

Nails, Box, 25 lb.,16d
Vacuum

Screwdriver, Phillips, Stubby

Rebar Cutter—Battery

Screwdriver, Slotted, Stubby

Animal Rescue Snare

Electrical Cord Reel, 10/3

Etriers—In High Angle Rescue Packs

Ram Set Boosters, Purple, Hilti

Battery, 12-volt lead Acid—
Spare for Exothermic Cutting Torch
and Other Similar-Powered Tools

Bars, Pry and Pinch Point, 60", 18 lb.

Rescue Resource List—With Contact Numbers

Wedges, Pr. 4 x 4 x 24—Cases

Swiftwater Helmets

Screw Jacks 1$\frac{1}{2}$" x 18", /Foot

Stick, Light, 12 hr.

Air Bag, Rescue, 31.8 ton, 13.1" lift

Electrical Adapter, Wye

Saw, Rotary, Air Filter

Label Machine Tape, 4"

Rotary Hammer Bit, 1$\frac{1}{4}$"

Awning Side Wall, EZ-UP—
For Rehab and Equipment Pool

Rebar Cutter Blades

Lineman Gloves, pr.

Cell Phones—One Per Post Position—
For Emergency Communications

Nails, box, 25 lb., 8d

Adapter Kit, Air Knife, and Air

Chalk Line Chalk, 8 oz.

Rebar Cutter—Battery—Spare Blades

Horse Rescue Harness—Small

Rotary Hammer Bit, $\frac{11}{16}$",
Spline Shank

Nails, Box, 25 lb., 16d Duplex

Crane Operators Manual

ICS Field Operations Guide (FOG)

Ventilation Blower Ducts, 8", 1 ea.
25', 15'

FEMA US&R National Response
System Field Operations Guide
(FOG)

Cribbing, 2 x 4 x 24—Cases

Generator, 5kw, portable, gas

Fireground Search Rope Pack

Blade, Carbide Tip, 14" fire rescue

Lumber, 6" x 6" x 20'

Petrogen Metal Burning Systems

Cribbing, 2 x 6 x 30

Osborn Aluminum Lockouts, 1" hasp

Ram Set Pins, box, 1⅞"

Anchor Kit, in ammo boxes, 1/size

Chock Block, Metal

Eight Plate, Rescue, Aluminum—
In High Angle Rescue Packs

Carpenter Belts—One Per Post Position

Square, Tri/Speed

Rope, Rescue, 100'

Lumber, 2" x 4" x 20'

Respirator Dust Pre-Filter

Light Saddle, Pelican

Diagonal Cutters,

Electrical Plug, L14P, Male Plug

Swiftwater Gloves, Pair—One Per Post Position

Swiftwater Fins, pair—One Per Post Position

Elbow Pads, Pr. Cordura w/ Plastic
Cover—One Per Post Position

Atmospheric Monitor, Battery Shell pack
Line

Electrical Plug, L14R, Female Receptacle

Economy Lockout Padlock w/ 2½" Shackle

Electrical Adapter

Ellis Jacks

Lumber, 2" x 12" x 12'

Cans—Gasoline—1 gallon—
for Petrogen Metal Burning Systems

Ram Set Pins, box, Steel, ⅞"

Osborn Aluminum Lockouts 1½" hasp

Ram Set Pins, Box, 2⅞"

Communicable Disease Kits—
One Per Post Position

Line, Chalk

Eight Plate, Rescue, Steel—
In High Angle Rescue Packs

Hydraulic Tool, Fluid, 3 gallons

Shovels, Folding

Chalk Line

Lumber, 4" x 4" x 16'

Haul Buckets, Metal, 5 gal.

Electrical Adapter

Swiftwater Dry Suit—
One Per Post Position

Swiftwater Goggles—
One Per Post Position

Spark Plug, for Rotary Saw

Kneepads, Pr. Cordura, w/ Plastic
Cover—One Per Post Position

Swiftwater Booties, Wet Suit, pair—
One Per Post Position

SABA Air Hose Reel, w/ 250' of Air

& Comm

Spark Plugs, for Stihl 044 Chainsaw

Electrical Adapter

Webbing, Blue, 12'

Floodlight, 500 wt., portable

Canvas Bag, Tool, Carpenters, Extra Strength—One Per Post Position

Ellis Clamps

Paddles, Carlson, for Inflatable Rescue Boat Operations

Exothermic Cutting Torch Collet Nuts, ¼"

Zip Ties, Medium Duty, 10"

Topographical Maps for Mountainous Areas

Webbing, Red, 22'—In High Angle Rescue Packs

Pulleys, Pussik minding—In High Angle Rescue Packs

Post Screw Jacks

Pipe, Steel Schedule 40, 1½" x 4'

Pulleys, 2"—In High Angle Rescue Packs

Atmospheric Monitor, Spare Batteries, Lithium

Saw, Rotary, Blade, 14", Metal Cutting

Filter, Water/Dust Stop

Radio Harness—Chest—One Per Post Position

Exothermic Torch Flash Arrestor

Radio Bone Microphones, for Motorola Saba radio—One Per Post Position

Pipe, Schedule 40, 1½" x 20'

SCBA Bottles, 60 Minute—One Primary and One Spare Per Post Position, With 5 Extra for Equipment Operations

Electrical Plug, L5P, Male Plug

Electrical Plug, L5R, Female Receptacle

Air Bag, Rescue, Regulators—In Hardened Case for Industrial and Street Operations

Air Bag, Rescue, Regulators—In Back Pack for Rapid Intervention Operations and Collapse Rescue Operations

APPENDIX 4

Fireground RIC equipment per each member:

- Two or three light sources
- Personal tools
- Self contained breathing apparatus: (1 hour bottle if available, or consider closed circuit systems)
- Drop bag
- Webbing, rapid intervention strap, or other capture and harnessing device

Equipment for RIC:

- Thermal imager(s)
- Extra SCBA with mask (or SABA) for trapped firefighter/rescuer
- Rotary rescue saw
- Sawzall with extra blades and batteries (or electric)
- Portable battery-powered hydraulic rescue tools
- Air bags in bag with regulators and hose
- SCBA bottle for air bags
- Hydraulic jack
- Hydraulic or pneumatic emergency shores
- Attic ladder
- Debris bag with handles to carry out firefighter/rescuer
- Axes and belts
- Sledge hammer
- Rabbit tool
- Irons
- Sounding tool

- Chain saw
- Other tools based on hazards at the scene
- Rescue RIC equipment pool

Per RIC member:

- Personal protective equipment
- Three sources of light
- Respiratory protection based on hazards
- Rescue harness for high angle situations
- Other equipment based on hazards

Per RIC team:

- SABA or Biopacks for victims as well as rescuers.
- Intrinsically safe monitors, 3 sources of light, and other standard tools for confined space entry.
- Air-powered tools for extrication
- Confined space litters and rescue harnesses.
- Rapid intervention straps
- High angle equipment

For water rescue:

- Throw bags and high angle gear
- Inflatable rescue boats or personal watercraft
- Helicopter-based swiftwater teams, or helicopters equipped with cinch harnesses and other self-rescue devices that can be deployed from a copter.
- Whistles, bullhorns, radios, and other signaling devices.

For trench/excavation rescue:

- Air knives and air vacuums
- Hydro Vac trucks
- Ladders and other escape devices
- Digging tools
- Rapid shoring devices
- High angle gear

Checklist for RIC standby during emergencies

- ❑ Monitor incident frequencies
- ❑ Observe fire behavior and other factors
- ❑ Observe effects of tactics and strategy
- ❑ Watch for signs of impending collapse, backdraft, etc.
- ❑ Watch for rescue-related hazards like secondary collapse, flood surges, rope operation problems, helicopter mishaps, etc.
- ❑ Think LCES

Procedure for when RIC situation occurs

1. Assess the event with incident commander: what just happened, to whom did it happen, where were they last seen, are you in contact with them, what is likely to happen next, what is your best strategy to get to the victim(s) and extract him/them without delay?
2. Assemble RIC and equipment, give a quick briefing and move in.
3. Ensure use of RIC frequency
4. Monitor other frequencies to communicate with trapped/lost personnel, to maintain awareness of other conditions, to request assistance as necessary.
5. Make access to victim, or to search area.
6. Assess situation and take appropriate action (search, extrication, etc.)

APPENDIX 5

Quake Magnitude Indications [1]

Mild (0–4.3 Magnitude): Monitoring instruments record vibrations; shaking felt indoors; hanging objects swing.

Moderate (4.3–4.8): Dishes rattle; standing cars rock (commonly setting off car alarms); trees shake noticeably; doors swing; liquid spills from glasses.

Intermediate (4.8–6.2): Windows break; pictures fall from walls; plaster, bricks, and tiles fall; church bells ring; people lose their balance; unreinforced masonry and other vulnerable buildings may collapse; some overpasses may be damaged or collapse; tsunamis possible from ocean floor displacement or underwater landslides.

Severe (6.2–7.3): Chimneys fall; tree branches break; trees fall; landslides occur; foundations damaged; mud bubbles from ground (signs of liquefaction); buildings can be destroyed; overpasses may collapse; tsunamis may occur form ocean floor displacement or underwater landslides.

Catastrophic (7.3–9.6): Railway tracks can bend; roads can break up; large cracks may appear in the ground; buildings may sink, shift, or overturn (liquefaction); rockfalls and landslides; "waving" of ground surface is seen during the quake; river courses may be altered; river flow may be reversed; lakes may drain, depressions may fill with water, creating new lakes; tsunamis are possible from ocean floor displacement or underwater landslides.

NUMERICAL EARTHQUAKE INTENSITY RATING[2]

This rating system is used by the County of Los Angeles and other fire/rescue agencies to rapidly quantify the observed level of damage in the moments after an earthquake has struck. The ratings are used by all fire/rescue companies conducting "windshield surveys" to determine the level of damage in their first-in districts, which is reported to their battalion headquarters, and quickly relayed to the department's headquarters, where it is used to deploy fire/rescue resources where to the areas with the worst damage.

0—Nothing felt

1—Earthquake felt, no damage

2—Windows broken

3—Block walls down

4—Structures shifted off foundations

5—Structural collapse

References

[1] U.S. Geological Survey, Southern California Earthquake Center, USC School of Engineering.

[2] County of Los Angeles Fire Department Earthquake Procedures—Engine Company Quick Reference Sheet.

APPENDIX 6

Getting the Helo/High-Rise Team Onto the Roof

Helo/high-rise team will generally deploy to the roof of a high-rise building by the following methods, in order of preference:

Helicopter landing insertion. The helicopter lands on the roof (if it has a designated helipad or a roof that's certified to accommodate the helicopter landing), and firefighters step out of helicopter with their equipment. If there is no helipad, or if the ability of the roof to accommodate the weight of a helicopter landing is uncertain, one of the other following methods will be used.

"One skid" insertion. The helicopter places one skid on the edge of the roof, while the pilot maintains power to held the copter steady at roof level. The helo/high-rise firefighters step off the copter, unload their equipment, and move out of the way.

Rappel insertion. The firefighters assigned to the helo/high-rise team rappel from the helicopter, which is hovering over roof. Their equipment is lowered down to them by helicopter crewman using the rappel rope; then the rope is dropped onto the roof for firefighters to use for potential rescues or searches.

Hoist insertion. The helo/high-rise team firefighters are lowered onto the roof using the hoist cable as the helicopter hovers over the roof. Their equipment is lowered to them in the same manner.

On the roof. Once on the roof, firefighters assigned to the helo/high-rise team can perform the following tasks:

- Remove rooftop obstructions to clear access for helicopter access (to facilitate additional firefighters and equipment being delivered to the roof, and to ensure rapid extraction of trapped victims and firefighters as necessary).
- In coordination with command, open stairwells to help facilitate vertical ventilation to reduce "stack effect," to improve the atmosphere above the fire (improving survivability for victims and visibility for firefighters).
- Perform other tasks as assigned to improve ventilation.
- Search upper floors for victims (using thermal imagers, night vision, etc.) and remove them to safe atmosphere or shelter in place as necessary.

- Conduct fire attack on upper floors to reduce "lapping" and other vertical extension, as well as horizontal extension and to protect stairwells and victims.
- Conduct crowd control on the roof and on upper floors (in the event victims are trapped above the fire floor(s); conduct "shelter in place" operations as directed; and perform evacuation/rescue of victims, either to the roof, or down the stairwells (with proper coordination with command, assuming the stairwells are secure, and assuming moving down is the best option at the time).
- Conduct helicopter evacuation of victims trapped on the roof if other avenues of escape are cut off and it is deemed necessary to undertake helicopter evacuation. In many cases (except in cases of potential terrorist attacks with possible secondary attacks on adjacent buildings), it will be advantageous to move victims from the rooftop of one high-rise to the rooftop of another high-rise. In this case, at least one member of the helo/high-rise team should be transferred to the roof of the second building *with or before* the first group of victims.
- Conduct Rapid Intervention Crew (RIC) operations from above, and facilitate (as necessary) the rooftop evacuation of downed firefighters.
- Conduct forcible entry to allow victims trapped in stairwells and offices that may be damaged by fire, collapse, explosion, or other forces.
- Effect rescue of victims trapped in elevators.
- Conduct high angle rescue of victims trapped in situations requiring those tactics.
- Coordinate the rooftop deployment of additional companies and equipment as deemed necessary by incident command.
- Conduct other tasks as directed by the incident commander or other officers.

Rooftop evacuation of victims by helicopter. In order of preference, the following rooftop evacuation tactics can be employed by the helo/high-rise team:

Roof landing extraction. As the helo/high-rise team controls the crowd and as the helo/high-rise team leader (the captain of the USAR unit, rescue company, truck company, etc) communicates with the pilot of the incoming helicopter (keep in mind that a number of fire/rescue helicopters may be operating at the fire), the copter lands on the roof. The helo/high-rise team members assist people into the cabin of the copter (the number of people on each flight to be determined by the respective pilot). An air crewman or a member of the helo/high-rise team secures the door before takeoff.

One skid extraction. As the crowd is controlled and kept at a safe distance, and at the signal from the helo/high-rise team leader, the copter approaches the building and performs a "one skid," while the helo/high-rise team members assist a prescribed number of people into the cabin, with the air crewman securing the door before the copter departs.

Short haul extraction. As the crowd is controlled, one or more citizens is placed in an evacuation harness (or, in the case of injured patients, an appropriate rescue litter). The copter is signaled in, and the pilot hovers overhead while a short-haul line is lowered. Members of the helo/high-rise team attach the victim(s) to the loop(s) in the rope and signal the air crewman (standing on the skid above) to go ahead and lift.

In turn, the air crewman, on "hot mic" inside the cabin, directs the pilot to ascend. As the victim(s) is short-hauled to the other building or to the ground, the air crewman maintains a close watch to communicate the victim's position to the pilot until the victim is set down at the designated point. At that time the remaining member of the helo/high-rise team assists the victim's "touch down," disconnects the rope from the victim(s), and signals the copter to depart back to the burning high-rise to evacuate more victims.

Back on the roof of the burning high-rise, the helo/high-rise team should already have harnessed up the next citizen to be extracted, using additional harnesses they carry to set up a "round robin" that will expedite the process of removing victims.

At the set-down point, the remaining member of the helo/high-rise team removes the harness from the victim that was just extracted. This harness is then clipped to the end of the short haul when the next person is dropped off, so the air crew can transfer the harness back to the roof of the burning high-rise, where it is placed on another citizen.

Hoist Extraction. As members of the helo/high-rise control the crowd, one or more citizens is placed in an evacuation harness (or, in the case of injured patients, an appropriate rescue litter). The copter is signaled in, and the pilot hovers overhead while the hoist cable is lowered. Members of the helo/high-rise team attach the victim to cable and signal the air crewman (standing on the skid above) to begin hoisting.

In turn, the air crewman, on "hot mic" inside the cabin, directs the pilot to very slowly begin to ascend until the power of the helicopter itself was lifting the victim off the rooftop (this is standard protocol for hoist operations).

Normally, once the victim is actually suspended in the air under power of the copter, the pilot tells the air crewman, "Okay, you've got the load," at which time the air crewman activates the hoist motor to begin lifting the victim toward the copter, eventually bringing them all the way to the level of the cabin, assisting them into the cabin and securing them inside.

However, in the case of a high-rise fire with multiple victims waiting to be evacuated, the hoist cable might be treated as a sort of "short haul," and the victim might simply be short hauled at the end of the extended cable directly to the roof of an adjacent high-rise or to the ground. This would have the effect of reducing the time between each "rescue cycle" by eliminating the time it takes to retract the hoist cable for each victim.

As the victim is set down, the designated member of the helo/high-rise team disconnects the victim's harness from the cable and signals the copter to return to the burning high-rise to evacuate more victims.

Back on the roof of the burning high-rise, the helo/high-rise team would already have harnessed up the next citizen to be extracted, using additional harnesses. Finally, back at the set-down point, the designated member of the helo/high-rise team would remove the harness from the victim. The harness is then connected to the cable after the next person is dropped off, and the cycle continues until all the victims—and possibly the remaining firefighters from the helo/high-rise team—are extracted from the roof of the burning high-rise.

Operational retreat. In all of these cases, the helo/high-rise team and the team leader must be cognizant of the potential for an "operational retreat" to be ordered by the incident commander. This would mean the helicopter would come back and evacuate all the firefighters off the roof…if the smoke and other conditions allow the copter to get back to the roof.

Every member of the helo/high-rise team must also recognize from the beginning that it is possible for conditions to deteriorate to the point where it becomes impossible for the helicopter to return to the rooftop at all, effectively stranding them above a raging high-rise fire with the potential for structural failure.

APPENDIX 7

About Soil for Trench/Excavation Rescue[1]

- 1ft.^{3} of soil weighs about 100 lb.–125 lb.
- Victims may suffocate with only 18 in. to 24 in. (700 lb.–1,000 lb. on their chest) covering them if they are horizontal.

Soil type A is cemented or cohesive (clay, silty clay, clay loam). It has an unconfined compressive strength of 1.5 tons per square foot (tsf) or greater. It is generally virgin, non-fissured soil. It clumps easily, and because of its compactness it can be very difficult to remove from around victims after it collapses and buries them.

Soil type B is cohesive (angular gravel, silt, sandy loam) with an unconfined strength between 0.5 tsf and 1.5 tsf. It also clumps easily, but also can surround a victim tightly, and therefore may be quite deadly in a collapse situation.

Soil type C is granular (gravel, sand, loamy sand) with an unconfined strength less than 0.5 tsf. It may also be called submerged or free seeping soil. It breaks apart easily, which makes it somewhat easier to remove, but also less stable and more apt to fill a victim's mouth and nose when collapse occurs.

Stable rock is a natural solid material that can be excavated with vertical sides and remain intact while exposed.

Thumb penetration test

Some soils may be tested for their typing by pushing one's thumb into them and observing the reaction. Type A soil allows the thumb to penetrate only with very great effort. Type B soil allows the thumb to penetrate ¾ in. with moderate effort. Type C soil allows thumb penetration with ease.

References

[1] County of Los Angeles Fire Department Trench Rescue Manual.

APPENDIX 8

Los Angeles County Marine Disaster Plan

In the 1980s, Federal Aviation Administration (FAA) regulations required a Marine Disaster Plan to be developed in case of an airliner crash in the coastal waters near Los Angeles International Airport (LAX). Two airliners had crashed into the ocean after takeoff from LAX during the 1960s (one of the crashes resulted in survivors being rescued at sea by the Coast Guard, the Los Angeles County Lifeguards, and the Los Angeles County Fire Department), but in recent times the safety record there had been free of air crashes in the ocean. Still, the FAA understood that with thousands of takeoffs a day and millions of passengers going through LAX each year, it was necessary to have an elaborate plan in place in case of another air/sea disaster.

In response to the FAA regulations, the U.S. Coast Guard, Los Angeles County and City Fire Departments, the Los Angeles County Sheriff Department, the Los Angeles County Lifeguards (now a division of the fire department), the Harbor Patrol, and other associated agencies developed the LAX Marine Disaster Plan, which has since been expanded to be capable of responding to incidents where victims are in need of search and rescue on the open sea as a result of small airplane crashes, capsized boats, boat collisions, boat fires, ship collisions (including passenger cruise lines), etc.

When the Marine Disaster Plan is implemented, the Los Angeles County Fire Department (LACoFD) automatically responds with other agencies (including the U.S. Coast Guard, the Los Angeles County Sheriff Department, the Harbor Patrol, the City of Los Angeles Fire Department, etc.) in a coordinated response of marine-based, air-based, and land-based fire/rescue/EMS resources, to effect timely rescue and recovery, provide rapid transportation to shore, triage, and treat victims through a "landslide multi-casualty incident system," and transport them to the appropriate hospital.

Ultimate responsibility for off-shore search and rescue operations along the coast of Los Angeles County is assigned to the commander of the Eleventh Coast Guard District as the Federal Search and Rescue Coordinator (SMC). However, it has been agreed by all agencies participating in the LAX Marine Disaster Plan that the first on-scene unit from any participating agency will establish the initial incident command. Upon arrival of the Coast Guard and other agencies, unified command will be established with the Coast Guard as the lead agency.

When a Marine Disaster occurs, the LACoFD immediately dispatches at least two USAR task force stations, whose personnel are assigned as rescue swimmers transported by the department's fire/rescue helicopters for heli-deployment directly into the ocean to inflate 10-person rescue platforms and inflatable rescue boats (IRBs) and begin rescuing victims.

To maintain constant readiness for marine disaster operations, members of the LACoFD USAR Task Force stations are expected to complete standard helicopter deployment training (and annual recertification), HEEDS/dunker training and annual recertification, and rescue swimmer training. Their personal protective equipment for marine disasters (*e.g.*, wetsuits, fins, snorkels and masks, etc.) is always kept aboard the USAR apparatus for immediate response from anywhere they may be—whether they are attending training, conducting some public education function, or simply returning from a rescue or multiple alarm fire in the 4,000 square mile county.

Although their assignment as rescue swimmers during marine disaster response is only one of many technical rescue roles they may be called upon to fill, it is one that is constantly in the back of the minds of USAR Task Force members because it could happen at any time, and when it does minutes and seconds will truly count for people struggling to stay on the surface of the sea. They must be constantly ready to assemble as a rescue swimmer team, board a fire/rescue helicopter for flight to the crash site, to assess the situation, to deploy out of the copter into the ocean using standard methods that prevent injury to rescuers, to inflate flotation devices for victims on the surface, and to begin the process of rescuing any survivors. They must also be ready to begin recovering non-survivors after the living have been rescued.

For anyone who survives a sea crash, there is a narrow window of survivability, and the window grows narrower each second. That's why the most critical thing is to get rescuers to the scene without delay, to deploy floatation for the victims, and to get them out of the water. That is why the USAR Task Forces go immediately to the scene by helicopter, while the boats respond from the coast.

And then there is the issue of Large Predatory Marine Animals (LPMAs), which will almost certainly be coming to investigate and perhaps to eat. It is just another hazard for which the rescue swimmers must be prepared.

At the same time as the USAR Task Forces and helicopters are dispatched as part of the marine disaster plan, the Los Angeles County Lifeguards (now a division of the LACoFD) launch their Baywatch boats from strategic locations, as well as a number of beach-based inflatable rescue boats and rescue-ready personal watercraft (PWCs) staffed with lifeguards, to join the search and rescue operations. While the USAR task force personnel are being deployed into the water from helicopters, the lifeguards will be arriving in the big white Baywatch boats, in IRBs, and on PWCs. They have the ability to take victims that the USAR personnel have provided floatation for and shuttle them to the bigger boats for transportation back to shore.

Simultaneously, the U.S. Coast Guard, the Los Angeles County Sheriff Department, and other agencies (based on the location of the disaster) are dispatched. All these agencies and units are trained and prepared to work together in a coordinated fashion under the LAX Marine Disaster Plan, which includes provisions for airliners crashing in the ocean, ships (including cruise liners) sinking or catching fire, small planes, and helicopters down in the ocean, boats capsizing, boats on the rocks, and ship collisions, and other marine disaster situations.

The FAA requires periodic multi-agency air sea disaster exercises to ensure that that system works and that problems are solved ahead of time. The concept of multi-agency marine disaster plans is now being evaluated for expansion to other southern California counties.

APPENDIX 9

SWIFTWATER RESCUE EQUIPMENT

Firefighters dispatched to river and flood (swiftwater) rescues should be provided a minimum of U.S. Coast Guard approved Type III or Type V personal flotation devices (PFDs). Attached to the PFDs should be a rescue knife (to cut away rope or debris in case of entanglement in moving water) and a whistle (for emergency communication). Other equipment should be suited to the level of exposure to danger. This equipment includes foot wear (wet suit booties, sneakers, or other light weight foot wear with hard rubber soles); head protection (rescue helmets); and occasionally wet suits or dry suits for thermal protection and reduction of abrasions. There are a wide variety of other equipment types and models on the market to deal with particular hazards.

Minimum rescue equipment includes throw bags (60 ft. to 80 ft. of polypropylene or nylon rope stuffed into a floating bag), rescue rope and hardware, lights, and tools for reaching assists (*i.e.,* pike poles). There is a wide variety of rescue equipment ranging from line throwing devices to inflatable rescue boats.

STANDARD OPERATING GUIDELINES FOR SWIFTWATER RESCUE

Fire department Standard Operating Guidelines (SOGs) for swiftwater rescue should include the following topics.

1. Mandatory use of appropriate personal protective equipment (PPE).
2. Prohibitions on use of equipment or apparel likely to endanger personnel if they enter the water to attempt a rescue, or if they accidentally fall into the water. Examples may include full turnouts, structure fire helmets, ropes tied around the body of rescuers in moving water, and other items that may adversely affect safety.
3. Upstream safety spotters to warn of debris approaching the rescue scene.
4. Downstream safety positions (personnel staged at strategic downstream points, ready to conduct a Rapid Intervention in case a rescuer or victim is swept away).

5. The use of low risk methods whenever possible, and consideration of increasingly higher-risk options as necessary.
6. Rules specifying that only properly trained and equipped personnel should operate in IDLH environments (such as low head dams and concrete flood channels with vertical walls).
7. Standard communications, techniques, and tactics. Mutual aid resources should be operating in the same mode.
8. Waterway rescue pre-plans should be utilized as the initial Incident Action Plan whenever possible. This provides a basis for the initial actions of all responders.

APPENDIX 10

The Three Phases of Suicide in Progress Rescue

Sergeant Barry Perrou of the Los Angeles County Sheriff Department's Crisis Negotiation Team describes a three-phase system that has proven effective in managing these difficult situations. In some cases, one of the two crisis negotiators is designated as the *shooter.* His job is to allow the *talker* to negotiate with the subject, but remain ready to shoot the subject in case he decides to pull a gun, attempt to take the negotiator hostage, or otherwise harm the *talker.* It is a tactic that may not immediately be evident to firefighters on the scene, but one they should be aware of so they remain out of the line of fire and so they wear Personal Protective Equipment (PPE) such as flak vests and helmets. Some of these tactics (*i.e.*, designating a *shooter,* for example) are intended as law enforcement actions, but fire/rescue personnel should be aware that these tactics are being considered by the police at the scene of a "jumper" or other crisis negotiation situation, because it can and will impact their operations. This emphasizes the need for a unified incident command system and good coordination and communication between law enforcement and the fire department.

Suicide in progress phase one

1. Stabilize the situation.
2. Assess the situation for weapons (indications the subject is intent on "suicide-by-cop").
3. Some so-called "jumpers" may also wish to harm police, firefighters, and others sent to help. They may wish to take someone with them.
4. Designate a shooter(s) in the event the subject prompts suicide-by-cop actions.
5. Do not rush to the rescue (remain wary of unseen weapons, explosives, etc.).
6. Request EMS, fire department rescue or USAR resources (and, if appropriate a truck company) to provide belay lines and other protection for the negotiators, and to be ready in case of an "adverse event."
7. Request a fall bag through the fire department (if applicable), with non-code response and set-up out of sight of the subject.

8. Determine if the incident is a truly valid threat or a public nuisance.
9. Observe the total situation.
10. Request highway patrol for freeways, and have the entire freeway shut down in both directions.
11. Shut down bridges in both directions.
12. Request crisis negotiators and appropriate mental health authorities.
13. Identify local emergency room hospitals, trauma centers, routes, etc.
14. Avoid/prevent heroic or independent acts until appropriate or necessary.

Suicide in progress phase two

1. Secure a larger inner perimeter (isolate, and deny entry).
2. Eliminate public spectators and other "audiences" as much as feasible.
3. If first responders are communicating effectively with subject, leave them in place until appropriate to relieve them.
4. Develop a plan between the fire and police departments to ensure everyone is on the same page. Avoid strategic and tactical surprises that can be prevented through good communication and planning.
5. Assure that crisis negotiators operate as a team: the first armed negotiator is the Primary (the *talker*), which the second armed negotiator is the Secondary (the *shooter*, or *non talker*).
6. Turn on microcassette to record the discussion with subject.
7. When appropriate, use the fire department rescue companies or US&R units and truck company to assist with high angle access, belay lines, and a tactical rescue plan.
8. Request family and friends of the subject to command post—keep them out of sight!
9. Deploy video camera to document crisis negotiations.
10. Gather subject profile information.
11. Complete data-bank search (CWS, RAPS, AWS, Psych, etc.—a negotiator function).
12. Assess subject for psychosis, paranoia, agitation, etc. (a negotiator function).
13. Assess the subject for rationality.
14. Assess for alcohol and recreational drug use.
15. Secure a safe working area (eliminate slip, fall, and other hazards).

a. US&R units/rescue companies assess hazards and advise remedies.
b. US&R units/rescue companies establish backup/safety for negotiators.
c. US&R units/rescue companies provide special equipment.
d. US&R units/rescue companies assist in tactical rescue plan.
e. US&R units/rescue companies to assist in tactical rescue?
f. US&R units/rescue companies provide safety for subject (as appropriate).

Suicide in progress phase three:

1. Remember: The clock starts *now,* not when the first units arrive!
2. Develop an incident action plan.
3. Develop a tactical plan.
    a. Verbal intervention (issues causing despair).
    b. Tactical intervention/rescue (when appropriate).
4. Select trained primary negotiator/talker (possibly female).
5. Select trained secondary/non-talker/shooter.
6. Keep containment officers advised.
7. Develop a surrender plan.
8. Have negotiators on separate, restricted or dedicated frequency.
9. Initiate crisis intervention/negotiation.
10. Negotiate in good faith.
    a. Compassion, respect, dignity, hope, and help.
    b. Be prepared for rage and anger.
11. If tactical rescue is initiated, use one command that everyone understands!
12. If tactical rescue is considered, ask "Why?" and "Why Now?"
13. Assess for pre-death behaviors during negotiations (if things are not going well).
    a. Rhythmic movements (bad sign!)
    b. Counting (bad sign!)
    c. Closing eyes as if to avoid watching the scene.
14. Never initiate independent or heroic acts that aren't 110% certain.
15. Prepare to assist with surrender.
16. Be respectful to subject as they surrender.
17. Assume negotiators should remain with subject to hospital, custody, etc.
18. Be prepared for possible death leap or other sudden act.
19. Assume mandatory critical incident stress debriefing for all personnel involved.

19

HAZARDOUS MATERIALS OPERATIONS

Gregory G. Noll

CHAPTER HIGHLIGHTS

- Key legislative, regulatory, and voluntary consensus standards make up the foundation of fire service hazardous materials (hazmat) planning and emergency response operations.
- Planning and prevention have resulted in a decline in the number and severity of working hazmat incidents.
- To be effectively managed and controlled, hazmat problems must be approached systematically, from a broad and coordinated perspective.
- The Eight-Step Process provides a flexible management system and consistent structure that enables an operation to expand if the scope of an incident changes.

INTRODUCTION

The decade of the 1980s saw the emergence, growth, and acceptance of hazardous materials (hazmat) and environmental issues within the fire service. These issues have impacted many aspects of our work, including training, community planning, fire code development and enforcement, live fire training, personal protective clothing and equipment, and the selection of fire extinguishing agents. In many respects, hazmat was to the 1980s what emergency medical services (EMS) was to the 1970s. As the emergency response community moves through the 21st century, the fire service will continue to see the maturation of hazmat as one of the primary public safety responsibilities, along with fire, rescue, and EMS.

Regardless of the response "level" to be provided within a community, fire service executives must recognize that hazmat and environmental planning, prevention, training, and response issues are inherent elements of a comprehensive public safety program. The absence of a hazmat response team does not alter the fact that there are still a number of hazmat operational and regulatory issues that must be continuously monitored and managed by fire service executives.

Traditionally, the fire service has been the primary local government agency responsible for hazmat emergency planning and response, although in some jurisdictions this responsibility has been delegated to other agencies, including local health departments, police departments, and emergency medical services. In addition, there are a number of other regional, state, and federal agencies with responsibilities regarding hazardous materials and the environment, including state and federal environmental agencies (*e.g.*, EPA, DER), the United States Coast Guard (USCG), state and federal emergency management agencies (*e.g.*, FEMA), and state and federal law enforcement agencies (*e.g.*, state police environmental crimes units, FBI, ATF). When combined with the growth of the terrorism threat, the need for intergovernmental communication and coordination is critical if hazmat problems are to be effectively managed.

This chapter has been written to provide chief fire officers and fire service executives with an overview of the fire service's role in managing hazmat problems. It is not meant to be a treatise on the subject, but merely an overview with thoughts and ideas on where additional information and assistance can be obtained. It should be noted that a key resource in managing a fire service hazmat program is the fire department's hazmat officer/coordinator or a senior fire officer charged with responsibilities in this area. In many respects, this individual serves as the fire department's "in-house environmental consultant."

Hazmat Laws, Regulations, and Standards

Operations involving the manufacture, transport, and use of hazardous materials, as well as the response to hazmat incidents, are impacted by a large body of laws, regulations, and voluntary consensus standards. These rules influence virtually every aspect regarding the handling of hazardous materials. Because of their importance to emergency planning and response operations, chief fire officers must have a working knowledge of how the regulatory system works. What is the difference between a law, a regulation, and a standard? The three terms are often used interchangeably, but they have distinctly different meanings.

Laws

Laws are created through an act of Congress or by individual state legislatures. Laws typically provide broad goals and objectives, mandatory dates for compliance, and established penalties for noncompliance. Federal and state laws enacted by legislative bodies usually delegate the details of implementation to a specific federal or state agency. For example, the United States Occupational Safety and Health Act delegates rulemaking and enforcement authority for worker health and safety issues to the Occupational Safety and Health Administration (OSHA).

Regulations

Sometimes called *rules*, federal and state regulations provide the guidelines for complying with laws enacted through legislative action. Regulations permit individual governmental agencies to enforce laws through audits and inspections conducted by federal and state officials. Sometimes agencies such as the EPA and OSHA adopt consensus standards that have the effect of regulations using a process known as the "general duty clause." This concept is discussed in greater detail later in this section.

Voluntary consensus standards

These standards are normally developed through professional organizations or trade associations as a means for improving the individual quality of a product or system. Within the fire service, the National Fire Protection Association (NFPA) is recognized for its role in developing consensus standards and recommended practices that impact fire safety and fire department operations. In the United States, standards are developed primarily through a democratic process whereby a committee of subject matter specialists representing varied interests writes the first draft of the standard. The document is then submitted to either a larger body of specialists or to the general public, who may then amend, vote on, and approve the standard for publication. Collectively, this process is known as the "consensus standards process."

When a consensus standard is completed, it may be voluntarily adopted by government agencies, individual corporations, or organizations. Many hazmat consensus standards are also adopted as references in regulations. In effect, when a federal, state, or municipal government adopts a consensus standard by reference, the document becomes a regulation. An example of this process is the adoption of NFPA 30 Flammable and Combustible Liquids Code and NFPA 58 Liquefied Petroleum Gas Code as laws in approximately 30 states.

General duty clause

Voluntary consensus standards may also have the effect of regulations when federal regulatory agencies such as OSHA and the EPA use a process known as the *general duty clause.* For example, Section (5)(a)(1), the general duty clause of the OSHA Act, requires that employers provide employees with work and a place of employment free from recognized hazards causing or likely to cause death or serious physical harm. If there are no specific regulations pertaining to a health or safety concern, OSHA can use the general duty clause to address the issue.

When the general duty clause is used, OSHA establishes what employers must do to protect their employees from hazards (*e.g.*, fire, hazardous materials, confined space) and determines how the agency will ensure that employers provide the required protection. In these situations, OSHA typically relies upon standards promulgated by other branches of the federal government (*e.g.*, EPA, USCG), along with guidelines developed by professional associations such as the NFPA or the American Petroleum Institute (API) that have in effect become industry practice. For example, when evaluating safety issues at facilities that manufacture, store, or use hazardous materials, OSHA often adopts by reference NFPA and American National Standards Institute (ANSI) standards, or related industry standards published by organizations such as the American Society of Mechanical Engineers (ASME), Compressed Gas Association (CGA), and American Conference of Governmental Industrial Hygienists (ACGIH).

Federal Hazmat Laws

Hazmat laws have been enacted by Congress to regulate everything from finished products to toxic waste. Congress has enacted many hazmat laws in response to specific environmental events, such as Love Canal, New York; Bhopal, India; and the Exxon Valdez oil spill. Because of their lengthy official titles, most people simply use abbreviations or acronyms when referring to these laws. The following summaries will help you understand some of the more important laws affecting the fire service.

Comprehensive Environmental Response Compensation and Liability Act (CERCLA). Known as "Superfund," this law addresses hazardous substance releases into the environment and cleanup of inactive hazardous waste disposal sites. It also requires that companies responsible for the release of hazmat notify the National Response Center (NRC) when a specified "reportable quantity" occurs. This center is the single point-of-contact for the federal government.

Resource Conservation and Recovery Act (RCRA). This law establishes a framework for the proper management and disposal of all waste materials (*i.e.*, solid, medical, hazardous), including hazardous waste treatment, storage, and disposal (TSD) facilities. The act also establishes installation and notification requirements for underground storage tanks.

Clean Air Act (CAA). This law establishes requirements for airborne emissions and the protection of the environment. CAA amendments in 1990 addressed emergency response and planning issues for certain facilities that use highly hazardous chemicals. Amendments established the National Chemical Safety and Hazard Investigation Board; the requirement of risk management and prevention plans (RMPPs) at the facilities; and OSHA's 29 CFR 1910.119: *Process Safety Management of Highly Hazardous Chemicals.* In addition, certain facilities are required to make information available to the general public regarding the manner in which chemical risks are handled within a facility.

Title III, Superfund Amendments and Reauthorization Act of 1986 (SARA). Also known as the *Emergency Planning and Community Right-to-Know Act.* This law requires chemical manufacturers to notify the local emergency planning committee (LEPC) and the state emergency response commission (SERC) of hazardous substance inventories that exceed the threshold planning quantity (TPQ). Also, certain chemical manufacturing facilities are required to notify the Environmental Protection Agency of normal toxic chemical releases annually.

Of these four laws, SARA Title III has had the greatest impact on fire service operations. As the name implies, SARA acted to both amend and reauthorize the *Comprehensive Environmental Response, Compensation, and Liability Act of 1980* (CERCLA or Superfund). While many of the amendments pertained to hazardous waste site cleanup, SARA added a number of new requirements that established a national baseline with regard to planning, response, management, and training for hazmat emergencies.

SARA Title III mandated the establishment of both state and local planning groups to review or develop hazmat response plans. The state planning groups are referred to as *state emergency response commissions* (SERCs). An SERC is responsible for developing and maintaining a state's emergency response plan. This includes ensuring that planning and training are taking place throughout the state, as well as providing assistance to local governments, as appropriate. States generally provide an important source of technical specialists, information, and coordination. However, they typically provide only limited operational support to local government and the fire service in the form of equipment, materials, and personnel.

The coordinating point for both planning and training activities at the local level is the local emergency planning committee (LEPC). In many communities, either the local emergency management agency (EMA) or the fire department staff and manage the activities of the LEPC. Among the LEPC membership are representatives from the following groups:

- Elected state and local officials
- Fire department

- Law enforcement
- Emergency management
- Public health officials
- Hospital
- Industry personnel, including facilities and carriers
- Media
- Community organizations

The LEPC is specifically responsible for developing or coordinating the local emergency response system and capabilities. A primary concern is the identification, coordination, and effective management of local resources. Among the primary responsibilities of the LEPC are:

- Develop, regularly test, and exercise the hazmat emergency response plan.
- Compile information on hazmat facilities within the community. This includes chemical inventories, material safety data sheets (MSDS) or chemical lists, and points of contact.
- Coordinate the community right-to-know aspects of SARA Title III.

In a number of communities, the LEPC has expanded it scope and responsibilities to adopt an "all hazards" approach to emergency planning and management. Chief fire officers desiring more information on both hazmat and "all hazards" planning should consult the following websites:

- EPA Chemical Emergency Preparedness and Prevention Office (CEPPO)
  http://www.epa.gov/swercepp/
- Federal Emergency Management Agency (FEMA)
  http://www.fema.gov/
- International Emergency Management Association (IAEM)
  http://www.iaem.com/
- U.S. National Response Team
  http://www.nrt.org/
- U.S. Chemical Safety and Hazard Investigation Board
  http://www.chemsafety.gov/

HAZARDOUS MATERIALS REGULATIONS

As previously noted, laws delegate certain details of implementation and enforcement to federal, state, or local agencies who are then responsible for writing the actual regulations for enforcing the legislative intent of the law. Regulations are detailed and either define the broad performance required to meet the letter of the law or provide very specific guidance. The following summaries include some of the more significant federal regulations that affect the fire service.

***Hazardous Waste Operations and Emergency Response* (29 CFR 1910.120).** Also known as HAZWOPER, this federal regulation was issued under the authority of *SARA Title I.* The regulation was written and is enforced by OSHA in the 23 states and two territories having their own OSHA-approved occupational safety and health plans. In the remaining 27 states and the District of Columbia, fire service personnel will be covered by a similar regulation enacted by the EPA (40 CFR Section 311). However, despite the common aspects of OSHA and EPA regulations, Congress did not provide EPA with any enforcement power for Section 311 issues. For a listing of the states covered by the OSHA and EPA regulations, see the Technical Resources section of this chapter.

This regulation establishes important requirements for both industry and public safety organizations that respond to hazmat or hazardous waste emergencies. This includes firefighters, EMS personnel, and hazmat response team (HMRT) members. Requirements cover the following areas:

- Hazmat emergency response plan
- Emergency response procedures, including the establishment of an incident command system, the use of a buddy system with backup personnel, and the establishment of a safety officer
- Specific training requirements covering instructors and both initial and refresher training
- Medical surveillance programs
- Post-emergency termination procedures

Of particular interest to the chief fire officer are the specific levels of competency and associated training requirements identified within OSHA 1910.120 (q)(6). These levels are outlined as follows:

First responder at the awareness level. These are individuals who are likely to witness or discover a hazardous substance release and who are trained to initiate an emergency response notification process. The focus of their hazmat responsibilities is on securing the incident site, recognizing and identifying the materials involved, and making the appropriate notifications. These individuals take no further action to control or mitigate the release. First responder awareness personnel shall have sufficient training or experience to objectively demonstrate the following competencies:

- An understanding of what hazardous materials are and the risks associated with them in an incident
- An understanding of the potential outcomes associated with a hazmat emergency
- The ability to recognize the presence of hazardous materials in an emergency, and, if possible, identify the materials involved
- An understanding of the role of the first responder awareness individual within the local emergency operations plan (This would include site safety, security and control, and the use of the *DOT Emergency Response Guidebook.*)
- The ability to recognize a need for additional resources and to make the appropriate notifications to the communication center

The most common examples of first responder awareness personnel are law enforcement and plant security personnel and some public works employees. There is no minimum hourly training requirement for this level; employees should have sufficient training to objectively demonstrate the required competencies.

First responder at the operational level. Most fire department suppression personnel fall into this category. These individuals respond to releases or potential releases of hazardous substances as part of the initial response team for the purpose of protecting nearby persons, property, or the environment from potential harmful effects. They are trained to respond in a defensive fashion without actually trying to stop the release. Their primary function is to contain the release from a safe distance, keeping it from spreading and damaging exposures. First responder operational personnel shall have sufficient training or experience to objectively demonstrate the following competencies:

- Knowledge of basic hazard and risk assessment techniques
- Knowledge of how to select and use proper personal protective clothing and equipment available to the operations-level responder
- An understanding of basic hazmat terms
- Know how to perform basic control, containment, or confinement operations within the capabilities of the resources and personal protective equipment (PPE) available
- Know how to implement basic decontamination measures
- Possess an understanding of the relevant standard operating procedures and termination procedures

First responders at the operational level shall receive at least eight hours of training or have sufficient experience to objectively demonstrate competency in the previously mentioned areas, as well as the established skill and knowledge levels for the awareness-level first responder.

Hazmat technicians. These individuals respond to releases or potential releases for the purpose of stopping it. Hazmat technicians generally assume a more aggressive role than individuals at the operations level. They are often able to approach the incident site to plug, patch, or otherwise stop the release of a hazardous substance.

Hazmat technicians are required to receive at least 24 hours of training equal to the operations-level first responder and have competency in the established skill and knowledge levels to:

- Implement the community emergency operations plan
- Classify, identify, and verify known and unknown materials using field survey instruments and equipment (direct reading instruments)
- Function within an assigned role in the incident command system
- Select and use the proper specialized chemical personal protective clothing and equipment provided to the hazmat technician
- Understand hazard and risk assessment techniques
- Perform advanced control, containment, and/or confinement operations within the capabilities of the resources and equipment available to the hazmat technician
- Understand and implement decontamination procedures
- Understand basic chemical and toxicological terminology and behavior

Many communities have personnel trained as emergency medical technicians (EMT), but these EMTs do not have the primary responsibility for providing basic or advanced life support medical care. Similarly, hazmat technicians may not be part of a hazmat response team. However, if they are part of a designated team as defined by OSHA, they must also meet the medical surveillance requirements within OSHA 1910.120.

Hazmat technician training courses generally range from 24 to 200+ hours. In evaluating where and how to initially train personnel to the technician-level, chief fire officers must consider these questions: (1) What are the local hazardous materials risks? (2) What are the expected tasks and duties of the personnel? and (3) What equipment and resources will be available for emergency response use? Industry emergency response teams are regularly trained to the hazmat technician level in a 24-hour course as they respond to a limited number of chemicals and response scenarios. In contrast, given the wide range of potential situations that could occur in a community, fire service technician-level training will likely require 40 to 200+ hours.

Hazmat specialists. These individuals respond with and provide support to hazmat technicians. While their duties parallel those of the technician, they require a more detailed or specific knowledge of the various substances they may be called upon to contain. These individuals may also act as the site liaison between federal, state, local, or other governmental authorities in regard to site activities.

Similar to technicians, hazmat specialists shall receive at least 24 hours of training equal to the technician level and have competency in the following established skill and knowledge to:

- Implement the community emergency operations plan
- Classify, identify, and verify known and unknown materials using advanced field survey instruments and equipment (direct reading instruments)
- Know the state's emergency response plan
- Select and use the proper specialized chemical personal protective clothing and equipment provided to the hazmat specialist
- Understand in-depth hazard and risk assessment techniques
- Perform advanced control, containment, and/or confinement operations within the capabilities of the resources and equipment available to the hazmat specialist
- Determine and implement decontamination procedures
- Develop a site safety and control plan
- Understand basic chemical, radiological, and toxicological terminology and behavior

Whereas the hazmat technician possesses an intermediate level of expertise and is often viewed as a "utility person" within the response community, the hazmat specialist possesses an advanced level of expertise. Within the fire service, the specialist often assumes the role of the assistant safety officer for the hazmat group or the hazmat group supervisor. Within industry however, an industrial hazmat specialist's expertise may be "product specific." All specialists must meet the medical surveillance requirements outlined within OSHA 1910.120.

On-scene incident commanders. Incident commanders (ICs), who assume control of the incident scene beyond the first responder awareness level, shall receive at least 24 hours of training equal to operations-level first responder training. In addition, employers must certify that incident commanders have competency in the following areas:

- Know and be able to implement the community's incident command system
- Know how to implement the local emergency operations plan
- Understand the hazards and risks associated with working in chemical protective clothing
- Know the state's emergency response plan and have awareness of the federal regional response team
- Know and understand the importance of decontamination procedures

Readers seeking further information on the application of OSHA standards to hazmat emergency response situations should consult the OSHA website at *http://www.osha-slc.gov/*. Specific attention should be paid to (1) OSHA interpretations of the HAZWOPER standard, and (2) OSHA Directive Number CPL 2-2/59A: *Inspection Procedures for the Hazardous Waste Operations and Emergency Response Standard*, 29 CFR 1910.120 and 1926.65, Paragraph (q): *Emergency Response to Hazardous Substance Releases* (April 4, 1998).

***Hazard Communication Regulation* (29 CFR 1910.1200).** HAZCOM is a federal regulation that requires hazardous materials manufacturers and handlers to develop written material safety data sheets (MSDS) on specific types of dangerous chemicals. MSDSs must be made available to employees who want information about a chemical in the workplace upon request. Examples of information on an MSDS include: known health hazards; the physical and chemical properties of the material; first aid, firefighting, and spill-control recommendations; protective clothing and equipment requirements; and emergency telephone contact numbers.

The chief fire officer should recognize that fire service personnel are also covered by state and federal HAZCOM and "right-to-know" laws. Depending on the jurisdiction, MSDS's must be on-file for fire department supplies and materials having hazardous ingredients, including cleaning supplies and fire-extinguishing agents.

Hazmat transportation regulations (49 CFR)

This series of regulations is issued and enforced by the U.S. Department of Transportation (DOT). The regulations govern container design, chemical compatibility, packaging and labeling requirements, shipping papers, transportation routes and restrictions, and so forth. The regulations are comprehensive and strictly govern how all hazardous materials are transported by highway, rail, pipeline, air, and water. Some fire departments have been trained and certified to conduct inspections of hazmat cargo tank trucks using DOT regulations.

State regulations

Each of the 50 states and the U.S. territories maintains an enforcement agency responsible for hazardous materials. The three key players in each state usually consist of the state fire marshal, the state OSHA, and the state Department of the Environment (sometimes known as Natural Resources or Environmental Quality). While there are many variations, the fire marshal is typically responsible for the regulation of flammable liquids and gases due to the close relationship between the flammability hazard and the fire prevention code, while the state environmental agency is responsible for the development and enforcement of environmental safety regulations.

While known by various titles, most states have a governmental equivalent of the federal OSHA. Approximately 23 states have adopted federal OSHA regulations as state law. This method of adoption has increased the level of enforcement of hazmat regulations such as the HAZWOPER previously described.

State governments also maintain an environmental enforcement agency and environmental crimes unit that usually enforces the federal RCRA, CERCLA, and CAA laws at the local level. Increased state involvement in hazardous waste regulatory enforcement has significantly increased the number of hazmat incidents reported. This increase is expected to continue into the future and will continue to generate more fire service activity at the local level.

Voluntary Consensus Standards

Standards developed through the voluntary consensus process play an important role in making the workplace and the public safe from both fire and hazmat releases. Historically, a voluntary standard improves over time as each revision reflects field experience and adds more detailed requirements. As users of the standard adopt it as a way of doing business, the level of safety gradually improves over time. Consensus standards can usually be developed more quickly than regulations to meet "issues of the day." For example, in response to potential terrorist incidents involving dual-use industrial chemicals, chemical, or biological agents, in 2001, the NFPA approved NFPA 1994 Protective Ensembles for Chemical/Biological Terrorism Incidents. This standard provides guidance for emergency response personnel in the selection and use of protective clothing and equipment during such incidents.

Consensus standards are updated on a more regular basis than governmental regulations. For example, since the OSHA 1910.120 interim regulation was released in December, 1986, NFPA 472 Standard for Professional Competence of Responders to

Hazardous Material Incidents (described below), has been revised four times, the most recent being the 2002 edition. NFPA consensus standards are typically revised on a five-year cycle.

In many respects, voluntary consensus standards provide a way for individual organizations and corporations to self-regulate their businesses or professions. Interestingly, all of the national fire codes in the United States are developed through the voluntary consensus standards process. Historically, the two key players have been the NFPA and Western Fire Chiefs Association. Many of the standards developed by these two organizations address hazmat storage and handling, personal protective clothing and equipment, and hazmat professional competencies. The next section describes the most important hazmat consensus standards used within the fire service.

NFPA 471 Recommended Practice for Responding to Hazardous Material Incidents. This document covers planning procedures, policies, and application of procedures for incident levels, personal protective clothing and equipment, decontamination, safety, and communications. The purpose of NFPA 471 is to outline the minimum requirements that should be considered when dealing with responses to hazmat incidents, and to specify operating guidelines.

It should be noted that NFPA 471 is a recommended practice and not a technical standard. A recommended practice is a document similar in content and structure to a code or standard, but that contains only recommended or non-mandatory provisions. This difference is usually illustrated by using the term "should" to indicate recommendations in the document (as compared to "shall" in a standard).

NFPA 472 Standard for Professional Competence of Responders to Hazardous Material Incidents. The purpose of NFPA 472 is to specify minimum competencies for those who will respond to hazmat incidents. The overall objective is to reduce the number of accidents, injuries, and illnesses during response to hazmat incidents, and to prevent exposure to hazmat to reduce the possibility of fatalities, illnesses, and disabilities affecting emergency responders.

It is important to recognize that NFPA 472 is not limited to fire service personnel (although this is a common misconception within the emergency response community), but covers all hazmat emergency responders from both the public and private sector. NFPA 472 provides competencies for the following levels of hazmat responders. These levels parallel those listed within OSHA 1910.120, with the exception that the hazmat specialist has been deleted and the private sector specialist employee has been expanded upon.

First responder at the awareness level. These are individuals who, in the course of their normal duties, may be the first on scene of an emergency involving hazardous materials. They are expected to recognize hazmat presence, protect themselves, call for trained personnel, and secure the area.

First responder at the operational level. These are individuals who respond to releases or potential releases of hazmat as part of the initial response to the incident for the purpose of protecting nearby persons, the environment or property from the effects of the release. They shall be trained to respond in a defensive fashion to control the release from a safe distance and keep it from spreading.

Hazmat technician. These are individuals who respond to releases or potential releases of hazmat for the purpose of controlling the release. Hazmat technicians are expected to use specialized chemical protective clothing and specialized control equipment.

Incident commander. The incident commander is responsible for directing and coordinating all aspects of a hazmat incident.

Private sector specialist employee. These are individuals who, in the course of their regular job duties, work with or are trained in the hazards of specific materials and/or containers. In response to incidents involving chemicals, they may be called upon to provide technical advice or assistance to the incident commander relative to their area of specialization. There are three levels of private sector specialist employee:

- Level C—Includes those persons who may respond to incidents involving chemicals or containers within their organization's area of specialization. They may be called upon to gather and record information, provide technical advice, and arrange for technical assistance consistent with his or her organization's emergency response plan and standard operating procedures. This individual is not expected to enter the hot or warm zone at an incident.
- Level B—Those persons who, in the course of their regular job duties, work with or are trained in the hazards of specific chemicals or containers within their organization's area of specialization. Because of their education, training, or work experience, they may be called upon to respond to incidents involving these chemicals or containers. The Level B employee may be used to gather and record information, provide technical advice, and provide technical assistance (including working within the hot zone) at the incident, consistent with his or her organization's emergency response plan and standard operating procedures, and the local emergency response plan.
- Level A—Includes personnel who are specifically trained to handle incidents involving specific chemicals and/or containers for chemicals used in their organization's area of specialization. Consistent with his or her organization's emergency response plan and standard operating procedures, the Level A employee shall be able to analyze an incident involving chemicals within his/her organization's area of specialization, plan a response to that incident, implement the planned response within the capabilities of the resources available, and evaluate the progress of the planned response.

Hazmat branch officer. This person is responsible for directing and coordinating all operations assigned to the hazmat branch (or group) by the incident commander. This branch officer is trained to the technician level, with additional competencies in the command and control area.

Hazmat branch safety officer. The branch safety officer works within an incident management system to ensure that recognized safe practices are followed within the hazmat branch (or group). This individual is trained to the technician level, with additional competencies in the safety area.

Hazmat technician with a specialty (cargo tank, tank car, intermodal tank). These individuals provide support to the hazmat technician, provide oversight for product removal and movement of damaged hazmat containers, and act as liaison between technicians and other outside resources.

It should be noted that the hazmat specialist level was originally included in NFPA 472 and was consistent with OSHA 1910.120, but was later dropped from the standard as a result of user comments. In simple terms, users noted that there was very little difference in the competencies between the hazmat technician and the hazmat specialist. In its place, the NFPA Technical Committee created the above position of hazmat technician with a "specialty."

NFPA 473 Standard for Professional Competence of Responders to Hazardous Material Incidents. The purpose of NFPA 473 is to specify minimum requirements of competence and to enhance the safety and protection of response personnel and all components of the emergency medical services system. The overall objective is to reduce the number of EMS personnel accidents, exposures and injuries, and illnesses resulting from hazmat incidents. There are two levels of EMS/HM responders.

- ***EMS/HM Level I.*** These are persons who in the course of their normal duties may be called on to perform patient care activities in the cold zone at a hazmat incident. EMS/HM Level I responders provide care to only those individuals who no longer pose a significant risk of secondary contamination. Level I requires varying competency for basic life support (BLS) and advanced life support (ALS) personnel.
- ***EMS/HM Level II.*** In the course of their normal duties, Level II responders may be called on to perform patient care activities in the warm zone at a hazmat incident. Level II responders may provide care to individuals who still pose a significant risk of secondary contamination. In addition, personnel at this level shall be able to coordinate EMS activities at a hazmat incident and provide medical support for hazmat response personnel. Level II requires varying competency for basic and advanced life support personnel.

NFPA technical committee on hazmat protective clothing and equipment (NFPA 1991, 1992, 1994). This technical committee is responsible for the development of standards and documents pertaining to the use of personal protective clothing and equipment (excluding respiratory protection) by emergency responders at hazmat incidents. The committee scope includes PPE selection, care, and maintenance. The following three standards have been developed by this committee:

- NFPA 1991 Standard on Vapor-Protective Ensembles for Hazmat Emergencies
- NFPA 1991 Standard on Liquid Splash-Protective Ensembles for Hazmat Emergencies
- NFPA 1994 Standard on Protective Ensembles for Chemical/Biological Terrorism Incidents

Other standards organizations

There are many other important standards writing bodies, including the American National Standards Institute (ANSI), the American Society for Testing and Materials (ASTM), the Compressed Gas Association (CGA), the Safety Equipment Institute (SEI), and the American Petroleum Institute (API). Each of these organizations approves or creates standards ranging from hazmat container design to personal protective clothing and equipment. For more information on these standards organizations, consult the following websites:

- American Petroleum Institute (API) – *http://api-ec.api.org/intro/index_noflash.htm*
- American Society for Testing and Materials (ASTM) – *http://www.astm.org*
- Compressed Gas Association (CGA) – *http://www.cganet.com*
- National Fire Protection Association (NFPA) – *http://www.nfpa.org/*
- Safety Equipment Institute (SEI) – *http://www.seinet.org/*

In addition, the Building Officials and Code Administrators International (BOCA), the International Conference of Building Officials (ICBO), and the Southern Building Code Congress International (SBCCI) have developed code provisions for the storage, handling, and use of hazardous materials. There are also a wide range of NFPA standards that address hazmat storage, handling, and use issues. These model codes also permit the local fire official to require facilities that fall within their hazmat requirements to submit one or both of the following:

- Hazmat inventory statement (HMIS)—Inventory of regulated materials in a form approved by the fire official. It includes: chemical name; hazardous ingredients; UN, NA, or CIS identification number; and maximum quantities stored or used onsite at any time.
- Hazmat management plan (HMMP)—A management and contingency plan in a form approved by the fire official that includes: site plan, floor plan, material compatibility, monitoring and security methods, hazard identification, inspection procedures, employee training, and emergency equipment available.

HAZMAT LIABILITY ISSUES

Since the passage of SARA Title III, there has been a tremendous increase in the interest level on the subject of liability within the hazmat field. Communication between the senior fire executive and their legal representative is critical in this area. Liability may be either civil or criminal. Civil liability results from an action or inaction that creates harm to a third party (*i.e.*, tort). In contrast, criminal liability results from an action or inaction that is deemed criminal by statute (*i.e.*, causing harm to people or the environment). Although fire service liability issues involve civil issues by a third party, there have been cases of criminal liability raised as a result of actions by fire service personnel during training, as well as while controlling or mitigating a hazmat release.

The concept of "standard of care" has frustrated many fire service managers and emergency planners. Standard of care is not static, but rather is constantly evolving as a result of new laws, regulations, standards, and court decisions. Within the hazmat field, the standard of care is defined by laws and regulations such as SARA Title III, , OSHA 1910.120 and EPA Section 311, and consensus standards such as NFPA 472 and 473. Unless fire department policies and procedures are updated on a regular basis to reflect the current standard of care, liability could be incurred.

Four key elements must be present to prove that negligence and liability exist. These elements are:

- ***A duty or standard to act or perform***—The existence of a duty establishes a standard of conduct. The duty may be statutory or common-law; for example, a fire marshal who is responsible for enforcing a building code. Duties can also be established by consensus standards organizations and bodies (*e.g.*, NFPA).
- ***A breach of that duty, which can be either an action or an omission***—There must be a breach of duty to prove negligence. This violation may be either intentional or, as in most cases, unintentional.
- ***A connection between the failure and the resulting harm***—A cause and effect relationship must be established. It must be demonstrated that the failure actually resulted in the harm being alleged. The harm may be direct, such as injury, or indirect, such as business disruption.
- ***An actual loss or harm to the parties involved***—There must be measurable harm, which can be in the form of injuries, damage to properties, or mental anguish.

Liability in the hazmat field may arise in a variety of areas. Among the most common are the following:

- ***Planning***—Poor or incomplete emergency response plans and their corresponding implementation standard operating procedures (SOPs) or guidelines are potential sources of liability. Remember, planning is an ongoing process and plans must be continuously improved and updated. Response plans and SOPs that are out-of-date or that reflect obsolete capabilities or unrealistic planning assumptions are liability hazards. Hazmat emergency response plans must be based on the hazards present and represent the realistic capabilities possessed by the community and the fire department.

- ***Training programs and exercises***—Failure to comply with the OSHA 1910.120 or EPA Section 311 hazmat training requirements may result in liability for the fire department from both the employee and members of the general public, if either is harmed. Training programs must be well documented and personnel and organizational capabilities tested with safe and regular exercises. Emphasis should be on ensuring that all personnel are trained to perform their expected duties and that competencies are documented in an objective manner.
- ***Failure to identify chemical hazards or prioritize chemical hazards from a planning perspective***—The hazards analysis and emergency planning processes are the foundations of an effective hazmat emergency response plan. Not only must a community and the fire service have an effective risk analysis process, but that process must be based upon accepted standards and practices, such as those outlined in NRT-1: *Hazmat Emergency Planning Guide* and the EPA's *Technical Guidance for Hazards Analysis* (see Bibliography and References).
- ***Failure to warn the public of impending hazards and threats***—In the event of a hazmat release, the proper warnings and directions must be given (*e.g.*, evacuate or shelter-in-place). If a warning message results in civilians moving toward or into a hazard, rather than away from it, the liability potential is clear. In addition, some communities have suffered liability problems because they failed to activate the community alerting system, or because the alerting systems failed to operate as designed, leaving areas without notification. The fire service should work with its emergency management agency to ensure that its community warning system meets the requirements for today's standard of care.
- ***Negligent construction or operation of emergency response systems***—The fire department must ensure that both the hazmat equipment and the department's services are safe, well maintained, and operated properly. The use of old or outdated practices and equipment can be a major liability concern. An example of a negligent operation is a fire department hazmat response team that is not trained, staffed, equipped, or operated on-scene to the standard of care established by both federal and state governmental regulations and voluntary consensus standards.

The Hazmat Management System

Fires in the United States have traditionally been managed by fire suppression operations at the expense of prevention activities. Fortunately, since the release of *America Burning* in 1973, there has been a growing emphasis on managing fire related problems from a systems perspective. Community master planning, public education, residential sprinklers, improved fire code enforcement, and fire protection engineering are some examples of this change in philosophy.

A similar situation exists with hazmat in the community. If it is to be effectively managed and controlled, the approach to hazmat issues must be broad, coordinated, and systematic. There are four key elements in a hazmat management system: planning and preparedness, prevention, response, and cleanup and recovery. Planning and prevention efforts during the last 15 years have resulted in a decline in the number and severity of "working" hazmat incidents.

Planning and preparedness

Planning is the first and most critical element of the hazmat management system. A community's ability to develop and implement an effective hazmat management plan depends upon two elements: hazards analysis and the development of a hazmat emergency operations plan. The fire service must play a major role in the development of both of these elements.

Hazards analysis is the analysis of the hazmats that exist in the community, including their location, quantity, specific physical (*i.e.*, how they behave) and chemical (*i.e.*, how they harm) properties, previous incident history, surrounding exposures, and risk of release. Hazards analysis is normally performed by either the fire department or the local emergency management agency as part of the LEPC's development of the overall emergency response plan.

Emergency planning is a comprehensive and coordinated response to the community hazmat problem. This response builds upon the hazards analysis and recognizes that no single public or private sector agency is capable of managing the hazmat problem by itself. The emergency planning process and the development of the actual hazmat emergency response plan is normally coordinated through the LEPC. However, the fire department must be a major player during this planning process, as the data and information generated by these activities will enable both the LEPC and the fire service manager to assess the potential risk to the community, the level of hazmat response to be provided, and the resources to be allocated.

Hazards analysis is the foundation of the planning process and can be viewed as a "refined" pre-incident planning process. Community hazards analysis is normally the responsibility of either the fire department or the local emergency management agency, with the information then channeled through the LEPC for inclusion into the local emergency response plan and procedures. Fire service pre-incident plans often serve as the starting point for the hazards analysis process. It should be conducted for every hazmat location designated as a moderate or high probability location for an incident. In addition to risk evaluation, vulnerability, or what in the community is susceptible to damage should a release occur, must also be examined. A hazards analysis program provides the following benefits:

- Lets firefighters know what to expect
- Provides planning for less frequent incidents
- Creates an awareness of new hazards
- May indicate a need for preventive actions, such as monitoring systems and facility modifications
- Increases the likelihood of successful emergency operations

An evaluation team familiar with the response area can facilitate the hazard analysis process. The team should include fire officers and members from each battalion or district, as well as representatives from prevention and the hazmat section. In addition, personnel from outside the fire department may sometimes be included, such as emergency management staff, local health or environmental department officials, and other technical specialists. The primary concern here is geographic, as most firefighters are very familiar with their "first due" area.

There are four steps involved in a hazard analysis program:

1. ***Identify the hazards***—Hazards identification typically provides specific information on situations that have the potential for causing injury to life or damage to property and the environment due to hazmat spills or releases. Hazards identification is initially based on a review of the community's history of incidents and evaluation of facilities that have submitted chemical lists and reporting forms under SARA Title III and related state and local right-to-know legislation. Information should include:
    - Chemical identification
    - Location of facilities that manufacture, store, use, or move hazardous materials
    - The type(s) and design of chemical containers or vessels
    - The quantity of material that could be involved in a release
    - The nature of the hazard associated with the hazmat release (*e.g.*, fire, explosion, toxicity, etc.)
    - The presence of any fixed suppression and/or detection systems
2. ***Perform a vulnerability analysis***—This procedure identifies areas in the community that may be affected or exposed, and what facilities, property, or environment

may be susceptible to damage should a hazmat release occur. A comprehensive vulnerability analysis provides information on:

- *The size/extent of vulnerable zones*—Specifically, the size of the area that may be significantly affected as a result of a spill or release of a known quantity of a specific hazardous material under defined conditions is determined. Computer dispersion models, such as CAMEO, ARCHIE, and other product-specific models, are extremely helpful tools for this process.
- *The population, in terms of numbers, density, and types*—For example, the number of facility employees, residents, special occupancies (*e.g.*, hospitals, nursing homes, schools, etc.) must be predetermined.
- *Private and public property that may be damaged*—This includes essential infrastructure and support systems (*e.g.*, water supply, power, communications) and transportation corridors and facilities.
- *The environment that may be affected*—This includes the impact of a release on sensitive environmental areas and wildlife.

3. ***Perform a risk analysis***—A risk analysis is an assessment of (1) the probability or likelihood of an accidental release, and (2) the actual consequences that might occur. The risk analysis is a emergency response resources—Based upon potential risks, consider the personnel, equipment judgment of incident probability and severity based upon the previous incident history, local experience, and the best available hazard and technological information.
4. ***Evaluate supplies necessary for firefighting, EMS, protective actions, traffic management and control, etc.***—Take an inventory of available equipment and supplies along with their ability to function. For example, are firefighting foam supplies adequate to control and suppress vapors from a gasoline tank truck rollover?

A completed hazard analysis should enable fire service managers to determine the level of response to emphasize, what resources will be required to achieve that response, and what type and quantity of mutual aid and other support services will be required. The hazards analysis process is also an integral element of both the OSHA process safety management (PSM) and EPA risk management program (RMP) regulations being implemented at industrial facilities that manufacture, store, and use hazardous chemicals. Targeted at facilities that pose the greatest risk to the public and the environment in the event of a release, PSM primarily pertains to onsite worker protection while the RMP is primarily concerned with offsite impact. However, the two systems share a number of commonalities including the need to evaluate worst-case and alternative worst-case scenarios using accepted hazards analysis methods.

Among the commonly used hazards analysis methods used by safety professionals are the following:

- ***What if analysis***—This method asks a series of questions, such as, "What if Pump X stops running?" or "What if an operator opens the wrong valve?" to explore possible hazard scenarios and consequences. The method is often used to examine proposed changes to a facility.
- ***HAZOP study***—This is the most popular method of hazard analysis used within the petroleum and chemical industries. The hazard and operability (HAZOP) study brings together a multidisciplinary team (usually five to seven people) to brainstorm and identify the consequences of deviations from design intent for various operations. Specific guidewords ("No," "More," "Less," "Reverse," etc.) are applied to parameters such as product flows and pressures in a systematic manner. The process requires the involvement of a number of people working with an experienced team leader.
- ***Failure modes, effects, and criticality analysis (FMECA)***—This method tabulates each system or unit of equipment, along with its failure modes, the effect of each failure on the system or unit, and how critical each failure is to the integrity of the system. Then the failure modes can be ranked according to criticality so as to determine which are the most likely to cause a serious accident.
- ***Fault tree analysis***—A formalized deductive technique that works backwards from a defined accident to identify and graphically display the combination of equipment failures and operational errors that led up to the accident. It can be used to estimate the quantitative likelihood of events.
- ***Event tree analysis***—A formalized deductive technique that works forward from specific events or sequences of events that could lead to an incident. It graphically displays the events that could result in hazards and can be used to calculate the likelihood of an accident sequence occurring. It is the reverse of fault tree analysis.

Originally, RMP results were to be made available to the public to ensure that the community was aware of the potential risks posed by specific facilities and chemicals. However, given the potential for this same information to be used as background and intelligence for a terrorist attack, its distribution and release to the general public is now being more closely controlled. Chief fire officers desiring the latest information should consult the EPA Chemical Emergency Preparedness and Prevention Office (CEPPO) website at *http://www.epa.gov/swercepp/*.

Emergency operations planning. Hazmat management is a multidisciplinary issue that goes beyond the resources and capabilities of any single agency or organization. As there will be a variety of "players" responding to a major hazmat emergency, the emergency operations plan and related procedures will establish the framework for carrying out the emergency response effort. Consequently, to effectively manage the overall hazmat problem within a community, a comprehensive planning process must be initiated. This effort is usually referred to as "contingency planning" or "emergency operations planning."

There are many federal, state, and local requirements that apply to emergency operations planning. The one that most directly affects the fire service is SARA Title III. Title III requires the establishment of state emergency response commissions and local emergency planning committees. Title III also outlines specific requirements covering factors such as extremely hazardous substances (EHS); threshold planning quantities; make-up of LEPCs; dissemination of plans, chemical lists, and MSDS information to the community and general public; facility inventories; and toxic chemical release reporting.

The LEPC is the coordinating point for the development of the emergency operations plan. The fire department, as a member of the LEPC and the initial responder to most hazmat emergencies, must play a leadership role in the overall development and implementation of all plans and procedures developed by the LEPC. However, experience shows that plans and procedures alone are insufficient, as procedures must reflect the ability to handle not only a major emergency, but also "day-to-day" operations. Remember these two points: (1) emergency response plans and procedures that are not "user friendly" do not get used, and (2) "compliance-oriented" emergency response plans are not necessarily operationally-effective.

Figure 19–1 provides an overview of the hazmat emergency planning process, including these steps:

1. ***Organize the planning team.*** Planning requires both fire service and community involvement throughout the process. Experience has shown that plans prepared by a single person or a single agency will likely fail. Remember, there is no single agency (public or private) that can effectively manage a major hazmat emergency alone.
2. ***Define and implement the major tasks of the planning team.*** These include reviewing any existing plans, identifying hazards, and analyzing and assessing current prevention and response capabilities.
3. ***Write the plan.*** There are two approaches to this step: (a) develop or revise a hazmat appendix or a section of a multi-hazard emergency operations plan, or (2) develop or revise a single-hazard plan specifically for hazardous materials. Once the plan is written, it must be approved by the LEPC and all the respective planning groups involved. Again, the fire department must be a major player during this development process.

   It should be noted that in recent years there has been a push toward the development of single integrated contingency plans (ICPs) or the "one-plan" approach. Readers desiring further information on the ICP concept should consult the EPA Chemical Emergency Preparedness and Prevention Office (CEPPO) website at *http://www.epa.gov/swercepp/.*
4. ***Revise, test, and maintain the plan.*** Every emergency plan must be evaluated and kept up-to-date through the review of actual responses, simulation exercises, and the regular collection of new data and information.

While community-level emergency planning is essential, the completion of a plan does not guarantee that the community is actually prepared for a hazmat incident. Planning is only one element of the total hazmat management system.

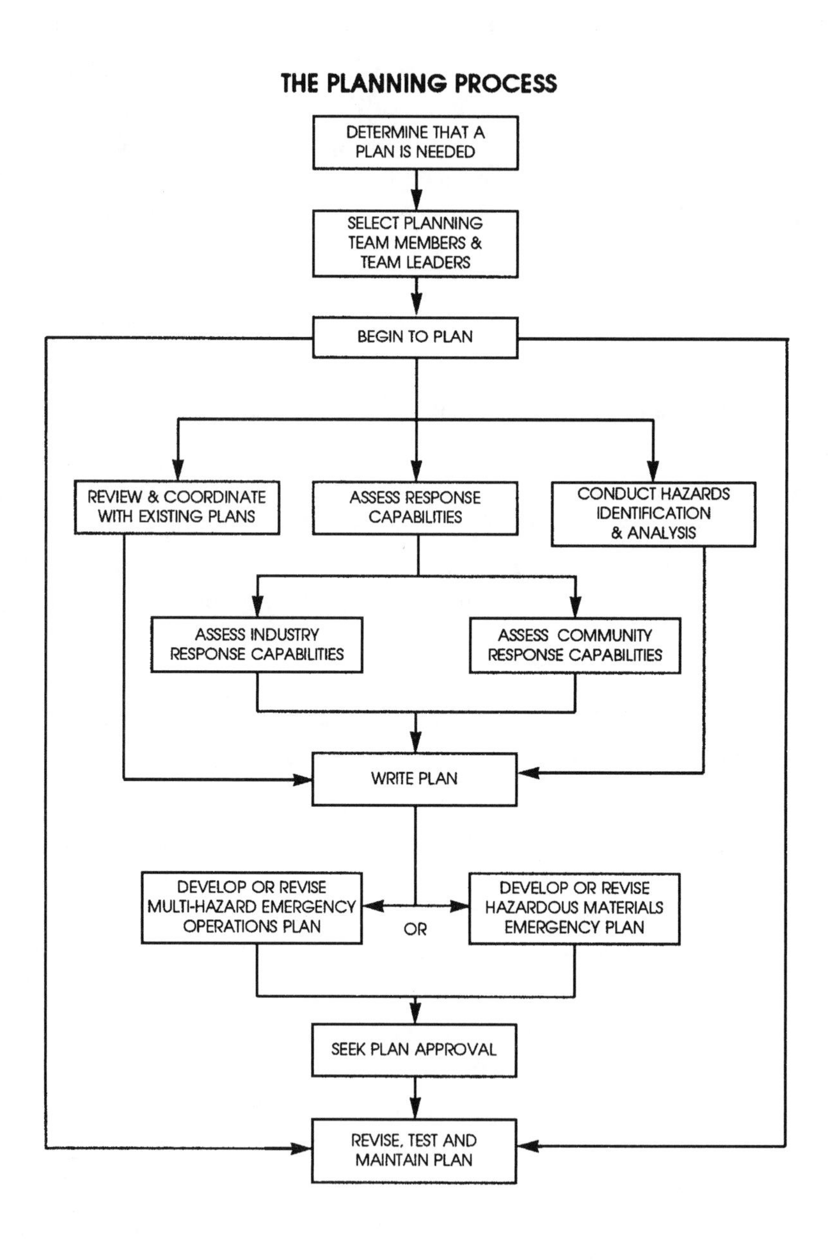

Figure 19–1 The planning process

Prevention

Responsibility for the prevention of hazmat releases is shared between the public and private sectors. Because of their regulatory and enforcement capabilities, however, public sector agencies generally receive the greatest attention and often "carry the biggest stick." However, chief fire officers should understand that the insurance and risk management industry could also be used as an advocate for making facility-level changes and modifications that can reduce the risk of an incident or release. Prevention activities often include the following resources and functions:

Hazmat process and container design/construction standards. Almost all hazmat facilities, containers, and processes are designed and constructed to some standard. This "standard of care" may be based upon voluntary consensus standards, such as those developed by the NFPA and the ASTM, or on government regulations. Many major petrochemical, hazmat companies, and industry trade associations (*e.g.*, Chlorine Institute, Compressed Gas Association) have also developed their own respective engineering standards and guidelines.

All containers used for the transportation of hazmat are designed and constructed to both specification and performance regulations established by the U.S. Department of Transportation. These regulations can be referenced in Title 49 of the Code of Federal Regulations (CFR). In certain situations, hazmat may be shipped in non-DOT specification containers that have received a DOT exemption.

Inspection and enforcement. Fixed facilities, transportation vehicles, and transportation containers are normally subject to some form of inspection process. These inspections can range from comprehensive and detailed inspections at regular intervals (*e.g.*, six months, one year, five years, etc.) to visual inspections each time a container is loaded. Fixed facilities will commonly be inspected by state and federal OSHA and EPA inspectors, in addition to state fire marshals and local fire departments. It should be recognized that many of these inspections focus upon fire safety and life safety issues and may not adequately address either the environmental or process safety issues.

Transportation vehicle inspection is generally based upon criteria established within Title 49 CFR. The enforcing agency is often the state police, but this will vary according to the individual state, the hazmat being transported, and the mode of transportation. Some local fire departments, such as Aurora, Colorado's, routinely perform inspections of hazmat cargo tank trucks. U.S. DOT agencies with hazmat regulatory responsibilities include those that follow.

- *Office of Hazmat Safety (HMS) of the Research and Special Programs Administration (RSPA).* HMS is responsible for all hazmat transportation regulations except bulk shipment by ship or barge. Includes designating and classifying hazardous materials, container safety standards, label and placard requirements, and handling, stowing, and other in-transit requirements. HMS serves as the DOT representative to the National Response Team, supports the National Response Center (NRC) operation in coordination with the U.S. Coast Guard, and serves as the DOT liaison with the Federal Emergency Management Agency (FEMA) on hazmat transportation issues

- *Office of Hazmat Enforcement (OHME) of the Office of Hazmat Safety (HMS).* OHME's inspection and enforcement staff determine compliance with safety standards by inspecting entities that offer hazmat for transportation; manufacturing, rebuilding, repairing, reconditioning, or retesting packaging used to transport hazardous materials; and handling intermodal transfers of hazardous materials. OHME is responsible for the management and coordination of RSPA's hazmat inspection and enforcement program.
- *Federal Railroad Administration (FRA).* The FRA is responsible for enforcement of regulations relating to hazmat carried by rail or held in depots and freight yards.
- *Federal Aviation Administration (FAA).* The FAA is responsible for the enforcement of regulations relating to hazmat shipments on domestic and foreign carriers operating at U.S. airports and in cargo handling areas.
- *U.S. Coast Guard (USCG).* The USCG is responsible for the inspection and enforcement of regulations relating to hazmat in port areas and on domestic and foreign ships and barges operating in the navigable waters of the United States. The USCG supports the operation of the National Response Center in Washington, DC

Public education. Hazmat are a concern for both industry and the general community. The average homeowner contributes to this problem by improperly disposing of substances such as used motor oil, paints, solvents, batteries, and other chemicals used in and around the home. As a result, many communities have initiated full-time household chemical waste awareness, education, and disposal programs. In other instances, communities have established used motor oil collection stations and chemical cleanup days in an effort to cope with the problem.

An example of a highly successful public education program is the Wally Wise Guy program developed by the Deer Park, Texas LEPC. An example of industry and government partnership, Wally Wise Guy is a turtle that teaches children and their parents how to shelter-in-place in case of a chemical emergency. Further information on the Wally Wise Guy Program can be obtained by consulting the Deer Park LEPC website at *http://www.wally.org/.*

Handling, notification, and reporting requirements. These guidelines actually act as a bridge between planning and prevention functions. There are many federal, state, and local regulations that require those who manufacture, store, or transport hazmat and hazardous wastes to comply with certain handling, notification, and reporting rules. Key federal regulations include CERCLA (Superfund), RCRA and SARA Title III. There are also many state regulations similar in scope that often exceed the federal standard requirements.

Response

When prevention and enforcement functions fail, response activities begin. Since it is impossible to eliminate all risks associated with the manufacture, storage, and use of hazardous materials, the need for a well trained, effective emergency response capability will always exist. Response activities should be based upon the information and probabilities identified during the planning process and an evaluation of local hazmat problems.

While every community should have access to a hazmat response capability, that capability does not always have to be provided by either local government or the fire service. Numerous states and regions have established both statewide and regional hazmat response-team systems that ensure the delivery of a competent and effective capability in a timely manner.

Levels of incidents. Fortunately, not every incident is a major emergency. Response to a hazmat release may range from a single engine company responding to a natural gas leak in the street, to a railroad derailment involving numerous government and private agencies, to a terrorism event involving chemical or biological weapons with mass casualties. In addition, most working hazmat incidents require a unified command organization that combines the key players from fire, law enforcement, and emergency medical service (EMS) agencies at all levels of government.

Incidents can be categorized based upon their severity and the resources they require. Figure 19-2 outlines these response levels.

A common refrain voiced by most incident commanders at "working" hazmat incidents is, "Why is it taking so long?" Among the factors that can lengthen the timeline of a hazmat incident are:

- Research must be conducted on the hazardous material(s) involved. When multiple chemicals are involved, the timeline will increase.
- There is a need to contact technical information sources remote from the scene, including CHEMTREC, chemical manufacturers, and other sources for information.
- The appropriate type of chemical protective clothing along with the associated donning and pre-entry activities must be determined.
- The decontamination corridor must be established.
- Getting "out-of-area" resources to the scene, such as product or container specialists, heavy equipment, product transfer, and uprighting equipment, etc. are subject to logistics and response times.

In addition, several entry operations may be required to control and stabilize the problem. Establishing a liaison officer to communicate with these outside groups and agencies can help facilitate the response operation, minimize political and community impacts associated with the incident, and reduce the incident timeline.

Response groups. The emergency response community consists of various agencies and individuals who respond to hazmat incidents. These response groups can be categorized based upon their knowledge, expertise, and resources, and include fire department, law enforcement, emergency management, emergency medical services (EMS), and environmental and industry personnel.

RESPONSE LEVEL	DESCRIPTION	RESOURCES	EXAMPLES
LEVEL I POTENTIAL EMERGENCY CONDITIONS	AN INCIDENT OR THREAT OF A RELEASE WHICH CAN BE CONTROLLED BY THE FIRST RESPONDER. IT DOES NOT REQUIRE EVACUATION, BEYOND THE INVOLVED STRUCTURE OR IMMEDIATE OUTSIDE AREA. THE INCIDENT IS CONFINED TO A SMALL AREA AND POSES NO IMMEDIATE THREAT TO LIFE AND PROPERTY.	ESSENTIALLY A LOCAL LEVEL RESPONSE WITH NOTIFICATION OF THE APPROPRIATE LOCAL, STATE AND FEDERAL AGENCIES. REQUIRED RESOURCES MAY INCLUDE: • FIRE DEPARTMENT • EMERGENCY MEDICAL SERVICES (EMS) • LAW ENFORCEMENT • PUBLIC INFORMATION OFFICER (PIO) • CHEMTREC • NATIONAL RESPONSE CENTER	EXAMPLES-500 GALLON FUEL OIL SPILL, INADVERTENT MIXTURE OF CHEMICALS, NATURAL GAS IN A BUILDING, ETC.
LEVEL II LIMITED EMERGENCY CONDITIONS	AN INCIDENT INVOLVING A GREATER HAZARD OR LARGER AREA THAN LEVEL I WHICH POSES A POTENTIAL THREAT TO LIFE AND PROPERTY. IT MAY REQUIRE A LIMITED EVACUATION OF THE SURROUNDING AREA.	REQUIRES RESOURCES BEYOND THE CAPABILITIES OF THE INITIAL LOCAL RESPONSE PERSONNEL. MAY REQUIRE A MUTUAL AID RESPONSE AND RESOURCES FROM OTHER LOCAL AND STATE ORGANIZATIONS. MAY INCLUDE: • ALL LEVEL I AGENCIES • HAZMAT RESPONSE TEAMS • PUBLIC WORKS DEPARTMENT • HEALTH DEPARTMENT • RED CROSS • REGIONAL EMERGENCY MANAGEMENT STAFF • STATE POLICE • PUBLIC UTILITIES	EXAMPLES-MINOR CHEMICAL RELEASE IN AN INDUSTRIAL FACILITY, A GASOLINE TANK TRUCK ROLLOVER, A CHLORINE LEAK AT A WATER TREATMENT FACILITY, ETC.
LEVEL III FULL EMERGENCY CONDITION	AN INCIDENT INVOLVING A SEVERE HAZARD OR A LARGE AREA WHICH POSES AN EXTREME THREAT TO LIFE AND PROPERTY WHICH MAY REQUIRE A LARGE SCALE EVACUATION.	REQUIRES RESOURCES BEYOND THOSE AVAILABLE IN THE COMMUNITY. MAY REQUIRE THE RESOURCES AND EXPERTISE OF REGIONAL, STATE, FEDERAL AND PRIVATE ORGANIZATIONS. MAY INCLUDE: • LEVEL I AND II AGENCIES • MUTUAL AID FIRE, LAW ENFORCEMENT AND EMS • STATE EMERGENCY MANAGEMENT STAFF • STATE DEPARTMENT OF ENVIRONMENTAL RESOURCES • STATE DEPARTMENT OF HEALTH • ENVIRONMENTAL PROTECTION AGENCY (EPA) • U.S. COAST GUARD • FEDERAL EMERGENCY MANAGEMENT AGENCY (FEMA)	MAJOR TRAIN DERAILMENT WITH FIRE, EXPLOSION OR TOXICITY HAZARD, A MIGRATING VAPOR CLOUD RELEASE FROM A PETROCHEMICAL PROCESSING FACILITY

Hazardous Materials - Noll. Hildebrand, Yvorra ©1988 Peake Productions

Figure 19–2 Levels of hazmat incidents

These responders can be compared to the levels of capability found within a typical EMS system. In that system, an injury such as a fractured arm can be effectively managed by a first responder or emergency medical technician-ambulance (EMT-A), while a cardiac emergency will require the services of an intermediate (EMT-I) or a paramedic (EMT-P).

In the same way, a diesel fuel spill can usually be contained by first responder operations-level personnel, such as a fire department's engine company using dispersants or absorbents. An accident involving a poison or reactive chemical, however, will require the on-scene expertise of a hazmat technician or hazmat response team. In simple terms, we try to match the nature of the problem with the capabilities of the responders. Figure 19-3 illustrates this comparison.

Hazmat response teams (HMRT). In order to respond to hazmat emergencies in a more effective and efficient manner, many communities and areas have established hazmat response teams (HMRT). NFPA 472 defines an HMRT as an organized group of trained response personnel operating under the emergency response plan and appropriate standard

HAZARDOUS MATERIALS VS EMS
COMPARING RESPONSE GROUPS

KNOWLEDGE LEVEL	EMERGENCY MEDICAL SERVICES	HAZARDOUS MATERIALS RESPONSE
BASIC	FIRST RESPONDER	FIRST RESPONDER
INTERMEDIATE	EMERGENCY MEDICAL TECHNICIAN	HAZARDOUS MATERIALS TECHNICIAN
ADVANCED	PARAMEDIC	HAZARDOUS MATERIALS SPECIALIST

Hazardous Materials - Noll. Hildebrand, Yvorra ©1988 Peake Productions

Figure 19–3 Comparison of HM vs. EMS response groups

operating procedures, who are expected to handle and control actual or potential leaks or spills of hazmat requiring close approach to the material. The HMRT members perform responses to releases for the purpose of control or stabilization of the incident.

HMRTs typically function as an incident command system (ICS) group or sector under the direct control of a hazmat group supervisor. Based upon their assessment of the hazmat problem, the HMRT, through the hazmat group supervisor, provides the incident commander with a list of options and a recommendation for mitigation of the hazmat problem. However, the final decision always remains with the incident commander.

This can be an interesting and dynamic political dilemma for both the incident commander and the hazmat group supervisor. When the concept of HMRTs became "mainstream" fire service thinking in the early 1980s, many ICs voiced their concern with a group of individuals responding and "taking over" the incident. Today, many ICs are only too willing to "give" overall incident responsibility to the HMRT because they may not possess a strong hazmat background. The reality of the situation is simple—the HMRT is essentially an "environmental consultant" to the incident commander and is present to provide both technical expertise and resources so that the "problem" can be managed in a safe and effective manner. At the end of the day, the incident commander cannot delegate overall responsibility for managing the incident.

Personnel on HMRTs are trained to a minimum of the OSHA hazmat technician level and must participate in a medical surveillance program based upon the requirements of 29 CFR 1910.120. Among the specialized equipment carried by an HMRT are technical reference libraries, computers and communications equipment, personal protective clothing and equipment, direct-reading monitoring and detection equipment, control and mitigation supplies and equipment, and decontamination supplies and equipment.

The need for an HMRT should be one of the conclusions determined by the hazards analysis and planning process. In evaluating this need, however, the chief fire officer should consider the following points:

- There is no single agency that can effectively manage the hazmat issue by itself. While the fire department may operate the HMRT, it must work closely with other local, state, and federal government agencies. A unified command organization is a necessity.
- Every community does not require an HMRT. However, every community should have access to an HMRT capability through either local, regional, state, or contractor resources.
- An HMRT will not necessarily solve the community hazmat problem. Remember the hazmat management system: planning and preparedness, prevention, response, and cleanup and recovery.

- There are numerous constraints and requirements associated with the delivery of an effective HMRT capability. These include legal, insurance and political issues, both initial and continuing funding sources, resource determination and acquisition, personnel and staffing, and initial and continuing training requirements.
- Successful HMRT response programs are those that truly understand what services an HMRT can provide at all emergencies, not just those involving hazardous materials. For example, no organization within the emergency response community better understands and routinely practices, (1) risk evaluation, (2) air monitoring/detection equipment and its interpretation, and (3) the fundamentals of safe operating practices in a field environment better than HMRT personnel. The HMRT is not a chemical resource; it is a health and safety resource with capabilities that can be used in a variety of response scenarios, including hazardous materials, confined space, structural collapse, aircraft accidents, and other significant fires and emergencies.

Types of hazmat response teams. Many areas of the United States have established levels of HMRTs based upon local risks and available resources. Although variations may be found, there are two basic types of HMRT used in the emergency response community.

Technical information HMRT. This is an HMRT that provides technical assistance and information to the incident commander, but has a limited hands-on capability. This type of HMRT would be useful in very rural areas where there is a need for hazmat technical expertise, but the volume of incidents and the resource and training requirements do not justify having an extensive "hands-on" offensive-oriented HMRT.

The technical information team is capable of accessing technical information and resources (*e.g.*, emergency response guidebooks, computer databases, cellular communication capability, etc.), and could provide both strategical and tactical control recommendations to the incident commander. This team would have a limited protective clothing and control equipment inventory, with its capabilities being primarily defensive-oriented. However, key HMRT personnel could be trained to the hazmat technician level, while support personnel could be trained to the first responder operations level and be provided additional training in specific areas (*e.g.*, hazard and risk analysis, air monitoring equipment, decontamination, and limited use of specialized PPE).

Hands-on HMRT. These HMRTs are trained to the hazmat technician level and are expected to handle and control actual or potential releases of hazardous materials. In some instances, there may be two levels of hands-on HMRTs based upon the level of personal protective clothing and equipment carried on the hazmat unit. A Level II-HMRT may be limited to chemical splash protective clothing (Level B), basic monitoring devices, and defensive and limited offensive control equipment and supplies. Level III-HMRTs might carry chemical vapor protective clothing (Level A), advanced monitoring equipment, and specialized control equipment to implement offensive control measures. In addition, a Level III-HMRT typically has a greater staffing level than a Level II-HMRT.

Consider the following example. Regional and statewide hazmat response systems have been developed in some areas of the country, including North Carolina, Massachusetts, Pennsylvania, and California. A number of these response systems have different levels of HMRTs based upon the HMRT's staffing and equipment inventory. For example, the commonwealth of Virginia is divided into eight regions, each region staffed by a regional hazmat officer from the Virginia Department of Emergency Management (DEM). These regional officers provide assistance to localities when the services of an HMRT are not required and serve as liaison between the state and all levels of government. Each region is also staffed by at least one Level-III HMRT who can respond to the most serious emergencies.

The tiered-response concept is also used in a number of metropolitan fire departments in which the HMRT is supported by a number of ladder or rescue companies designated as hazmat support companies. These companies are trained to handle the lower risk incidents, as well as provide support to the HMRT for a "working" hazmat response.

HMRT needs assessment. In addition to the local risk levels and actual hazmat incident experience identified during the planning process, a number of other criteria must be assessed during the HMRT needs assessment process. These criteria apply equally for both new and established HMRTs.

- **Legal issues.** Legal and related issues should include:
    a) Legislative authority—Within the local or regional government, is there enabling legislation that assigns responsibility for hazmat response to a specific agency or organization? If so, is the fire department assigned as lead agency? Is hazmat response an element within the fire department's charter or mission statement? If the fire department is not the lead agency, who is? What is this agency's relationship to the fire department?

    b) Potential liabilities—Although one can never totally eliminate the potential for being the subject of a regulatory citation or liability suit, the fire service manager can enact measures to minimize the potential of a successful suit or regulatory citation. These include the development and revision of the emergency response plan and SOPs, updating mutual aid pacts to ensure that hazmat response activities are included, procedures for cost reimbursement for hazmat supplies and equipment, and the development of "Good Samaritan" or similar legislation to facilitate the utilization of private sector and outside specialists for technical expertise.

    c) Insurance coverage—Does the present insurance policy cover HMRT activities? If the HMRT consists of personnel from several different fire departments or public organizations, how is the insurance policy structured? What is the HMRT's responsibility after the emergency phase is stabilized?

- **Political factors.** For better or worse, chief fire officers must live and operate in a political environment. In the event of a major emergency, political leaders will take "public heat" for major hazmat emergencies. Likewise, political leaders may also use emergency responders and the fire department as scapegoats for emergency response problems and failures.

  Major hazmat emergencies are often tragedies for fire department personnel and the community, but they also represent opportunities for resources, as well as policy and management action. One fire service colleague referred to this as "crisis diplomacy." Take every opportunity to educate your leaders about the need for an effective hazmat response program and the inherent problems associated with delivering a timely, professional, well-trained and well-equipped hazmat response capability.

- **Economic factors.** Regardless of the type of HMRT being evaluated or already in place, recognize that an HMRT requires a tremendous financial commitment to both initiate and continue. While costs can vary greatly, initial costs have ranged from $100,000 to $1 million, depending on equipment, supply and vehicle requirements, as well as the method of HMRT staffing.

  A critical economic issue is cost recovery for the expenses associated with delivering the emergency response. In general, recent court cases nationwide have shown that in the absence of state or local legislation authorizing cost recovery for emergency responders, it is difficult to successfully recoup associated expenses from the responsible party. If you are fortunate to have cost reimbursement legislation in place, documentation of all "billable" expenses is critical. Today, most responsible parties (*i.e.*, the spiller) have an independent third party review both fire service and emergency response hazmat response invoices to ensure that all expenses were related to the emergency response effort.

  When searching for alternative funding sources, consider industry, public or private sector grants, and public interest groups. In some areas, fines assessed for environmental crimes and chemical discharges have been given to the LEPC and local emergency response units to assist in HMRT operations and training. Some recommendations to consider when approaching private sector organizations include:

  a) Determine your needs and list them

  b) Make your requests specific

  c) Have several options

  d) In-house items are generally easier to access

  e) Don't forget skills and services as an option

- **Personnel and staffing.** The staffing and operation of the HMRT will be dependent on the HMRT's scope of operations. Options will vary, and range from career staffing of a dedicated hazmat unit, to career staffing on a shared function unit (*e.g.*, engine company/hazmat unit). In other instances, staffing may be entirely through volunteer personnel or some combination of career and volunteer.

In initially selecting personnel for a fire department HMRT, consider several points. First, just because a person is a good firefighter does not mean they will automatically be a good hazmat responder. This is particularly critical in organizations that may rely on career personnel for dedicated or shared-function HMRTs. Second, technical smarts or "book smarts" does not necessarily equal "street smarts." One does not have to be a chemist to be an effective hazmat responder; however, one must have an understanding of chemistry and be able to ask the right questions and interpret the responses. Third, depending on how long the HMRT has been in existence, the selection process can be viewed as the fire department equivalent of the NFL draft. When HMRTs start out, look for the best players regardless of their specific field of expertise. Emphasis should be on team players with an established reputation. As time progresses however, you will identify areas or gaps within your HMRT structure and should select specific types of personnel. Examples include personnel with strong technical backgrounds, mechanical skills, command and control expertise, and EMS support personnel.

Upon selection to the HMRT, personnel must be provided a medical surveillance program. Components of the medical surveillance program include medical examinations at the following intervals:

- Baseline examination prior to assignment on the HMRT
- Periodic medical examinations, typically 12 months, but not greater than 24 months based upon a physician's recommendations
- At termination of employment or upon reassignment from the HMRT
- After any exposure to a hazardous material
- At such times as the physician deems necessary

Finally, an HMRT must be properly trained. The two primary information sources for establishing an HMRT training program are OSHA 1910.120 and NFPA 472. While OSHA 1910.120 outlines the regulatory requirements (what you have to do), NFPA 472 spells out the specific training and educational competencies for the training program (how you can do it).

Both OSHA 1910.120 and NFPA 472 recommend that HMRT personnel be trained to the hazmat technician level. According to OSHA 1910.120 (q), the hazmat technician requires a minimum 24 hours of initial training at the first responder operations level. However, hazmat emergency response operations require personnel to have a broader, more comprehensive background. It is extremely doubtful that any personnel receiving only 24 hours of hazmat technician training will become viable and functioning HMRT members. As a point of reference, most fire department HMRT programs require 40 to 200 hours of initial hazmat training before being classified as HMRT members. In addition, there is the need for both annual refresher courses and continuing education.

Tools, resources, and equipment. Specialized supplies and equipment will be required for personal protection, monitoring and detection, and mitigation and control of agents or supplies. In addition, there is also the need for a vehicle. The selection of these items must be based upon the scope of the HMRT's operations. Many fire service managers falsely believe that they possess an HMRT capability, when in reality they lack the necessary supplies and equipment to perform the job safety and effectively.

While no single source exists, there are a number of references the fire service manager can consult in this area. These include the *NFPA Hazmat Response Handbook* and *EPA Hazmat Team Planning Guidance.* In addition, a number of states have established equipment lists for HMRTs that are either certified by the state or receive state funding.

Cleanup and recovery

Cleanup-and-recovery operations are designed to restore a facility and/or community back to normal as soon as possible. In many instances, chemicals involved in a hazmat release will be eventually classified as hazardous wastes. At a minimum, cleanup-and-recovery operations will fall under the guidelines of OSHA, *CERCLA*, and *RCRA* federal regulations.

Cleanup-and-recovery operations can be classified on the basis of short-term or long-term operations. Short-term activities are actions immediately following a hazmat release. They are primarily directed toward removal of any immediate hazards and restoration of vital support services (*i.e.*, reopening transportation systems, drinking water systems, etc.) to minimum operating standards. Short-term activities may last up to several weeks.

In contrast, long-term activities are remedial actions that return vital support systems back to normal or improved operating levels. Examples include groundwater treatment operations, the mitigation of both aboveground and underground spills, and the monitoring of flammable and toxic contaminants. These activities may not be directly related to a specific hazmat incident, but are often the result of abandoned industrial or hazardous waste sites. These operations may extend over months or years.

Fire service role during cleanup operations. Fire service personnel are usually not directly responsible for the cleanup and recovery of hazmat releases. Depending on the nature of the incident, however, they may continue to be responsible for site safety until all risks are stabilized and the emergency phase is terminated. After the emergency phase is terminated, there must still be a formal transfer of command from the lead response agency (*e.g.*, fire department) to the lead agency responsible for post-emergency response operations.

At short-term operations immediately following an incident, the incident commander should ensure that the work area is closely controlled, that the general public is denied entry, and that the safety of emergency responders and the public is maintained during cleanup-and-recovery operations. When interfacing with both industry responders and contractors, the incident commander should ensure that these groups are trained to meet the requirements of OSHA 1910.120.

Long-term cleanup-and-recovery operations do not normally require the continuous presence of the fire service. Depending on the size and scope of the cleanup, a contractor or government official (remedial project manager, or RPM) will be the central contact point. The fire department should be familiar with the cleanup operation, including its organizational structure, the RPM, work plan, time schedule, and site safety plan. Cleanup operations should conform to the general health and safety requirements of both state and federal EPA and OSHA standards.

Although the fire service may not have the regulatory authority to conduct inspections or issue citations at cleanup operations, they can bring specific concerns to the attention of the state or federal regulatory agency having jurisdiction.

Hazmat Group Operations

Depending on the scope and complexity of an incident, special operations (*e.g.*, hazmat, bomb squad, technical rescue, etc.) may be managed as either a branch or separate group within the incident command system (ICS). This section provides a brief overview of the application and use of a hazmat group at a hazmat incident. Readers desiring additional information on hazmat group operations should consult the *IMS Model Procedures Guide for Hazmat Incidents.* The book is available through Oklahoma State University, Fire Protection Publications at 1-800-654-4055.

The hazmat group is normally under the command of a senior hazmat officer (known as the hazmat group supervisor) who, in turn, reports to the operations section chief or the incident commander. The hazmat group is directly responsible for all tactical hazmat operations that occur in the hot and warm zones of an incident. Tactical operations outside of these hazard control zones (*e.g.*, public protective actions) are not the responsibility of the hazmat group.

Although a number of resources may be assigned to the hazmat group, it is typically made up of the hazmat response team with its complement of personnel and resources. The scope and nature of the problem will determine which roles are staffed. Primary functions and tasks assigned to the hazmat group are as follows:

- **Safety function.** Primarily the responsibility of the hazmat group safety officer (may also be referred to as the assistant safety officer - hazmat). Responsible for ensuring that safe and accepted practices and procedures are followed throughout the course of the incident. Possesses the authority and responsibility to stop any unsafe actions and correct unsafe practices.

- **Entry/backup function.** Responsible for all entry and backup operations within the hot zone, including reconnaissance, monitoring, sampling, and mitigation.
- **Decontamination function.** Responsible for research and development of the "decon" plan, and the setup and operation of an effective decontamination area capable of handling all potential exposures, including entry personnel, contaminated patients, and equipment. If necessary, will include the coordination of a safe refuge area.
- **Site access control function.** Responsible for establishing hazard control zones, monitoring egress routes at the incident site, and ensuring that contaminants are not being spread. Monitors the movement of all personnel and equipment between the hazard control zones. Manages the safe refuge area, if established.
- **Information/research function.** Responsible for gathering, compiling, coordinating, and disseminating all data and information relative to the incident. This data and information is used within the hazmat group to assess hazard and evaluate risks, evaluate public protective options, the selection of PPE, and development of the incident action plan.
- **Secondary support functions.** Secondary support functions and tasks that may also be assigned to the hazmat group, including medical and resource functions.
- **Medical function.** Responsible for pre- and post-entry medical monitoring and evaluation of all entry personnel and provides technical medical guidance to the hazmat group as requested.
- **Resource function.** Responsible for control and tracking of all supplies and equipment used by the hazmat group during the course of an emergency, including documenting the use of expendable supplies and materials. Coordinates, as necessary, with the logistics section chief (if activated).

Figure 19-4 illustrates the standard ICS positions found at a typical hazmat incident. Remember, ICS is a modular organization. Only those functions necessary for the control and mitigation of the incident would be activated, as appropriate.

Hazmat group staffing

Hazmat group supervisor. This individual is responsible for the management and coordination of all functional responsibilities assigned to the hazmat group, including: safety, site control, research, entry, and decontamination. The hazmat group supervisor must have a high level of technical knowledge and be knowledgeable of both the strategic and tactical aspects of hazmat response.

The hazmat group supervisor is trained to the hazmat technician level. This position is normally filled by the HMRT team leader or HMRT officer. Depending on the scope and nature of the incident, the hazmat group supervisor will report to either the operations section chief or the incident commander. Based on the IC's strategic goals, the hazmat group supervisor develops the tactical options to fulfill the hazmat portion of the incident action plan (IAP), and is responsible for ensuring that the following tasks are completed:

- Hazard control zones are established and monitored
- Site monitoring is conducted to determine the presence and concentration of contaminants
- A site safety plan is developed and implemented

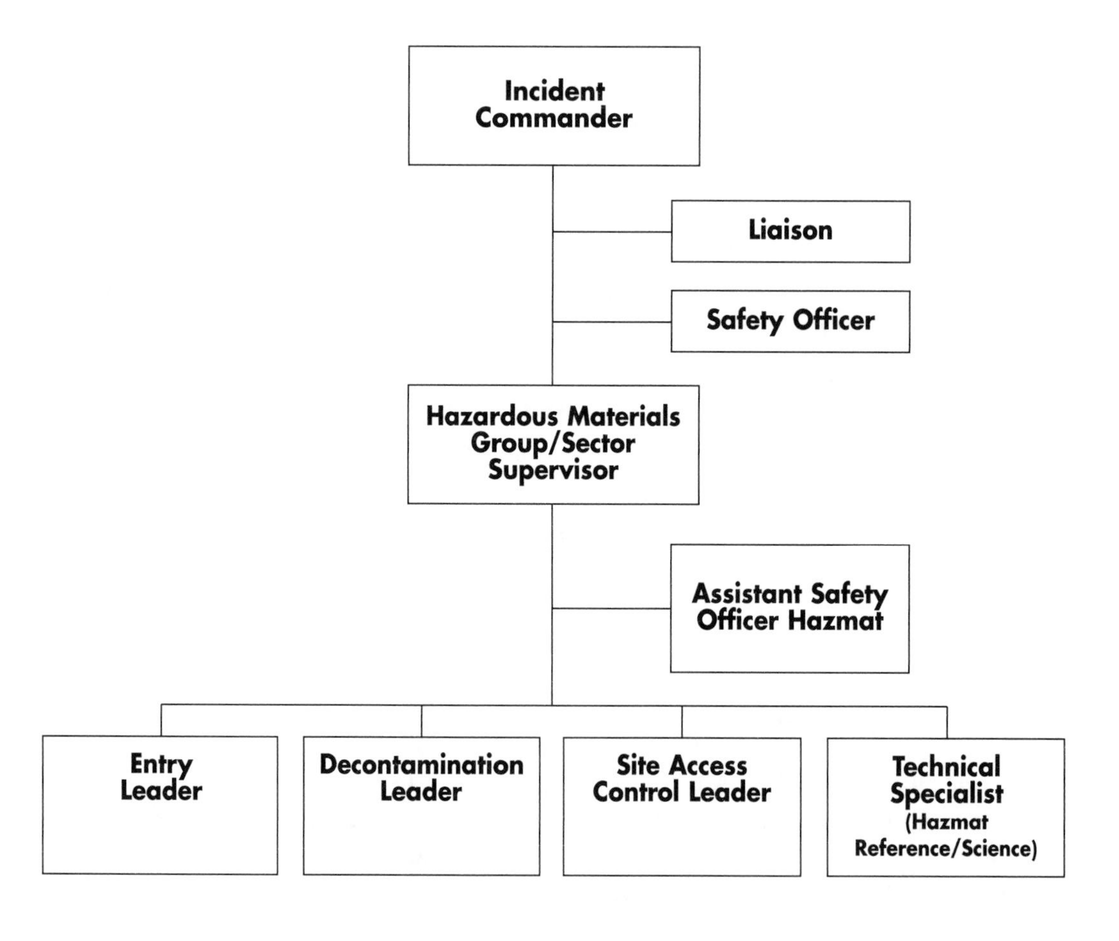

Figure 19–4 Hazmat group organization chart

- Tactical objectives are established for the hazmat entry team within the limits of the team's training and equipment limitations
- All hot zone operations are coordinated with the operations section chief or incident commander to ensure tactical goals are being met

Hazmat group safety officer (i.e., assistant safety officer—hazmat). The hazmat group safety officer reports to the hazmat group supervisor and is subordinate to the incident safety officer. This individual is responsible for coordinating safety activities within the hazmat group, but also has certain responsibilities that may circumvent the normal chain of command. The incident safety officer is responsible for the safety of all personnel operating at the emergency, while the hazmat group safety officer is responsible for all operations within the hazmat group and within the hot and warm zones. This includes having the authority to stop or prevent unsafe actions and procedures during the course of the incident.

The hazmat group safety officer must have a high level of technical knowledge to anticipate a wide range of safety hazards. This should include being hazmat-trained, preferably to the hazmat technician level, and being knowledgeable of both the strategical and tactical aspects of hazmat response. This position is typically filled by a senior HMRT officer or member. While it is not the hazmat group safety officer's job to make tactical decisions or to set goals and objectives, it is their responsibility to ensure that operations are implemented in a safe manner.

Specific functions and responsibilities of the hazmat group safety officer include:

- Advising the hazmat group supervisor of all aspects of health and safety, including work/rest cycles for the entry team
- Coordinating site safety activities with the incident safety officer, as appropriate
- Possessing the authority to alter, suspend, or terminate any activity that may be judged unsafe
- Participating in the development and implementation of the site safety plan
- Ensuring the protection of all hazmat group personnel from physical, chemical, and/or environmental hazards and exposures
- Identifying and monitoring personnel operating within the hot zone, including documenting and confirming both "stay times" (*i.e.*, time using air supply) and "work times" (*i.e.*, time within the hot or warm zone performing work) for all entry and decon personnel
- Ensuring that EMS personnel and/or units are provided, and coordinating with the hazmat medical leader
- Ensuring that health exposure logs and records are maintained for all hazmat group personnel, as necessary

Entry team. The entry team is managed by the entry leader. This individual is responsible for all entry operations within the hot zone and should be in constant communication with the entry team. The entry team and the entry leader are responsible for the following:

- Recommending actions to the hazmat group supervisor to control the emergency situation within the hot zone
- Implementing all offensive and defensive actions, as directed by the hazmat group supervisor, to control and mitigate the actual or potential hazmat release
- Directing rescue operations within the hot zone, as necessary
- Coordinating all entry operations with the decon, hazmat information, site access, and hazmat medical units

Personnel assigned to the entry team include the entry and backup teams and personnel assigned for entry support. The entry team consists of all personnel who will enter and operate within the hot zone to accomplish the tactical objectives specified within the incident action plan. Entry teams always operate using a buddy system.

The backup team is the safety team that extracts the entry team in the event of an emergency. It may also be referred to as the rapid intervention team (RIT). The backup team must be in place and ready whenever entry personnel are operating within the hot zone. Entry support personnel (also known as the dressing team) are responsible for the proper donning and outfitting of both the entry and backup teams.

Decontamination team. The decontamination (decon) team is managed by the decon leader. The decon team and the decon leader are responsible for the following:

- Determining the appropriate level of decontamination to be provided
- Ensuring that proper decon procedures are used by the decon team, including decon area setup, decon methods and procedures, staffing, and protective clothing requirements
- Coordinating decon operations with the entry leader, site access control, and other personnel within the hazmat group
- Coordinating the transfer of decontaminated patients requiring medical treatment and transportation with the hazmat medical group
- Ensuring that the decon area is established before any entry personnel are allowed to enter the hot zone
- Monitoring the effectiveness of decon operations
- Controlling all personnel entering and operating within the decon area

Site access control. The site access control unit is managed by the site access control leader and is responsible for the following:

- Monitoring the control and movement of all people and equipment through appropriate access routes at the incident scene to ensure that the spread of contaminants is controlled
- Overseeing the placement of the hazard control zone lines based upon recommendations from the entry, decon, and info/research units
- As necessary, establishing a safe refuge area and appointing a safe refuge area manager (This would include coordinating with decon concerning decon and medical priorities for contaminated persons.)
- Ensuring that injured or exposed individuals are decontaminated prior to departure from the incident scene

Hazmat information/research team. The hazmat information/research team is managed by the information leader (also known as research or science). Depending on the level of the incident and the number of hazmat involved, the information team may consist of several persons or teams. The hazmat information team and the information leader are responsible for the following:

- Providing technical support to the hazmat group
- Researching, gathering, and compiling technical information and assistance from both public and private agencies
- Providing and interpreting environmental-monitoring information, including the analysis of hazmat samples and the classification and/or identification of unknown substances
- Providing recommendations for the selection and use of protective clothing and equipment
- Projecting the potential environmental impact of the hazmat release

Hazmat medical unit. Medical support services may be provided by either a hazmat medical unit or a medical group within the ICS organization. Hazmat medical unit personnel will be located in the entry team dressing area and in the rehabilitation area. The hazmat medical unit and hazmat medical leader are responsible for the following:

- Providing pre-entry and post-entry medical monitoring of all entry and backup personnel
- Providing technical assistance for all EMS-related activities during the course of the incident
- Providing emergency medical treatment and recommendations for ill, injured, or chemically-contaminated civilians or emergency response personnel
- Providing EMS support for the rehab area

The hazmat medical unit will conduct post-entry medical monitoring, cooling, and rehydration of entry and backup personnel in the rehab area. All operating personnel should not be given anything to eat or drink unless approved by hazmat medical personnel. Medical findings and personal exposure forms should be forwarded to the hazmat group safety officer and/or the entry team leader.

Hazmat resource unit. At some "working" incidents, a hazmat resource function may be established to support hazmat group activities. Directed by the hazmat resource leader, this unit is located in the cold zone and will be responsible for acquiring all supplies and equipment required for hazmat group operations, including protective clothing, monitoring instruments, leak-control kits, etc. In addition, the hazmat resources leader will also be responsible for documenting all supplies and equipment expended as part of the emergency response effort. The hazmat resources unit and leader must work closely with the logistics section chief.

Incident Management Operations: The Eight-Step Process

An effective hazmat response system differs little from a computer. With computers, the software package drives or directs the computer's hardware or machinery. Similarly, in hazmat response, standard operating procedures (SOPs) are the software that drive and give direction to emergency response personnel and the use of their equipment.

On-scene response operations must always be based on a structured and standardized system of protocols and procedures. Regardless of the nature of the incident, the nature of the response, and the personnel involved, reliance upon standardized procedures will bring consistency to the tactical operation. If the situation potentially involves hazardous materials, this reliance upon standardized tactical response procedures will help minimize the risk of exposure to all responders.

The eight-step process outlines the basic tactical functions to be evaluated and implemented at incidents involving hazardous materials. Like all SOPs, the eight-step process should be viewed as a flexible guideline and not as a set of rigid rules. Individual departments and agencies should decide what works best for them.

The eight-step process offers several benefits. First, it recognizes that most hazmat incidents are minor in nature and generally involve limited quantities. It also builds on the action of first responding units and identifies the roles and responsibilities of each level of response. The eight-step process provides a flexible management system that expands as the scope and magnitude of the incident grows. Finally, it provides a consistent management structure, regardless of the classifications of hazmat materials involved.

Essentially, there are eight basic functions that must be evaluated at emergencies potentially involving hazardous materials. These eight functions typically follow an implementation timeline at the incident. The functions are:

1. Managing and controlling the site
2. Identifying the problem
3. Evaluating the hazard and risk
4. Selecting personal protective clothing and equipment
5. Managing information and coordinating resources
6. Implementing response objectives
7. Implementing decontamination and cleanup operations
8. Terminating the incident

Step 1: Site management and control

Site management and control refers to managing the physical layout of an incident. It is one of the most critical areas of managing an emergency. Experience has shown that incidents that are poorly managed in the initial stages become increasingly difficult to control as the emergency progresses in both time and complexity. You cannot safely and effectively manage the incident if you do not have control of the scene. As a result, site management and control is a critical benchmark and is the foundation on which all subsequent response functions and tactics build. Specific tasks associated with site management include the following:

- *Approaching and positioning emergency response units*—Focus should be on approaching the emergency scene in such a manner that responders do not become part of the problem (*e.g.*, uphill, upwind, based on nature/location of the incident)
- *Establishing command and implementing the incident command system*—To be effective, ICS must be implemented upon the arrival of the first on-scene emergency response unit
- *Staging other responding emergency units to minimize safety concerns*
- *Isolating and denying entry and establishing hazard control zones*—Failure to isolate the area during the initial stages of an incident will lead to an increased potential for exposure and injury. At a minimum, a hot zone must be established. See the additional information on hazard control zones that follows
- *Implementing public protective actions for the protection of the community*—This will be evacuation, protection-in-place, or some combination of the two options

Hazard-control zones. The establishment and maintenance of hazard-control zones is a critical element for safe operations. Hazard-control zones must be established by the incident commander at all hazmat incidents. If on-scene, the hazmat group supervisor will normally provide the necessary technical information and recommendations to assist the incident commander in determining the control-zone size and boundaries. The shape and dimensions of the hazard-control zones shall depend on such factors as the size and nature of the release (*e.g.*, liquid vs. vapor, instantaneous vs. continuous release), chemical concentrations present, visual observations or air-monitoring results, wind direction and velocity, surrounding topography, or adjacent exposures, etc.

Hazard-control zones are broken into three categories: hot zone, warm zone, and cold zone (see Figure 19–5).

- *Hot zone (restricted, high hazard area)*—Immediate hazard area surrounding the problem or release site that extends far enough to prevent adverse effects to personnel outside the zone. Only hazmat-trained personnel or individuals may enter the hot zone under monitored conditions, and these individuals (minimum of two) must possess particular knowledge of the problem or situation. During both entry and reconnaissance operations, a backup team (minimum of two individuals) with appropriate protection should be stationed at the edge of the hot zone or in a location where they can quickly gain access to the entry team in an emergency situation.
- *Warm zone (limited-access area, decontamination zone)*—Area surrounding the hot zone and bounded by the cold zone where entry support and decontamination operations take place. It includes a corridor with access-control points to assist in reducing the spread of contamination. Entry is restricted to emergency response personnel, as well as those assigned by the incident commander. Individuals entering the warm zone must be wearing appropriate personal protective clothing.
- *Cold zone (support area)*—Area surrounding the warm zone that presents no hazard to emergency response personnel and equipment. Reserved for emergency services functions only, such as the incident command post and other support functions deemed necessary to control the incident. Support personnel without the proper level of personal protective clothing should be limited to only the cold zone.

The outer boundaries of the hazard-control zones are normally identified with safety tape, traffic cones, or some other identification method.

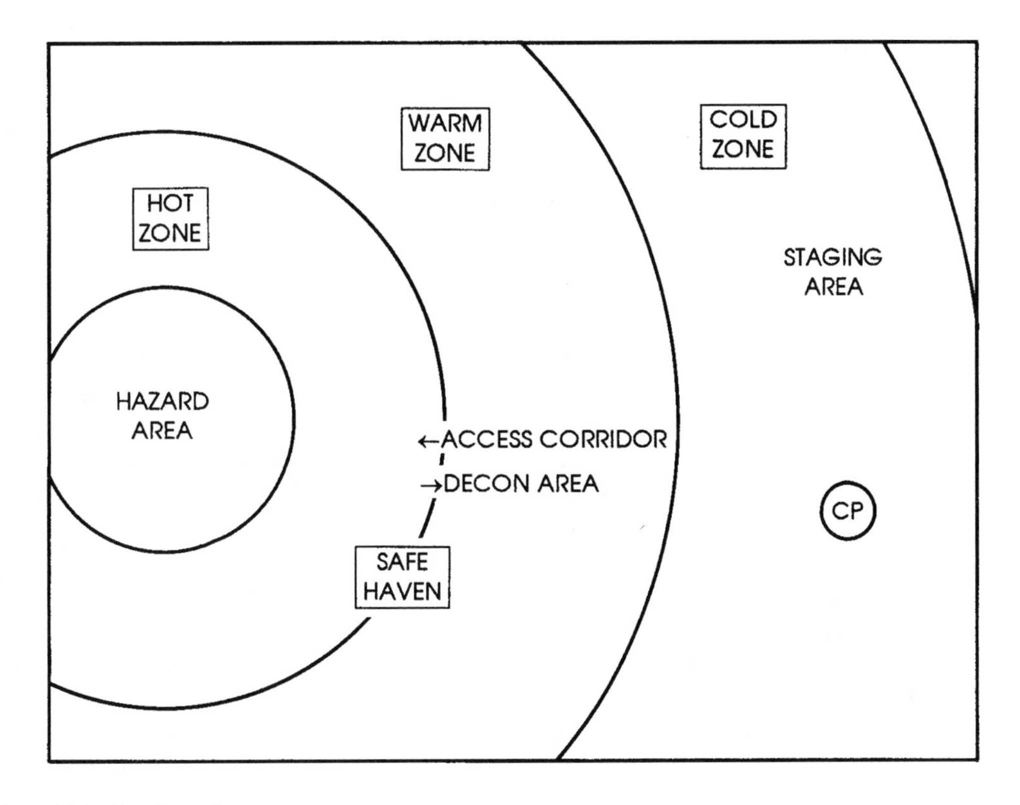

Figure 19–5 Hazard control zones

Step 2: Identify the problem

Once the hazard area is isolated, proper identification of the materials involved becomes the cornerstone of all further decision making. Basic principles of identification in an emergency focus on recognition, identification, classification, and verification of hazardous materials. The steps are:

1. Recognize that presence of hazardous materials.
2. Identify the hazardous material(s) involved.
3. If it is not possible to specifically identify the materials involved, try to classify and determine the hazard classification (*e.g.*, corrosive, poison gas, etc.).
4. Upon arrival at the scene, always verify the information you are initially provided. Never assume that the initial information you receive is correct.

Common methods of recognition and identification of hazmat include:

- Occupancy and location

- Container shapes
- Markings and colors
- Placards and labels
- Shipping papers and facility documents
- Monitoring and detection instruments
- Senses

Most critical for the fire officer are situations where emergency operations are already underway when it is discovered that hazmat are involved. Possible scenarios include terrorism events involving hazmat or WMD agents where large numbers of civilians may be impacted. In these instances, efforts must be directed toward establishing an effective site management-and-control policy, and limiting the spread of contaminants. Personnel who have been exposed or contaminated must be isolated to ensure that contaminants are not carried beyond the immediate control area.

At some incidents, it may first be necessary to conduct reconnaissance (*i.e.*, recon) operations to determine the scope and nature of the problem. Depending on the nature of the incident, recon operations may be carried out by either HMRT personnel or first responders. Recon operations may be classified as either defensive or offensive. In defensive recon, the objective is to obtain information on site layout, physical hazards, access, and other related conditions from beyond the hot zone. Recon is normally obtained through threat assessments, interviews, physical observations, etc. In an offensive recon operation, the objective is to obtain incident information by physically entering the hot zone. Entry tasks can include air monitoring, sampling, and video or photo documentation for analysis.

Step 3: Evaluate hazards and risks

The evaluation of hazard information and the assessment of the relative risks is a critical decision-making point in the successful management of a hazmat incident. The decision to intervene—or more often, not to intervene—is not easy. While most individuals recognize the initial need for isolating the area, denying entry, and identifying the materials involved, many overlook the need for developing effective analytical skills.

Hazard-and-risk evaluation is the most critical function that emergency responders perform. The primary objective of the risk evaluation process is to determine whether or not responders should intervene, and what strategical objectives and tactical options should be pursued to control the problem at hand. You absolutely cannot afford to get this wrong. If you lack the expertise to perform this function, get help from someone who can provide assistance, such as HMRTs or chemical or container specialists.

Hazards. A hazard refers to a danger or peril. In hazmat response, it usually refers to the physical or chemical properties of a material, such as flammability properties, exposure values, and chemical protective clothing compatibility. At an emergency, all of these elements are regarded as constant values (*i.e.*, they do not change regardless of the location of the emergency) and can be referenced from sources such as emergency response guidebooks and material safety data sheets (MSDS).

Risks. A risk is the probability of suffering a harm or loss. Risks are intangibles that are different at every hazmat incident and must be evaluated by knowledgeable emergency response personnel. Among the factors that can influence the level of risk are:

- *Hazard class and quantity of material(s) involved*—Risks are generally greater when dealing with bulk quantities of hazmat in storage or transportation than when the substances are in limited quantity, individual containers. However, this must be balanced against the specific materials or class of materials involved (*e.g.*, toluene—an aromatic hydrocarbon, flammable liquid—as opposed to phosphine—a flammable and poisonous gas).
- *Container type, the type of "stress" applied, and its ability to adapt to the stressor*—As long as the hazardous material remains within its container, there are no problems. In simple terms, it remains in a "controlled" state. Therefore, emergency responders must be able to analyze the type of hazmat container and pertinent safety features, and the type of stress applied to the container (*e.g.*, thermal, mechanical, chemical or combination). They must also be able to estimate what will occur once the hazardous material escapes from its container.
- *Proximity of exposures, including the size, distance, and rate of dispersion of a chemical release*—Exposures can affect emergency responders, the community, property, and the environment. These exposures can consequently produce systems disruptions (*e.g.*, shutdown of transportation corridors, loss of community water supplies and utilities, etc.).
- *Level of available resources*—This includes their training and expertise in dealing with the hazmat involved and associated response times. For example, fire departments have a great deal of experience in handling flammable liquid and gas emergencies, but do not have a commensurate level of training and experience in handling fires and emergencies that involve clandestine lab operations.

Although the risks associated with hazmat response never will be eliminated, they can be managed successfully. Hazmat responders must see their role as risk evaluators, not risk takers.

Hazard-and-risk evaluation. Hazard-and-risk evaluation can be divided into three tasks: (1) assessing the hazards present; (2) evaluating the level of risk; and (3) establishing an incident action plan to make the problem go away. These tasks are explained below.

1. *Assessing the hazards present*—Hazards are the physical (*i.e.*, how the material behaves) and/or chemical (*i.e.*, how the material harms) properties of a material. Examples include flash point, boiling point, exposure vales, etc. Some sources of hazard data and information are provided below.
2. *Emergency response guidebooks*—Most responders rely on three to five basic guidebooks as primary sources. Several operational considerations should be kept in mind when using guidebooks:
   - You must know how to use guidebooks to use them effectively.
   - Remember the "Rule of Three's." Evaluate a minimum of three independent hazard information sources (*e.g.*, reference guidebooks, specialists, CHEMTREC, etc.) before permitting personnel to operate within a hostile environment.
   - If there is conflicting information among different sources, always select the most conservative recommendations or values.
   - Be realistic in your evaluation of the data contained in guidebooks. For example, many first responders do not carry any air-monitoring equipment. However, a continuous release from a bulk chemical container of a chemical with relatively low TLV-TWA and IDLH values will easily exceed safe exposure levels and require the use of personal protective clothing and equipment.
3. *Emergency communication centers*—A number of private and public sector hazmat telephone centers exist. Their functions include providing immediate chemical hazard information, accessing secondary forms of expertise for additional action and information, and acting as a clearinghouse for spill notifications.

Among the most widely used chemical information and industry access hotlines are CHEMTREC (1-800-424-9300) and CANUTEC in Canada (613-996-6666). CHEMTREC's purpose is two-fold. First, it provides immediate advice to callers anywhere on how to cope with chemicals involved in a transportation or fixed facility emergency. Secondly, it can access shippers or other forms of expertise for additional and appropriate follow-up action and information. CHEMTREC also has the ability to download emergency information to the field via fax and computer modem. CANUTEC is the Canadian counterpart of CHEMTREC, and provides assistance in identifying and establishing contact with shippers and manufacturers that originate in Canada.

For emergency medical issues, the Agency for Toxic Substances and Disease Registry (ATSDR) provides a 24-hour link with medical professionals who can provide appropriate advice (404-639-0615). Within 10 minutes of a call, ATSDR will link the caller with an emergency response coordinator to give advice on immediate actions. Within 20 minutes, the agency can provide access to toxicologists, environmental health scientists, chemists, and others.

Hazmat transportation incidents that cause death, serious injury, or property damage in excess of $50,000, or a continuing threat to life and property must be reported to the National Response Center (1-800-424-8802, 202-426-2675, or, if calling from within the Washington, DC area, 267-2675). Likewise, when hazmat releases exceed the reportable quantity provisions of CERCLA, the responsible party (*i.e.*, the spiller) must also notify the NRC. The NRC can supply assistance in identification, technical information, and initial response actions, and serves as the federal government's single point-of-contact in the event of a suspected terrorist incident.

Databases. Portable computers and personal desk assistants (PDAs) can provide rapid field access to databases and technical information during an incident. In many respects, these devices have revolutionized our ability to access information in a rapid manner. Other examples include CAMEO (computer-assisted management of emergency operations), Tomes™ and Tomes Plus™, and RTECS (registry of toxic effects of chemical substances).

Technical information specialists. These are generally product or container specialists from industry and allied professions who can provide responders with technical information to control and mitigate problems. The group also includes individuals who may possess some specialized knowledge, such as chemistry, container design, toxicology, or environmental sciences.

MSDS/right-to-know information. Many state and local worker right-to-know laws have been enacted across the country. In addition, OSHA has issued its hazard communication standard for chemical markings and worker exposures to chemicals in the workplace and the EPA has implemented the *Emergency Planning and Community Right-to-Know Act of 1986* (SARA Title III). These laws have provided responders with substantial access to information, as well as the burden of how to handle it effectively.

Monitoring and detection equipment. Air monitoring and detection are critical in implementing a risk-based response. Monitoring and detection results should be used to:

- Identify or classify materials involved
- Verify hazard information
- Determine appropriate levels of personal protective clothing and equipment
- Determine the size and location of the hazard-control zones
- Develop public protective action recommendations

Risk evaluation. The primary objective of risk evaluation is to determine whether or not emergency response personnel should intervene, and what strategical objectives and tactical options should be pursued. To make these determinations, responders must first estimate the likely harm that will occur without intervention. The question of the day becomes: "What will happen if we do nothing?" Using non-intervention as the baseline, responders can then evaluate other available strategical options in terms of their effect on the final outcome of the emergency.

This evaluation and comparison process is built around two interrelated actions. First, the incident commander must visualize the likely behavior of the hazmat and its container ("How will the juice behave when it gets out of the can?"). Second, he or she must describe the likely outcome of that behavior ("How will it harm me?"). As a risk evaluator, the incident commander should be able to assess the following conditions:

- Previous and current status of the incident
- Overall condition of the hazmat container
- Environmental conditions, including runoff, wind, precipitation, topography, etc.
- Exposures, including people (facility personnel, emergency responders, and the public), property, environment, and systems disruption
- Comparison of resources available with the level required to respond to the problem: Evaluate the risks of personnel intervening directly in the emergency. Consider the limitations of the people involved and their equipment
- Estimation of likely harm without active emergency response intervention and development of response objectives

Establish an incident action plan. Based on the answer to these questions, an incident action plan will be developed. Strategic goals are usually very broad in nature and are determined at the command level. Several strategic goals may be pursued simultaneously during an incident. Examples of common strategic goals include the following:

- Rescue
- Spill control (*i.e.*, confinement)
- Leak control (*i.e.*, containment)
- Fire control
- Public protective actions
- Recovery

Hazmat strategic goals and tactical objectives can be implemented from three distinct operational modes.

Offensive mode. Offensive-mode objectives commit resources to aggressive leak, spill, and fire control objectives. Offensive strategies/tactics are achieved by implementing specific types of offensive operations designed to quickly control or mitigate the problem. Although offensive operations increase the risk to emergency responders, the risk can be justified if rescue operations can be accomplished quickly, if the spill can be confined or contained rapidly, or the fire can be extinguished quickly.

Defensive mode. Defensive-mode objectives commit resources (people, equipment, and supplies) to less aggressive objectives. Defensive strategies/tactics are achieved using specific types of defensive tactics directed toward keeping the hazmat release confined to a specific area. A defensive plan may require "conceding" certain areas to the emergency, while directing response efforts toward limiting the overall size or spread of the problem (*e.g.*, concentrating the majority of response efforts on constructing dikes or dams in advance of a toxic spill to prevent contamination of a fresh-water supply).

Non-intervention mode. Non-intervention means taking no action. The basic premise of non-intervention is that responders will allow the hazmat incident to follow its natural sequence of events until the risk of intervening has been reduced to an acceptable level. Non-intervention is typically used when the incident commander believes that implementing offensive or defensive strategies will expose responders to unacceptable risk. Non-intervention would be a viable strategy when dealing with a potential boiling liquid expanding vapor explosion (BLEVE) scenario.

Step 4: Select personal protective clothing and equipment

The necessary level of personal protective clothing and equipment will be selected based on the assessment of the materials involved, the relative hazards and risks, and the selection of response objectives. No one type of protective clothing will satisfy personal protection needs under all circumstances. Chief fire officers should be familiar with the types and levels available.

The selection of personal protective clothing will depend on the hazards and properties of the materials involved, and the response objectives to be implemented (*i.e.*, offensive, defensive, and non-intervention). In evaluating the use of protective clothing at hazmat incidents, the following factors must be considered:

- The hazard to be encountered, including the specific tasks to be performed
- The level and type of specialized protective clothing to be utilized
- The user, or individual(s) who will use the PPE in a hostile environment. Remember, chemical protective clothing (CPC) places a great deal of both physiological and psychological stress on an individual. Experience has shown that HMRT responders are more likely to be injured as a result of heat stress than from chemical exposure

Protective clothing used at hazmat emergencies can be categorized in several methods. For our purposes, it will be broken into three generic levels: structural, chemical, and high-temperature.

Structural firefighting clothing. Although its hazmat applications are limited, structural firefighting clothing (SFC) can still offer sufficient protection to the user who is aware of the hazards encountered and the limitations of the clothing. It is typically worn in conjunction with self-contained breathing apparatus (SCBA). Structural clothing may be used when there is not likely to be any contact with splashes of extremely hazardous substances, or when total atmospheric concentrations do not contain high levels of chemicals toxic to the skin. In other words, when there will be no adverse effects from chemical exposure to small areas of unprotected skin.

With the recent growth of the terrorism threat, the use of SFC at incidents involving chemical or biological weapons has taken on greater significance. The U.S. Army's Soldier and Biological Chemical Command (SBCCOM) conducted extensive testing on protection factors provided by SFC in certain chemical agent environments. Readers should review the report, *Guidelines for Incident Commander's Use of Firefighter Protective Ensemble (FFPE) with SCBA for Rescue Operations During a Terrorist Chemical Agent Incident,* prepared by the U.S. Army SBCCOM Domestic Preparedness Chemical Team (August, 1999). The report can be downloaded at *http://www2.sbccom.army.mil/hld/.*

Structural firefighting clothing is not specifically designed to offer personal protection when operating in a chemical environment. It would certainly not be the PPE of first choice when operating in a toxic environment, such as chlorine or sarin. However, it may provide *limited* protection in certain response scenarios. When evaluating this issue at the local level, chief fire officers should consult with their hazmat officer or coordinator to develop a local protocol.

Chemical protective clothing. Chemical protective clothing (CPC) is designed to protect the skin and eyes from direct contact with the chemical(s) involved. Respiratory protection is worn in conjunction with the clothing. CPC is divided into two groups: chemical-splash and chemical-vapor protective clothing. However, the EPA also has developed a classification system for the various levels of CPC.

Chemical splash-protective clothing (EPA levels B and C). Chemical splash-protective clothing consists of several pieces of clothing and equipment designed to provide skin and eye protection from chemical splashes. It does not provide total body protection from gases, airborne dusts, and vapors. For hazmat operations, SCBA or airline hose units must also be provided for respiratory protection. However, if responders are conducting air monitoring, respiratory protection may be downgraded to air purifying respirators (APRs). Chemical splash protective clothing would be used under the following conditions:

- The vapors or gases present are not suspected of containing high concentrations of chemicals that are harmful to, or can be absorbed by, the skin.
- It is highly unlikely that the user will be exposed to high concentrations of vapors, gases, particulates, or splashes that will affect any exposed skin areas.
- Operations will be conducted in a flammable atmosphere. However, in some situations, it is possible to wear chemical-splash protective clothing with structural firefighting clothing to combine chemical and thermal protection.

Chemical-vapor protective clothing (EPA level A). Chemical-vapor protective clothing offers full-body protection against a hostile chemical environment when used with air-supplied respiratory devices. Unlike chemical-splash protective clothing, it provides a sealed, integral system of protection. Chemical-vapor protective clothing would be used under the following conditions:

- Extremely hazardous substances are known or suspected to be present, and skin contact is possible (*i.e.*, chlorine, cyanide compounds, toxic and infectious substances)
- Potential contact with substances that harm or destroy skin (*e.g.*, corrosives)
- Anticipated operations involve a potential for splash or exposure to vapors, gases, or particulates capable of being absorbed through the skin
- Anticipated operations involve unknown or unidentified substances and require intervention by ERP (emergency response plan or procedures)

CPC is primarily designed to provide chemical-splash or vapor protection. When operating in a combined flammable and toxic environment, recognize that CPC may not provide any protection against the flammability risk. Although fire-retardant garments or covers may be used in conjunction with CPC, there are still inherent risks. Remember, personal protective clothing is your last line of defense!

High-temperature protective clothing. High-temperature clothing is designed for protection from short-term exposures to high temperatures. There are two types, proximity and fire entry suits. These ensembles are used to function in high-heat environments that exceed the protection capabilities of structural firefighting gear. Flammable-liquid and gas firefighting are good examples. They are not designed to provide any chemical protection.

Proximity suits are constructed for short duration, close proximity to flame, and radiant heat. They will also withstand exposures to steam, liquids, and weak chemicals. They are commonly used for airport rescue and firefighting (ARFF) applications.

Fire entry suits offer complete protection for short-duration entries (maximum 60 seconds total suit exposure) into total-flame environments. They may be found in petroleum facilities where flammable gases, liquids, and other reactive chemicals are processed.

Chemical resistance. Two primary concerns when using CPC are chemical resistance and clothing integrity. CPC resistance is described in terms of chemical degradation, penetration, and permeation. Degradation is the physical destruction or decomposition of a clothing material due to exposure to chemicals, use, or ambient conditions (*e.g.*, storage in sunlight). Signs of degradation include charring, shrinking, and dissolving, or loss of tensile strength. Penetration is the flow of a hazardous liquid chemical through zippers, stitched seams, pinholes, or other imperfections in the material (*e.g.*, suit fasteners, exhalation valves, etc.).

In contrast, permeation is the process by which a liquid chemical moves through a given material on a molecular level. Unlike both degradation and penetration, there are often no obvious physical indications that the chemical has passed through the CPC. Permeation is a significant concern when evaluating protective clothing contamination and decontamination. Permeation through an impervious barrier is a three-step process.

1. Adsorption of the chemical into the outer surfaces of the material, generally not detectable by the wearer
2. Diffusion of the chemical through the material
3. Desorption of the chemical from the inner surface of the material, usually the first time the user will detect the chemical while inside the protective clothing

When evaluating chemical permeation charts supplied by a CPC manufacturer, the term "breakthrough time" will be listed. Breakthrough time is defined as the time from the initial chemical attack on the outside of the material until its desorption and detection inside. Obtaining the breakthrough time will enable responders to estimate the duration of maximum protection under a worst-case scenario of continuous chemical contact.

Heat stress. Heat stress is a critical concern to personnel wearing CPC. Both chemical-splash and chemical-vapor protective clothing are designed to protect the user from a hostile environment and prevent the passage of contaminants into the "protective envelope." Unfortunately, they also reduce the body's ability to discard excess heat and perform natural body ventilation.

A key indicator of body-heat levels is the body core temperature. If body heat cannot be eliminated, it will accumulate and elevate the core temperature. Physical reactions include heat rash, heat cramps, heat exhaustion, and heat stroke. Experience has shown that HMRT responders are more likely to be injured as a result of heat stress when wearing CPC than from a chemical exposure. Guidelines for reducing heat stress include the following:

- Responders must maintain an optimal level of physical fitness
- Provide plenty of liquids. To replace body fluids (water and electrolytes), use electrolyte mixes
- Body cooling devices (*e.g.*, cooling vests) can be used to aid natural body ventilation. These devices add weight and must be carefully balanced against worker efficiency
- Install mobile showers and portable hose-down facilities to reduce body temperatures and cool protective clothing
- Rotate personnel on a shift basis

- Provide shelter or shaded areas to protect personnel during rest periods. Entry personnel should be placed in cool areas before and after entry operations (*e.g.*, air-conditioned ambulances)

In summary, the evaluation, selection, and use of protective clothing at a hazmat emergency should be viewed from a "systems" perspective consisting of: (1) the hazard to be encountered, including the specific tasks to be performed, (2) the level and type of specialized protective clothing to be utilized, and (3) the user, or the individual(s) who will use the PPE in a hostile environment.

Step 5: Information management and resource coordination

Information management and resource coordination refers to the process of ensuring the timely and effective management, coordination, and dissemination of all pertinent data, information, and resources among all the players. The success of this coordination effort is directly related to the implementation of the local incident command system (ICS) procedure. If the ICS elements identified in site management and control have not been implemented, it will be extremely difficult for the various agencies and units represented on-scene to operate in a safe and effective manner. These ICS elements include the establishment and transfer of command, the establishment of an incident command post (ICP), and the development of a unified command organization when multiple response agencies are involved.

Coordination of information and resources at a hazmat incident can be a complex, time-consuming task. While it is difficult to operate safely when little or no information is available, it can be even more harmful to operate when overwhelmed with people and data that cannot be effectively organized for evaluation and decision making. Information coordination also plays a role after the fact, as the documentation arm of the incident commander. As with any major incident, it is important to establish a "paper trail" that follows the events leading up to and occurring during and after the incident. The following are some guidelines to consider:

- Confirm emergency orders and follow through to ensure that they are fully understood and correctly implemented. Maintain strict control of the situation.
- Make sure that there is continuing progress toward solving the problem in a timely manner. Do not delay in calling for additional assistance if conditions appear to be deteriorating.
- Make sure that all of the key players understand the incident action plan and have "bought into" the process. Unified command is critical!
- Bad news doesn't get better with time. If there is a problem, the earlier you know about it, the sooner you can start to fix it!

- Don't allow external resources to "free-lance" or do an "end run" around unified command.
- If activated, provide regular updates to the local emergency operations center (EOC).

Step 6: Implement response objectives

Based on the hazard-and-risk assessment, the incident commander determines the overall response strategic goals, tactical objectives, and incident action plan (IAP) for the emergency. Remember that strategic goals are very broad in nature, are determined at the command level, and may be pursued simultaneously during an incident. Examples of common strategic goals include the following:

- Rescue
- Spill control (*i.e.*, confinement)
- Leak control (*i.e.*, containment)
- Fire control
- Public protective actions
- Recovery

In contrast, tactical objectives are the specific objectives used to achieve strategic goals. Tactics are normally determined at the section level or lower in the command structure. For example, tactical objectives to achieve the strategic goal of spill control would include absorption, diking, damming, and diversion.

If the incident commander hopes to have strategic goals understood and implemented, the goals must be packaged and communicated in simple terms. If strategic goals are unclear, tactical objectives will become equally muddied. Hazmat strategic goals and tactical objectives can be implemented from three distinct operational modes.

Defensive mode. Defensive-mode objectives commit resources (people, equipment, and supplies) to less aggressive objectives. As a general rule, defensive strategies should be implemented to the fullest extent before attempting more aggressive, offensive strategies.

Defensive operations are typically directed toward keeping the released hazmat confined to a specific area. This may require "conceding" certain areas to the emergency, while directing response efforts to limit the overall size or spread of the problem. Advantages of defensive operations include:

- The problem may be controlled without direct exposure of emergency responders to the released hazardous materials

- In some cases, operations can be performed without the need for specialized protective clothing or equipment
- Special hazmat leak-control equipment may not be required
- First responder operations-level personnel can usually perform most spill control operations with minimal supervision

Offensive mode. In situations where defensive options have not produced the desired results or the public is at great risk from exposure, aggressive, well-planned offensive operations may be required. Offensive strategies and tactics are directed toward keeping the hazmat *contained* within its container. This process typically requires personnel to operate within the hot zone and at a higher level of potential exposure. Examples would include plugging and patching leaking containers or applying specialized leak-control kits, etc.

It is important to recognize that unless specialized training has been provided, such as with flammable liquid fuels and gases, firefighters at the first responder operations level will be unable to perform most offensive tactical options. While offensive options subject responders to a potential higher level of risk, they have several advantages over defensive options, including:

- Environmental damage is minimized. This is especially true when dealing with liquids that may be released and enter into drainage systems or water supplies, thereby creating major hazards and pollution problems.
- On-scene operating time is usually reduced. Leaks controlled and stopped at the container can limit the spread of the material and eliminate the need for evacuation. This is particularly true when the hazardous material is a vapor or gas and/or toxic.
- Cleanup costs are usually reduced when contaminants are limited to small areas or have not contaminated either ground or surface waters.

Realistically, defensive and offensive tactical options can be used at the same time. For example, first responders can begin downstream spill-control operations while HMRT members prepare for entry operations to control the release at the container.

Non-intervention mode. Non-intervention means taking no action. In this mode, operations focus on maintaining a passive position in a safe location, while allowing the hazmat incident to follow its "natural course" without direct emergency response intervention. An example is a potential BLEVE or reactive chemicals scenario, or a situation where emergency responders are awaiting the arrival of additional responders or resources on-scene before initiating control operations.

Prior to initiating any entry operations, the incident commander must ensure that properly equipped backup personnel wearing the appropriate level of personal protective clothing are in place. In addition, the entry teams should be monitored by EMS personnel and be briefed prior to entering the hot zone. At a minimum, this briefing should include the following:

- Removal of all watches, jewelry, and personal valuables
- Reviewing of the objectives of the entry operation
- Performing radio communications, SCBA, and CPC checks
- Identifying emergency escape signals
- Reviewing decontamination setup, location, and procedures

CAUTION: Decon should be set up prior to initiating entry operations. For situations where civilians are down and chemical exposures are suspected, emergency decon must be established as soon as possible.

Finally, while cleanup-and-recovery operations are typically not performed by emergency responders, they will often have control and overall responsibility of the emergency scene. Experience has shown that many accidents and injuries occur during this phase of the emergency. The combination of incident duration, personnel becoming tired and wanting to "wrap up" the emergency, and a new cast of "players" now arriving on the incident scene (*e.g.*, contractors, industry representatives, government officials, etc.) often set up a scenario in which safety can be compromised if responders do not remain alert.

The incident commander should ensure that the work area is closely controlled, that the general public is denied entry, and that the safety of emergency responders and the public is maintained during cleanup-and-recovery operations. When interfacing with both industry responders and contractors, the incident commander should ensure that these individuals are trained to meet the requirements of OSHA 1910.120.

Step 7: Decontamination and cleanup operations

Decontamination is the process of reducing or eliminating harmful substances (making "safe") from personnel, equipment, and supplies when entering and working in contaminated areas (*i.e.*, hot zone). Although decon is commonly addressed in terms of "cleaning" personnel and equipment after entry operations, command personnel should recognize that in some instances decontamination of clothing and equipment may not be possible, and these items may require disposal. Ultimately, the success of decontamination is directly tied into how well the incident commander can control on-scene personnel and operations. The following section discusses the basic concepts of decontamination and cleanup.

Contamination prevention. If contact can be controlled and minimized, the potential for contamination can be reduced or eliminated. Try to stress work practices that minimize contact with hazardous substances. Don't walk through areas of obvious contamination, and as much as possible (and practical), don't touch potentially hazardous substances. If contact is made, remove the contaminant as soon as possible.

Types of contamination. There are two types of contamination—surface and permeation. Surface contaminants have not been absorbed into the surface and are normally easier to detect and remove than permeated contaminants. Typical examples include dusts, powders, and asbestos fibers.

Permeated contaminants are absorbed into the material and are often difficult or impossible to detect and remove. If the contaminants permeating a material are not removed by decontamination, they may continue to permeate through the fabric and cause an exposure on the "inside" of a protective clothing material. Permeation is affected by the following factors:

- *Contact time*—The longer a contaminant is in contact with an object, the greater the probability and extent of permeation.
- *Concentration*—Molecules flow from areas of high concentration to areas of low concentration. The greater the concentration, the greater the potential for permeation.
- *Temperature*—Increased temperatures generally increase the rate of permeation.
- *Physical state*—As a rule, gases, vapors, and low-viscosity liquids tend to permeate more readily than high-viscosity liquids or solids.

Methods of decontamination. While decontamination is performed to protect health and safety, it can also create hazards in certain situations. For example, decon methods may be incompatible with the hazmat involved and thereby cause a violent reaction. Further, they may pose a direct health hazard to personnel from the inhalation of hazardous vapors.

Decon methods, solutions and techniques may be obtained from shippers, manufacturers, and medical facilities, including poison control centers, and other technical resources, such as CHEMTREC and other hazmat databases.

The physical and chemical compatibility of the decon solutions must be determined before they are used. Any decon method that permeates, degrades, damages, or otherwise impairs the safe function of PPE should not be used unless unusual circumstances dictate such. For example, sodium hypochlorite (*i.e.*, bleach) may be used as a decon agent when dealing with chemical and biological materials. However, bleach will also degrade common fire clothing materials such as Nomex® and Kevlar®, and will shorten the life of the material.

CAUTION: Beware of decon methods that may pose a direct health hazard to emergency response personnel. Measures must be taken to protect both decon workers and the personnel being decontaminated.

Methods of decontamination can be divided into physical and chemical. Physical techniques include:

- Brushing or scraping
- Dilution
- Absorption
- Heat
- Use of low or high air pressure (Note: Pressurized air may cause contaminants to become airborne resulting in an inhalation hazard and spreading the contaminant. OSHA regulations restrict the use of pressurized air for use on people.)
- Vacuuming
- Disposal

Chemical methods for decontaminating include:

- Chemical degradation (use of bleach, solvents, surfactants, cleaners, etc.)
- Neutralization
- Solidification
- Disinfection and sterilization

Personal protection of decon workers. Response personnel serving as the decon crew must be properly protected. In some situations, decon personnel should wear the same level of PPE as personnel operating in the hot zone (*i.e.*, in incidents involving poison liquids and gases). In most situations, however, decon personnel are sufficiently protected by wearing PPE one level below that of the entry crews. Air monitoring may allow the level of respiratory protection to be downgraded to air purifying respirators. In addition, when using a multi-station decon system, decon personnel who initially come into contact with personnel and equipment leaving the hot zone often require more protection than those workers assigned to the final station in the decon corridor.

All decon personnel must also be decontaminated before leaving the decon area (warm zone). The extent of their decontamination process will be determined by the types of contaminants involved in the emergency and the type of work performed within the decon operation.

Testing for decontamination effectiveness. Decon methods vary in their effectiveness for removing different substances. The effectiveness of any decon method should be assessed at the beginning of the decon operation and periodically throughout the operation. If contaminated materials are not being removed or are permeating through protective clothing, the decon operation must be revised.

In assessing the effectiveness of decontamination, visual inspections can show stains, discolorations, corrosive effects, etc. Monitoring devices such as photo-ionization detectors (PIDs) and detector tubes can show that contamination levels are at least below the device's detection limit. Wipe-sampling can provide after-the-fact information about the effectiveness of decon. After a wipe-swab is taken, it is analyzed in a laboratory. Protective clothing, equipment, and skin may be tested using wipe-samples.

Post-incident decon concerns. Following the termination of the emergency, it may be necessary to collect and contain all disposable clothing, plastic sheeting, and other discarded materials. All containers containing contaminated materials should be sealed, marked, and isolated. In addition, determine whether any equipment requires isolation for further analysis or decontamination.

Types of decontamination. Chief fire officers should be familiar with the types of decon operations available. This section will provide a basic overview of each type, including emergency decon, technical decon, patient decon, and mass decon.

Emergency decontamination. Situations can occur when personnel become unknowingly or accidentally contaminated. Emergency decon is required when someone who is not wearing PPE is contaminated or a responder's PPE is breached and there is body contact with the contaminant. This problem is normally associated with routine events in which hazardous materials are not suspected.

When responders have been accidentally contaminated, emergency decontamination must be performed in an expedient manner. Remember, this is not a controlled situation. Typically, little or no chemical protective clothing is in place, and there is usually no specialized equipment or expertise on-scene to assist. Some key steps to remember in these situations are as follows:

1. Establish an area of refuge within the hot zone as soon as possible.
2. Establish a gross decon area using some form of water.
    - Hose lines are best, even a garden hose.
    - If time allows, construct a basin to collect runoff, or use a place that will hold the runoff, such as a depression in the ground.
    - Emergency decon operations are graded on speed, not neatness. The sooner you decontaminate, the better.
3. Provide EMS for the victims.

Emergency decon can be innovative, but the bottom line is to clean the contaminated person as soon as possible. Soap and water is a universal solution and should be applied in large quantities. When acids or bases are on bare skin, the minimum amount of time for a water flush is at 20 minutes.

Technical decontamination. This is a more formal setup and in some cases may be called formal decontamination. It is normally implemented by HMRTs operating at a working incident. Technical decontamination is a multi-step process that goes from dirty to clean, with assistance being provided to do the cleaning. The process has changed over the years from a long, cumbersome process to a more practical approach. If the entry team has accomplished its task and will not be using the same PPE, the process can be very simple. If the PPE is to be used on another entry, more steps and care must be taken. As stated previously, responders must use safe work practices to avoid contamination. Disposable overgarments, such as gloves and boots, should also be used to help minimize the amount of decon that is required.

In previous sections we have made reference to the use of soap and water as a universal decon solution. For most materials this is the case. However, with some materials, bleach and a wetting agent may need to be added. Nerve agents and pesticides are neutralized faster when soap, alcohol, and bleach are part of the decon mixture. The chemical determines the type of decon solution required. When determining the type of decon solution needed, we recommend always consulting with a chemist.

The process of technical decon is made up of several stages that can vary locally. The basic process is the same, but the order and the mechanism used to accomplish the steps may vary. To perform a gross decon, some localities use hoses, while others may use a shower setup or even a shower in a tent. The intent is the same—to remove the big chunks or gross contaminants—but the mechanism varies. The basic steps in technical decon are as follows:

1. Tool drop—Where tools are placed that may be reused or will require further cleaning.

2. Overglove/overboot drop—Efficient removal of the overgloves and overboots can minimize the potential for secondary contamination. Overgloves and overboots are used to minimize the amount of contamination on PPE. This is also where the majority of personal contamination occurs.

3. Gross or primary decon—The initial step where the entry crew is rinsed off, paying attention to the hands and feet. A soap and water wash may also be used, followed by a rinse.

4. Formal or secondary decon—Additional washing and rinsing steps designed to further reduce the level of contamination. Special attention is given to the hands and feet.

5. PPE removal—PPE should be removed in a manner that minimizes the potential for the decon crew touching or coming in contact with those being decontaminated. Large trash bags are a good tool for making handling and disposal easier. If the materials require additional post-incident handling (*e.g.*, evidence), clear plastic bags should be used.

6. Respiratory protection removal—Should always be the last item removed. If individuals being deconned are wearing coveralls, the user should continue to wear the face piece until the PPE is removed.

7. Clothing removal—If necessary, change out of the undergarments. In this case, it will also be necessary to provide an additional change of clothing for personnel.
8. Body wash—If a breach in a suit occurs, the whole body, or at least the potentially contaminated area, should be washed. This may be done offsite, but responders should always shower prior to going home.
9. Medical evaluation—Responders should always be medically evaluated after an entry. They should sit, rehab, and have their vital signs taken 10-15 minutes later to ensure that they are recovering from the stress of entry. Responders who are not recovering should be further evaluated, hydrated, and possibly transported for further evaluation.

All equipment, including vehicles, monitoring instruments, hand tools, etc. should be decontaminated onsite prior to crossing the hot zone and leaving the site. Fire hoses should be cleaned according to the manufacturer's recommendations. For most materials, detergents will perform adequately. However, certain detergents and cleaning agents may damage fire hose fibers. The fire hose should be thoroughly rinsed to prevent any fiber weakening. The hose should be marked and pressure-tested before being placed back into service.

Hand tools may be cleaned for reuse or disposed. Cleaning methods include hand cleaning or more commonly, pressure washing or steam cleaning. One must weigh the cost of the item against the cost of decontamination and the probability that it can be completely cleaned.

For vehicle decon, it may be necessary to construct a decon pad. The pad may be a concrete slab or a pool liner covered with gravel. Engines exposed to toxic dusts or vapors should have their air filters replaced. If the engine has been exposed to corrosive atmospheres, a mechanic should inspect it. Permeable materials, such as seats, floorboards, and steering wheels may have to be removed and disposed.

Patient decontamination. Emergencies can occur in which personnel are exposed to hazardous materials. If the patient is treated and transported to a hospital before decontamination, the patient, emergency medical responders, and hospital staff will be exposed to the chemical threat. The potential for completely shutting down the hospital emergency room is a real and distinct possibility.

Individuals exposed to hazmat should be decontaminated as much as possible prior to being transported to a medical facility. Proper and immediate field decon procedures will limit patient exposure, as well as protect EMS and hospital personnel. The extent of field decon will depend on correctly determining whether the patient has been exposed or contaminated, and the possibility of secondary contamination.

Mass decontamination. A "worst-case scenario" is one in which a large number of civilians have been contaminated and require decontamination or medical treatment. With the growth of the terrorism problem, the likelihood of dealing with tens, hundreds, and perhaps even thousands of contaminated people is possible.

Remember that the removal of clothes is a part of decon and can often remove most of the contaminants. U.S. Army SBCCOM tests with simulated chemical agents show that this simple action can remove up to 80% of the contaminant. Mass decon can done in a variety of methods, including:

- Establishment of mass decon corridors using either hand lines or soft hose streams from fire engines or aerial devices—As more units arrive on-scene, they can be added to the corridor.
- Use of portable shower setups (both commercial and available systems)
- Use of indoor or outdoor swimming pools in which the victims wade into the water
- The use of sprinkler heads in a building with a fire protection sprinkler system
- Multiple bathroom showers, such as in a high school locker room

In cold weather environments, victims can be washed off in cold weather and survive hypothermia if they are moved to a warm building or vehicle as soon as possible. The rule of thumb is simple—if the victims may die from contamination, wash them off, no matter how cold. You do need to move them to a covered environment of at least 70°F (preferably higher) as soon as possible, but you will not cause permanent damage to them by washing them.

Nothing in the preceding statements should suggest that mass decon will be easy!

Step 8: Terminate the incident

Step 8 is the termination of emergency response activities and the initiation of post-emergency response (PERO) operations, including investigation, restoration, and recovery activities. Terminating the incident includes the transfer of command to the agency responsible for coordinating all post-emergency activities. Termination activities are divided into three phases: debriefing emergency response personnel and the incident staff, post-incident analysis, and incident critique.

Incident debriefing. This activity may be held during the final phases of terminating the emergency or after emergency responders have been released. A debriefing should meet the following objectives:

- Inform emergency response personnel of possible hazmat exposures and associated signs and symptoms
- Identify equipment damage and unsafe conditions requiring immediate attention or isolation for further evaluation

- Assign information-gathering responsibilities for a post-incident analysis and critique
- Summarize the activities performed by each section or sector within the incident command system
- Reinforce positive aspects of the emergency response

The debriefing is most effective when one individual is selected to lead it. The incident commander may not be the best facilitator, but he should be present to summarize the incident from the perspective of the command staff and to reinforce the performance of the command staff and response personnel. The debriefing session should be concise, cover only the major aspects of the incident, and be limited to no more than 30 minutes. The following subjects are recommended in the order listed:

1. *Health information*—Exact materials and potential stresses to which personnel have been exposed, including exposure signs and symptoms. In addition, need for any followup medical evaluations and the documentation of exposure levels.
2. *Equipment and apparatus exposure review*—Identification of equipment and apparatus potentially exposed and plans for special cleaning or disposal. Identification of personnel and procedures to decontaminate or dispose of equipment.
3. *Problems requiring immediate action*—Equipment or procedural failures, major personnel problems, or legal implications of response and recovery operations.
4. *Reinforcement of things that went correctly and appreciation by command for a job well done.*

Post-incident analysis. The post-incident analysis is a reconstruction of the emergency response to establish a clear picture of the events that occurred. The primary objective of the post-incident analysis is the improvement of future emergency response operations.

An individual or task force is selected to collect information pertaining to the emergency response and recovery operations, as well as issues raised at the debriefing session. This will guarantee that sensitive or unverified information is not improperly released. A checklist of key data and documentation should include information on the cause of the incident and contributing factors, chemical hazard information from available resources, and records on levels of exposure and decontamination.

Additional information can be acquired from interviews with department personnel, mutual aid units, and any photographs or videotapes made of the emergency response effort. This material will also serve as documentation for the post-incident investigation and potential cost recovery efforts.

When all data is assembled and a rough-draft report is prepared, the report should be reviewed by the key players at the emergency to verify the contents. Once completed, the analysis can begin. Post-incident analysis should focus on five key topics:

- Command and control
- Tactical operations
- Resources
- Support services
- Plans and planning

When the post-incident analysis is completed, it should be forwarded to management for review and then distributed to those responsible for appropriate action. Conclusions and recommendations should be incorporated into the existing emergency response plan (ERP) and procedures, or used as the basis for developing a new or revised ERP.

Critique. An effective incident critique or self-evaluation supported by senior management is a positive way to outline and discuss lessons learned. A commitment to critique emergency response operations will improve performance and planning by increasing efficiency through the detection of deficiencies.

The purpose of a critique is to develop recommendations for improving the emergency response system, and *not* to find fault with the performance of ERT personnel. The crucial player in the critique is the facilitator who leads the process. A facilitator can be any individual who is comfortable and effective working in front of a group, knowledgeable about the ERP and standard operating procedures (SOPs), and experienced in emergency response. This individual need not have been part of the response effort. The following is a recommended critique format for large-scale or working emergency responses:

1. *Begin with a critique by the participants*—Each individual should make a statement relevant to their performance and what they feel are the major issues. Depending on available time, more detail may be added. There should be no interruptions during this phase.
2. *Critique of the operation*—Participants should comment on the strengths and weaknesses of each section or sector's actions and contributions. Through a spokesperson, each section or sector should present problems encountered, unanticipated events, and lessons learned. Each presentation should not exceed five minutes.
3. *Conclude with a session critique*—At the end of the critique, participants should focus on the problems that need to be addressed by each group. The facilitator should encourage discussion, reinforce constructive comments, and record important points.

Following the critique, the facilitator should forward written comments to management. These comments should emphasize suggestions for improving emergency response capabilities and for revising/upgrading the emergency response program. A final report should then be circulated within the emergency response organization for all personnel to review.

Conclusion

This chapter has provided an overview of hazmat emergency response operations within the fire service. Topics included a review of the key legislative, regulatory, and consensus standards that impact fire service hazmat planning and emergency response operations, an overview of the hazmat management system for managing the community hazmat problem, and the introduction of the eight-step process as a methodology for the on-scene management of a hazmat emergency.

Technical Resources

Trade and business associations

American Petroleum Institute
1220 L Street, NW
Washington, DC 20005
202/682-8100
Website – *www.api.org*

American Chemistry Society
CHEMTREC
1300 Wilson Blvd.
Arlington, VA 22209
703/741-5100
Website – *www.americanchemistry.com*
CHEMTREC Website – *www.chemtrec.org*

Association of American Railroads
50 F Street, NW
Washington, DC 20001-1564
202/639-2100
Website – *www.aar.org*

American Trucking Associations
2200 Mill Road
Alexandria, VA 22314
703/838-1770
Website – *www.truckline.com*

Center for Chemical Process Safety
American Institute of Chemical Engineers
3 Park Ave
New York, NY 10016-5991
212/591-7319
Website – *www.aiche.org/ccps*

Chemical Producers and Distributors Association
1220 19th Street, NW, Suite 202
Washington, DC 20006
202/872-8110
Website – *www.cpda.com*

The Chlorine Institute
2001 L Street, NW, Suite 506
Washington, DC 20036
202/872-4729
Website – *www.cl2.com*

Compressed Gas Association
4221 Walney Road, 5th Floor
Chantilly, VA 20151
703/788-2700
Website – *www.cganet.com*

Dangerous Goods Advisory Council
1101 Vermont Avenue, NW
Washington DC 20005-3521
202/289-4550
Website – *www.dgac.org*

The Fertilizer Institute
Union Center Plaza
820 First Street, NE, Suite 430
Washington, DC 20002
202/675-8250
Website – *www.tfi.org*

Institute of Makers of Explosives
1120 19th Street, NW, Suite 310
Washington, DC 20036
202/429-9280
Website – *www.ime.org*

National Association of Chemical Distributors
1560 Wilson Boulevard, Suite 1250
Arlington, VA 22209
703/527-6223
Website – *www.nacd.com*

National Propane Gas Association
1600 Eisenhower Lane, Suite 100
Lisle, IL 60532
630/515-0600
Website – *www.npga.org*

National Tank Truck Carriers
2200 Mill Road
Alexandria, VA 22314
703/838-1960
Website – *www.tanktransport.com*

Propane Education and Research Council
1776 K Street, NW, Suite 204
Washington, DC 20006
202/452-8975
Website – *www.propanesafety.com*

Synthetic Organic Chemical Manufacturers Association
1850 M Street, NW, Suite 700
Washington, DC 20036
202/721-4100
Website – *www.socma.com*

Fire service/hazmat professional organizations

International Association of Fire Chiefs (IAFC)
Hazmat Committee
4025 Fair Ridge Drive
Fairfax, VA 22033-2868
703/273-0911
Website – *www.ichiefs.org/*

National Fire Protection Association (NFPA)
Hazmat Committee Staff Liaison
1 Batterymarch Park
Quincy, MA 02269
617/770-3000
Website – *www.nfpa.org/*

International Association of Fire Fighters (IAFF)
Hazmat Training Office
1750 New York Avenue, NW
Washington, DC 20006
202/737-8484
Website – *www.iaff.org/academy/content/hazmat.html*

Federal government agencies

Agency for Toxic Substances and Disease Registry
Emergency Response and Consultation Branch
Division of Health Assessment and Consultation
1600 Clifton Road, NE
Atlanta, GA 30333
404/498-0110
404/498-0120 (24-hour emergency response assistance)
Website – *www.atsdr.cdc.gov/atsdrhome.html*

U.S. Department of Transportation
Research and Special Programs Administration
400 7th Street, SW
Washington, DC 20590
800/467-4922 (Hazmat Information Center)
Website – *http://hazmat.dot.gov* (Hazmat Safety Homepage)

Environmental Protection Agency
Chemical Emergency Preparedness and Prevention Office
OS-120
401 M Street, SW
Washington, DC 20460
202/260-8600
Website – *www.epa.gov/swercepp/*

Occupational Safety and Health Administration
Office of Information and Consumer Affairs
U.S. Department of Labor
Room N3647
200 Constitution Ave, NW
Washington, DC 20210
202/523-8151
Website – *www.osha-slc.gov/*

List of OSHA and Non-OSHA states

The following list includes states and territories currently having a delegated OSHA-enforcement program and state/local government employers covered by OSHA 1910.120.

Alaska	Arizona	California
Connecticut	Hawaii	Indiana
Iowa	Kentucky	Maryland
Michigan	Minnesota	Nevada
New Mexico	New York	North Carolina
Oregon	Puerto Rico	South Carolina
Tennessee	Utah	Vermont
Virginia	Virgin Islands	Washington
Wyoming		

The following list includes states that currently do not have a delegated OSHA-enforcement program, nor state/local government employers covered by EPA Section 311 regulations.

Alabama	Arkansas	Colorado
District of Columbia	Delaware	Florida
Georgia	Idaho	Illinois
Kansas	Louisiana	Maine
Massachusetts	Mississippi	Missouri
Montana	Nebraska	New Hampshire
New Jersey	North Dakota	Ohio
Oklahoma	Pennsylvania	Rhode Island
South Dakota	Texas	West Virginia
Wisconsin		

References

Agency for Toxic Substances and Disease Registry (ATSDR). *Managing Hazmat Incidents: A Planning Guide for the Management of Contaminated Patients.* Vols. I and II, Atlanta, GA: ATSDR , 1992.

Bevelacqua, Armando. *Hazmat Chemistry.* Albany, NY: Delmar-Thomson Learning, 2001.

Bronstein, Alvin C., and Phillip L. Currance. *Emergency Care for Hazmat Exposure.* 2nd ed. St. Louis: The C.V. Mosby Company, 1994.

Callan, Michael. *Street Smart Hazmat Response.* Chester, MD: Red Hat Publishing, 2002.

Fire, Frank L. *The Common Sense Approach to Hazardous Materials.* 2nd ed. New York: Fire Engineering, 1996.

Hawley, Chris. *Hazmat Response and Operations.* Albany, NY: Delmar-Thomson Learning, 2000.

Hawley, Chris; Gregory G. Noll; and Michael S. Hildebrand. *Special Operations for Terrorism and Hazmat Response.* Chester, MD: Red Hat Publishing, 2002.

Hildebrand, Michael S., and Gregory G. Noll. *Hazmat for Fire and Explosion Investigators.* Stillwater, OK: Fire Protection Publications, Oklahoma State University, 1998.

Hildebrand, Michael S., and Gregory G. Noll. *Storage Tank Emergencies: Guidelines and Procedures.* Stillwater, OK: Fire Protection Publications, Oklahoma State University, 1997.

International Fire Service Training Association (IFSTA). *Hazmat for First Responder.* 2nd ed. Stillwater, OK: IFSTA, Oklahoma State University, 1994.

Isman, Warren E., and Gene P. Carlson. *Hazardous Materials.* Encino, CA: Glencoe Publishing Company, 1980.

Lesak, David M. *Hazmat Strategies and Tactics.* Upper Saddle River, NJ: Brady/Prentice Hall, 1999.

National Fire Protection Association. *Hazmat Response Handbook.* 4th ed. Quincy, MA: National Fire Protection Association, 2002.

National Fire Service Incident Management System Consortium Model Procedures Committee. *IMS Model Procedures Guide for Hazmat Incidents.* Stillwater, OK: Fire Protection Publications, Oklahoma State University, 2000.

National Institute for Occupational Safety and Health (NIOSH). *Occupational Safety and Health Guidance Manual for Hazardous Waste Site Activities.* Washington, DC: NIOSH, OSHA, USCG, EPA, 1985.

National Response Team. *Hazmat Emergency Planning Guide (NRT-1).* Washington, DC: National Response Team, 1987.

Noll, Gregory G. and Michael S. Hildebrand. *Gasoline Tank Truck Emergencies: Guidelines and Procedures.* 2nd ed. Stillwater, OK: Fire Protection Publications, Oklahoma State University, 1998.

Noll, Gregory G., Michael S. Hildebrand, and James G. Yvorra. *Hazardous Materials: Managing the Incident.* 3rd ed. Stillwater, OK: Fire Protection Publications, Oklahoma State University, 2002.

Noll, Gregory G., Michael S. Hildebrand, and Michael L. Donahue. *Intermodal Containers: Guidelines and Procedures.* Stillwater, OK: Fire Protection Publications, Oklahoma State University, 1995.

Stilp, Richard, and Armando Bevelacqua. *Emergency Medical Response to Hazmat Incidents.* Albany, NY: Delmar-Thomson Learning, 1997.

Stringfield, William H. *A Fire Department's Guide to Implementing Sara, Title III and the OSHA Hazmat Standard.* Ashland, MA: International Society of Fire Service Instructors, 1987.

Stringfield, William H. *Emergency Planning and Management: Ensuring Your Company's Survival in the Event of a Disaster.* 2nd ed. Rockville, MD: Government Institutes, 2000.

Stutz, Douglas R., and Stanley J. Janusz. *Hazardous Materials Injuries: A Handbook for Pre-Hospital Care.* 2nd ed. Beltsville, MD: Bradford Communications Corp., 1988.

U.S. Environmental Protection Agency. *Hazmat Planning Guide.* Washington, DC: EPA, 2001.

U.S. Environmental Protection Agency, et al. *Technical Guidance for Hazards Analysis: Emergency Planning for Extremely Hazardous Substances.* Washington, DC: EPA, FEMA, DOT, 1987.

York, Kenneth J., and Gerald L. Grey. *Hazardous Materials/Waste Handling for the Emergency Responder.* New York: Fire Engineering, 1989.

Glossary

absorbent material. A material designed to pick up and hold liquid hazardous material to prevent contamination spread. Materials include sawdust, clay, charcoal, and polyolefin-type fibers.

absorption. 1) The process of absorbing or "picking up" a liquid hazardous material to prevent enlargement of the contaminated area. 2) Movement of a toxicant into the circulatory system by oral, dermal, or inhalation exposure. 3) Process of adhering to a surface.

air monitoring. To measure, record, and/or detect contaminants in ambient air.

air purifying respirators (APR). Personal protective equipment item. A breathing mask with chemical cartridges designed to either filter particulates or absorb contaminants before they enter a worker's breathing zone. They are intended for use only in atmospheres where the chemical hazards and concentrations are known.

American Chemistry Council. Parent organization that operates CHEMTREC.

American National Standards Institute (ANSI). Serves as a clearinghouse for nationally coordinated voluntary safety, engineering, and industrial consensus standards developed by trade associations, industrial firms, technical societies, consumer organizations, and government agencies.

American Petroleum Institute (API). Professional trade association of the United States petroleum industry. Publishes technical standards and information for all areas of the industry, including exploration, production, refining, marketing, transportation, and fire and safety.

area of refuge. A "holding area" within the hot zone where exposed or contaminated personnel are protected from further contact and/or exposure and controlled until they can be safely decontaminated or treated.

BLEVE. Boiling liquid expanding vapor explosion.

boiling liquid expanding vapor explosion (BLEVE). A container failure with a release of energy, often rapid and violent, accompanied by a release of gas to the atmosphere and propulsion of the container or container pieces due to an overpressure rupture.

boom. A floating physical barrier serving as a continuous obstruction to the spread of a contaminant.

breakthrough time. The elapsed time between initial contact of a hazardous chemical with the outside surface of a barrier, such as protective clothing material, and the time at which the chemical can be detected at the inside surface of the material.

branch. That organizational level within the incident command system having functional/geographic responsibility for major segments of incident operations (*e.g.*, hazmat branch). The branch level is organizationally between section and division, sector, or group.

buddy system. A system of organizing employees into work groups in such a manner that each employee of the work group is designated to be observed by at least one other employee in the work group (per OSHA 1910.120 (a)(3)).

Canadian Transport Emergency Center (CANUTEC). A 24-hour, government-sponsored hotline for chemical emergencies. It is the Canadian version of CHEMTREC.

Center for Disease Control (CDC). The federally-funded research organization tasked with disease control and research.

chemical protective clothing (CPC). Single or multi-piece garment constructed of chemical protective clothing materials designed and configured to protect the wearer's torso, head, arms, legs, hands, and feet. Can be constructed as a single or multi-piece garment. The garment may completely enclose the wearer either by itself or in combination with the wearer's respiratory protection, attached or detachable hood, gloves, and boots.

chemical resistance. The ability to resist chemical attack. The attack is dependent on the method of test and its severity is measured by determining the changes in physical properties. Time, temperature, stress, and reagent may all be factors that affect the chemical resistance of a material.

chemical resistant materials. Materials specifically designed to inhibit or resist the passage of chemicals into and through them by the processes of penetration, permeation, or degradation.

Chemical Transportation Emergency Center (CHEMTREC). The chemical transportation center operated by the American Chemistry Council. Provides information and technical assistance to emergency responders (Phone 1-800-424-9300).

chemnet. A mutual aid network of chemical shippers and contractors, activated when a member shipper cannot respond promptly to an incident involving chemicals (contact is made through CHEMTREC).

clandestine laboratory. An operation consisting of a sufficient combination of apparatus and chemicals that either have been or could be used in the illegal manufacture or synthesis of controlled substances or weapons of mass destruction.

Clean Air Act (CAA). National legislation that resulted in EPA regulations and standards governing airborne emissions and ambient air quality.

cleanup. Incident scene activities directed toward removing hazardous materials, contamination, debris, damaged containers, tools, dirt, water, and road surfaces in accordance with proper and legal standards, and returning the site to as near a normal state as existed prior to the incident.

Code of Federal Regulations (CFR). A collection of regulations established by federal law. Contact with the agency that issues the regulation is recommended for both details and interpretation.

cold zone. The hazard-control zone of a hazmat incident that contains the incident command post and other support functions as deemed necessary to control the incident. May also be referred to as the clean zone or support zone.

colorimetric tubes. Glass tubes containing a chemically-treated substrate that reacts with specific airborne chemicals to produce a distinctive color. The tubes are calibrated to indicate approximate concentrations in air.

combustible gas indicator (CGI) detector. Measures the presence of a combustible gas or vapor in air.

Community Awareness and Emergency Response (CAER). A program developed by the American Chemistry Council to assist chemical plant managers in developing integrated hazardous materials emergency response plans between the plant and the community.

compatibility. The matching of protective chemical clothing to the hazardous material involved to provide the best protection for the worker.

compatibility charts. Permeation and penetration data supplied by manufacturers of chemical protective clothing to indicate chemical resistance and breakthrough times of various garment materials as tested against a battery of chemicals. This test data should be in accordance with ASTM and NFPA standards.

Comprehensive Environmental Response, Compensation and Liability Act (CERCLA). Known as CERCLA or Superfund, it addresses hazardous substance releases into the environment and the cleanup of inactive hazardous waste sites. It requires those who release hazardous substances, greater than certain levels (known as "reportable quantities" as defined by the EPA) to notify the National Response Center.

computer-aided management of emergency operations (CAMEO). A computer database storage-retrieval system of preplanning and emergency data for on-scene use at hazardous materials incidents. Developed and maintained by the EPA.

confinement. Procedures taken to keep a material in a defined or localized area, once released.

contact. Being exposed to an undesirable or unknown substance that may pose a threat to health and safety.

container. Any vessel or receptacle that holds a material, including storage vessels, pipelines, and packaging. Includes both bulk and non-bulk packaging and fixed containers.

containment. Actions necessary to keep a material in its container (*e.g.*, stopping a release of the material or reduce the amount being released).

contamination. An uncontained substance or process that poses a threat to life, health, or the environment.

control. The offensive or defensive procedures, techniques, and methods used in the mitigation of a hazardous materials incident, including containment, extinguishment, and confinement.

corrosivity (pH) detector. A meter or paper that indicates the relative acidity or alkalinity of a substance, generally using an international scale of 0 (acid) through 14 (alkali-caustic). (See also pH.)

critique. An element of incident termination that examines the overall effectiveness of the emergency response effort and develops recommendations for improving the organization's emergency response system.

dam. A defensive confinement procedure consisting of constructing a dike or embankment to totally immobilize a flowing waterway contaminated with a liquid or solid hazardous substance.

debriefing. An element of incident termination that focuses on the following factors: 1) Informing responders of exactly what hazmats they were (possibly) exposed to, and the signs and symptoms of exposure. 2) Identifying damaged equipment requiring replacement or repair. 3) Identifying equipment or supplies requiring specialized decontamination or disposal. 4) Identifying unsafe work conditions. 5) Assessing information-gathering responsibilities for a post-incident analysis.

decon. Popular abbreviation referring to the process of decontamination.

decontamination. The physical and/or chemical process of reducing and preventing the spread of contamination from persons and equipment used at a hazardous materials incident (also referred to as "contamination reduction").

decontamination corridor. A distinct area within the "warm zone" that functions as a protective buffer and bridge between the "hot zone" and the "cold zone," where decontamination stations and personnel are located to conduct decontamination procedures.

degradation. The physical destruction or decomposition of a clothing material due to exposure to chemicals, use, or ambient conditions (*i.e.*, storage in sunlight). Degradation is noted by visible signs such as charring, shrinking, swelling, color change or dissolving, or by testing the material for weight changes, loss of fabric tensile strength, etc.

dike. A defensive confinement procedure consisting of an embankment or ridge on ground used to control the movement of liquids, sludge, solids, or other materials. Barrier that prevents passage of a hazmat into an area where it will produce more harm.

dike overflow. A dike constructed in a manner that allows uncontaminated water to flow unobstructed over the dike while keeping the contaminant behind the dike.

dike underflow. A dike constructed in a manner that allows uncontaminated water to flow unobstructed under the dike while keeping the contaminant behind the dike.

dilution. Application of water to water-miscible hazmats to reduce to safe levels the hazard they represent. Can increase the total volume of liquid having to be disposed of. In decon applications, it is the use of water to flush a hazmat from protective clothing and equipment and is the most common method of decon.

dispersion. To spread, scatter, or diffuse through air, soil, surface or ground water.

diversion. A defensive confinement procedure to intentionally control the movement of a hazardous material to an area where it will pose less harm to the community and environment.

doublegloving. Involves the use of gloves under work gloves. Enables wearing of work gloves without compromising exposure protection. Also provides an additional barrier for hand protection and reduces the potential for hand contamination when removing protective clothing during decon procedures.

emergency decontamination. The physical process of immediately reducing contamination of individuals in potentially life-threatening situations without the formal establishment of a decontamination (or contamination-reduction) corridor.

emergency medical services (EMS). Provides emergency medical care for ill or injured persons by trained personnel.

emergency operations center (EOC). The secured site where government or facility officials exercise centralized direction and control in an emergency. Serves as a resource center and coordination point for additional field assistance. Also provides executive directives to and liaison for government and other external representatives and considers and mandates protective actions.

emergency response. Response to any occurrence that has or could result in a release of a hazardous substance.

emergency response organization. Organization that utilizes personnel trained in emergency response. Includes fire, law enforcement, EMS, and industrial emergency response teams.

emergency response personnel. Personnel assigned to organizations having the responsibility for responding to different types of emergency situations.

emergency response plan. Plan that establishes guidelines for handling hazmat incidents as required by regulations such as SARA Title III and HAZWOPER (29 CFR 1910.120).

Environmental Protection Agency (EPA). The purpose of the EPA is to protect and enhance our environment today and for future generations to the fullest extent possible under laws enacted by Congress. The agency's mission is to control and abate pollution in the areas of water, air, solid waste, pesticides, noise, and radiation. The EPA's mandate is to mount an integrated, coordinated attack on environmental pollution in cooperation with state and local governments.

EPA. (See Environmental Protection Agency)

EPA levels of protection. EPA system for classifying levels of chemical protective clothing. Level A is chemical-vapor protective suit, level B is chemical liquid-splash protective suit with SCBA, and level C is chemical liquid-splash protective suit with air purifying respirator.

evacuation. A public protective option that results in the removal of fixed facility personnel and the public from a threatened area to a safer location. Typically regarded as the controlled relocation of people from an area of known danger or unacceptable risk to a safer area, or one in which the risk is considered acceptable.

exposure. The subjection of a person to a toxic substance or harmful physical agent through any route of entry (*e.g.*, inhalation, ingestion, skin absorption, or direct contact).

first responder. The first trained person(s) to arrive at the scene of a hazardous materials incident. May be from the public or private sector of emergency services.

first responder, awareness level. Individuals likely to witness or discover a hazardous substance release and trained to initiate an emergency response sequence by notifying the proper authorities. They take no further action beyond notifying the authorities.

first responder, operations level. Individuals who respond to releases or potential releases of hazardous substances as part of the initial response to the site for the purpose of protecting nearby persons, property, or the environment from the effects of the release. Trained to respond in a defensive fashion without actually trying to stop the release. Their function is to contain the release from a safe distance, keep it from spreading, and prevent exposures.

flame ionization detector (FID). Device used to determine the presence of hydrocarbons in air.

full protective clothing. Protective clothing worn primarily by firefighters. Includes helmet, fire retardant hood, coat, pants, boots, gloves, PASS device, and self-contained breathing apparatus designed for structural fire fighting. Does not provide specialized chemical-splash or vapor protection.

gas chromatograph/mass spectrometer detector. Instrument used for identifying and analyzing organics.

gross decontamination. Initial phase of the decontamination process during which the amount of surface contaminant is significantly reduced. This phase may include mechanical removing and initial rinsing.

hazard. Refers to a danger or peril. In hazmat operations, usually refers to the physical or chemical properties of a material.

hazard analysis. Part of the planning process. The analysis of hazmats that exist in a facility or community. Elements include hazards identification, vulnerability analysis, risk analysis, and evaluation of emergency response resources. Hazard analysis methods used as part of process safety management (PSM) include HAZOP studies, fault tree analysis, and "what if?" analysis.

hazard and risk evaluation. Evaluation of hazard information and the assessment of the relative risks of a hazmat incident. The evaluation process leads to the development of the incident action plan.

hazard class. The hazard classification designation for a material as found in the Department of Transportation (DOT) regulations, 49 CFR. There are currently 9 DOT hazard classes, divided into 22 divisions.

Hazard Communication (HAZCOM). OSHA regulation (29 CFR 1910.1200) that requires hazmat manufacturers to develop material safety data sheet (MSDSs) on specific types of hazardous chemicals, and provide hazmat health information to employees and emergency responders.

hazard-control zones. The designation of areas at a hazardous materials incident based on safety and the degree of hazard. Many terms are used to describe these zones; however, for the purposes of this text, these zones are defined as the hot, warm, and cold zones.

hazardous materials. Any substance or material in any form or quantity capable of posing an unreasonable risk to safety, health, and property when transported in commerce (U.S. Department of Transportation, 40 CFR 171).

hazardous materials response team (HMRT). An organized group of employees, designated by the employer and expected to handle and control actual or potential leaks or spills of hazardous substances requiring possible close approach to the substance. A hazmat team may be a separate component of a fire brigade or a fire department or other appropriately trained and equipped units from public or private agencies.

hazardous materials specialists. Individuals who respond and provide support to hazardous materials technicians. While their duties parallel those of the technician, they require a more detailed or specific knowledge of the various substances they may be called upon to contain. Also act as liaisons with federal, state, local and other governmental authorities in regard to site activities.

hazardous materials technicians. Individuals who respond to releases or potential releases of hazardous materials for the purposes of stopping the leak. They generally assume a more aggressive role in that they are able to approach the point of a release to plug, patch, or otherwise stop release of a hazardous substance.

hazmat. Acronym used for hazardous materials.

HAZWOPER. Acronym used for the OSHA *Hazardous Wastes Operations and Emergency Response* regulation (29 CFR 1910.120).

heat detector. An instrument used to detect heat by sensing infrared waves.

high-temperature protective clothing. Protective clothing designed to protect the wearer against short-term high temperature exposures. Includes both proximity suits and fire entry suits. This type of clothing is usually of limited use in dealing with chemical exposures.

HMRT. (See hazardous materials response team)

hot zone. An area immediately surrounding a hazardous materials incident that extends far enough to prevent adverse effects from hazardous materials releases to personnel outside the zone. Also referred to as the "exclusion zone," the "red zone," and the "restricted zone" in some documents.

immediately dangerous to life or health (IDLH). An atmospheric concentration of any toxic, corrosive, or asphyxiant substance that poses an immediate threat to life, would cause irreversible or delayed adverse health effects, or would interfere with an individual's ability to escape from a dangerous atmosphere.

incident. A release or potential release of a hazardous material from its container into the environment.

incident action plan. The strategic goals, tactical objectives, and support requirements for an incident. All incidents require an action plan. For simple incidents (level I), the action plan is not usually in written form. Large or complex incidents (levels II or III) require that the action plan be documented in writing.

incident commander (IC). The IC is in charge of the incident site and is responsible for the management of all incident operations. May also be referred to as the on-scene incident commander, as defined in 29 CFR 1910.120.

incident command post. The location from which all incident operations are directed and planning functions are performed. The communications center is often incorporated into the incident command post.

incident command system (ICS). An organized system of roles, responsibilities, and standard operating procedures used to manage and direct emergency operations.

isolating the scene. The process of preventing persons and equipment from becoming exposed to an actual or potential hazmat release. Includes establishing isolation perimeter and control zones.

isolation perimeter. Designated crowd-control line surrounding the hazard-control zones. The isolation perimeter is always the line between the general public and the cold zone.

leak. The uncontrolled release of a hazardous material that could pose a threat to health, safety, and/or the environment.

LEPC. (See local emergency planning committee)

local emergency planning committee (LEPC). A committee appointed by a state emergency response commission, as required by SARA Title III to formulate a comprehensive emergency plan for its corresponding local government or mutual aid region.

material safety data sheet (MSDS). A document containing information regarding the chemical composition, physical and chemical properties, health and safety hazards, emergency response, and waste disposal of a material as required by 29 CFR 1910.1200.

MSDS. (See material safety data sheet)

National Fire Protection Association (NFPA). An international voluntary membership organization to promote improved fire protection and prevention and to establish safeguards against loss-of-life and property by fire. Writes and publishes national voluntary consensus standards (*e.g.*, NFPA 472 Professional Competence of Responders to Hazardous Materials Incidents).

National Response Center (NRC). Communications center operated by the U.S. Coast Guard in Washington, DC Provides information on suggested technical emergency actions and is the federal spill notification point. The NRC must be notified within 24 hours of any spill of a reportable quantity of a hazardous substance by the spiller. (Phone 1-800-424-8802).

National Response Team (NRT). The National Oil and Hazardous Materials Response Team consists of 14 federal government agencies that carry out the provisions of the national contingency plan at the federal level. The NRT is chaired by the EPA, while the vice-chairperson represents the U.S. Coast Guard.

Occupational Safety and Health Administration (OSHA). Component of the U.S. Department of Labor. An agency with safety and health regulatory and enforcement authority for most U.S. industries, businesses, and states.

offensive tactics. Aggressive leak, spill, and fire-control tactics designed to quickly control or mitigate a problem. While increasing risks to emergency responders, offensive tactics may be justified if rescue operations can be quickly accomplished, if the spill can be rapidly confined or contained, or the fire can be quickly extinguished.

penetration. The flow or movement of a hazardous chemical through closures, seams, porous materials, pinholes, or other imperfections in a material. While liquids are most common, solid materials (*e.g.*, asbestos) can also penetrate through protective clothing materials.

permeation. The process by which a hazardous chemical moves through a given material on the molecular level. Permeation differs from penetration in that permeation occurs through the clothing material itself rather than through the openings in the clothing material.

personal protective equipment (PPE). Equipment to shield or isolate a person from the chemical, physical, and thermal hazards that may be encountered at a hazardous materials incident. Adequate PPE should protect the respiratory system, skin, eyes, face, hands, feet, head, body, and hearing. Personal protective equipment includes: personal protective clothing, self-contained positive pressure breathing apparatus, and air purifying respirators.

photo-ionization detector (PID). A device used to determine the presence of gases or vapors in low concentrations in air.

post-emergency response operations (PERO). That portion of an emergency response performed after the immediate threat of a release has been stabilized or eliminated and the cleanup of the site has begun.

post-incident analysis. An element of incident termination that includes completion of the required incident reporting forms, determining the level of financial responsibility, and assembling documentation for conducting a critique.

process safety management (PSM). The application of management principles, methods, and practices to prevent and control releases of hazardous chemicals or energy. The focus of both OSHA 1910.119 Process Safety Management of Highly Hazardous Chemicals, Explosives and Blasting Agents and EPA Part 68 Risk Management Programs for Chemical Accidental Release Prevention.

protection in place. Directing fixed facility personnel and the general public to go inside a building or structure and remain indoors until the danger from a hazardous materials release has passed. May also be referred to as in-place protection, sheltering-in-place, sheltering, and taking refuge.

public protective actions. Strategy used by the incident commander (IC) to protect unexposed people from a hazardous materials release by evacuating or protecting-in-place. This strategy is usually implemented after the IC has established an isolation perimeter and defined the hazard-control zones for emergency responders.

radiation beta survey detector. An instrument used to detect beta radiation.

radiation dosimeter detector. An instrument that measures the amount of radiation to which a person has been exposed.

radiation gamma survey detector. An instrument used for the detection of ionizing radiation, principally gamma radiation, by means of a gas-filled tube.

reportable quantity (RQ). Designated amount of a hazardous substance that, if spilled or released, requires immediate notification to the National Response Center (NRC).

risks. The probability of suffering harm or a loss. Risks are variable and change with every incident.

risk analysis. A process to analyze the probability that harm may occur to life, property, and the environment and to note the risks to be taken to identify the incident objectives.

risk-management programs. Required under EPA's proposed 40 CFR Part 68. Risk-management programs consist of three elements: (1) hazard assessment of the facility, (2) prevention program, and (3) emergency response considerations.

safety officer. Officer responsible for monitoring and assessing safety hazards and unsafe conditions, and developing measures for ensuring personnel safety. Member of the command staff required at a hazmat incident, based on the requirements of OSHA 1910.120.

SARA. (See *Superfund Amendments and Reauthorization Act*)

scene. The location impacted or potentially impacted by a hazard.

self-contained breathing apparatus (SCBA). A positive-pressure, self-contained breathing apparatus or combination SCBA/supplied air-breathing apparatus certified by the National Institute for Occupational Safety and Health (NIOSH) and the Mine Safety and Health Administration (MSHA), or the appropriate approval agency for use in atmospheres that are immediately dangerous to life or health (IDLH).

site management and control. The management and control of the physical site of a hazmat incident. Includes initially establishing the following: command, approach, positioning, staging, initial perimeter and hazard-control zones, and public protective actions.

spill. The release of a liquid, powder, or solid hazardous material in a manner that poses a threat to air, water, ground, or the environment.

Superfund Amendments and Reauthorization Act (SARA). Created for the purpose of establishing federal statutes for right-to-know standards, emergency response to hazardous materials incidents, reauthorization of the federal superfund program, and mandates for states to implement equivalent regulations and/or requirements.

structural fire-fighting clothing (SFC). Protective clothing normally worn by firefighters during structural fire-fighting operations. Includes helmet, coat, pants, boots, gloves, PASS device, and hood to cover parts of the head not protected by the helmet. Structural fire-fighting clothing provides limited protection from heat, but may not provide adequate protection from harmful liquids, gases, vapors, or dusts encountered during hazmat incidents. May also be referred to as turnout or bunker clothing.

technical information specialists. Individuals who provide specific expertise to the incident commander or the HMRT, either in person, by telephone, or through other electronic means. May represent the shipper or manufacturer or be otherwise familiar with the hazmats or problems involved.

temperature detector. An instrument, either mechanical or electronic, used to determine the temperature of ambient air, liquids, or surfaces.

termination. That portion of incident management where personnel are involved in documenting safety procedures, site operations, hazards faced, and lessons learned from the incident. Termination is divided into three phases: debriefing, post-incident analysis, and critique.

warm zone. The area where personnel and equipment decontamination and hot-zone support takes place. Includes control points for the access corridor and thus assists in reducing the spread of contamination. Also referred to as the "decontamination zone," "contamination-reduction zone," "yellow zone," "support zone," or "limited-access zone" in some documents.

20

AIRCRAFT CRASH RESCUE AND FIRE FIGHTING

James Goodbread
Don Hilderbrand
James W. Hotell

CHAPTER HIGHLIGHTS

- Unique aspects of the airport terminal, which include aircraft and aircraft movement areas, large numbers of people who are generally unfamiliar with its design, egress, and security, and its overall configuration require specialized personal protective equipment and specialized training by the fire services.
- The two agencies that set the standards for aircraft fire fighting are the Federal Aviation Administration (FAA) and the National Fire Protection Association (NFPA).
- Aircraft today present many unique challenges since they can be constructed of unique materials such as carbon fibers and associated glues, combustible or potentially explosive metals, and other materials that when subjected to impact fragmentation or fire can threaten both the firefighter and passengers.

INTRODUCTION

Aircraft rescue fire fighting (ARFF) is a widely diverse and dynamic field. It covers a full spectrum from small commuter–type airports providing limited services, to private and commuter aircraft, to providing a full line of emergency response services for major airports supporting thousands of commercial flights a day. Some areas of an ARFF fire department are similar to conventional departments and some are very different. However, one thing is consistent; most ARFF departments operate with considerably fewer resources than are available for immediate deployment compared to their conventional counterparts.

The Federal Aviation Administration (FAA) provides mandatory minimum requirements for the various categories of airports. These requirements deal mainly with firefighting equipment, agent gallonage, vehicles, and response times. From a tactical standpoint the FAA requires only that a department be able to create a survivable path for passengers to self-egress. So, an airport department could staff one member on each piece of required equipment arriving within the specified time and be in full compliance. Beyond those very minimal requirements the level of emergency services provided by the individual airport is determined locally and varies greatly. Some ARFF departments operate with very limited resources providing only limited services.

Airport Facilities

Airport facilities can range from single buildings to complexes the size of cities. At medium to large airports the infrastructure is similar to those of cities. Typically the airport grounds contain large hotels, various business occupancies, and even large tank farms with complex underground piping to support refueling operations and aircraft maintenance facilities. Most of these types of airport occupancies are common to facilities found in the city environment and can be dealt with in the same way you would deal with their city counterparts.

The thing that sets the airport environment apart from that of a city is the aircraft, the aircraft movement areas, and the airport terminal. The terminal is a unique place. It sometimes houses thousands of people, most of whom are relatively unfamiliar with its design, egress, security, and other aspects of its overall configuration. Add to those considerations the fact that there are sometimes hundreds of aircraft loaded with thousands of people sitting directly adjacent and linked to the terminal via jet ways. These conditions have always been challenging to emergency response organizations but the recent terrorist threat has greatly complicated the task of providing protection for these facilities. They are now considered as prime targets for terrorist activity. An attack on a large airport terminal could cost hundreds of lives and disrupt air travel worldwide. The possibility now exists for the use of weapons of mass destruction resulting in hundreds of people injured and requiring the movement of thousands of victims to pre-designated areas for treatment and refuge. Airport security and community emergency plans also have been strengthened to prevent the placement of secondary terrorist devices that could cause even more destruction. Planning has never before been more important than it is today.

The airport environment

The area of operations in an airport can be an intimidating, dangerous, and sometimes alien place for someone who is not familiar with its unique function. The FAA mandates minimum requirements for personnel operating on airport grounds. The local authority having jurisdiction may adopt other requirements or regulations, such as those outlined by the National Fire Protection Association (NFPA) as long as they meet or exceed the FAA requirements. Runway, taxiway markings and procedures, radio and communication protocols, and general operating procedures are usually complicated. Most airport fire departments require their personnel to be certified through extensive training programs to ensure they can safely perform before they are allowed to operate on the airport grounds. This certification is usually a part of the local fire department training program and coordinated with whatever authority has jurisdiction. Part of this training program should include identification of the various airport operational agencies, along with their function, as well as planned and coordinated procedures for communicating within and traversing through the various aircraft movement areas. Procedures for the various types of emergency responses as defined in the department's mission statement should be documented, well coordinated, and included in the training program.

Aircraft-related emergencies evolve quickly and on short notice, thus the execution of ARFF operations can differ from conventional fire and rescue services because the events occurring and the associated fire scene changes so rapidly. Response time elements for outside support agencies can also create critical manpower and resource shortages. Events involving fire will likely represent severe challenges with high heat release rates. Large numbers of people must be kept off active runways and taxiways. There are significant aircraft entry and extrication challenges. There may be inordinate response distances, over challenging terrain features, through multiple security control points, and during continuing aircraft movement. There is also the command, control, and communications challenges found in most disaster management scenarios. Many jurisdictions and different communications channels can further compromise command capability. The response situation may also require the logistics for a mass casualty situation with many severe trauma and burn victims. If the operation is to be successful, the situation must be dealt with within this extreme set of constraints.

Design improvements in the structural integrity of aircraft and increased usage of fire-resistant materials aboard aircraft have decreased the extent of fatal injury to occupants. Since more can survive, more will need help. These factors, along with the increased size of aircraft and a greater frequency of flights, justify the need for properly staffed and equipped ARFF services. Incidents in the airline industry have always represented unique challenges for emergency response forces. That has never more apparent than during the September 11, 2001, terrorist attacks, wherein commercial aircraft were used as weapons of mass destruction. For the first time in airline history, the public began to comprehend the potentially destructive capabilities and risks that a large, fully-fueled aircraft represents. Since that time, the potential hazards associated with the airline industry have been emphasized to the public as never before.

The Mission

One of the first things that the ARFF fire department manager must accomplish is to develop a mission statement that describes what emergency services his ARFF fire department is going to provide. The FAA dictates basic ARFF requirements and the training that is required for each ARFF firefighter. Each fire department member must be trained to these standards. These FAA requirements are very basic and if an ARFF fire department commits to perform other emergency services such as EMS, technical rescue, or hazardous materials response, these services should be included in the department's mission statement along with those mandated by the FAA. Once the mission statement has been established, an associated risk assessment can be accomplished to identify exactly how these services are to be provided and the necessary resources to be allocated. Pertinent NFPA guidelines such as Standards 1500, 1710, and 403 should be used to outline these services so that they exceed the mandatory FAA requirements, thus ensuring that the efficiency and safety of both the firefighter and his customer are not compromised. This process identifies the type and level of additional training that is going to be required for the airport fire department members. Along with the mandatory FAA requirements, each member should be trained, certified, and equipped to respond to each additional mission element. This process is critical to provide a full understanding of the department's emergency response capabilities to both the ARFF department members and the community it serves.

Resource Management of Personnel

Without question, the most important resource that any fire department has are its members. No matter how well-equipped a department is, if an adequate number of human resources are not deployed to an emergency and if the welfare of these human resources are not provided for, they cannot be expected to perform effectively. If the proper numbers of firefighters are deployed and if they are properly trained, equipped, and supported, they should be able to provide for the maximum efficiency and effectiveness in support of the mission. NFPA 1500, 1710, and 403, provide standards for both conventional and ARFF fire departments. These standards directly address the unique situations that are encountered during ARFF incidences. As mentioned earlier, the size and complexity of the mission of ARFF departments vary greatly. Each ARFF department must address their unique mission requirements and apply these standards to ensure that all parties know exactly what emergency services are to be provided, the level of those services, and the resources are that are required to support the mission. Only when this has been accomplished can the manager measure the effectiveness of his department.

Risk Management of Firefighter Safety and Health

Once the mission requirements have been established, the associated risks can be identified and department operating procedures established to manage that risk. Many ARFF departments operate with limited manpower compared to their community counterparts and depend heavily on support from outside agencies. In such cases, a department's operating procedures should outline that the ARFF should not implement aircraft interior fire fighting or rescue until the emergency response forces from these outside agencies are on the scene and the required support has been established. This also will help the ARFF with limited manpower to comply with the OSHA breathing respiratory standards in such situations.

Specialized ARFF Vehicles

The ARFF vehicle is a complicated, unique, and relatively fragile piece of firefighting equipment. Major pieces of ARFF equipment are usually larger, heavier, more powerful, and capable of reaching much higher speeds than conventional structural equipment. (The FAA's Federal Aviation Regulations [FAR] Part 139 dictates basic requirements for the various types of ARFF vehicles. Part 414 of NFPA provides more detailed information on construction specifications and performance.) These vehicles usually carry large quantities of water and agents that can create stability problems. Vehicle rollovers are not uncommon and the repair costs can run as high as 75% of the cost of the original vehicle. Before purchasing an ARFF vehicle, the department needs to not only pay attention to the vehicle specification, but if possible, past performance at other locations. There have been situations where it has been well established that certain models of ARFF vehicles have had a much higher incidence of rollover and other performance problems.

Hence, the most important factor in dealing with the aforementioned problems is driver training. The unique performance capabilities and handling qualities of an ARFF vehicle require that special emphasis be placed on training personnel to operate it correctly. The driver must be fully aware of all of the ARFF vehicle capabilities and limitations and trained to recognize when he or she is reaching the performance limitations of the ARFF vehicle. This type of quality training is not easy to provide since high usage of the vehicle equates to higher maintenance costs. Plus, operating these vehicles to their limits, in many cases, is not practical due to safety considerations.

One recent answer to this problem is the ARFF driver-training simulator. These high-tech simulators give the driver an opportunity to reach the performance limitations of ARFF equipment without the associated maintenance and safety problems. However, at this time these devices are relatively rare and expensive, and since these devices are relatively new, performance data is not available. However, it has been well documented that simulator training in other applications can significantly reduce accidents. When taking into consideration the high number of ARFF vehicle rollovers each year, the associated costs of simulator training may become more acceptable when compared to the alternatives. This training is available through companies that specialize in simulator training or the simulators can be purchased by individual departments for local or regional use. There are also other safety devices that may be purchased to warn the driver when he is approaching the vehicle limitations, such as rollover warning devices that alert the driver when the angle of lean and G forces are becoming unsafe.

Personal Protective Equipment (PPE)

FAR Part 139 addresses general requirements for ARFF personal protective equipment (PPE) and basically states that protective clothing shall be provided for each firefighter. It does not specify that the clothing meet any other standards of protection. NFPA 1500 provides more specific guidance that substantially exceeds that of the FAA for the various types of emergency responses that may be included in the airport fire department mission statement. NFPA 1500 requires that ARFF use certified proximity firefighting PPE, NFPA 1976. Proximity PPE differs from the conventional structural firefighting PPE. This proximity PPE has heat-reflecting qualities and is designed to withstand substantially higher temperatures than the standard structural clothing, providing for a higher level of protection for the firefighter. NFPA also requires that the firefighter be provided with the type of PPE that is certified for the particular mission or emergency response assignment.

There can be some confusion when both structural and ARFF equipment are housed in the same station, such as when military fire departments routinely keep structural and ARFF equipment in the same building or area. This situation can raise the question of which type of PPE should be used. The PPE for aircraft incidents is generally more restrictive and presents considerably more heat stress to the firefighter than PPE approved for structural fire fighting. It is also more costly and does not meet some of the requirements of structural PPE such as reflective markings. Before the NFPA proximity standard for PPE was introduced, some departments used PPE designed for aircraft incidents for both ARFF and structural assignments. During that time the military acquired an OSHA decision that allowed the firefighter to wear the type of PPE certified for his primary assignment. So, PPE designed for aircraft incidents was used for both ARFF and structural assignments. With the introduction of the NFPA proximity standard, however, the requirements were clarified. If the primary firefighting assignment was ARFF related, the PPE had to be proximity compliant, and if structural the PPE must be structural compliant. This translated into the issuing of both types of PPE when there is a possibility of a dual assignment for a firefighter.

Agents

FAR Part 139 also dictates the minimum numbers of vehicles and the types and quantities of firefighting agents. They are generally determined by the size or index of an airport. Agent requirements for ARFF vehicles are addressed as "on board" water capacity, plus the amount of foam required for the proper mix ratios, plus pounds of the dry chemical extinguishing agent. (Alternative agents such as Halon can be substituted for dry chemical agents.)

Aqueous film forming foam (AFFF)

AFFF has been the standard agent for more than 30 years, and is used by most ARFF firefighting agencies worldwide. It is by far the most economical and effective agent on the market for extinguishing pool fires of hydrocarbon fuel. Originally designed for use in mass application on large pool-type fires, it is also an excellent wetting agent. AFFF mix ratios provide excellent on-board supply capacity and, as a mixed agent, meets any of the FAA or NFPA distance requirements for "agent throw." As good as AFFF is, there are some environmental considerations that, over the past 10 years, have motivated the firefighting agent industry to look for suitable replacements. However, they have not been able to find anything that can match AFFF. The agent provides good knockdown and cooling, and its ability to prevent burn back is excellent. AFFF is mixed with water and can be obtained in 1%, 3%, and 6% mix ratios. Over the years the 3% has become standard and is required by FAA.

Dry chemical agents

Both FAA and NFPA require certain quantities of dry chemical agents to supplement AFFF. The strong points of these agents are their knockdown capabilities and their effectiveness on three-dimensional fires. Their drawbacks are the inability to provide cooling and limited throw range.

Compressed air foam

A more recently introduced agent for ARFF fire fighting is compressed air foam. This technology relies on the expansion characteristics of foaming agents to deliver large quantities from relatively small units. These units typically consist of a foam tank and a pump or compressed air vessel. The compressed air is used for both the method of delivery and the expansion of the foaming agent. When the system is charged the air is injected into the premixed foam creating high expansion ratios. The result is a highly effective and durable foam. The advantage of these systems is they can be packaged into a relatively small and inexpensive unit that is highly effective. The disadvantage is that the recharge time is too lengthy for the unit to be used more than once at an incident.

Dual agents

The ARFF firefighting community has long realized the effectiveness of dry chemical agents on hydrocarbon fuel fires. However, as mentioned above, the limited throw range of these dry chemical agents and their inability to provide any effective cooling limits their applications. To better take advantage of these agents the firefighting industry has introduced a dual agent concept. This technology allows foam or dry chemical agents to be used independently or together. When the agents are used together, the dry chemical is discharged into the center of a foam master stream using the foam stream to extend its throw range, or the dry chemical is mixed with the foam. The latter has no effect on the extinguishing properties of the dry chemical agent. This solves the problem of the limited range of the dry chemical and effectively takes advantage of both agents extinguishing properties. The combination of the two is much more effective than either one individually. The dry chemical provides for a more effective initial knockdown and three-dimensional fire control, and the foam agent prevents any flashback and provides its inherent cooling effect. However, they still have the inherent problem of excessive re-supply time in the field. Since the combination of the two agents provides for more capability there has been a great deal of interest generated in smaller ARFF vehicles with essentially the same extinguishing capabilities of some larger ones. Even though required gallonage requirements must still be maintained, the airport community is taking a close look at these smaller, less expensive vehicles that provide comparable capability of the larger, more expensive units for the rapid intervention mission.

Halogenated agents

Halogenated agents have been widely used in the past by the ARFF industry for an alternative agent. Its primary advantage over other agents is its clean agent properties. However, these agents have been identified as ozone depleting agents. Since that designation, their use has diminished greatly. Industry has tried to find an environmentally acceptable substitute but to this date none has been found. Nevertheless, the use of halogenated agents for ARFF continues to decline.

Agent application

ARFF firefighting agent requirements are determined from the size of the aircraft being protected and anticipated fuel pool fires. The FAA and NFPA provide the requirements for the amounts of agent for ARFF fire fighting. FAA requirements are the lesser of the two. These agent amounts are mandatory minimum requirements for all commercial airports. NFPA 403 requires more agent and since that amount exceeds that of the FAA, these national consensus standards may be used when adopted by the local authority. There are some other requirements dealing with turrets and the distance they are able to throw the agent, but basically both standards only require the ability to deliver the required amount of agent to predetermined locations within a given amount of time.

Agent re-supply

In the event of an aircraft incident, the responding emergency response force will most certainly be dealing with not only pooling fuel, but also the possibility of both three-dimensional fuel fires and fire in the aircraft interior compartments or other inaccessible areas. Studies of aircraft incidences have repeatedly shown that the agent amounts required by both FAA and NFPA are substantially exceeded during suppression operations. U.S. Air Force studies in the early 1990s on fire growth and decay during aircraft incidents indicate that if sustained re-supply operations are not initiated immediately upon the arrival of emergency forces that the aircraft will be lost. These studies indicated that a well-coordinated agent re-supply plan must be developed. The studies also found that large quantities of agent were wasted because of the inability to apply agent to areas that were not easily accessible. The fire can also re-ignite if it is three-dimensional or extends into these inaccessible areas of the aircraft and the fire fighting agent becomes depleted. To address this problem, the industry has called upon technology to help provide the firefighter with more efficient tools in order to have the ability to apply these agents in a more efficient and precise manner, thereby maximizing the effectiveness of the agent that is carried to the scene. However, it is still essential that a predetermined plan be developed for immediate agent re-supply to ensure that the required quantities are available.

Roof turrets

The standard application method for AFFF is the high flow roof turret. These devices are designed for mass application of agent on pool fires. On most units the discharge gallonage can be varied from 500 to 1,200 gallons a minute. These high flow turrets are very effective when used as they were designed—to extinguish large pool-type hydrocarbon fuel fires. The disadvantages are that if the fire is three-dimensional or is inaccessible, large quantities of agent can be wasted. The high angle discharge of these devices can also obscure the vision of the operator, further hampering the ability to place the agent where it is needed.

Bumper turrets

Bumper turrets typically are mounted on the front of the ARFF vehicle directly below the driver and operator. They generally are low flow devices designed to protect the area directly in front of the ARFF vehicle during the initial mass application attack, in support of hand line operations and general mop-up. The low angle of application of these devices provides excellent visibility to the operator, which allows for more accurate placement of the agent.

Elevated waterways

Though these devices have been around for a while, only in the last few years have they become widely accepted. They add a great deal of flexibility when compared to the standard roof turret by providing the ability to precisely place large amounts of agent virtually anywhere it is needed. This allows for a much more efficient and effective use of the on-board agent. These devices may be equipped with bayonet or drill type nozzles that can effectively penetrate passenger decks to support evacuation, or inaccessible areas in an aircraft, such as baggage or utility compartments. They also can apply large amounts of agent in close proximity to a fire, such as a fully involved flight deck, without subjecting firefighters to the inherent dangers involved. Elevated waterways are relatively complicated pieces of equipment that require a rather high degree of training. Departments that purchase these devices should include training in the acquisition process. Recently, computerized simulators have been made available to the persons in the field to help familiarize operators with the waterway's basic functions before actually operating the device. These PC-based simulators are highly effective, relatively inexpensive, and can also be used to maintain competency without operating the actual waterway. Departments should also make available to their members realistic field training props to simulate the various waterway operations so that operators can initially acquire and maintain competency. Elevated waterway training should be integrated into the department training program and members should be licensed to operate the waterways the same as any other type of aerial fire fighting device. Pre-fire plans should also be developed on the various types of aircraft that these devices are to be used on. From these plans, tactics and strategy should be developed to pre-identify situations that could occur where the waterways are most likely to be used.

Low angle/high flow turrets

When AFFF was originally introduced, it was thought that the most efficient way to apply it was from a roof-mounted turret at a high angle to achieve the so-called *raindrop effect*. Extensive studies by the U.S. Air Force, however, have demonstrated that low angle, high volume agent application is much more efficient and effective on pool fires than that of the traditional rain drop application from a roof mounted turret. The reason is simple: The heat thermals from the fire tend to break up, disrupt and even evaporate the agent droplets before they can reach the surface area of the fire. Roof turrets also block the vision of the operator limiting his ability to efficiently and effectively apply the agent. To take advantage of these findings, the industry has introduced the high flow, articulating bumper type turret. Like the roof turret the high flow bumper turret can flow up to 1,200 gpm. The low angle of discharge from these devices provides for much more effective agent application when compared to the standard roof turret while giving the operator an unobstructed view allowing for more precise agent placement. Although not as versatile as the waterway, these devices are comparatively inexpensive and are simple to operate and maintain.

ARFF Fireground Operations

Today's modern aircraft present a number of formidable hazards to both the emergency responders and the passengers. Of primary concern is the large quantity of volatile fuel, but there are many other hazards that must be dealt with. These hazards include oxygen vessels, combustible metals, batteries, composite materials, and in some cases such as on military aircraft, munitions. In the past, military aircraft very rarely made transit flights fully loaded with live munitions. Usually the worst situation that could be expected on a military aircraft would be that it was armed for training maneuvers only, which presented much less of a hazard than standard munitions. Since the war on terrorism, however, that is not the case. The U.S. Air Force has the capability of launching fully loaded aircraft from bases in the continental U.S. in order to deliver them anywhere in the world. So, great care must be taken when responding to one of these aircraft, especially if the department is not familiar with what they are dealing with. The key to being able to deal with any hazard associated with aircraft is planning. Recognizing what hazards are present, where these hazards are located in the aircraft and how to cope with them is critical. In any aircraft incident there is the potential for multiple emergency situations, such as proximity fire, structural fire, mass-casualty, hazardous materials, and environmental contamination. As in structural fire fighting, pre-fire plans for each model of aircraft that utilizes airport facilities provides critical information to all airport emergency response forces.

Aircraft hazards can be both common and unique. Strategies for common hazards such as wheel fires, oxygen systems, and auxiliary power unit fires can be pre-planned to standardize the response. The aircraft industry has made great improvements in the combustibility and design of passenger compartments. However, aircraft today present many unique challenges. They can be constructed of unique materials such as carbon fibers and associated glues, combustible or potentially explosive metals, and other materials that when subjected to impact fragmentation or fire can threaten both the firefighter and passengers. Composite fibers, when fragmented, can not only present respiratory hazards, but studies conducted by the U.S. Air Force have suggested that they may also penetrate unprotected skin and may become lodged in the organs of the body.

It should be noted that there have been many cases of people who are unprotected and operating in the immediate area of an aircraft incident who then later become ill. These illnesses are hard to diagnose and there still are questions as to how dangerous the exposure actually is. There is one thing that is obvious, however. You need to provide protection for and limit the exposure of all personnel involved at an aircraft incident.

During any aircraft emergency one of the first things that must be accomplished by the incident commander is to identify the area that is immediately dangerous to life and health (IDLH). Only then can the associated risks be managed. Just as with structural fire fighting, anyone entering into this area should be protected with full PPE and associated support functions. NFPA 1500 dictates that initially the area within 75 feet of the skin of the aircraft will be identified as the area immediately hazardous to life and health. The requirement establishes an area of immediate concern for all emergency responders to recognize. The standard then allows the incident commander to reestablish the IDLH as the situation dictates. After an incident where firefighters and their PPE have been exposed, full decontamination procedures should be completed to minimize the risk of post-incident exposure.

Aircraft Construction Factors

ARFF operational success is affected by the ability to understand and manage emergencies in an environment alien to the structural firefighter's domain. The ARFF firefighter is challenged by his limited capacity to access the interior of an aircraft to attack a fire or assist in passenger removal while the fuselage remains pressurized.

When comparing aircraft structure to standard building construction, forced entry to aircraft is difficult due to the use of high strength, light-weight construction which utilizes heat-treated metals and composite materials for construction. These elements are not easily breached without specialized rescue tools and equipment, and having the benefit of aerial equipment in place and activated within the time of the initial response. The ability to forcibly enter the aircraft is further frustrated by the reality that manpower is limited and that exterior pool fire suppression and crew-initiated passenger evacuation are priorities in an initial response to an aircraft incident.

Disaster planning

Both the FAA and NFPA require that the airport have a disaster plan much the same as a community. If there is a difference between the two it is that the airport plan needs to be more detailed. During an aircraft incident the situation has the potential to change so rapidly that the plan must provide the capability for emergency forces to react to these rapidly changing situations. Both airport and community resources should be considered during the planning process. NFPA 424 Airport-Community Disaster Plan is the most comprehensive NFPA document for ARFF operations. NFPA 1600 Disaster Planning provides guidance for the development of disaster plans.

Incident Management

A critical part of the airport disaster plan is the airport incident management system (IMS). Airport incident command structures can vary greatly. It is critical to identify command echelon's to ensure that all agencies are familiar with the various levels of authority. A well-coordinated IMS system establishes the critical chain of command required to utilize the resources of not only the fire department, but also other airport agencies and the adjacent community. NFPA 1561 FD Incident Management System provides guidance for developing an IMS system.

Communications

Airport fire departments' communication capabilities can vary widely, usually being dictated by the complexity of the airport's activities. They can be as basic as a single frequency or as complicated as a multiple frequency, multiple agency system. FAA requirements are basic and only require that the fire department be able to communicate as a unit. There has been a great deal of discussion on exactly who the fire department should be able to communicate with. Some departments have communications with only airport operations. They have to depend on emergency information that is relayed from air traffic control through airport operations. Some have communications with airport operations and can additionally monitor emergency information broadcast from air traffic control, but cannot converse with them. The incident commander still has to request clarifications or updates through airport operations. The obvious problems are that the fire department incident commander is receiving second- or third-hand emergency information and the excessive time involved the relaying of information. That's risky, to say the least, when so much is at stake. Generally, pilots don't want to converse with anyone except air traffic control while they are airborne because of the emergency requiring their full attention. Air traffic controllers are sometimes hesitant to converse with more than one agency because of the hectic situation they are dealing with. The answer to this problem is to get all interested agencies together, voice each one's concern, and develop operating procedures that provide for the best solution. Many departments have settled on the following procedure: While the aircraft is airborne the fire department incident commander can communicate with airport operations, air traffic control, and with the pilot of the emergency aircraft only at his request. Once the aircraft lands, air traffic control notifies the fire department incident commander of the ground frequency assigned to the emergency aircraft. At that time he may freely communicate with the aircraft commander. During this procedure communications are limited to information inquiries that have been predetermined. This procedure addresses all of the concerns of each agency and provides for much more effective direct voice communications, thus greatly reducing the possibility of inaccurate or untimely information.

COMPLIANCE COMPARISON

Airport ARFF requirements are dictated by the FAA. FAA requirements are mandatory, generally considered basic, and major airports are inspected periodically to assure compliance. The FAA has the authority to shut down operations at airports if their compliance requirements are not met. NFPA standards are locally adopted and usually exceed, or in some instances differ, from those of FAA. NFPA standards are generally used when the airport fire department identifies the need to improve the basic requirements of the FAA. Combining the requirements of these two agencies can be confusing. Below is a comparison of some of the major compliance areas addressed by both agencies.

Primary ARFF response

Re-supply/reserve fire suppression agent supply for sustained operations

- **FAA**: FAR Part 139 requires ARFF vehicles to carry a reserve of AFFF foam concentrate on board adequate for one re-supply of the ARFF vehicle water capacity. There is no FAA requirement for water re-supply to ARFF vehicles.
- **NFPA**: NFPA 403 requires a 100% on board reserve of AFFF foam concentrate and a 100% reserve of water supply to be delivered by tankers or other rapid re-supply means for sustained operations.

ARFF Vehicles (Reserve)

- **FAA**: FAR Part 139 does not require reserve ARFF vehicles. The airport is allowed to operate for up to 48 hours after an ARFF vehicle is removed to out-of- service status resulting in reduced ARFF capacity. When an ARFF vehicle is out of service for more than 48 hours, the airport fire/rescue index is reduced reflecting the actual ARFF capability.

- **NFPA**: Reserve ARFF vehicles are recommended for replacement of out-of-service primary ARFF vehicles.

ARFF vehicle construction specifications and testing

- **FAA**: FAR Part 139 specifies ARFF vehicle flow rates and turret nozzle discharge criteria to be a function of the water capability of the vehicle. Dry chemical discharge rate is specified as a function of the on-board capacity.
- **NFPA**: ARFF vehicles are to be constructed and equipped in accordance with NFPA 414 criteria and tested annually in accordance with the criteria of NFPA 412.

ARFF vehicle emergency response times

- **FAA**: FAR Part 139 requires that the first responding ARFF vehicle to arrive at the midpoint of the farthest runway within 3 minutes of alarm notification, with all remaining vehicles arriving within 4 minutes of emergency notification.
- **NFPA**: NFPA 403 requires that the fire station to be located so that the first responding ARFF vehicle will arrive at any point on an operational runway within 2 minutes, and to any remaining portion of the on airport rapid response area within 2 minutes. The remaining required ARFF vehicles are to arrive at intervals not exceeding 30 seconds.

ARFF vehicle staffing requirements

- **FAA**: FAR Part 139 does not require specific staffing requirements. The minimum number of firefighters is determined by the number of personnel necessary to operate the required ARFF vehicles. This requirement may be satisfied with one person for each required vehicle.
- **NFPA**: NFPA 403 requires there be three firefighters on each piece of major ARFF equipment. Additional staffing requirements should be based on guidance provided in NFPA 1500 and NFPA 1710. These standards require that a sufficient number of personnel be on the scene to perform the tactical assignments that are made utilizing guidance provided. NFPA 1500 recommends three firefighters for each piece of major ARFF equipment.

Auxiliary fire suppression agents

- **FAA**: FAR Part 139 requires various airport indexes to provide sodium bicarbonate dry chemical capability that varies with the index rating of the airport.
- **NFPA**: Potassium bicarbonate dry chemical is required for some airport categories as an auxiliary suppression agent.

Firefighting personnel professional qualifications

- **FAA**: FAR Part 139 requires that firefighting personnel be trained on a variety of subjects listed in 319(j)(2) and that they have an annual live fire certification.
- **NFPA**: Requires firefighters to conform with NFPA 1003, 6-1.4. Airport firefighters are required to be trained and equipped to fight interior aircraft fires.

Firefighting training and personnel protective clothing requirements

- **FAA**: FAR Part 139 requires proper firefighter protective clothing.
- **NFPA**: Proximity protective clothing is required in accordance with NFPA 1976.

Operations

- **FAA**: FAR Part 139 requires the fire department to be on duty during aircraft operations.
- **NFPA**: NFPA 403 requires the fire department to be on duty during flight operations.

Phased ARFF vehicle response

- **FAA**: FAR Part 139 requires that the on-duty fire department capability be a function of the aircraft type operating at the time.
- **NFPA**: Requires that the number of ARFF vehicles required to respond to an incident be based upon the agent discharge rate contained in Table 3-3.1(a), 7-1.3. This requires that an adequate number of vehicles respond and meet the NFPA 403 agent discharge requirements for the airport category.

Emergency communications

- **FAA**: FAR Part 139 requires two-way emergency communications in ARFF vehicles. Additionally, a means to alert responding firefighters is required.
- **NFPA**: Emergency communications are required to meet the functional needs of the facility. NFPA requires fire departments to have backup power and to be tested every 24 hours.

Disaster plan

- **FAA**: FAR Part 139 requires an emergency/disaster plan.
- **NFPA**: A disaster plan is required and practiced on a biannual and annual basis.

21

WILDLAND FIRE FIGHTING

John R. Hawkins
James L. McFadden

CHAPTER HIGHLIGHTS

- A failure to understand the unique problems of wildland fires will have negative consequences for fire departments who deal with them.
- Suppression operations for wildland fires should be proactive, not reactive, and based on long-term, prioritized objectives.
- Conditions, such as the weather, topography, fuels, and exposures must influence the strategy and tactics used to control a wildland fire.
- Defensible space and built-in fire-protection design will save structures at I-zone fires.

INTRODUCTION

Although large-scale wildland fires do not occur on a regular basis in all parts of the country, there are few areas that do not have the potential for them. Analyses of wildland fires in the United States over the past several years indicate that wildland fires are occurring more frequently and, in many cases, in places where they seldom occurred in the past.

The recent increase in the number and severity of large-scale wildland fires can probably be attributed to two factors. The first is a change in national weather patterns that has created drought-like conditions and erratic winds in many areas, both of which are conducive to the development and spread of large and damaging wildland fires. A second reason is the expanding development of many communities to the point that there is direct encroachment into wildland areas, thus causing an interface problem. These areas are often called structural-wildland interface or I-zone areas.

I-zone fires present unique problems in terms of command and effective control. The failure to recognize this fact can have a negative impact on the ability of fire department officers to deal with these problems. The management of a large-scale wildland fire can be very demanding and complex because of the large number of potential problems that can occur. Conditions may exist where the fire starts, or may develop before it can be extinguished, that can rapidly change it from an incipient blaze to a major emergency. Many existing conditions, such as the weather, topography, fuels, and exposures can compound fire suppression actions and result in fires that are extremely fast moving and totally unpredictable. In these situations, the control capability of even the largest fire departments can be stretched to the limit due to the amount of firefighting resources and degree of logistical support required, and the fact that the incident may continue to burn over a prolonged period of time and cover great distances.

Since the possibility of these types of fires exists to some degree in most communities, fire department officers must be prepared to deal with them by understanding the types of problems that may be encountered and identifying possible solutions to these problems prior to a fire occurring. The information that follows focuses on problems associated with I-zone and wildland fires and offers strategic and tactical considerations that fire officers can use to reduce or eliminate problems.

Wildland fire terms

Figure 21–1 shows a typical burned area at a wildland fire. The diagram contains graphical examples of primary fire terminology including the fire head, rear, flanks, fingers, pockets, spot fires, and islands.

- *Head*–The fastest spreading and advancing part of the fire.
- *Rear*–Generally, the point where the fire started.
- *Flanks*–The sides of the fire. The right flank is always described as that side on the right when looking from the rear to the head. (The opposite is true for the left flank.)
- *Fingers*–Develop when a wildland fire meets an obstacle such as a rock outcropping, a retardant drop, wind change, etc. A finger is nothing more than a part of the fire that is spreading away from the previous perimeter.
- *Pockets*–Develop when fingers form and enlarge. Pockets can be very dangerous, particularly if firefighters are trapped inside a pocket with no escape route or safety zone.
- *Spot fires and new fire perimeters*–Often start outside and beyond the fire head. They are a nemesis of the firefighter and often lead to serious fire behavior.
- *Islands*–Can occur inside the main perimeter when wildland fire fuels do not burn for various reasons including some of the reasons listed for the formation of fingers.

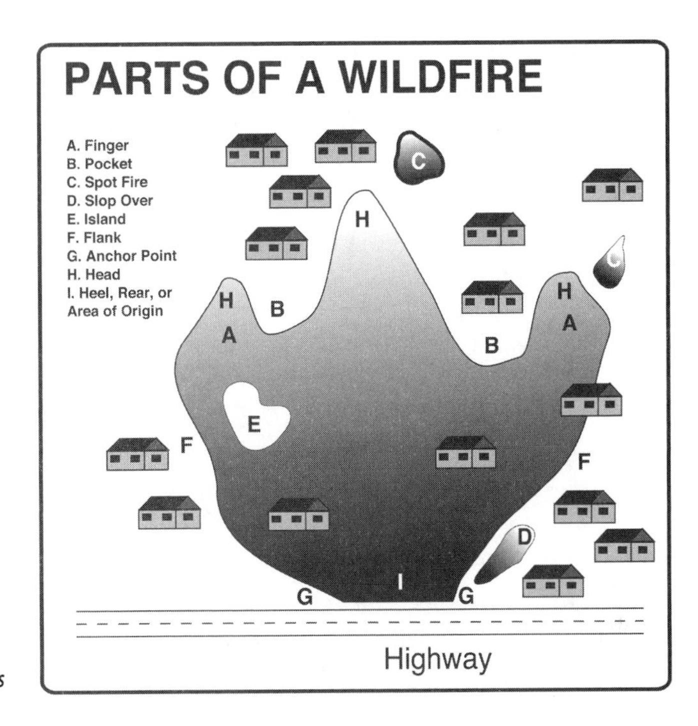

Figure 21–1 Wildland fire terms

Major Fire Spread Factors

Three major spread factors affect wildland fire movement: weather, topography, and fuel. Primary weather factors include: air mass stability, temperature, relative humidity and precipitation, and wind. Topography factors include: the steepness of the slope; the aspect (direction a slope faces); position of the fire on the slope whether at the bottom, mid-slope, or at the ridge; shape of the land; and topographic fire funnels called chimneys or chutes. Fuel factors include: fuel moisture, fuel loading, fuel characteristics, fuel continuity, and vertical arrangement and fuel position.

WEATHER

Weather is the most important and variable factor affecting wildland fire spread. Therefore, the importance of weather and its impact on chief officers' knowledge and their application of knowledge at actual emergencies becomes extremely important. Anyone acquainted with the behavior of a wildland fire recognizes the vital and dynamic influence of weather. Likewise, anyone who has studied weather is impressed with the complexity of the science. At every atmospheric level, there are great fundamental laws constantly at work making and remaking the weather.

When experienced officers are given a general prediction of the weather, they automatically convert the prediction into anticipated fuel conditions and fire behavior. The long established habit of associating brisk, dry winds with severe burning conditions is inclined to make anyone forget that the first is cause and the second is the effect.

Three master forces govern the making of weather upon the earth: heat from the sun (called solar energy), the force of gravity, and the tendency of all dynamic elements to seek a state of balance or equilibrium.

Air mass stability

Stability and instability refer to a relationship between the vertical temperature distribution within an air mass and a vertically moving parcel of air. If air mass temperature decreases sharply with altitude (generally, about 5 °F/1,000 ft. of change in altitude), conditions are favorable for air currents to rise vertically through the air mass. Thus, an unstable condition exists, because vertically rising warmer air can continue to move with little or no restriction. Under this condition, calm fires may suddenly and violently explode causing erratic fire behavior. Unstable air above the ground surface is responsible for the worsening fire behavior. A simple example of increased fire activity due to the vertical air mass profile regularly exists in a fireplace. When the damper is opened, the fire increases in intensity. The same phenomenon occurs when the "atmosphere damper" opens because of an unstable air mass.

Hot air rises, with the rising air mass getting cooler with increased altitude. In an unstable condition, an air mass, which is warmer nearer the ground, rises and keeps rising—higher and higher, faster and faster—creating unstable air aloft. As the hot air ascends, it pushes aside colder, heavier air. Displaced, this heavier air descends toward the ground and flattens out on reaching the earth's surface. Firefighters get a sudden "unexpected" wind. The fire is suddenly intensified and may be pushed in various directions, perhaps across a newly constructed fire line.

In contrast, when the air temperature near the ground is more approximate to the temperature aloft, warm air travels slowly upward (3.5 °F/1,000 ft.), and upon reaching cooler air, is itself cooled and stops rising. This is a stable condition. Vertical air movement is limited, and temperature equalization occurs. The resulting fire behavior during a stable air mass is usually much less severe than during unstable air conditions.

Temperature

Wood ignites between 400 °F and 700 °F. Normally woody wildland fuels will burst into flame at approximately 540 °F (assuming sufficient oxygen is present). Of course, the time required to produce fire at this temperature will vary with the amount of moisture in the fuel that first must be driven out by the heat.

The highest temperature that the ground could be expected to reach on a wind-sheltered surface is 150 °F or 160 °F. This temperature is far below the point of self-ignition of wildland fuels. However, solar heating is significant, aside from the drying effect of such a high temperature, in that the fuel that gains a boost of 100 °F or more from the sun is well along toward its ignition temperature before the igniting source is ever applied.

Of course, a fire officer may take advantage of the same rule operating in reverse, especially at night or during periods of relatively cool air. Heavy burning fuel may be separated and turned over to cause it to give up accumulated heat into the air. This leads to temperature reductions and mimics the effects of nighttime cooling.

Another variable is that of daily, or diurnal, temperatures. Diurnal temperature variation is dependent on the same factors as those for seasonal temperatures. Typically, daily temperatures are the coolest at daybreak with warming continuing until mid-afternoon when temperatures are at their highest, followed by evening cooling. The angle at which the sun's radiation strikes the earth is important. Topographic features can modify the process, with southern-facing slopes receiving more radiation than northern-facing slopes.

Smoldering fuels along a fire line will begin to support flame production as the sun moves higher in the morning sky. With the arrival of mid-afternoon, the sun will have brought fuel temperatures to their peak, and the fire will probably make concerted runs into new, formerly unburned fuels. A general cooling will take place as the sun falls closer to the horizon and flame production eventually may die back as nightfall signals a return to a smoldering-fire state. An important departure from this scenario occurs when fire burns as fiercely at night as in the daytime, due to yet another weather influence—wind.

Temperature inversions

As discussed earlier, under most conditions warmer air is nearer the earth's surface, with cooler air at higher elevations. However, there also are normal conditions under which quite the opposite is true. Such an "opposite" is called a temperature inversion. With an inversion, cooler air is found near the surface or ground, and a warm band of air is found aloft with cooler air above the warm band.

Along coastlines, marine inversions develop when cool, moist ocean air flows onshore across lowlands or through hill and mountain passes, to settle heavily within land depressions. This cooler, heavier air slides under warmer air; thus, a temperature inversion develops. Likewise on land, when a mass of air is cooled by nighttime temperatures and becomes sufficiently heavy with moisture, it flows downslope to collect in land depressions, such as into bowl-shaped valleys. A covering layer of warmer air above, sometimes at a point along a canyon wall, creates a temperature inversion.

It is well known that wildfire spread slows during cooler nighttime hours. Many large fires are contained during cooler weather periods. Certainly, fire burning in the cooler environment of an inversion can be expected to subside, and becomes dormant and sluggish as higher relative humidity and higher fuel moistures become controlling factors. Firefighters working a fire in a warmer band of air above the inversion will probably be faced with fire activities quite different from their counterparts in the cooler area below. While nighttime temperatures generally can be expected to drop at all elevations, temperatures will drop less in the warm band of an inversion and fire behavior may remain very active. The area along a slope where fire activity remains active above an inversion is called the thermal zone or thermal belt.

Inversions become most dangerous when a lower-elevation fire suddenly burns through the bottom, cool inversion lid into the warmer air above. With that change, firefighters will find intense burning conditions, accelerated fire spread, and higher flame lengths, all of which can place unprepared firefighters in immediate jeopardy.

Relative humidity

Water vapor, which is simply water in a gaseous state, is a very important weather factor. An air mass that passes over water picks up vapor. The oceans are a primary source of water vapor for the atmosphere. Lakes, rivers, moist soil, snow, and vegetation are also water sources for vapor, but furnish smaller amounts.

The term *relative humidity* compares the ratio of water vapor actually present in the air to the maximum amount of water vapor that the air could potentially hold at a particular temperature and air pressure. If the relative humidity is 100%, the air is completely saturated; that is, it can hold no more water vapor.

In contrast, if the relative humidity is 50%, the air is only half-saturated and can potentially hold 50% more water vapor. When the relative humidity drops below about 30%, the situation becomes favorable for intense wildland fire spread. The dryer air is better able to pick up water vapor from the fuel; consequently, less time is required for heat to bring about combustion. Fuels absorb and give up moisture at different rates; in fact, larger fuels such as logs are very slowly affected by humidity changes. Grass, a lighter fuel, is affected quite readily and may not burn at all when a relative humidity of 30% or higher is present for an hour or longer.

Firefighters can take advantage of relative humidity in many ways. Sometimes it is desirable to use low humidities to more easily burn away undesired fuels. Other times, fire officers must wait for rising humidities of night to more directly attack a formerly hot burning fire that is subsiding. Failing to recognize the effects of humidity can work against fire officers. Many firing or backfiring operations have outright failed because late afternoon or nighttime humidities prevented even grasses from burning cleanly, and fire officers found themselves faced with an incomplete burn, which more than likely roared to life during the heat of the next day.

Wind direction and velocity

Wind is simply the movement of air. The average person thinks of wind as generally horizontal air movement fast enough to be felt. That is because a person is much less aware of the causes of wind than of the horizontal air movement one feels and observes as a mover of light objects.

Over the face of the earth, several major forces work unceasingly to stimulate air movement. First, there are the vast areas of heated earth surface that produce rising air currents that return to earth in the cooler regions. Next, is the gravitational effect of the turning earth on these tremendous churning currents. The changing seasons alter the pattern of wind movement because the hot and cold regions of the earth seasonally shift about. Water bodies and land masses modify wind (and with it all weather behavior) effects.

Wind can be compressed under pressure, expanded and contracted with heat and cold, and made moist or dry. It may pause unmoving and then spring in any direction with violent gusts. Wind is like a fluid, except that it is capable of being compressed.

Wind movement is critically important to the fire officer. Perhaps the best approach to mastering such a flexible, untamed natural element is to become acquainted with the air movement habits of your own region.

If fuel and topographic conditions do not adversely influence fire spread, a strong wind from one prevailing direction will cause a long wedge or elliptically-shaped fire. This results not only from the driving force of the wind, but also because intensified combustion will demand an in-draft of wind toward the fire from the flanks and thus provide more oxygen for the combustion process.

Winds may cause fires to jump prepared fire-control lines or natural barriers. Winds can drive a crown fire through the tops of trees when normally a lack of surface fire heat would prevent a crown fire from developing.

Large fires make their own local weather, especially with respect to air movement. Large convection updrafts cause air currents along the ground to move toward the fire and sometimes cause downdrafts of importance reaching out beyond the fire perimeter. Smoke columns may shade the sun and alter the temporary radiation of solar heat toward and away from the earth, resulting in changed winds and fire behavior.

Over broad areas such as plains or long, wide valleys, the prevailing direction of the wind can be predicted quite easily. But in irregular topography, local wind movement may be quite different from the major prevailing conditions and also more changeable from time to time. Irregular topographic objects and vegetation act as a friction source on ground wind movements.

Winds change direction and intensity throughout the day and night. The particular change in each locality will depend on the temperature changes in and around local topography. Large bodies of water, such as the ocean and lakes, usually cause winds to blow inland (onshore wind) as the sun warms the land area about mid-day. The wind blows seaward (offshore wind) at night when the land cools more rapidly than the water area. Isolated mountains draw air upward as in a chimney when the mountain slopes warm under the warming sun.

The local effects of topography on wind are as varied as the shapes of the topographic features. The time of day, the aspect (exposure of the slope or the direction that the slope faces) with respect to the moving sun, and many other influencing factors control the direction and magnitude of the wind. Gradient (general) winds blowing above the surface are the predominant element much of the time. However, when these winds weaken in the presence of strong daytime heating and nighttime cooling, convective winds of local origin become important features of weather in areas of broken topography.

The formation of cumulus clouds directly over peaks or ridges can have a marked influence on wind velocity and direction. As cumulus clouds grow higher, strong in-drafts are created that can increase upslope winds on the higher land surfaces. When a thunderstorm cloud (cell) is in its most active form (mature stage), large volumes of cold air may cascade to the ground as a strong downdraft. Although wind downdrafts usually last only a few minutes, resulting gusty winds can strike suddenly and violently with speeds up to 50 miles per hour and faster. Meteorologists refer to this type of wind downdraft as a microburst.

The passage of a cold front will invariably affect wind velocity and, more often than not, cause a shift in wind direction. Cold fronts will often give visible evidence of their presence in the form of high clouds. As these fronts pass, there usually will be a marked increase in wind velocity, followed by an abrupt clockwise shift in wind direction of 45° to 180°. These sudden changes in wind direction will cause dramatic changes in fire spread and potentially place firefighters in the path of a rapidly approaching fire front.

Orientation of topography is an important factor governing strength and timing of wind flow. The upflow of wind begins first on east-facing slopes as the sun rises and slopes begin to warm. The intensity of upslope wind increases as daytime heating continues. South and southwest slopes heat the most and therefore have the strongest upslope winds. Often velocities are considerably greater on those slopes than on opposite north-aspect slopes. Morning upslope winds flow straight upslope. The increased velocity of canyon winds later in the day turns the direction of upslope winds diagonally up-canyon. Therefore, first expect upslope winds with initial morning heating, followed by up-canyon winds with maximum afternoon warming. Generally, upslope winds will occur as the result of surface heating in the daytime, and downslope winds will occur as the result of surface cooling at night. There is an exception when a particular type of wind (Foehn wind) blows around the clock and dilutes the diurnal upslope and downslope wind process.

The change from an upslope to a downslope wind will usually begin on those areas first shaded from the sun. Initially, the upslope wind will gradually diminish. There will be a period of calm and then a gentle downslope wind movement will begin. A large drainage can easily have varying degrees of this transition in process at the same time, depending on exposure or decreasing exposure to the heating rays of the sun. When all areas are in the shadows, the downward movement of air strengthens until winds are moving in a 180° change of direction from daytime topographic wind flows.

Strong up-canyon or upslope winds can be quite turbulent and can form large eddies at topographic bends and tributary junctions. Fires burning in these locations will behave very erratically and may spread alternately one way and then another, but generally will move upslope and up-canyon. Daytime heating often produces rising convective winds that are capable of holding gradient winds aloft. In contrast, strong gradient winds can completely obliterate topographic local wind patterns. The gradient wind effect will vary with the stability of the lower atmosphere. Stable layers, of which inversions are an extreme type, tend to "insulate" the local wind patterns from the gradient wind, thus minimizing its effect. When the air is unstable, expect more mixing of the gradient wind and local topographic winds.

Along the western coastal mountain ranges of the U.S., marine air intrusion complicates the fire officer's ability to forecast wind movement. All of the preceding discussion of convective winds would lead one to believe that downslope winds in the heat of summer day are either nonexistent or rare. Such is not the case; in fact, they do occur regularly and frequently. On the west coast, marine air moving inland over the coastal mountains can often spill into east-facing canyons or draws, and flow beneath the locally created warmer upslope winds. The heavier marine air will flow first through saddles and over low points and closely follow drainages and slopes, thus reversing ground wind direction from upslope to downslope. The change in direction can occur rather quickly. Cases of reversals occurring within a few minutes have been documented, with the downslope wind velocity blowing considerably greater than the preceding upslope wind speed. Obviously, this sudden shift in wind direction can adversely affect fire behavior in the areas involved.

A gradient wind blowing toward the sea (offshore wind) will reduce the sea breeze and, if strong enough, may block the sea breeze entirely.

Wind of any kind has a marked influence on fire intensity and fire behavior. Foehn winds, however, have the most devastating and adverse effect on fires. Foehn winds are known as north winds in northern California, mono winds in the central Sierras, and Santa Ana (also called Santana) winds in southern California. In the Rockies, they are called chinook winds. Foehn winds are capable of reaching extremely high velocities, 80 to 90 mph, across ridge tops and peaks. Foehn winds blow from high atmospheric pressure and elevation to low atmospheric pressure and elevation. They are characterized by blowing downhill while hugging the land profile, being warm (at least warmer than they were at the same elevation on the windward side of the mountain), and becoming progressively more desiccating (drying) as they descend. These winds can blow unabated around the clock for several days. When a Foehn wind reaches a canyon, its speed is accelerated, much like the speed of water when forced through a small diameter nozzle. This funneling effect is devastating on fire behavior.

A fire occurring under extreme Foehn wind conditions can spread with such violence that control forces will be temporarily powerless to take control action, except at the rear and on the flanks of the fire. Foehn winds cause significant air and fuel drying. Spot fires can occur in dried fuels a mile or more in advance of the main fire and become raging infernos on their own before being joined by the original fire front.

Such fires are capable of crossing both natural and manmade barriers, not only by long-distance spotting, but also by direct contact with new fuels when winds literally flatten flame sheets to a horizontal position. Flame lengths of more than 100 to 200 ft. are common.

Wind direction generally is responsive to the direction from which the greater atmospheric pressure bears, as local pressures and topographic wind channels modify that dominant pressure. Wind speed is responsive entirely to the strength of the dominant force that causes a pressure to bear on an air mass from any single direction at any single instant. The rules are rather simple. Wind velocity is disturbed by vertical obstructions and will not return to its velocity previous to the obstruction for a distance of seven times the height of the obstruction. The result can be most complex, especially in the broken topography where so many wildland fires occur.

Knowledge of local climate changes that occur during the day and night is vitally important to the fire officer. The long, quiet flank of a fire at dawn may suddenly become the fire's head when the usual afternoon wind shifts direction, and the fire officer may become unnecessarily surprised. Ignorance of wind changes can cause an unnecessary disaster. Fire officers must understand and be able to predict weather events and fire behavior. They must know and use the *Ten Standard Fire Orders*, *18 Situations that Shout Watch Out!*, the common denominators of fatal and near-miss fires, the *Downhill and Indirect Fire Guidelines* and lookouts, communications, escape routes, and safety zones (LCES).

Wind not only affects fire behavior, but also creates safety problems for fire officers. Perhaps the most common is eye injury from wind-blown material. In addition, specialized equipment such as aircraft will find limited use in high winds. Air tankers and helicopters lose effectiveness when wind speeds reach 20 to 30 mph. Above 30 mph, for safety reasons, air tankers will normally be grounded because of their ineffectiveness.

Fuel moisture

We have considered water vapor in the air (measured as relative humidity) and the effect of temperature on fire behavior. Both relative humidity and temperature are important, but what really concerns the fire officer is the condition of the fuel that will feed the fire. That condition is termed *fuel moisture* and is defined as the amount of moisture in the fuel.

Whether that fuel will be available to burn depends on its moisture content. Moisture in the fuel will not burn and must be converted to steam by heat and driven away before effective combustion can take place. Air moisture is absorbed into dry fuel or taken away from the fuel by moist or dry air during nature's eternal quest to reach equilibrium.

While many parts of the country are subject to considerable precipitation or rainfall, it is almost a forgotten element during most fire behavior discussions. By delivering water to the soil reservoir, precipitation sets the length of the vegetation growing season, thus controlling the moisture content of living fuels and, often, the severity of the fire season.

As the fire season, or dry season, progresses, stored rain moisture is dissipated into the air through growing leaves or by evaporation from logs and litter. Cumulative fuel moisture loss can be measured. This moisture loss can easily be measured as a loss in weight of the fuel particles. Eventually, the lighter dry fuels will lose their stored winter moisture and then reflect only a change in moisture content as relative humidity changes cause water vapor to move into and away from the body of the fuels.

Heavy fuels such as logs slowly release and absorb moisture. Dry leaves and grass respond quickly to relative humidity changes in the air; their moisture levels will vary from hour to hour and from day to day. Green, living leaves naturally respond in accordance with their complex transpiration habits particular to each species.

A wildland fuel that will not burn in a midnight or early morning backfire will probably be ready to roar when the decreasing relative humidity of mid-morning or afternoon arrives.

Topography

Of the three major factors influencing wildland fires (weather, topography, and fuel) topography is the most static; nevertheless, it is still a major fire behavior factor. The word topography refers to the shape of the earth's surface. It is the land elevation, or relief. Simply stated, topography is the lay of the land.

Elevation

Elevation changes affect fire behavior. This is true both at a local fire and from an overall change in vertical position from sea level to the top of the highest peaks. On a summer's day, going from a valley floor with an elevation of 1,500 ft. above sea level to a mountain meadow at a 5,000-ft. elevation, results in some dramatic weather and fuel changes.

Generally, the temperature drops as the elevation increases. Changes in elevation cause dramatic fuel changes, and as elevation increases, precipitation increases as well. The mountain snow pack also has a distinct influence on fire burning conditions, length of fire season, and fuel types.

Elevation has a strong influence on the length of the fire season. Lower elevations have longer fire seasons than higher elevations. Higher elevations also have a strong influence on the movement of air between the valleys and the mountains. The elevation at which a fire is burning in relation to the surrounding topography is important. Depending on the fire's location, it may be influenced by strong local wind effects under stable overall weather conditions. It is important for the fire officer to understand the topographic influences on local weather as it changes from day to night. Under settled, mid-summer atmospheric conditions, a daily interchange of airflow occurs between the mountaintops and the valley bottoms.

As previously stated, local winds result from several factors, many of which are also affected by topography, particularly elevation.

Some species of vegetation require significant solar heating that occurs on hotter, southern exposures, whereas others cannot survive in high heat. However, the requirements for moisture are probably more important than heat and sunlight in determining the type of plant species, growth rate, and size of each species. Higher temperatures and generally prevailing southwest winds remove moisture more rapidly from southern-exposure fuels, both affecting soil moisture and ground litter fuel moisture and leading to more plant transpiration (sweating). The result is that vegetation that adapts to these southern exposures is dryer, sparser, and more flammable than the vegetation that grows on opposite northern-facing slopes. Rainfall on both north and south slopes may even be equal, but the deeper soil on the north slope, coupled with less sunlight and heat, produces a remarkably different and wetter micro or highly localized climate. This climate and soil difference produces a heavier vegetative cover on northern slopes.

Because of the drier, sparser flash-fuel-type vegetation that grows on southern exposures, the potential for fire ignition and rapid fire spread is greater than on the north slopes. Also, spot fires generally occur more readily on southern exposures. Weather and fuel conditions combined with the interactions of people (people tend to be outdoors, *e.g.*, camping in warmer weather) have caused a unique cyclical fuel-fire effect on the southern exposures.

The exception regarding fuels and burning conditions occurs after prolonged periods without measurable rainfall. When this happens, the north/east aspects may burn as intensely as the south/west aspects due to the dry fuel conditions.

Slope

Slope (steepness) is basically the change in elevation divided by the corresponding change in horizontal distance, expressed as a percentage. Slope influences fire spread through impacts of the type of vegetation present, fire behavior, and fire rate of spread. There is a strong relationship between slope and the other physical factors affecting fire behavior. A 5% slope means a 5-ft. rise in elevation for each 100-ft. of horizontal distance.

To calculate slope, simply divide the change in elevation (rise) by the horizontal distance (run) and multiply the result by 100%. For example, if the elevation rose 40 ft. in a distance of 200 ft., dividing 40 by 200 and multiplying by 100% equals a slope of 20%.

It is most important to remember that slope is expressed as a percentage and not in degrees of angle. A 45% angle is a slope of 100%, because for each elevational increase of one ft., the horizontal distance also increases by one ft.

It is a basic rule in wildland firefighting that if all things remain unchanged, a fire will burn faster uphill and more slowly downhill. Yet why do relatively minor changes in slope have such a strong effect on the rate of spread? Changes in slope have a distinct effect on two methods of heat transfer, convection and radiation, and two of the important elements of combustion—heat and oxygen. As a fire burns up a steep slope, the angle of the flame is closer to the fuel than it is during a fire burning on a lesser slope or level ground. The radiation of these flames on the increased slope removes fuel moisture and preheats unburned fuel.

The convective effects of heat transfer cause hot air and gases to also rise, thus providing additional heat to the unburned fuel. The convective heat of a fire, as evidenced by the size and shape of the smoke column, also lifts and distributes firebrands that, in turn, may fall ahead of the main fire, causing spot fires. Spot fires, like the main fire, require large amounts of oxygen. Oxygen is supplied from the area near the fire and may be preheated, further increasing the rate of combustion. Because of the in-draft effect, spot fires and the main fire will tend to draw toward each other at an increased rate. Extreme examples of adverse fire behavior and spread can take place under such conditions.

As previously stated, winds generally blow upslope during the day and downslope at night. If downslope winds are strong enough, their effect may completely offset the factors governing the upslope spread of fire and may actually cause a fire to burn downhill at an accelerated rate. Knowledge of local winds and microclimates is of utmost importance to the incident commander and all officers on the fire line.

In areas of broken topography or tall aerial fuels, a fire of low intensity may cross over the top of a ridge and slowly burn downhill. As the fire burns downhill, its pattern is usually highly irregular. The irregular pattern is caused by the rolling of burning material downslope generally causing a slower rate of spread. This happens quite often at night, and the fire leaves an unburned canopy of dried-out, preheated fuel. Under such conditions a dangerous reburn of the same area can take place. Strong safety precautions, including the clear identification of LCES, must be taken to ensure firefighter safety.

Several distinct topographical features have a strong effect on fire behavior and control methods.

Wide canyons. The prevailing wind direction will not be affected by the general orientation of a wide canyon. Since there are few sharp river bends in large canyons, the wind will generally not be deflected. Cross-canyon fallout of spot fires is not common, except during high-wind periods. Within large canyons, aspect has a strong effect on fuel and fire behavior conditions.

Narrow canyons. Narrow canyons have many sharp bends and more rugged, irregular topographic features. As a result, narrow canyons normally have more independent wind currents than wide canyons. The wind usually follows the direction of the canyon. Sharp breaks and forks in the canyon may cause strong eddies and turbulent drafts. Because of the more turbulent wind conditions found in narrow canyons, spot fires are common. The bends in the bottom of narrow canyons lead to more spot fires crossing the changes in canyon direction. Aspect has little effect on changes in fuel conditions in the bottom of narrow canyons. In fact, the fuels in the very bottom of very steep, narrow canyons are much more like northern-aspect fuels.

Steep slopes. The steeper the slope, the more likely it is that a fire will drive upward in a wedge shape, forming a narrow head. The rapid movement of the fire may cause strong in-drafts on the flanks. The occurrence of spot fires in front of the main fire is likely.

Ridges. Quite often, as a fire reaches a ridge top, it meets an opposing airflow from the other side of the ridge. Such a condition will slow the spread of the fire, but also may cause adverse fire behavior. Under the same conditions, fire will generally burn 17 times faster upslope than down the same slope. The topography and opposing winds will cause eddies and turbulence. A fire will often slow as it burns downhill on the lee side of a ridge. The change in rate of spread from the fire running up the windward side of the ridge to a slower spread on the lee side is called a *slope reversal.* Another type of slope reversal occurs when the fire backs to the bottom of the drainage, crosses it, and races up the other side.

Chimneys. A chimney is a narrow draw or chute that extends from the top of a ridge at a topographical feature called a saddle to the bottom of the hillside or canyon. It carries water down the hill during the wet season. Chimneys draw the fire upward as does an actual chimney flue. Up-canyon winds are drawn into the topographic features of a chimney, convected upward, and confined within the chimney by its steep side slopes. As the fire moves up the slope, the radiant heat from the steep side slopes intensifies the rate of spread in the chimney. Firefighters have been trapped in and above chimneys with resulting loss of life, injuries, and damaged equipment. All firefighters must be able to identify a chimney. Even a shallow chimney 5-ft. deep and less than 100-ft. long can cause an increase in the rate of spread and intensify a fire's behavior.

Physical barriers. Barriers can have a significant influence on the spread of wildfires. Barriers, either natural or person-made, often help contain portions of the fire. Natural barriers include rockslides and other barren areas, lakes, rivers, dry streams and riverbeds, and the ocean. Areas that contain wet vegetation, such as some agricultural zones and riverbed vegetation, may slow or stop the spread of a fire. In addition, certain types of fuel may not burn at different times of the year, such as grassland during wet times of the year.

Artificial barriers. Topography has a strong influence both on how a fire burns and on the type of fire attack methods and equipment used to suppress a fire. Artificial barriers are person-made changes in topography, such as major highways and other road systems, fire breaks, fuel breaks, power-line clearances, reservoirs, housing developments, and cleared land areas. Changes of natural vegetation to agricultural use may help or hinder the fire containment. Each of these barriers may affect the spread of the fire directly, through the absence of fuels, or indirectly, through the modification of relative humidity, local winds, and other fire weather conditions.

Fuel

Fuel is the third major element affecting fire behavior. Fuel is one of the three sides of the fire triangle and is a necessary element for combustion. Fuel is any substance that will ignite and burn. Most wildland fuels are nothing more than different species of vegetation or types of structures. There are many different biological classifications of wildland vegetation. Here, however, we will generally identify and classify fuels only as they relate to wildland firefighting. The fuel types will be discussed in terms of their flammability and burning characteristics.

With wildland firefighting, fuels are not limited strictly to either living or dead vegetative types. In urbanized areas, contiguous to or surrounded by wildland fuels, structures and improvements provide fuel for wildland fires, affecting firefighting tactics and actual fire behavior. For purposes of classification, wildland fuels are grouped with respect to:

- Whether they are living or dead
- Their position on the ground or in the air
- Their size
- The relative rate at which they burn
- Their compactness, continuity, volume, and moisture content

One of the first classifications of fuels must be in terms of whether the fuel is alive or dead. Obviously, dead fuels ignite and burn much more easily, although living fuels are a definite part of fuel loading. Dead fuels include dry grass, material that has fallen from the vegetation and remains suspended on live branches or that rests on the ground, and limbs that have died and remain affixed to the vegetation's skeletal structure. Live fuels are still a part of the growing plant and contain some amount of sap or moisture. With the exception of annual grass, which dies each year, most wildland fuels enjoy a spring growing period when the moisture content often reaches 250% of normal. With the advent of the dry season, fuels begin to lose live fuel moisture and reach a dormancy period when the moisture content may bottom out at 50%. This concept directly affects fuel quantity and availability.

Fuel quantity and availability

Fuel quantity, also called fuel loading, often is expressed in tons per acre and refers to the amount of combustible fuel available. Quantities can range from less than one ton/acre for some grass types, to 200 or more tons/acre for heavy slash (forest tree refuse on the ground, often the result of logging). In actual fire situations, estimations are usually made by trained observers in order to estimate fire spread and control factors. From time to time, actual samples are measured to verify casual observations.

In consideration of fuel size, fuels are divided into three main categories: light, medium, and heavy. This broad classification is generally sufficient for wildland firefighting. *Light fuels* usually consist of grass and mixed light brush, up to 3-ft. high. *Medium fuels* consist of brush up to 6-ft. in height that grows in fairly thick stands. *Heavy fuel* refers to thick brush taller than 6-ft., such as timber slash, and standing conifer and hardwood trees.

The fuel that is actually burning determines the fuel-quantity classification. A pine needle or grass fire of low intensity, burning on a forest floor, would be considered a fire in light fuel even though it is burning in a timber-covered area. It is generally better to use relative terms rather than numerical designations when dealing with fuel quantities.

Fuel availability refers to the proportion of fuel, usually finer fuel that will burn in a wildland fire. Fuel availability varies with the time of year. A dry grass fuel that is available to burn during the dry season may well be green and unavailable to burn during the wet season. In a grass fire, it is quite easy to determine fuel availability, as most of the available fuel burns during a fire. In brush and forest fuel types, fuel availability varies widely with fuel moisture conditions and with the thickness of the fuel itself. Fuel consumption also varies with the duration and intensity of the fire. Therefore, fuel availability is determined by the time of year and the amount and type of fuel consumed under a specific set of burning conditions.

Fuel position

Wildland fire fuels occupy three general vertical positions. *Ground fuels* are right against the ground and include peat, roots, buried trees, duff (decomposing vegetation), and humus. *Surface fuels* are located above ground fuels and include examples such as grass, pine, or fir needles, leaves, twigs, down wood, brush, and slash (dead limbs, dead tree tops, or other remnants of logging or a natural act that knocked down trees). *Aerial fuels* are those fuels located above surface fuels including trees, snags (dead trees), and large brush species.

The distinction between surface fuels and aerial fuels is probably more difficult to define because some of the most prominent fuels fit both classifications. Intermediate fuels are given the broad name of brush or chaparral. Brush is a generalized fuel term, whereas chaparral refers to a dry Mediterranean-climate type of vegetation. Mature brush species range in height from only a few inches to more than 20-ft. above the ground. The differences depend on the individual species and growing site. The major value in any definition of fuels is the creation of a general reference term for fuel types that have similar burning tendencies. The compactness of fuel particles and their size regulate the transfer of heat and the availability of oxygen to the fuel itself. These two factors allow for great differences in the burning characteristics of various brush-type fuels.

Low aerial fuels, or surface fuels, are such that they often ignite easily, burn rapidly, and have relatively complete combustion; hence, they are called flash fuels. Among the different vegetative types identified as surface fuels are grass, sage, and other perennials. Any low-brush growth that does not allow for the easy movement of air through the foliage is considered a surface fuel, including small trees growing as reproduction (plantation) timber.

Nonliving surface fuels include downed logs, heavy limbs, and smaller twigs, leaves, needles, bark, duff, and cones. Timber slash is a prime example of a surface fuel. In some areas, slash can be 4-ft. to 8-ft. deep with fuel loading exceeding 200 tons/acre. Fires in slash pose unique and very difficult suppression problems.

Some types of surface fuels pose other problems because of their ease of ignition. Punky logs (old, dead, and decaying logs) or any type of pulverized wood material allows quick ignition when exposed to sufficient radiant heat, sparks, or firebrands. Large logs, particularly if their surface is not punky or splintered, may resist ignition, and once burning, may not aid the fire's spread. However, if a group of logs at a lumbermill site or a lumbermill log deck catch fire, the close proximity of the burning logs to each other can cause a tremendous heat buildup and spread fire through radiation, convection, and by spot fires caused from firebrands.

Large limbs burn more easily when dead needles are attached. Pine slash poses this hazard for several years while fir limbs usually drop their needles after the first year. Several types of needles and leaves become ready firebrands when they are lifted into the fire's convection column. Eucalyptus, because of its oil content in the bark and leaves, and the aerodynamics of each, can cause a considerable fire-spotting problem. In parts of Australia where burning and fuel conditions resemble those in the United States, eucalyptus firebrands have caused spotting for a distance of several miles ahead of the main fire.

In its cured state, grass is a fast-burning surface fuel. Its density and thickness allow a maximum amount of oxygen to be available for flame production. In other than early-season fires, when certain grasses still maintain some moisture, combustion is usually complete. Grass fires are relatively easy to suppress and, once out, provide a safe area for firefighters to retreat if necessary. Grass is also an easy fuel in which to construct a fire line and a good fuel to use in backfiring operations.

Regardless of the size of the fire, never overlook the fact that a fast-moving grass fire can be extremely dangerous to firefighter safety and that fatalities are most likely to occur at small fires or during docile periods of fire behavior under what appear to be innocent overall conditions. Hence, a small fire is not necessarily safer for firefighters to work than a large fire. A change in wind can can cause a rapid change in the fire's direction and rate of spread and can, as has happened in the past, overrun engine companies and personnel.

The results to firefighters are serious burns or even death. Firefighters have been killed and injured while fighting grass fires because they underestimated the potential of this light flashy fuel.

Unlike grass, in which changes in seasonal rainfall may affect fuel quantity quite dramatically from year to year, the quantity of fine fuels (vegetation-type fuel with a diameter less than 2-in. to a height less than 6-in., *i.e.*, brush) in many mature brush fields and forest-covered areas remains relatively stable. This is true unless temporarily reduced by fire, or increased by such factors as a severe snow (down and dead fuel), insects, windstorms, drought, or logging operations (slash).

The rate at which leaves, bark, twigs, and other fine fuel material are deposited on the ground varies according to the type and density of the parent fuel. A litter bed in the process of breaking down and decomposing is referred to as *duff*. Fires in duff (ground fuel), although not spectacular, can cause many fire-control problems if fire lines are not properly constructed and the fire is not completely suppressed. A fire in duff and peat can creep along for long periods of time and escape established control lines. Therefore, duff or peat fires must be suppressed completely to prevent "rekindles."

Sawdust is a ground fuel that produces a slow-burning, smoldering type of fire with little or no active flame production. Sawdust fires are very difficult to extinguish and the extensive use of water to suppress this type of fire is seldom effective. Sawdust fires must be separated and allowed to burn out or be extinguished by extensive handwork and mop-up operations. Some sawdust piles have been known to burn for months or longer. The safety hazard in sawdust fire operations is that the fire will burn underneath the surface and cause burned-out areas that are not visible. These burned-out areas produce large caverns and sinkholes that contain hot material posing a distinct safety hazard for firefighters.

Living and dead leaves and needles are considered aerial fuels while located above the ground, as are bark, lichens, moss, and vines such as poison oak. The needles of most evergreens and the leaves of some hardwoods are highly flammable because of the availability of air, exposure to higher-level winds, and high oil or sap content.

Dead tree limbs near the ground or in the proximity of hot, burning ground fuels can provide a ladder (called ladder fuels) for a fire to reach the aerial fuel or tree crowns. Once the fire is established in the crowns of a few trees, radiation and convection can sustain a crown fire. Homeowners can help protect their property by removing ladder fuels. Generally one should remove the lower branches on trees that are in contact with the ground and up to at least 6-ft. to 8-ft. above the ground.

Snags are dead, standing trees that cause a multitude of problems for the fire officer. Snags are caused by the death of a tree from drought, disease, insects, animals and, quite often, previous fires. Burning embers can ignite a snag quickly, and once a snag is burning, it can spread fire for some distance. Extinguishing a fire in a snag is difficult because in most cases the snag must be cut down before the fire can be extinguished. Obviously, felling burning and weakened snags can pose serious safety problems for suppression personnel.

There are several fuel types that fall into both classifications: under certain conditions will be classified as a surface fuel, and under more lush growing conditions might be considered an aerial fuel. The important point is to learn the burning characteristics of surface and aerial fuels under various weather patterns and topographic features.

Structural fuels

Many profound textbooks and training manuals have been written on the burning characteristics and behavior of structural fires. This chapter will deal only with structural fuels as they directly relate to wildland fires. Structural development and encroachment into previously undeveloped wildland areas are taking place at alarming rates throughout the world. Fire officers must learn how to protect structures from wildland fires, how to extinguish structure fires once they ignite and how structural fire fuels interrelate with wildland fires. Every fire department should accept that its mission includes protecting both structural and wildland fire environments.

Structural fuels are high-volume fuels that produce large amounts of radiant and convective heat. Once a structure becomes involved, it can spread quickly to other nearby structures and to uninvolved wildland fuels. The heat generated from a structural fire tends to remove the moisture from fuels for some distance. Because an involved structure will burn for some time, it provides a long-term source of ignition for other fuels.

The biggest problem with interface structures is their construction, particularly wood shingle roofs. Many homes, cabins, and resorts in wildland areas have roofs constructed of wood or shake shingles. These roofs provide a ready location for spot fires to start, and once ignited, these fires are difficult to extinguish.

Burning shingles are rapidly uplifted into the convection column and can travel great distances before landing on the ground to start new spot fires. During the Paint Fire (Santa Barbara, California, June 1990), which destroyed 450 homes, much of the fire's rapid spread (3.5 miles downhill-spread in 81 minutes) was attributed to shake-shingle roofs. Shingles can easily cause spot fires across fire lines, endangering firefighters.

Fuel arrangement

The two major factors that affect fuel arrangement are the continuity of the fuel and its density. Continuity refers to the distribution of fuel particles with respect to each other and their relative distances that may or may not allow the transfer of heat from one fuel particle to another. The rate of a fire's spread and the total heat generated over any area are dependent on the continuity of the fuel. Areas that have some distance between fuel particles, such as in deserts or high-alpine areas are termed patchy fuel areas. Rocks, plowed land, wet drainages, green herbs, and other noncombustible surfaces also produce areas of patchy fuel. Generally, areas of patchy fuel favor firefighting efforts to fight a fire. However, some areas of patchy fuel can cause problems for the fire officer because sparse fuel may cause difficult or totally ineffective backfiring operations.

Areas of uniform fuel consist of fuel stands that evenly and continuously cover a sizable location. This uniformity relates to the horizontal continuity of the fuel rather than to the pureness of the stand of a specific species. If other fire behavior variables remain constant, it is fairly easy to determine rates of spread in a uniform fuel cover.

Compactness refers to the proximity of fuel particles to one another with respect to the unobstructed flow of air around the particles. Compactness is an opposite characteristic from continuity in that close continuity produces faster and greater heat spread. Extreme compactness could mean less heat generation and the discouragement of combustion because of lack of sufficient oxygen to sustain an active combustion process.

A grain field fire is a good example of a fire in fuel of uniform horizontal continuity. Usually these fires burn very hot and produce a relatively rapid rate of spread. In contrast, a fire burning in leaf mold (a fuel of great compactness) tends to burn with little flame production and produces a relatively slow rate of spread.

Both the continuity and compactness of fuels vary from high- to low-combustion rates as local weather conditions take effect. For example, an area may be so thinly covered with scattered fuel that, normally, its poor continuity (patchy fuel) will reduce both the chances of a fire starting and the possible spread of a fire once it starts. Desert brush growing in a mountain rain shadow is a good example of such a fuel, yet a strong, dry Foehn wind could produce a hot-burning fire that would increase the rate of combustion and tend to bridge the gaps between the available fuels. The expanding fire would produce more intense flame production, and at the same time, the strong wind would be pushing the convective heat through the available sparse desert fuel in a leeward direction.

Suspended pine needles on timber slash may be loose enough that fire would not move along the clumps on a high-humidity day, but the same fuel may be adequately compacted to carry fire with a slight breeze on a dry day.

Fuel moisture

Fuel moisture is an important factor in determining the burning capability of different wildland fuels. There are two fuel-moisture classifications—dead fuel moisture for dead fuels and live fuel moisture for living fuels. Dead fuel moisture is the measured amount of moisture available in dead fuels such as cured or dry grass, dead but still attached brush and tree limbs, dead and down limbs and treetops, and litter and duff. It is measured by weighing a sample fuel stick that is exposed in the open air. Weighing the stick indicates how much moisture it has absorbed or released. The moisture is expressed as a percentage of its dry weight.

Dead fuel moisture readings below 10% indicate that the dead fuels are in a combustible state. Remember, a kiln-dried piece of lumber has a fuel moisture of 12% to 14%. During Foehn wind conditions, dead fuel moisture readings for fine fuels have been recorded as low as 0.5%. When fuel moisture is low, fires easily start because the ignition source has very little moisture to drive from the fuel. The applied heat quickly causes flammable gases to vaporize and the fuel to ignite. When fuel moisture is high, many ignition sources such as sparks, smoking materials such as tobacco, exhaust, carbon, etc., may cool down before the water vapor is removed from the fuel and flame-producing gas starts to vaporize. Generally, as fuel moisture increases, rate of spread decreases, and conversely, as fuel moisture decreases, rate of spread increases. However, if fuel moisture is already low, a further reduction will cause the rate of spread to increase five to six times.

Live fuel moisture is the amount of sap or moisture in a live wildland fuel. It is controlled by two major factors, weather and the curing stage of the vegetation. Weather considerations that influence live fuel moisture include: temperature, relative humidity, wind speed, and precipitation. The curing stage factors that affect live fuel moisture are the type of vegetation, time of year or season, and days since new growth. These factors, combined

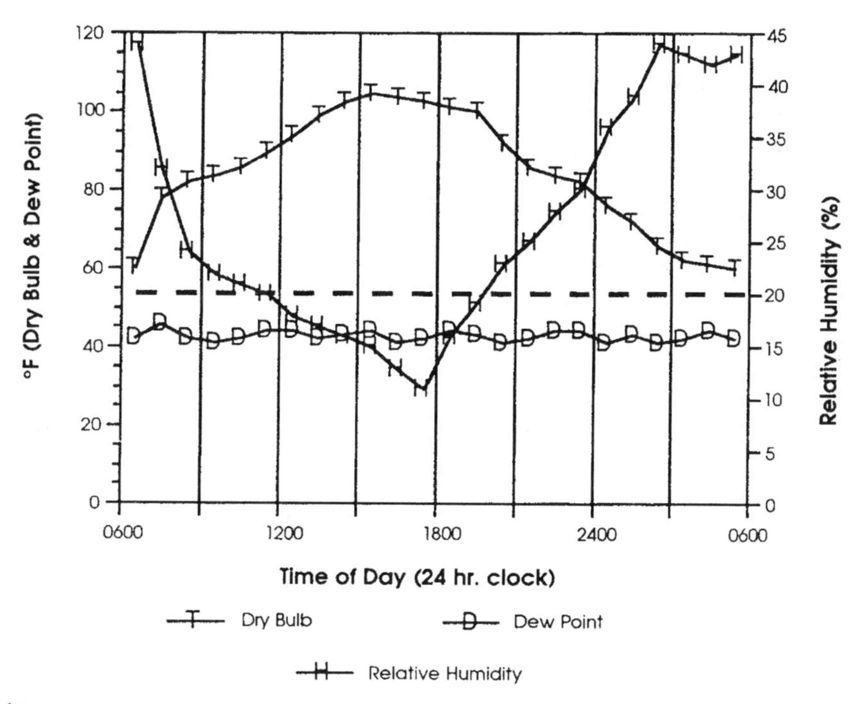

Figure 21–2 Fuel moisture

with seasonal weather, determine whether the fuel is green, curing, or cured. Fires burn well in live fuels when the live fuel moisture drops below 100%. During the spring, live fuel moistures may exceed 250% of normal. During the driest parts of the year, live fuel moistures may drop to below 60%.

Fuel temperature

Fuel temperature influences both the probability of a fire and its rate of spread. Air temperature and direct sunlight are responsible for fuel temperatures. As fuel temperature rises, less heat is needed from an ignition source to start a fire. Most wildland fuels will ignite within a temperature range of 400 °F to 700 °F. During the night, the temperature of a fuel might drop to 50 °F or 60 °F; on a hot day, this same fuel could rise to an afternoon temperature in excess of 150 °F. Most of us have experienced this "higher-than-air" temperature factor when we have touched the hot steering wheel of our vehicle that has been parked in the sun for some length of time.

The rate of spread will increase at a greater rate than a proportioned increase in the fuel temperature. For example, if a fuel temperature of 61 °F causes a rate of spread factor of 1, a rise in fuel temperature to 100 °F will result in a spread factor of 2, or twice as fast if all other variables remain constant. The effect of temperature is quite evident when we realize that most wildland fires start and spread the fastest during the heat of the day, usually from 10:00 a.m. to 4:00 p.m.

Because it is impractical to measure fuel temperatures while fighting a fire, try to keep these general guidelines in mind:

- The rate of spread factor is doubled long before the fuel temperature is doubled.
- Fine fuels are heated more quickly and easily by air temperature and direct sunlight than are heavy and compacted fuels.
- During the hottest part of the day, all fuels on south- and west-facing exposures will have higher fuel temperatures than those on north- and east-facing slopes.
- Heavy fuels will usually have a lower fuel temperature than the surrounding fine fuels in the daytime, and the reverse is true at night.
- When fuels are exposed to direct sunlight, surface fuels will usually have a higher temperature than aerial fuels.

The behavior of a wildland fire will vary as it consumes different types and amounts of fuel under varying weather and topographic conditions. An understanding of how a fire will react to all fuel types and the other major factors affecting wildland fire behavior must influence the strategy and tactics used to control a wildland fire.

FIRE BEHAVIOR

General fire behavior

Fuel, topography, and weather are found to have varying influences on fire behavior. Any one of these may predominantly influence what any individual fire will do, but it is usually the combined effect of all three that dictates the fire's behavior. The particular topography will not change of course, but wind movements are influenced by the orientation of topography and, hence, the direction that the fire will burn. The actual shape of the topography will similarly have an effect. For example, a saddle or low point in a ridge will act as a funnel, tending to draw a fire in its direction. Remember that below every saddle there is a chimney.

Severe funneling action often occurs on steeper slopes, as rising air currents are forced upward through narrow, chimney-like ravines or gullies. Fire is also drawn upward in the chimneys. Even with very sparse vegetative cover, fire is capable of moving long distances in an extremely short time. Fatalities have occurred within chimneys when the superheated air alone, rising from fire below, actually seared the lungs and depleted the oxygen supplies of firefighters. There is long-established understanding that it is dangerous to work above wildland fires. Working above a wildland fire is no different from working on a floor directly above a basement fire or descending into a cellar fire. The danger factors must be multiplied many times when personnel are at the top of, or within, a chimney.

Fires will burn very rapidly upslope, but in the absence of other strong influences, will usually burn slowly downslope. Fires burning in the upper portions of steep slopes or backing downhill often spread themselves rather quickly by means of rolling pieces of flaming fuel. Pinecones are noted for this because of their rounded shape. Pinecones lying on the ground are loosened as they burn and then roll downslope into unburned fuel.

Fires spreading upslope or with the wind assume a wedge-like shape. The point of the wedge is the fire head where the most rapid spread occurs. The flanks of the fire will gradually widen with the passage of time. That portion of both flanks immediately behind the head will unquestionably be influenced by an in-draft caused by the convection column at the head. A shift in wind direction will, of course, cause a greater movement of the former leeward flank and probably change the direction of the head.

Should the head of this theoretical fire be stopped suddenly and no action occur on the remainder of the fire, the rate of spread of the flanks will increase. If the wind is strong or the slope is steep, the original forward movement of the fire will remain unchanged. It is probable that the fire will now progress with two new heads (one on either side of the old head) until they join, or until one becomes dominant.

Depending on fuel type and condition, spot fires can occur ahead of both small and large fires. Burning embers are carried with the wind in the convective updrafts and dropped in unburned fuel in advance of the main fire. Embers will nearly always come from the underside of the heaviest smoke concentration. Flying embers can also originate from burning snags or trees with punky (dead or rotten wood) fire pockets and, again, the spots will occur downwind. The rapid spread of a fire up a steep slope to a ridge crest often results in spot fires on the opposite downslope side or even on the upslope of the next ridge. Spotting also can be expected downwind when individual trees "torch out" or when large piles of dead material such as slash are burning at peak intensities.

A small fire is not necessarily safer for firefighters to work than a large fire. Small grass fires have overrun many engine companies and personnel have sustained serious burns or died. Regardless of the size of the fire, fatalities are most likely to occur at small fires or during docile periods of fire behavior and under what appear to be innocent overall conditions.

When the relative humidity drops below about 30%, the situation is favorable for wildfires. For every 20° increase in temperature, the relative humidity is reduced by one half. As the relative humidity drops, the chances become greater for a smaller fire to grow unless an adequate suppression force is available.

Spotting

Spot fires are fires that start from embers given off by the main fire. Initially, they have separate fire perimeters ahead of the main fire, but they burn together with the main fire. One of the characteristics of an intense, small fire is the development of spot fires. Spotting is an important eventuality of many wildland fires. Spotting occurs when wind and convection columns broadcast hot "firebrands" to an unburned (green) fuel ahead of the main fire. The use of lookouts can sometimes expedite quick attacks on spot fires. In many instances, the burning embers evade discovery until a later time, when sufficient smoke rises above the foliage canopy alerting firefighters to the spot fire. Patrolling officers often are the first to detect hot material or spot fires outside the main fire edge.

Although burning grass seems often to be fully consumed by fire, spots can and do occur in unburned grass fuel well ahead of the main fire, such as across several lanes of a freeway. Spotting occurs more easily across narrow ravines than wider ones, and given favorable conditions, spots quickly spread to become fire problems themselves. Documented examples of fire spotting include spot fires starting up to three miles ahead of the main fire. Long-distance spotting will more than likely occur on the right flank of an advancing fire head because of the tendency of the smoke column under the influence of wind to rotate in a clockwise direction with increased height.

Although spotting is often related to such firebrands as wood shakes and ash from burning brush and timber, there are other causes of fire spread as well. Small animals, such as rabbits, have been observed carrying fire into unburned fuel.

Small fires

Work should begin on the flank that has the greatest potential, keeping in mind the necessity of protecting important exposures and holding the fire to a small size. Many small fires, and resulting larger ones, originate along roadways from a number of causes. The roadway serves as an excellent anchor point from which to begin work along the chosen flank.

Smaller fires in brush may dictate the use of a "hot-spotting" team, whose job it will be to pass up dormant or slower burning portions of line in order to knock down portions with greater activity. Hot-spotting can be carried out by either an engine company or a fire crew as a delaying tactic until additional forces arrive. Hot-spotting, if it is successful (and it has been on thousands of small fires), can save the day in terms of avoiding a long duration battle with an extended or major fire. This tactic also can be used on large fires.

Small fires can be quite damaging to improvements such as fences, outbuildings, and homes because a small fire burns, generally untouched, from the time of its origin until the first-in company arrives. Thus, even a small, one-acre grass fire could have sufficient time to move into an area of improvements, and barring an attack by local residents, do considerable damage. On many such fires near residential areas, local residents can and often do knock down many potentially dangerous fires. Because many fires start near or in populated areas, first-in companies should expect traffic congestion, dense smoke, confusion, and access problems.

An alert, well-trained company can successfully knock down small, hot, running fires if access to the fire is known and available. When fence lines and other security measures are present, access can be very difficult.

"Small" does not mean "easy" when it comes to firefighting a fire. Because of the intense aggressiveness needed to hold such a fire to a small size, almost every nozzleperson will tell you that it is hard, very hot work along the fire line. Most will say that although the action may be short-lived, it is usually nip-and-tuck for a while.

Large fires

Statistics show that 95% of reported wildland fires are contained within the first burning period (10:00 a.m. the first day to 10:00 a.m. the following day). The other 5% could have been reduced by more pre-suppression efforts, more suppression effort at the proper place, availability of sufficient firefighting resources to fight the fire, or by less human carelessness. Most large fires result from a combination of adverse conditions of fuel, weather, and topography.

As a wildland fire approaches and enters a populated area, such as a subdivision, fire officers must turn their attention to saving homes and other improvements while still actively pursuing perimeter fire containment. Protecting structures draws suppression resources away from perimeter containment efforts. If personnel reinforcements are inadequate, wildland fire control may be greatly hindered. Therefore, it is critical that initial attack incident commanders order reinforcements early to ensure an integrated attack on both the wildland and structural problems.

If weather is to blame for a small fire's growth, the odds against the fire officer increase as the weather becomes more adverse. A doubled wind speed can quadruple fire spread when winds are blowing at only 10 mph. Reducing fuel moisture by half, when it is already low, may cause fire spread to move not twice, but five or six times faster.

Added to whatever adversity may exist because of the shape of topography or the weather, large fires have unique characteristics that mean trouble for the fire officer. These fires have the tendency to crown and spot ahead because of the towering convective column. Also, the effect of upslope radiation is increased as the intensity of the fire increases.

On the other hand, unless unusually high-velocity and dry winds from one general direction overwhelm the effect of local broken topography, a large fire can be expected to subside considerably when it reaches a ridge top. This will result from the heavy in-draft demanded by the large fire running uphill.

The development of towering convection columns over large fires, with smoke rising tens of thousands of feet into the air, is one of the most striking differences between small and large fires. The essential cause of this difference is, quite naturally, found in the difference in the concentrated heat mass and the atmospheric stability.

Consider the cause-and-effect stages with respect to the single matter of spot fires caused by firebrands originating from the main fire. If the air mass over the fire is stable, that air will resist the development of a strong convective updraft in the form of a heavy smoke column and the mass transport of burning embers downwind. Consequently, the fire will burn less intensely than it would under a strong updraft. This does not mean that there will not be spot fires. Rather, spot fires that develop under stable air are likely to be fewer in number and to start closer downwind from the fire's head. Unstable air is more conducive to the development of a strong convection column and a more intense fire. The stronger updraft naturally has more burning ember carrying capacity. However, the longer that firebrands are suspended, the less potential there will be for burning firebrands to reach downwind spot fires.

A convective smoke column carries considerable heat and burning embers. The columns grow to very high levels when there is atmospheric instability and little wind. Under this condition, downwind ember fallout will not extend as far as when the wind bends over a column and drops the embers much farther downwind. Also, with the wind-driven column, the embers will spend less time aloft and better hold their heat resulting in a higher percentage of embers causing spot fires.

When a large fire burns during unstable weather conditions, the smoke column can tower over the fire. The towering, convective updraft effect can lead to a thunderstorm forming over the fire. If the storm becomes fully developed, downdraft winds may occur. These winds can adversely affect the already spreading fire. Sometimes, downdraft winds can reach 50 mph. These winds may compound existing winds and burning conditions.

Another important difference between small and large fires is the relative unimportance of size, distribution, and arrangement of fuel particles in favor of total fuel volume in the high-intensity fires. Larger-sized fuels will burn faster and hotter, producing temperatures up to 2,650 °F. Such intense heat results in the rapid and violent consumption of large areas and nearly total consumption of all combustible material.

Area ignition

Area ignition is the ignition of numerous individual fires throughout an area, either simultaneously or in quick succession, and so spaced that they soon influence and support each other to cover the entire area. Experienced fire officers have long appreciated this principle of augmented ignition and combustion from an adjacent source of heat. However, the full significance of area ignition from the standpoint of potential destruction seems to have been overlooked from a scientific perspective, until the occurrence of "fire storms" in large cities created by wartime bombing. Area ignition in the wildland fire setting is very similar to flashover in the structural fire setting with the physical combustion processes are quite similar.

A large fire tends to beget a larger fire principally because adverse conditions around a fire act to make bad things become even worse. Part of the subsequent trouble may be due to the effect of area ignition as it occurs in a natural manner during the progress of the wildfire. However, this phenomenon is much more likely to occur outside than within the perimeter of a large fire. This would happen when a number of spot fires outside the fire area flare up in unison all at once in fuel that is highly combustible. This, of course, constitutes an added hazard for the fire officer.

When a number of fires are ignited geographically so that the heat of each one affects the others, the following situation develops. Radiant heat prepares large amounts of unburned vegetation for easy ignition at approximately the same time. If we could look down on such a condition of multiple fires at an early stage, we might observe perhaps one-tenth of the area in flames. This would leave some nine times the flaming area in a state rapidly approaching a readiness for ignition. As this scenario develops, individual convection currents thrust hot air upward, creating in-drafts at their bases. Soon the convection updrafts begin to mingle and multiply the dimensions of the invisible "chimney." A constantly and dynamically increasing supply of air generates numerous fires into one massive flame front until a blowup occurs. Whether intentionally created or otherwise, the blowup is a fire of such intensity that it is difficult to hold within prescribed control lines.

It is obvious that this condition should not normally be expected from a fire moving outward from a central source, actually flaming for the most part at the perimeter and leaving behind only the ashes of the fuel it has consumed. If quantities of fuel are bypassed, either unburned or smoldering, the powerful forces of multiple ignitions may suddenly be unleashed at one critical moment when heat, air, and the crucial spark of ignition are ready.

Area influence

The effects of fire burning large areas in a relatively short period of time can extend for a considerable distance from the actual flames. This may influence fire behavior in other areas, sometimes with adverse effects. It is, therefore, quite important for fire officers to be aware of rapid fire occurrence on any part of the fire and be alert for possible dynamic changes resulting in their own areas. The intensity of this type of fire is demonstrated by the "Highway 41 Fire" in San Luis Obispo County, California, in 1994. At its peak this fire burned two acres per second for four hours, completely destroyed everything in its path.

Fire whirls

A fire whirl can be described as a violent, noisy tornado of fire, shaped like an elongated inverted funnel. A fire whirl spins at an extremely high velocity and emits a loud roar that is best compared to the sound of an aircraft engine. Its size can vary from a few feet to several hundred feet in diameter and from a few feet high to 4,000-ft. in height.

Fire whirls are usually associated with large fires, although they have occurred at small fires. This is probably because some of the conditions (such as unstable air) conducive to whirlwind formation are also factors at small fires. The cyclonic action (rapidly whirling fire around the outer perimeter of a center cone of air or gas) can pick up debris, sometimes including small logs, and raise them to great heights. A central "tube" is present, whether or not it is always visible.

Topography plays an important part in fire whirl occurrence. Although fire whirls have been known to happen on flat terrain, by far the majority occur in mountainous areas. Most of the whirlwinds observed by fire officers have, generally, occurred on the leeward sides of ridges, near the top. The shearing action of wind flowing over the abrupt edge of a ridge can cause the formation of dust devils.

All of the theories about whirlwind formation include air stability as an important factor. Some people, however, advance the belief that it is local thermal instability caused by the fire and not the degree of upper air instability that has the greatest effect. Still others point to the extremely unstable upper air conditions known to exist when large, destructive fire whirls have occurred. Unfortunately, fire officers in the past have paid too little attention to upper air instability as an indicator of potentially violent fire behavior.

Heat supplied by the fire provides the "trigger" to set the whirl in motion when all of the other factors are present. Sometimes a mass of fire is required; this usually results in a rather large whirl. At other times, relatively small whirls have formed along a moderately burning fire edge with no noticeable increase in fire intensity prior to their formation.

Strategy and Tactics

To successfully combat a wildland fire, the incident commander must have a working understanding of several critical command concepts. The concepts are as applicable at a small grass fire requiring only three or four engines as they are at a major fire that requires hundreds of suppression units and thousands of firefighters supported by complex command and general staff functions. The concepts include the development and execution of effective strategies commensurate with the fire problem, conducting an effective size-up, integrating the management cycle components with command concepts, executing practical suppression and rescue tactics and methods, implementing the incident command system, and dealing with complex I-zone fires.

Often, firefighters start suppression operations based on what they determine to be the immediate requirements of a situation. They augment their resources based on what the fire is doing. Their organization grows with little concern for long-term, prioritized objectives, and the command and suppression organization becomes chaotic, producing poor overall results. Such leadership, more of a happening than a commanded effort, results from a reactive rather than a proactive approach to proper and cognitive fire command.

Strategy and tactics concepts

The first course of action at any emergency is to clearly define strategy and tactics. Through the years, fire officers have termed the subject tactics and strategy, which is unfortunate since strategy must always precede tactics.

Strategy is the overall incident plan. It is a function of the incident commander or command. Strategy can simply be defined as *what* is to be accomplished. Management objectives are a major component of a strategy, as are incident objectives. Both types of objectives are important and must be prioritized. Management objectives might include: not having any firefighter injuries, protecting life and property, and not ruining the environment with the suppression efforts. Incident objectives are more operationally oriented and include objectives such as: protecting structures, holding the fire north of Sunset Boulevard, holding the fire east of the San Diego Freeway, holding the fire west of Laurel Canyon Drive, and keeping the fire south of Mulholland Drive. Creating concise incident objectives enables fire officers to clearly judge their suppression efforts against planned activities.

Tactics are *how* the incident commander will implement the incident objectives. At a wildland fire, particularly an I-zone fire, the incident commander must integrate ground and air attacks, while being sure to provide firefighters necessary logistical support. Tactics are a function of the operations section chief (OSC) unless the incident commander has not staffed the OSC position. The following concepts are basic to all command operations including wildland fire.

Size-up concepts

An officer's initial decisions are usually irreversible and the consequences of errors can quickly compound and become disastrous. Unlike business executives, fireground officers must base all-important decisions on hastily obtained information gathered under the most stressful conditions. This process of hastily gathering important information is called size-up. Size-up is the basic foundation for subsequent firefighting decisions and operations. The old saying that a building is no stronger than the foundation on which it rests has a direct application to fire size-up. Build a house with square corners and plumb walls and adding onto it later is simple; the same applies to building an emergency organization.

Size-up should begin long before the fire actually starts. Fire officers should be familiar with the fuel types, topographical features, exposure problems, and daily weather patterns in their areas of responsibility. Having this basic information can provide a foundation for future observations. When the alarm sounds, each firefighter should know the general fire behavior that can be expected, the location of the fire, and the types of physical factors that are present. These factors include: fuel, topography, road access, structural exposure, water supplies, naturally and unnaturally occurring person-made barriers, and the number and types of suppression resources dispatched.

In his 1941 book, *Tactics and Strategy*[1], Lloyd Layman developed the components of a size-up that are still valid to this day. Layman determined that an effective size-up must include: (1) facts, (2) probabilities, (3) own situation, (4) decision, and (5) plan of operation. The first three components are information inputs that lead to the fourth component, which is the strategy that leads to the last component, the tactics. The first two components, facts and probabilities, relate to situation status. The third component, own situation, is really a measure of the resource status. Together, facts, probabilities, and own situation and resource status are the three input additives that create the decision or strategy. From there, we can take the strategic or incident objectives and develop a plan to implement the strategy, or plan of operation (tactics).

The following list provides some examples of facts needed during the decision-making process:

- Available pre-plans
- Weather, current and predicted
- Date and time of alarm
- Location of fire start
- Life hazard
- Topography and fuel conditions
- Fire history of area
- Exposure problems
- Available water systems or water tenders

Examples of probabilities include:

- Likelihood that firefighters will be able to contain the fire relatively quickly
- Threat that fire may become a major fire
- Fire spread projection
- Loss of life or serious injuries to firefighters or civilians
- Water system long-term viability
- Need for evacuation

Examples of own situation or resource status include:

- Availability of fire engines
- Availability of fire crews and dozers
- Chief officer availability
- Aircraft availability
- Ability to staff a major incident with operational and support personnel
- Other fires that will compete for resource availability
- Support public safety organization availability including law enforcement, public health, welfare, etc.

These input examples lead to a strategic decision and the formulation of prioritized incident objectives. From there, tactics evolve as previously discussed. Table 21–1 illustrates the Two x Two Status Matrix, a great aid for gathering information.

Table 21–1 Two x Two Status Matrix with Status Examples

	Current Status	Projected Status
Situation Status	Fire location? Fire size? Fuel, topography, and weather? Exposures threatened or burning? Suppression effectiveness?	Where will the fire be? What will the size of the fire be? What will the fire involve? Will the fire burn more structures or communities? What is the predicted weather? What is the future plan?
Resource Status	Engines, crews, and dozers? Air tankers and copters?	What will the availability of engines, crews, and dozers be? Are aircraft available? Support personnel? Incident command team?

Remember that no fire is static, and, therefore, your size-up should not be static. Many fires have escaped control because, despite the fact that the incident commander made a good initial size-up, he or she failed to reappraise the situation. Continue your size-up throughout the confinement, extinguishment, and mop-up process. As mentioned before, many firefighters have been killed or injured because sudden changes in weather affected fire behavior.

The management cycle

Fire command consists of the proactive consideration and intelligent use of six basic management elements: planning, organizing, staffing, directing, controlling, and evaluating. The incident commander must formulate a plan of action, build and staff an organization to accomplish the objectives and strategies outlined in the plan, direct the organization by communicating the strategies and tactics to the necessary members, control the identified fire organization with adequate systems so the incident command staff can meet the objectives as outlined, and, finally, evaluate actions to compare planned efforts vs. actual accomplishments. Figure 21–3 illustrates the management cycle.

Planning

A plan is an ordered sequence of events over a specified time to accomplish a specific objective(s). Plans are formulated based on considerations of the past, present, and future factors that are or may affect the fire. The fire suppression planning process consists of three separate elements—an information-gathering system, an information evaluation and prediction system, and a constant re-evaluation system. These elements provide the information for the facts, probability, and own situation processes discussed under size-up.

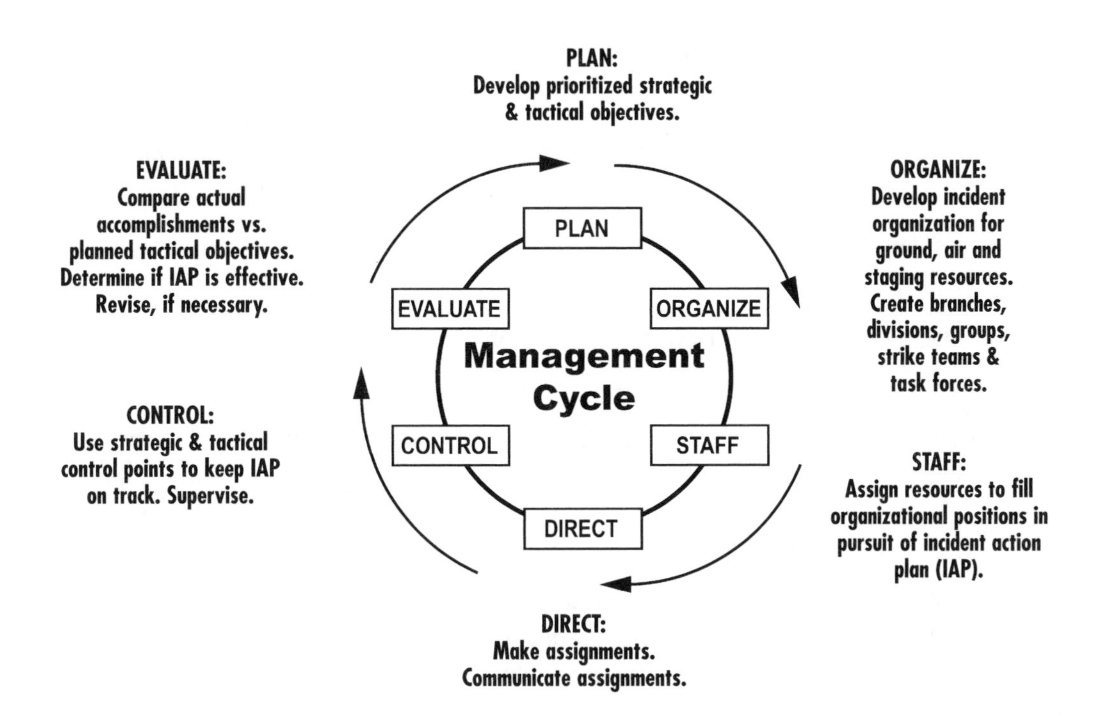

Figure 21–3 The management cycle

At wildland fire situations, the information-gathering system may consist of facts gathered from pre-fire plans, observations and radio traffic from first-in companies, information received from the dispatch center, and communicated observations from aircraft, such as the ATGS (air tactics group supervisor) and helicopters. Information also is gathered from reconnaissance personnel and the personal observations of command officers including yourself. The information evaluation and prediction system consists of the incident commander and various staff planning actions. Under size-up, it is the decision (strategy) and plan of operation (tactics) formulation efforts.

The re-evaluation system is a never-ending process. Good firefighters and command officers make effective planning a dynamic and ongoing process. A good plan is constantly adjusted based on the fire status and the proactiveness of the suppression forces.

Ideally, fire command incident action plans (IAP) should be based on facts alone. However, if the officer developing the action plan does not have the training and experience necessary to evaluate the facts and probabilities, the plan may fail. Attitudes, sentiments, values, and, therefore, planning conclusions, are not always based solely on facts. Successful firefighters objectively, rather than subjectively, review data in order to develop effective control plans.

Organization and staffing

A fire organization, simply stated, is a group of fire officers, firefighters, and emergency responders working together to achieve a common objective(s) or the incident objective(s). To be effective, the organization must include the following elements:

- A system of authority and responsibility
- A system of direction and communication
- A consolidated action toward identified objectives
- A system of maintaining acceptable norms and standards

The system of authority and responsibility is usually outlined with a legal or an operational basis in mind. Statutes, policies, and procedures set the direction for fire department actions. Much of the United States has adopted the incident command system and, as such, each department has developed specific operational policies and procedures.

ICS is communicated through an organizational chart and supplemental position descriptions. The organization must maintain a system of direction and communication. This system can consist of face-to-face contacts, written messages, and radio and telephone communications, including electronic mail and facsimile transmissions. An important key to a successful organization is coordinated action toward specific objectives. This coordinated action is based on how well the forces are organized and how realistically the objectives have been identified. The important elements to consider in a fire suppression organization are span of control, unity of command, clear definition of authority and responsibility, and unity of objective.

Span of control relates to the number of people or units a manager can effectively direct or supervise. For example, a management unit may be an individual (firefighter), or a unit such as a fire crew, that reports directly to one supervisor. Generally, in a fire organization, the span of control number ranges from three to seven units with five units being the optimum.

Proper *unity of command* dictates that each firefighter should have only one supervisor. Although almost all fire officers agree with this concept, it was once the most violated organizational principle on the fireground. Today, with the incident command system and accountability procedures, it is less frequently violated.

A clear definition of *authority and responsibility* is paramount to fire suppression efforts. Each person should be given clear authority to carry out the responsibilities assigned to them. Again, the ICS provides position definitions and outlined duties.

Unity of objective must be clearly understood and practically pursued to accomplish the overall goal of suppressing the fire. All fire officers and firefighters must play a supporting role in achieving a single purpose or objective(s).

Staffing is simply the delegation of personnel or resources to planned organizational assignments. Incident commanders should always assign the very best personnel to critical assignments where the risk is high. Strong players personally discipline themselves to perform to the highest level during the greatest periods of stress and disorder. Wisely and effectively utilize them. Sometimes, you will create organizational positions for which you do not have human resources. In that case, you will assign a person or unit to act in those positions whenever they become available.

Organizing and staffing a wildland fire is much easier than it may seem. This is particularly true during the early stages of an initial action fire when no earlier organization exists. Division of labor, or dividing the fire, is critical even at smaller fires. Often, the best initial division is to split the fire with the left flank being division A ("Alpha") and the right flank being division Z ("Zebra"). The incident command system specifies that fires should be divided by starting at the origin and progressing clockwise around the fire, identifying divisions as you proceed. In reality, there will probably never be a fire where the incident commander can create all divisions that are perfectly arranged clockwise around the fire and are all created at the same time. In spite of this, by naming the left flank division A and the right flank division Z, the incident commander can easily add divisions as necessary.

In addition to creating two divisions, the incident commander could create a staging area. Until additional fire officers arrive, it is perfectly acceptable to assign company officers to staff the two created divisions. If the fire continues to expand, the incident commander should add and staff additionally required divisions or structural protection groups. Figure 21–4 illustrates how to split a fire into two divisions and create a staging area.

There are two schools of thought on using divisions or structure protection groups for I-zone fire protection. One school advocates that all actions within an area are assigned to a division, thereby ensuring better coordination and communications within one organization. With this approach, divisions must be small enough to not overload the span of control of the division supervisor. The other school of thought suggests that perimeter and structure protection actions should not be combined into a division, but rather should be separated as a division and as a group. Under the ICS, groups can move across division or branch boundaries as necessary. Sometimes, structure-protection group assignments lead to engine strike teams sitting on houses when they could be better utilized doing both structures protection and extinguishing the fire edge or perimeter.

Some fires start as major fires. At those infrequent wildland fires, it is perfectly acceptable to initially start the organization by creating branches. Instead of the two initial divisions discussed earlier, create two branches; make Branch I the left flank and Branch II the right flank. Divisions must be created within the branches as needed. Again, consider establishing a staging area. In fact, the incident commander may need to down-staff an engine company and use the company to staff and manage the staging area. The company officer could act as the staging area manager.

Multiple staging areas can be created at a fire. Consider critical transportation response nodes near the fire when locating multiple staging areas. One staging area might be established at the north side of the fire near a particular access intersection and a second area established near the south end of the fire near another critical transportation route. Often a

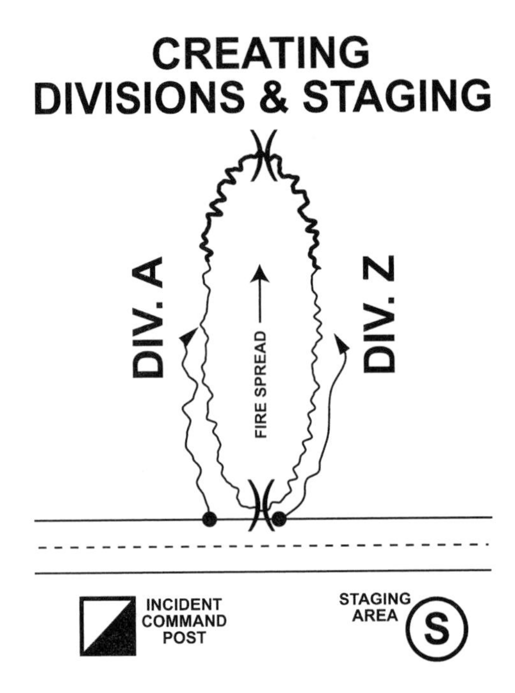

Figure 21–4 Creating divisions & staging

surface street near the fire can be blocked and used as a staging area. This works very well for funneling resources into one end of the staging area and assigning them out the exit.

The incident commander must locate near the incident. Locate means establishing a fixed incident command post and staying at the post. Roving incident commanders cannot adequately plan and direct a major wildland fire. This is particularly true once the incident becomes a running fire or during unified command situations. Assisting and cooperating agency representatives must be able to easily locate the incident commander.

If the fire becomes large with literally hundreds of firefighters assigned, the incident commander must establish an incident base where the six logistical functions—food, medical unit, communications, supply, ground support, and facilities—are provided. At major incidents where logistical support requires the establishment of an incident base, the incident commander should assign at least one-fourth of the assigned incident workers to support roles and the remaining three-fourths to operations activities.

Direction

The fourth element of the management cycle involves direction, which is the issuing of directives during the fighting of a moving wildland fire. Direction requires a clear understanding of the process of human communications under stressful conditions and the necessary parts of a proper and effective directive.

Put simply, a directive communicates some specific information toward the accomplishment of a task. For a directive to be clear and effective, it should answer the following questions (the five Ws):

- *Who* is to perform the work?
- *What* is to be done?
- *When* is it to be done?
- *Where* is it to be done?
- *Why* is it to be done?

Sometimes the *why* is implied or understood and does not have to be fully explained in the directive.

The following is an example of a proper incident directive that meets the requirement of providing the "who, what, when, where, and why." "Engines 28 and 72, upon arrival at the fire, start a progressive hose lay on the right flank. The hose lay is to tie in with the tractor plow line that is progressing on the left flank."

After a directive is issued, it is important to ask, "Do you have any questions?" Good directives involve two-way communication, are acknowledged or confirmed, and are best handled on a face-to-face basis.

Controls

Controls are methods or systems designed to inform the incident commander of operations progress and breakdowns so the system can be adjusted and corrected to meet the objectives as identified in the IAP. To ensure a proper set of controls, a fire organization must contain several important elements.

One of these elements is visual control, or actually observing the progress taking place. Periodic reporting from different suppression units and fire officers can pass along pertinent information. Monitoring radio traffic and holding face-to-face meetings with responsible members of the incident organization can provide timely controls. Also, pilots of aircraft, both fixed and rotary-wing, can quickly provide important and timely information to field commanders.

Consideration should be given to establishing strategic control points that evaluate individual objectives that are part of the overall IAP. If an incident strategic objective dictates that the wildland fire not cross the Mississippi River, then a strategic control point exception would be any extension of the fire over the river. From a command standpoint, the use of strategic control points greatly helps an incident commander clearly determine the need to adjust strategic operations. The concept of control points also can be extended to tactics through the use of tactical control points.

Evaluating

Evaluating is the simple process of comparing actual accomplishments against planned actions. It is an extension of controls, but happens on a more global basis and often validates or requires the amendment of incident objectives. If the test question validates the incident objectives, the incident commander is on the way to a successful outcome. If the test question shows that the strategic plan is not working, the plan must be revised, possibly leading to another entire trip around the management cycle. Using this as a mental check enables incident commanders to better plan, execute, and evaluate an incident operation.

The chapter thus far has covered the concepts of fire management and the development of strategic objectives. From this point on, fire ground tactics and suppression methods, the deployment of fire attack resources, and I-zone firefighting will be addressed.

TACTICS AND WILDLAND FIREFIGHTING METHODS

Firefighters have two options for combating wildland fires, direct attack and indirect attack. Quite often, the method chosen is based on fuel types, exposure problems, fire behavior, and availability of suppression resources. The safety requirements for the personnel involved should be considered first when choosing a method or combination of methods.

Direct attack

The direct attack method involves working directly at the fire's edge. This method has several tactical advantages. Suppression resources instantly produce results with the amount of fire controlled. This type of suppression action might involve the following actions:

- Laying fire hose and spraying water on the fire's edge
- Constructing a handline along the immediate edge of the fire

- Making air tanker or helicopter drops to support firefighters working on the perimeter
- Having bulldozers or tractor plows construct a containment line along the edge of the fire

The direct attack method requires a continuous and finished fire line (line cut and scraped to mineral soil to prevent future fire escapes). The fire line and fire edge are coterminous. The line should be started from an anchor point (starting point from which the fire cannot fishhook behind and trap firefighters), such as a road, plowed field, or other unnaturally occurring or natural barrier. The direct attack will limit the fire's ability to build momentum and eliminates the need for backfiring. It is generally used on fires in the initial attack phase and can be used at the flanks and head. The main objective is to stop the spread of the fire as soon as possible.

The advantages of the direct method are that it holds the size of the fire to minimum acreage; takes advantage of portions of the line that are not burning because of a lack of fuel or poor burning conditions; eliminates fuels at the fire's edge; provides direct operational control and higher levels of firefighter safety; and reduces loss of resources caused by backfiring or allowing the fire to run to predetermined control lines.

The disadvantages of the direct attack are that it might not be effective against an intensely hot or fast-moving fire; results in an irregular, and thus, long fire line to construct; exposes firefighters to direct flame and smoke; normally does not take advantage of naturally or unnaturally occurring manmade barriers in the fire's path; and requires additional mop-up. Additionally, direct fire attack requires close coordination of the suppression forces involved.

The following is a simple explanation and example of a direct attack using the *box-it-in* concept. With this method, the incident commander superimposes a box over the wildland fire and then prioritizes the sides of the box to determine the necessary incident objectives. When present, structures are always considered the first priority. After structures, stopping the head is frequently the second priority. Next, the incident commander must decide which flank is most threatening, and that decision will then determine the third priority. The fourth priority is usually the other flank with the fifth priority being the fire rear. The incident commander then spatially relates each side of the box to a containment road, naturally or unnaturally occurring human-made barrier. The box easily enables the incident commander to visualize where the fire should be contained. Naturally, some fires may require several different box analyses as the fire expands beyond containment priorities. Figure 21–5 shows an example of the box concept.

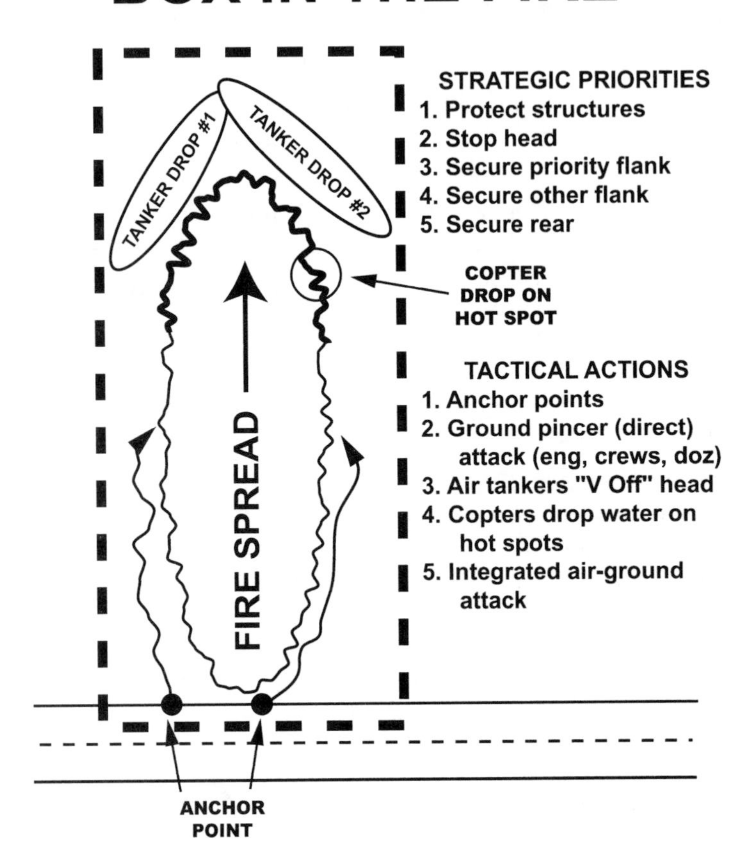

Figure 21–5 The box concept

Indirect attack

This attack method is mostly used on large fires. The indirect attack consists of fighting a wildland fire by constructing or using control lines at a considerable distance ahead of or away from the main fire edge. It is used when the fire is burning too hotly and rapidly to use direct attack methods or when too few suppression forces are available to use direct attack methods. Pre-constructed lines placed ahead of the fire are either fired or held by ground and air resources and become the final control line. The indirect attack establishes barriers or fire lines in advance of the fire edge and either allows the fire to burn out to the breaks or removes (burns out) the unburned fuel between the fire edge and the fire line. The indirect method puts firefighters some distance ahead of the main fire edge.

The advantages of the indirect attack method are that a larger fire can be contained by using breaks in advance of the main fire edge with fewer suppression resources; a simpler fire line can be constructed that does not follow irregular perimeters as happens with a direct attack; firefighters can work away from large amounts of heat and smoke; and mop-up is limited.

The disadvantages of the indirect method are that the fire cannot be extinguished until the control lines are identified, constructed and staffed, by which point additional acreage is sacrificed; portions of the fire line that may have gone out aren't used; and firefighters are placed in front of an advancing fire. In addition, secondary lines may have to be constructed and a frontal stand by firefighters and a successful backfiring operation, or both, may be required. In terms of firefighter safety, the indirect attack is the most dangerous. More firefighters are killed or injured with indirect attacks than with direct attacks.

Application of attack methods

No one method must be followed exclusively throughout fire suppression operations. More often, on any sizable fire, both methods are used. On one flank it may be possible to use dozers or tractor plows and engines with a direct attack and use handline construction and hose lays on the other flank with an indirect attack.

Backfiring and burning-out actions

There is a distinct difference between backfiring and burning out. These two terms have been used interchangeably with resulting confusion for many years. The term *burning out*, sometimes called *firing out*, refers to the removal by fire of residual fuel between a constructed line and the edge of a fire. The term also refers to the burning of fuel to protect structures or other improvements. Sometimes burning out is described as a defensive measure when timing is not critical and the main fire will not affect the action.

Backfiring is an indirect fire-control action typically used against a rapidly spreading fire. The fuel between the preplanned control line and the active fire's edge is intentionally fired to eliminate fuel in advance of the fire. This widens the control line to possibly change the fire's direction and to slow the fire's progress. Backfiring may be described as an offensive measure because of effects on the main fire and the critical timing necessary for success.

Backfiring often involves considerable planning, organizing, and physical preparation. The decision to backfire should be made by the command function, communicated through line channels, and executed at the division level by strike teams or single resources. Under special circumstances or during true emergencies, line personnel may be forced to backfire and concurrently notify higher command.

Whenever backfiring or burning out fuels, safety considerations must be the first priority. No backfiring operation, regardless of the importance of the strategic values involved, is worth risking one human life. The probabilities of both success and failure must be calculated and considered by command personnel because any backfire could result in losing the fire and subsequent compounded injuries, deaths, or property damages.

Knowing when and how to properly conduct a backfiring operation is essential because when done properly such operations can control a major fire rapidly with relatively limited suppression resources. Overall fire strategy and resulting tactics must be understood by all personnel involved in the backfiring operation as the fire behavior on other portions of the control line is likely to be affected.

Backfiring should be directly related to the behavior of the fire. If the main fire is burning intensely and spreading rapidly, backfire quickly enough to prevent the fire from jumping prepared or proposed control lines.

When weather, fuel, topography, and sufficient suppression forces are available for a backfiring operation, it should proceed without delay. Many fires have been lost because of indecision, while other backfires have created additional problems because time was wasted and burning conditions changed to less than favorable, causing the backfire to *lay down* or go out. At times, the same indecision and burning conditions can cause backfires to escape, resulting in additional problems. Fire officers must evaluate the decision carefully. The tactic should be performed only by trained and experienced firefighters. Active fires (either the main fire or intentionally set backfires) produce conditions that draw other fires. The larger and more intense a fire becomes, the more it affects local weather conditions, thus greatly influencing other fires at increasingly greater and greater distances. Noticeable wind direction and intensity changes have taken place more than a mile away from a fast-moving fire in heavy fuel.

An officer who understands fire behavior and knows how to take advantage of favorable topography, fuel, and weather conditions should always direct backfiring operations. The officer in charge of the backfire must know the overall fire strategy and be kept constantly informed about the progress of the main fire. The officer should have considerable experience based on other previous experiences with firing operations.

Once a decision to backfire is made, the following important guidelines should be followed:

- When fire danger is extreme, backfiring is most hazardous and may fail. Suppression action on the flanks may prove to be more effective.
- A backfire should be started as close as possible to the main fire, balancing the time needed to establish an effective line against the rate of spread of the main fire. Always allow a margin of time.

- Once backfiring has started, it is essential that all fuel between the backfire and the main fire be burned.
- Backfiring should be conducted by a group of highly experienced firefighters under experienced leadership.
- Sufficient forces should be committed to hold the line created by the backfire.
- The main fire and the backfire should meet a safe distance from the control line as intense, erratic fire behavior is common when the two fires meet.
- Weather conditions, both current and predicted, should be noted at all times.
- Never start more fire than can be controlled by personnel assigned to the holding operation. However, once started, the backfire should gain maximum depth in the shortest amount of time without endangering firefighters.

Remember that there will always be a calculated risk with backfiring operations!

INTERFACE FIRES (I-ZONE)

One of the greatest challenges facing today's firefighter is combating a wildfire that burns in, around, and past structures. This type of fire is frequently called the structure-wildland interface or I-zone fire. Throughout the United States, I-zone fires continue to destroy structures, and sometimes, result in injuries or worse to firefighters. Few fire departments or communities are exempt from the incidence of I-zone fires.

Most I-zone problems are the result of lack of fire protection planning. In fact, many I-zone problem areas developed long before fire chiefs recognized the problem of structures intermixed in a wildland setting. One of the earliest I-zone fire incidents occurred in 1956 at Malibu, California during Christmas. Today, there are destructive I-zone fires almost every year in the United States, and some communities have significantly greater problems than others. Not only is the combination of structures and wildland fuels a dangerous mix, but when the community is located in an area prone to strong winds, the result can be catastrophic.

On one hand, the public wants to live in a natural area, possess their piece of America away from the rush of city life, and live in a private, often uncontrolled way. On the other hand, the public also wants and expects a certain level of protection from harm including wild fires. Fire chiefs and officers must ensure that the public receives effective fire protection wherever possible. To that end, many I-zone fire problems can be reduced by comprehensive fire protection planning measures, including actions to ensure that residents provide and maintain clearance of vegetation around structures. Community-based water systems should be installed (if not already in place) and maintained. Ingress and egress roadways should be well thought out, well designed and maintained. Finally, building design features need to include fire-safe structure characteristics such as suitable structure location, non-combustible roofs and siding, and automatic fire sprinklers.

Typical I-zone fire situations experience the following obstacles:

- Many structures are potentially threatened, but few firefighting resources are available.
- There is limited time to plan or organize firefighting efforts to fight fire.
- Limited water supplies exist.
- A subsequent wildfire drives all activities.
- Non-wildfire problems develop incidental to the fire, such as law enforcement, evacuation, civilian welfare, etc.

As with all wild fires, fuel, weather, and topography significantly impact I-zone fires. The type and quantity of fuel directly affects flame length, which thereafter has an impact on structure survivability, particularly as it relates to defensible space around the structure. Critical weather factors include dry bulb temperature, relative humidity, fuel moisture, spot fire potential, and wind—the most variable (direction and speed) and dangerous spread factor. Topographic characteristics, such as aspect (direction slope faces), incline steepness of the slope, and shape of the topography (particularly the presence of chimneys or chutes) are critical factors when attempting to protect endangered buildings threatened by a wild fire.

Structure access is always an issue when fighting an I-zone fire. Specific access problems include the following:

- Location of the fire in relation to roadways that can be negotiated by fire apparatus
- Roadway limitations such as road width, road surface, and bridge capabilities
- Roadway hazards such as falling or rolling rocks
- Lengthy travel or response times
- Civilian vehicle traffic, particularly during panic situations
- Lack of posted street addresses and road signs
- Poor visibility due to smoke or fire conditions

The bottom line on access is to safely reach the affected structures and acquire a map of the local area where the fire is burning.

When triaging threatened structures, consider the entire structure environment, including the area surrounding the structure and the building construction. Density of structures, yard debris (firewood piles, scrap lumber, lawn furniture, and general junk), and typical structure clearance of natural and landscape fuels are the most critical environmental factors. Lack of clearance is arguably the most dangerous environmental factor, although yard junk, ornamental shrubbery, and leaves or needles in rain gutters contribute to fire severity. LPG (liquid propane gas) tanks may also present a risk. Clearance of vegetation from around LPG tanks is critical to prevent events related to heated tanks.

Exterior building construction factors include the type of roof covering, the type of siding, attic and foundation openings, the size and location of windows, and the position of the structure in regard to the topography. Combustible roofs particularly wood-shingled roofs, combustible siding or deteriorated siding that exposes wood to flames, large windows facing wildland areas, and cantilevered homes located on slopes overhanging areas of heavy vegetation pose the greatest exterior threat risks.

Interior construction factors include the type of window covering, particularly flammable drapes, HVAC systems that enable fire or heat to spread within the building, doors that allow the structure to be compartmentalized (thwarting fire extension within the structure), and the control of electrical and gas utility services. Generally, firefighters should leave the electrical service activated to energize local water system pumps and for lights to operate (particularly the front porch light) so that firefighters will know that the structure has been searched. Gas service should be shut off to avoid an unnecessary fire hazard.

I-zone structure triage will consist of the following steps:

1. Eliminate hopelessly threatened structures.
2. Ignore those structures not immediately threatened.
3. Deal with the remaining structures.

When making triage decisions, consider (1) the range of possible outcomes, (2) the probabilities, and (3) without fail, be decisive in developing an incident action or tactical plan.

An I-zone fire will hit a community in one of two ways. The first way is called a broadside hit where a fire front, sometimes a wide fire front, will strike a developed area. The front may be one or more miles in width. It will almost always be wind-driven, but may be slope-driven and have a narrow head when wind is not the primary driving factor. If wind is the primary driving factor of a broadside hit, the fire may well burn straight through the community causing massive destruction. If wind is not the primary driving factor, expect the fire to burn into the community some distance, but not through the community.

The second type of hit is a progressive fire, which enters a community one area at a time. Progressive fires hit communities because of the oblique angle of the fire vector movement in relation to the general cardinal direction layout of the community. A progressive hit may also be the flank of a rapidly spreading fire. Progressive hits can be very destructive depending on the primary spread factor driving the fire. Whole communities can incur severe destruction from progressive hit fires.

At I-zone fires, incident commanders will find it hopeless to save structures when:

- The fire is making sustained runs in live fuels with poor clearance.
- Newly developing spot fires occur faster than suppression efforts can extinguish them.
- The water supply will not last as long as the threat will be present.
- You cannot remain, and your escape route is becoming unusable.
- The structure roof is more than one-half involved, and the day is windy.
- There is a fire in the structure, windows are broken, and the day is windy.

The incident commander must develop incident objectives (IAP) for the wild fire threatening or burning structures. The plan should include efforts to provide both structure protection and perimeter control. If lack of resources only allows fire apparatus to protect structures, the main fire will continue to spread, involving more and more structures, increasing the I-zone life and property threat, and increasing losses. The incident commander must order sufficient resources to staff all tactical positions where structures are threatened and at locations limiting fire perimeter spread. The action plan must include the following information:

- Current and projected situation status
- Escape routes and safety zones
- Prioritized incident objectives
- Tactical assignments
- Firefighter safety and accountability
- Water source considerations
- Arrival times vs exposed structure threats
- Comparison of actual burning vs predicted fire behavior conditions
- Planning, logistical, and financial considerations for extended operational periods
- Public information dissemination, particularly as it applies to involved and threatened communities

An example of initial, prioritized incident objectives (strategy) for an I-zone fire might include:

- Protect life and property, including structures
- Hold the fire north of Sunset Boulevard
- Hold the fire east of the San Diego Freeway
- Hold the fire west of Laurel Canyon Drive
- Keep the fire south of Ventura Boulevard

When fire conditions change, incident objectives must also change and reflect the change in fire spread and structure involvement. The change may make the situation worse or better. Incident commanders must always be ready to change plans whether that means increasing the incident resource commitment, decreasing the incident resource commitment, or modifying the incident objectives.

Strategy situations at I-zone fires will typically be offensive, defensive, or a combination of both. Offensive situations are possible when the ability to directly attack the fire exists, when the main fire can be controlled, and when sufficient suppression resources are present or available. Defensive strategies are more likely in situations in which the main fire cannot be controlled before it burns through structures and when efforts are concentrated mostly on saving buildings. A combination strategy exists when some structures must be defended, but at least a part of the main fire can be controlled.

Some critically important, but simple rules-of-thumb guidelines for I-zone fire operations include:

- For isolated residences, assign one engine per residence.
- Where common access (structures less than 50-ft. apart) exists, assign one engine per two residences (may need only one hose line).
- For multifamily structures, assign two or three engines per building.
- Try to maintain a floater or uncommitted engine per engine strike team.

Expected commitment-time rules-of-thumb guidelines include:

- Allow 20–40 minutes per protected site.
- Allow additional time with heavy fuels where residual burnout times are lengthy.
- As long as there is a threat from firebrands, you must stay.
- To ensure a safe exit, allow time for the fire to settle sufficiently.

Expected water-usage rule-of-thumb guidelines are:

- Allow approximately 200 gal. per residence.
- Allow approximately 400-800 gal./hr. per engine.
- When possible, assign a water tender for each strike team.
- Allow greater water usage with heavy fuels.

Required defensible-space (fuel clearance) rule-of-thumb guidelines are:

- Always observe the existing fire behavior, particularly the flame length.
- Flame lengths in live fuels will be two to three times the height of the fuel.
- Defensible space clearance requirements in heavy fuels will be two to three times the flame length.
- Defensible space clearance requirements in light fuels will be less than three times the flame length.

Engine placement at I-zone fires is critically important. There are some very basic, but highly important, directions including:

- When responding, note landmarks and hazards.
- Consider the possibility of exiting through a hot and smoky escape route.
- Locate effective safety zones.
- Back the apparatus in against the threatened structure to create a heat shield from the encroaching fire.
- When parking the fire apparatus, do not block other traffic movement.
- Do not park near or over flammable materials or objects.
- Park away from heat sources and sealed vessels.
- Do not park under power lines.
- Park close enough for effective hose line deployment.
- Close apparatus windows and doors.

The basic process for engine companies protecting structures includes the following concepts:

- Let the fire come to the structure unless the fire can be extinguished, as in an offensive strategy situation.
- Do not leave the safe refuge that the structure provides to chase the fire remotely from the structure unless you can be offensive and extinguish the fire.
- If sufficient engine company personnel are available, stretch two hose lines. If not, stretch one hose line.
- Use 1½-ft. or 1-ft. hose lines with combination nozzles.
- Hose lines should be able to wrap around the threatened structure, but must be able to stay where a heat barrier exists until the fire stabilizes. Often a stem wall or building corner provides such a heat barrier.
- Keep hose lines as short as possible.
- Lay hose lines outside yard obstacles.
- Don't lay hose lines in front of the engine's exit path.
- Hose lines should come off the same side of the fire engine to facilitate rapid departure, if necessary.
- Be able to quickly disconnect hose lines should a rapid exit be necessary.

- Put a flowing garden hose in the fire engine tank filler and attempt to keep the tank filled.
- For personal protection, always try to keep a minimum of 100 gals. of water in the tank at all times.
- Try to use the resident's ladder if it is serviceable.
- Send a hose line to the roof if it is combustible or there is a roof fire that can be stopped.
- As a rule, don't lay long supply-hose lines. However, sometimes 100-ft. of large-diameter hose (LDH) laid to a hydrant will really enable a hard and effective hit with a deck gun on the approaching fire. Consideration must be given to putting a firefighter on the deck gun in an exposed position and to water supply consumption when using a deck gun. It is a rapid and effective method of knocking down an approaching fire.

Figure 21–6 illustrates a typical I-zone engine deployment.

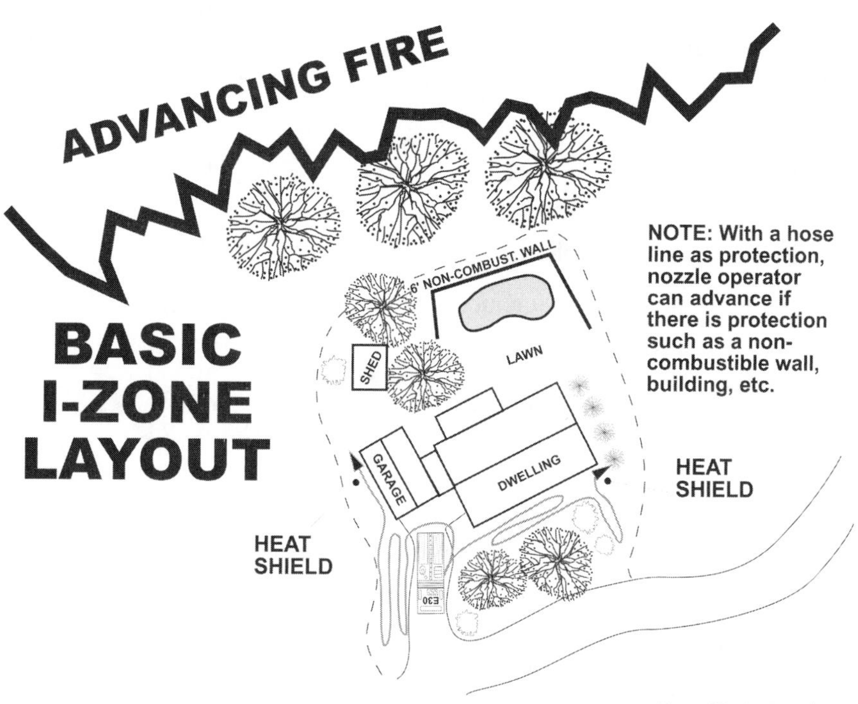

Figure 21–6 Typical I-zone engine deployment

There are basically three types of outcomes at I-zone fires. Table 21–2 shows several fire situations and their possible outcomes.

Table 21–2 Three Basic Types of Outcomes for I-Zone Fires

Type of Attack	Fire Situation	Possible Actions
Full Control	• Fire is not burning hot and can be. • Sufficient resources are available.	• Directly attack the fire at the edge of the yard. • Be sure to stop and hold the fire. • Combine structure protection with perimeter, direct attack efforts. • Check structure for hidden or small fires.
Partial Control	• Lack of time to prepare the structure or fire intensity prevents full control. • Fire will burn around structure. • Should be little structure damage.	• Knock down the fire front that is moving at the structure. • Lead fire around the structure. • Check the structure for hidden or small fires.
Damage Control	• Fire has or will pass around structure. • Fire is too hot to stop. • Firefighters locate and utilize heat shields until fire passes. • May need to stay in structure or engine until fire passes. • Could be significant structure damage.	• First consideration is firefighter and civilian safety. • Get inside the structure or apparatus, or use heat barrier for protection until the fire passes. • When fire passes, extinguish fire in and around structure and residual vegetation fire. • Be ready to bump ahead to another threatened structure.

Water application is very important during I-zone fires. Some important water-use factors include:

- Avoid wetting down ahead of the fire arrival.
- Keep the fire out of heavy fuels, including wood piles.
- Extinguish the fire at its lowest intensity level.
- Avoid wasting water during the peak heat wave.
- Apply water directly to the structure to avoid ignition.
- Avoid putting water on window glass.
- Apply water only if it controls the fire spread or reduces excessive heating of the threatened structure.

Additives for water application really increase the suppressing capability of firefighters' universal extinguishing agent, water. The additives include educted class A foam, compressed-air agitated class A foam, and a relatively new fire blocking gel, Barricade. The most commonly applied additive is class A foam, which is added to hose streams via a normal inboard or external eductor. Class A foam is mixed with water using a 0.3% and 1.0% mixture. It is very effective with lighter fuels, reduces knockdown problems, and helps prevent rekindles. Compressed-air class A foam systems utilize an air compressor to agitate the foam. They are called CAFS for compressed-air foam systems. More and more fire departments are using CAFS. When CAFS-generated foam is applied to a structure, it tends to stick to the structure for several hours and really reduces structure ignition and suppression problems. Barricade is a new product made from the same materials as that found in baby diapers. Barricade is probably the only additive that allows a structure to be gelled and then left without further guarding by firefighters. It is still not widely used in the fire service, but is rapidly becoming recognized for its fire suppression value. Barricade is also available to, and can be applied by, the homeowner with a garden hose.

Obstacles around the structure impede hose line operations. Many obstacles are additional fuels to feed the fire and may include buildings, firewood piles, shrubbery, yard furniture, and junk. All act as rub points on hose lines, creating additional friction and difficulty in advancing the hose lines. Stretching hose lines becomes an important issue for a three-person or smaller engine company: advancing only one hose line is often the best option as it allows the company officer to act as a safety lookout.

Should firefighters at a threatened structure need to retreat, there are simple guidelines for their departure:

- Beware of fire along the exit route.
- Proceed carefully in the smoke.
- Watch for falling trees or snags.
- Watch for debris on the road such as rocks or burned debris.
- Watch for downed power lines.
- Check for burned bridges or cattle guards.

The following are guidelines for preparing structures threatened by an advancing fire:

- Structure exterior—cover openings and windows exposed to fire using:
  - Plywood, scrap lumber
  - Outdoor tables
  - Tarps (not plastic)
- Structure roof
  - Remove needles
  - Cover roof vents and coolers
  - Remove combustible debris from rain gutters

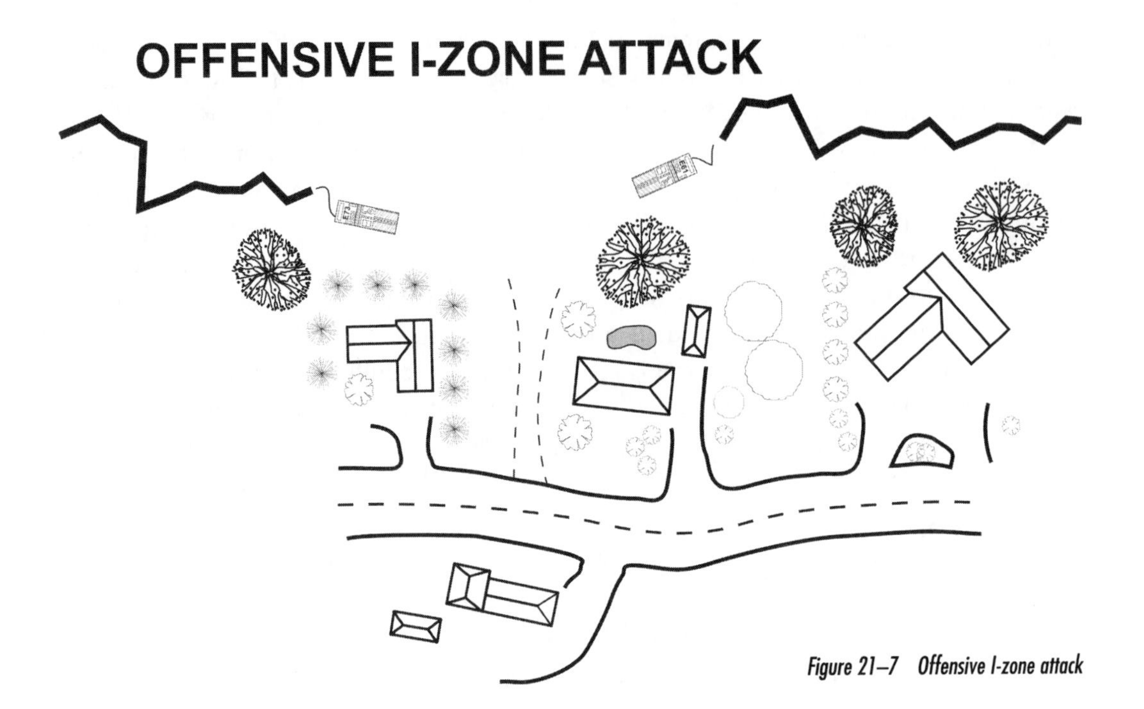

Figure 21–7 Offensive I-zone attack

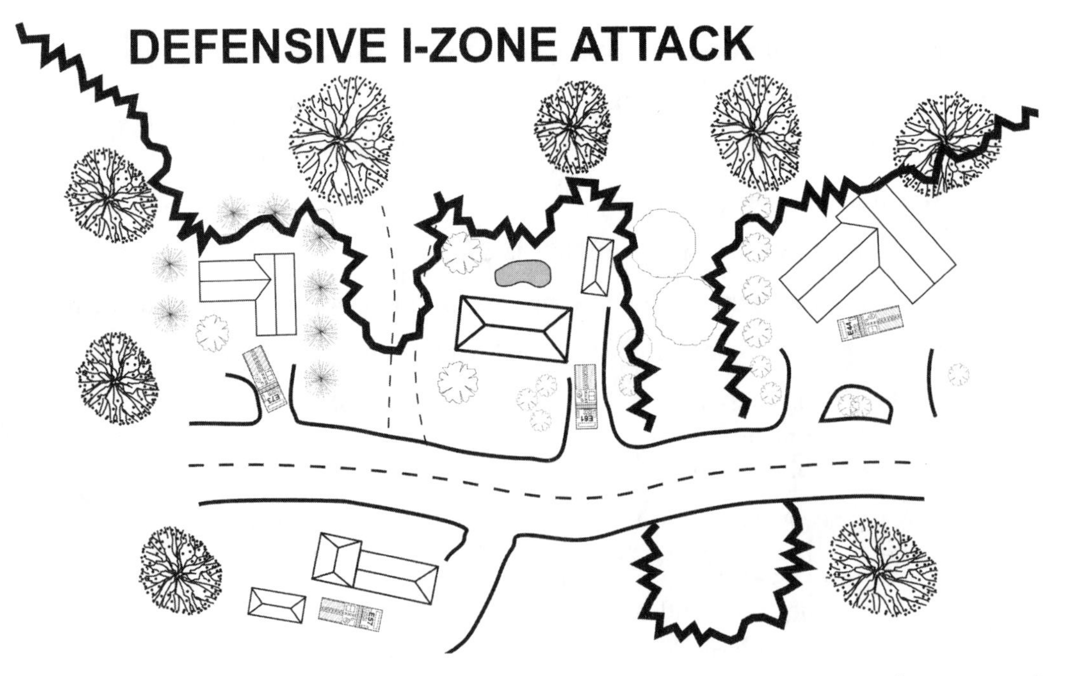

Figure 21–8 Defensive I-zone attack

- Utilities
  - Shut off LPG
  - Leave electricity on
  - Leave exterior light on
- Interior windows
  - Close all windows
  - Remove flammable drapes
  - Close non-flammable drapes (insulated drapes)
- Doors
  - Close interior doors (compartmentalize the structure)
  - Close, but do not lock, exterior doors
- Ventilation systems (HVAC)
  - Shutdown the system
  - Close vents and registers
  - Leave pump energized to keep swamp cooler pads wet
- Leave note for homeowner

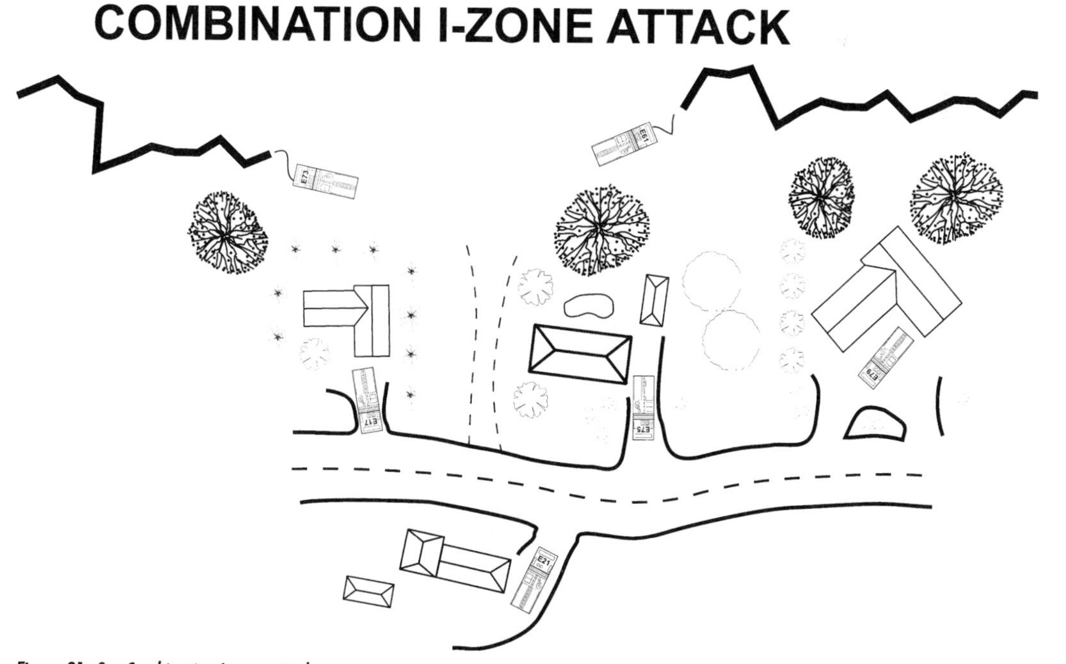

Figure 21–9 Combination I-zone attack

When making the critical decision to attack or retreat, the company officer may logically analyze the situation in the following possible ways:

- The engine drives up, the company officer assesses the situation from the engine's cab, determines that the situation is either too dangerous or there is no hope to save the structure, and drives past to a safe refuge.
- The engine drives up, backs in against the structure, the officer does an assessment (hot lap), decides that the situation is marginal, and directs the engine company to stay in the engine and ready for an apparatus retreat.
- The engine drives up, backs in against the structure, the officer does an assessment (hot lap), decides that there is a chance to save the structure and pulls hose lines, but does not advance hose lines beyond a safe point that acts as an effective heat shield.
- The engine drives up, backs in against the structure, the officer does an assessment (hot lap), decides that there is a chance to save the structure, pulls the hose lines and advances hose lines, safely attacks the fire, and saves structure.

The suppression actions of firefighters along with necessary defensible space and built-in fire protection design will save structures at I-zone fires. But, firefighters cannot risk their lives just to save property. Two very good safety thoughts are: (1) Risk a life to save a life, but risk no life to save property, and (2) Protect yourself, your company, and the public in that order. Fighting fire in the I-zone will be dangerous and very challenging. Firefighters will be tested to their extreme, but using good judgment along with practicing well-known and accepted firefighting practices will help ensure that firefighters survive and that structures are saved. Often, firefighters will save more than one structure at a fire. In fact, on November 2, 1993, at the Old Topanga Fire near Malibu, California, one engine strike team saved over 21 structures in a little less than six hours. Their actions were outstanding during a very severe wind-driven fire that burned more than 18,000 acres and in excess of 300 structures.

AIR ATTACK

Aircraft generalities

Wildland firefighting aircraft are termed either fixed (planes with wings) or rotary-wing (helicopters). Examples of fixed-wing aircraft used in firefighting include planes used to transport the air tactics group supervisor who directs all airborne firefighting operations, air tankers that drop retardant or slurry on fires, personnel transport airplanes, and general reconnaissance airplanes.

Examples of rotary-wing aircraft include copters (helicopters) that are used to drop water or water-foam mixtures on fires, copters used for reconnaissance, copters that transport personnel or equipment to and from remote fire locations, and those used for medevac (medical evacuation copters). These aircraft drop either plain water or a mixture of class A foam and water.

Fixed-wing aircraft that drop retardant on fires are called air tankers. The retardant they drop may be a slurry solution of a fire-suppressing chemical (that is also a fertilizer), a gum thickener, a fugitive dye and water, or possibly a mixture of class A foam and water. Fire retardant is no longer called borate, as it was termed in the 1960s when a sterilizing chemical was used as an aerial retardant.

Each fire has a different set of circumstances that can only be evaluated effectively by the fire officers in command. The incident commander and the ATGS must work as a team to make the most effective, economical, and integrated use of fire-fighting aircraft and ground resources. The ATGS and incident commander continually evaluate the need for additional air support, whether by air tanker or helicopter drop. Generally, aircraft will be diverted from large fires to new, small fires to ensure that a new fire does not become another large and damaging fire.

Wildland fire agencies certify both an aircraft and its pilot for particular missions. The pilot must have a certification "card" in his or her possession at all times of employment. There must also be a card within the aircraft for that particular unit. Possible certifications include: personnel transport, air reconnaissance, personnel transport, water drops, air tanker drops, external loads (copters only), short-haul rescues, rappelling, etc.

Fixed-wing aircraft

Air tankers are used for two purposes. They are most successful at holding small fires in check until ground fire-fighting resources can arrive and contain the fire. They are also used for containing small fires and creating fire lines using fire retardant. They either drop their retardant along the fire edge to try and knock out the heat and flames, or drop it ahead of the fire to pretreat wildland fuels to resist ignition. Air tankers generally do not extinguish wildland fires, but rather help hold the fire until ground resources can contain and control it. For safety reasons, air tankers must drop no lower than 150 feet above the highest obstacle.

An air tanker is under the command of the pilot, and in some cases, assisted by a co-pilot. The pilot has the final say on aircraft safety and mission approval. The pilot receives his tactical assignments from the ATGS unless an air tanker coordinator (commonly called a lead plane that leads air tanker pilots through a drop run) is working for the ATGS and directing air tanker supervision. The air tanker coordinator is assigned when visibility or air space congestion develops at a fire. The incident commander provides the overall incident objectives to the ATGS, who then tactically decides how many air tankers are needed at the particular fire. On the average, an ATGS can safely direct about 10 to 12 air tanker drops per hour. Remember, many air tankers are capable of making more than one drop before reloading. Knowing this rule-of-thumb fact and also the turnaround time for an air tanker to arrive at the fire, drop his load of fire retardant, and return to the air attack base to reload with retardant, the ATGS determines how many aircraft can be worked at a maximum-effort fire. If the turnaround time is 30 minutes, the air tanker can make two trips per hour. That means that the ATGS will need six air tankers for an expanding fire.

Initial attack air tanker drops at a fire will often create an inverted V-pattern across the head of the fire. This will slow the forward spread of the fire until ground resources can effect a direct or pincer attack on the fire. In some cases, the air tanker might drop around a structure to buy time and allow fire engines time to protect the structure from the encroaching fire. Figure 21–10 illustrates typical air tanker drops at the head of a fire. The goal is to either stop the head of the fire or protect the exposure(s).

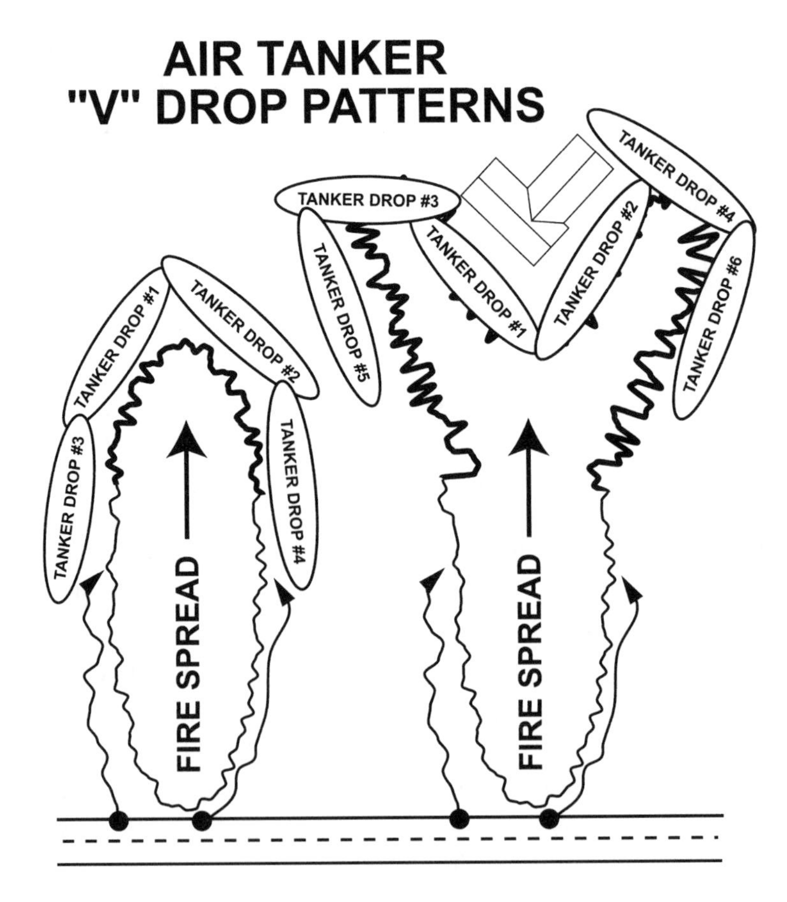

Figure 21–10 "V" drop patterns

Even air tankers have a difficult time stopping a mid-slope fire. Sometimes mid-slope attacks work, but that is usually only during very moderate burning conditions. Often, airdrops will hold the head or pinch the fire flanks and attempt to herd the fire toward a barrier. Success holding a fire at a mid-slope position should be expected only if the frequency of drops can dominate a rate of fire spread that could cause the fire to outflank the retardant line.

A typical air tactical assignment might proceed as follows:

1. The incident commander has a need for an airdrop on a portion of the fire that cannot be reached by fire engines. Tthe ATGS is called on the air-to-ground tactical radio frequency and describes what the air tanker is to accomplish, *i.e.*, "Have one air tanker drop on the right flank."
2. The ATGS acknowledges and directs the type of drop to the air tanker.
3. After the drop, the ATGS evaluates the effectiveness of the drop and converses with the incident commander about additional needs.
4. Fire crews or engine companies will be immediately sent to the right flank to work the fire edge, which, it is hoped, is burning with less intensity following the retardant application.

As with all firefighting resources, air tankers are capability typed under the incident command system. Table 21–3 lists the ICS typing configuration for air tankers.

Table 21–3 ICS Air Tanker Typing Chart

Resource	Components	Minimum Standards for Type 1	2	3	4
	Minimum tank capacity (gal)	3,000	1,800	600	100
	Examples	C-130 Hercules P-3 Orion DC-7	DC-4 SP2H P2V Neptune	S-2	Thrush

At many large fires, the most logical strategy with airdrops is to pretreat a ridge or bench where both retardant material and the natural decrease in fire intensity would provide the most beneficial result in holding the fire.

The effectiveness of air tankers increases as the following conditions are approached:

- The fire or fires are still small fires.
- Grass or light brush fuel predominates.
- Wind impact decreases its influence on the fire.

- Topography becomes less steep.
- The time of fire start occurs after mid-afternoon.
- The distance to the fire from the air attack base decreases within the 20-minute maximum ideal first-attack striking limit.

Finally, all retardant drops must be supported by ground efforts in order to support the firefighting actions. Fires do burn through drops, and therefore, drops must never be considered the final action until fire crews or engine companies have secured the fire. During critical backfiring operations, an orbiting air tanker may be desirable to provide immediate action on spot fires. Finally, air tankers can, normally, be used most successfully when a series of separate fires occur almost simultaneously in the same general area.

Aircraft limitations

Certain conditions can seriously limit the use of air tankers. Incident commanders should recognize that:

- Steep topography seriously reduces air tanker effectiveness. Deep canyons may rule out use entirely for certain fire targets.
- Winds in excess of 20 mph sharply reduce air tanker effectiveness. Shifting and high-velocity winds (> 30 mph) and turbulent air may restrict or exclude air tanker use.
- During early morning and late afternoon, air tankers may be less effective. Deep shadows cast by the sun over certain aspects of topography make it difficult for pilots to see fire targets or ground obstructions.
- Dense smoke may make air tanker operations both hazardous and ineffective on part or all of the fire area.
- Air tankers cannot be used at night when a fire is normally expected to become less active.
- Tall, dense timber and snags may require air tankers to make drops at higher-than-desirable altitudes and may intercept most of the retardant before it reaches the fire.

Older air tankers generally drop fire retardant in three patterns or configurations. They have from 1 to 12 doors that the pilot can configure to drop all at once, in sequence, or one at a time. The following is a description of these three patterns:

- *Salvo*–Total load at one time and place. All tanks are open.
- *Trail*–Overlapping series of from two to eight tanks in tandem.
- *Split*–Single drop from one-tank-at-a-time at widely-spread intervals or two to eight times on the same place or on separate small fires.

Newer air tankers have the latest tank configuration, which provides for constant-flow doors. Based on the fire's intensity, the pilot can select the desired retardant coverage level and the doors on the tank will open according to the selection made by the pilot. After the air tanker passes over the drop zone, the doors are closed and another pass can be made. Air tankers with constant-flow tanks deliver more consistent drops at a lower cost per gallon of retardant.

Often, airdrops are made on a fire before ground forces arrive. Such a "delaying tactic" pays big dividends when the drops are successful in holding the fire to initial attack size.

Alertness at the fire helps ground firefighters avoid possible serious injuries from the force of the falling retardant. While aircraft do fly at sufficient altitudes to permit the falling retardant to disperse in the form of a rain or heavy mist, occasionally the payload does not break up, and personnel have been hurt and equipment damaged. When the retardant hits in a mass, it is capable of tearing brush out of the ground, kicking rocks into the air, or knocking personnel off their feet.

Sometimes fire conditions along the fire line become untenable or pose a potential threat to firefighters. In such cases, it is not uncommon for fire officers to ask for a protective retardant drop. The problem with such a request is that, if asked, any ATGS would probably say that the very conditions threatening the firefighters will likewise obscure the pilot's visibility. The ATGS will probably not be able to find firefighters on the ground asking for assistance. The aircraft may be unable to locate the firefighters in time to help them. Even with the brightly-colored, fire-resistant clothing worn by ground firefighters, it may be difficult to pinpoint the exact area of need and to establish a proper drop pattern because of dense smoke and topographic hazards. Therefore, the dependence on aircraft to provide safety relief from a hazardous situation should not take the place of implementing LCES. Air tanker movement adjacent to the fire line can fan the fire, setting a dormant section ablaze, or can force a change in spread direction. Both contingencies are detrimental to firefighting efforts to fight the fire and accentuate the relative immobility of ground personnel.

The more heavily loaded the aircraft and the lower and slower it flies, the stronger the vortex turbulence will be and the more likely the vortex action will reach the ground. The vortex may strike the ground with velocities up to 25 mph, sufficient to cause sudden and violent changes in fire behavior on calm days in patchy fuels. On the other hand, wind gustiness and surrounding high vegetation tend to break up or diminish vortex intensity.

Firefighters should be alert in air tanker drop zones when:

- The air is still and calm
- The fire is burning in open brush or scattered timber
- The air tanker is large or heavily loaded
- The air tanker is flying low and slow

Rotary wing aircraft

One of the principal advantages of a helicopter is its ability to operate from locations near and on the fire line. Its vertical takeoff and landing characteristics make it a valuable piece of equipment for close support action of ground-based firefighting operations.

Copters perform many functions at wildland fires. They transport personnel and equipment, provide an aerial platform to recon the fire, provide medical evacuation of injured firefighters, and drop water or foam-water mixtures on the fire.

Water-dropping copters have two types of water containers. The first is a fixed tank under the belly of the copter, much the same as that affixed to an air tanker. Since the copter does not return to an air tanker base to refill with water, the fixed tank must either be filled by landing at a helibase and reloading via an attached fire hose or by dipping from a static water source and using the attached snorkel and pump to draft water. Many copters today use a snorkel-pump configuration to take on water. The other current system for copter drops involves using a bucket suspended under the copter by cables. Both the fixed tank and the bucket have the ability to inject class A foam concentrate into the water. Flying to the fire from the water source agitates the foam-water mixture.

Copters support air tanker drops by working difficult-to-reach hot spots and dropping between air tanker drops. The ATGS will direct water-dropping copters to hold out of the air-tanker-drop zone until the tanker clears the area. Then, the ATGS will direct the copter to hit a hot spot with a load of water or water-foam mixture. When more than about three copters are assigned to a fire, the ATGS will often commission a copter to perform as the copter coordinator. That position then directs and coordinates all incident-assigned copters under the immediate supervision of the ATGS.

Because of their ability to hover, copters can be precisely directed via radio communication with ground personnel. Some helicopters carry a collapsible water tank that can be strategically placed on the fire line. A portable pump is placed in the tank, which can be filled by helicopters using water drops. Firefighters then are afforded some of the benefits of having stored water and a pump at locations to which fire engines cannot drive. The portable pump includes a complement of fire hose, adapters, nozzles, and hose clamps.

At major fires, copters principally work from a main base termed a helibase. Helibases should be accessible to vehicles for copter fuel and equipment service. Helispots are locations temporarily on or near the fire perimeter for the purpose of delivering or returning personnel and equipment. Helispots can be constructed at meadows, on mountain peaks, and on ridge tops. During some long duration fires, helispots may be used to stockpile sleeping bags, lunches, drinking water, back pumps, and extra tools. Fire crews are flown to helispots to work a shift, then perhaps return to the helispots for supplies and to spend the next shift sleeping in a designated area adjacent to the helispots. A miniature ICS camp is thus created that can be advantageous in reducing the number of ferrying trips to and from the fire line. The procedure also allows for a longer period of fire line construction because copter ferry times are not involved. For easy identification by personnel, helispots are identified by name or number on maps. Pilots approaching helispots are able to confirm identification from the air because the helispots are marked with a large plastic sign or by some other means.

Table 21–4 ICS Helicopter Typing Chart

		Minimum Standards for Type			
Resource	Components	1	2	3	4
	Seats, including pilot (minimum)	16	10	5	3
	Card weight capacity (lb)	5000	2500	1200	600
	Tank, gals. of retardant (minimum)	700	300	100	75
	Examples	Bell 214 Vertol 134 Chinook Firehawk Skycrane	Bell 204 Bell 205 Bell 212 Sikorsky S58T	Bell 206 Hughes 500	Bell 47

Helicopters are useful for transporting injured firefighters, including burn victims, who are flown to pre-designated hospital burn wards. Helicopters have often served dramatically to evacuate persons stranded or threatened by fire. These medical rescues are called *medevacs.*

Most agencies that have and coordinate helicopters require certain documented information for each load of personnel and equipment. Written manifest forms are routinely completed prior to each flight assignment. Operationally, for accountability reasons, it is desirable to keep track of a crew once it has been dropped on the fire line at a helispot. The fire crew scheduled to return to the pickup point may end up in quite a different location. There are several reasons why this might happen, including reassignment to another section of fire line or the necessity (for various reasons) to walk the crew out rather than fly, or easier access to another helispot, etc. As is the case in so many fire operations, good communication can help prevent lost personnel and equipment, even on remote sections of fire line.

All fire control personnel are subject to work with or around helicopters. Personnel should have at least a general familiarity with the type of work that can be accomplished as well as the limitations. All personnel who may be working near, or transported by, helicopters should be trained in safety practices. They must observe such precautions at all times. Copter safety includes the following rules:

- Keep the helibase or helispot free of debris
- Do not smoke within 100 ft. of copter landing area

- Stay back at least 100 ft. until directed to approach the copter
- Stay well clear of main and rear rotors
- Stay forward of the rear door
- Approach or leave a copter in a crouched position
- Approach or leave a copter from low ground
- Approach or leave a copter in plain view of the pilot
- Carry tools in a horizontal position
- Wear the required PPE (personal protective equipment)
- Either hold or wear a helmet as directed by the copter crew chief
- Fasten your seat belt in the copter
- Do not throw anything from the copter
- Do not slam copter doors
- Do not touch any controls in the copter
- Do not distract the pilot
- Listen closely to the pre-briefing given by the copter crew chief

Firefighter Fire Safety

Firefighter safety must be the "number one" priority for the fire service. Unfortunately, annual reports do not show the dramatic reduction in injuries and fatalities that should be taking place across the country. Certainly the fire service has improved training standards, PPE, equipment, and incident scene safety requirements. Nonetheless, the wildland fire service has experienced several years of being at the top of the list of line-of-duty firefighter deaths. Like public safety organizations everywhere, wildland fire organizations subscribe to the "law of catastrophic reform." In other words, whenever a firefighter dies in the line of duty, we make changes and develop checklists or training standards to hopefully prevent further injuries and fatalities.

What follows are some of the common-sense lists frequently ignored and reported as violated in after-action reports. Some have been modified over many years, while others are fairly recent, but be assured they all were developed after firefighters have died in the line of duty. In addition, there is a description of the minimum standard for wildland PPE and personal safety equipment that every fire officer and firefighter must have before any suppression action is taken.

As stated before LCES stands for lookouts, communications, escape routes, and safety zones. This system for operational safety originated with Fire Management Officer Paul Gleason, who witnessed six fatalities in the Dude Fire of 1990. He recognized that multiple checklists had become a source of confusion during wildland safety training for all federal firefighters. In an effort to make these checklists usable and easily understood by everyone, he developed LCES. This is a "system" and only works when all four elements are in place.

Elements of LCES

1. Lookouts:
    a. must be experienced.
    b. must be able to see the fire (hazard) and the firefighters (risk) and be able to recognize risk to firefighters.
2. Communications between lookouts and firefighters must be effective.
3. Escape routes:
    a. work best with at least two routes, which is difficult on some wildland fires.
    b. must lead to safety zones.
    c. are most elusive for fire crews.
4. Safety zones are locations where fire shelters are not required.

When these four operational safety practices are in place, all of what follows is covered. LCES information should be a part of every IAP on major incidents and posted for everyone to read. The application of LCES has its place at every incident and in every type of emergency to which the fire service responds on a daily basis. For example, consider the ventilation of a roof on a supermarket. The *lookout* observes for any signs of weakening and *communicates* with the firefighters. *Escape* routes are identified to the nearest wall, then to the corner and down the ladder. The *safety* zone is the engine company.

Common denominators

Carl C. Wilson and James C. Sorenson, researchers for the USFS (U.S. Forest Service), studied wildland fire fatalities and near-miss incidents. They found the following common denominators on 67 federal and state fires where one or more fatalities or near-misses were present:

- Most incidents happen on small fires or on isolated sections of large fires.
- Flare-ups generally occur in deceptively light fuels, such as grass, herbs, and light brush.
- Most fires are innocent in appearance before unexpected shifts in wind direction and/or speed result in flare-ups. In some cases, tragedies occur during the mop-up stage.
- Fires respond to large- and small-scale topographical conditions, running uphill surprisingly fast in chimneys, gullies, and on steep slopes.
- Some suppression tools, such as helicopters and air tankers, can adversely affect fire behavior. The blasts of air from low flying aircraft have been known to cause flare-ups.

Firefighters must constantly be vigilant before taking suppression action when one or more of these common denominators is present. Some are very obvious, but others, such as chimneys, can be hidden by fuel or so slight that they are not readily observed.

Ten Standard Fire Orders. These standard fire orders started with the Mann Gulch Fire of August 5, 1949, which resulted in 13 fatalities. The USFS commissioned a task force to analyze why the firefighters died and to develop some general orders. The task force analyzed four other major fires in the 50s before developing the *Ten Standard Fire Orders* using the U.S. Marine Corp's general orders. The original orders follow and are subdivided by impacts, fire behavior, safety, communications, and suppression.

Fire behavior

1. Keep informed on fire weather conditions and forecasts.
2. Know what your fire is doing at all times. Observe personally and use scouts.
3. Base all action on current and expected behavior of the fire.

Safety

4. Have escape routes and make them known.
5. Post a lookout when there is possible danger.
6. Stay alert. Keep calm. Think clearly. Act decisively.

Communications

7. Give clear instructions and be sure they are understood.
8. Maintain prompt communications with your firefighters, your supervisor, and adjoining forces.
9. Maintain control of your forces at all times.

Suppression

10. Fight the fire aggressively, having provided for safety first.

Fire orders

In the mid 80s, the *Ten Standard Fire Orders* were revised to enable better memorization. They were renamed *Fire Orders* and organized so that the first letter of each order follows the letters, f-i-r-e-o-r-d-e-r-s.

Fight fire aggressively, but provide for safety first

Initiate all action on current and expected fire behavior

Recognize current weather conditions and obtain forecasts

Ensure instructions are given and understood

Obtain current information on fire status

Remain in communication with crew members, your supervisor, and adjoining forces

Determine safety zones and escape routes

Establish lookouts in potentially hazardous situations

Remain in control at all times

Stay alert, keep calm, think clearly, act decisively

Situations that Shout Watch Out! The USFS developed the original *13 Situations that Shout Watch Out!* after the *Ten Standard Fire Orders.* The original 13 have now been expanded to 18 situations. Both the orders and 18 situations are referred to in almost every after-action report in which firefighters have died or a near-miss incident has been investigated.

The *18 Situations that Shout Watch Out!* are:

1. Fire in the country not seen in daylight
2. Fire not scouted or sized-up
3. Safety zones and escape routes not identified
4. Unfamiliar with weather and local factors influencing fire behavior
5. Uninformed on strategy, tactics, and hazards
6. Instructions and assignments not clear
7. No communications link with crew members or supervisor
8. Constructing a line without a safe anchor point
9. Building fire line downhill with fire below you
10. Attempting a frontal assault on the fire

11. Unburned fuel between you and the fire
12. Cannot see main fire and not in contact with anyone who can
13. On a hillside where rolling fire can ignite fuel below you
14. Weather getting hotter and drier
15. Wind increases or changes direction
16. Getting frequent spot fires across line
17. Terrain and fuels make escape routes to safety zones difficult
18. You feel like taking a nap near the fire line

Downhill and Indirect Fireline Guidelines

The *Downhill and Indirect Fireline Guidelines* were developed in 1967 by the USFS after yet another series of fatalities in the November 1, 1966 Loop Fire in the Angeles National Forest. These guidelines are obviously directed toward areas in which there is considerable change in elevation, when firefighters may be working downhill or using an indirect attack on the fire. Firefighters must recognize the danger when there is fire below their position on a slope. Just as firefighters would not attack a third story fire from the fourth floor without considering all available options, wildland firefighters should use extreme caution when there is fire below them.

Direct attack methods should be used whenever possible, and the following guidelines should be followed:

1. The decision for direct attack is made by a competent firefighter after thorough scouting.
2. Downhill/indirect line construction in steep terrain and fast burning fuels should be done with extreme caution.
3. Downhill line construction should not be attempted when fire is present directly below the proposed starting point.
4. The fire line should not be in or adjacent to a chimney or chute that could burn out while a crew is in the vicinity.
5. Communication is established between the crew working downhill and the crews working toward them from below. When neither crew can adequately observe the fire, communications will be established between the crews, supervising overhead, and a lookout posted where the fire's behavior can be continuously observed.

6. The crew will be able to rapidly reach a zone of safety from any point along the line if the fire unexpectedly crosses below them.
7. A downhill line should be securely anchored at the top.
8. Avoid under-slung line if at all practical.
9. Line firing should be done as the line progresses, beginning from the anchor point at the top. The burned out area provides a continuous safety zone for the crew and reduces the likelihood of a fire crossing the line.
10. Be aware of and avoid the *watch out situations!*
11. Full compliance with *the standard fire orders* is assured.

In all the "lists" referred to in nearly all after-action and investigation reports of firefighter fatalities or near misses, there is no mention of the two most important items. These two items are training and personal protective equipment (PPE). Training and PPE are referred to in all investigative reports. Every firefighter's goal should be to prevent the law of catastrophic reform leading to the development of another checklist.

General safety and PPE

No writing on wildland firefighting safety would be complete without mentioning PPE and fire shelters. Just as a firefighter would not consider entering a burning building without wearing proper turnouts and an SCBA, no firefighter should consider attacking a wildland fire without wearing approved wildland protective clothing and a fire shelter. To expect one set of protective clothing to be suitable for both missions is no more acceptable than asking a firefighter in turnouts to handle a hazardous materials incident.

At a minimum, PPE for everyone on a wildland fire should include protective pants over uniform pants, long-sleeved shirt with long-sleeved cotton T-shirt, protective over-jacket, lace-up boots, helmet with a shroud for neck and face protection, gloves, and goggles for eye protection. All of this clothing must be approved for fighting wildland firefighting. There are several manufacturers that meet these standards and sell various styles of PPE. Additional equipment is available for respiratory protection with new advances being tested yearly.

The one item that cannot be overlooked is a fire shelter assigned to every firefighter. Fire shelters have and will continue to save firefighters' lives. Every fire officer and firefighter must be trained in the use of, and carry along with them, a fire shelter. It is recommended that extra shelters be carried in vehicles so they can be placed against the vehicle's glass to block radiant heat if firefighters have to use a completely enclosed cab for protection. Information and training are available from most wildland firefighting agencies. Drills should be conducted at least annually deploying shelters in a prone position and in a cleared area.

References

[1] Layman, Lloyd. *Fire Fighting Tactics.* NFPA, Boston, MA. 1953

22

EMERGENCY MEDICAL SERVICES

Judy Janing, Ph.D., R.N., EMT-P
Gordon M. Sachs, MPA, EFO

CHAPTER HIGHLIGHTS

- The current definition of an emergency medical services (EMS) system is a comprehensive, coordinated arrangement of resources and functions which are organized to respond in a timely, staged manner to targeted medical emergencies, regardless of their cause and the patient's ability to pay, and to minimize their physical and emotional impact.
- Fire departments are typically involved in the first seven stages of EMS. These stages are: prevention, detection, notification, dispatch, pre-arrival (instructions en route), on-scene care, and transport and facility notification.
- The non-clinical aspects of EMS are human resource management, management of non-human resources, customer service, interactions with external agencies, training and certification or licensure, information management, and an understanding of current issues affecting EMS.
- The six steps of the EMS Leadership and Management Process are to adopt a vision, analyze the system, establish a plan, direct implementation, monitor effectiveness, and revise appropriately.

INTRODUCTION

An emergency medical services (EMS) system is essentially a comprehensive, coordinated arrangement of health and safety resources that serves to provide timely and effective care to victims of sudden illness and injury. As an integral part of this system, pre-hospital EMS plays a critical role in the effectiveness, efficiency, safety, and quality of the total system. Every fire department is a key component of their EMS system, whether or not they provide ambulance transportation, or are even routinely dispatched on EMS calls.

The role in EMS is continually expanding for the American fire service. As a result, many fire departments are expanding their services to meet the needs of the citizens, or customers, they serve. Many departments are even changing their names to "fire and EMS department" or "emergency services." This is a sign of the future of the American fire service.

Fire departments play a critical role in providing EMS services to their communities

History and Background

In many ways, the growth of EMS across the nation is the result of public demand. Prior to the 1960s, the victim of a medical emergency did not receive much medical assistance, other than transportation to the hospital. Often staffed by only a driver, ambulance services until that time offered little in the way of lifesaving care.

The first recorded ambulances—carts used to remove wounded soldiers from the battlefield—date back to the Crusades. The first pre-hospital care was provided during the Napoleonic Wars in the late eighteenth and early nineteenth centuries, when a vehicle staffed with a medical attendant went directly onto the battlefield to remove the wounded. The attendant would treat the wounded soldiers then transport them to a field hospital where surgeons awaited.

A battlefield ambulance from the Civil War.
(Photo courtesy Library of Congress,
Prints & Photographs Division)

America's first ambulance service was reportedly started by the U.S. Army in 1865. This was followed closely by city ambulance services in many parts of the country, typically established by hospitals. This type of system—hospitals providing a vehicle staffed with a nurse and/or intern, sent to treat and transport patients when necessary—continued in most urban areas of the country through the 1930s, while rural areas had no pre-hospital care system other than local doctors making "house calls." Some cities, or parts of cities, developed "city ambulance services" when services were not provided adequately by hospitals or when there were no hospitals nearby.

An ambulance from the 1940s.
(Photo courtesy Library of Congress, Prints & Photographs Division)

During World War II, the loss of hospital personnel due to the war effort caused many hospitals to turn their ambulance services over to volunteer groups and agencies capable of operating a motor vehicle for this type of service. Typically, this was either a fire department, police department, or funeral home. In some places, local citizens joined together to provide the service, forming independent ambulance corps or rescue squads. At this time, ambulance services were little more than "horizontal taxi" operations, as nurses, interns, and physicians were no longer available to staff the ambulances, and civilian medical training, if any, was rudimentary.

The evolution of EMS within the fire service is sketchy and there is no record of the first fire department to provide EMS as a routine service. Most fire departments began providing the service as a reaction to a crisis—absorbing the service during World War II, following a major disaster where EMS was not available in sufficient quantity, or when the quality or timeliness of EMS was in question. Like the hospital, the local fire department operated 24 hours each day for emergency response and was an obvious candidate as a "first on the scene" EMS provider. Often, the fire fighter had to rescue and administer aid to victims of a fire. It did not take long for many communities to recognize this potential source of assistance during all types of medical emergencies, and add ambulance services to the fire department functions.

During the 1960s, the medical community focused on the problem of heart attacks. Technical developments led to knowledge about electrical defibrillation and closed-chest pulmonary resuscitation. While these positive steps were reducing the death rate of patients who reached the hospital, any major reduction in overall mortality was still limited because many persons died at the scene or en route to the hospital. Consequently, an emphasis was placed on upgrading ambulance technicians nationwide to advanced first aid and cardiopulmonary resuscitation (CPR) training.

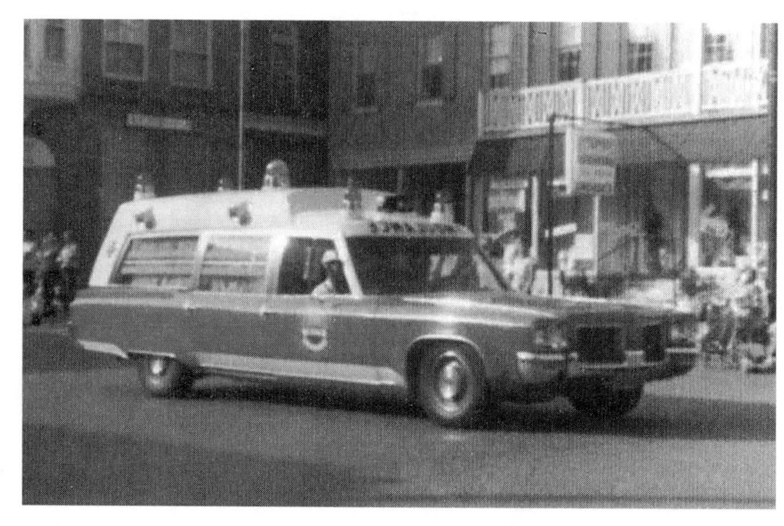

An ambulance from the 1960s. (Photo courtesy Fairfield, PA Fire & EMS)

As a result of increased attention to the medical community's goal of reducing the overall mortality rate through increased preparedness, emergency medical care began to receive added attention throughout the country. The long-range effects of quality emergency medical care on mortality were discussed in a 1966 report by the National Research Council's National Science Foundation, entitled "Accidental Death and Disability, the Neglected Disease of Modern Society." The report concluded that, of the mobile emergency teams studied, most had an average response time of more than 40 minutes, and had inadequately equipped and trained crews. This study, along with others, resulted in the development of pre-hospital EMS, although it would require nearly a decade for this reformation to take shape.

Primarily as a result of concern over highway safety, the Highway Safety Act of 1966 (Standard 11) addressed EMS issues by developing specifications covering ambulance attendant training, equipment requirements, and the design of the emergency vehicle itself. A series of national training courses were created and implemented in many communities across the country. Many fire departments, while not providing ambulance services, sought to meet these emergency medical training requirements and began responding to medical emergencies to provide "first response" medical care until an ambulance from another agency arrived.

The Emergency Medical Services Systems Act of 1973 designated federal funding for improved EMS across the nation through the development of regional EMS systems. Specific requirements outlined by this act—training and certification, interagency cooperation, equipment development, communications, and public education—are still recognized today as key elements of effective EMS. Fire departments provided a ready model for the development of regional EMS systems, as they offered an example of regionalized training and certification programs, and used communications technology and strategically located stations to reach anywhere in a community within minutes.

In the mid-1980s, much of the federal funding for EMS programs was reduced or eliminated. As a result, EMS systems became more dependent on state and local funding. Many differences in levels of service that persist today—the lack of universal 9-1-1 coverage, for example—can be traced to the decline in federal funding.

Since the mid-1960s, and particularly since the mid-1980s, the number of fire departments providing EMS as a significant part of their service has risen. Many departments already providing EMS have expanded the level of service they provide, or have started providing transport services in addition to first response. The demand for increased services may be one reason, as well as a decrease in the number of fires due to increased fire prevention and protection efforts. Another major reason is because the fire chief of today has come up through the ranks with knowledge and understanding of EMS and perhaps was even an EMS provider. This "hands-on" perspective can go a long way toward developing the relationships that are necessary for a fire department to be a successful part of an emergency medical services system.

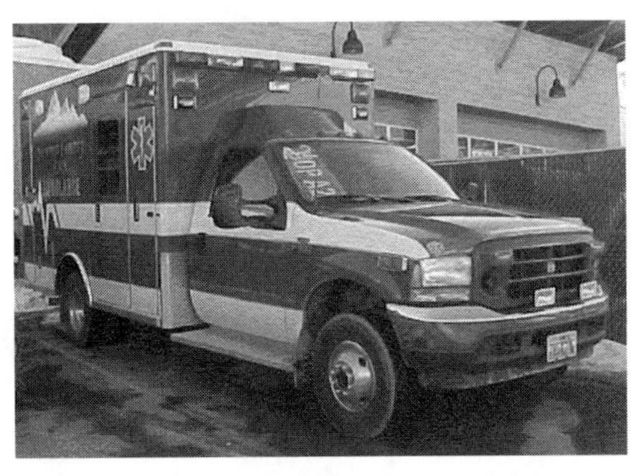

A 2002 "Type 1" ambulance. (Photo by Gordon M. Sachs)

The Emergency Medical Services System

Terminology is always important when dealing with a critical issue like EMS, as confusion or inaccurate perceptions can lead to organizational differences or disputes and ultimately to ineffective service or harm to patients. One often misinterpreted term is "EMS system." It is important that fire departments realize that pre-hospital EMS is only one small, yet important part, of an EMS system.

The Emergency Medical Services Systems Act of 1973 (Public Law 93-154) defined an EMS system and the components that such a system must have. The program regulations for this law referred to this definition of an EMS System:

> *[An EMS system] provides for the arrangement of personnel, facilities, and equipment for the effective and coordinated delivery of health care services in an appropriate geographical area under emergency conditions (occurring either as a result of the patient's condition or of natural disasters or similar situations) and which is administered by a public or nonprofit private entity which has the authority and the resources to provide effective administration of the system.*

This original definition of EMS system listed 15 component parts, which included:

- Manpower
- Training
- Communications
- Transportation
- Facilities
- Access to specialized care facilities
- Coordination with other emergency services
- Citizen involvement in policy-making
- Provision of services without regard to ability to pay
- Follow-up care and rehabilitation
- Standardized record keeping system
- Public information and education
- Review and evaluation
- Mass casualty and disaster plan
- Linkages with other agencies

In January 1993, the National Association of State EMS Directors (NASEMSD) and the National Association of EMS Physicians (NAEMSP) ratified a joint position statement that, in effect, revised the 20-year-old definition of EMS systems to be more applicable today. This new definition, widely recognized by national fire service and EMS organizations, states that an EMS system is:

> *A comprehensive, coordinated arrangement of resources and functions which are organized to respond in a timely, staged manner to targeted medical emergencies, regardless of their cause and the patient's ability to pay, and to minimize their physical and emotional impact.*

The NASEMSD/NAEMSP position paper lists the resources of an EMS system as currently including (but expanding beyond):

- Professional, occupational, and lay disciplines, such as pre-hospital EMS providers and other public safety personnel, and includes emergency medical and public safety dispatchers
- Facilities, agencies, and organizations, including fire departments, rescue squads, ambulance companies, law enforcement agencies, as well as hospitals, government agencies, and EMS professional organizations
- Equipment, such as ambulances and rescue vehicles, medical equipment and supplies, extrication devices, communications equipment, and personal protective equipment
- Funding, whether from various government sources, fees and other revenue sources, reimbursement mechanisms, or donations

Functions of a comprehensive EMS system include:

- System organization and management
- Medical direction
- Human resources and education
- Communications
- Transportation
- Definitive care (facilities)
- Quality assurance and improvement, evaluation, and data collection
- Public information and education
- Disaster medical services
- Research
- Care of special-needs patients

The resources and functions of an EMS system are coordinated through specific stages of EMS response to medical emergencies. These stages define the scope of the total system. According to the NASEMSD/NAEMSP position paper, the stages of EMS response are (in order):

1. Prevention (injuries and illnesses that don't happen)
2. Detection (realizing that a situation requiring medical attention has occurred)
3. Notification (calling EMS)
4. Dispatch (trained EMS personnel)
5. Pre-arrival (providing instructions on basic care to the caller while EMS personnel are en route)
6. On-scene (care provided by EMS personnel)
7. Transport and facility notification (by EMS personnel)
8. Emergency department/receiving facility (care provided at the emergency facility the patient is taken to)
9. Inter-facility transport (if the patient is taken to another facility after evaluation and/or stabilization)
10. Critical care (specialized care for seriously injured or ill patients)
11. In-patient care (care provided after being admitted to the hospital)
12. Rehabilitation (to return the patient to the appropriate "quality of life")
13. Follow-up (as required, based on the type of injury or illness)

These stages are applicable to all types of medical emergencies. Fire departments are typically involved in the first seven stages; many departments also provide inter-facility transportation.

It is clear that pre-hospital EMS is only part of the EMS system. The fire department, as a part of its regular functions, is involved in the EMS system from public education and prevention activities, through dispatch and on-scene activities, and often through transport. Similarly, hospitals may be involved in prevention, medical control, and

Fire and injury prevention programs are important parts of an EMS system, and should be an important part of all fire departments, not just those that provide EMS. (Photo by Lisa H. Sachs)

receiving, through critical care and rehabilitation. Because an EMS system is pluralistic, no single agency can take credit for managing and operating the entire EMS system.

The NASEMSD/NAEMSP position paper also addresses the concern that fire departments providing EMS often put a greater emphasis on fire fighting than on EMS, while the majority of their emergency response activity is EMS-related. There are similar concerns about specialty hospitals, health care agencies, and other related services. The position paper states:

> *When EMS, at any response stage, is provided by an agency or institution that also provides non-EMS services, the role and responsibilities of that agency or institution as a sub-component of the EMS system must not be jeopardized by its non-EMS role(s) and responsibilities. Quality patient care will depend upon total commitment to the development and operation of an integrated and comprehensive EMS system.*

If the fire service is to provide efficient, effective pre-hospital EMS, they must operate as part of the total EMS system. This may be an adjustment to some traditional viewpoints, as the fire service has often looked at itself as an untouchable island or monopoly in the area of providing public safety services. Many fire departments, however, have been successful at being an integral part of the overall "EMS team."

Dispatchers are an important component of an EMS system, as they are the first link between the patient and emergency medical care. (Photo by Gordon M. Sachs)

EMS in the Fire Service

Today, as many as 85% of the approximately 34,000 fire departments in the United States are routinely dispatched on emergency medical calls. Personnel in these departments are trained to the level of "First Responder," "Emergency Medical Technician (EMT) Basic," "EMT-Intermediate," or "Paramedic." Those trained at the First Responder and EMT-Basic levels provide basic life support (BLS) care, while those trained at the EMT-Intermediate and Paramedic level provide advanced life support (ALS) care. Modern technology has led to the availability of the semi-automatic defibrillators. Personnel at all levels can receive training and become certified to use this piece of life-saving equipment.

While emergency medical care and transportation are provided by many fire departments, many communities use a system where the fire department responds to EMS calls in an engine or ladder truck as a "first response" unit. Personnel on that unit may be trained at the BLS or ALS level, and provide care until an ambulance arrives. Further care and transportation to a medical facility may be provided by a private, for-profit ambulance company, separate municipal EMS agency, or hospital-based ambulance service. Whether fire service or not, EMS providers may be career or volunteer, and may respond in BLS ambulances, ALS ambulances, or aeromedical ambulances (such as helicopters), in addition to the first response unit.

New designs in apparatus for EMS response include ambulances that also have a small pump, water tank, and one or two 1½" hoselines, and full size pumpers with enclosed cabs that have an ambulance cot and EMS equipment for transporting patients when necessary. The future will probably see totally redesigned ambulances with side loading capabilities, where the cot is secured across the width of the ambulance, and the personnel ride facing only forward or backward (for safety). First response units in the future may range from specially equipped sedans to sport utility vehicles to other small response units.

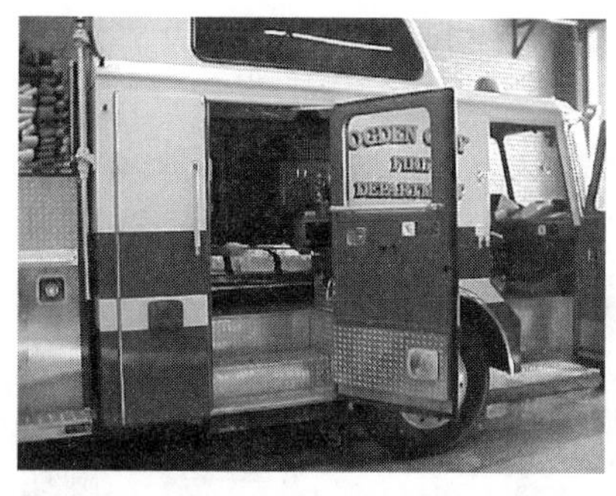

Many departments have experimented with transport-capable engines. Some have found it beneficial, others haven't. (Photo by Gordon M. Sachs)

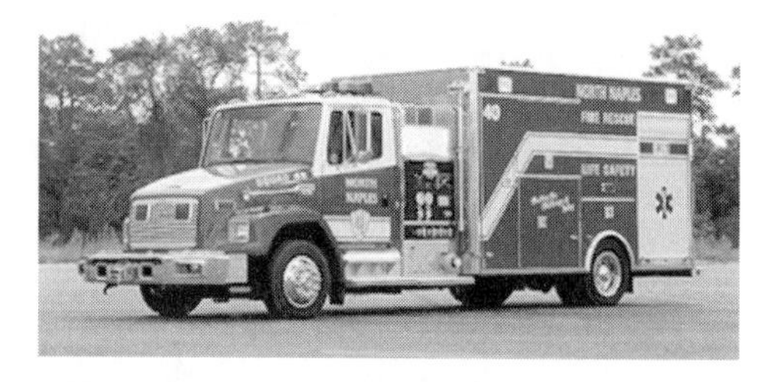

A more popular option is the use of a suppression-capable ambulance. Usually equipped with water, foam, a pump, one or two hoselines, and a crew of two, these versatile vehicles provide a valuable "first punch" on vehicle fires, brush fires, and smaller structure fires. (Photo courtesy Freightliner/Medic Master)

The need for clarity in terminology has not been more evident in EMS than when referring to "first responders." First responder systems are often confused with the level of EMS training known as "First Responder." These are often confused with the level of hazardous materials emergency response training.

Specifically, first responder systems refer to those pre-hospital EMS systems where the closest emergency response unit with EMS-trained personnel is dispatched along with a transporting EMS unit. The personnel may be trained at the first responder level; however, more often they are trained at the EMT, EMT-I, or paramedic level. The first response unit is usually a non-transporting unit—typically an engine company, but often a rescue or squad company or even a ladder company. Some fire departments use ambulances as first response units, but have arrangements for the transport of non-critical patients with a private ambulance company.

Many fire departments are putting one or more firefighter/paramedics along with advanced life support equipment on engines. This often reduces response times and increases the level of care provided. (Photo by Gordon M. Sachs)

The benefit of a first response system is the rapid arrival of EMS providers, without the financial burden on a department to have ambulance transportation capabilities at every station. Similarly, private EMS companies can increase their efficiency without jeopardizing patient care as well. It is up to the citizens to determine the level of service they want; however, the highest level of care provided at the scene must be continued while transporting. Thus, if ALS care is provided by first responders at the scene, an ALS provider must accompany the patient to the hospital. Most fire departments that provide first response at the ALS level also provide ALS transportation or work in a system that utilizes ALS transportation. Others operate under an agreement where the ALS firefighter accompanies the patient to the hospital in the ambulance; in this case, the department must determine whether the fire company has adequate staffing to remain in service, and must have arrangements for the return of the ALS firefighter to the station.

Many fire departments and non-fire EMS departments use "tiered" EMS response systems, which are related to first response systems. There are three primary types of tiered systems:

1. ALS ambulances respond to all calls, but can turn a patient over to a BLS unit for transport.
2. A BLS ambulance is dispatched on all calls, while an ALS ambulance is only dispatched on ALS calls.
3. A non-transporting ALS unit is dispatched with a transporting BLS unit.

These tiered systems work very effectively, whether or not the fire department provides the transport services, as they allow for a given number of ALS units to serve a large population, while maintaining a rapid response time.

Non-clinical Aspects of EMS

There are many aspects of EMS, related to EMS management and leadership, that have little to do with the provision of medical care. They are as follows:

- Human resource management
- Management of non-human resources (such as vehicles, equipment, and facilities)
- Customer service
- Interactions with external agencies (including the medical director–an integral part of a fire department's EMS activities)
- Training and certification or licensure
- Information management, as medico-legal aspects of record keeping must be addressed with confidentiality
- An understanding of current issues affecting EMS (such as health care reform and violence in the workplace)

While all of these are important, special emphasis must be given to three specific areas: public information, education, and relations; risk management; and quality management.

Public information, education, and relations

EMS public information, education, and relations (PIER) programs provide the means for promoting an EMS system, developing positive public attitudes, and informing citizens on specific EMS issues and techniques. Public information deals with providing facts, typically incident-related, to the public. Public education programs teach functional knowledge and/or skills, as a means to modify behavior. Public relations programs are designed to create an attitude or general impression, generally positive, rather than to convey specific information.

Letting the public know what you do and how you do it can be important for continued support and funding. Here, EMS equipment is displayed for a public event. (FEMA photo)

EMS PIER is vital to a successful EMS service. Public support, system abuse, public awareness, and education are all areas which a PIER campaign can address. Such a program is very similar to a public fire education program, but specifically addresses EMS issues such as injury prevention, calling for medical help appropriately, and citizen CPR. Together, public fire education and public EMS education programs are often referred to as "life safety education" or "fire and injury prevention."

A good example of a comprehensive EMS PIER campaign is "Make the Right Call–EMS" designed by the United States Fire Administration (USFA) and the National Highway Traffic Safety Administration. "Make the Right Call–EMS" provides information and materials to local fire and EMS departments to teach their citizens:

- What EMS is
- How to recognize a medical emergency
- When and when not to call EMS
- How to call for emergency help
- What to do until help arrives

The campaign includes posters, pamphlets, classroom presentation and workshop guidelines, and two videotapes—one for general audiences and one for children. (Campaign materials are available at no cost from USFA.)

Risk management

Risk management in any healthcare setting is not just about preventing monetary losses; it is about preventing disability, loss of life, and/or irreparable business damage as a result of the provision of patient care. Risk management involves direct "hands-on" patient care as well as various indirect aspects of patient care, including the development of effective training programs and the selection of qualified personnel. Regardless of its specific focus, however, the overall goal of risk management is to reduce the frequency and severity of preventable, adverse events that create losses.

Risk management includes the development and application of an occupational safety and health program designed to reduce the risks to EMS personnel in all aspects of their work environment. Measures must be implemented to make the EMS workplace as safe as possible—on physical, emotional, and mental terms. This is done through the identification of potential risks; the measurement of risks to determine their probability; the development of strategies to lessen risks; the implementation of these strategies; and the monitoring of these strategies and activities to ensure their effectiveness. An additional step that can be added or incorporated into this process is prioritization of risks, which is typically inserted between the risk identification and risk control techniques. Infection control and hazardous materials response are areas where such risk management measures have been mandated by law.

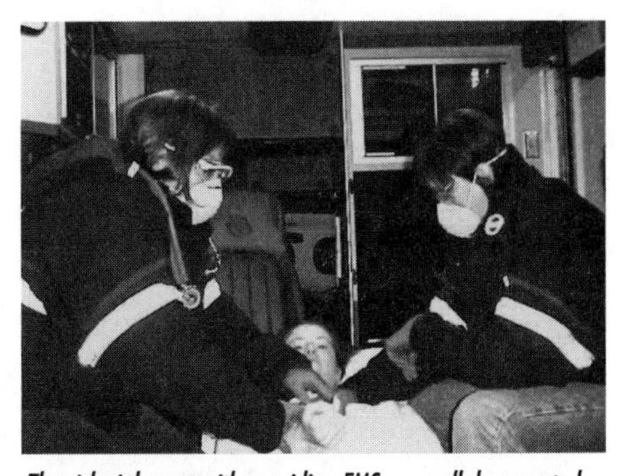

The risks inherent with providing EMS are well documented. A solid risk management program will address these risks through SOPs, training, and personal protective equipment. (Photo by Judy Janing)

For risk management efforts to be successful, they require support from personnel at all levels of the organization. Problem identification often requires considerable fact-finding and information-gathering activities. Those performing risk management functions must be seen as trustworthy in order for personnel to be forthcoming regarding risk management issues. To be successful, it is important that organizational members understand and feel that they are valuable contributors to the risk management process, not that the process is being forced upon them.

Quality management

Quality management differs from the traditional quality assurance program. Quality management is meeting the needs and expectations of the customer based on a quality strategic plan, objectives, and performance indicators. Quality assurance is carried out by post-incident review and audit to determine the compliance with performance indicators. Customers are both external and internal. In this case, external customers include patients and their families, physicians, nurses, taxpayers, visitors to the area, and suppliers; internal customers are the employees of the organization. Each customer has different needs and expectations regarding quality which must be balanced to provide the best result for the ultimate consumer—the patient. In this context, "quality" includes both the clinical quality of medical care (how well the care is provided, how the care affects the patient's outcome, and if the proper type of care is being provided) and the customer's perception of that care and the responders' satisfaction with the work environment. This extends the concept of quality beyond the traditional focus on clinical proficiency to include all aspects of care and work satisfaction.

The goal of every EMS system should be to continuously evaluate itself and constantly strive for performance improvement. In order to do this, it is first necessary to establish difficult but achievable goals for every behavior inherent in the system. Examples of such goals would include a response time of six minutes or less to 90% of all ALS calls and an on-scene time of no more than ten minutes on 95% of all trauma calls. The role of quality management is then to measure the system's actual performance against those goals. If data analysis reveals that an EMS system is meeting performance goals, new goals should be set which encourage higher levels of performance. If however, it appears that a system is continually falling below standards, the system should be analyzed to determine the cause(s) of the problem. Once the cause has been identified, a plan should be developed to correct the problem and monitored for effectiveness after it is implemented.

The benefits of adopting a quality management process are many, including improved patient care, employee morale, patient satisfaction and outcomes, and decreased cost. For the process to succeed, however, everyone in the EMS organization, particularly those at the top, must be involved and committed to performance improvement.

Fire chiefs often look at EMS as strictly the provision of emergency care, and overlook these other important aspects of EMS: public information, education, and relations; risk management; and quality management.Unfortunately, this can lead to a misrepresentation of the departmental needs related to EMS at budget time, in public forums, or during times of crisis. There are many aspects of EMS that are not clinical in nature, but should be fully integrated with other related department activities or programs.

Challenges and Issues Facing Fire Service EMS

There have always been challenges to fire service EMS, and the future holds more of these than ever before. To avoid being discouraged, however, it is important to point out that past challenges have been overcome by those with a vision of making EMS an integral part of the fire service.

The biggest challenge EMS in the fire service has faced has been acceptance. Almost a moot point now, there was a time when many firefighters and fire chiefs wanted nothing to do with EMS, as it was a new service and they had joined the fire department "to fight fires." Often, it was a lack of understanding about EMS, insecurity about learning EMS skills, or a similar attitude that was actually the cause of this negativity. Some areas still find a lack of acceptance of fire department EMS providers by medical personnel, such as physicians and nurses. Again, these attitudes may be due to job insecurity or a lack of understanding.

As may be expected, funding of fire service EMS has been, is, and probably always will be a challenge. In the past, EMS had to compete with fire suppression forces, fire prevention, and even equipment maintenance for funding. (Now, with EMS making up the majority of a fire department's emergency responses, a more complicated type of competition is being seen—from the private sector.) Financial concerns—costs, budgeting, and even revenue—are important EMS considerations. Many fire departments find EMS to be "keeping them in business," because the fire problem is decreasing and EMS responses are increasing; plus, they can bill for EMS service.

There are many other issues facing EMS today and an even larger laundry list will be facing fire department EMS in the future. Some of today's issues include:

- *Privatization.* Private companies are sometimes contracted to provide EMS for municipalities.
- *Legislation and standards.* Requirements, often in the name of responder safety, put fiscal and policy requirements on fire and EMS departments.
- *Health care reform.* Changes in the way health care is administered may affect the fire department's role in the EMS system.

- *Changes in regulations.* Changes relating to Medicare reimbursement for care administered, for example.
- *Expanded roles for EMS providers.* Many departments are now offering services beyond emergency response, including vaccinations, home healthcare, and episodic care.
- *The changing workforce.* Workplace diversity is a very important management issue in the emergency services.
- *Increased educational and training demands.* Requirements such as an expanded basic and paramedic primary training curriculum and continuing education to maintain certification and licensure increase demands on the department for personnel, time, and funding.
- *Accreditation.* All EMS departments may be required to be "accredited" by a national accreditation agency in the future. Fire departments must begin preparing now.
- *Overcoming tradition.* This is the "square wheel" which has kept the fire service from forging ahead in the delivery of services such as EMS.

Additional issues that will be facing fire service EMS in the future include:

- Review of response procedures as a result of the terrorist attacks of September 11, 2001
- Potential regulatory changes related to safety
- Improvements in technology affecting communications, equipment, and vehicles
- A demand for an increasing scope of practice for EMS providers, including providing primary care and public health services
- Bidding for services, with the potential for losing EMS transportation services
- A better educated, more savvy public (due, in part, to television), which will result in higher expectations
- A changing and aging population, changing public education needs and increasing responses
- The need to manage fire service EMS competently, like a business
- Increased research in EMS, necessitating a close working relationship with medical directors, hospitals, and research institutions
- Regionalization or consolidation of EMS services
- The need to market the organization and services provided in the same manner that a private business would

FIRE SERVICE EMS LEADERSHIP AND MANAGEMENT

Fire chiefs and other fire service EMS managers need to be ready to face the challenges and issues confronting fire service EMS today and in the future. They need the knowledge, skills, and abilities to move forward, to affect change, and to break traditional paradigms that have held the fire service back for so many years. These leaders and managers must ensure an efficient and dynamic EMS system, given current EMS requirements and future EMS trends. One way to do this is by following the "EMS Leadership and Management Process" as taught by the National Fire Academy.

The EMS Leadership and Management Process involves analyzing a problem or issue, setting appropriate goals, and establishing and maintaining an effective approach to the challenge or opportunity, while focusing on the department's vision for EMS. Ideally, EMS leaders and managers should cycle the entire process with virtually every issue and topic they confront. The six steps of the process are:

Step 1: Adopt a vision. Adopt a vision for EMS in order to develop an effective, viable EMS system consistent with that vision. Rationale: Leaders must possess a vision of the future of the constantly evolving field of EMS and must identify their organizational and personnel mission, consistent with that broader vision for EMS. *(Note: This step is different from specific goal setting that is done as part of step 3.)*

Step 2: Analyze the system. Analyze performance and characteristics of current EMS system. Rationale: Leaders and managers must understand the current system—its level of performance, functioning and local needs—as a baseline from which to effectively lead that system toward fulfillment of its mission.

Step 3: Establish a plan. Set and establish the plans and policies necessary to accomplish the vision. Rationale: Once the current system is evaluated and understood, plans and policies which lead that system from its current status toward fulfillment of its mission must be developed.

Step 4: Direct Implementation. Direct, motivate, and inspire the implementation of plans and policies required to achieve the vision. Rationale: In order to succeed, leaders and managers must ensure that organizational policies are embraced and plans are followed. Often, they must "sell" and/or justify policies and programs both inside and outside the organization. They must exercise the savvy and interpersonal skills necessary to achieve the vision and the judgment to delegate the operational and technical aspects of this step, as appropriate.

Step 5: Monitor effectiveness. Monitor the effectiveness of changes through sound analysis techniques. Rationale: Leaders and managers must understand the effects of their actions by systematically monitoring system performance. Effective documentation systems and procedures must be in place to ensure accurate data analysis and interpretation.

Step 6: Revise appropriately. Generate solutions to refine and improve the system and continue to monitor. Rationale: The EMS Leadership and Management Process is cyclical; leaders and managers must constantly develop, implement, and evaluate refinements. Effective leaders must continually modify their vision and develop new creative strategies and policies in light of future opportunities and challenges.

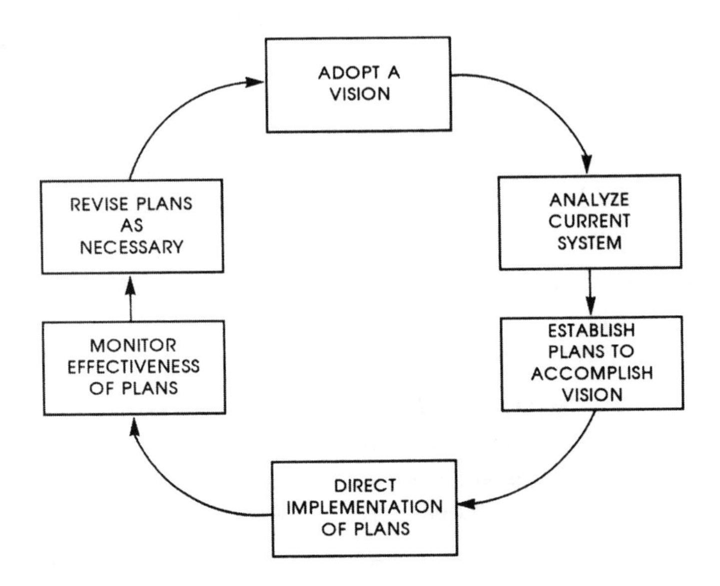

EMS Management Resources

There are many local, state, national, and federal resources that can help fire service EMS leaders and managers apply the EMS Leadership and Management Process and meet the challenges of the future. Rather than trying to reinvent the wheel, fire departments should look to other emergency services (even those outside the fire service) for ideas and advice. Regional EMS councils, public health agencies, health care entities, and hospitals (especially emergency departments) are also available to provide information and, possibly, other resources.

At the state level, many fire departments already have a relationship with the state fire marshal's office, state fire academy, and other fire-related organizations. These groups may have EMS information or EMS management-related courses available. State agencies that do have this type of information, however, are also available to fire departments. These agencies include the state EMS director and State EMS training coordinator. These state EMS offices are often in the state health department, but deal in particular with EMS system issues, including issues related to pre-hospital EMS.

At the national level, there are many fire- and EMS-related organizations. Most fire departments are familiar with the fire organizations, but often are not aware of the EMS activities of these organizations. For example, the International Association of Fire Chiefs (IAFC) has an EMS Section that deals with fire service EMS issues on a national level, and the National Volunteer Fire Council (NVFC) now represents the voice of both fire service and non-fire service volunteer EMS personnel across the country. While not specifically a fire-related organization, NAEMSP has, as members, many fire department medical directors. Another important national organization is the National EMS Alliance (NEMSA), an "organization of EMS organizations" with a mission to promote cooperative working relationships between and among EMS organizations. With the support of groups like the American College of Emergency Physicians, the American Ambulance Association, and NAEMSP, NEMSA will be a powerful EMS ally. Additionally, the fire service, through the IAFC, International Association of Fire Fighters, and National Fire Protection Association, is a part of a cooperative EMS network.

There are several federal agencies that are programmatically involved in EMS. Some agencies conduct research, others compile statistics, and still others develop or provide training materials or information to assist fire and EMS departments and providers across the country. The agencies most likely to have information useful to fire service EMS include: the U.S. Department of Agriculture and Rural Development Administration (low interest loans for rural EMS facilities and equipment); the U.S. Department of Transportation and National Highway Traffic Safety Administration/EMS Division (national standard EMS training curricula, EMS PIER information, and some grant programs); the Federal Emergency Management Agency/United States Fire Administration (EMS publications and information, the FIRE Grant program, and EMS management training); and the General Services Administration (specifications for ambulances). These agencies coordinate their activities through the Federal Interagency Committee on EMS, which is chaired by USFA.

One federal agency of special note is USFA's National Fire Academy (NFA), which has a comprehensive EMS management curriculum. EMS courses at NFA include: Management of EMS; Advanced Leadership Issues in EMS; EMS Special Operations; EMS Administration for Volunteers (web-based course); Incident Command System for EMS; two EMS hazardous materials courses; and health and safety courses related to EMS. There are other EMS-related courses at NFA taught outside the EMS Management curriculum as well, including an EMS-specific course in the Emergency Response to Terrorism curriculum.

Conclusion

EMS is a primary service provided by the majority of fire departments across the country. Within these departments, EMS typically accounts for 70% to 80% of the emergency activity. Life safety education programs are now encompassing both public fire and public EMS education, as a part of an overall PIER program. What used to be fire fighter safety programs are now department risk management programs and what used to be EMS quality assurance programs have become department-wide quality management programs. In other words, it is almost impossible to segregate EMS from the other functions of a fire department. Many departments are now called "fire and EMS" or "emergency services."

EMS has changed dramatically since the Emergency Medical Services Systems Act of 1973 was promulgated and it continues to change at a rapid rate. To keep abreast of these changes and remain as a viable EMS agency, fire departments must have a vision for the future of EMS, and must use the EMS Leadership and Management Process as they confront different issues and challenges. Fire departments must also tap into the many local, state, national, and federal resources that are available to them.

The pre-hospital component of an EMS system is recognized as "a public safety entity delivering a public health service." As always, the main purpose of EMS is patient care. However, if EMS in the fire service is not managed appropriately, it won't remain in the fire service for very long. Another more viable entity will be providing the patient care—and reaping the revenue and public relations benefits—from the citizens the fire department is sworn to serve.

The National Fire Academy, part of FEMA's U. S. Fire Administration, provides EMS management and leadership courses as well as other courses applicable to fire service EMS. (FEMA photo)

REFERENCES

Advanced Leadership Issues in Emergency Medical Services (Student Manual). Emmitsburg, Maryland: United States Fire Administration/National Fire Academy, 1999.

Braun, O., McCallion, R., and Fazackerley, J. "Characteristics of Midsized Urban EMS Systems." *Annals of Emergency Medicine,* May 1990, pp. 536-546.

Compton, Dennis. *When In Doubt, Lead! The Leader's Guide to Enhanced Employee Relations in the Fire Service.* Fire Protection Publications, Oklahoma State University, 1999.

Compton, Dennis. *When In Doubt, Lead...Part 2! The Leaders Guide to Personal and Organizational Development in the Fire Service.* Fire Protection Publications, Oklahoma State University, 2000.

Compton, Dennis. *When In Doubt, Lead...Part 3! The Leaders Guide to a Focused and Empowered Workforce.* Fire Protection Publications, Oklahoma State University, 2001.

Edwards, Steven T. *Fire Service Personnel Management.* Prentice-Hall, Inc., 2000, pp. 30-31.

Emergency Medical Services: 1990 and Beyond. Washington, DC: National Highway Traffic Safety Administration, 1990.

Emergency Medical Services Agenda for the Future. Washington, DC: National Highway Traffic Safety Administration, 1996.

Emergency Medical Services Agenda for the Future Implementation Guide. Washington, DC: National Highway Traffic Safety Administration, 1997.

Fire Service EMS Planning Guide. Washington, DC: United States Fire Administration, 1982.

Fire Service EMS Program Management Guide. Washington, DC: United States Fire Administration, 1981.

Fitch, J., Keller, R., Raynor, D. and Zalar, C. *EMS Management: Beyond the Street, 2nd ed.* Carlsbad, California: JEMS Communications, 1993.

Fitch, J. *Prehospital Care Administration.* JEMS Communications, 1995.

Goldbach, George. *The Fire Chief's Handbook.* "Chapter 6." Saddlebrook, New Jersey: Fire Engineering Books & Videos, 1995, pp. 225-228.

Implementation of EMS in the Fire Service. Washington, DC: U. S. Fire Administration, 1997.

"Joint Position Statement on Emergency Medical Services and Emergency Medical Services Systems." National Association of State EMS Directors and National Association of EMS Physicians, 1993.

Leadership Guide to Quality Improvement for Emergency Medical Services Systems. Washington, DC: National Highway Traffic Safety Administration, 1997.

"Make the Right Call–EMS" public EMS education campaign. Washington, DC: United States Fire Administration, 1994.

Management of Emergency Medical Services (Student Manual). Emmitsburg, Maryland: United States Fire Administration/National Fire Academy, 1999.

Managing Fire Services. Washington, DC: International City Management Association, 1988.

Marquardt, Michael J. and Berger, Nancy O. *Global Leaders for the 21st Century.* State University of New York Press, Albany, 2000. pp. 20-21.

Page, James O. *Emergency Medical Services for Fire Departments.* Boston: National Fire Protection Association, 1975.

Sachs, Gordon M. *Fire & EMS Department Safety Officer.* Brady Publishers, 2000.

Sachs, Gordon M. *Officer's Guide to Fire Service EMS.* Fire Engineering Books & Videos, 1999.

GLOSSARY

advanced life support (ALS). All basic life support measures, plus invasive medical procedures, including intravenous therapy; administration of anti-arrhythmic medications and other specified drugs, medications, and solutions; use of adjunctive ventilation devices, including endotracheal intubation; and other procedures that may be authorized by state law and performed under medical control. ALS is typically provided by EMS personnel certified at the EMT intermediate or paramedic level.

basic life support (BLS). Generally limited to airway maintenance, ventilatory (breathing) support, CPR, hemorrhage control, splinting of suspected fractures, management of spinal injury, protection and transportation of the patient in accordance with accepted procedures. BLS providers with special training can use automatic or semi-automatic defibrillators for cardiac defibrillation as well. BLS is typically provided by EMS personnel certified at the first responder or EMT basic level.

clinical. Relating specifically to the treatment of injured or ill persons.

customers. Anyone who may use, support, or participate in the delivery of services, including patients and their families, firefighters and EMS providers, physicians, nurses, taxpayers, visitors to the area, and suppliers.

emergency medical services. The provision of service to patients with medical emergencies; in particular, the pre-hospital delivery of these services.

emergency medical services system. A comprehensive, coordinated arrangement of resources and functions that are organized to respond in a timely, staged manner to targeted medical emergencies, regardless of their cause and the patient's ability to pay, and to minimize their physical and emotional impact.

emergency services. Agencies that provide the essential public safety services of fire suppression, emergency medical services, rescue, and hazardous materials response.

EMS. Emergency medical services.

first responder. 1) The basic level of medical training for emergency response personnel. 2) personnel on a first response unit (may be trained at any level of EMS).

first responder system. A tiered EMS system wherein the closest emergency response unit with trained EMS providers responds to medical emergencies.

first response unit. A non-transport emergency response unit with personnel trained at any EMS level, dispatched to provide medical care prior to the arrival of an EMS transport unit.

life safety education. Public education programs that are designed to teach citizens about fire safety, injury prevention, traffic safety, and other aspects of rural and urban survival.

medical director. A physician who advises the fire chief on EMS issues and who approves the training and protocol development related to EMS for a department. In many departments, pre-hospital EMS providers actually operate under the medical director's license.

PIER. Public information, education, and relations. EMS PIER programs provide the means for promoting an EMS system, developing positive public attitudes, and informing citizens on specific EMS issues and techniques.

pre-hospital. The aspect of the EMS system that occurs prior to the delivery of an injured or ill patient to a medical facility.

primary tiered systems. 1) ALS ambulances respond to all calls, but can turn a patient over to a BLS unit for transport; 2) BLS ambulance is dispatched on all calls, while an ALS ambulance is only dispatched on ALS calls; and 3) A non-transporting ALS unit is dispatched with a transporting BLS unit.

quality management. Actions taken to meet the needs and expectations of the customer; management based on the quality strategic plan, objectives, and performance indicators.

risk management. Actions taken to prevent disability, loss of life, and/or irreparable business damage as a result of the provision of patient care.

tiered response system. An EMS delivery system designed to ensure that the appropriate level of EMS care and transportation is provided to all patients, while ensuring the response of the closest EMS unit.

23

VOLUNTEER, PAID ON-CALL, AND COMBINATION DEPARTMENTS

Richard A. Marinucci

CHAPTER HIGHLIGHTS

- Recruitment and retention remain the most critical issues for volunteer, on-call, and combination departments.
- There is a strong correlation between successful recruitment and successful retention.
- Training is essential to compensate for limited experience and has become more demanding as more responsibilities are added to departments.
- Management and administration of volunteer, on-call, and combination departments require creativity, innovation, and a passion for the job.

INTRODUCTION

According to the National Fire Protection Association (NFPA), an estimated 7% of the more than 26,000 fire departments in the United States are career and another 5% are mostly career. Approximately 93% of the fire departments use the services of volunteer members (firefighters that are not considered career or full-time employees, regardless of whether or not they receive compensation), with 88% using mostly or all volunteers

(*United States Fire Department Profile Through December, 2001, NFPA*). Volunteer firefighters provide fire protection and other emergency service in a variety of communities either solely as volunteers or as part of a combination department (departments that utilize both career and volunteer firefighters, regardless of the percentage of make-up). Though many communities use volunteers out of necessity (cost, political climate, growth of community, or the transition of the department), others elect this method of fire protection not only for its cost savings, but because it provides competent professional service that meets the needs of the community.

There are benefits and drawbacks to any system of providing service. The use of volunteers and on-call firefighters is popular because of cost, but can also be advantageous because it involves the community more in the fire department; encourages creativity, innovation, and dialogue; generates adequate personnel in a labor-intensive business; and addresses the specific needs of the community. On the other hand, response times may be longer, available time of the members may limit the abilities of the department, and conflicts between career and volunteer members can distract the fire department management and leadership from their mission. Another consideration is the limited experience that may be the result of rapid turnover. Regardless, you need to know both the positive and negative aspects of each form of fire department service.

Good management and administrative practices are important in volunteer and combination departments. Fire chiefs of these departments are challenged to hone their management and administrative style in order to properly discharge their duties and provide leadership. They are expected to perform, to be held accountable for their actions, and to be faced with ever increasing responsibilities. They require knowledge of laws and regulations. Clearly, chiefs of volunteer and combination departments need to possess the knowledge and skills identified in all administrators, but they also need to possess special abilities to deal with the unique nature of these organizations. Often the financial constraints of a combination or volunteer department require that the chief seek better or more innovative ways to tackle problems. Traditions must be examined and evaluated, allowing alternate methods to be investigated. While innovative leadership cannot be taught, you must recognize the need to develop and expand upon current practices.

Recruitment

No fire department can survive or function without adequate staffing. The most important responsibility of volunteer, on-call, and combination departments is recruiting and retaining sufficient numbers of qualified firefighters to provide the required services of the department. There is a direct correlation between recruiting and retention. The reasons that existing members remain with the fire department are very often the same benefits that attract potential recruits. Members join the fire department for excitement, the opportunity to save lives, the challenge, the recognition, a chance to contribute to the community, social involvement, camaraderie, and a chance to learn new skills. An added reason for some to consider working with a volunteer department is to gain experience and training to help in the pursuit of a career position. This may or may not be an issue. Often a member will remain in a department even after obtaining a career position. They may serve your department for a few years before successfully obtaining a career firefighter job. Regardless, this has become an unpleasant fact in some areas. Recognize it and understand that it may be a valid reason to some candidates. These reasons remain after an individual joins and continues their career with the department. If you provide for these needs, you address both recruitment and retention. As a fire chief, you must view the firefighters as your customers who then provide service to their customers, the citizens.

Recruitment is the process of attracting and evaluating potential candidates for the department. It includes all steps of the hiring process, beginning with advertisement (both formal and informal), application, testing, and selection. Retention is the ability to maintain personnel once they have been hired or enlisted. You need to establish a reasonable goal with respect to longevity (how many years of service can be expected from each successful recruit?) before you can address the issue of retention. While establishing a goal of greatly increasing retention longevity, there are some things that are out of your control. Today's society is much more mobile. Personnel may relocate because of their full-time employment even though you are doing all the right things. If you attract young candidates who may be seeking a career in firefighting, there is not much you can do to retain them after they get their dream job.

While statistics indicate that the number of active volunteers in the United States is on the decline, there are many departments that are able to meet their personnel needs through successful recruitment and retention. If you select the right people, they will stay. This chapter provides some suggestions for recruitment and retention, but does not neglect

Professional organizations and associations, and associations like the IAFC, NFPA, and NVFC can offer assistance.

Whether paid, part-time, or volunteer, the job of firefighter requires physical activity. Develop a test to be included as part of your selection process.

to investigate other successful fire departments. The departments that have active enlistment processes provide many great examples of how to recruit, as well as establishing effective programs designed to retain existing members. Much can be learned through networking with other fire chiefs and similar departments and professional organizations, such as the International Association of Fire Chiefs [IAFC], NFPA, and National Volunteer Fire Council [NVFC]. Also, consider what can be learned from other volunteer organizations or agencies that rely on volunteers. Charitable organizations, such as the Red Cross, or other governmental agencies (police auxiliaries) need to aggressively recruit and retain. Find out what is successful for them and determine if these ideas and concepts can be adapted to your organization. While there are many ways to generate applications and interest in a fire department, failure to consider all aspects of the entire system is an exercise in futility. This system must also include the impact of the department on families, jobs, and social lives. You are asking people to make a major commitment to the department, which will affect their life style. If the department does not compensate for the disruptions it causes or does not meet the needs of the individual, recruitment and retention will suffer. The department must be well organized, professional, and competent. It must also be fun. You are trying to provide something important to the individual, whether it is social, educational, developmental, providing a sense of belonging, or satisfying a community need.

Prior to bringing any additional members into the department, there must be an evaluation of the organization. The department must have a mission with established goals and objectives. Without direction, the organization will flounder, creating confusion for both existing members and recruits. A most important step in this process is to develop a fire department that is effective and efficient. Ineffective, mismanaged, or incompetent organizations cannot attract members and will lose any quality fire fighters within the organization. No competent individual will remain in a voluntary situation if their needs are not met or the organization is not effective. Remember why people join, to meet specific needs of achievement, recognition, ability to do the job, and so on. If you do not meet these needs, you will lose good people.

Developing an effective organization requires skills in the areas of administration and management, regardless of whether it is for the fire service or any other type of business. The ideas in this text will help whether the department is paid, combination, or volunteer. The point to remember is that few people will consider joining an organization that is perceived to be mismanaged, poorly administrated, or that breeds discontent. Further, if they do decide to join, quality people will not remain in a disorganized group. The "house" must be "in order" and relatively free of political turmoil and internal strife. Even the best recruitment plans are useless until this is accomplished. A healthy organization supports its members and allows them to perform their duties without distractions. As a result, each member has assigned responsibilities and is committed to the improvement and future development of the department.

To begin, a plan must be established. This plan must be developed by and involve the entire department, as recruitment is not just the chief's job. There is a vested interest for every department member to recruit quality people to maintain the organization. This "buy-in" cannot be overemphasized. A recruitment committee should be organized to develop the plan, offer suggestions, and get the commitment of the department as a whole.

The first task is to determine personnel needs. This would include the number needed, skills required, and the type of person desired. It is also based upon services provided, operating activity levels, and the ancillary services offered or required.

One question that is seldom asked is, "Exactly how many fire fighters are needed to properly discharge the duties of the fire department?" Because this question is seldom asked, the actual roster size of many departments is arbitrary, or has a historical significance of "always being that number." Remember that having too many members for the workload can cause personnel, management, and administrative problems. The department should establish its rosters based upon level of service, response load, services provided, and the history of the department. By knowing the appropriate number, you have a better idea of the task before you. If you only need to add a few people, you do not need to expend as much effort as someone who is looking to greatly increase their roster size.

In order to sell, you need to know your product. Develop a list of the benefits of membership. You need to know what you are offering potential members. Recruiting for the fire department should not be that difficult. The fire service has a lot to offer: excitement, friendship, respect, the chance to save lives, and a whole host of other positive reasons. You can convince people you have a great product (or opportunity) for them if you know exactly what it is. The actual benefits can be listed by talking to those who already have them—the existing membership. Ask the current members what they like and don't like, what works for them, and for any other suggestions that they may have. You then have answers for most of the items considered by potential recruits. By identifying the privileges of membership, you answer questions from people who want to know "what's in it for me?"

Next, establish the criteria for the position of firefighter. You need a job description. There are many examples to help get you started; check with other departments. Based upon the job description, there should be some physical as well as academic requirements. These standards may vary based upon services provided. For example, departments heavily involved in EMS will require people capable of handling the rigors of intense medical training. As a minimum, consider a high school diploma as a prerequisite. Also require an agility test and physical examination. Obviously, whether career or volunteer, the job can be rigorous. Statistics show that the job is dangerous and requires a physically fit individual. Do not be tempted to take anyone that walks in the door. There are too many volunteer firefighters injured and killed each year trying to do the job. Heart attacks happen too frequently for you to disregard the fitness of your candidates. The NFPA has guidelines for this and a sample is also provided in the technical resource section.

Though many people consider volunteer fire departments different from career, most departments are covered by a variety of federal and state laws including the Americans with Disabilities Act, the Civil Rights Act, and other anti-discrimination laws. The physical and intellectual requirements must be job related and not biased toward any particular group of people or protected class. The laws in this field change rapidly, so you are advised to consult with an attorney to reduce the risk of violating labor laws. Also, assistance can be provided by a variety of state and federal organizations such as municipal leagues, risk management agencies, and professional associations.

Another big issue that must be considered is the time needed by the individual volunteer to be successful. Determine how much training time will actually be required in order to be considered an active volunteer. This is done so that you can be up-front with the people you recruit and not waste anyone's time—either yours or theirs. Also, review the special needs of your department. For example, you may be looking for daytime personnel to fill certain spots on the roster.

Outline the criteria that will be used on a basis of selection. Consider the skills, knowledge, and ability necessary for the job and list the prerequisites needed to meet these requirements. Ask what the elements are that indicate the possible success as a volunteer firefighter. This can include time available for the job, stability, physical skills, ability, willingness to train and learn, being a team player, basic communication skills, and other related issues.

The next step is to identify your target group or audience from which you will seek to solicit potential recruits. It may help to develop a list of exactly who you are looking for. This list should be very broad so as not to eliminate any particular group. While there may be some "typical" members, leaving the target group as large as possible enhances your

Conducting an informal meeting can help explain the work of volunteer firefighters and the needs of the department.

selection field. You need to eliminate your stereotypes and begin to look outside the traditional volunteers. Typically, volunteers have been predominately white males. Investigate ways to attract females, minorities, or others typically not associated with volunteer fire departments. There should be no predetermined mindset. A variety of members enhances your service to the community, gives you a better perspective of the citizens being served, and provides a better understanding of and relationships with the fire department. Volunteerism *by* the community is *for* the community and must reflect the community to be truly successful.

The next step is to develop a strategy that will include where to look and ways to advertise. There are many good ways to reach the public and advertise your needs. This can be done through the media, information meetings, signs and posters, targeted mailings, and newsletters. All this is very helpful, but the most effective recruitment tool is personal contact between department members and the potential recruit. Existing department members should ask other people to join. This is a very effective and proven method of recruitment. If your members are doing the recruiting, they will be looking for potential successful candidates and become part of the process. This helps when the candidate joins, as the existing members will provide the necessary assistance for the recruit to succeed and helps with social inclusion. The personal touch also "shows off your product," and if your quality people are the sales staff, the product sells itself. Proud members are your best sales people. People need to be coaxed into joining because they are often unsure of their interest or what the job entails. Show the members of your department how to recruit. This would include the following information to be included when talking to potential recruits:

- Introductions
- Their background
- History of department

- Benefits of membership
- How to apply

It is a continuous process and requires diligence and commitment. Always be on the lookout for prospective members. If you were to check with other departments and chiefs, you would be amazed at the various locations that volunteers were discovered. Check your own department membership; you may be surprised at the various backgrounds of your members.

A good tool to use to get your message out and explain your department is a "recruit night." The recruit night is an opportunity to invite all potential candidates to the fire station for an informational meeting. The department operation can be explained and questions can be answered. Your department can be "sold" to interested parties and you can discourage candidates unable to meet your predetermined requirements. Formal presentations, tours of the station, and personal contact with firefighters can deliver your message.

After the applications are collected, continue the selection process. This can include a written test, agility test, oral interviews, and background checks (among others). Cost may be a factor in determining the final method and should be considered in establishing the order in which tests are used. Do the least expensive tests first to eliminate unsuccessful candidates before more costly tests are conducted. A good selection process can help weed out unacceptable candidates before you waste a lot of time, energy, and money on training. Even though they are volunteers, a mistake in hiring can be very costly—both financially and organizationally.

Retention and Motivation

The retention and motivation of volunteer and on-call firefighters is not different from any other employee. You must understand why the fire fighter joined and why they stay with your department. They want recognition for a job well done and a voice in the operation. Much can be done to show appreciation. Depending on the background of each department, some examples to consider are:

- *Awards programs*—Recognize the accomplishments of your members (a sample is provided in the technical resource section).
- *Recognition banquets*—Reward your members and their significant other with a night out. Present their awards and certificates in front of their family and friends.
- *Tax incentives*—Some states allow municipalities to offer this. Check your local laws.
- *Length of service awards*—Reward longevity.
- *Retirement benefits*—Invest in your members' future.

- *Health insurance.*
- *Employee assistance programs*—Take care of your employees.
- *Social activities*—Involve entire families. Try anything (picnics, holiday parties, golf, bowling, roller-skating, etc.).
- *Tuition reimbursement*—Help with higher education.
- *Free use of public facilities*—Your local YMCA, etc.
- *Longevity pay or bonuses.*
- *Disability insurance.*
- *Newsletters*—For information and recognition.

All of the above may be considered, but the most effective incentive is to treat people fairly and with respect. With this in mind, it is important to train managers, chiefs, and officers about this need (as many have never had any training in this area). How the firefighters are treated will have more impact on retention than any other programs or "gimmicks" if you are sincere. Members ultimately stay because they have a passion for the work and/or enjoy the friendships they have in the organization.

Training

Training is extremely important in all fire departments to ensure the response readiness of all of the members; however, it has become more complex as a result of the changing demands placed upon the fire service. Time and legislation have become big constraints. Changes in legislation at the state and federal level have had an impact by requiring specific numbers of hours to become certified, licensed, or just to meet the law. In addition, many volunteer departments are being held accountable in the legal arena if things go wrong. People expect a competent response to their emergency regardless of pay status.

Recognition is part of any successful retention program.

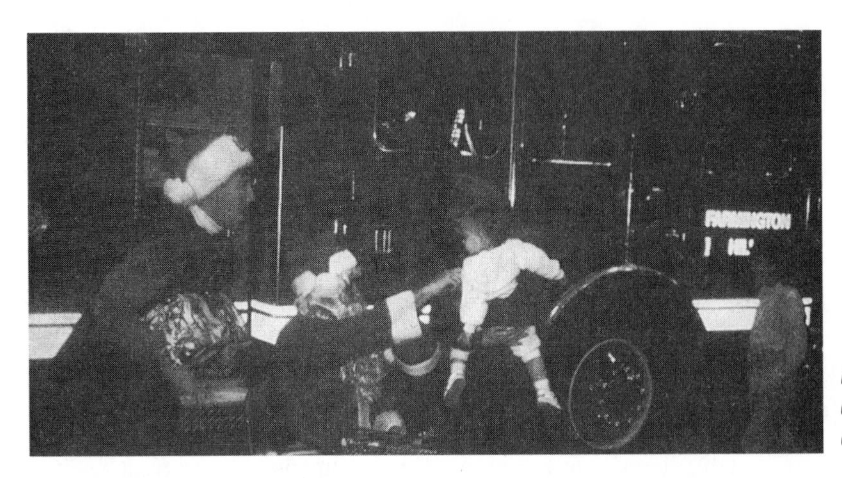

Involve families in department-sponsored activities.

Training is needed for firefighting, hazardous material response, EMS, rescue, special incidents, and terrorism—a new addition to the responsibilities of fire departments. This variety can create a barrier to quality training in all areas because of the time requirements and the need for knowledge of special equipment. The volume of emergency activity, competition for leisure time, family life, and work all become factors. These barriers can be overcome through proper planning and the implementation of worthwhile training.

Evaluate training needs. Initially, evaluate the needs of the department with respect to training to determine the type and amount of training necessary to prepare personnel for their job responsibilities. This is based upon the legislative mandates, desired service levels of the community, and the desire of the individual department members and chief to commit to training while working within the above mentioned time constraints.

Establish the training schedule. Next, establish a schedule based upon these needs. When scheduling, bear in mind the time and availability of volunteers and on-call fire fighters. Remember that they can't get all the training at one time; you must consider the time commitment required in responding to emergencies as well as time for their personal life and their full-time job. This can be accomplished by extending the usual time scheduled to obtain the training. Perhaps two or three sessions spread over a month's period would be better than committing an entire weekend. You may wish to take a multi-year approach to obtain the necessary training, allowing personnel a few years to obtain the certifications and licenses required by your department. If there is no sense of urgency, allow members to proceed at a reasonable pace, one that does not overwhelm the individual. Whatever you choose, prioritize it based upon your needs (*i.e.*, types of fires, EMS, rescue, etc.) Also, ask your members what would work best for them. New members typically want all the training they can get early in their career. They then pace themselves as they gain experience. Do not neglect to consider the impact on your training officer(s); they, too, can spend only so much time.

Establish the position of training officer. There should be at least one department training officer responsible for the coordination of all training activities. Depending on the size of the department, additional instructors may be needed. One of the advantages of volunteers is the talent that they bring to the organization from their primary employment. Utilize as much of this talent as you can by developing their instructional capabilities. Some states have certification requirements for instructors. Check your state requirements. If you do not have the expertise, contract with people outside your organization.

A training program must include the basics for everyone, and some variety based upon the members' levels of experience. You can't expect all members to require the same training all of the time. For example, entry-level training is not essential for senior personnel and officers. Recruit firefighters require basic skills and senior firefighters require training that is more advanced. There needs to be specialized training for recruits, firefighters, and officers. There should be officer classes for existing officers, as well as courses designed to prepare the future officers. Topics for officers should include incident command, fireground strategy, personnel management, and leadership. This is important to make up for a possible lack of experience.

The department may also require specific training to maintain licensure and certification. For example, some state laws require continuing education credits to maintain emergency medical technician licensure. In addition, OSHA and other federal laws require specific training in the area of hazardous materials for personnel required to respond to hazardous materials incidents. Training should begin with "need to know" information and expand to "nice to know" information as time allows.

Training is extremely important in preparing personnel to properly discharge their duties as a firefighter. Quality training is not only essential, but when done properly it enhances

Live fire training is essential in developing and maintaining firefighting skills.

the public's perception of the department's members. Members also feel better if the training is of good quality. Continue the training to keep their skill levels high. Training members is most important; it will motivate members and set or change policy. Consider it an investment in the future.

Safety

Safety remains a critical issue within the fire service. According to the United States Fire Administration (USFA), approximately 100 firefighters per year die doing their job. While many of the safety issues are applicable to all who respond, whether paid or not, there are causes that affect volunteers more. Of the approximate 100 deaths, approximately 70% are volunteer or part-time firefighters. Two areas affecting volunteers more than career firefighters are responding to alarms, returning from alarms, and heart attacks. As these are fairly consistent from year to year, it is imperative for departments to work towards the reduction of these numbers. Strong policies, good training, enforcement of policies, and attention to physical fitness must become part of every department. History shows that there is reluctance to address these issues (as indicated by the fact that the statistics have not changed in recent years). Members must be both mentally and physically fit for duty, and trained to the appropriate level. Fire chiefs and the entire organization must commit to improve upon the unacceptable loss of firefighters' lives.

Promotions

For immediate and long-term development of any fire department, there must be a fair and equitable system of promotion that elevates competent, qualified personnel to appropriate positions in the organization. This is possible by establishing a promotional system that considers the individuals, the positions needed, and the political realities of the department. There are three steps that should be taken (examples are provided in the technical resources section):

- Establish the officer's position with corresponding job description
- Establish criteria for officers (prerequisites for the positions—knowledge, skills, abilities, and training)
- Establish a selection process

There are two reasons to have officer positions in volunteer and on-call departments. The first is that some positions are needed to efficiently run the organization and lend support for command and control operations. The second reason is that some positions may be needed as a motivational tool for volunteer firefighters. It creates a career ladder. Having officer positions available allows the firefighter to establish goals and strive to accomplish something within the organization. Many people seek promotion and are extremely proud when they finally earn their "stripes." Having multiple officer ranks also allows for the development of the future leaders of your organization.

Various prerequisites can be used for officer positions. These can include experience, previous training, evaluations, and past contributions to the department. Establishing training requirements for promotional opportunities motivates members to train, as well as improves the status for each position. It also improves public perception and public confidence in the department's ability.

Probably the most difficult decision to make is to choose the appropriate method for selection of officers. Three methods that are commonly used are:

- Voting by membership
- Appointment by the chief
- A competitive testing process

All of these methods have their advantages and disadvantages and have also been successful within various departments.

Voting can become a popularity contest and restrict officers when tough decisions need to be made. It also may expose the department to additional liability if qualifications are not used as part of the promotion process. On the positive side, the elected officers may receive more support from the membership.

Appointment by the chief may be perceived as favoritism, but can allow the chief to better develop his own team. Since the chief has the responsibility for the entire organization, proper choices can be made. However, personalities can cloud the selection.

Competitive testing is perceived to be the fairest, but should not be construed as foolproof. There is always the human factor to consider. Competitive testing probably offers the best long-term solution to the promotion process.

The department history and culture play an important role in selecting the appropriate promotional process. Regardless of the selection method, the most important part of promotion is to establish prerequisites for the position. If this is done, only qualified people will compete for the position. Training requirements must be part of any promotional procedure.

Quality control

One of the greatest challenges faced by volunteer and on-call departments is the maintenance of quality service to the community. There is a general decline in the number of fires, lack of experience, and lack of adequate training facilities to expose members to live burn situations. There is no question that these issues affect and impact the quality of service delivery. This issue of quality control is directly related to training and the constant evaluation of the physical and mental preparedness of the department members. To have a prepared firefighter, there is a need for:

- *Training*—Growth in firefighting skills (and EMS, Haz-Mat, Terrorism, etc.)
- *Skills Maintenance*—Those skills required of all members
- *Practice*—To ensure efficient, effective delivery
- *Testing*—To ensure that all members are prepared at all times

A call for assistance can occur at any time and, therefore, all members must always be capable of performing their assigned duties. With fewer fires to practice on, members must be motivated to maintain their skill level, as the department is unable to monitor members' skill levels on a daily basis. Members must be motivated to do the best job they possibly can and prepare themselves. The department can monitor this through random periodic testing and on a scheduled annual basis. In addition, tactical simulation training with slides or simulators are an effective alternative to actual incident command in maintaining command skills, and expose senior firefighters and junior officers to command training. The department must define the skills, knowledge, and abilities required to perform the job and take the necessary steps to ensure that each member is capable.

Policies and procedures

Every organization needs a set of rules to govern its members. These rules should include guidelines for the various functions and operations within the department. This lends control as well as informs the membership of the expected behavior and performance requirements. The whole package includes rules and regulations, special orders, policies, and procedures.

- Rules and regulations are the basic internal "laws" that govern the department. These serve as your "constitution" and it should not be easy to change them.
- Special orders are emergency directives needed to address concerns on a short-term basis.
- Policies and procedures offer guidelines on actions and expected behavior in situations.

Drafting rules and policies is one of the most difficult things for a department to do, as it is time consuming and not one of the jobs that many people like to do. Fortunately there are places to look for help. Consider other fire departments and professional organizations, such as the IAFC or NVFC. Procedures can also be adapted from other non-fire service organizations (police agencies, private sector, etc.). In addition, the committee process lends itself well to the development of rules and regulations, as well as certain policies and procedures. By using department members to draft the rules, there will be more acceptance and compliance. The committee should be representative of the department make-up, including fire fighters, company officers, and command staff. The size of the committee will be large enough to include adequate representation, but not too large that it impedes its ability to operate.

Rules and regulations should be developed as a simple document outlining acceptable behavior. They also define the discipline process including filing charges as well as grievance and appeal steps. Rules and regulations must include the organization chart, which establishes the vertical and horizontal relationships and responsibilities.

Policies and procedures should be developed to provide guidelines for the operation of the department. Possible subjects include administration; fireground operation; maintenance of apparatus, equipment, and stations; safety; hazardous materials; training; membership benefits; and communication. There should be a formal adoption process that allows for review before implementation. This would include an attorney and the governing body (fire board, city council, commissioners, etc.) where appropriate.

Emergency scene operations

As with any other function of a fire department, the emergency scene needs to be managed and organized to produce efficient and effective service. The incident command system was developed for this purpose. It provides a good basis for operation. A formal incident command system is required as part of federal legislation (SARA, Title III, CFR 1910.120) and could be required by the various states or localities. In addition, national standards, such as those issued by the NFPA, recommend the use of a formal incident command system.

There are actual incident command systems already in print. However, volunteer, on-call, and combination departments present special circumstances within the system. There are some questions that need to be accounted for within each volunteer and combination department.

- Do you know who will be responding on each and every run?
- In what order will they be arriving?
- Will they be arriving on fire apparatus as part of a team, or individually in privately owned vehicles?
- Given the time of day and day of week, how many people will be responding?

Since a fire or other emergency can occur at any time and there is no guarantee that all members of a volunteer department will always be around, the incident command system must be developed so that it can operate in all situations. There must be flexibility in the system.

The number of firefighters responding and their method of response (on fire trucks or private vehicles) affect the organization, command, and control of the fireground.

Of particular concern in combination departments is the issue of chain of command and who has the ultimate authority on the scene of an emergency. This must be resolved and be perfectly clear to all members of the department. Departments have had success both in designating career members as senior officers and cases where volunteers are designated to be in charge. This is certainly a local decision. A suggestion is to allow for a transition process. As new career personnel are hired, they probably do not have the necessary training and experience. As they put in time and attend classes, this will change. A gradual, well-planned phase-in program may address the concerns of the department and allow for the most competent command on the scene of an emergency.

Accountability

The tracking of emergency personnel on the scene is extremely important and can become complex. The incident commander must be able to track both the location and identity of each responder. The nature of volunteer fire departments does not guarantee who will respond or necessarily when or if they might arrive. They may or may not arrive with fire apparatus. Each department must develop a system to track personnel from arrival at the station and/or on the scene and as they conduct various fire scene operations. Various systems have been developed and implemented including nametags on the personnel, which are delivered to the incident commander or are left with appropriate fire apparatus. Your system and training must instill discipline in your personnel to eliminate the risk of freelancing. Regardless of what is used, all personnel must report to the officer in charge for assignment. This provides the officer information on all the resources available.

Volunteer and combination departments are faced with an added incident command system challenge in that any department member may be faced with making initial decisions on the fireground and implementing the incident command system. They must be able to start the incident command process and have it operating properly until the first arriving officer can assume command. Therefore, all members must receive training, be offered practice, and be familiar with the transfer of command. Training is extremely important to overcome the lack of experience of some of the department members and to compensate for the nature of the volunteer and combination services that make the use of an incident command system more difficult than in a traditional career department.

To summarize, all departments must develop an incident command procedure for use on the scenes of emergencies. The system must be custom fit to address the nuances of each individual department, and all department members must be adequately trained in its use and operation.

Support and auxiliary services

Fire prevention. Fire departments wishing to provide a totally comprehensive service package to the community must include fire prevention programs. This is an added challenge in that most fire prevention activities are not glamorous nor are they traditionally part of a volunteer fire department. Combination departments have a distinct advantage in this area in that career personnel can be utilized to provide these services. All volunteer departments face the difficulty of finding the necessary time and interested personnel for these services. To be successful, these departments need to incorporate fire prevention as a requirement of the job of all department members.

When we speak of fire prevention activities it includes fire safety inspections, public fire safety education, and fire investigation. The ability of a volunteer department to perform inspections is based upon two factors: the number of buildings that require inspection, and the availability of personnel to receive the proper training. If there are relatively few buildings, personnel can be trained to perform these services on a part-time basis. An added consideration is the level and type of inspections to be performed. This can vary from simple safety inspections (exit lights, extinguishers, and smoke detectors) to complete code enforcement with legal authority to obtain compliance. If the department is unable to do so then they must look to other agencies to support their services. Local building departments, county agencies, or agents of the state may be able to provide this service. In cases of combination departments, career personnel are the logical choice to perform these duties. This promotes productivity as down time waiting for alarms to occur is utilized to enhance fire safety.

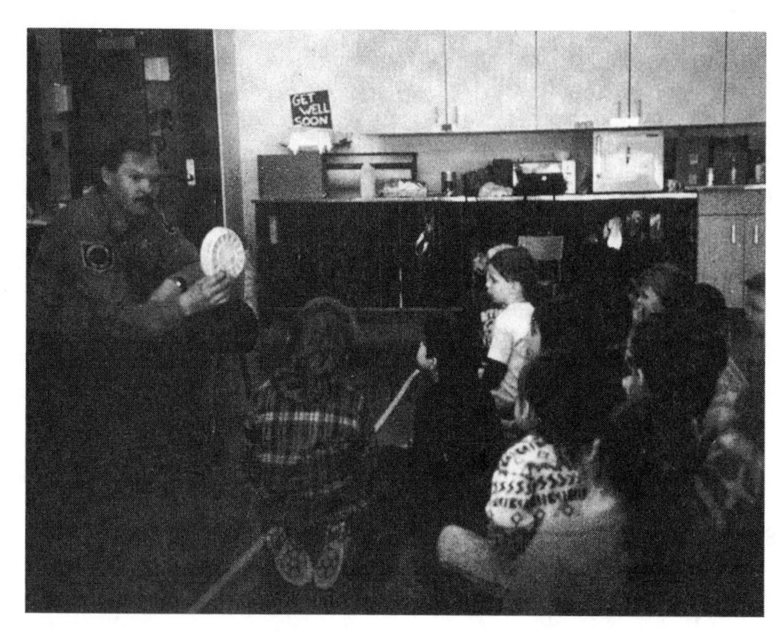

Public fire safety education can provide added benefits of improving public relations.

Public fire safety education is becoming an important tool in reducing the number of fires, fire fatalities, and injuries. Any department and its members can become involved in public fire safety education to different degrees. There must be the desire and interest on the part of some of the department members. Usually one or two "champions" can promote the cause of public fire safety education. If a member or members are interested, support from the fire chief's office will enable the program to succeed. The benefits to the department are two-fold:

- They educate the public.
- Public fire safety education has become a tremendous public relations tool.

Through this, departments can gain support for other programs including suppression operations, station improvements, additional apparatus or replacement, personnel benefits, uniforms, safety equipment, etc. As with other activities, this requires planning that would include identifying the target audience, the message to be delivered, and the method to be used. It can be done through schools, the media, open houses, presentations, or involvement in other community activities such as parades, civic involvement, and honor guards at festivals. Quite often there are volunteers who have talent or experience in this area that can prove to be very beneficial.

Pre-fire planning. Gathering information before an emergency is extremely important for the safety of firefighters as well as giving the fire department the best opportunity to perform properly at the scene of an emergency. This information is gained from commercial, industrial, institutional, recreational, and residential (apartments, townhouses, hotels, and motels) properties within the community. County, state, or federal laws may require some activities. Regardless, this reconnaissance information gathering should be mandatory for all departments, whether they are paid, combination, or volunteer. The format may be simple and computers are not necessary. A single sheet of paper can be used to provide basic information. This can be put in a loose-leaf notebook, which can be carried on all fire apparatus. If the dispatch center has the resources to provide the information, it can be kept there, which would keep it out of the weather. Either way, it gives the first responders the information they need to begin operation at the emergency scene.

Fire investigation. Fire departments have the responsibility to determine the origin and cause of fires to which they respond. Gaining the necessary experience and training is difficult. It is sometimes easier to request that county or state agencies provide this function. Some mutual-aid organizations have pooled their resources to create investigation teams or task groups. This approach can minimize the impact on any one department. However, larger departments may have personnel interested in performing this service. If so, they must receive the necessary training and acquire the appropriate credentials. If personnel are willing, the department will benefit. Training personnel and getting them the appropriate credentials is important, but time consuming and costly. Consider the other options of utilizing police agencies and other county or state services.

EMS. One of the biggest decisions that any fire department must make is whether or not to be involved in EMS and to what degree. The trend certainly is to become involved as a majority of fire departments in the United States provide EMS service to their communities. In the case of volunteer, on-call, and combination departments, the municipality or community must decide if they wish to become involved. Once they have done that, they have to answer other questions regarding the type and level of service to be provided—whether it is basic, advanced, or as first responders (or something in between).

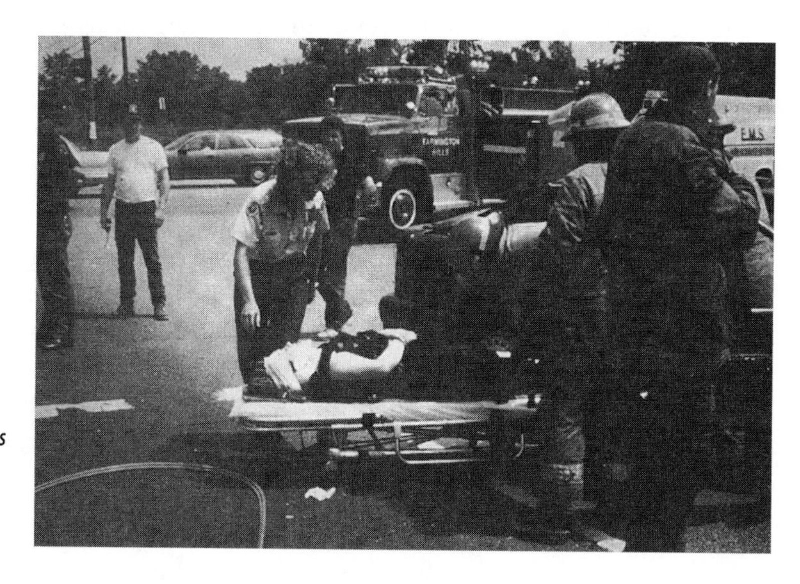

The addition of EMS responsibilities significantly increases the number of emergencies and the training requirements.

Further adding to the complex nature of EMS would be special rescue situations. Some departments may be involved in these even if they are not directly involved in EMS. These would include dive rescue, high angle rescue, extrication, confined space operations, as well as search and rescue.

The biggest concern for volunteer departments has to do with the time necessary to complete the mandatory training for EMS, including initial training and continuing education credits necessary to maintain licensure. Typically, medical calls account for anywhere from 50% to 80% of a fire department's activities. This added run response, in addition to the time necessary for training, can put a tremendous burden on the fire department and its personnel.

Hazardous materials. The advent of hazardous material responsibilities for fire departments has also placed an added burden on volunteer fire departments. All too often, the fire department is the only choice to respond to these incidents. These volunteer departments must try to decide how involved they wish to be and what costs they are willing to absorb. Most volunteer departments would be advised to seek other alternatives. This could involve the use of mutual aid, regional hazardous material teams, or private contracts. If the departments wish to get involved they should investigate all state and federal laws, as they would be required to comply.

Terrorism. The 1995 terrorist events in Oklahoma City, the 1993 World Trade Center bombing, and the attacks on September 11, 2001, have raised the awareness level of all first responders to the threat of more incidents. All firefighters, whether paid, combination, or volunteer must prepare. It can be an explosion, fire, or the newer threats of chemical, biological, or nuclear releases. What does this mean? It means more training time, more equipment, and more safety issues to consider. One approach is to view these incidents as either a hazardous materials, mass casualty, or special rescue incident. The preparation for response is the same with some specific exceptions related to the criminal act. There are federal and state resources available to help locals prepare. Investigate all the options available to you. You may be able to find financial assistance, but the biggest impact on your organization will be the time commitment necessary to train and prepare.

Fundraising. Many volunteer departments are not supported by tax revenue and must raise money so that the department can remain in operation. The fire service is the only government agency that is forced to do this. Though it may be a part of the department's duties, many times the department elects this method of funding to limit the control of government agencies on the fire department. If fundraising is required, determine your needs. How much money is needed? Involve the entire department and community in this effort. Form a steering committee and check the laws of your state and municipality. Some successful events used have been carnivals, bake sales, raffles, bingo, and dinners—get creative. If you don't like fundraising, make an appeal for funding through taxes.

CONCLUSION

The use of volunteers, on-call, and combination departments can be an efficient and effective means to provide emergency service to a community. They typically utilize innovation and creativity of their members to solve problems. Administrators and managers of these departments must know and understand the advantages as well as the history of the department. Good management and administrative practices are essential.

Resources Available

Organizations

International Association of Fire Chiefs
Volunteer Committee
Management Information Center

National Volunteer Fire Council

National Fire Protection Association

Federal Emergency Management Agency
Emergency Management Institute

United States Fire Administration
National Fire Academy
- Learning Resource Center

Women in the Fire Service

Publications and literature

Trade Journals
Fire Engineering
Fire Chief
Fire Journal
Fire House
FireRescue Magazine
Various other fire/rescue publications

State Fire and Emergency Associations

State Fire Marshals and State Training Directors

Snook, Jack W., et. al. *Recruiting, Training, and Maintainng Volunteer Firefighters.* Fire Protection Publications, Oklahoma, 1998.

APPENDIX 1

________________ FIRE DEPARTMENT

SAMPLE PERSONNEL PROCEDURE

Awards and Recognition **No: ______**

Rescinds: ________ **Effective: ________** **Page: 1 of 5**

During the course of the year members of the ________________ Fire Department often do things that deserve special recognition. The Department intends to identify and recognize these individuals because of their various achievements.

These approved awards shall be presented during the annual fire department banquet:

- Medal of Valor
- Medal of Bravery
- Fire Fighter of the Year
- Life Saving Award
- Chain of Survival Award
- Station Fire Fighter of the Year
- Meritorious Service Award
- Unit Citation
- Recruit of the Year

Criteria for awards are as follows:

1. Medal of Valor

The fire chief shall present this award to members who have, under especially hazardous conditions, courageously risked their own life to save another. The intention of this is to reward the truly outstanding performances under times of duress and shall be considered for emergencies only. Members receiving this award shall be nominated for the International Fire Chiefs Association Benjamin Franklin Fire Service Award.

2. Medal of Bravery

This award shall be second only to the Medal of Valor and will be presented to a member for an act that exhibited disregard for personal safety in an effort to save another. This will generally be considered for members acting above and beyond the call of duty and within safe operating policies and procedures of the Department.

3. Fire Fighter of the Year (Department)

This award is intended for the Department member who, over the course of the year, has continually put forth an effort of the highest degree. This may involve fire suppression, emergency medical service, fire prevention, training, or any combination of the above. Further, it may involve an individual event or a collection of exceptional performances. Any current member of the Department may nominate another for whatever reasons they feel appropriate. The fire chief at the annual awards banquet will present the award, and the recipient will also serve as the official representative at the City Service Awards Program.

4. Life Saving Award

To be awarded to an individual for the saving of a human life. Intended for an individual *directly* responsible for the saving of a human life and shall be issued to members of the Department for the saving of a life through various actions such as the application of pre-hospital emergency medical care or public safety measures.

5. Chain of Survival Award

This award will be presented to an individual whose pre-hospital diagnosis and treatment is directly attributed to stabilizing the patient's condition prior to arrival at the hospital emergency room.

6. Station Fire Fighter of the Year

The district chief at each of the four stations shall present this award to the member who, as nominated by his fellow firefighters at his station, is most deserving. All members are entitled to nominate one member from his station. District chiefs are not entitled for nomination.

7. Meritorious Service Award

This shall be awarded to members of the Department whose actions have distinguished them from standard performance expected of the position. This award may apply to any phase of the Department.

8. Unit Citation

This award may be presented to members of the Department that participated in an action that contributed to the overall professionalism of the _______________ Fire Department. This award may apply to any phase of the Department.

9. Recruit of the Year

This award shall be given to the recruit fire fighter that best exemplifies the conduct required of a _______________ firefighter and continually demonstrates readiness, performance, and excellence in completing the recruit training program. The recruit shall display maturity and leadership potential before fellow recruits and, through dedication and commitment to duty, has made a significant contribution to advancing the goals of the Department.

The following are additional awards that may be presented at any time during the year or during Department functions.

- Certificate of Appreciation
- Certificate of Training
- V.F.W./Optimists Club Fire Fighter of the Year
- Letter of Appreciation
- Highest Response Record

Nominations

Any member of the Department may nominate another at the time of the incident or occurrence. Submittals should be made within 60 days of the incident. Nominations shall be forwarded to the station representative of the Awards Committee, who will submit the nomination to the Awards Committee for review.

For the selection of Department Fire Fighter of the Year, the Awards Committee will accept nominations. A minimum of three nominations will be forwarded to the fire chief for his review and selection. The Awards Committee may select an individual to provide a minimum of three nominees.

The Awards Committee will accept nominations for V.F.W./Optimists Club Fire Fighter of the Year. Nominations will be reviewed and the Awards Committee will recommend the recipient to the Fire Chief for approval.

Ribbon Placement

Ribbons will be placed in order from highest to lowest as follows.

Medal of Valor	metal gold bar
Medal of Bravery	metal bronze bar
Fire Fighter of the Year	solid blue
Life Saving Award	white/blue
V.F.W./Optimist Fire Fighter of the Year	solid red
Station Fire Fighter of the Year	red/white/red
Meritorious Service Award	red/white/blue
Unit Citation	solid white
Recruit of the Year	n/a

The Medal of Valor and the Medal of Honor will always be placed above and centered on any ribbons displayed on the uniform. The following are examples of ribbon placement:

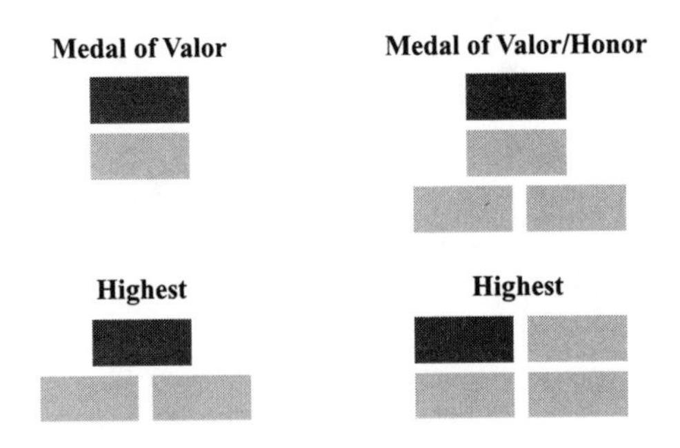

APPENDIX 2

_______________ FIRE DEPARTMENT

SAMPLE FIRE PREVENTION PROCEDURE

Pre-emergency Information Sheets **No: _____**

Rescinds: _______ **Effective: _______** **Page: 1 of 2**

Department personnel shall complete and disperse a pre-emergency information sheet for all addresses in the city with the exception of one- and two-family dwellings. The purpose will be to provide information to responding personnel that will be beneficial during times of emergencies.

The information shall be gathered any time Department personnel completely tour any building. Examples of this would include annual inspections of occupancies, station tours of buildings, or as part of assignments for training programs. Though these can be completed by anyone at anytime, the vast majority will be conducted during the duty day in conjunction with the inspection program. Personnel working during the day will be responsible for the completion of the forms in addition to the paperwork associated with the inspection. Indicate on the inspection form who is responsible for submitting the preplan.

If a sheet has been completed for an occupancy, any subsequent contact with the building shall be used to update the existing information sheet. The information shall be recorded on the forms provided and will be distributed to the first-due station, squad, and district chief; Department staff officers; dispatch; and headquarters for file purposes. This information shall be carried on the squad. If the person completing the information sheet feels that more information is required and the occupant does not offer it, he may utilize the "Right-to-know" law to request this information. Members completing the form shall submit it to Headquarters for review and copying for distribution. The member who prepares the preplan shall be indicated on the inspection form submitted by the full-time fire fighter.

In order to maximize the benefit of these sheets, time at each station drill shall be set aside for the review of any recently completed forms. Normally, a mention of the pre-emergency information sheets to the personnel will be sufficient, with members reviewing them on their own. Certain occupancies, because of the nature of the hazards present, may require a more in-depth review of the information

APPENDIX 3

________________ FIRE DEPARTMENT

SAMPLE PERSONNEL PROCEDURE

Promotions **No: ______**

Rescinds: ________ **Effective: ________** **Page: 1 of 3**

I. SCOPE

This procedure shall apply to members seeking promotion.

II. PURPOSE

To outline the policy and procedure along with minimum requirements to be acquired before being considered for promotion.

III. PROCEDURE

The promotion to the position of officer within the Department will be accomplished through a consistent, organized, and fair process. When a vacancy in the officers cadre of one of the fire stations occurs and a replacement is desired, the position will be posted at all stations for a period of two weeks. Those members interested shall apply, in writing, to their district chief within the specified timeframe.

Candidates that meet the minimum qualifications for the position as outlined in these procedures will be evaluated based upon written and oral examinations, seniority, and a review by the district chief responsible for the promotion. The written test will consist of questions developed specifically for the vacated position and will address fire suppression knowledge, Department procedures and directives, emergency medical service, hazardous materials, fire prevention practices (*i.e.*, preplans), and station policies. For the oral examination, a panel will be established consisting of Department officers not affiliated with the station requesting the promotions. A single interview board will be used to evaluate candidates seeking a position. The district chief will provide input into the selection process by reviewing and evaluating each candidate. The final score will be calculated with the following percentages:

Written Examination:	30%
Oral Examination:	30%
District Chief Evaluation:	35%
Seniority:	1% per year, up to 5 years

(Seniority points shall accrue from the point a member is eligible for a promotion.)

The candidate with the highest score will then be recommended to the fire chief for promotion. Individuals promoted shall serve a one-year probationary period from the date of appointment.

Prerequisites for officers' positions

In order to be considered for promotion to the position of sergeant, lieutenant, or captain, Department members must meet the following requirements (the fire chief will evaluate equivalency):

I. Sergeant

  A. Three years on the Department
  B. Fire Fighter II
  C. Fire Officer I
  D. State licensed EMT
  E. Good history of community service
  F. Satisfactory evaluations and service record
  G. Run Percentage greater than 45% during the past 12 months

II. Lieutenant

  A. Sergeant's qualifications

  B. One year as a sergeant

  C. Fire Officer I certification

  D. Fire Officer II

  E. Satisfactory sergeant evaluation and service record

III. Captain

  A. Lieutenant's qualifications

  B. Lieutenant for two years

  C. Fire Officer II certification

  D. Fire Officer III

  E. Satisfactory lieutenant evaluation and service record

APPENDIX 4

SAFETY PROCEDURE

Driver Safety Standards **No: ______**

Effective: ________ **Page: 1 of 5**

Purpose

To provide guidelines for the safe and effective operation of emergency vehicles and related equipment.

Procedure

All personnel shall be charged with the responsibility of operating personal vehicles (authorized emergency vehicles) and apparatus in a safe manner so as to prevent accidents and injury during the performance of their duties.

I. Introduction

A. The driver of each Department vehicle has the responsibility to drive safely and prudently at all times. Vehicles shall be operated in compliance with the State Motor Vehicle Code. This code provides specific legal exceptions to regular traffic regulations, which apply to fire department vehicles only when responding to an emergency incident or when transporting a patient to a medical facility.

B. Emergency response (Code 3) does not absolve the driver of any responsibility to drive with due caution. The driver of the emergency vehicle is responsible for its safe operation at all times. The officer in charge of the vehicle has the responsibility for the safety of all operations.

II. Warning lights

A. When responding Code 3, warning lights must be on and sirens must be sounded to warn drivers of other vehicles, as required by the State Motor Vehicle Code. The use of sirens and warning lights does not automatically give the right-of-way to the emergency vehicle. This requests the right-of-way from other drivers, based on their awareness of the emergency vehicle. Emergency vehicle drivers must make every possible effort to make their presence and intended actions known to other drivers, and must drive defensively to be prepared for the unexpected inappropriate actions of others.

III. Speed limit

A. Public Act allows emergency vehicles to exceed the posted speed limits. However, Department vehicles are authorized to exceed posted speed limits only when responding Code 3 under favorable or ideal conditions. This applies only with light traffic, good roads, good visibility, and dry pavement. While the Act establishes no limit, under these conditions, the maximum of 10mph over the posted speed limit is authorized. Under less favorable conditions, the posted speed limit is the absolute maximum permissible. Drivers are responsible for operating in a safe and prudent speed at all times. Special caution should be made when entering and driving through subdivisions. At no time should a vehicle exceed the speed limit while driving inside a subdivision. Further, the driver must be alert for any pedestrian traffic within the subdivision.

IV. Intersections

A. Intersections present the greatest potential danger to any emergency vehicle. When approaching and crossing an intersection with the right-of-way, drivers shall not exceed the posted speed limit. When approaching a negative right-of-way intersection (red light, stop sign), the vehicle shall come to a complete stop. When approaching a yield sign, the driver shall slow to a speed not greater than 10mph and shall proceed through the intersection only when it is determined to be safe. The emergency vehicle may proceed only when the driver can account for all oncoming traffic in all lanes yielding the right-of-way. If there is any doubt, the emergency vehicle shall wait until the intersection can be crossed safely.

V. School buses

A. When responding to alarms, vehicles shall stop for school buses loading and unloading as required by law. If the bus driver gains control of all students, turns off warning signals, and acknowledges the presence of the Department vehicle, the driver may continue to the incident.

VI. General safety precautions

A. Code 3 response is authorized only in conjunction with emergency incidents. Unnecessary emergency response shall be avoided. In order to avoid this, the following rules shall apply:

1. When the first unit reports on the scene with nothing showing, or an equivalent report, any additional units shall continue Code 3, but shall not exceed the posted speed limit.
2. The first arriving unit will advise additional units to respond Code 1 whenever possible or appropriate. The term "go easy" shall be synonymous with Code 1.
3. Drivers shall avoid backing up whenever possible. When this is unavoidable, guides shall be used at all times. Guides shall remain in visual/radio contact with the driver, preferably at the right rear corner of the apparatus. If no guide is available, the driver shall dismount and walk completely around the apparatus before backing.
4. All employees are required to use seat belts at all times when they are operating a department vehicle equipped with seat belts. Anyone riding as a passenger in a department vehicle is also required to use seat belts.
5. Compartment doors shall not be left open and unattended unless drying in winter months.
6. Transmission by radio is prohibited while the apparatus is inside the station unless the vehicle is not running and is parked inside of the station.
7. All personnel shall ride only in regular seats provided with seat belts. Riding on tailboards or other exposed positions is not permitted on any vehicle at any time.
8. During an emergency response, fire vehicles are not to pass other emergency vehicles. If passing is absolutely necessary, arrangements should be conducted through radio communications. Privately owned vehicles (POVs) shall not pass Department vehicles with the exception of district chiefs who have radio contact with the emergency vehicle.

VII. Driving on the fireground

A. The unique hazards of driving on or adjacent to the fireground require the driver to use extreme caution and to be alert and prepared to react to the unexpected. When driving apparatus or personal vehicles on the fireground, drivers must resist the tendency to drive hastily or imprudently due to the urgent nature of the fireground operations. Drivers must consider the dangers their moving vehicles pose to fireground personnel and spectators who may be preoccupied with the emergency, and may inadvertently step in front of or behind a moving vehicle. Personnel shall not dismount fire apparatus until they have come to a complete stop and have been instructed to do so by the officer in charge. When stopped at the scene of an incident, vehicles should be placed to protect personnel who may be working in the street, and warning lights shall be used to make approaching traffic aware of the incident. At night, vehicle mounted floodlights and any other lighting available shall be used to illuminate the scene. Bright lights on oncoming traffic should be avoided whenever possible. All personnel working in or near traffic lanes shall wear their turnout gear with the reflective striping.

VIII. Privately owned vehicles

A. Personnel responding to alarms in authorized privately owned vehicles shall comply with this procedure. Privately owned vehicles shall yield the right-of-way to all Department vehicles responding Code 3. Drivers shall use due caution when approaching intersections opposite the travel of Department apparatus. Personnel responding to the station in privately owned vehicles shall respond Code 3 only until appropriate apparatus is en route to the incident. Department members are responsible for maintaining their vehicles in a safe operating condition.

IX. Emergency response policy

A. Department vehicles shall be operated in a manner that provides for the safety of all persons and property. Safe arrival shall always have priority when driving en route to an emergency incident.

1. Prompt, safe response shall be attained by

    a. Leaving the station in a standard manner.

        (1) Quickly mounting apparatus.

        (2) All personnel on board, seated, and seat belts on.

        (3) Station doors fully opened.

(4) Compartment doors closed and positively latched.

(5) Use of radios in the station is prohibited.

b. Driving defensively and professionally at safe speeds.

c. Knowing where you are going before leaving the station. Plan your route taking into account traffic, type of road, construction, weather, etc.

d. Using warning devices to move around traffic and to request the right-of-way in a safe and predictable manner.

e. Approach the scene with caution.

f. Ensure the vehicle has stopped before dismounting.

2. Fast response shall not be attained by

a. Leaving quarters before crew has mounted safely and before apparatus doors are fully opened.

b. Driving too fast for conditions.

c. Driving recklessly without regard for safety.

d. Taking unnecessary chances with negative right-of-way intersections.

3. Definitions

a. Code 1—Driving without emergency lights or sirens and within normal driving limits.

b. Code 3—Driving with emergency lights on and audible warning devices activated.

Attachment:

Reprint of State Vehicle Code

APPENDIX 5

TRAINING PROCEDURE

Minimum Department Training Requirements **No: ______**

Effective: ________ **Page: 1 of 2**

Purpose

To list the minimum levels of training, license, or certification to be obtained during the member's probationary period and maintained throughout the member's employment with the Department.

Procedure

I. All members of the Department are required to obtain and maintain the following minimum training requirements during their probationary period (with the exception of an Emergency Medical Technician license that shall be obtained within 24 months following station assignment).

A. State Department of Public Health, Emergency Medical Services: First Responder license. This remains the minimum requirement for employees with seniority dates before October 01, 1988.

B. American Red Cross Professional Rescuer Basic Life Support certification.

C. Fire Fighter I certification.

D. Hazardous Materials First Responder: Operation level training.

E. Apparatus driver certification.

F. State Department of Public Health, Emergency Medical Technician license must be obtained within twenty-four months of station assignment.

II. Failure to obtain an EMT license will result in disciplinary action returning the member to the rank of Probationary Fire Fighter.

III. Once successfully obtaining the required license or certification, all Department members are required to maintain these as the minimum training requirements of the Department.

IV. Newly hired members may request to be granted equivalency under this procedure. When presented with proper credentials, the training division, with the approval of the fire chief, may grant equivalency or require that the newly hired members complete elements of the Department training program. The training division shall respond to all equivalency requests in writing.

APPENDIX 6

JOB DESCRIPTION

Position Title: ____________________________

Supervised By: ___________________________

Division: _______________________________ **No:** ______

General purposes

Working under the general guidance and direction of a sergeant or other Department officer, he/she will prepare for and respond to hostile fires, medical emergencies, personal injury accidents, and hazard material spill/leaks and provide assistance to the general public. He/she will be required to participate in public fire/safety programs and activities.

Essential duties

1. Ensures that all assigned station apparatus and equipment are in a state of readiness.
2. Carries out their duties in a safe and efficient manner in accordance with department policy and recognized safe practices.
3. Assists with the supervision of Probationary Paid-on-Call personnel in their duties as directed.
4. Performs tasks during emergency scene operations including: locating incidents using available resources; driving and operation of apparatus; the proficient use of equipment, tools, and supplies to suppress hostile fires; rescue persons; provide pre-hospital emergency medical treatment; and contain hazardous material spills or leaks.

5. Performs tasks during non-emergency operations including: maintenance of facilities, apparatus and equipment to include cleaning and minor repairs; inventory and stocking of medical supplies; updating resource materials; delivering public education programs; and assisting with the company inspections/surveys.
6. Actively participates in departmental activities.
7. Ensures that their conduct and performance conform to department policy and community standards.
8. Reports violations of Department policy/procedures or illegal acts to supervisor.
9. Attends and actively participates in company and available outside training opportunities.
10. Prepares incident and associated reports that are accurate and complete.
11. Maintains a positive working relationship with all department members.
12. Coordinates and exchanges information with sergeants, other officers, and firefighters.

Peripheral duties

1. Performs the duties of subordinate personnel in their absence, and fills in for staff members, as assigned by the fire chief.
2. Attends conferences and meetings to keep informed of the activities of the Department, the city, and the fire service.

Minimum education and experience qualifications

- High school diploma or equivalent
- Fire Fighter I certification (within first year of service)
- Fire Fighter II certification (within third year of service)
- Medical First Responder training (within first year of service)
- Basic Emergency Medical Technician license (within second year of service)
- One year of experience in the Department (probationary year)
- Any combination of equivalent education and/or experience

Necessary knowledge, skills, and abilities

Extensive knowledge of:

- Department policy, procedures, personnel, facilities, apparatus, equipment, and organizational philosophy

Thorough knowledge of:

- Fire behavior and characteristics
- Firefighting techniques, practices, and standards
- Basic arson detection techniques
- City and Department policy and procedures
- Station apparatus and equipment capability
- EMS techniques, practices, and standards

Working knowledge of:

- Department operational procedures
- Department policies
- Skill in operation of listed equipment and apparatus

Ability to:

- Work effectively with other staff, supervisors, and the public
- Follow verbal and written instructions
- Establish and maintain effective working relationships
- Handle the physical requirements of the job
- Analyze situations quickly and correctly as well as make decisions
- Analyze situations correctly and make decisions regarding the management of assigned personnel
- Work effectively as part of a management team
- Prepare incident reports and correspondence
- Communicate effectively, both orally and in writing
- Effectively analyze situations and provide solutions to problems

Desirable knowledge, skills, and abilities

- Basic Emergency Medical Technician license

- Hazardous Material Operations course
- Confined Space Operations course
- Hazardous Materials Technician training
- Experience working with career and paid-on-call personnel
- Familiarity with the area

Special requirements

- Must maintain CPR certification for the Professional Rescuer
- Must maintain Automatic External Defibrillator certification
- Must maintain driver certification for all apparatus
- Must possess a valid state driver license
- Must be able to read, write, and speak the English language
- No felony convictions or criminal history

Selection guidelines

May include any or all of the following: formal request for promotion, review of education, training, and experience; written examination; psychological examination; oral board; background/driver's license check; offer of employment; post physical examination, including drug screen.

Apparatus and equipment used

All vehicles/apparatus, two-way radios, pagers, personal computers, telephones, calculators, tape recorders, photo equipment, miscellaneous EMS equipment and supplies, operational tools, and related equipment.

Physical demands

The physical demands described herein are representative of those that must be met by any member to successfully perform the essential functions of this job. Reasonable accommodations may be made to enable individuals with disabilities to perform the essential functions.

While performing the duties of this job, the member is frequently required to stand; sit; walk; talk; listen; use hands or fingers to handle or operate objects, tools, or controls; and reach with hands and arms. The member is frequently required to climb or balance; stoop, kneel, crouch, or crawl; and taste or smell. The member must frequently lift and/or move heavy objects. Specific vision abilities required by this job include close, distance, color, peripheral vision, depth perception, and the ability to focus.

Work environment

The work environment characteristics described herein are representative of those a member may encounter while performing the essential functions of this job. Reasonable accommodations may be made to enable individuals with disabilities to perform the essential functions.

Work is performed primarily in office, vehicle, and outdoor settings, in all weather conditions including temperature extremes during day or night shifts. Work is often performed in emergency and stressful situations. Individual is exposed to sirens and hazards associated with fighting fires and rendering emergency medical assistance including infectious substances, smoke, noxious odors, fumes, chemicals, solvents, and oils.

The member occasionally works near moving mechanical parts, in high, precarious places and is occasionally exposed to wet and/or humid conditions, fumes or airborne particles, toxic or caustic chemicals, radiation, risk of electrical shock, and vibration. The noise level in the work environment is usually quiet in office settings, moderate during the daily work routine, and loud at the emergency scene.

The duties listed above are intended only as illustrations of the various types of work that may be performed. The omission of specific statements of duties does not exclude them from the position if the work is similar, related, or a logical assignment to the position.

The job description does not constitute an employment agreement between the employer and member and is subject to change by the employer as the needs of the employer and requirements of the job change.

APPENDIX 7

Personnel Procedure

Probationary evaluation process/requirements : **No: ______**

Effective: ________________________________ **Page 1 of 3**

Purpose

To outline the requirements and expectations of a member hired by the Department and serving during the recruit school and probationary period. Each member is encouraged to become completely familiar with this procedure utilizing its provisions and objectives to learn his or her responsibilities as a measure of personal growth with the Department.

Procedure

I. Definitions—For purposes of this procedure, the following definitions are established.

   A. *Applicant/Candidate*—Any person who is seeking employment as a paid-on-call fire fighter during the selection process.

   B. *Recruit Member*—Any person selected by the Department for employment and assigned to the Department recruit school having been appointed by the fire chief and taken the probationary oath.

   C. *Probationary Paid-On-Call Fire Fighter*—Any new member under review who performs basic functions while learning, having satisfactorily met the requirements of recruit member (see A-415.1), and assigned to station response for a period of 12 months. This shall be known as the probationary period.

D. *Probationary Paid-On-Call Fire Fighter Review*—The final process of review and evaluation conducted at the end of the 12-month probationary period prior to the recommendation for promotion to the rank of Fire Fighter I.

II. Evaluation process

A. Recruit and probationary members of the Department shall be evaluated at 3-, 6-, and 12-month intervals.

B. Evaluations shall be completed by the station's officer cadre to which the member is assigned and department training division utilizing the appropriate Departmental forms and shall be submitted to Fire headquarters for approval and filing with the member's employment records. More frequent evaluations should be conducted with employees whose performance is below an acceptable level. This is particularly important if you contemplate releasing a probationer for unsatisfactory performance. It is necessary that releases of this type be fully documented with performance ratings, which show written evidence of attempts to motivate the employee to perform satisfactorily. Please contact the chief's office should you desire to make such a review.

C. The training division and an officer from the probationer's station shall complete the 3-, 6-, and 12-month evaluations on the appropriate Probationary Member Progress Evaluation form. After the review form has been prepared, it should be signed by the preparer and submitted to the district chief for review and approval prior to discussion with the employee.

D. Following the district chief's approval, an interview shall be conducted in private with the probationary member and the training officer or his designee. This interview should be more of a counseling session whereby strengths and weaknesses can openly be discussed and a satisfactorily agreed upon course of improvement established. Schedule the interview in advance. The member should be told that the purpose of the meeting will be to discuss his/her performance. The meeting should be held in private, and it should not be interrupted, barring emergency response.

E. After the interview, the employee should read over the review form, write his/her comments, if any, and sign the form.

F. The completed evaluation shall then be forwarded to the fire chief's office by the established due date.

G. The 12-month evaluation shall include a probationary fire fighter review and an annual performance review form. The annual evaluation and review process should be conducted as indicated in the paid-callback member evaluation procedure. All probationary fire fighters shall be required to complete the probationary fire fighter review during the twelfth month of the probationary period. This shall be the sole means by which the station officer cadre and training division may recommend promotion to the rank of Fire Fighter I.

Attachments

3-Month Review Form

6-Month Review Form

12-Month Review Form

APPENDIX 8

THREE-MONTH PROGRESS EVALUATION

Member: ______________________ **Station No.** ______________________

Date of employment: _____________ **Station assignment date:** ___________

Return evaluation to training division before: ____________________________

To the satisfaction of the Training Division and its evaluators within the established procedure (No. 600.2), the probationary employee shall have completed the following objectives at the time of the three-month progress evaluation.

Requirements	Satisfactory	Unsatisfactory
1. Shall have read and be able to demonstrate familiarity with Department Rules and Regulations, Special Orders, and Procedures.	____	____
2. Be able to identify and name the Chief of Department and all Department staff officers and clerical personnel.	____	____
3. Know the names and ranks of all station officers.	____	____
4. Demonstrate an understanding of proper Departmental radio communication procedures and be able to state the call signals of all apparatus, apparatus operators, officers, sector commanders, incident commanders, district chiefs, fire chief, deputy chief, fire marshal, staff officers and investigators.	____	____
5. Know the location of all fire stations.	____	____
6. Know the apparatus response assignments by type of incident to all alarms as stated in Departmental procedures.	____	____
7. Know the basic addressing scheme in the City and list same for each mile road, north to south, and each cross road, east to west. Must be able to name target occupancies within first alarm area when given either address or occupancy name.	____	____

8.	State the requirements of private vehicle response to emergency incident location.	_____	_____
9.	Be familiar with and demonstrate understanding of the Department's safety procedures relative to protective clothing and the wearing of self-contained breathing apparatus.	_____	_____
10.	Be familiar with and demonstrate knowledge for servicing or obtaining service for self-contained breathing apparatus according to Department procedure.	_____	_____
11.	Be able to demonstrate the safe and proper operation of all breathing air compressors and cascade systems for the refilling of self-contained breathing apparatus cylinders.	_____	_____
12.	Be able to demonstrate the proper procedure for refilling pressure water and Light Water stored pressure water extinguishers as well as the procedure for requesting the recharging of any dry chemical, carbon dioxide or other special agent extinguishers located on apparatus or within his/her station.	_____	_____
13.	Be able to demonstrate the correct procedure for refilling both portable "D" cylinder oxygen bottles and the servicing of the station oxygen cascade.	_____	_____
14.	Properly demonstrate making a hydrant and hose connection as well as any signaling involved.	_____	_____
15.	Know how to operate and release a standard cot stretcher from medics and private ambulances.	_____	_____

As of this date, the probationary employee's performance and progress with regard to the above listed objectives has been:

❑ Satisfactory. Recommend continuation of employment.

❑ Improvement required. Continued employment contingent upon satisfactory development in areas identified above.

Recommend special review be held on ______________________________ (date).

❑ Unsatisfactory. Minimum objectives for job not met. Recommend termination or reassignment (specify):

____

____

____

____

____

____

____

Training Division   Date

To the satisfaction of the station officer cadre and within an established Departmental procedure, the probationary employee shall have completed the following objectives at the time of the probationary member's three-month progress evaluation.

Requirements	Satisfactory	Unsatisfactory
16. Be familiar with and demonstrate knowledge of any special station procedures, *i.e.*, maintenance and cleanup assignments, vehicle or station responsibilities, etc.	____	____
17. Be able to state the purpose of the "hot line" and demonstrate its use.	____	____
18. Be familiar with and demonstrate the use of emergency power generators and other emergency backup systems within the station (*i.e.*, manually operate overhead bay doors).	____	____
19. Be able to name the proper storage location of all equipment carried on apparatus or locate it within his/her assigned fire station.	____	____
20. Be able to demonstrate the proper and safe operation of all equipment carried on the apparatus or assigned to his/her fire station.	____	____
21. Be able to demonstrate the safe and proper operation of all power equipment located on apparatus or assigned to his/her station.	____	____
22. State what constitutes a proper uniform according to Departmental rules and regulations and where and when it is to be worn.	____	____
23. Maintain a minimum response percentage as defined by Departmental procedures of 40% of all tone-alerted emergencies within the member's response district.	____	____
24. Maintain an 80% attendance rate of all regularly scheduled drills at his/her station.	____	____

Area of Performance	Below Minimum	Needs Improvement	Meets Requirements	Exceeds Requirements
(Place your comments regarding each area directly under the category.)				
Technical or functional proficiency	_____	_____	____	_____
Contact with public	_____	_____	____	_____
Communication with employees	_____	_____	____	_____
Work performance	_____	_____	____	_____
Participation/attendance	_____	_____	____	_____
Punctuality/response	_____	_____	____	_____
Progress	_____	_____	____	_____
Decisiveness and Judgment	_____	_____	____	_____

As of this date, the probationary employee's performance and progress with regard to the above listed objectives has been:

❑ Satisfactory. Recommend continuation of employment.

❑ Improvement required. Continued employment contingent upon satisfactory development in areas identified above.

Recommend special review be held on ____ (date).

❑ Unsatisfactory. Minimum objectives for job not met. Recommend termination or reassignment (specify):

____

____

____

____

____

____

Station Officer Date

I have seen and discussed this performance with the evaluator.

❑ I have no comments to make

❑ I request a discussion of my review with (select one) ❑ District Chief ❑ Fire Chief

❑ I have the following comments. (Use additional paper if necessary.)

____

____

____

____

____

____

____

____

____

Member Date

District Chief's Comments:

____

____

____

____

____

____

____

____

____

District Chief Date

Approved ❑ Yes ❑ No

____

Date

SIX-MONTH PROGRESS EVALUATION

Member: ______________________ **Station No.** ______________________

Date of employment: ____________ **Station assignment date:** __________

Return evaluation to training division before: ____________________________

The following are specific objectives the probationary member shall have completed by the sixth month of station assignment. Since the last probationary review, the probationary fire fighter must have maintained the 45% response percentage to all alarms as defined in Department procedure. the probationary fire fighter must have maintained an 80% attendance of all assigned station drills at his/her station.

Requirements	Satisfactory	Unsatisfactory
1. Demonstrate knowledge of the location, capability, and proper use of all medical equipment assigned to his/her station and apparatus.	____	____
2. Be familiar with the differing medical apparatus and equipment assigned to all other fire stations.	____	____
3. Demonstrate knowledge of the medical capabilities of the fire departments and private ambulance companies within the immediate surrounding communities.	____	____
4. Demonstrate the proper operation and release of a standard cot and stretcher.	____	____
5. Be able to name the location of area hospitals and state the location by closest mile roads and the best routes to be used to reach them from the member's response district.	____	____
6. Have satisfactorily completed the Department's driver training program, being able to safely drive all vehicles assigned to his/her station and how to properly activate and utilize all emergency warning signals (see Driver's Training procedure).	____	____
7. Demonstrate the proper operation of all vehicle-mounted and portable radios assigned to his/her station.	____	____

8. Be able to state the proper channel assignment for all frequencies utilized in vehicle-mounted and portable radios assigned at his/her station. _____ _____

9. Be able to demonstrate the safe operation of all vehicle-mounted pumps and appliances on apparatus assigned to his/her station and have completed the Department's pump operator's certification course. _____ _____

10. Identify the hazardous material transportation identification resources provided on Department apparatus and demonstrate correct usage of the Department of Transportation Guidebook. _____ _____

11. Have obtained certification in public safety responder CPR. _____ _____

Area of Performance	Below Minimum	Needs Improvement	Meets Requirements	Exceeds Requirements
(Place your comments regarding each area directly under the category.)				
Knowledge of work	_____	_____	_____	_____
Technical or functional proficiency	_____	_____	_____	_____
Contact with public	_____	_____	_____	_____
Communication with employees	_____	_____	_____	_____
Work performance	_____	_____	_____	_____
Participation / attendance	_____	_____	_____	_____
Punctuality / response	_____	_____	_____	_____
Progress	_____	_____	_____	_____
Decisiveness and judgment				

As of this date, the probationary employee's performance and progress with regard to the above listed objectives has been:

❑ Satisfactory. Recommend continuation of employment.

❑ Improvement required. Continued employment contingent upon satisfactory development in areas identified above.

Recommend special review be held on ____ (date).

❑ Unsatisfactory. Minimum objectives for job not met. Recommend termination or reassignment (specify):

____

____

____

____

____

____

Station Officer Date

I have seen and discussed this performance with the evaluator.

❑ I have no comments to make

❑ I request a discussion of my review with (select one) ❑ District Chief ❑ Fire Chief

❑ I have the following comments. (Use additional paper if necessary.)

____
____
____
____
____
____
____
____

____
Member Date

District Chief's Comments:

____
____
____
____
____
____
____
____

____
District Chief Date

Approved ❑ Yes ❑ No

____
Date

PROBATIONARY PROGRESS EVALUATION

Member: ____________________ **Station No.** ____________________

Date of employment: ____________ **Station assignment date:** ___________

Return evaluation to training division before: ____________________________

In accordance with Fire Department Procedure, all probationary fire fighters shall be required to complete the probationary fire fighter review during the twelfth month of the probationary period. This shall be the sole means by which the station officer cadre may recommend promotion to the rank of Fire Fighter I.

The following are specific objectives the probationary member shall have completed by the twelve-month probationary review of station assignment. Since the last probationary review, the probationary fire fighter must have maintained the 45% response percentage to all alarms as defined in Departmental procedure. The probationary fire fighter must have maintained an 80% attendance of all assigned station drills at his/her station.

Requirements	Satisfactory	Unsatisfactory
1. Satisfactorily completed the three-month probationary evaluation.	_____	_____
2. Satisfactorily completed the six-month probationary evaluation.	_____	_____
3. Obtained FF-I Certificate.	_____	_____
4. Demonstrates the ability to accurately complete a fire report.	_____	_____
5. Obtained EMS First Responder training course certificate.	_____	_____

DATE ISSUED: ______________________________

AND

American Red Cross and/or American Heart Association Public Safety Rescuer CPR/BLS Certification.

DATE ISSUED: ______________________________

OR

Obtained a valid State Department of Public Health EMT-B, EMT-A, OR EMT-INT license. (Provide copy)

LICENSE NUMBER: ___________________________

EXPIRATION DATE: __________________________

		Yes	No
6.	Completed 6 hours of public fire education activity.	____	____
7.	Participated in the annual Fire Prevention Open House during Fire Prevention Week at Fire Headquarters.	____	____

	Requirements	Yes	No
8.	Optional requirement for career fill-in assignments. In order to be eligible for future fill-in assignment (for full-time personnel), the probationary member must complete 20 hours of fill-in training, working as a third member at the training and service pay rate. Record training dates here and indicate if member satisfactorily completed this requirement.	____	____

Dates:

____

9.	Has the probationary member been unable to or restricted from participating in Department activities and programs for any length of time due to sickness, injury, personal leave of absence, disciplinary action, etc. during the probationary period? If yes, please explain:	____	____

____

____

____

____

____

____

10.	Has the member received any Department, civilian, or personal commendations (*i.e.*, Recruit Fire Fighter of the Year, Medal of Valor, distinguished Service Award, Certificate of Merit, Letter of Appreciation) during the probationary period? If yes, please explain:	____	____

____

____

____

____

____

____

Topic	Rating 1	Rating 2	Rating 3
Safety Using equipment properly and safely; promoting safe operations.	❑Unsafe practices or fails to follow safety standards and regulations (Specify below)	❑Consistently performs in safe manner of operation and follows safety standards/ regulations.	❑Uses intuition in eliminating potential safety hazards; encourages others to work safely. (Specify below)
Dependability Following instructions and carrying out assigned tasks.	❑Instructions and procedures not adequately followed; does not respond in a timely manner to requests from supervisor. (Specify below)	❑Carries out assigned duties in a timely manner.	❑Exceptional reliability; exceeds supervisor's expectations in carrying out requests. (Specify below)
Housekeeping Keeping work area and equipment clean and maintained.	❑Neglects routine station housekeeping and equipment/apparatus maintenance. (Specify below)	❑Performs required station housekeeping and equipment/apparatus maintenance.	❑Self-motivated to perform routine housekeeping and maintenance and uses purposeful task time efficiently. (Specify below)

Contact with Public Manner in which employee interacts with public.	❑ Occasionally exhibits ineffective communication skills with the public; may exhibit lack of composure or tactlessness. (Specify below)	❑ Exhibits poise, tact, self-control, and businesslike manner.	❑ Projects optimism, confidence, and enthusiastic attitude. (Specify below)

____

____

____

____

____

Working with Others Maintaining positive working relationship with peers and superiors; accepting instructions and assignments; assisting others to accomplish work group activities.	❑ Indifferent toward work objectives and assisting others; may place work group objectives behind personal or social objectives; may be a source of conflict. (Specify below)	❑ Gets along well with supervisors and peers; strives to achieve work group objectives.	❑ Respects and is respected by others, regularly provides assistance, reinforcement, and support to others. (Specify below)

____

____

____

____

____

Job Knowledge Application of technical and procedural know-how to get the job done.	❑ Knowledge is limited to certain areas; insufficient to handle most tasks. (Specify below)	❑ General knowledge is sufficient to handle most tasks.	❑ Broad general knowledge and in depth expertise in most areas; can handle advanced and unusual tasks. (Specify below)

____

____

____

____

____

Motivation to achieve Being results-oriented desire to excel on the job; working steadily and actively.	❑ Takes action only when instructed; must be prodded to keep working. (Specify below)	❑ Does assigned tasks diligently; accepts responsibility for work beyond regular duties when necessary.	❑ Accepts responsibility for getting the job done; initiates independent action. (Specify below)

____

____

____

____

____

Functional proficiency Basic skills and precision in working with apparatus/equipment.	❑ Employee is unable to use job-required equipment and required skills. (Specify below)	❑ Employee is proficient in the use of all equipment and required skills; provides all the functional proficiency expected of the job.	❑ Employee has exceptional skills and versatility in working with equipment (Specify below)

____

____

____

____

____

Decisiveness and Judgment Ability to make productive decisions based on all available facts.	❑ Employee is unable to make required job decisions or decisions are incorrect. (Specify below)	❑ Employee can be consistently relied upon to make decisions, which require minimum review, amendment or correction.	❑ Employee makes independent decisions within scope of responsibility. Decisions are based on sound judgment and evaluation of pertinent data. Decisions are timely and firm. (Specify below)

____

____

____

____

____

Run Percentage Annual run percentage of all toned-out runs.	❑ Below 35% (Specify below)	❑ Between 35%–55%	❑ Above 55% (Specify below)

____

____

____

____

____

____

Training Involvement Attendance of station drills and other available training.	❑ Below 80% drill attendance. (Specify below)	❑ 80% drill attendance; participates in required Department training and maintains required licenses.	❑ Exceeds required drill attendance and participates in additional training opportunities. (Specify below)

____

____

____

____

____

____

Personal Appearance Uniform usage and appearance to the public.	❑ Consistently out of uniform and has unkempt appearance. (Specify below)	❑ Meets requirement for uniform usage and has generally neat appearance.	❑ Exceeds department requirement for uniform usage; well groomed and has pride in his appearance. (Specify below)

____

____

____

____

____

____

Attitude Feeling toward work assignments and overall Department.	❑ Indifferent toward work assignments, generally negative opinion of overall Department or is disruptive to others. (Specify below)	❑ Good overall attitude toward Department and its activities; has balance between good sense of humor and the need to take job seriously.	❑ Performs assignments diligently; shows exceptional interest and commitment, is a positive motivating force to others. (Specify below)

____

____

____

____

____

____

____

As of this date, the probationary employee's performance and progress with regard to the above listed objectives has been:

❑ Satisfactory. Recommend member be promoted to the rank of Fire Fighter I effective ____/____/____.

❑ Improvement required. Continued employment contingent upon satisfactory development in areas identified above.

❑ Recommend that probationary period be extended until ____/____/____. (Any extension of the probationary period must be accompanied by a complete explanation including terms and conditions.)

❑ Unsatisfactory. Recommend the probationary member be terminated at this time. (Any termination recommendation must be accompanied by a full and detailed explanation.

____

____

____

____

____

____

____

____

Station Officer Date

I have seen and discussed this performance with the evaluator.

❑ I have no comments to make

❑ I request a discussion of my review with (select one) ❑ District Chief ❑ Fire Chief

❑ I have the following comments. (Use additional paper if necessary.)

Member Date

District Chief's Comments:

District Chief Date

Approved ❑ Yes ❑ No

Date

APPENDIX 9

Fire Department
Rules and Regulations

(Sample)

Effective: ______________________________, ____, 200__

APPROVAL

Recommendation and approval:

I, Fire Chief of the City, in accordance with the powers and authority as vested in me, do hereby recommend and approve the attached Fire Department Rules and Regulations, this __________ day of ____________________, 200__.

____

Chief

I, City Manager, in accordance with Section 2.54 of the City Code and other powers vested in me, approve the attached Fire Department Rules and Regulations, this __________ day of ____________________, 200__.

____

City Manager

PURPOSE AND SCOPE

1-1. The Fire Department shall be administered in conformance with the powers and duties prescribed by the laws of the United States, the state, municipal code, city manager directives, and the policies and procedures authorized by the fire chief, and be governed by the following Rules and Regulations.

1-2. In case of any conflict between these Rules and Regulations and the provisions of the city charter, state, or laws of the United States, then these rules and regulations shall be subordinate thereto.

1-3. The fire department shall be responsible to prevent fires from starting; to save life and property, protect the citizens should fire, medical emergency, or other threatening situation occur; to confine fire to its place of origin and to extinguish fire by the most effective means possible; and to provide mutual aid to surrounding communities in accordance with established policy.

1-4. The purpose of these Rules and Regulations is to promote efficiency and maintain discipline within the fire department in order that it may function most effectively for the protection of life and property in the city.

1-5. The fire chief retains the right to add or to amend the Rules and Regulations with the approval of the city manager.

1-6. These Rules and Regulations cover either in a specific or general way the obligations and duties of the members of the fire department. These Rules and Regulations are not designed or intended to limit any member in the exercise of good judgment or initiative in taking the action a reasonable person would take in extraordinary situations.

1-7. Members shall be familiar with and see to it that the Rules and Regulations are adhered to by all members.

1-8. Members not in compliance with these Rules and Regulations and fire department special orders, policies, procedures, and inter-office correspondence may be subject to disciplinary action.

1-9. The fire department is an equal opportunity employer. The fire department does not discriminate against any member or applicant for membership because of race, religion, color, national origin, age, sex, height, weight, or the presence of a non-job related medical condition or handicap.

1-10. The use of masculine pronouns is not meant to be gender specific, nor is it meant to be offensive to female members of the fire department. The use of masculine pronouns shall imply that both female and male members are referenced in this document.

DEFINITIONS

2-1. BOARD OF CHIEFS—shall mean the fire chief, deputy fire chief, fire marshal, district chiefs, or their designee, and staff lieutenants and shall serve in an advisory capacity to the fire chief.

2-2. CITY—shall mean the City of _________________, a municipal corporation.

2-3 DEPARTMENT—shall mean the fire department.

2-4. DEPARTMENT OFFICERS—shall mean all members of the rank of sergeant and above, including the station training officer.

2-5. DIRECTIVES—shall mean any oral or written orders and directions to include but not be limited to special orders, policies and procedures, and interoffice correspondence to be followed by all members.

2-6. DISTRICT CHIEF—shall mean the paid-on-call chief located at each station and appointed by the fire chief with the approval of the city manager.

2-7. FIRE CHIEF—shall mean the executive head of the fire department appointed by the city manager.

2-8. DEPUTY CHIEF—shall mean the executive head of the fire department in the absence of the fire chief and appointed by the fire chief with approval of the city manager.

2-9. FIRE MARSHAL—shall mean the executive head of the Department in the absence of the fire chief and deputy fire chief and shall be responsible for the functions related to the Fire Prevention Division.

2-10. GRIEVANCE—a written complaint pertaining to discipline and/or working conditions.

2-11. GRIEVANCE BOARD OF APPEALS (GBA)—shall mean a panel comprised of a paid-on-call member from each station and a staff member, appointed by the fire chief.

2-13. MEMBER—shall mean any employee of the fire department including probationary fire fighters.

2-14. PROBATIONARY FIRE FIGHTER—shall mean a new member under review who performs basic functions while learning for a period of time, or as a demoted fire fighter under review for a specific period of time as a result of disciplinary action.

2-15. PROCEDURES—shall mean those specific directives that outline the standard operating policies affecting Department operations and personnel.

2-16. SENIORITY—shall refer to time in grade beginning with the last date of hire or promotion not to include leaves of 30 days or more. Seniority with respect to chain of command shall mean the time accumulated at a rank level.

ORGANIZATION AND CHAIN OF COMMAND

3-1. The fire department shall be organized as a combination paid-on-call and full-time Department with primary fire fighting and emergency response capability.

3-2. The organizational chart shall be used in matters involving the administration of the fire department except as otherwise provided in these Rules and Regulations.

3-3. The Department shall conform to the chain of command as outlined in policies and procedures.

3-4. Promotions within the system shall be in accordance with policies and procedures.

FIRE CHIEF

4-1. The fire chief shall be appointed by and responsible to the city manager and shall be in charge of the fire department.

4-2. He shall be responsible for planning and implementing all phases of Department operation-administration, including but not limited to fire control, hazardous materials incident, planning and mitigation, human rescue, fire investigations, fire prevention, and training. All members of the Department are subject to his orders. He shall be responsible for the enforcement of the Department Rules and Regulations and directives.

4-3. He shall be responsible for the care and management of all buildings, apparatus, equipment supplies, and any and all other property assigned to the fire department.

4-4. He shall be responsible to see that the officers and members of the Department are properly trained to perform their duties and responsibilities.

4-5. He shall be responsible for the preparation and management of budget, as directed by the city manager, including control of expenditures and payroll, and all other matters pertaining to financial operations.

4-6. He shall submit a report to the city manager covering Department activities for each month.

4-7. He shall have developed and maintained a policy and procedure manual for the Department. Such manual shall establish standards for the operation and control of the Department within the framework of these Rules and Regulations.

4-8. He shall determine that proper command and emergency response capability is maintained at all times.

4-9. He may appoint a member to the position of district chief for each station with the approval of the city manager.

4-10. He shall make promotions within the Department pursuant to procedures established by the Department or bargaining agreement between the city and the union.

4-11. He may issue directives supplementing these Rules and Regulations for the efficient and proper operation of the Department.

4-12. Nothing contained in these Rules and Regulations shall be construed so as to limit the power and authority granted the fire chief by the city, state, and federal laws.

4-13. He shall cause to be done, a personnel evaluation of all employees as needed. Members failing to meet the minimum requirements may be subject to demotion or dismissal at the discretion of the fire chief.

NEW MEMBERS' OATH

5-1. All new probationary members shall be deemed to be under the selection process during the time of probation and may be dismissed at will by the fire chief and shall have no right of appeal.

5-2. Upon initial acceptance into the Department, the member must have fulfilled minimum entrance and qualification requirements and appear before the fire chief to take the following oath: "I, ________________(name)________________, do solemnly swear that I will support the Constitution of the United States, the constitution of the state, and the city charter and ordinances. That I will comply with the Rules and Regulations of the fire department and will discharge the duties of fire fighter to the best of my ability. I understand that I must satisfactorily complete a probationary period which is part of the selection process."

CONFIRMED MEMBERS' OATH

6-1. Upon satisfactory completion of the initial probationary period, the member must have completed the qualification requirements and appear before the fire chief to take the following oath: "I, ______________(name)______________, do solemnly swear that I will support the Constitution of the United States, the constitution of the state, and the city charter and ordinances. That I have read, understand and will comply with the Rules and Regulations, special orders, procedures, and interoffice correspondence of the fire department and will discharge the duties of fire fighter to the best of my ability."

6-2. A signed and dated copy of the Department's oath will be obtained from each member and placed into each member's personnel file.

APPARATUS, EQUIPMENT, AND FACILITIES

7-1. Members shall be responsible for all Department apparatus, equipment, and facilities.

7-2. Apparatus, equipment, and facilities shall be used for fire department use only, unless prior approval by the fire chief, deputy chief, fire marshal, and/or district chief shall be obtained to do otherwise.

7-3. No fire department equipment shall be removed from the station for personal use unless prior approval is given by the fire chief and/or district chief.

7-4. Any new equipment or change in location of equipment on the apparatus unit must first be approved by the fire chief, deputy chief, and/or district chief. It shall be the responsibility of the district chiefs to notify all members at their respective station of any and all changes.

7-5. Broken, lost, damaged, or unsafe equipment shall be reported immediately by all members to their senior officer who shall forward such information to headquarters. All stolen or suspected stolen Department property shall be reported to the police department and the fire chief.

EQUIPMENT ISSUE AND UNIFORMS

8-1. The member is responsible for all equipment and uniforms issued to them and shall treat such with care. Repairs, replacement, and/or return of any equipment shall be facilitated through the member's supervisor. The member, at the member's expense shall replace lost, stolen or damaged issued equipment or uniforms if, in the opinion of the fire chief, the member was negligent.

8-2. When personnel leave the employ of this Department, all issued equipment shall be returned to the Department. To cover the costs for all equipment not returned, a deduction may be made from the employee's final check.

8-3. No equipment, except Department issued lights and sirens, shall be carried in private vehicles, except when on Fire Department business or unless otherwise approved by the fire chief. Only Department issued or members shall use approved lights and sirens. Department issued lights are not to be kept on top of the vehicles unless responding to an emergency. No emergency lights shall be permanently mounted on the roof of a privately owned authorized vehicle.

8-4. Uniforms of all members shall conform with and only be worn in accordance with regulated specifications prescribed by the Department's policy and procedures.

8-5. Members shall wear, when reasonably expected, the departmental uniform when representing the Department.

8-6. No combination of departmental uniform and civilian dress is permitted—with the exception of the wearing of Department jacket, coat, t-shirt, and/or cap when representing the Department.

8-7. The personal appearance of Department members is a reflection on both the city and the fire department. When wearing Department issued uniforms, the uniform shall be clean and members shall be well groomed and neat in appearance.

8-8. No member shall bring discredit or unfavorable attention to the fire department or the city, whether are on or off duty. The unauthorized use of the Department's name, uniforms, or insignias is strictly prohibited.

CHARGE PROCEDURE AND DISCIPLINARY ACTION

9-1. Any violation of the provisions of these Rules and Regulations or directives of the fire department or the neglect or evasion of duties prescribed herein, shall be subject to disciplinary action.

9-2. Charges preferred against a member of the Department shall be in writing and submitted within seven days of the alleged violation. Copies shall be jointly delivered to the member charged, the member's district chief or supervisor, and the chief of the department. charges, which are mailed and registered, to addressee only within seven days of the alleged violation shall be considered delivered to the member in accordance with these Rules and Regulations.

9-3. The fire chief, deputy fire chief, fire marshal, or a district chief may suspend any officer or member of the Department from duty when, in his or her judgment, the circumstances warrant such action.

9-4. Probationary employees may be subject to dismissal without cause.

9-5. All members of the Department may be subject to reprimand, suspension, demotion, or dismissal from the Department for violating any of these Rules and Regulations or directives governing the Department.

9-6. Notwithstanding an appeal, actions against a member may include a warning, reprimand, suspension, discharge, demotion, or loss of privileges.

9-7. Members of the Department shall answer any and all questions truthfully and directly as it may relate to any Departmental investigation. Members may have representation during any phase of the disciplinary process. Any expenses for such representation if any shall be at the member's expense.

9-8. No member shall obstruct, hinder, or impede any departmental investigation.

9-9. In recognition of the fact that each instance may differ in many respects for somewhat similar situations, the Department retains the right to treat each occurrence upon its individual merits without creating any precedent for the treatment of any other situation that may arise in the future and under extreme or unusual circumstance. The member's previous record may be considered and more or less severe action may be taken depending upon the circumstances and the individual's service record.

9-10. Members shall have the right to present grievances relative to the interpretation or application of these rules and disciplinary action.

9-11. The informal resolution of differences or grievances is encouraged at the lowest possible level of supervision. If the problem is not resolved, the following steps shall be followed:

Step 1—If a member feels he or she has a grievance, they shall submit through their district chief or supervisor a written request to bring the grievance to the Grievance Board of Appeals (GBA). Such requests shall be made within seven days of the date of notice of discipline and shall state the facts giving rise to the grievance, shall identify the provisions of the Rules and Regulations or directives allegedly violated, and shall indicate relief sought. The request shall be signed and forwarded to the chief of the department through the district chief or supervisor. The chief or his designee shall call a meeting of the GBA within seven days to review the grievance and hear from any or all parties with information pertinent to the case before them. The GBA shall have authority to:

1. Overturn the disciplinary decision
2. Concur with the disciplinary decision
3. Modify the disciplinary decision

The GBA shall reduce the decision to writing and shall forward same to the chief, district chief, supervisor, and member who requested the hearing. The GBA shall render a decision within seven days of the hearing.

Step 2—If the member does not concur with the decision of the GBA, he may request in writing a meeting with the fire chief. Such a request must be made within seven days of the GBA decision. The chief shall review the decision of the GBA and may at his discretion seek information from other members of the Department regarding the matter at hand. The fire chief may:

1. Overturn the decision
2. Concur with the decision
3. Modify the decision

The chief shall render a decision within seven days.

Step 3—If the member does not concur with the decision of the fire chief, he may request in writing a meeting with the city manager. Such a request must be made within seven days of the fire chief's decision and may at his discretion seek information from other members of the Department regarding the matter at hand. The city manager may:

1. Overturn the decision
2. Concur with the decision
3. Modify the decision

The decision of the city manager shall be final not withstanding any other rights afforded by law to an employee. The city manager shall render a decision within ten days.

Step 4—If the member does not concur with the decision of the city manager, he may submit the grievance to the American Arbitration Association with written notice delivered to the city manager within 10 days of receipt of the city manager's answer in Step 3 or the date such answer was due. The arbitrator shall be empowered, except as his powers are limited below, after proper hearing to make a decision in cases of alleged violations of the Rules and Regulations, policies and procedures. He shall have no power to subtract from, disregard, alter, or modify any portion of the Rules and Regulations, policies, or procedures. If either party disputes the arbitrability of any grievance under the terms of this agreement, the arbitrator shall first determine the question of arbitrability. In the event that a case is appealed to an arbitrator on which he has no power to rule, it shall be referred back to the parties without decision or recommendation on its merits. If the arbitrator's decision is within the scope of his authority as set forth above, it shall be final and binding on the member and the city.

The city and the member shall share the fees and expenses of the arbitrator equally. All other expenses shall be borne by the party incurring them.

1. All claims for back wages shall be limited to the amount of wages that the employee would otherwise have earned, less any compensation that he may have received from any source during the period of back pay.
2. No decision in any one case shall require a retroactive wage adjustment in any other case. By accepting arbitration, the member waives the right of any future litigation against the city on matters before the arbitrator. At the time of the arbitration hearing, both the city and the member shall have the right to call any employee as a witness and to examine and cross-examine witnesses. Each party shall be responsible for the expenses of the witnesses that they may call. Upon request of either the city or the member, or the arbitrator, a transcript of the hearing shall be made and furnished to the arbitrator with the city and the member having an opportunity to purchase their own copy. At the close of the hearing, the arbitrator shall afford the city and the member a reasonable opportunity to furnish briefs. The arbitrator will render his decision within thirty days from the date the hearing is closed or the date the parties submit their briefs—whichever date is later.

GENERAL RULES

10-1. The rules of general behavior apply to all members of the fire department on- and off-duty, unless otherwise denoted in these general rules. These rules are intended as examples of impermissible conduct, which is not intended to serve as an all-inclusive list.

10-2. Members shall be courteous in their dealings with the public and shall give them assistance when it is in their power to do so.

10-3. Members shall not engage in altercations or disruptive activities while on duty at Department/city functions or when on city property.

10-4. Members shall not threaten, intimidate, coerce, or interfere with their superiors, other members, or the public.

10-5. Members shall not use indecent, profane, sexist, ethnic slurs, or abusive language.

10-6. Members shall not become intoxicated (alcohol/controlled substances) while on duty, when in uniform, or respond to duty when under the influence of intoxicants.

10-7. Any member convicted of a felony or misdemeanor may be subject to immediate discipline up to and including discharge.

10-8. Members of the Department shall not destructively criticize the Department or its policies, programs, actions, or officers or make any written or oral statements, which tend to bring discredit to the Department.

10-9. Any member who has been arrested for a serious crime may be subject to suspension pending final determination.

10-10. Members shall accord all officers of the Department the reasonable respect due their rank and shall address and refer to all officers by proper title in the presence of the public and other Department members.

10-11. Members shall maintain a valid state driver's license and all other required licenses and certificates, and shall immediately submit a written report through the chain of command to the fire chief if such license is suspended, revoked, or restricted.

10-12. Members shall familiarize themselves with the laws governing the response of emergency vehicles and exercise due care according to the law.

10-13. Members are subject to being contacted in case of necessity during their off-duty time. They shall at all times maintain a telephone where they can normally be contacted and shall be sure that the Department knows the number of such telephone. Any change of residence or telephone number shall be reported to the Department in writing within 48 hours.

10-14. Members shall not present themselves as representing the Department before the city council, the state legislature, or any legislative body or media on any matter relative to the Department without the prior knowledge of the fire chief or his designee.

10-15. Members, while in uniform or while functioning as an employee of the fire department or presenting themselves to be an employee of the fire department, shall neither directly or indirectly, in any manner solicit, suggest, or request contributions, subscriptions, or donations or engage or take part in any scheme or enterprise intended or likely to induce any person or make presence of gifts of money or goods of any description without the sanction of the fire chief.

10-16. No member shall use the uniform badge or prestige of the Department for the purpose of personal gain or soliciting gifts.

10-17. Members shall not use official fire department channels of communication including stationery or bulletin boards, without prior approval of the fire chief or district chief.

10-18. Members shall not make maliciously false reports or gossip concerning the character and conduct of any other member or of the business of the Department to the discredit or detriment thereof.

10-19. No member shall publicly express any opinions on any litigation which the Department or city is involved or the circumstances surrounding same without prior permission from the fire chief.

10-20. Department provided or privately owned property on premise, shall be subject to entry, search, and inspection with the approval of the fire chief or his designee without notice.

10-21. The city shall not be held responsible for any privately owned items belonging to members.

10-22. Members shall not knowingly falsify or alter any personal records, pay records, or other records/reports.

10-23. Members shall report to their superior officers any complaint regarding unsafe practices, which might endanger the public or other members of the Department.

10-24. Members shall not litter or otherwise contribute to poor housekeeping, unsanitary, or unsafe conditions on any Department premises.

10-25. Members authorized to use private vehicles for emergency response are required to keep their vehicle in safe mechanical condition and appearance, and properly license and insure in accordance with Public Act ________________, (state) Vehicle Code.

10-26. Members planning to be unavailable for duty for more than 48 hours must notify their district chief or fire chief 24 hours in advance. Absence from the Department for more than 30 days shall require a written request and prior approval of the district chief and/or fire chief—failure to obtain approval may be grounds for termination.

10-27. Members flouting authority of any superior officer by obvious disrespect or by disputing or ignoring such officer's orders shall be guilty of insubordination.

10-28. Members shall refrain from exceeding their authority in giving orders.

10-29. Members who, by lack of good judgment, neglect, or carelessness cause or contribute to the damage of city property, shall be held responsible for their actions and may be subject to disciplinary action and/or financial restitution.

10-30. Members receiving orders or directions from other members, when the authority of the directing officer is delegated, shall regard such orders as coming directly from the designated officer and shall promptly execute said orders.

10-31. Members of the Department shall not furnish information pertaining to the business of the Department or about its members or past members to anyone not connected with the Department or city unless as authorized by the fire chief or his designee.

10-32. The growing of a beard by members of the fire department is strictly prohibited. Mustaches and sideburns shall be allowed when the same does not interfere with the safe use of self-contained breathing apparatus.

10-33. Members shall be medically fit commensurate with the duties and requirements of the position that a member occupies.

10-34. Members must maintain a 35% response record to tone-alerted incidents within their district during the course of a one-year period. This is subject to review by the district chief and fire chief.

10-35. Members must maintain an 80% attendance record at required Department training drills unless excused in writing by the fire chief, district chief, or his designee for reasonable cause.

10-36. In order to maintain personal physical fitness commensurate with the duties and requirements of those positions a member occupies, certain conditions and standards must be met. This shall include demonstrating proper conditioning through actual job performance and evaluations by medical examination and/or skills testing. If a member fails to perform, the member shall follow a corrective physical and/or training program and shall be retested within a reasonable period of time. If any member fails to follow the prescribed corrective program, he shall be subject to review, which may result in suspension or termination.

10-37. Members shall maintain a neat and professional appearance when representing the Department.

10-38. Jewelry shall not be worn that may be considered offensive or pose a safety hazard.

10-39. Members shall bring to the attention of the city any condition that inhibits that member's ability to perform prescribed fire fighting tasks.

10-40. The Department retains the right to require a member to have a physical exam at the city's expense at anytime. If, in the opinion of the city's doctor, an employee is not medically fit, he or she may be suspended or terminated.

10-41. Ignorance of Rules and Regulations or directives will not be accepted as an excuse for any violation thereof.

BOARD OF CHIEFS

11-1. The Board of Chiefs shall act in an advisory capacity to the fire chief and shall determine a regular schedule of meetings.

11-2. The Board of Chiefs shall exercise the following responsibilities.

1. Make recommendations on policies and procedures affecting the Department in its operations.
2. Make recommendations and assist in the preparation and review of the annual budget for the Department.
3. Make recommendations on the administrative and operating procedures of the Department including such items as staffing, equipment purchases and maintenance, medical emergency and fireground tactics, response policies, and other similar items affecting the efficient operation of the Department.
4. Make recommendations on programs, special projects, and all internal and community activities affecting personnel and the image of the Department to the public.
5. Be responsible for clear transmittal and dissemination of all Department policies, Rules and Regulations, and directives to all members of the Department.
6. Provide all other assistance and advice as may be requested by the fire chief.

OFFICERS

12-1. All officers shall be responsible for discipline of their subordinates and shall inspire and motivate members under their command as to promote and maintain efficiency and morale while accomplishing Department goals.

12-2. Officers shall be just, dignified, and firm in their dealings with subordinates and shall refrain from violent, abusive, or immoderate language in giving orders and directions.

12-3. Officers shall be responsible for making certain that all Rules and Regulations, policies and procedures, and directives are known by their subordinates and are carried out and obeyed and shall set an example to their subordinates in showing due regard for such regulations and orders.

12-4. Officers shall properly report well substantiated charges in writing. Any violations of the Department Rules and Regulations or other conduct discrediting to the interest of the Department shall be reported to their superior.

12-5. Officers in disagreement with an order or directive shall first obey the order and then express themselves to their superior officer at the appropriate time. Their opinions will be received and shall be given due consideration. They shall refrain from expressing dissatisfaction with Department policy or directives to or in the presence of their subordinates.

12-6. Officers shall relay all written communications through proper channels and no intervening officer shall suppress or disregard any such communication.

12-7. Officers of the Department shall be included in the definition of the general term member and shall be governed by all Rules and Regulations pertaining to such members.

AMENDMENTS, ADDITIONS, AND DELETIONS

13-1. When a member of the ____________ Fire Department believes an amendment, addition, or deletion should be made to these Rules and Regulations, the member shall submit in writing the same through the chain of command to the fire chief. The fire chief will review such proposed amendments with the Board of Chiefs and may deny or recommend approval to the city manager. If approved, the fire chief will forward such action to the city manager for his review and decision. The decision of the city manager shall be final. The Department shall ensure that all members are kept informed of any change affecting these Rules and Regulations, procedures, and directives.

Section V

FIRE PREVENTION AND LOSS REDUCTION

24

PLANNING FOR COMMUNITY FIRE AND EMERGENCY SERVICES

Jack A. Bennett
Douglas P. Forsman

CHAPTER HIGHLIGHTS

- The concept of master planning and the importance of the "safety element" in the general plan.
- Implementing "planning teams" within the department for the purpose of preparing the fire and emergency services plan.
- The importance of the fire chief's role in the master planning process and in political and legislative activities.
- How to use the methodologies described to produce a fire risk analysis.

THE CONCEPT OF MASTER PLANNING

Since the colonial days in the United States, the fire service has evolved into the first line of defense from unfriendly fires, emergency medical incidents, and a continually growing list of other related emergencies.

How did Benjamin Franklin and the pioneers after him decide where to locate the fire station in the village? How did they evaluate the needs of the "community at risk"? Whatever they did, most of us in this century are still living with their decisions or doing something to change their system to meet today's needs. The heart of master planning for fire and emergency services is found in the following questions, questions that must be asked by all fire chiefs.

- What are the greatest risks in the community?
- What does the community expect from the fire service?
- What level of service does the community get from the fire service?
- Can the level of service be improved?
- If improvements are necessary, what will they cost?
- What can the community afford?

This chapter will discuss master planning concepts, the role of the fire service in the planning and implementation phases, the methodologies used to complete the planning and implementation phases, and the future of planning community fire emergency services, with an emphasis on environmental concerns.

The Comprehensive Plan

The responsibility for master planning is normally delegated to local government. This planning process sets authority for regulating land use through zoning requirements, regulating or restricting the amount of growth, and establishing local controls over all types of new and existing developments. Guiding this process is a document called the *comprehensive plan.* In most cases these comprehensive plan documents are legally mandated, and the local governing body usually passes ordinances to put the comprehensive plan in place. This process usually requires public hearings and, sometimes, voter approval before the plans are enacted as statutes by local government. These comprehensive plan documents also contain data and analysis sections where the conclusions are supported and quantified.

Many states require comprehensive plans to be adopted, and then reviewed periodically. The components of a comprehensive plan are shown in Figure 24–1.

The safety element of the comprehensive plan will contain information about fire and emergency services in the community. Too often these "safety elements" are incomplete and have not had the input of the local fire department. The information in this chapter will help you to be a "facilitator and motivator" in the preparation of the safety element of the comprehensive plan.

The information on fire protection may include project-specific plans, building construction fire protection ordinances, interior and exterior life safety measures, wildland/urban interface, and special ordinances dealing with the fire problems of the community.

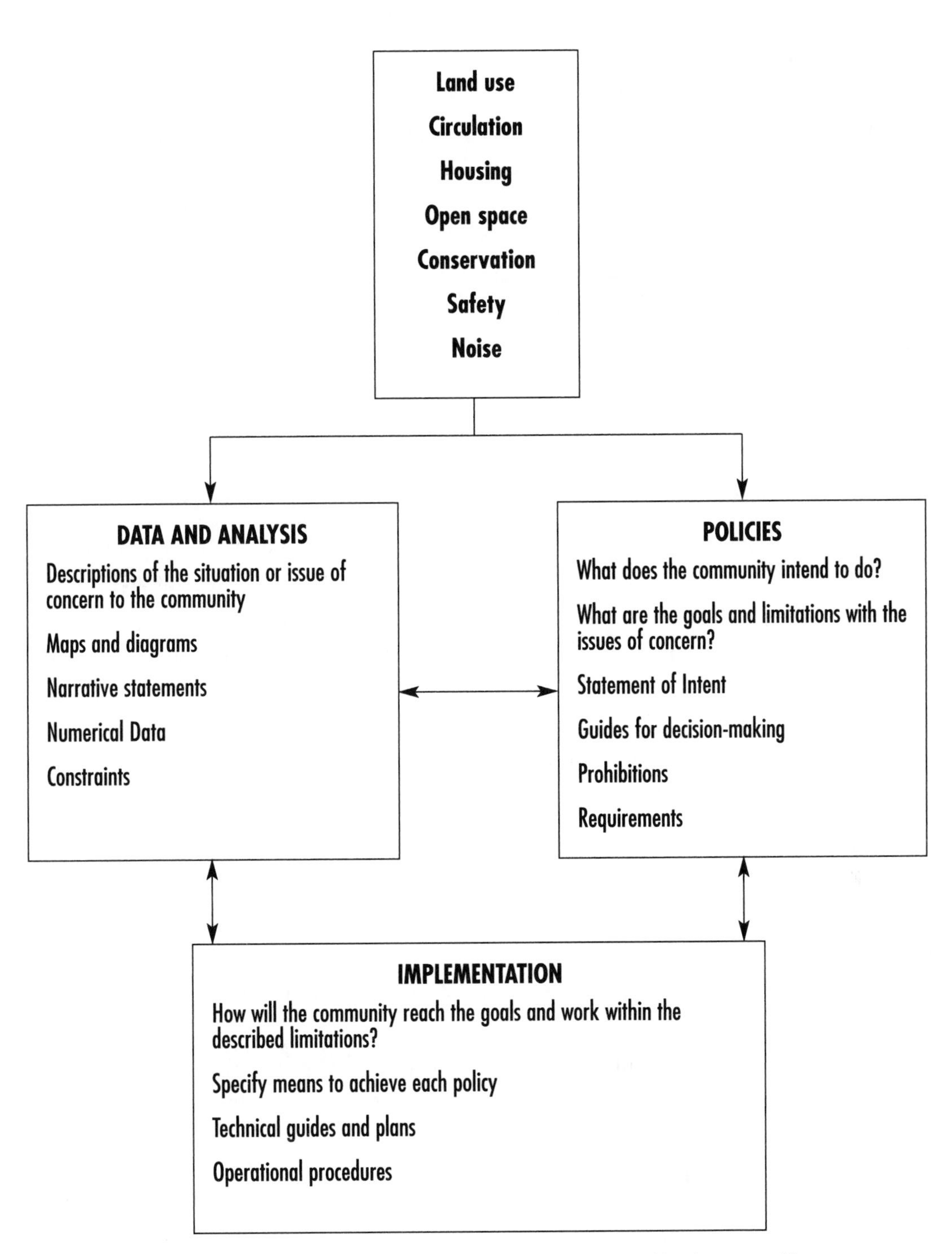

NOTE: "Implementation...within the described limitations," refers to the sphere of influence of the plan in terms of being too restrictive in environmental, zoning, or other aspects. The comprehensive plan must not supersede or replace specific laws and ordinances in the community.

Figure 24–1 Components of a comprehensive plan

The *National Fire Protection Association (NFPA) Handbook* contains an excellent reference to master planning and fire prevention:

> *Master planning is a participative process that should result in the establishment of a fire prevention and control system that is goal-oriented, long-term, comprehensive, provides known cost/loss performance, and adapts continually to the changing needs of your community.*

Fire and emergency services master planning should involve all community governmental agencies, specifically as each affects or in any way supports the fire and emergency services system. There also must be an opportunity to interact with private fire protection and health care systems, appropriate citizen groups, and with national, state, and regional fire service organizations. Planning involves the participation, cooperation, and commitment of all parties interested in the development of a defined cost-loss relationship for community fire protection. It allows you to systematically analyze fire prevention and control and other services through common-sense procedures. Master planning has three phases:

1. *Preplanning.* During the preplanning phase necessary commitments, committees, estimates, and schedules are assembled, and go-ahead approvals are received.
2. *Planning.* The planning phase is a time of gathering and analyzing data, setting goals and objectives, determining an acceptable level of fire and emergency service, identifying alternatives, and constructing the plan.
3. *Implementation.* The implementation phase never ends because the plan is ongoing and is always being revised and updated.

The planning process involves the following elements:

- The *mission statement* of the municipality, and especially of the fire department. Mission statements tend to be very broad in concept and are not related to individual objectives, for example:

  To protect the citizens of the fire district from emergencies and disasters through fire suppression, emergency medical services, hazardous materials mitigation, fire prevention, and public education program.

  The mission statement can be amplified through specific objectives, tasks, and programs. Consider taking each element of the mission statement and establishing a list of objectives for it.

  For example, fire suppression objectives would include the ability to respond to 90% of emergencies with a travel time of four minutes anywhere in the district; the capability to maintain all equipment and apparatus in a high state of readiness; and the ability to implement training programs to meet the demands of the public in fire fighting and emergency medical services.

- An *environmental evaluation* should be made of the community to identify the factors that influence the welfare of the citizens. These factors can include weather, water, transportation, use of hazardous materials in the workplace, noise, and air pollution. A key environmental concern is the impact of the wildland/urban interface problems of some communities.
- *Organizational factors* play an important role in this process. Especially important is the difference between a municipality, a fire district, and a non-profit fire department district organization. Generally speaking, a city government includes a city council, a mayor (and/or city manager), and various department heads. In a municipality there are layers of functional organization and resources. The resources are associated with the various departments. In a municipality, the fire department uses the resources of the city government to provide public works, personnel, administration, payroll, and other services.

  In a fire district the organization is entirely different. The fire district must structure its organization to include all of those services and resources internally. The office staff provides personnel, payroll, accounts receivable and payable, employee benefits, and coordination of Workmen's Compensation.

  The district's various officers may be assigned to oversee facilities and supplies, communications, training, safety, and apparatus and equipment. Because of these differences, the budgets of the municipal fire department and the fire district are very different. The third organization is the non-profit fire department organization. Here you will usually find "line officers" for the suppression side of the organization, and support personnel or "administrative officers" who are responsible for business and finance. The planning process must have a high degree of accountability to be effective in these organizations.
- The *objectives of the organization* must be developed internally by the members, and externally with community input from outside of the organization. These objectives must pertain to specific programs that will influence the overall fire protection plan. Examples include automatic sprinkler requirements for businesses and residential occupancies, life safety requirements in places of public assembly, and brush clearance requirements for areas that are in the wildland/urban interface.
- The *implementation of the objectives* has to be accomplished with careful understanding of the impact on the community. This implementation phase is done in concert with the annual budget and the interaction of the community. There must be an "action plan." The action plan spells out what, how, when, where, and who will perform tasks. Time lines are a critical part of any action plan. Time lines spell out specifically when the task is *expected* to be completed and, later, can show also when the task actually *was* completed.

- The final element is the *continuous review and evaluation* of the fire emergency services plan. This becomes perhaps the most important ongoing element. The department must schedule routine discussions of the fire plan. It may be necessary to rewrite certain portions of the plan at intervals, depending on the rate of change in the technological environment. In the evaluation and review phase, do not overlook community involvement and any external factors that may have changed or been modified.

The Planning Methodologies

The field of fire and emergency service is already very complex and will continue to be complicated in our technological society. To carry out their responsibilities and duties, fire chiefs will expend great effort to cope in this environment. Individual skills can make or break a fire chief's survival. One of the skills that the fire chief must use is *planning.* In the past, in fact even today, planning skills have not always received enough attention from city managers and directors as they search for the best fire chief. This section will introduce planning concepts and ideas to assist both new and experienced chief officers.

The methodologies used to develop the fire and emergency services plan rely on the following factors:

- *Human resources.* These can be from the fire department organization and the community.
- *Planning experience.* Consider outside assistance.
- *Management control.* This can be by a city council, board of supervisors and directors, or commissioners of the district or non-profit organization.
- *Internal and external pressures.* These will be evident and may result from affirmative action, valuing differences, and management diversity.
- *Quality inputs.* The parties responsible for fire, building, health, emergency medicine, planning, water, public works, police, education, private industry (fire brigades), as well as federal, state, county, and local government should be part of the input process. There should also be several opportunities to receive citizen input.
- *Levels of planning.* There are essentially three levels of planning: long range, operational, and tactical. Long range is typically 3 to 5 years, whereas operational and tactical planning tends to address near term issues at a level that is not as broad as those addressed in the long range plan.

There are other planning elements that will be used during the overall process. These include prioritization of subjects, finance considerations, employer/employee relations and, most importantly, the documentation of the plan and approval by the city council or board of directors.

There are some "change factors" that are also important to consider in the planning process:

- Personal future—Career opportunities, growth, and upward mobility
- Organization future—Grows out of individual effort and loyalty
- Societal future—Economics, social values, and public perceptions

The entire planning process is most interesting and challenging. One challenge is the long-range plan, which describes the next five years of the department's work. This is called "visioning" by some managers. In today's world of dynamic changes, a useful and valid five-year plan requires excellent staff work and many difficult and thoughtful decisions. It is, however, a worthwhile and profitable management exercise. To remain current yet still be a five-year plan, it must be updated annually; to do so, drop the year just ending and add a new "fifth" year. A more accurate and currently applicable approach to planning is a two-year plan that is updated every year.

There are two more important factors in this planning process. One is the input and coordination of the decision-makers (city council and/or board of directors) during the planning phases. This involves the policy makers in the overall plans of the department. Then, when support for the budget becomes necessary, these people have already given their tacit approval for the programs.

The other element is the input from the firefighters' representatives (labor union or association). The fire chief's world can become extremely difficult if they do not cooperate and coordinate with the labor element of the department regardless of its career or volunteer status. Remember, while the fire chief does not always have to agree, it is essential to meet and confer with firefighters' representatives.

The "Community Fingerprint"

Establishing the "community fingerprint" is the next step in planning for fire protection. It is very important to start the planning process with a good understanding of the community and its variables.

These variables include:

- Demographics
- Economics
- Environment
- Weather
- Culture
- Ethnic influence

Demographics

Demography is the study of a community's characteristics, for example, size, growth, density, distribution, age, and vital statistics. These data frequently can be found in the planning department or census bureau, or in the nearest public library or city hall. It is important to collect four or five years' worth of data to determine whether there are changes in the demographic information. Perhaps there is a trend that may have a serious impact on future fire and emergency service planning.

For instance, in the past 20 years in the San Fernando Valley area of Los Angeles a trend has been emerging. The Valley once was considered a bedroom community of the city with housing scattered among ranches and agricultural activities. These agricultural areas were replaced with residential developments and shopping centers. Later the housing stock changed from single-family dwellings to multiple-family dwellings, condominiums, and town houses.

When these apartment houses and condos were newly constructed, there were few reports of fires. Today, in many areas of the San Fernando Valley, fires occur more frequently in those aging buildings and there is a definite change in the living environment.

Similar changes in demographics are occurring in almost every community. The planner must be aware of and plan to meet the changes.

Economics

Economics can drive the fire protection planning situation in a positive or a negative direction. The financial condition of federal, state, county, and local government at any given time can clearly have an impact on a community. In the effort to resolve a crisis, legislative process can significantly change matters important to the financing of local government. In some past cases this meant closing fire stations, reducing company staffing, and making overall reductions in normal services. The critical question is, "What can the community afford for fire protection?" The economic situation, good or bad, affects our ability to provide fire and emergency services.

Environment

The environment is one of the newest and most important factors to be considered in a community evaluation. The agendas of many environmental groups will affect your planning and ability to provide fire protection. For instance, the impact of these groups can be felt as a fire chief plans to build a new fire station.

Have any environmental impact reports been filed? Have you considered the existence of the brown beetle or the yellow butterfly in your plans to build the new training facility?

Have you considered the environmental impact of flowing large quantities of water during a heavy-stream training exercise? Is it still possible to conduct a live fire training exercise and meet the requirements of the air quality? All of these questions will arise in fire and emergency service planning.

High among environmental considerations in some areas of the country is the impact of wildland/urban interface fires and their threat to the community. Several fire protection planning models are available for these wildland areas from the NFPA and the U.S. National Forest Service. They encourage the participation of many agencies, private citizens, and businesses in the community.

Other environmental factors that impact service type and demand, and that must be considered are:

- Residential structures—Types of construction and occupancies
- Mobile environment—Types of transportation in the community
- Community—Public assembly occupancies, offices, hospitals, schools, churches, businesses, and hotels.
- Industrial environment—Factories, warehouses, and industrial manufacturing
- Outdoors—Wildlands, forest, timber, and crops
- Other structures—Vacant, under construction, bridges, tunnels, etc.
- Other—All other and unclassified properties.

Weather

Weather conditions affect all fire departments across the United States in their efforts to control emergency incidents. Western states experience droughts, dry winds, and hot temperatures, while in the Midwest rain, floods, and tornadoes are common. Eastern states suffer cold weather, hurricanes, and ice storms that have an effect on fire and emergency service operations. Probably the largest focus of the U.S. Federal Emergency Management Agency for many years has been the effort to plan for flood disasters. Their planning includes not only the normal disaster planning modes of preparedness, response, and recovery, but also a significant effort to mitigate potential losses in predictably flood-prone areas.

What are the weather conditions in your community? Are there seasons? How does the wind affect fire conditions? It has been said that fire without wind is a piece of cake, but a fire with high winds can be very dangerous. This holds true for both structure and wildland fires. As the pieces of the fire protection plan fit together, weather will play an important role.

Culture

Webster's Dictionary defines culture as "a particular form of civilization; the beliefs, customs, arts, and institutions of a society at a given time; a refinement in intellectual and artistic taste; and, the art of developing the intellectual faculties through education."

The planner must have a feel for this community "culture." What are the customs? What are the beliefs? What institutions are prevalent in the community? How will these institutions affect the fire department in its quest for an effective fire and emergency services plan?

Understanding the community can be accomplished by attending city council meetings. Listen to the people who speak. Listen also to the elected officials. Are they pro-growth or anti-growth? Are they fiscal conservatives or liberal spenders? Are they more interested in staff reports and surveys than in direct action? All of these observations will assist the planner.

Ethnic influence

What is the ethnic makeup of the community? Is there any one group that is particularly affected by the emergency problems in the community? Is this group active in the affairs of the community? Are there social and economic problems associated with any particular group? Is any one ethnic group suffering an unusually high proportion of service demand within its neighborhoods? Can these problems be mitigated through education or other means? The planner must have this kind of information to be able to prepare the fire and emergency service plan.

The Fire Department Role

It is very important for the fire department personnel involved to understand their role in the planning process. This is the time for the fire department to take a leadership position. The following are roles that are useful in moving the plan forward.

- Facilitator
- Motivator
- Communicator
- Salesperson
- Politician
- Legislator (and/or legislative advocate)

Facilitator

The facilitator keeps the group on track. The following checklist can help the fire chief serve in the role of facilitator:

- Keep a focus on the group's progress
- Keep a tempo going with daily objectives
- Develop a design (an outline to follow)
- Develop a purpose statement and focus on that statement daily
- Be committed to the objectives
- Be receptive to ideas
- Care about the project
- Understand conflict resolution
- Be flexible
- Be a good listener
- Be a participant
- Show enthusiasm
- Make every effort to be at each session
- Pay attention to details.

Being a facilitator can be interesting and demanding. The fire chief shows interest by facilitating the planning process.

Motivator

The fire chief's role as motivator is very closely aligned with the "leadership" factor in the process. The motivation factor will be present if the chief expresses ideas, supports other ideas, sets the pace, and establishes a "vision" for the group.

Also, a motivator:

- Focuses on a better, more positive future
- Encourages hopes and dreams
- Appeals to common values and interests
- States positive outcomes
- Emphasizes the strength of the unified group
- Uses word pictures, images, and metaphors
- Communicates enthusiasm
- Kindles excitement

Communicator

The art of communication is not all talking. It also has to do with being a good listener. The key is connecting with the other person or group. There are five basic activities that pertain to good listening:

1. Clarifying
2. Restating
3. Maintaining a neutral position.
4. Reflecting
5. Summarizing

Although the best form of communication is face to face, the fire department role in the planning process also means that written communications originate in the department and must be followed up with personal contact. This is very important in scheduling meetings and events.

Good communication sets the stage for the leadership role of the fire department. It reinforces the impression that the fire department can and will play an active, intelligent role in the planning process.

Legislator/legislative advocate

Over the years the fire service has learned some bitter lessons regarding legislation. It has been the victim of too many pieces of legislation that have interfered with day-to-day operations, finances, service delivery options, and the safety of the public. Thanks to the efforts of Pennsylvania Congressman Curt Weldon and others, the national fire service is enjoying a higher degree of legislative interaction. The Congressional Fire Service Caucus, one of the largest on Capitol Hill, is a prime example of efforts to be more involved in the rulemaking process. We should pick up that "gauntlet" and continue to run with it, especially locally.

The fire service today is affected by various federal, state, and local laws and ordinances that require operational changes, as well as changes in education and training that exceed the training levels of yesterday. Hazardous materials, weapons of mass destruction, and EMS are just three of the areas that require additional effort, funds, and programs. Disaster preparedness has also become significant for fire departments. Many fire departments are responsible for the entire city disaster preparedness program.

Planning efforts may have to be tempered with the fact that some jurisdiction may be subject to restrictions on the breadth of regulatory actions in relation to state or provincial codes and standards. This concept, known as a "mini-maxi code," precludes local jurisdictions from enforcing codes and standards that are either more or less restrictive than documents adopted by the state or province.

The fire service needs good advocates who are active in federal, state, and local legislative programs. Monitoring and influencing the legislative process is extremely important to fire and emergency service planning.

Politician

The fire department must interact within the community in a political sense. The fire chief must interact with the various members of the city council and various boards and commissions. This can be a real learning experience for a new fire chief.

A good starting point is regular attendance at city council meetings and other government functions. Attending local chamber of commerce meetings and affairs also is very helpful. The fire chief should consider joining the chamber of commerce and such other community organizations as the Kiwanis and Rotary clubs, and other high-profile community service organizations. These political connections certainly help the fire chief locally.

Membership in statewide organizations such as the state firefighters association, or fire chiefs associations is a must for the department and the chief. It is with the help of these organizations that good legislation is promoted and bad legislation is defeated. It also helps to identify planning resources and to achieve community buy-in to the plan.

Salesperson

Advocating or marketing your fire department is a subject that needs to be high on your priority list. Telling the world how good you are, through visual or written media, will make a big difference when you are trying to sell a new fire protection plan or accomplish anything that requires communication with and understanding by others. Some fire departments use the medium of television for public education announcements and to communicate to the public stories about recent fires and how the fire department saved property and lives. Broadcasting radio announcements over local stations is another method. Publishing a newsletter periodically about the department and its employees, and including public education announcements is easy to do with a personal computer and a publication software program. The biggest cost is mailing the newsletter. Cut mailing costs by using the local markets and stores for distribution or by working with the chamber of commerce to get the message out to the citizens.

Risk Analysis

Risk analysis is a systems approach that involves the following steps:

1. The community fire and emergency service system
2. The community at risk
3. The response capability
4. The unprotected risk
5. The strategies to consider

The first step is to form a team to conduct the study. Place one person in charge. The team may consist of a chief officer, company officer(s), and members. The project research then can be delegated and divided among the members of the team.

The balance of this section consists of checklists for each of the main steps with explanations for each item on the checklist. The example given relates to fire risk. However, the same principles apply to other perils that are addressed by the fire and emergency response system in a community.

Step 1—The community fire protection system

Evaluate your community's fire and emergency services system. Begin by developing a description of your community's resources and systems. Efficient and effective use of public and private fire protection resources is the anticipated result of good planning. In today's environment, the service demands of the fire department include fire prevention, fire safety education, emergency medical services, disaster management, and hazardous materials controls. The community fire protection system includes *all* public and private services that are available to protect people and property from fire, explosion, and other hazardous situations.

The primary elements of the community fire and emergency services system are the fire department, private fire protection entities, federal, state, and county governments, public utilities, health care organizations, building departments (building inspections, permits), planning departments (zoning regulations), and schools (both public and private).

A general statement of built-in fire protection systems. This statement includes the requirements for automatic fire sprinklers in various types of occupancies. Describe these requirements in detail. Is there a certain part of the community that is not affected by these requirements? The ordinances are not always enacted retroactively and, therefore, buildings constructed before a certain date may not require sprinklers.

The fire prevention code requirements other than sprinklers systems. There are many fire prevention code requirements that have an effect on life safety and property damage.

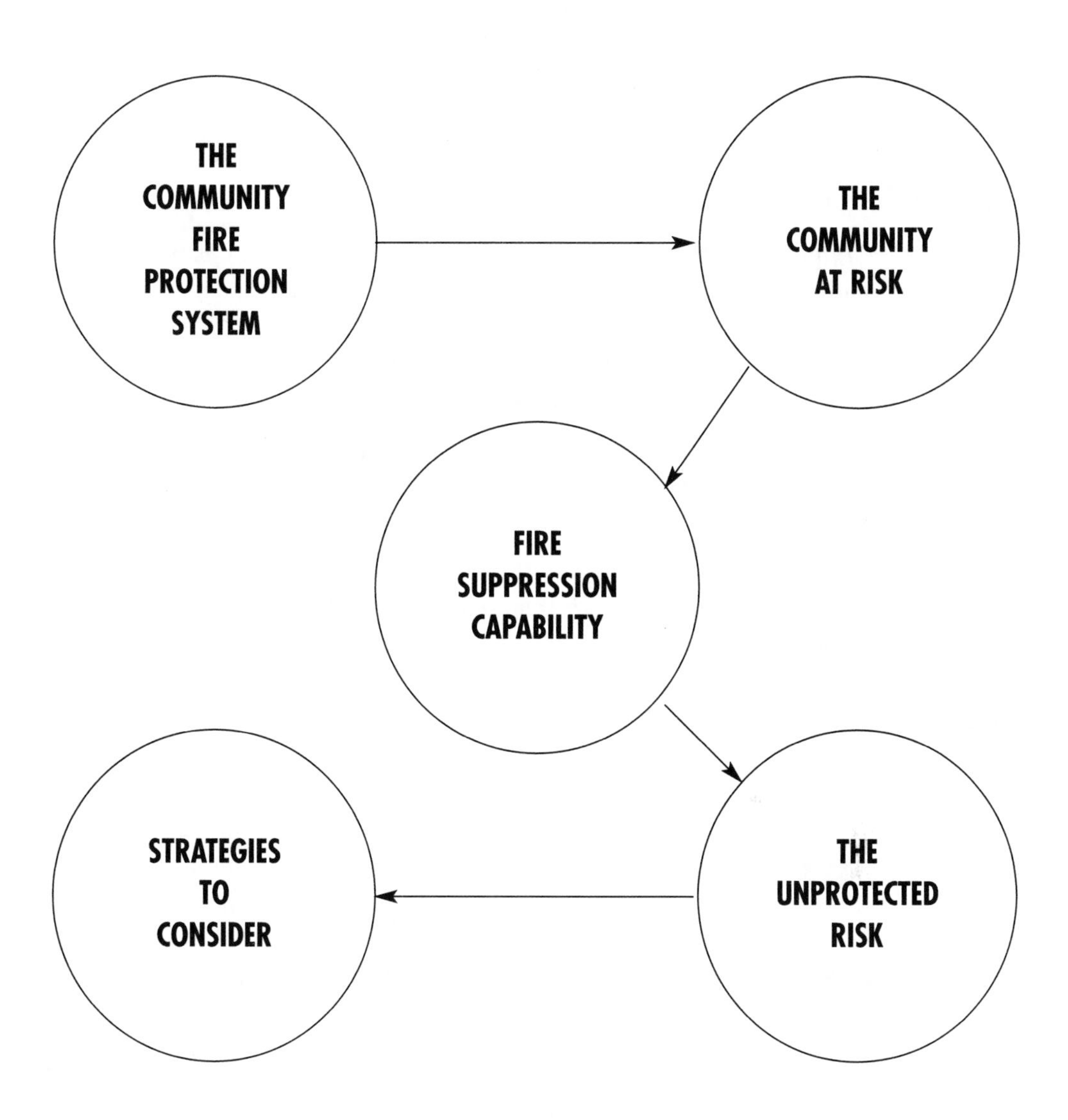

Figure 24–2 Fire risk analysis—a systems approach

These requirements should be described in this section. They may include the requirements for automatic fire doors and windows, fire separations (to separate hazardous areas), occupancy load requirements for public assembly type occupancies, flame spread requirements for decorations, and exit signs and exit door hardware. All of these requirements are a part of the community fire protection system.

Step 2—The community at risk

Describe community demographics. This is where statistical information about the present population and future population growth is described. This information is available from the planning department, census bureau, public library, the chamber of commerce, or the county's tax records. Is growth expected in a geographic area of your community? New housing and shopping malls? Industrial parks and new commercial areas?

Describe what has been burning in the community. Use statistical information that describes the fire activity in the community. What types of occupancies were involved? Has the number of fires been decreasing or increasing? A good way to display this type of information is to compare local statistics with state, county, and national statistics.

Describe what is available to burn in the community. This is the section that describes the target hazards in the community. A target hazard is a building or occupancy that is unusually dangerous in terms of life loss or has a high potential for property damage. It may be that old hotel that houses many senior citizens, or the chemical plant near the river that has little built-in fire protection. It also may be an area of the community that borders a wildland fire hazard area.

Occupancies that have life loss potential are a priority in this analysis. You should ask the following questions when evaluating a life loss target hazard.

- What is the occupancy type? Crowded occupancies present a potential problem.
- Are there sprinklers or other fire protection systems? Is there some type of early warning system (*e.g.*, smoke and heat detectors, or a fire alarm system)?
- What is the condition of the building? What is its age? Is the housekeeping good? What is the condition of the structural members? Are there poke-throughs in walls and ceilings? Is there a chronic storage problem in the hallways or fire escapes?
- What is the occupant load? How many occupants are in the building during the day and at night?
- What are the ages of the occupants? The very young and the very old are high-risk groups.
- What is the physical condition of occupants (ambulatory or non-ambulatory)?

- What are the means of egress from the building? Are there exterior or interior fire escapes? Are there adequate exit signs? Is there adequate lighting in the exit hallways and stairways?
- Have these occupants received any evacuation training? Are they supervised, as in nursing homes and similar institutions?

Identification of target hazards may also be by other hazards associated with the building or occupancy. A large-loss fire in a "one-industry" town will have a big economic impact on the entire community.

The remaining factor in "what is available to burn" and the identification of "target hazards" is the application of the fire flow requirements for the community. Fire flow, which at first seems more appropriately included under step 3, Fire Suppression Capability, is introduced at this point under "what is available to burn" because structures are evaluated in terms of fire flow requirements. Once the flow requirements have been determined, they can be used to assist in determining manpower and equipment needs.

Fire flow requirements. The fire flow requirement is the amount of water needed to extinguish a fire in a given occupancy. The flow is always stated in gallons per minute (gpm) and may be required for a specific period of time.

The needed fire flow allows planners to study resource needs within the fire department. How many pumping engines will deliver how many gallons per minute? How many firefighters will be needed to handle the hoselines and nozzles or to operate large stream appliances? Hand-held lines will flow from 100 gpm to 250 gpm, requiring one to three firefighters. It follows that the higher the required fire flow, the more firefighters will be required. Large stream appliances will deliver 300 gpm and more, usually with fewer personnel.

There should have been a water flow test performed within the previous three years. This often is done during hydrant testing, with flow test information recorded by geographical location. The information may well be available from the Insurance Services Office if a recent rating study has been conducted by that organization for the community in question. The type of occupancies and what fire flow is required also must be listed. This information shows areas in the community that have good water supplies and those that have inadequate supplies.

It also should be noted that the water flow requirements are based on the occupancy being fully involved with fire and requiring the maximum gpms. It is obvious also that fixed fire protection systems will reduce the required flow.

Another factor to this equation is that if the maximum fire flow is 3,000 gpm for your community, and a certain occupancy exceeds that demand, you will need to plan additional resources for the probable fire. This could mean implementing automatic or mutual aid agreements.

The identification of "fire management areas." One of the last requirements in the Community at Risk section is the identification of "fire management areas" (FMAs). Fire management areas are designated when the community is divided into manageable pieces. These are specifically identified on a community map. It may be advantageous to divide the community into FMAs by using natural geographical separations. Natural separations include rivers, freeways, main streets, railroad tracks, and open space areas.

Another way to determine FMAs is to place a grid overlay on the map of your community. Be careful not to make too many FMAs by using a grid system that is too small. Some fire departments have made their FMAs consistent with their first-in districts (those districts where an engine company normally should be the first in when responding from the fire station). If you have six fire stations, then you should have six "first-in districts" and six FMAs.

Step 3—The fire suppression capability

Evaluate your department and its organization. This is the place to describe the organization of the fire department. An organization chart should show the chain of command and responsibilities for various functional activities in the department. Record the number and location of fire stations, the number of personnel (both uniformed and civilian), and the number, type, and location of apparatus.

Describe your staffing level per company and any special staffing plans that you have adopted. Special staffing plans may include additional firefighters during the busier times of the day or week, or special overtime plans to staff companies during personnel vacancies. Charts would be helpful in describing this type of information.

The service time for emergency and non-emergency responses should be made available in this section. This information usually can be found in the department's annual statistical report and in the dispatch records.

The life loss and fire injury statistics for the last five years. This information also is available from incident reports, fire prevention records, or the annual report.

Your communications system. Describe how your companies are dispatched. Do you have your own dispatch center? Are you contracting with another agency for the dispatch service? What types of emergencies are being received at the dispatch center? Is the 911 system or enhanced 911 being used?

What are you using for a radio system? How many frequencies or talk groups are available? Does the system have good interoperability with other responder agencies? Do you share a frequency with other agencies? Do you have a mutual aid frequency that is used for larger emergencies? You should describe that system and how it is activated.

Describe your mutual aid and automatic aid resources. Mutual aid resources are those from other fire departments and agencies not in your own geographic area. Typically a cooperatively established system is designed to activate these mutual aid companies. The dispatch center is charged with this responsibility and implements the mutual aid plan when it is needed. Automatic aid resources are preplanned with adjoining fire departments and agencies.

These may include an engine or truck company that is automatically dispatched to certain areas of your city or to special occupancies that represent a special hazard (life loss and property damage).

Step 4—The unprotected risk

Determine the unprotected risk. Unprotected risk is the degree of imbalance that exists between risk and suppression capability. If suppression forces available to respond to any location are inadequate to deal with the fire situation, that particular situation must be considered part of the community's unprotected risk.

Unprotected risk can be reduced or eliminated by decreasing the risk or increasing the suppression the capability. Risk can be reduced by the classic three "E's."

- *Education*—Educating the public, particularly the owners and operators of properties at risk, creating in them an understanding of why the public fire forces cannot protect their property effectively, and how they can reduce the imbalance.
- *Engineering*—Requiring built-in structural fire protection measures, retrofitted if necessary, and fire protection systems (automatic sprinklers, special hazard systems, alarm systems) through codes and ordinances.
- *Enforcement*—Assuring, through frequent and competent inspection, that all statutory fire protection and fire prevention measures are in full compliance.

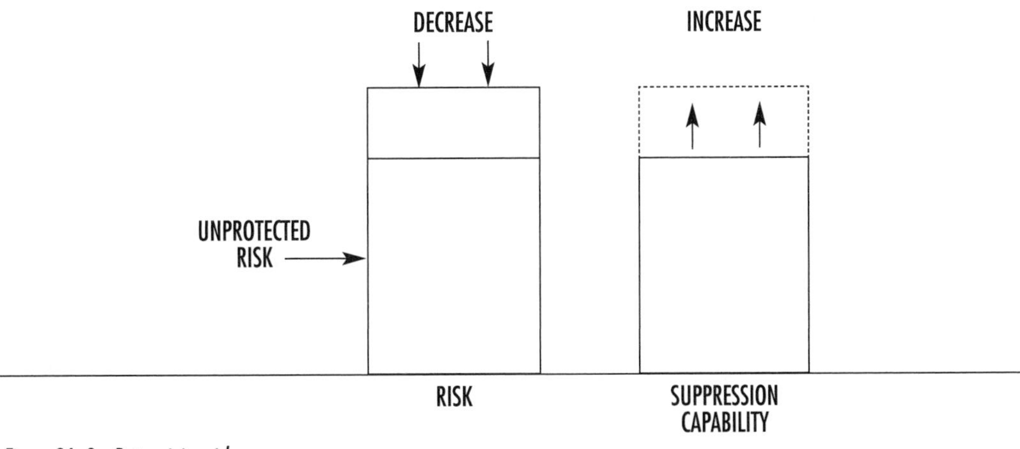

Figure 24–3 Determining risk

Suppression capability can be increased by upgrading the water supply system; adding or relocating fire stations, apparatus, and personnel; improving training, strategy, and tactics; and improving communications and mutual aid.

While improving suppression capability has been the traditional fire service response to risk imbalance, we have never been able to fully close the gap. We will always need suppression forces and they should be the best we can provide, but the final answer lies in risk reduction through education, engineering, and enforcement. Make sure both are a part of your fire protection plan.

Establish the concept of acceptable risk. "Acceptable risk" is the level of unprotected risk with which the community is willing to live. Unfortunately, few citizens have any understanding that they have, by default, established a level of acceptable risk. When a fire occurs, they fully expect the fire department to arrive with red lights, sirens, and the equipment and personnel to suppress the situation. Sometimes we win, sometimes we lose.

A significant part of "classic" master planning is to give the community a clear understanding of what the fire department can do and what it cannot do. We should make every effort to explain to the community the purpose and goals of higher taxes for improved fire suppression, or the need for more stringent fire protection and fire prevention codes, or for private investment in improved fire safety. The final decisions will be made in the councils of government and in the marketplace, where dollars drive decisions. It is the responsibility of the fire chief to see that those decisions are made with all the necessary information and the most complete understanding that we can provide. Only then will "acceptable risk" be a meaningful concept.

In the late 1990s the NFPA developed two standards that address deployment of public fire services in career and volunteer systems, respectively. These controversial documents (NFPA Standards 1710 and 1720) describe standard levels of service in terms of response time in 90% of calls and the number and training of resources. As is the case with all NFPA standards, they are voluntary consensus documents that may or may not be adopted by a jurisdiction.

Step 5—Strategies to consider

The closing stage is the time to evaluate all of the information and establish some final plans or strategies.

One of the final planning actions is examining the location of the existing fire stations. Fire station location should be determined based on the factors discussed here.

Fire stations should be located near major streets. Traffic patterns and use must be determined. However, the new fire station should not be located on the busiest street in town. Surrounding land uses must be considered. The environmental concerns include apparatus noise and other station activities. Future land use requiring additional fire suppression efforts also must be examined. A new development with 1,000 homes, a new industrial center with hazardous materials, or a new baseball stadium, are all examples of land use considerations which affect the fire protection system.

The most important factor is apparatus response times. Can the community's standards for response (*i.e.*, four minutes in 90% of the emergency calls) be met? Can maximum coverage of the entire area be accomplished from that location? Consider greater alarms. Consider freeways and railroads that might be impassable.

Station location analysis also requires three basic types of data:

1. The nature of the incidents experienced
2. The locations of most fires (areas of response concentration)
3. The response times to these locations

Prime considerations in the "nature of the incident" data category are the number of structure fires and the number of emergency medical incidents.

The response times throughout the community can be measured by actual incidents, the history of those calls, or by driving the district and measuring the times for a specific area. When describing the response times to the city council, consider mentioning the "time-temperature curve" and explaining how important a rapid response is in saving lives and property.

Comparisons

It may be very helpful to compare your fire department with those of neighboring agencies of similar size. You might be able to profitably incorporate the answers to these questions; comparing your statistics with those of other communities.

- What is the fire protection cost per capita?
- What is the fire loss per capita?
- What were the fire loss trends in the last five years?
- What are the total alarms annually?
- What types of structural fires are occurring?
- What is the total number of emergency medical calls?
- How many hazardous materials responses occur annually?

Insurance ratings

Another area is the impact of the insurance industry on the community. What is the recognized authority for setting insurance rates in your area? Is it the Insurance Services Office (ISO)? Are rates based on ISO's information and individual insurance company analyses?

A hypothetical fire department that has just been rated by the ISO and has received a rating of III should ask what it could do to improve the rating. ISO rating personnel can assist in identifying these factors. It may be that the fire department would have to establish additional fire stations, place additional personnel on all companies, increase the number of dispatchers on duty at all times, and improve the city water supply system.

It would be difficult for a fire department to complete all of those recommendations at once, but they should be in the department's long-range plan, and methods to fulfill those recommendations should be established as resources permit.

Additional considerations

The plan should include a section on additional considerations that addresses recommendations in such areas as:

- Employee preparedness training
- Citizen training and preparedness
- Marketing the department's role in the community
- Consolidation of services within and with other agencies
- Water supply contingencies during disasters
- Prefire planning and inspection by all suppression and prevention personnel
- Employee physical fitness levels with programs to improve and maintain fitness
- Employee skill levels and formal education

Final recommendations

The closing portion of the final risk analysis contains the final recommendations. These are developed with the fire chief, the staff, other employees, the labor union or association, and most importantly, citizens who have an interest in the subject. The question of community fire protection must involve the entire community. When that large fire loss occurs, citizens must understand that the loss was because the "acceptable loss" levels were exceeded. In this way, citizens will better understand fire protection in the community. Data analysis information should be displayed in public meetings and other forums to inform the citizens of the concepts of "unprotected risk," "protected risk," and "acceptable risk."

Once the recommendations have been developed, discussed, prioritized, and finalized, they should be presented to the decision-making authority in the city or fire district for consideration.

Suggested format for the final report

- *A well-designed, colorful cover.* The title must be highly visible.
- *An inside title page.* Include date of report, name of official sender (the fire chief), and the names of those who participated in the preparation of the report.
- *Table of contents.* Do this last to incorporate any changes made in the report as it is put together.
- *A chart or graphic symbol that communicates the essence of the report.*
- *An executive overview of the report.* A summary of the goals and objectives. A purpose statement. Keep this to no more than two pages.
- *The contents of the report divided into sections or divisions.* Each section should have a cover page with a short description of that section. It is helpful if the cover pages of each section are a different color.
- *A glossary of terms and a bibliography or reference section at the end of the report.*

Once the final report has been completed, it will need to be "marketed" throughout the community. Because you have involved the community in the planning process, this selling job may not be too difficult. Be ready to share your hard work with other fire departments and government agencies. They may want to use your methods to determine their fire protection plan for the future.

Conclusion

The concept of master planning involves participation by many players. The state, city, county, and special districts all participate in this process of planning for the future.

This chapter encourages a high level of participation by the fire chief in this process. It requires a high degree of communication between the fire chief and the other city planners and politicians. The concept of "planning teams" within the fire department, managed by the fire chief, is the suggested method to use. These planning teams are capable of producing significant results with a high degree of initiative and vision. The planning teams must understand the "fingerprint" of the community (*i.e.*, demographics, economics, environment, culture, and ethnic influence). These questions must be answered at the beginning of the planning process.

Fire department personnel and, most importantly, the fire chief, must play an active role in the development of the fire protection plan. The roles of facilitating, motivating, and communicating will all bring results. The fire chief must be active in the legislative process within his own organization and at city, state, and federal levels. There have been many occasions when the fire service has not been active in the legislative process and it has been the beneficiary of poor legislation. The political battleground is another area that needs the full attention of the fire chief. The fire chief should develop a good understanding of the political atmosphere in the community. The five steps of the fire risk analysis system will provide a methodology for the fire chief to follow in developing the fire protection plan.

25

FIRE PREVENTION AND CODE ENFORCEMENT ORGANIZATION

Glenn P. Corbett
Ronald R. Farr

CHAPTER HIGHLIGHTS

- The historical development of fire prevention activities in the United States, the current thrust of fire prevention efforts, and the future influence of fire prevention goals within the fire service
- Fire prevention organization in different types of departments
- The members of a fire prevention organization, their duties, and how they are selected for appointment to the organization, the positive traits of an effective chief of fire prevention
- How a well-organized organization operates efficiently
- Codes enforced by fire prevention organization and fire companies, the types of plans reviewed and inspections performed, and the legal issues involved therein
- Organizations that promulgate model codes as well as the techniques used to adopt codes at the local level

The focal point of a fire department's efforts to minimize fire losses in a community is the fire prevention organization. Fire prevention is considered to be the most important non-fire suppression activity a fire department can be involved in. The main goal of the fire prevention organization is improving the safety and quality of life for the citizens it serves. In most fire department mission statements, fire prevention usually is included as one of their primary goals.

Fire prevention efforts will continue to play an important role in fire department activities as we move forward in the 21st century, primarily because of the tremendous fiscal constraints being placed on the fire service. With fewer firefighters and fewer fire stations, a shift to mandated installations of fixed suppression systems and increased public education, both key fire prevention organization activities, is likely.

The chief of the fire prevention organization plays a pivotal role in leading a department to meet this newly enhanced role. The position requires technical and legal competency, political awareness, and persistence. Changes in a fire prevention organization do not occur overnight; long-range planning and goal-setting require a fire prevention chief to persevere.

This chapter will discuss the organization and operation of the fire prevention organization. It also will discuss a fire prevention organization's three basic activities: code enforcement, public education, and fire investigation. Code enforcement is reviewed in detail; public education and fire investigation are given an overview treatment since they are the topics of chapters 26 and 27, respectively.

The fire prevention organization can be considered symbolic of a fire department's mission—to prevent fires and minimize losses. Over the years, the public has developed certain expectations of fire prevention bureaus—to identify and abate fire hazards in the community. The public also expects the fire prevention organization, as well as the fire service as a whole, to be their fire protection experts.

It is important then that chief officers have a working knowledge of how a organization functions in order to meet this critical goal of proactively protecting lives and property from fire. With these expectations in mind, this chapter strives to provide a detailed overview of fire prevention organizations and how they can be used to meet public expectations.

The chapter opens with a historical look at fire prevention activities and concludes with a glimpse into the future. The intervening detailed text describes the organization and management of a fire prevention organization, the organization's personnel, as well as the codes and other tools used to carry out its charge of preventing fires.

Historical Perspective

Fire prevention activities actually predate the creation of organized fire departments in the U.S. Fire was a constant threat to early European settlements in America. Major fires struck in Jamestown, Virginia, in 1608 and Plymouth, Massachusetts, in 1623.

Fire regulations in the early years primarily affected the construction, height, and maintenance of chimneys. The carrying of hot coals from fireplace to fireplace was also a constant source of large blazes. Wooden building construction aggravated the situation.

Boston banned smoking paraphernalia in 1638. Ten years later, Peter Stuyvesant forbade the use of wood or plaster chimneys in New Amsterdam (New York City).

Stuyvesant also appointed four fire wardens who served without pay. They inspected chimneys and patrolled for fire hazards. They also assessed large fines for carelessness. The fines were used to buy fire fighting buckets, ladders, and hooks.

Arson was a problem then as it is now. A rash of arson fires prompted Boston to pass legislation in 1652 that imposed punishments of flogging or death for convicted arsonists.

Besides inspections of chimneys, fire wardens also performed fire watch duties in many cities. Patrolling the streets at night, they would use large "rattles" to summon assistance and notify the city of a fire.

Ben Franklin, the father of the organized volunteer fire service, also was an early supporter of fire prevention. His words, "an ounce of prevention is worth a pound of cure," are famous. In addition to inventing the "Franklin stove," he also was a proponent of chimney cleaning and the proper handling of burning materials.

Even when rudimentary fire apparatus were introduced in major U.S. cities during the first half of the 18th century, fires were still a major threat. Major conflagrations, some larger than even today's biggest fires, continued to plague our cities well into the 20th century.

These conflagrations were the impetus for the fire service to begin to take an active role in fire prevention activities. The National Association of Fire Engineers (later to become the International Association of Fire Chiefs) during its first annual meeting in 1873 developed a list of eight fire safety concerns:

1. The limitation or disuse of combustible materials in the structures of buildings, the reduction of excessive height in buildings, and restriction of the dangers of elevators, passages, hatchways, and mansards
2. The isolation of each apartment in a building from other apartments and of every building from those adjacent by high party walls
3. The safe construction of heating apparatus
4. The presence and care of trustworthy watchmen in warehouses, factories, and theaters—especially during the night
5. The regulation of the storage of inflammable materials and use of same for heating or illumination; also the exclusion of rubbish liable to spontaneously ignite
6. The most available method for the repression of incendiarism
7. A system of minute and impartial inspection after the occurrence of every fire, and rigid inquiry into the causes—with reference to future avoidance
8. Fire escapes actually serviceable for invalids, women, and children

Fire wardens were the predecessors of today's fire inspectors. They continued to play an important role through the 19th century, as evidenced by this New York City helmet from c. 1855. In 1850, the fire wardens (appointed from the rank of firefighter) made inspections at construction sites throughout the city looking for deficiencies, inspected other buildings for improper storage, and strived to "inquire and examine into any and every violation of any of the provisions of the acts previously passed for the prevention of fires in the city of New York...."

It is interesting that many of the recommendations outlined above were the basis for a number of current fire and building code requirements. The early recognition of the importance of fire investigations is an insight as well. An example is the reference to "fire escapes actually serviceable of invalids," which came to the fore in the form of the Americans with Disabilities Act.

A few years later, in 1884, former Chief Damrell of the Boston Fire Department presented a committee report to the National Association of Fire Engineers concerning the formulation of a building code. This pioneering work was important in the development of the first model building code in the United States.

The insurance industry, also seeing the obvious need for a set of standardized building regulations, especially after having to pay the extraordinary claims that resulted from conflagrations, promulgated the "National Building Code" in 1906. This first model code (a uniform set of construction requirements to be adopted by the cities) published by the National Board of Fire Underwriters was patterned after Damrell's report and the New York State Building Code that had recently been published.

Life safety also became a major code concern during the early 20th century, primarily because of the Triangle Shirtwaist factory fire in New York in 1911. The fire killed 146 textile workers. It led the National Fire Protection Association (NFPA) to form the Committee on Safety to Life, which eventually developed what was to become NFPA 101 Code for Safety to Life from Fire in Buildings and Structures.

In retrospect, it seems likely that this fire also prompted many fire departments to form fire prevention bureaus. New York City, in fact, formed its bureau shortly after the Triangle fire in 1911, although some fire prevention bureaus were in existence before 1911.

Unfortunately, the formulation and actual adoption of fire safety code requirements have, in many cases, come only after the occurrence of tragic fires. For example, consider the following "occupancy-specific" fires. Each of these major fires led to legislative and code changes targeted to their particular occupancy types:

- 1903 Iroquois Theater fire, Chicago, Ilinois—602 dead
- 1942 Coconut Grove Nightclub fire, Boston, Massachusetts—492 dead
- 1944 Ringling Bros. Circus tent fire, Hartford, Connecticut—162 dead
- 1946 Winecoff Hotel fire, Atlanta, Georgia—119 dead
- 1958 Our Lady of Angels school fire, Chicago Illinois—95 dead
- 1977 Beverly Hills Supper Club fire, Southgate, Kentucky—165 dead
- 1980 MGM (high-rise) Hotel fire, Las Vegas, Nevada—85 dead
- 1982 K-Mart warehouse fire, Bucks County, Pennsylvania—100-million dollar loss

There is no doubt the April 19, 1995 bombing of the Alfred P. Murrah Federal Building in Oklahoma City, Oklahoma, (168 dead) and the September 11, 2001 terrorist attacks on the Wold Trade Center and the Pentagon (more that 2,500 lives lost) will be added to this list of historic fires. The fire service, in all likelihood, will study these incidents and develop code changes as a result.

Another fire prevention effort, public education, has its roots in the early part of this century. In October 1922, President Harding initiated the first Fire Prevention Week in remembrance of the great Chicago fire of October 1871.

The latter half of this century has seen the development of fire codes, codes developed specifically to address hazardous activities in and around buildings, and maintenance of fire safety features of buildings themselves. Although a fire code is not considered a building construction code, current fire codes do address issues of importance pertaining to specific occupancies relative to fire safety.

Several organizations were originally involved in the development of codes (both building and fire) in the United States: the National (formerly the Basic) Fire Prevention Code developed in 1969, the Uniform Fire Code developed in 1971, and the Standard Fire Prevention Code developed in 1974. In 1971 the National Fire Protection Association (NFPA) also began to develop a fire prevention code, NFPA 1, covering all aspects of fire prevention and would reference the other codes and standards developed by NFPA.

These codes were generally adopted by jurisdictions on a regional basis. The National Fire Prevention Code was used primarily in the northeast and midwest, the Uniform Fire Code in the west and southwest, and the Standard Fire Prevention Code in the southeast. NFPA 1 (NFPA's Fire Prevention Code) was not affected by the regional use concept and could be used in any area desired.

In 1994 the International Code Council (ICC) was formed with the goal of beginning to develop a single set of national model codes. The founding organizations of the ICC were Building Officials and Code Administrators International, Inc. (BOCA), International Conference of Building Officials (ICBO), and Southern Building Code Congress International, Inc. (SBCCI). The end result was the development and publication of a single set of codes (instead of three sets). In 2000, the International Code Council published the International Fire Code along with companion codes (International Building Code, Residential Code, Mechanical Code, Plumbing Code, and others).

In 2002, the NFPA partnered with the Western Fire Chiefs Association and brought both the Uniform Fire Code and NFPA 1/Fire Prevention Code into a single document. Also in 2002, NFPA published their first building code (NFPA 5000) to work with their family of codes.

Although we have seen a reduction in fire fatalities, approximately 3,700 people still die annually in fires. Today, true conflagrations are rare resulting in the public's fear of fire being somewhat diminished. Overall, the number of fires per capita since the beginning of the 20th century has decreased. This is the direct result of improved firefighting and fire protection equipment, better training, enhanced public fire prevention education activities, better building and fire codes, and better origin and cause determination for prosecution of fires considered suspicious.

There is still work to be done to lower the number of annual fire fatalities. We remain suppression-oriented—a reactive mode—but must do all we can to move toward a more pro-active mode and prevent fire from occurring.

Organizational Structure

Fire prevention organizations, also known as "fire marshal's offices" and "fire prevention bureaus," range in size from a single individual to more than 100 people. Volunteer departments may have only one or two part-time inspectors, while large paid departments may have a full complement of inspectors, fire protection engineers, plan reviewers, investigators, educators, administrators, and clerical staff.

The number of organization personnel is a function of both the size of the community served and the number of specific duties carried out by the organization. For example, some larger departments assign individuals exclusively to high-rise structures and others to institutional occupancies. Some smaller departments neither investigate fires nor have a public education specialist.

Reporting to the chief of the department, the head of the organization may be known as the Chief of Fire Prevention, Division Chief, or Fire Marshal. It is the duty of the chief of fire prevention to administer the overall activities of the organization, assist other fire department division heads, and interact with other city agencies that affect fire prevention.

As the manger of fire prevention activities within the department, the chief of fire prevention is a key individual within the department as a whole. The fire prevention manager needs to be an effective leader who is able to direct the activities of those who follow so that the goals and objectives sought are achieved. For the most part, this individual is second only to the chief of department in terms of responsibility and exposure.

The following qualities are characteristic of most successful fire prevention chiefs:

- ***Level-headed.*** The chief must remain calm under pressure. Contrary to popular belief, an active organization can be a high stress position. Great demands are placed on the individual in charge.
- ***Technically agile.*** The fire prevention chief must act quickly when confronted with a technical problem, be able to break down and understand technical issues, and be prepared to convey such information to higher authorities (including the fire chief and city management).
- ***In control.*** The chief is not a dictator, but a person who exudes an air of authority and leadership and must be able to get others to follow in pursuit of agency goals.
- ***Non-wavering.*** When under pressure, the chief must stand by his personal convictions, as well as back those of subordinates (assuming, of course, such convictions are legitimate and legal).
- ***Good communicator.*** "People" skills and writing skills are important elements of this position. The fire prevention chief's professionalism reflect both on him and the department as a whole.
- ***Politically astute.*** This is a highly political position. Enforcing regulations does not turn land developers into friends. Fire prevention chiefs must know how to navigate "shark-infested waters" and still meet their objectives (see discussion below).
- ***Energetic.*** Directing an organization is a mentally strenuous job, one that can take its toll eventually. Without enough personal motivation, being willing to continue to learn will start to fade.

New fire prevention chiefs often experience culture shock when they come into the organization after having been in the field. The workload, myriad problems addressed by the organization, and the public's attitude toward the organization all contribute to the culture shock.

A fire prevention organization's progress is measured in years, not months. New fire prevention chiefs must expect to spend time understanding a city's unique fire problems as well as invest time in developing and implementing their own programs, hence the need to "stick it out." For meaningful results, a commitment of at least five years is probably necessary.

Politics play a critical part in fire prevention organization activities. Before individuals accept a fire prevention chief position, they must be assured of a commitment to fire safety from the top city management and the fire chief, a commitment that will be backed up. Without such commitments, an individual should be prepared to spend time spinning his wheels.

This does not mean that city management will not occasionally influence organization decisions. The fire prevention chief is expected to defend the organization's position, and to do so with technical accuracy. Compromises sometimes have to be made. A fire prevention chief who feels uncomfortable with compromise should have no qualms about passing it to the chief of the department.

The chief of fire prevention in volunteer departments may be the fire chief or an officer appointed by the chief to serve in this role to oversee fire prevention activities. In smaller volunteer departments, this individual typically works part-time and may or may not be compensated on an annual basis. Large volunteer departments may employ two or three full-time employees, one of which should be the fire prevention officer.

In either case, part-time inspectors who work a few hours a week may be used. Since volunteer departments vary widely in size and fire prevention duties, other organizational structures may be encountered.

In paid departments, the chief of fire prevention is most often promoted into the position from within the department. In larger paid departments, the chief of fire prevention oversees three distinct sections that correspond to the organization's primary activities: code enforcement, public education, and investigations. Each of these sections is generally managed by a support officer.

At the section level, the captains and lieutenants are the "backbone" of the organization. It is these individuals who carry out the code enforcement program (which is subdivided into inspections and plan review in some larger departments), investigations, and public education activities.

One group of personnel often overlooked in the organizational structure is civilian clerical staff. They play a critical role in keeping the organization running smoothly, especially since they handle the paperwork that is the lifeblood of the organization.

Because of the complexity and magnitude of issues that must be reviewed by fire departments, specially trained individuals (such as fire protection engineers) with technical expertise to enhance the organization's effectiveness and capabilities are being employed.

Personnel and basic training and performance requirements

As mentioned above, paid departments traditionally use lieutenants and captains to staff task-oriented positions. It is advantageous to have officers from the firefighting ranks in the organization since firefighting experience can be useful in code enforcement and fire investigations. Such officers also tend to learn fire prevention techniques and skills more readily than civilians with no fire protection experience. These officers possess a level of fire protection expertise that is necessary for successful fire prevention effort.

This is especially true with respect to model fire prevention codes, which have become more complex in recent years. For example, more technical and diverse industrial processes have led to greatly expanded code provisions for the use and storage of hazardous materials. These code changes, as well as increasing litigation against fire departments, demand that bureaus be staffed by highly trained and technically competent inspectors. As mentioned, some departments have hired fire protection engineers to assist them in meeting these technical demands.

More recently, there have been sweeping changes in the field of fire investigation, changes that have made this area of fire prevention more technical. Findings from laboratory research on fire dynamics are being applied in fire investigation and investigations themselves are being conducted with greater scientific scrutiny. Some of the old fire investigation "principles" and accepted practices have been discarded as fire investigation has become more scientifically based. Many of these changes can be attributed to NFPA 921 Guide for Fire and Explosion Investigations, first published by the NFPA in 1992.

One frustration many fiscally strained departments face is having to look to the fire prevention organization to cut back expenditures, sometimes "civilianizing" the organization by hiring civilians as inspectors and public educators instead of uniformed personnel. In some cases financially strapped departments have had to eliminate positions or even the entire division for a period of time.

These staffing changes have underscored the demand for personnel standards, such as state-developed standards or national standards of practice—like those published by the NFPA.

To be certified and re-certified under state standards an individual must participate in a set number of hours of training as well as meet specific areas of proficiency. States that enforce their own standards apply them primarily to fire inspectors; some of these states also have standards for fire investigators, including requirements for peace officer status.

Some states adopt the NFPA standards directly as their statewide minimum performance requirement. NFPA 1031 Standard for Professional Qualifications for Fire Inspector, NFPA 1033 Standard for Professional Qualifications for Fire Investigator, and NFPA 1035 Standard for Professional Qualifications for Public Fire Educator are the three minimum standards applicable to fire prevention bureaus.

Officers of fire prevention bureaus must meet their state's requirements. In the absence of state requirements, a fire prevention chief must, at a minimum, strive to have his staff meet the NFPA standards. An untrained fire prevention staff is dangerous both to the public and to firefighters, as well as a waste of a city's financial resources. An untrained fire prevention person can result in litigation being taken against the agency.

Selection of personnel for the organization

Placing the proper individuals in the organization is an important goal for a fire prevention chief. In the past, fire prevention bureaus were a dumping ground in some fire departments. The lame, lazy, troublemakers and those who wanted to ride out their last days in the department found their way into the fire prevention organization. Fortunately, those days are slowly coming to an end. Fire prevention activities have become much more technical and as a result have demanded the inspector to be qualified.

It is critical that prospective personnel for the fire prevention organization be selected based on their desire to work in the organization, make a difference, be willing and able to learn, commit to the job for an extended period of time, and be willing to take on a new set of responsibilities that are much different than those they had in a fire company.

It is becoming apparent that the practice of forced transfers into a fire prevention organization, including those of promotional transfers, can be a problem—manifested by high turnover and dissatisfaction among personnel.

Some departments have had success in requiring a tour of duty in the organization prior to promotion to the senior ranks. This policy can be beneficial, especially if new personnel are exposed to the realities of fire prevention work and can acquire an appreciation for it. On the other hand, requiring these individuals to stay in the organization for an extended period of time (forced transfers) can have a detrimental effect.

Organization personnel encounter the public on a different plane than do their fire fighting counterparts. Members of the organization must often respond to citizen complaints. They rarely receive the accolades given to firefighters during emergency situations. Many times fire inspectors are looked at as "the enforcer," someone who will find problems that will cause a business to expend money. Fire prevention is often a thankless job.

One problem in retaining trained fire prevention personnel is that very often individuals must leave the organization to be promoted. Thus, an experienced person leaves the organization, increasing turnover, and necessitating the training of yet another person. In today's technically complicated and demanding fire prevention organization, a personnel problem can prove disastrous. A smooth-running and competent fire prevention organization needs a cadre of experienced, well-trained, and self-motivated individuals to maintain consistency and continuity.

As mentioned earlier, the use of civilians is on the rise in fire prevention bureaus and it is an advantageous practice based on:

- Lower salaries and less costly benefits
- Lower overtime costs
- Overall lower turnover rates
- The special talents some civilians offer (previous work as a fire protection system installer/designer or firefighter in another jurisdiction, for example)

The disadvantages of using civilian personnel include:

- Civilians who lack fire and life safety protection experience start much lower on the fire protection "learning curve."
- Civilians without fire service experience may find understanding fire and life safety code provisions concerning fire department operation more difficult.
- Civilians are often considered "outsiders" by uniformed personnel.
- Civilians may find few or nonexistent promotional possibilities.
- The pay and benefits disparity between uniformed inspectors and civilians may lead to animosity among the civilians.

By exploring personnel options, the fire prevention chief can put the "best team on the field." Without the proper personnel, a fire prevention organization is bound to fail in its objectives. Since clerical staff often deal with the public, their ability to work well under pressure and to maintain a positive demeanor are important qualities. Since they route numerous phone calls and queries from the public, they are often the first contact the public has with the organization. Overall, they must understand the organization operations and objectives.

Today, fire protection engineers and technicians are slowly finding their way into fire prevention bureaus across the country in increasing numbers either as a full time employee or on a contractual basis. The fire protection engineers are graduates of the University of Maryland or Worcester Polytechnic Institute. Fire protection engineering technicians are very often graduates of Oklahoma State University.

Fire protection engineers and technicians bring substantial technical expertise to the fire prevention organization. This is especially helpful in the area of plan review and inspections of complicated projects, where they can use their training to ensure that codes are enforced properly. As fire codes become more performance based (described below), the use of fire protection engineers and technicians in fire departments will increase.

Other fire protection engineering and technician duties include fire investigation, where the engineers' and technicians' knowledge of fire dynamics can help investigators to determine the origin and cause of a fire. They also can assist the incident commander at major fires by providing information on fire protection systems, water supplies, etc.

Facilities and Equipment

The facilities and equipment of the organization are important. The offices and associated areas must provide a good working environment, an atmosphere of professionalism, and project the seriousness with which the organization carries out its fire prevention duties. A fire prevention office located at the back of an apparatus bay or in a dormitory area is certainly inappropriate.

The fire prevention chief and other fire prevention administrators such as battalion chiefs responsible for the three "sections" of code enforcement, investigations, and public education as well as a fire protection engineering administrator need individual offices equipped with a desk, chair, bookcase, and small conference table with a few chairs.

Inspectors and public educators need individual desks, chairs, and telephones. Sharing of desks, chairs, and telephones creates serious distractions and is problematic, resulting in the loss of valuable time. Space dividers can be used here, because inspectors and educators spend more of their time in the field. They are usually in the office early in the morning and late in the afternoon.

Plan reviewers and fire protection engineers need individual offices. Space dividers provide neither privacy nor a noise-free atmosphere where they can concentrate to review a set of plans. Each should have a desk, chairs, bookcases, drafting table, and the necessary equipment for plan review.

In addition to individual offices, investigators need access to an interrogation room, a secure and protected evidence room, and possibly a laboratory. Should it be necessary to detain arson suspects in a holding cell, arraignments can be made with the appropriate law enforcement agency.

Other rooms needed by all organization personnel include a library, conference rooms(s), and a training room. (A large conference room and a couple of smaller rooms may be needed in larger departments.) Rooms for dispatch, clerical, and other support staff, as well as a reception area, also are important and should be considered.

Reliable communication equipment (radios, cell phones, and pagers) are needed to ensure constant contact with prevention personnel and to maintain busy schedules. Each investigator, inspector, and educator should have a portable radio or cell phone with an automobile charger. This is important especially for operations that involve more than one organization person, for example, a stakeout by investigators or a fire alarm test of a large building by several inspectors. Separate radio frequencies must be provided for inspectors and investigators, both of them different from the fire fighting frequencies.

Automobiles also are important for inspectors, investigators, and educators. These personnel are highly mobile; they need individually assigned and comfortable automobiles, since quite a bit of time is spent in the car.

The organization's telephone system is a critical piece of equipment. The heavy volume of calls justifies the use of a quality system, one that minimizes the frustrations of the public and fire prevention personnel alike.

An automated system, with a quick default to a "human," is probably the best type, because people become frustrated quickly when they must deal with a completely automated system. The system must have an adequate number of incoming phone lines—a constant busy signal also generates numerous complaints. The message-taking capability of an automated system is perhaps its best feature. This eliminates the avalanche of paper messages and avoids missed messages.

Inspector-specific equipment

Equipment needed for public educators often includes a variety of technical aids, while investigators need specialized detection and laboratory equipment, photography equipment, surveillance equipment, etc.

For fire inspectors, however, a basic list of equipment includes:

- Metal clipboard with cover
- Hardhat
- Flashlight
- Street mapbook
- Personal copy of fire code, building code and amendments
- Full complement of inspection forms, traffic, and misdemeanor ticket books
- Tape measure

Additional equipment may include:

- Hydrant wrench
- Pitot tube and gauge and static pressure gauge
- Coveralls and boots
- Camera

A piece of equipment finding more use by some fire prevention bureaus is the hand-held electronic clipboard or PDA. These devices record the results of inspections electronically, save the information, and then "download" the stored information into the organization's record system when desired. With this type of equipment, paperwork and time is greatly reduced.

Infrequently used equipment—water flow test kits, binoculars, distance measuring wheel, explosimeters/gas detectors, photo equipment, or blast monitor (to measure particle velocities and frequencies in blasting operations) may be stored at the organization office. Such equipment is typically signed out before it is used.

Record keeping

Most paper-intensive organizations, including fire prevention bureaus, have benefited from computers. A computer's capability to store tremendous amounts of information and to generate data instantly assists fire prevention bureaus tremendously. Such data is especially helpful when a fire prevention chief is studying organization statistics for a report or checking up on progress made by specific programs that have been implemented. Each of the two general types of computers—mainframes and personal computers—has its own advantages and disadvantages for a fire prevention organization. It is up to the fire prevention chief to research what type is best for the particular organization.

Records typically kept in computer files include inspections, plan reviews (including a tracking system to follow projects through the approval process), licenses and permits issued by the organization, variances granted to an occupancy, fee records, violations, court orders, etc. In addition, the computer can be used for its word processing capabilities.

Records that are stored in the computer can be retrieved by address, project name, or a number of other ways that reference project-specific information. The computer can also perform such calculations as how many fire alarm system permits were issued during a given period of time. You also can create information files for specific topics such as fireworks or blasting operations.

A "tickler" paper file is a good idea. This is a file of documents flagged for specific action at a future date; for example, a manufacturing company commits to providing sprinklers in its building in five years. The tickler file is referred to periodically to review what outstanding actions must be taken by specific parties.

An important related issue is "forms." Your organization needs to use standardized forms so that your personnel can become familiar with them and can scan them quickly for pertinent information.

Code enforcement

Code enforcement personnel typically comprise the largest contingent within the organization. It is in this area that fire prevention chiefs expend the greatest resources. Code enforcement can be broken down into two distinct subsections: plans review and inspections (although not all fire prevention bureaus perform all of these types of inspections and plan reviews).

Personnel may review one or more of the following types of plans:

- Subdivision plans
- Site plans
- Water supply plans, both public right-of-way and private water supply plans
- Building permit plans, including architectural and mechanical, electrical, and plumbing plans
- Automatic sprinkler system and standpipe system plans
- Fire alarm system plans
- Specialized fire protection systems plans, including CO_2 systems, wet or dry chemical systems, and "clean agent " systems

Obviously, not every aspect of these different types of plans is fire-related. The fire department looks only at the relevant parts of the plans for which they have been designated as the review authority. The fire department may, in fact, be only one of several city departments that review these plans. After the plan review, the fire prevention organization renders comments, and either disapproves or approves the set of plans. After approval a permit is issued by either the fire prevention organization or another plan review coordinating agency, such as the city's building department. There are also municipalities who have created partnerships between fire and building departments and have established office space for a fire plan examiner within the building division.

Plan review personnel sometimes provide such other services as pre-construction meetings or preliminary plan reviews designed to help architects, contractors, and developers meet fire code requirements. Fees may be assessed for these services.

Types of inspections

- *Site* inspections during construction to check fire department access and water supply for firefighting.
- *Building shell* inspections before a building is occupied.
- *Certificate of occupancy* inspections to ensure that an existing or new building is ready for a specific type of occupancy.
- *Licensing* inspections of specific occupancies, such as liquor stores or daycare centers, for which city or state-issued permit requires a fire inspection.

- *Complaint* inspections to follow up on complaints from the public or fire companies about overcrowding in public assemblies or locked exit doors, etc.
- *Nighttime* inspections of such building as nightclubs where operating hours preclude their inspection during daytime hours.
- *Re-inspections* of buildings that fail initial inspections.
- *Substandard or vacant building* inspections of buildings targeted for demolition by the city.
- *Fire protection system* inspections of systems in new or remodeled buildings to verify repairs to existing systems that were not in compliance.
- *Routine* inspections of all "inspectable" occupancies, essentially all occupancies except one-and two-family dwellings in the jurisdiction on at least an annual basis. A routine inspection is the type carried out by nearly all fire prevention bureaus.

Most medium and large cities assign inspectors to a specific district. This way the workload can be handled more easily and the inspectors within a district can become more familiar with their area. Inspectors can be given routine inspection sheets for their districts that they complete when they have time.

In large cities it is advantageous to assign specific inspectors to particular types of target hazards—for example, handling all institutional occupancies, all educational occupancies, all high-rise buildings, or all occupancies with hazardous materials present. This makes it possible to turn inspectors into specialists.

Code enforcement inspectors sometimes perform other duties, for instance, fire watches or crow control details in buildings for sporting events or concerts, or in other places of assembly. In most cases, the promoter or other party that requests the fire watch detail pays the fee for organization's services, on an overtime or flat-rate basis. Inspections of outdoor festivals and carnivals can include inspecting cooking appliances and electrical wiring for adequate fire safety. Fire watches are necessary in buildings where fire protection systems are temporarily disabled.

A fire code specialist should be assigned in the organization to answer telephone questions too technical to be handled by clerical personnel. Designate a specific telephone for code questions and use a machine to record messages for the technical specialist to return quickly. By assigning a specific person who is knowledgeable and has good research skills, you can avoid interrupting or tying up other organization personnel.

Bureaus also issue permits, and, in some cases, license individuals for particular occupations. For example, permits are issued for welding operations and fireworks displays, and tests are given to individuals employed in fire-related jobs (such as blasters who use explosives where obtaining the license is contingent on passing the test).

Legal issues

An important consideration for fire prevention officers is the legal authority with which the organization performs its duties. Organization personnel need to understand the breadth of this authority and its limitations.

The legal authority to perform fire code enforcement may come from state laws as well as from city or county ordinances, including provisions in an adopted model fire code. State laws usually outline proper due process and warrant procedures, while local ordinances specify what constitutes a fire code violation and the corresponding type of infraction. A jurisdiction must identify if their fire code is to be a misdemeanor violation of law or a civil infraction. More and more municipalities are establishing their fire code as a civil infraction.

Since the laws that regulate fire code enforcement vary from jurisdiction to jurisdiction, the following general concerns should be researched to customize a legal program to a particular fire prevention organization.

- What legislation created your fire prevention organization? Does the legislation specify the organization's duties?
- Who is authorized to conduct fire inspections? Are firefighters included?
- What are the proper legal procedures for conducting an inspection? Must the fire inspector demand entry? Is the fire inspector prohibited from inspecting specific occupancies such as one-and two-family dwellings?
- What is the legal recourse for a fire inspector who has been denied entry? Are search warrants required and where can they be obtained?
- What type of paperwork must be completed for a fire code violation and to whom must copies be given? May the fire inspector issue citations for violations? How are the citations processed?
- How is the fire code violator's case decided—by a Board of Appeals or the court system? What are the violator's appeal rights?
- What is the fire inspector's authority in the case of an imminent hazard (an especially hazardous condition that must be abated immediately)? May the inspector order a building to be vacated in such a situation? May the fire inspector seek a court injunction to abate a hazardous condition? Is the fire code a misdemeanor violation or a civil infraction?

These questions show that the legal ramifications of fire code enforcement on the fire prevention organization are significant. It is important that the fire prevention organization's chief officer meet with the jurisdiction's attorney first, to establish a proper legal process for code enforcement, and then periodically to keep the attorney apprised of organization activities and to seek legal advice.

Fire company inspections

Some paid departments use in-service fire companies to carry out routine inspections. These company inspections can be very useful in reducing a jurisdiction's overall fire hazards and making fire companies more productive.

Fire companies should not undertake facility preplanning (or pre-fire survey) activities at the same time they are conducting an inspection. Inspecting and preplanning have different objectives. In addition, permission granted to conduct a preplan does not extend to conducting an inspection—entry for an inspection may have been denied.

Normally, firefighters are trained to conduct basic inspections, looking for the fire hazards typically found in various occupancies. These inspections are usually not as thorough as one conducted by a fire inspector. There are several training programs for fire company officers who supervise company inspections. The department may elect to choose one of these programs or develop one of their own. In either case it is important to provide the level of training necessary to accomplish the tasks expected of the company inspectors.

One or more members of the fire prevention organization must be assigned to coordinate the results of these inspections, review the inspection reports, and follow up on noted violations or any difficult problems.

Fire Codes and Fire-Related Codes/Standards

At the heart of any code enforcement program is an adopted fire code and related standards. It is from this adopted fire code and the adopted standards that the fire prevention organization derives its authority. Codes and standards establish minimum requirements, but should not prevent or discourage someone from exceeding the minimum requirement. (See the discussion on mini-maxi codes.)

The terms code and standard have specific meanings within the code text. Basically, a code dictates *when* a specific requirement must be met, while a standard dictates how to meet a specific requirement. This distinction is important, especially when you must decide which regulation to apply to a particular situation.

The text of specific codes and standards may be considered either prescriptive or performance. For years we have been very "prescriptive" oriented when designing structures; however, there is an increasing demand to design buildings based on performance. The difference is very important in determining how much leeway there is to comply with a particular requirement.

Prescriptive codes or standards explicitly spell out requirements (*i.e.*, "the enclosure wall of a stairwell in a five-story building shall be of eight-inch-wide concrete block [cells filled] construction and have a fire-resistive rating of two hours"). A performance-type requirement might read "the enclosure wall of a stairwell in a five-story building must be capable of withstanding a five-megawatt fire for the full duration of egress from the building."

A prescriptive example is very specific—it prescribes only one way to meet the requirement, whereas a performance-type requirement allows latitude for compliance. A rice paper wall might comply if it met the egress time requirements! To date, most fire codes and standards have been exclusively prescriptive in nature, but this is changing—performance-type requirements are slowly finding their way into the codes.

Model fire codes and standards are developed primarily by two national code-writing organizations. The NFPA and the International Code Council (ICC). While these code-writing organizations have different membership rules, development procedures, and voting rules, they do follow a basic pattern.

Committees of organization members are formed to develop and prepare a particular code or standard. Under the NFPA's structure, technical committee members are selected based on their interest group, so that the NFPA can balance the committee by appointing representatives from each interest group. Once a subcommittee has prepared a draft document, it is voted on by the full committee. In some organizations, such as the NFPA, the document is then made available for public comment. Documents for which public comments are received may be revised to reflect the comments. The revised document is voted on again by the committee. The document then is presented to the organization's general membership for a vote of acceptance. These voting sessions are held at annual or semi-annual meetings.

The NFPA's membership essentially votes on an entire code or standard at a time, and may vote on 30 different codes or standards at one meeting. (The NFPA has more than 300 fire-related codes and standards.)

This is in contrast to organizations that prepare the three regional model fire prevention codes. These organizations vote on their one code, section by section. Also, because membership is limited in these organizations, only fire officials (and other code officials in the case of the National Fire Prevention Code) may vote on the fire codes.

Outlined below are the original code-writing organizations and the codes they published that now comprise the International Code Council (ICC). They are:

- The Building Officials and Code Administrators, International (BOCA) published the National Code.
- The International Fire Code Institute, in conjunction with Western Fire Chiefs and the International Conference of Building Officials (ICBO), prepared the Uniform Fire Code.
- The Standard Code was prepared by the Southern Building Code Congress, International (SBCCI).

• NFPA 1 Fire Prevention Code was prepared by the National Fire Protection Association, International.

There are currently four fire prevention codes in use in the United States; however, with the development and publication of the International Fire Code in 2000 by the ICC, the desired plan is to reduce the number, leaving the International Fire Code and NFPA 1.

As previously indicated, NFPA 1 Fire Prevention Code was recently expanded, with the development of a partnership with the Western Fire Chiefs Association and NFPA and the joining of NFPA 1 Fire Prevention Code and the Uniform Fire Code. It has similarities to the other fire code, NFPA's companion building code NFPA 5000 Building Construction and Safety Code, first published as the 2003 edition.

Both NFPA 1 Fire Prevention Code and the International Fire Code are published every three years. Supplements and addendums are prepared in intervening years when necessary.

These fire prevention codes are considered "maintenance" codes. They are not building codes—except for certain provisions within the documents as described—and apply to the hazardous processes and activities in and around a building. They also contain requirements for maintaining building fire protection features. Relevant areas include general fire safety provisions such as water supplies for firefighting and fire department access, requirements for such equipment as mechanical refrigeration systems, and particular hazards such as flammable/combustible liquids and high-piled stock.

Although the fire codes apply primarily to the building's occupancy, the fire prevention organization should not wait until a certificate of occupancy is issued by the building department. Many of the fire code's provisions will affect the building itself. The NFPA publishes a variety of fire protection codes and standards relating to all facets of fire protection. These include installation standards for most fire protection systems, standards that cover various hazardous materials and processes, and personnel standards.

Included in the NFPA documents is NFPA 101 Code for Safety to Life from Fire in Buildings and Structures, which has elements of both a fire and building code. The Life Safety Code has been adopted by some jurisdictions, particularly at the state level, and is often used for institutional occupancies such as hospitals, nursing homes, and schools.

Building codes

Building codes also have a great impact on fire safety. A large majority of the model building code provisions are fire-related. Building codes cover such particulars as building area and height limitations based on building construction type, firewalls, shaft requirements, egress requirements, fire-rated separations, etc. It is imperative that fire inspectors and plan reviewers be well versed in both the building code and the fire code.

Two building codes correspond to the fire prevention codes described above: the Building Construction and Safety Code, prepared by the National Fire Protection Association and the International Building Code, prepared by the International Code Council.

Mechanical and plumbing codes

There are companion mechanical and plumbing codes that complement the two building and fire codes mentioned above: The International Mechanical Code and the International Plumbing Code used with the ICC codes and IAPMO's Uniform Mechanical Code and Uniform Plumbing Code.

The mechanical codes have implications for the fire inspector because it is in the codes that heating appliance requirements, as well as requirement for smoke detectors in air-handling systems, are found.

The plumbing code also has implications for the fire inspector and plan reviewer (for example, the regulations that govern combined domestic and fire protection water supplies used in single-family home fire sprinkler systems).

Electrical code

The National Electrical Code, used for nearly all electrical installations in the U.S., is published by the NFPA in NFPA 70. Within its many pages, the fire inspector or plan reviewer can find requirements for fire alarm system wiring and wiring used in hazardous environments.

Zoning and subdivision regulations

These regulations are developed by local municipalities or counties. They can be described as "quality of life" regulations, intended to bring order and minimum infrastructure requirements to the various neighborhoods and sectors of a jurisdiction. These regulations do have some impact on fire prevention activities. For example, zoning regulation often prohibit dangerous industrial facilities from being constructed in residential areas.

Subdivision regulations usually specify minimum public roadway widths that affect fire apparatus access. They sometimes specify water supply requirements, including fire flow and hydrant placement. These water regulations also may be found in the regulations of the local water provider or even the fire code itself. The jurisdiction's planning department normally has jurisdiction over these documents and may ask the fire prevention organization for assistance on technical issues involving these documents.

Organizations with an interest in fire prevention

Two organizations, Underwriters Laboratories (UL) and Factory Mutual (FM), prepare documents that influence fire prevention activities. UL has been conducting tests and developing standards of safety for nearly 100 years. These standards and tests apply to such diverse products as fire extinguishers, fire doors, foam concentrate, and sprinkler heads.

Equipment that meets UL's standards is listed in one of UL's listing directories. Fire codes often require that equipment used for fire protection service be listed. Care should be taken not to confuse the term "listed" and "approved" as they are not the same. Approved means acceptable to the Authority Having Jurisdiction, in this case the fire prevention organization.

Factory Mutual also conducts tests on fire protection equipment, particularly equipment found in industrial facilities and publishes a list of "approved" equipment. In addition, FM publishes a set of Loss Prevention Data Sheets that are particularly useful for determining appropriate fire protection measure for industrial processes.

Adoption of Fire Codes and Standards

To be enforced, a fire code or standard must be adopted or referenced within a statute enacted by a city, county, state, or other political subdivision. Otherwise it is not enforceable. The authority to adopt fire regulations at the local level comes from the state. The state empowers the jurisdictions within the state to enact laws not reserved for the state itself. This distinction is important since states with a state-wide fire code often limit individual cities in the adoption of their own fire codes. Progressive state code statutes do not restrict subdivisions from enacting more stringent provisions, if they so desire.

States with a so-called "mini-maxi" code allow individual cities to adopt their own fire codes; however, the city's adopted code must meet the minimum requirement of the state code and cannot exceed the state code's requirement. Such a situation leaves little room to handle particular problems within an individual city. For this reason many fire prevention chief's look unfavorably on state mini-maxi codes.

Once a jurisdiction has selected a particular fire code for adoption, it usually develops a set of amendments to meet local conditions. These amendments serve to address the unique fire hazards of the community.

Since the fire code normally will be adopted along with the jurisdiction's other codes, such as the building code and plumbing code, it is best to meet with representatives of the other city departments that enforce those codes. This is a good way to resolve conflicts among amendments.

Once a working draft of an ordinance has been prepared, the fire prevention chief should meet with groups affected by the regulations. These groups could include the Chamber of Commerce, architects' and engineering associations, developers, contractors, and other interested parties. Hold working sessions with these groups to determine points of disagreement. Additionally, public hearings are typically held to allow the general public to review the proposed code, with any amendments, and provide input.

The fire prevention chief should consider all suggestions from these groups, review these suggestions and, where acceptable, implement these suggestions in the code. Partnerships built during this process will assist in developing a more positive atmosphere between the local business community and the fire service with the end result being improved fire prevention.

The governing board (City Council/Commission, Township Board) of a community normally has the final say on the code and any amendments. Having the board on your side makes the adoption of codes and amendments easier. Take the time to explain the code to the adopting body. Highlight its benefits, how it compares with the present code, and the implications of its adoption. Do not hide any significant implications, but do not overemphasize them either. With these efforts and with your information on the table, you have the best chance at adoption.

Application and interpretation of fire codes and standards

Well-trained inspectors and plan reviewers are expected to enforce the fire codes and standards uniformly. There is no room for selective enforcement and doing so can invite litigation. They are expected to prepare complete inspection and plan review reports, documenting all deficiencies noted during inspections or plan reviews. In the case of inspections, reasonable compliance dates should be set and follow up inspections conducted. Follow up (or re-inspections) are one of the most important inspections that is performed. Violations that are allowed to go uncorrected increase a fire department's liability. An advantage to standardized basic and advanced training for fire prevention personnel is that all of them will handle similar situations in the same manner. This avoids multiple interpretations.

While it is desirable to have perfectly clear codes and standards, the reality is most fire codes and standards are not written with perfect clarity. This means that different people render different interpretations. In some cases, what a fire inspector or plan reviewer stipulates may be construed by a citizen as a misinterpretation or non-requirement of the code or standard.

Two safety nets help rectify this situation. The fire prevention code organizations and the NFPA all have staff to assist in interpretations. Keep in mind, however, that these interpretations are the opinion of the individual staff member. The NFPA has a special formal interpretation process. Written inquiries that pose specific questions on a particular code or standard section are reviewed by the committee that prepared the document and a formal interpretation is issued.

Neither an informal interpretation by a code-writing organization staff member nor a formal interpretation from the NFPA has the effect of law within a given jurisdiction where the code in question has been legally adopted. They merely attempt to resolve a problem before it is taken before the Board of Appeals, the legal means of conflict resolution. The Board of Appeals is a board appointed by the governing body of the jurisdiction, such as the city council. The composition of the board varies from jurisdiction to jurisdiction, but often includes members who represent various interest groups in the city (*i.e.*, developers, architects, engineers, and contractors, as well as citizens at large). Many board charters require that some of the members have specific fire protection experience.

The board can vote on interpretations of the code as well as the acceptability of alternative means of code compliance. Its duties tend to be narrowly defined; unless specifically stated, the board cannot waive code requirements for an appellant.

The Future of Fire Prevention

Fire prevention activities will begin to accelerate as we move forward in the 21st century. Departments still will be expected to do more with less, necessitating creative approaches to reducing fire losses through fire prevention. Fire prevention organization personnel will be expected to have greater technical capabilities. As a result it will be imperative that fire prevention personnel receive the best and most up-to-date training possible. Only constant training will keep them proficient in technological advances.

As more and more fire protection systems are installed, the demands on fire prevention bureaus to oversee proper maintenance of the systems surely will increase. In addition, since more reliance will be placed on these systems because of diminishing fire fighting resources, there is little room for error. New fire fighters must be trained differently when it comes to fire prevention. Their basic fire academy training must include instruction on effective fire prevention techniques. They must be made increasingly aware of their fire prevention responsibilities and why fire prevention must be the departments'—and their—top priority.

If the fire services do not start making fire prevention a priority, who will?

References

Cannon, Donald J., *Heritage of Flames,* Pound Ridge, NY, Artisan Books, 1977.

Ditzel, Paul C., *Fire Engines, Firefighters,* New York, NY, Crown Publishers, 1976.

Costello, Augustine E., *Our Firemen, A History of the New York Fire Department, Volunteer and Paid,* New York, NY, Augustine E. Costello, 1887.

Brayley, Arthur W., *A Complete History of the Boston Fire Department, Including the Fire Alarm Service and the Protective Department, from 1630 to 1888,* Boston, MA, John P. Dale and Company, 1889.

O'Brien, Donald M., *A Century of Progress Through Service: The Centennial History of the International Association of Fire Chiefs,* 1873–1973, Washington, DC. IAFC, 1972.

26

PUBLIC SAFETY EDUCATION

Robert C. Barr

CHAPTER HIGHLIGHTS

- Reviews the various types of Public Safety Education programs available
- Reviews how to identify PSE programs that are needed or required, and methods for delivering
- Identifying individuals or groups that can deliver PSE programs and reviews resources for delivering PSE programs
- What to consider when evaluating PSE programs

Education: A science dealing with the principles and practice of teaching and learning.

INTRODUCTION

Public Safety Education (PSE) programs are part of the essential services that are or should be delivered by every fire and rescue organization. These programs are the proactive programs that serve to instruct citizens in actions that will prevent loss or injury. If effective, they should decrease the demand for other types of emergency services.

Many activities that are described as "education" are not education, but "show and tell." Show and tell activities are those where the fire and rescue organization is called to present a program at a school or public gathering on short notice with the result being the closest company tasked to take their apparatus to a school and present a program. Quite often the program content is left to the discretion of the company officer. The result is that the program consists of showing the fire apparatus and equipment, having a firefighter don protective equipment, and then squirting a little water.

These types of activities are nice for public relations and serve to make the children familiar with fire apparatus and the people who respond on them. However, in many cases there is very little educational content in the visit. Perhaps with a little prior planning these visits could be turned into educational events. In order to make the visit more effective all it would take would be a written operating procedure or guideline that contains a simple lesson plan, with learning objectives that anyone on the company could use to convey some educational material in addition to the show and tell part. Prior planning could also equip each company with some literature that could be left with each individual. Remember that public safety education is more than just petting the spotted dog.

Effective programs should be well planned and supported in order to achieve a desired result. In addition, the program should have some means of measuring the effectiveness. If you can't measure it, you can't improve it.

What Is PSE?

Public safety education programs are those that seek to prevent injury or loss through programs that are educational, instructional, or informative on a wide range of issues that the fire rescue organization is responsible for.

It is often said that good fire protection consists of three elements: engineering, education, and enforcement. This chapter provides information on public fire safety education programs.

Objectives of PSE

Some of the objectives of a public safety program should be to:

- *Educate*—Educate target audiences in specific subjects in order to change behavior. One example would be to have people check their smoke detector on a regular schedule, or for children to react positively to the sound of a smoke detector in the home.
- *Instruct*—Instructing target audiences in how to do things such as operate fire extinguishers, or "stop, drop, and roll," etc.
- *Inform*—Inform large groups of people about public safety issues. This could be public safety announcements on a variety of issues.
- *Distribute*—Distribute information on timely subjects to target audiences.

Perhaps the greatest objective is to change the behavior of individuals through various educational programs in order to prevent injury, death, or loss due to fire or other types of incidents.

PSE and the Fire Department Mission

Public safety education programs should be fully compatible with the fire and rescue organization mission and other programs. One way to examine what is being done or not being done is to use the Fire Prevention Effectiveness Model that was developed by the Ontario, Canada, Office of the Fire Marshal. A graphic representation of the Model is shown below.

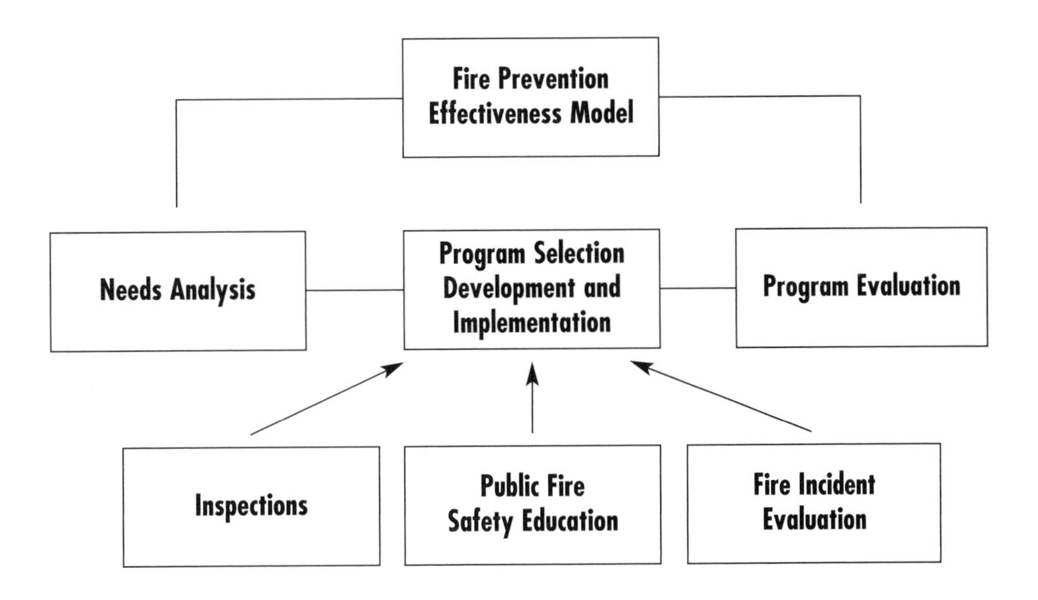

Fire Prevention Effectiveness Model

The model is explained as follows.

Purpose: To assist fire service managers to thoroughly plan fire prevention activities. It stresses the need for programs to be effective and properly evaluated. It also defines fire prevention activities as including inspections, public fire safety education, and fire incident evaluation.

Introduction: The Fire Prevention Effectiveness Model is a planning aid that focuses on one of the eight components of the Comprehensive Fire Safety Effectiveness Model. The Fire Prevention Effectiveness Model is a tool that ensures that all issues are identified and addressed when considering any fire prevention programs or activities or when reviewing existing programs.

Process: The process promoted in the model involves three distinct components:

1. A needs analysis to identify the significant (fire) risks to the community. *Note: The needs analysis can also be used to identify other significant risks. For example, the Phoenix Fire Department has identified child drownings as a significant risk.*
2. The selection, development, and implementation of appropriate programs to address identified risks.
3. An evaluation of the effectiveness of fire prevention programs (Public Safety Education Programs).

Needs Analysis: Needs analysis for fire prevention programming involves assessing the relative fire risks in a community and identifying the significant risks that should be addressed. It also involves compiling adequate information to conduct the analysis and for appropriate program selection, development and implementation. Information including fire losses, implications of fire occurrences, building stock, EMS related data, and demographics of the community have to be gathered and assessed.

Types and Kinds of Public Safety Education Programs

The following list describes various public safety education programs subject areas that can be presented to the community. This does not mean that all of the programs listed should be used in any particular community. And, the list does not contain a list of all subjects that could be presented. The fire and rescue organization should undertake a risk analysis and/or a needs analysis to determine which programs are needed or required.

It should also be noted that not all of the programs listed relate to fire. Some of the non-fire related programs relate to services that the fire and rescue organization deliver such as EMS. Any related programs that decrease the potential for injury or death should be explored and delivered, if the analysis substantiates the need.

- Home escape planning
- Smoke detectors
- Burn and scald
- Home fire prevention

- Natural disaster
- Man-made disaster
- Smoke detectors
- Carbon monoxide
- Children's car seat
- Fire drills
- Fire extinguishers
- Fireworks
- Hazardous materials
- Bicycle safety
- Stop, drop, and roll
- Babysitting
- CPR and AED

Identifying Specific Public Safety Education Program Needs

Every organization should undertake some type of needs analysis to determine the types of public safety education programs that are needed in the community. The needs analysis can be any method that produces results. Given the fact that there are more than 30,000 fire and rescue organizations in North America and that they service all types of communities, this section will not recommend any specific method but will make some general suggestions on how a community determines what their needs are. Suggested methods for identifying program areas are reviewed next.

Awareness

Most fire department members, regardless of organization size, will have some continuing general awareness of what types of fire and rescue service demands that are created by the community. This is probably most apparent in smaller organizations and communities because of a lesser number of incidents. Asking a chief or company officer about the types of incidents that they respond to most often will most likely generate an immediate answer. And, if asked what types of education programs should be implemented they can list those.

This means of identifying program needs is not very scientific, but is valid based on actual experience. In some cases there may be an increase in bicycle accidents or an increase in school fires. These trends will be readily apparent and indicate a need for increased educational efforts in these areas.

Anyone who is proposing new programs should interview those who deliver services on a regular basis in order to determine what types of services are in demand, and match the programs to the demands.

Fire cause data

Fire cause data gleaned from fire investigation reports is another important source of determining what types of education programs are needed in the community. This data or the actual reports will provide a wealth of information of related to trends in the community. One or two incidents of a specific type may not justify the need for a formal program, but may indicate the need for a targeted program for specific individuals.

Emergency medical data

The examination and analysis of emergency medical call data should provide some specific indicators on what types of prevention programs should be developed and delivered in order to decrease the demands for emergency medical services.

The ongoing examination of emergency medical data should also show trends in the community of specific periods of time.

One example of what this data may show is that given the current configuration of the community the response data indicates that the survival rate of heart attack victims is not what the community desires, and that some type of intervention strategy should be implemented. One strategy might be the increased emphasis on CPR to citizens of the community in order to improve the survival rate of heart attack victims. Or this may be an opportunity to implement a program that advocates the use of automatic external defibrillators (AED) in certain occupancies within the community.

Another example would be a trend in children ingesting household poisons. An increase in calls in specific parts of the community may indicate a need for increased awareness on the part of parents.

Request for programs

The request for programs from the community is another indicator of what the needs are. Fire and rescue organizations receive many requests from various community organizations for them to present programs at meetings or other types of gatherings. In most cases the request is not specific as to the content of the program. They just want someone to present a program on something related to the fire department. This is a great opportunity of present a program or sessions this is both informative and educational. The fire and rescue organization should be prepared with one of the "hot" topics that they want to present.

Again, this would be an opportunity of encourage the use of AEDs and CPR as a method of increasing the survival rate of heart attack victims. However, this would not be the best program for a group of children. Therefore there must be a menu of "hot" items ready for presentation depending upon the age and interest of the group making the request. All requests for programs should be viewed as opportunities to present a targeted education message for that specific group.

Fire report data

Fire report data is an excellent source of information that can identify trends in the demand for fire services, provided the reports are accurate and the data can be analyzed for specific time periods. This data can then be used to identify public safety education programs for specific audiences. This data can be used in conjunction with awareness to zero in on the larger issues that require immediate attention.

Developing PSE Programs

The development of public safety education programs requires more than writing an outline and a couple of lesson plans. However, not all programs require a large staff of education specialists.

One of the first steps is to identify the target audience.

- General audience—This would include the general public throughout the community.
- School age children—Includes grades K through 12.
- High risk—Includes people such as the elderly or socio-economic groups.
- Business/Industry.

The second step after the audience has been identified is to develop a set of learning objectives using the subject area that is to be presented. These objectives should be standard learning or behavioral objectives that simply state what the individuals are expected to know or do at the end of the presentation. The range of objectives should not be so extensive that it will take two weeks to convey the information. Experience shows that programs presented by fire and rescue personnel are limited in time. Therefore the list of objectives should be limited to the average amount of time available for the material to be presented.

This second step, however, may include long term programs such as those that may be presented in public and private school settings and integrated into the curriculum, as well as include a number of subjects. In this case, the program will require numerous learning objectives.

The third step requires the development of lesson plans or teaching outlines. The learning objectives and the lesson plans serve to standardize the programs so that the same message on a particular subject is delivered to everyone within the specified audience.

This step may require more effort than some of the other development steps because of the detail involved and the need to put the subject matter in a form that will be understood by the target audience. Fire and rescue organizations that lack trained specialists may wish to enlist the support of personnel from educational institutions in the community to help with this step.

After the lesson plans have been developed they must be tested to make sure that they work and do what they are intended to do. One way to test the lesson plans are to do a simple pre-test before presenting the material. This test may be designed as an ice breaker or warm up depending on the audience. Then, after the lesson or subject matter is presented, another simple test of some type may be given in order to find out what was learned.

The tests and presentation of the material will no doubt give the presenter some sense of what, if anything, needs to be changed. After the material has been presented several times and all the bugs are worked out, the lesson plan can then be distributed for use by the qualified presenters.

The fourth step should be the development of resources that support the lesson or lessons. Resources include handout materials that reinforce the subject, memory aids for the participants, audio visual materials for use by the presenter, and reference materials that can be used to prepare for the lesson.

This is a brief overview of a four-step development process for PSE materials. The development of effective materials takes time and effort. The approaches don't have to be dull lecture materials or presentations. During the last 30 years, fire and rescue organizations have used their imaginations and talents to develop materials that are both educational, effective, and fun. Perhaps one of the requirements for all PSE materials is that they should be informative, interesting, and a pleasure to learn—without removing the serious intent.

Delivering PSE Programs

There are as many methods of delivering PSE programs as one can imagine or create. The actual specifics on what methods to use may be better left to experts. It is recommended that each organization match the delivery of programs to the subject of the programs to be delivered. If you don't have the experts on staff seek out people who are in the business of delivering educational and instructional programs to the public. These people can be individuals involved in education, public relations, advertising, etc. These people all have skills that can be used to get the message out.

In addition to presentations by instructors or public safety education personnel to small groups, there are other means that can be used. Perhaps some of the newest additions to the public safety education arsenal are the fire and rescue organization websites. A little web surfing will reveal that quite a number of fire and rescue organizations throughout North America are providing a wide range of public safety education materials that can be accessed through their websites.

In addition to the methods mentioned above, some others are the use of radio spot messages, television notices, newspaper notices, neighborhood newsletters, religious groups, neighborhood associations, and other city, county, and regional groups. Many of these organizations are always looking for material. The fire and rescue organization, if they have prepared material and can make it available, will be received favorably by these groups. The key is preparation, and making the material available on a timely basis.

Again, there are no set rules. Use whatever means are available to deliver the PSE message.

Who delivers PSE programs?

When the first wave of public safety education programs were being delivered in the 1970s, fire departments sought elementary school teachers to manage and deliver the programs. This was a well founded and successful approach because the individuals understood the educational process, and were able to acquire the necessary fire knowledge from the fire community and convey the information to the target audiences. Many individuals currently managing public safety education programs have an education background. In addition, the individuals also possess a large amount of enthusiasm for the work that they do. The work of public safety educators has now progressed to the point that there are currently professional qualifications standards for those who deliver programs. These standards are part of the fire service professional qualifications standards that are published by the National Fire Protection Association (NFPA).

The standard is NFPA 1035 Standard for Professional Qualifications for Public Fire and Life Safety Educator. The standard defines and specifies the minimum requirements for the three levels of public fire and life safety educator. The three levels are:

- *Public Fire and Life Safety Educator I.* Certification at this level requires that the individual demonstrate the ability to coordinate and deliver existing educational programs and information as specified in the standard.
- *Public Fire and Life Safety Educator II.* Certification at this level requires that the individual demonstrate the ability to prepare educational programs and information to meet identified needs as specified in the standard.

- *Public Fire and Life Safety Educator III.* Certification at this level requires that the individual demonstrate the ability to create, administer, and evaluate education programs and information as specified in the standard.

The delivery of PSE Programs should not be limited to those who are certified as Public Fire and Life Safety Educators, but should include anyone in the fire and rescue organization who wants to be involved. There are a great number of fire and rescue personnel throughout North America who have used their talents to spread the public safety message. These individuals should be cultivated and supported so that they are delivering the message that the department wants delivered.

Resources outside the fire and rescue organization should also be recruited and encouraged. There are many civilians with a wide range of talent who can be used when properly supervised and supported to assist in delivering PSE programs.

Evaluating PSE Programs

The evaluation or measurement of the effectiveness or success of PSE programs may be somewhat difficult to determine without a great deal of effort, but it needs to be done. This falls into the same category of other types of education or training where specific learning objectives are used. This means that if you want to fix something, or make changes, or make it better you must have some means to measure it. Therefore, there must be some means of measuring the effectiveness of the PSE programs that are delivered.

Perhaps the simplest means is to monitor the activities within the community to determine if specific types of incidents that match PSE programs are increasing or decreasing. An increase in incidents of a specific type would indicate that additional effort or a change in method of focus is required. One thing that is necessary for measuring all programs is a set of clearly recorded objectives for each program. The objectives for the programs must state in some form the type of change in behavior or other type of change necessary as a result of the program.

For those organizations with ample resources there are more sophisticated methods for evaluating the effectiveness of PSE programs. These are the same methods that are used by educators or analysts to determine the outcome of educational programs. If additional information on more sophisticated methods of evaluation is desired, it is recommended that the NFPA *Fire Protection Handbook* be consulted. The *Handbook* contains a chapter on Evaluation Techniques for Fire and Life Safety Education.

The reason for evaluating the programs is to find out which programs are effective and which programs are not effective in order to capitalize on available fire and rescue resources. Ineffective programs waste time and effort.

PSE Resources

Programs and materials are available from a number of different sources. Some of the sources have comprehensive sets of materials that that require very little preparation time prior to presentation. Some have handout materials and some have programs that can be integrated into elementary school curriculums.

- **Federal Emergency Management Agency (FEMA).** FEMA has a number of items that are available to support the PSE effort. Most of these are available without charge. The most rapid way to access information on the items is to explore the FEMA website. Another valuable source of information is the *Emergency Preparedness Materials Catalog.* This catalog can be obtained by contacting FEMA. Their website address is: *www.fema.org*.
- **United States Fire Administration (USFA).** The USFA has a number of items available. The most expedient means is to access the website and look at the publications that are available. One publication that is useful is the *Resources On Fire.* Their website address is: *www.usfa.fema.org*.
- **State and provincial fire marshals.** Many state and provincial fire marshal organizations have some type of resource available to support PSE Programs. It is recommended that individuals who are seeking additional information on programs access their respective fire marshal's offices via the Internet.
- **State and provincial fire training organizations.** Most state and provincial fire training organizations have material that is available for public safety education. Some of these agencies are attached or part of the fire marshal's office, however some are independent state agencies.
- **National Fire Protection Association (NFPA).** The NFPA is a private, non-profit membership association that publishes a wide variety of print and audio visual materials that can be used in PSE programs. The Association publishes a catalog of all materials offered for sale. Call 800-344-3555 to request a catalog or membership information or access their website: *www.nfpacatalog.org*.
- **American Red Cross.** The Red Cross has materials and resources available to assist with public safety education, as well. For additional information contact your local Red Cross chapter or access their website: *www.redcross.org*.

27

INVESTIGATING FIRES

Jon C. Jones

CHAPTER HIGHLIGHTS

- Provides a basic understanding of the importance of investigating fires and explosions and to provide information that will assist in making decisions regarding this important function.
- According to NFPA 921, Guide for Fire and Explosion Investigations, a fire investigation is a process that determines the origin, cause, and development of a fire or explosion.
- To make sure that the investigator conducts a complete scene examination that addresses the information available at the fire scene, it is recommended that a systematic process be used to guide the investigation.

INTRODUCTION

The investigation of fires and the determination of their cause is normally one of the responsibilities of the fire chief. This responsibility, and the authority to conduct investigations, is often delegated to the fire chief by the fire prevention code adopted by their jurisdiction. The primary reason for this authority is for the collection of information regarding the origin and cause of fires in order to prevent similar occurrences in the future. If it is determined that the fire was intentionally set, the investigation will be expanded to collect evidence related to the crime.

In many jurisdictions, the fire chief delegates this responsibility to a fire investigations unit that may be part of the fire department or another agency with the responsibility at the local or state level. When a fire department responds to a fire incident, some form of investigation almost always takes place. The investigation may be very simple in order to complete a company report or NFIRS (National Fire Incident Reporting System) form (see appendix 1). If there is a question about the circumstances surrounding the incident, or an injury or fatality is involved, a more detailed investigation may be required. At the simplest level, the investigation may be conducted by a company officer assigned to the incident. When the complexity of the investigation increases, a fire investigator with specific training and experience in the field should be requested. At the very least, officers at the company and incident command levels should be trained in the basics of fire investigation and scene preservation so that they will be able to recognize conditions requiring additional support and prevent the destruction of critical evidence at the fire scene. The fire chief should consider state or national investigator certification for department personnel assigned as "origin and cause" investigators. Certification, discussed later in this chapter, adds to the credibility of the investigator should they be required to provide court testimony in criminal or civil cases involving fires they investigate.

A factor that has had a significant influence on the way fire departments conduct fire investigations today has been the adoption of NFPA 921 Guide for Fire and Explosion Investigations, by the National Fire Protection Association.[1] This document, first issued in 1992 (and revised several times since), has helped to define the practice of fire investigation both in the public and private sector. While the document is a "guide" in the NFPA system (and thus not written in mandatory language), it is widely accepted in the legal system as an authoritative reference.

Another NFPA document that has helped to shape the field of practice for fire investigations is NFPA 1033, Standard for Professional Qualifications for Fire Investigators.[2] This document defines the job performance requirements for the fire investigator. These requirements are used by several national organizations, as well as a number of states and provinces, as the basis for certification programs for fire investigators. Together, the two documents help to define the field of fire investigation. NFPA 1033 defines what the investigator does on the job and NFPA 921 provides the background material that the investigator uses to perform the job. Every fire chief responsible for the investigation of fires and explosions should be familiar with these documents and make sure that all personnel assigned to fire investigation duties are knowledgeable of their contents and properly trained to perform assigned tasks.

The Fire Investigation

According to NFPA 921 Guide for Fire and Explosion Investigations, a fire investigation is a process that determines the origin, cause, and development of a fire or explosion. The *origin* of a fire is the location where the first material or fuel involved in ignition is exposed to a heat source that is sufficient to cause the fuel/material to ignite. In fire investigation, the *area of origin* is the general location—a room or defined area where the fire began. The term *point of origin* is used to describe the exact point where ignition occurs. *Fire Cause* is the term used to describe the circumstances or conditions that allow a fuel and heat source to come together at the point of origin to result in ignition. Most fires are caused by either an accident or unintentional act, or they are intentionally set. *Fire development* is a term used to describe the factors that influence the growth and travel of a fire beyond the point of origin after the ignition sequence.

The NFIRS reporting system for fires tracks each of the elements previously discussed for evaluation of fires at a national level. The accurate determination of the information is directly related to the complexity of the fire and the amount of damage done prior to extinguishment. For many of the fires that departments respond to, the determination is fairly simple. Take, for example, a fire involving an electrical appliance in a residential kitchen where the occupant observes the ignition and unplugs the appliance prior to the arrival of the fire department. Damage is limited to the appliance and a small area of cabinets adjacent to the unit. The fire is extinguished prior to the arrival of the fire department and there is no extension beyond the area of origin.

Based on an interview with the occupant and observations at the scene, the company officer determines that the area of origin is the kitchen. The point of origin is the involved appliance. A careful examination of the appliance involved shows that the fire ignited in the rear of the unit adjacent to the point of entry for the power cord. Combustible material in this area is heavily charred and adjacent metal surfaces are discolored. The material first ignited appears to be combustibles in the appliance. The most probable source of the heat of ignition is electrical energy. The cause of the ignition, or what brought the heat and fuel together, is determined to be an equipment failure. The development of this fire was stopped by the removal of the energy source—unplugging the appliance. As a result, there is no additional information to collect regarding fire development. In this case, the fire officer should be able to complete the fire report and accurately document the essential information regarding the origin and cause without additional assistance from an investigator.

Now consider the same ignition sequence without an occupant in the home. The ignition occurs in the appliance and spreads to the nearby cabinets, which also begin to burn. Other fuels in the area reach their ignition temperatures and begin to burn. Within a few minutes the kitchen reaches flashover and is completely involved. The fire extends to the adjacent living room through a door opening and upward through a vent to the attic space of the home. A passerby observes smoke coming from the eaves of the house and calls 9-1-1 to report the fire. On arrival, the fire officer observes heavy smoke and orders a handline extended into the home for extinguishment. A ladder company is directed to ventilate the roof and a second engine provides a backup handline and conducts a primary search. After the fire is extinguished, a significant amount of the home is damaged and the suppression officers decide that the determination of the origin and cause is beyond their capacity and request an investigator.

While the ignition sequence for both fires is the same, the determination of the origin and cause will be much more difficult for the second scenario due to the amount of destruction caused. The fire burned for a longer period and resulted in much more damage to the structure and its contents. It may take the investigator several hours of work to properly document the scene and collect enough information to make an accurate origin and cause determination, if one can be made.

While the determination of the origin and cause of a fire is not the prime concern of the company officer or incident commander at the scene of a fire, it is an important issue. At some point in the incident, the incident commander will have to make a determination regarding the need for a detailed investigation at the fire scene. If it is determined that an investigator is not required, the assigned suppression personnel may be the only fire department representatives available to document the incident (as was the case in the first scenario discussed). The information they collect may be of importance in the prevention of similar incidents or in civil litigation that might take place as a result of the incident. If no formal investigation is to take place, the responsible officer should document the scene and write a report that details these observations.

If it is determined that a more detailed investigation is required (as in the second scenario), fire suppression personnel can assist in the investigation by maintaining control of the scene until the investigator arrives and by protecting the scene from unnecessary foot traffic, overhaul operations, and other activities that could destroy valuable physical evidence. The importance of coordination between fire suppression and investigative personnel is discussed in more detail later in this chapter.

The Investigative Process

The investigation of the origin and cause of a fire can be a very complex process. To make sure that the investigator conducts a complete scene examination that addresses the information available at the fire scene, it is recommended that a systematic process be used to guide the investigation. The fire department should adopt a standard operating procedure (SOP) and train all personnel who conduct investigations in the procedure. In general, the process should involve the identification of the origin of the fire and then move to determine the cause (including the ignition source and circumstances that resulted in the ignition and growth). NFPA 921 recommends the use of the scientific method as the model for a systematic process. Whatever the model selected by the fire department, the outcome should be the careful and complete review of all information available regarding the origin and cause of the fire.

An example of a systematic approach includes the following steps:

- Assignment/incident
- Scene examination
  - Exterior examination
  - Interior examination

- Document and collect evidence
- Analyze observations and findings
- Determine the origin and cause based on available evidence
- Report findings

This process can be used by fire officers involved in suppression procedures or by fire investigators. Each element of this systematic approach is discussed in more detail in the following section.

Assignment/incident

The investigation begins with an incident or assignment. As soon as a company or fire investigator is dispatched, the collection of information should begin. The time of dispatch and weather conditions should be noted. Companies should begin making observations as they arrive on the scene. These observations include fire conditions on arrival, individuals who are on-scene, vehicles on-scene and leaving the scene, signs of forced entry prior to the fire department's arrival, and signs of potential criminal activity.

An important consideration that must be made early in the investigative phase of the incident is the *right of entry* into the involved structure. Where fire conditions are present, the fire department has the right to enter a property without permission. This is commonly called "exigent circumstances." Where an emergency exists, it is assumed that the property owner gives permission to responders to enter a property and take emergency action. U.S. Supreme Court cases including *Michigan v Tyler*[3] and *Michigan v Clifford*[4] define the conditions surrounding the right of entry. In general, it is accepted that evidence in plain view of firefighters operating at a fire scene can be seized and used as evidence in a criminal case.

The issue of the right of entry and the collection of evidence for criminal cases come into play once the exigent circumstance no longer exists. This is typically an issue for the investigator who arrives on-scene after fire suppression units have completed their operations and cleared the area. Where a fire is suspicious, it is recommended that a fire company remain on the scene to retain control until the investigator arrives. Should access to a scene be required after the suppression and subsequent investigation is completed, the investigator must obtain permission to enter the property from the owner or get a warrant that allows entry for investigative purposes. Since the requirements for warrants and seizure of evidence vary from jurisdiction to jurisdiction, the fire chief should get clarification of the requirements from the city attorney or local prosecutor. An illegal entry into a property could result in the loss of a criminal case by excluding critical evidence obtained during the entry and possibly expose the department to civil action for trespassing.

Scene security is another issue that must be addressed early in the incident. The area surrounding the structure, or location of the fire, should be secured and only authorized personnel allowed to enter. Once the fire is controlled and overhaul operations begin, the number of firefighters in the fire area should be limited to those with a specific task to perform. Large numbers of firefighters traveling through the scene may result in unnecessary damage to very fragile fire evidence in the structure. Occupants and building owners should not be allowed entry into the building without an escort prior to the investigation. The purpose of scene security is to prevent the destruction or removal of physical evidence in the building prior to the arrival of the investigator.

Scene safety is also a major consideration for the assigned investigator. Fire investigators should be provided with protective equipment that is appropriate for the hazards they will be exposed to at the fire scene. Fire investigators should at least have head protection, eye protection, gloves for hand protection, and safety shoes or boots. At structural fires where there is significant damage, the investigator may require full structural firefighting equipment. Respiratory protection including SCBA should be available for scenes where products of combustion are still present. Fire investigators should also be trained to the appropriate level of hazardous materials response so that they can identify potential hazards that might be found at the fire scene, and operate using the appropriate protective equipment to match the potential hazard. Operations in hazardous locations may require that the investigators entering the scene wear full encapsulation suits for their protection. Departments where this potential exists may determine that specialized training in hazardous materials operations for some or all of its investigators is warranted.

Specific hazards for which the fire investigator should be alert for during the examination of the fire include:

- *Structural hazards*—weakened structural components that increase the potential for collapse and holes in floors or roofs.
- *Hazardous atmospheres*—ranges from products of combustion or oxygen depleted atmospheres to toxic materials released as a result of the fire or suppression operations. These could include pesticides in barns, hazardous materials in drug labs, or materials stored or used in industrial and storage occupancies.
- *Standing water*—accumulated water from suppression operations or broken pipes can add significant weight to a building and lead to collapse, or create deep pools that could possibly trap an investigator.
- *Building utilities*—utilities such as electricity and gas that are not secured can present a significant hazard to the investigator working in a structure.

The fire investigator should also be alert for other potential hazards that could be encountered in a structure such as animals or even booby traps left by occupants. Awareness of the potential hazards and the use of appropriate protective equipment during the scene examination, are critical to the safety of the investigator. Due to the potential hazards of the scene examination the fire chief should also ensure that investigators are not assigned to individually operate in hazardous conditions. When investigators do not work in pairs, a fire suppression company may be assigned to stand by and monitor the safety of the investigator while the scene examination is conducted. Whenever investigators are assigned to operate alone they should be provided with a means of communication, and dispatch should track their location and assignment.

The investigator may decide to conduct preliminary interviews prior to beginning the scene survey. Interviews would include any occupants in the building involved, firefighters involved in suppression, police officers that arrived on the scene early in the incident, and other witnesses. The preliminary interviews provide the investigator with basic information regarding the incident and the observations made early in the incident. This information may help to direct the investigator in the examination of the scene and determination of the fire cause. The investigator must, however, remember to remain as objective as possible during the examination of the scene. The examination should serve to verify the information provided during the preliminary interviews or show that the information was incorrect or deceptive.

Fire Scene Examination

The examination of the fire scene should be a systematic process that begins on the outside of the structure and moves into the building from the area of least damage. The examination of the fire scene is conducted to locate the origin of the fire and attempt to determine its cause. To that end, the investigator is looking for physical evidence that will help in that determination. Typical observations include signs of criminal activity (such as forced entry); physical evidence that might be related to the ignition (such as gas cans, broken glass, incendiary devices, appliances, heating devices, etc.); and patterns that provide the investigator with indicators of fire growth, development, and travel. For fire incidents that are confined to small areas, the scene examination will be relatively simple and require less time than incidents that involve multiple rooms or entire structures.

Most investigators begin by examining the outside of the building before they conduct an interior survey. The exterior survey allows the investigator to observe the conditions of the structure, including the location of the most visible damage. This is important when attempting to locate the area of origin and also gives the investigator an indication of how sound the building is before entering. The exterior survey provides information regarding the types of utilities the investigator should expect to find in the structure. For fire suppression personnel, the exterior survey may be part of the size-up conducted prior to fire attack. If possible, indicators of criminal activity or other physical evidence should be noted and protected from damage during fire suppression operations. Firefighters should also note the condition of doors and windows prior to performing forcible entry. Finding physical evidence such as broken glass, incendiary devices, or containers of flammable liquids should be an indicator that an investigator will be required at the scene.

Accelerant seeped into the wood surface of table and flames burned downward, leaving unmistakable flammable liquid burn pattern.

Horizontal burning along baseboards and a hole burned through floor where flammable liquid had been poured.

Heavy attack of flames on flooring at location where flammable liquid was poured. Note charring under the floor, where the accelerant has seeped through the floorboards and burned downward into the cross beam.

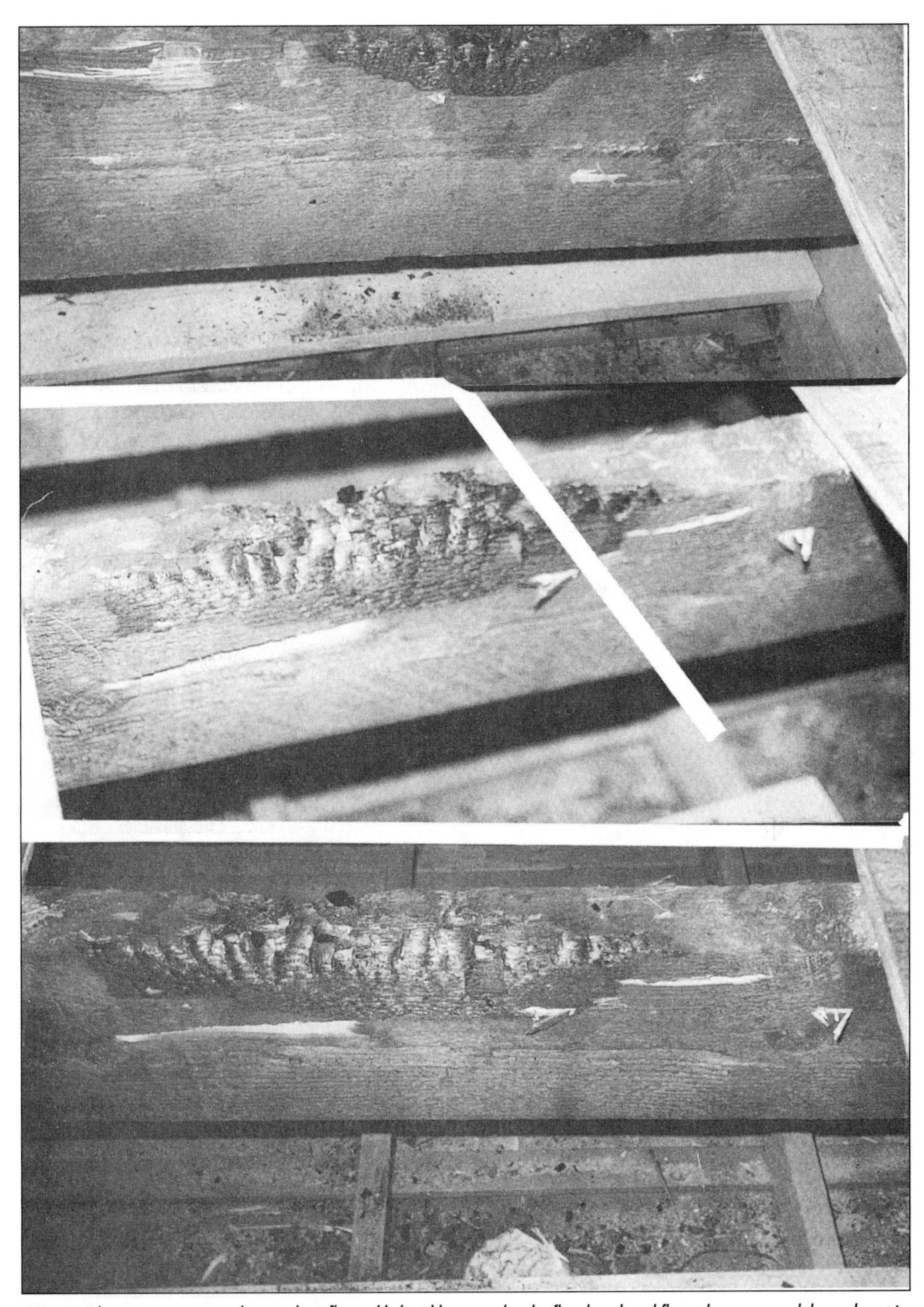

"Weeping" burn patterns on cross beams where flammable liquid has seeped under floor boards and flames have consumed the accelerant in downward paths.

The exterior survey is also an opportunity to evaluate the security of the fire scene. A perimeter should be established so that the scene will not be disturbed until all potential evidence is documented and collected (if necessary). From the perspective of the investigation the perimeter around a fire should extend beyond any potential evidence observed on the scene. For example, in the first scenario discussed the fire was confined to the kitchen. In this case, it would only be necessary to secure the kitchen while the company officer conducted the investigation. Security in this case could be accomplished by assigning a firefighter to monitor the room when the officer was not able to be in it. For the second scenario, the fire involves much of the building so the perimeter would most likely be established at the entry doors to the building. If physical evidence is observed outside the perimeter of a building, the perimeter must be expanded to encompass the evidence. For explosions, it is recommended that the perimeter be at least 1.5 times the distance of the farthest piece of evidence observed.[5]

The interior survey is the portion of the scene examination that will normally require the most time. Typically, investigators begin at the point of least damage to begin the interior survey. Starting in the less damaged area provides the investigator an idea of how the building looked before the fire. If the fire is intentional, incendiary devices or other points where ignition was unsuccessfully attempted may also be located. The investigator should also be looking for indicators of intentionally set fires, such as the lack of personal items in an occupied building or materials that are placed in locations that would impede firefighting operations.

As the interior survey continues into areas where there is more damage, the investigator should begin to observe fire patterns in the damaged areas. Fire patterns are defined in NFPA 921 as "the visible or measurable effects that remain after a fire." Patterns are usually lines of demarcation of damage caused by the fire on a surface. The patterns may be char, smoke or soot deposits, changes in color or character, destruction of material, as well as melting and distortion. The patterns left on surfaces such as ceilings, walls, floors, and contents of a building, help to tell the story of the fire including the movement and intensity of heat and flame in the room or building.

During the interior examination the investigator will look for burn patterns close to the ground that may help to locate the point of origin. Since the gases produced by a fire will rise until they reach an obstruction, such as a ceiling or other horizontal surface, a low burn pattern can indicate the location where the fire originated. Caution should be used in interpreting low burn patterns as they can also result from falling burning materials or the ignition of a fuel package close to the floor later in the development sequence of the fire. The classic pattern used in locating the point of origin is the V-pattern. The V-pattern is formed by the hot gas plume passing over a vertical surface (such as a wall). As the plume rises, the gases in it slow down and expand causing damage in the shape of a "V" to the surface it is in contact with. The bottom of the V will often point to the location of the initial flame in the area. The investigator should be aware that if a room reaches flashover, or full room involvement, low burn patterns generated early in the development by the fire will be destroyed.

For most fires the investigator will be able to determine the area or areas of origin. This determination allows the investigator to more closely examine that area in an attempt to locate the point of origin and the source of ignition. Depending on the conditions in the area, debris may have to be removed to allow for the examination of low burn patterns on surfaces. Debris removal should be done carefully and in layers so that potential evidence is not destroyed in the process. After the removal of debris, the investigator may also need to reconstruct the room by placing the remains of furniture and other contents in their pre-fire locations. This allows the investigator to observe any patterns on room contents that might point to a potential source of ignition.

The reconstruction process may require the investigator to conduct additional interviews of occupants to collect information on pre-fire conditions and locations of contents. If the scene is safe, investigators may decide to bring an occupant into the reconstructed area to obtain additional details on the pre-fire conditions. If safety is a concern, an occupant may be asked to provide a sketch of the area for the investigator to use. The reconstruction and examination of contents will be easier if overhaul operations are limited and any debris removed during overhaul is easily located for examination by the investigator.

Once the area and possible point of origin is located, the investigator begins the process of determining if there are sources of heat in the location that could have ignited combustibles. This is a process of elimination involving careful observation, where the investigator attempts to find evidence of a device, appliance, or other source of heat that could have provided the energy to ignite combustibles at the suspected point of origin. If there is no viable source of heat in the location, the investigator must consider what could have caused the ignition in that location or continue the examination of the area of origin for other possible points of origin.

Where the possibility of an intentionally set fire exists, the investigator should consider the possible use of flammable liquids or an incendiary device at the suspected point of origin. Patterns left by flammable liquids on floors can be one source of evidence. Research done by the National Institute of Standards and Technology (NIST) for the National Institute of Justice provides excellent information on the patterns that an investigator could expect to find on flooring after an ignitable liquid is burned.[6] If the investigator suspects that ignitable liquids were used in the ignition of the fire, it is important that samples for laboratory analysis be taken at the scene. Many department and investigative agencies use ignitable liquid canine/handler teams to assist investigators in locating areas where ignitable liquids may be present. These teams should be used to assist the investigator in identifying locations where samples for laboratory analysis should be taken. An alert by an ignitable liquid detection canine team should not be considered as a substitute for a laboratory analysis of ignitable liquid evidence. The collection of samples for laboratory analysis will be discussed later in this chapter.

The interior survey continues until the investigator has collected enough information to support a conclusion regarding the origin and cause of the fire. It is important to remember that the availability of the scene, and the potential evidence it contains, may be limited once the investigator leaves. The scene examination may be the only opportunity for the investigator to gather information regarding the fire. It is important to complete the process while the opportunity exists.

Document and collect evidence

The documentation and collection of physical evidence of a fire scene will begin as soon as the fire department arrives on the scene. This portion of the process is often completed in conjunction with the scene examination process. It is discussed separately here so that the importance of this step in the process can be emphasized.

NFPA 921 defines physical evidence as "any physical or tangible item that tends to prove or disprove a particular fact or issue." At a fire scene, physical evidence may assist the investigator in the determination of where the fire started, its cause and spread, as well as the factors related to ignition. The evidence found at a fire scene may be in many forms (such as fire patterns, devices or appliances, containers, samples of materials taken for laboratory analysis, broken glass or debris from the fire or explosion, etc.). During the investigation the investigator must identify potential evidence and then make a decision on whether it is relevant to the investigation and how to preserve it properly for use in the determination of the origin and cause. The fire investigator makes the decision to collect physical evidence at a fire scene. When it is decided that an item is of evidentiary value, the item must be properly preserved for use in the investigation and potential legal proceedings.

During suppression operations the preservation of fire scene evidence is not a prime concern. Suppression personnel can assist by protecting obvious items that may be of interest to the fire investigator, limiting the destruction of potential evidence in the area of origin, and protecting the scene from additional contamination by isolating the area of origin after the fire is controlled. Suppression personnel can also limit potential contamination by controlling the use and refueling of gasoline/diesel powered equipment on the scene and by limiting foot traffic within the scene after operations are completed. Overhaul and salvage operations in the area of origin may also need to be restricted until the arrival of the investigator to prevent the destruction or contamination of important evidence. The incident commander will have to weigh the importance of these operations against the potential damage they may cause to the evidence in the area.

Evidence that is to be removed from a fire scene should be protected in place prior to collection, and the location should be identified on sketches and documented with photographs. The actual removal should be accomplished with as little damage as possible to the item being collected. The item must then be packaged so that it can be transported and stored without additional damage. In criminal cases the services of specialized evidence technicians may be used.

Items or materials related to the presence of ignitable liquids that are collected for laboratory analysis must be handled, collected, and packaged with care to prevent contamination. The laboratory conducting the analysis should be asked to provide guidance in this process. *A Pocket Guide to Accelerant Evidence Collection,* developed by the Massachusetts Chapter of the International Association of Arson Investigators (IAAI), is an excellent source of information on the collection and preservation of accelerant evidence.[7]

There are times when an item such as an appliance may be identified as evidence in a fire. This may be to rule it out as a possible ignition source in an intentionally set fire or where the item is thought to be the source of ignition (such as in the previously mentioned example). If the item is removed from the scene by the fire department, the department then takes on the responsibility for its preservation. Any testing or examination that may result in the destruction of the evidence should be done in the presence of any potentially interested parties. If it is not possible to determine or locate all parties, the testing should be well documented. The issue here is the potential spoliation of the evidence.[8] Spoliation is defined by NFPA 921 as the "loss, destruction, or material alteration of an object or document that is evidence or potential evidence in a legal proceeding by one who has the responsibility for its preservation."

Primarily an issue in civil lawsuits, spoliation has become an issue in the area of evidence collection, preservation, and testing. A 1990 Florida case, *Continental Insurance Co. v Herman,* 576 So.2d 313 (Fla. 3ed Dist. Ct. App. 1990) listed several elements regarding spoliation:

- Existence of a potential civil action
- A legal or contractual duty to preserve evidence that is relevant to the potential civil action
- Destruction of physical evidence
- Significant impairment in the ability to prove the lawsuits
- Damages

Since the determination to collect physical evidence is the duty of the fire investigator, it becomes the responsibility of the fire department the investigator represents to protect items of evidence collected in the line of that duty. Procedures for the proper collection, storage and handling of physical evidence, whether for a criminal or potential civil action, must be in place to protect the department from the risk of a spoliation lawsuit. If physical evidence is collected, it must be stored and handled by the department so that it is protected from damage or destruction, and can be located when it is needed to support the investigation or as an exhibit in court. Once the item of evidence is determined to be of no use to the investigation, the owner of the property should be contacted prior to the disposal of the item. Evidence disposal and the permission to do so should be documented in the investigation file.

The right of entry is an issue directly related to the collection of physical evidence from a fire scene. The investigator must be certain that any evidence collected as part of a potential criminal investigation is going to be admissible in court. In some jurisdictions, when an investigation of a fire incident moves from an origin and cause determination to one that may involve criminal activity, the investigator may be required to obtain a search warrant prior to continuing with the examination and collection of evidence. As discussed previously, the department should work with local prosecutors and law enforcement officials to develop policies regarding search warrants and the collection of evidence indicating criminal activity.

Proper documentation of the collection process, procedures to the preservation and security of evidence, and preventing spoliation is the responsibility of the investigator and the fire department. Strict adherence to departmental policies and procedures related to evidence collection and preservation will protect the department from civil action and prevent the loss of critical evidence for criminal proceedings.

Analysis of investigative information

Once the fire scene has been examined and potential evidence collected, the fire investigator switches to the analytical side of the investigation. This is the process of reviewing the available factual information regarding the incident and using it to make an origin and cause determination. The scientific method uses the term *inductive reasoning* for this step. The analysis of the information is based on the training, knowledge, and experience of the investigator. During this process, the investigator should discount any speculative data and only use the hard facts obtained by observation or experiment. The analysis phase may be completed on the fire scene where no special analysis by a laboratory or product expert is required. Where special testing or laboratory analysis is necessary, the analysis of the data will be delayed until the results are made available to the fire investigator.

Determination of origin and cause

Based on the factual evidence available, the investigator will develop an opinion that explains the origin and cause of the fire. The scientific method refers to this as *developing a hypothesis*. Once these opinions are developed, the investigator uses inductive reasoning to evaluate the opinion in light of the information available. If the opinion cannot pass this test, the investigator should develop another hypothesis or opinion to evaluate. Only when the fire investigator can identify a sequence of events that is supported by the available evidence can an origin and cause be determined for the fire. Typical classifications used to describe fire causes include:

- Accidental—Fires that do not involve an intentional human act
- Natural—Fires that result without any human intervention (such as lightning, earthquake, flood etc.)
- Incendiary—Deliberately set fires
- Undetermined—Fires where the available evidence does not allow the investigator to classify the factors that resulted in the ignition

In many cases the origin of a fire may be determined, but due to the lack of physical evidence obtained during the investigation, no cause can be identified. In these cases the fire investigator should list the cause as undetermined. The use of the term "suspicious" as a cause should be avoided—as it implies that the fire was intentionally set, but there is insufficient evidence to support a classification of incendiary. Classifying this fire as undetermined allows the investigator to keep the file open and change the classification if additional information becomes available at a later date.

The report

One of the most important functions of the fire investigator is the accurate reporting of the investigative findings. The fire and investigation report and support materials (such as photographs) may be the only documentation of an incident. These reports may be used years after the incident to assist in civil litigation or in criminal prosecution. The report may be the only document that the fire department has to support testimony of officers or investigators involved in the incident.

The incident or investigation report should be in the format used by the department. Reports are legal documents and may be used for a number of reasons. As such, they should be legible and accurately reflect the findings of the investigation. At a minimum the findings of the origin and cause investigation should be reported with documentation of how the determination was made.

A key issue with investigative reports is the ability to store the information (including photographs and other support material used to document the scene) in a way that it can be easily located when required. The availability of computers and digital cameras makes the completion, storage and retrieval of reports much easier than ever before. Departments should develop a records storage system that allows for the retrieval and use of the information contained in the investigation report.

Fatal Fires

Fires that result in fatalities are a special case for both the firefighters involved in suppression and the fire investigator. The death of a civilian in a fire is a situation that will almost always result in an expanded investigation involving additional agencies (including the police and the medical examiner). When a body is found during suppression operations it should be left in place (if beyond medical assistance). The incident commander should modify operations to support the investigation and should take steps to protect the fatality's location so that it can be properly documented prior to removal.

Fires with fatalities have an increased potential for the involvement of the legal system (criminal, civil, or both). In light of this potential, the investigation and resulting report should be as detailed as possible. In addition to the normal origin and cause determination, the location and condition of the victim should be documented while the body is still in the location where it was found. Once the victim is removed an autopsy should be conducted to provide details regarding the mechanism of death. This information will be included in the investigation report and should be used in the development of the report and determination of the cause of the fatality.

When a search must be conducted for possible victims in a building, it should be conducted in such a way as to protect the remains and any potential evidence near the located victims. If there is considerable debris, the removal process should be completed as carefully as possible to avoid disturbing the remains and associated evidence. A fire investigator trained in debris removal should oversee the process. Using hooks, rakes, or other tools to move the debris should be done with extreme care out of respect to the victim, and also to preserve as much evidence as possible.

The investigation of a fire fatality should address the following:

- Identification of the victim
- Determining the cause of death
- Circumstance resulting in the death
- The activity of the victim prior to, during, and after the ignition of the fire

In some cases the victims are directly involved in the origin and cause of the fire that results in death. In other cases the fire develops without the interaction of the victim, and the products of combustion that result from the developing fire cause the death. The victim may have also died prior to the fire, and the fire was ignited in an attempt to destroy evidence. Using the systematic approach to the investigation, the fire investigator will develop opinions based on the facts available and make a determination regarding the fatality based on the available evidence. Any fire or building code violations that may have contributed to the fatality should be documented and forwarded to the appropriate authority for possible action.

Incidents involving a firefighter fatality are extremely stressful events for the personnel at the scene and the fire department personnel who respond to assist with the investigation. The incident commander must carefully control the actions of personnel at the scene during search and recovery efforts. The scene should be immediately secured and guarded to prevent the destruction of evidence. Just as with civilian casualties, the remains of the victim should be left undisturbed until the area surrounding it can be documented and potential evidence (such as protective equipment, clothing, and tools) is collected and preserved. There are special protocols available from the United States Fire Administration for the post-mortem examination of firefighters involved in line-of-duty deaths.[9] Properly following these protocols will assist in the development of critical information regarding the circumstances that resulted in the death, as well as provide information required for the Public Safety Officers' Benefits Program. The investigation of the incident must address all aspects of the incident, including the origin and cause as well as the circumstances related to the fatality. The resulting complete report should explore methods of preventing similar occurrences in the future.

When there are multiple fatalities in a fire, the incident commander should seek assistance to secure the scene. The assistance of the medical examiner should be obtained in planning for the documentation and proper removal of the victims. The location of each body should be identified and documented prior to removal. The incident commander will also have to make arrangements for handling the remains of the victims at the scene. This could involve the designation of a temporary morgue or another designated location. Transportation of the remains will also have to be coordinated. Arrangements for post-mortem examinations and victim identification must also be made. These considerations will be made in conjunction with the medical examiner and other involved agencies. As with other incidents involving fatalities, the investigation will be expanded to address not only the origin and cause, but the factors that led to the fatalities. Attention should be paid to potential fire code violations that may be related to the deaths. Multiple fatality incidents will also require additional investigative resources and a coordinated effort between the fire department, fire investigators, and other involved agencies.

TRAINING

The determination of the origin and cause of fires in a jurisdiction, and the accurate reporting of this information, requires training of personnel assigned to fire suppression duties as well as specialized training of fire investigators. The provision of policies and procedures that regulate the investigation of fires by department personnel along with proper training are ultimately the responsibility of the fire chief.

Members assigned to suppression operations should be capable of completing a basic origin and cause determination for minor incidents. They should also be trained to identify conditions that will require requesting a detailed origin and cause investigation. Additionally, all suppression personnel should be trained to recognize potential evidence and fire patterns that may be important to the investigation. Training in the importance of scene security, preserving evidence, and proper overhaul operations in areas where evidence may be located should be provided on a regular basis. Basic requirements for fire investigation, scene preservation, and security are found in the requirements for fire officers in NFPA 1021, Standard for Fire Officer Professional Qualifications, and in NFPA 1001, Standard for Fire Fighter Professional Qualifications, for fire fighters. The National Fire Academy has developed several training programs for fire suppression personnel. These programs are typically available through the state fire service training agency or directly from the National Fire Academy.

Personnel assigned as fire investigators for the department require specialized training in the field. NFPA 1033, Standard for Professional Qualifications for Fire Investigators, provides the job performance requirements for the investigator. These requirements can be met in many ways, including formal training programs designed to address these requirements, on-the-job training, and attendance at seminars and other in-service educational programs. Based on the requirements of NFPA 1033, certification to the level of fire investigator should be a goal of the department for all investigators. Certification is available from national organizations such as the IAAI and the National Association of Fire Investigators (NAFI). Some state fire service training agencies also offer fire investigator training and certification.[10]

It is essential that personnel assigned to fire investigation duties have the training and expertise to properly perform the assigned duties. Fire investigators must be familiar with the recommendations included in NFPA 921, Guide for Fire and Explosion Investigations, and reference books related to the subject of fire investigation. Fire investigators must also stay current with research and publications related to fire investigations and fire growth and development. Certification, coupled with in-service training and education, will provide the fire investigator with the skills and knowledge required to perform the assigned tasks, and increase the credibility of their findings when challenged in court.

Conclusion

The investigation of fires and explosions is one of the functions of most fire departments. To accomplish this function, members must assist in preserving the scene during suppression operations by observing conditions on arrival at the incident, establishing scene security early in the incident, and limiting overhaul operations in the area of origin until the investigation is completed. Fire suppression personnel should be aware of the need to preserve evidence to support the origin and cause investigation. Where no investigator is required at a fire scene, suppression personnel should complete a report that accurately details the observations at the fire scene so that the origin and cause information is available for analysis and use in future litigation. Fire investigators must be properly trained to perform assigned functions and be aware of resources such as NFPA 921 and current research in the field of fire and explosion investigation.

To be effective, fire investigation units must have the materials and equipment necessary to conduct investigations, document scenes, develop reports, evaluate and secure evidence, and analyze findings so that trends can be identified. The fire chief is responsible for ensuring that personnel are trained to perform and support the investigative process. The fire chief should ensure that policies and procedures are in place for the investigation of incidents, personnel safety during these activities, complete reporting of findings, and the preservation of physical evidence that investigators may decide to collect at fire scenes. Proper preparation, training, and support will enhance the investigation of fires and result in a safer community.

References

[1] NFPA 921 Guide for Fire and Explosion Investigations, 2001 edition, National Fire Protection Association, Quincy, Massachusetts.

[2] NFPA 1033 Standard for Professional Qualifications for Fire Investigators, National Fire Protection Association, Quincy, Massachusetts.

[3] *Michigan v Tyler,* 436 U.S. 499 (1978).

[4] *Michigan v Clifford,* 464 U.S. 287, 294, 104S. Ct. 641 (1984).

[5] *Guide for Fire and Explosion Investigations,* 2001 edition, Section 18.13.1, NFPA 921, National Fire Protection Association, Quincy, Massachusetts.

[6] *Flammable and Combustible Liquid Spill/Burn Patterns,* NIJ Report 604-00, National Institute of Justice, Washington, D.C., March 2001.

[7] *A Pocket Guide to Accelerant Evidence Collection,* Second Edition, Massachusetts Chapter International Association of Arson Investigators, 2000.

[8] Jones, Jon C. "Handle With Care," *Fire Chief,* July 1998.

[9] *Firefighter Autopsy Protocol,* United States Fire Administration, 1991.

[10] Jones, Jon C. "Facing Up To 921." *Fire Chief,* PRIMEDIA Business Magazines & Media, Inc., Kansas, July 1996.

28

THE NEXT GENERATION

John Granito

CHAPTER HIGHLIGHTS

- The flow of societal direction and the level of budgets still appear to be the major drivers of change in the operational characteristics of the fire service.
- Positive indicators such as additional training, a growing body of professional literature, the formation of the Congressional Fire Caucus and the Congressional Fire Services Institute, and the wide acceptance of newer operating methods and devices, reflect an internal change the fire service has experienced over the past several years.
- September 11, 2001 is a pivotal date wherein the fire service gained universal recognition for its overall great value.
- Intensive self-examination plus recognition of acceptable industry standards allow departments to set goals, measure goal attainment, and, if necessary, establish a phased-in program to reach the goals set.
- The next generation will need to exercise more imagination and foresight in disaster planning as it faces the new challenge of comprehensive preplanning for high consequence, low frequency events such as "weapons of mass destruction" incidents.

INTRODUCTION

The last edition of the *The Fire Chief's Handbook,* published in 1995, also ends with a chapter entitled "The Next Generation." The first sentence of that chapter, written some years ago, proclaims that "The future of the fire service has become a topic of great interest..."

That bit of wisdom certainly has been borne out, although not in ways clearly anticipated when it was written. The second paragraph of this same chapter in the fifth edition expresses the judgment that "...the driving forces behind fire service change have more to do with finances and the general direction of society than with the tools of our trade." That concept, as well, appears to reflect what has transpired in recent years. A good deal has changed, both in our society and in the fire service.

While thinking about what things will be like in the future, predictions concerning "the tools of the trade" and improvements in operations are more fun to speculate about and probably more exciting to read about. Yet, the flow of societal direction and the level of budgets still appear to be the major drivers of change. Of course, predictions surface more clearly when the forces and events of the past are analyzed first. The old adage says that if we don't study history we'll be forced to repeat it. There's another old saying: "If you don't know where you're going, you'll never be lost!" Fortunately, there's been more goal-directed thinking and forward-looking leadership in the fire service during the past several years than ever before, and we are clearly moving ahead. But forward movement has not been easy, even though fire departments are consistently rated tops by citizens who have been asked to judge their municipal services and the honesty and ethical standards of people working in various fields.

Recent Changes

A review of the many positive signs and events relative to the fire service that have occurred over the past several years, but prior to September 2001, would include the following:

- A high number of members are involved in obtaining education and training, ranging across a broad spectrum of topics from the Federal Executive Fire Officer Program to the widespread emphasis on hazmat response and firefighter safety.
- There is a growing body of professional literature, as evidenced by the vastly increased number of significant articles and monthly columns appearing in fire/rescue and EMS magazines, as well as the many books now available.
- There has been a general acceptance of NFPA standards and recommended practices relating to firefighter safety, health, and general wellness.
- Several national and international professional organizations have been re-vitalized and coalitions have been formed among them.
- The Congressional Fire Caucus and the Congressional Fire Services Institute have been formed.
- A carefully structured fire department accreditation program has been implemented.

- Additional experienced fire service professionals have been appointed to the National Fire Administration.
- Beginning in 2001, the federal government has given aid to local departments (with almost 2,000 recipients in that first year).
- Fire service and related professionals have composed several position papers in order to present a unified agenda to improve fire departments' abilities to serve the public.
- Attention has been focused on issues of initial response staffing and response time, as well as the development of two NFPA Standards directed at those and related issues.
- Departments have willingly expanded service offerings to include not only EMS and technical rescue specialties, but sophisticated public education programs.
- Operating methods and devices such as incident command, integrated emergency management, PASS units, compressed air foam, rapid intervention teams, two in–two out staffing, infrared aids, wildland/urban interface tactics, project management techniques, labor-management shared goals, leadership development programs, virtual symposia via the Internet, closed circuit and interactive video learning, and so on, have been widely accepted.
- Several model departments—career, volunteer, and combination—have emerged, whose programs and organizational techniques are described and available for replication or modification by other departments.
- There has been more leadership activity that includes several strong, influential, and professional organizations, plus influential individual leaders.
- There is a growing awareness within the service of the need for flexibility, imaginative problem solving, and a unified voice.
- There has been an increased emphasis on diversity of membership.

Each of these positive indicators is important in its own right and, taken together, they illustrate the significant volume of internal change the fire service has experienced over a relatively short time. When analyzed, these changes indicate both significant internal improvements and the resulting increased ability to serve communities.

Unfortunately, there are no guarantees that a department that has gained "internally" will improve in its ability to provide upgraded service delivery. Without widespread acceptance of performance standards and continuing quality assurance programs—as are commonplace in emergency medical service operations—communities may benefit only marginally, or not at all. This is a strong reason for departments to seek accreditation, where organizational status, operating protocols, and goal attainment are judged every several years using a standardized format. Despite the challenges associated with meeting nationally promulgated standards for emergency response time and capability statistics, these also provide a uniform yardstick for judging local performance adequacy.

New challenges

During the time that this "new generation" section was being written, the American fire service encountered the greatest challenge in its history. That challenge has continued and will be, without a doubt, with us for years to come. To reverse an old literary phrase, it has been the "worst of times" and the "best of times" with the death of hundreds of heroic firefighters and other responders in a horrendous terrorist attack. Ironically, this attack was followed immediately by universal recognition of the great value of the fire service in general and the courageous, unselfish character of individual firefighters.

The need for resources

No matter how firefighters were esteemed prior to September 11, 2001, on that date they became the great American (and worldwide) heroes. American fire service leaders and friends moved quickly to honor the fallen and to work for vastly increased federal support for additional resources. In a very real sense, the foreseeable future of American fire departments, and to some extent those of allied nations, will be measured from the infamous 9-11 date, when the meaning of "first responder" became abundantly clear.

Yet despite the overwhelming national focus of attention on firefighters, the contribution of millions of dollars from all sources for the families of those killed, and the increased height of the firefighter pedestal, relatively little has been done to strengthen the fire service as a whole since September of 2001. Let us hope that *sic transit gloria mundi* (thus passeth the glory of the world) will not apply. However, the best way to predict the future is to play a major role in shaping it, and the best way to avoid a replay of history is to keep in mind what has already transpired. So this writing turns now to a review of certain fire service events occurring over the past few years and a look at what the future might hold. What will the next generation encounter?

Emerging Key Questions

The concept that fire service change would come mostly from "finances and the general direction of society..." appeared to be the case during the decade of the 90s and for a few months into the new century. The congressional Fire Service Caucus, led by a few congressmen who are especially cognizant of the needs of the fire service, was successful in having $100 million in 2001 distributed to fire departments that had applied for help. Some new positions were established and experienced personnel were brought into the National Fire Administration. The growing concern over weapons of mass destruction and readiness for terrorist attacks prompted the fire service and several of its key organizations to stimulate planning and instruction aimed at moving fire departments into even more expanded roles in public safety. From a broad perspective, fire service organizations in the United States, Canada, Great Britain, Germany, Australia, South Africa, and other countries as well, were soul searching—and in some places, such as Great Britain, they were being scrutinized by governmental agencies—to determine more accurately the answers to several key questions:

- What level and types of resources are needed to provide adequate service delivery to various types of municipalities?
- What are the characteristics and indicators of an adequate or better fire department?
- Can career, volunteer, and part-time personnel organize and work together cooperatively and productively?
- Will sufficient numbers of volunteer firefighters be available to staff their departments?
- Which service items, such as EMS, are best delivered by fire departments rather than other organizations and agencies?
- What should the source or sources of funding for local fire departments be?
- What resources are needed to provide baseline community fire/rescue service?
- What should be the roles and responsibilities of the various national and state fire service organizations and the several fire and safety focused federal agencies (especially in the U.S.)?

In response to the need for at least partial answers to these questions, various fire related organizations, governmental agencies, and other interested groups initiated projects, launched studies, conducted reviews, and promulgated laws, standards, recommended practices, and accreditation procedures in the interest of improving community protection. Additionally, these actions were begun to increase firefighter safety, upgrade individual fire departments, foster resource sharing, and elevate the professional status of fire personnel. Because most of these efforts are either still developing or yet to be time-tested, they should be examined from two perspectives: impact as "recent history," and the stimulus for future conditions and events.

Other very important questions deal with fire suppression resource quantities and deployment. Around these two issues a heated and polarizing debate had grown which was set aside only because of the September 2001 terrorist attacks and ensuing military conflict. This debate started long before September, 2001, and will continue into the foreseeable future.

Performance Standards

It might be said that National Fire Protection Association (NFPA) Standard 1500, Health and Safety Programs for Fire Fighters, promulgated in 1987, and the federal Occupational Safety and Health Agency's "two in-two out" ruling of 1988—both of which call for the presence of four firefighters prior to structural interior attack—highlighted the concern over what the number of initial attack personnel required for safe operations at various types of incidents should be. Several studies had already taken place, ranging from the early Dallas study to the Centaur effort; from the Ohio State University study to the NFPA's Urban Fire Forum/Fire Department Analysis (FireDAP) project; and from the Ontario, Canada Fire Marshal's task analysis study to the very early NFPA *Fire Attack* books. None of these efforts, however, nor the additional experiments and writings on the subject, were able to convince all interested parties as to the initial number of attack personnel required for safe operations. All the groups mentioned who conducted these studies were, and are, both internal and external to the fire service.

Beginning in 1995, a formal and organized effort to set a national standard covering the organization, deployment, and operation of fire suppression and rescue services was launched by the NFPA Standards Council. The NFPA 1201 Technical Committee (Developing Fire Protection Services for the Public) however, was saddled with dealing with both volunteer and career departments–and that proved too formidable a task. Unable to see a consensus reached during the committee's two years of deliberation, the Standards Council appointed two separate technical committees, NFPA 1710 for "career" departments and NFPA 1720 for "volunteer" departments, thus recognizing that, at least initially, differing approaches were needed.

The 1720 Standard (Standard for the Organization and Deployment of Fire Suppression Operations, Emergency Medical Operations, and Special Operations to the Public by Volunteer Fire Departments) generated only tangential deliberation around such issues as "why should there be two different standards?" and "what is the precise definition of a volunteer department?" It easily passed the required NFPA general membership vote in 2001 because there were no staffing and service delivery requirements.

NFPA 1710, which contains both staffing and service delivery requirements, was controversial from the beginning and generated much debate before being passed at the same NFPA meeting as NFPA 1720. NFPA 1710 was appealed by a number of organizations, but the appeals were rejected by the NFPA Standards Council and the standard was issued by the Council in August, 2001

In the latter part of 2001, additional members were added to the Technical Committee on Fire and Emergency Service Organization and Deployment—Career. The new members represent some of the organizations who appealed NFPA 1710. The debate on the staffing and deployment will continue within the Technical Committee.

During the latter part of 2001 and into 2002, guidelines for understanding the 1710 Standard were issued by the International Association of Fire Chiefs (IAFC) and the International Association of Fire Fighters (IAFF), which is the pre-eminent labor organization of career firefighters in the United States and Canada.

There were discrepancies between the guidelines and some speculation that the IAFC interpretation was designed to "soften the blow" of additional cost to municipalities, while the IAFF document presented additional arguments in favor of the 1710 Standard. The IAFF had, of course, been exceedingly in favor of the new technical standard and was unwavering in its stance. At the extreme ends of the sometimes nasty debate were the opposing statements that the standard had as its purpose the creation of additional firefighter jobs, and that municipal officials put a low price on the personal safety and effectiveness of firefighters. In the midst of the controversy were the important questions that asked which standard, if any, applies clearly to the many combination departments, and which should be used as an operational template for those volunteer (or "substantially" volunteer) departments that protect urbanized areas, in some places more densely populated that those protected by some small career departments.

Also aiming at the goal of upgraded service delivery is the program of fire department accreditation, cosponsored by the IAFC and the International City and County Management Association (ICCMA) and managed by the Commission of Fire Accreditation International (CFAI). By early 2002, approximately 60 fire departments had successfully undergone the intensive process necessary and had been granted accreditation. Many others were attending the required process familiarization workshops so that they could begin the effort.

As with the accreditation of colleges and universities, the concept and process of fire department accreditation calls for a very detailed "self-study" using a template that forces the review of numerous topics and areas of importance, with described acceptable standards used as benchmarks. The immediate result of a successful process is twofold: the department now knows itself in minute detail as it has never before and the department now conforms within certain generally acceptable industry standards.

This combination of intensive self-examination plus recognition of generally acceptable industry standards allows departments to set goals, measure goal attainment, and, if necessary, establish a phased-in program to reach the goal. As an example, one department that achieved accreditation in 2001 examined its ability to respond in its area within certain time frames as specified by the "Standards of Cover" section of the accreditation process.

The department adapted the goal for career departments called for by NFPA Standard 1710, which states that the first due engine should arrive within 240 seconds running time and/or the entire first alarm assignment should arrive within 480 seconds running time, for a minimum of 90% of the annual incidents.

In this case, the department calculated that additional stations would need to be constructed and staffed if the department were to achieve its standards of cover goals. From this realization emerged a plan to fund and construct the required additional stations over a period of several years.

Thus the accreditation process not only insists on a detailed self-study, with results compared to recognized parameters, but also on the formulation of goal attainment plans. It also requires continued periodic reexaminations.

Hazard and Risk Analysis

To assist departments and their municipalities in determining the type, extent, and level of community fire risk, a computerized program called Risk Hazard and Value Evaluation (RHAVE) has been devised, and distribution began by the Federal Emergency Management Agency (FEMA) in early 2002. Following the September 2001 terrorist attacks on mainland targets, of course, a vastly increased focus on hazards identification and risk assessment emerged. Where the spotlight of attention primarily had been on high frequency-low consequence events, the World Trade Center and Pentagon incidents, coupled with continuing terrorist alerts mentioning municipal water supply systems and nuclear generating stations among others, caused greater concern over low frequency-high consequence possibilities.

In other words, while we had always known that certain hazards existed, we rated the chance that an incident involving them would occur as very low. Understandably, community officials will often limit spending if they judge the chance of a stated type of incident occurring as slim. This might be termed the low frequency-low preparation-low spending syndrome: "If it probably isn't ever going to happen, let's not spend much on a worst case scenario!" Emergency responders, on the other hand, often hypothesize worst case scenarios just in case the worst should actually happen.

What helps emergency management planners in their work is that if a community preplans for those incidents that will probably happen (mid to high frequency), most of those that might possibly happen (low or "never" frequency) will be mitigated at least somewhat by those same planning and action steps.

Multi-Organization Operations

It might be speculated, therefore, that one distinguishing earmark of the next generation will continue to be the understanding that low frequency-high consequence incidents can occur. Further, their consequences can be horrendous, and that whatever might prevent them or mitigate their impact must be accomplished. It is at this juncture that fire, rescue, emergency medical, police, and other first responders loom exceedingly large in the formula, with fire, rescue, and EMS likely in the forefront. Fire department planning will need to recognize such challenges.

If history does repeat itself, the possibility that national attention paid to fire departments and the intention to increase their resources, which began on September 12, 2001, will dissipate with time exists. The effort to increase by 75,000 the number of firefighters in the United States, paid for partially by federal funds, had not been successful by February, 2003, although the concept was still alive. Many communities at the time of this writing are on high alert and are aware of the paramount and dangerous role assigned to firefighter and EMS personnel. Should these homeland security alerts continue into future years, the next generation should benefit from increased resources, possibly at federal expense, with emphasis on numbers, training, equipment, and planning.

It is possible, also, that some first responder rescue and emergency medical tasks might be assigned to the National Guard, just as it took on homeland security assignments at airports and other high risk facilities. Although fire departments currently have a near monopoly on initial response and urban search and rescue teams (USAR), it's reasonable to assume that other organizations could easily be factored into the equation. The next generation of fire personnel, therefore, could operate in a more extensive co-responder environment necessitating shared resources.

Shared incident command is already commonplace in incidents ranging from wildland-urban interface to weapons of mass destruction (WMD) and more of this type of sharing needs to be anticipated. One problem associated with more complex incidents, including those involving WMD such as radioactive materials or bacteriological agents is that very sophisticated tactics, methodologies, and equipment are needed to mitigate their effects. Just as "overhead teams" of experts are called upon to handle incident command at wildland fires, so will specialized, mobile WMD overhead teams be needed to handle some events following local initial response. If fire departments are intending to protect communities from WMD and other smaller but relatively unique incidents, the types and levels of preplanning and training for the next generation will likely exceed those currently conducted. Additional training requirements will probably impact the ability of some communities to recruit and retain volunteer firefighters.

Considering that most departments in the United States and Canada are relatively small, and that the highest percentage by far is volunteer and time limited, the training demands as well as specialized equipment needs may prove too heavy of a time and cost burden. While vastly increased federal funding for equipment and training would help, it would seem that special regional multi-purpose response teams, very similar to the existing urban search and rescue (USAR) teams, will be necessary. The concept of mutual aid for fire, rescue, EMS, and related incidents will need to be recast and greatly expanded.

It may be that, just as some departments today run task force responses to local fire calls, task force responses composed of units from several or many departments will be dispatched to WMD and related incidents within a large geographic response area. Because of the possibility of wide spread disruption of services, possibly including wireless communication, disaster planning will have to include very different alternate methods of alerting, dispatch, and command.

Larger and More Complex Incidents

A worst case scenario for the next generation might well need to include such challenges as large area contamination and the disruption of common forms of wireless communication in that area. Response might need to include, or be composed entirely of, units totally unfamiliar with that locale. The fire/rescue service is already familiar with this type of operation through the USAR team program, which includes "self-contained" response groups. The September, 2001 attack at New York's World Trade Center illustrates the complexity of a WMD incident, but the area impacted, while large in scope, was geographically limited and was contained within a much larger city whose infrastructure and resources were otherwise intact. This might not always be the case.

Next generation challenges could include WMD incidents, such as those involving the release of toxic substances, where much larger geographic areas are impacted and where immediate nearby aid is not available. Although we understand that emergency medical services will be in great demand, and although federal medical evacuation and emergency treatment plans exist, the burden on local and nearby first responders who survive to operate could be overpowering. Just triage alone, assuming the release of a radioactive, chemical, or bacteriological substance, without entrapment, would present a huge, complex, and dangerous workload requiring very extensive resources.

The particularly new challenge of comprehensive preplanning for WMD incidents faces the next generation, but disaster planning, in which many communities already engage, provides a reasonable springboard for WMD preplanning. The major difference may be in the necessity to exercise more imagination and foresight in considering low frequency-high consequence events. We know the consequences of a hurricane because many have already occurred. We did not realize the consequences of two heavily fueled jet liners flying into the top stories of a high-rise complex. Next generation planners will need greater imagination and the ability to think "outside the box" in order to prepare for possible incidents.

Another crucial need will be increased ability to identify the resources needed to handle large scale incidents, to preplan their use, and to arrange beforehand for their availability. Currently we enjoy three types of mutual aid for fire/rescue incidents. The first is commonly termed "regular" mutual aid, which may be on-demand or automatic, and which generally is local in nature. The second is regional in nature and typically is controlled by a county, regional, or state dispatch or emergency operations center (EOC). The third type is multi-regional, multi-state, national, or even international in scope. USAR teams, Red Cross and Salvation Army disaster teams, hostage rescue teams (HRT), and some special weapons and tactic teams (SWAT) are examples of these broader scale operations.

It is this latter category of aid that most likely will become commonplace for the next generation, if WMD incidents and large-scale natural disasters become more frequent. Notice that as this type of aid, which may not be at all "mutual" in nature, is described, the terminology becomes more and more related to what is called "emergency management" (such as EOC) rather than the narrower fire department designations (such as alarm and dispatch center).

What may happen, of course, is that fire, rescue, and EMS personnel will not play the major roles in certain WMD and other types of large scale disasters. As the need for multidiscipline response groups becomes more evident, the template of USAR teams (that already have some specialized non-fire personnel) will become more encompassing of specialists in such disciplines as public health, engineering, demolition, radiation, environmental protection, toxic sampling, weather prediction, legal, crime scene investigation, martial law, evacuation, and public welfare. These experts will become full team members. In certain incidents, one or more will be lead players and firefighters will be members of the working team. Integrated emergency management systems (IEMS) will provide the framework for these large and complex operations, and IEMS will be as familiar to emergency responders as incident command is now to fire personnel.

Once again, existing USAR team operating protocols point toward future general operational modes. USAR teams often come from afar in their role as search and rescue specialists. They operate very effectively as a sub-system not typically in overall command, and they function alongside other types of specialists, all (hopefully) within an IEMS environment and a unified command structure.

Fire/rescue departments obviously will continue to function at typical incidents, and at almost every incident, just as they now do. They play the lead role, have the most specialized resources by far, and they establish incident command as a matter of course. But the next generation will need to develop and maintain the operational flexibility and psychological mind set which will enable responding departments—or even local departments out of their field of expertise—to become one of several operating partners in a more complex operation. The implications are here for more flexible leadership styles, less sensitive ego structures, and organizational cultures more receptive to transient change.

Most fire departments judge themselves under-resourced for normal operations, and much less for the additional challenges we are already seeing. For example, the anthrax scare alarms of the last months of 2001 and the early part of 2002 overtaxed many fire departments to the extent that the calls were queued and dispatched a minimal response, similar to the handling of routine non-emergency police calls. Of course, the possibility continues to exist that the next anthrax alarm can be a real one, just as the next automatic school or warehouse alarm can be an actual fire. With limited response units and tight budgets there is no completely satisfactory response protocol to handle this type of challenge.

However, should the next generation have to face and deal with the challenges of terrorist attacks against homeland security, as well as provide the basic fire suppression, fire prevention, EMS, technical rescue, hazmat, and public safety education services all communities need, the resource problem could become even more severe unless massive federal aid flows to almost all departments, no matter what type or location.

New Versions of Old Challenges

In addition to these new and complex issues, it appears very likely that the next generation will still have to battle some old, familiar challenges. However, there are several positive fire service trends that—if they continue—will help considerably.

Positive trends, indicators, and predictions

It seems quite clear that the fire and rescue business has spiraled upward to a point where it is viewed by some as a profession, in addition to being an occupational specialty. After all, it consists of many interwoven and complex sub-areas; it has developed a conceptual framework of sorts; it has generated a broad knowledge base; it contains a hierarchy of positions, each requiring advanced training and education; there is a body of technical literature; and both unit accreditation and, very recently, chief fire officer certification are possible and desired.

True, the fire service is still trying to regulate itself, and its internal problems and "dirty laundry" are sometimes featured in the media, but this is true of other occupational specialties purporting to be professions as well. So it is likely that this movement toward professionalism will continue, with all that it implies for younger people getting into the work, especially regarding training and education. The inclusion of emergency medical work has, of course, added considerable impetus.

One area of sometimes noticeable friction does exist and can be quite visible. The problems between "career" firefighters and "volunteer" or part-time firefighters continue to plague departments in several parts of the country. Friction is very minimal—or even nonexistent—in many places, but in others it often looms large. In one area for example, mutual aid into a larger "career" city would not be called from well-equipped and highly trained nearby volunteers, but instead is summoned from tiny, understaffed career departments some distance away.

National statistics indicate a reduction in the number of volunteer firefighters and an increase in the number of career/full time firefighters in the United States, but the underlying reasons for this appear not to be the desire of municipalities to spend more money on fire departments. Rather, the lifestyle of Americans and the growing time requirements to maintain active volunteer status seem to mitigate against volunteerism. It is interesting, however, that countless well-led and active volunteer departments in every state do not lack for dedicated members.

Recognizing this, it is evident that a focus of attention on those factors necessary to attract and retain volunteer firefighters will be necessary during the next generation, since literally thousands of communities cannot afford to fund small departments composed solely of full-time personnel. The following are variables to look for:

- A pronounced focus on improved volunteer department leadership.
- A broad "menu" approach to volunteer retention magnets, where a list of benefits may be selected, with a dollar allocation assigned to each active member.
- A reduced workload for volunteer firefighters in some communities through the removal of EMS runs from their assignments, and through a significant reduction in unnecessary automatic alarms created by a penalty system applied to property owners.
- The alleviation of volunteer officer administrative workload through the employment of administrative aids shared by several volunteer departments.
- The creation of on-duty flying squads of responders stationed centrally in geographic areas. These may be full-time firefighters or on-duty volunteers.
- An increase in the functional consolidation of several nearby departments, where organizational autonomy is retained, but operating and support areas are combined.
- The standardization of area volunteer firefighters and officers and the annual issuance of photo certification cards, enabling firefighters from other area departments to work at incidents without mutual aid call-up.
- The use of volunteer citizens to conduct support services.
- An increase in the "closest station response" concept and automatic mutual aid without regard to district boundaries.
- Increased use of part-time firefighters to cover heavy workload time periods.
- The creation of more combination departments.

It is important to apply teaching/learning technology to fire training requirements, since more volunteers will be retained if the necessary training can be self-paced, at home, and conducted at convenient times. Obviously, typical group drills and team exercises will continue to be necessary, but a great deal of training material can be presented through personal, interactive packages rather than at group sessions at the volunteer station. Sadly, in 2002, at least one state program had some training money for volunteers unexpended because of the necessary group time commitment that large numbers of volunteers were unable to make.

An important factor in attracting more volunteers is the recognition that, in many communities, pools of potential non-traditional members are never sought. Departments that work to attract non-traditional members will find that they are available and can fulfill requirements. Non-traditional recruits certainly include such people as nearby college students, women, and daytime workers who are non-residents.

Challenges for the Next Generation

It came as no surprise that less than one year after the September, 2001, terrorist attacks in the United States, and even as the media reported continuing threats, many communities in the United States, including New York City, were calling for large budget cuts. In many instances, fire departments—while recognized as a first line of defense—were assigned reductions or warned that reductions could be imposed. The challenge of achieving and maintaining an acceptable resource base doesn't go away.

Additionally, the recognition that some degree and volume of risk must be assumed by communities is growing. This concept of assumed risk runs counter to the historical "worst-case scenario" favored by fire personnel when laying out budget needs. The next generation will need to conduct all-hazard assessments, negotiate with officials and citizen groups, and be prepared to think innovatively to acquire needed resources. And the "needed resource" list will need defending, no matter the esteem in which fire departments are held.

An unusual challenge began to emerge in 2002, and may continue through the decade. In a number of cities, a typically high numbers of firefighters and officers will retire due to:

- General demographics
- Special provisions in some retirement systems encouraging early departure of higher paid personnel
- The positive impact of overtime payments on pension checks
- Plans that enable retirement and a type of time-limited reemployment by special contract available in some communities

- The focus on general health and physical fitness
- The growing realization concerning the type and span of required duties and responsibilities

What this means is that those departments where a high percentage are older members will not only need to recruit new members, but will need to fast-track some officer positions. The good news is that additional promotions will be necessary. The bad news is that decades of operational experience and mentor-class personnel will be lost.

The challenge of fostering continued cooperation among the major fire service organizations will continue and the emergence of active new specialty organizations will complicate matters.

Of course, the challenges of mounting comprehensive prevention and public safety education programs will continue as well, with public education programs using the all-hazard approach. Continued building in areas prone to natural disasters ranging from hurricanes and floods to wildland interface fires will require continuing and massive public education programs, with fire departments playing a major role.

The broadest and possibly most important challenge facing the next generation of fire service people is that of continuously modifying the list of important services that fire departments should be providing. While some redefining of departments will continue to be necessary, it appears that these three questions will be most important:

1. What are the services that fire departments should be providing their communities in addition to the traditional ones?
2. What constitutes adequate service levels?
3. How should adequate service levels be paid for?

It would be great to be able to say that the challenges and problems described in the last edition's chapter on the next generation have been taken care of. Unfortunately, most of those challenges are still with us and new ones have been added. But progress has been made since the last edition, and the fire service will continue to push ahead, meeting both the old and the new issues. It always has and it always will.

INDEX

A

C

D

I

J-K

L

N

T

U

V

Z